42nd Riso International Symposium on Materials Science: Microstructural Variability - Processing, Analysis, Mechanisms and Properties

IOP Conference Series: Materials Science and Engineering
Volume 1249

Roskilde, Denmark
5 - 9 September 2022

ISBN: 978-1-7138-6015-0
ISSN: 1757-8981

Printed from e-media with permission by:

Curran Associates, Inc.
57 Morehouse Lane
Red Hook, NY 12571

Some format issues inherent in the e-media version may also appear in this print version.

Institute of Physics
Dirac House, Temple Back
Bristol BS1 6BE UK

Phone: 44 1 17 929 7481
Fax: 44 1 17 920 0979

techtracking@iop.org

Additional copies of this publication are available from:

Curran Associates, Inc.
57 Morehouse Lane
Red Hook, NY 12571 USA
Phone: 845-758-0400
Fax: 845-758-2633
Email: curran@proceedings.com
Web: www.proceedings.com

TABLE OF CONTENTS

IOP Conf. Series: Materials Science and Engineering 1249 (2022) 011001 doi:10.1088/1757-899X/1249/1/011001

Preface

Today's industrially produced metallic materials generally are very advanced, and as such future breakthroughs will need to rely on detailed novel microstructural understandings. As a reflection of this, the 42nd Risø International Symposium will focus on the importance of a paradigm change from microstructural characterization largely based only on determination of overall average parameters to a new level, where microstructural variability is carefully analysed and included in simulations. The microstructural variability includes studies of the heterogeneity in microstructure, as well as effect of variations on a more local and distributed length scale. The whole range of metals and alloys from single phase conventional metals to advanced microstructurally designed ones are in focus. The microstructural parameters to be considered are not limited only to those for classic features, such as dislocations, cells, subgrains and grains but include local orientation relationships, local strain, etc. We welcome presentations that cover:

- Detailed characterization of microstructural variability in metallic materials on any length scale
- Analysis and simulations of microstructural evolution, including local variations, during processing and operation
- Effect of microstructural variability on mechanical and physical properties, as well as performance
- Concepts of microstructural design and enabling manufacturing techniques
- Linking local experimental findings to analytical and numerical models
- Next generation metals and alloys for engineering applications

These Proceedings contain 9 keynote, 12 invited, and 46 contributed papers dealing with the topics mentioned above. It is our hope that the Proceedings can inspire the future development of this important scientific field.

Peer Review Statement

All papers published in this volume have been reviewed through processes administered by the Editors. Reviews were conducted by expert referees to the professional and scientific standards expected of a proceedings journal published by IOP Publishing Publishing.

- **Type of peer review:** Single Anonymous
- **Conference submission management system:** Morressier
- **Number of submissions received:** 75
- **Number of submissions sent for review:** 75
- **Number of submissions accepted:** 67
- **Acceptance Rate (Submissions Accepted / Submissions Received × 100):** 89.3
- **Average number of reviews per paper:** 2
- **Total number of reviewers involved:** 8
- **Contact person for queries:**
  Name: Dorte Juul Jensen
  Email: doje@mek.dtu.dk
  Affiliation: Technical University of Denmark

IOP Conf. Series: Materials Science and Engineering 1249 (2022) 012001 doi:10.1088/1757-899X/1249/1/012001

Microstructural evolution during recovery of deformed aluminium—Effect of deformation strain

Tianbo Yu

Department of Civil and Mechanical Engineering, Technical University of Denmark, DK-2800, Kgs. Lyngby, Denmark

Email: tiyu@mek.dtu.dk

Abstract. Recovery takes place during annealing of a deformed metal, releasing some of the stored energy. Many recovery mechanisms have been proposed in the literature and these mechanisms strongly depend on the deformation microstructure. In this work, recent progress in the study of recovery in aluminium is reviewed. In lightly deformed aluminium, subgrain growth (through boundary migration and coalescence) is found to be an important recovery mechanism, whereas in heavily deformed samples, uniform coarsening through Y-junction motion is found to be the dominant recovery mechanism. The kinetics of recovery follows a universal coarsening model for aluminium deformed to both low and high strains, in which the apparent activation energy increases in the course of recovery. Furthermore, a new definition of recovery is suggested to be compatible with observations at both low and high strains.

1. Introduction

When a deformed metal is heated, restoration processes take place, releasing the deformation induced stored energy. The restoration processes are complex and influenced by a number of factors, e.g. the material and its purity, the deformation strain, and the annealing temperature [1]. The restoration processes are generally classified into three categories: recovery, recrystallization, and grain growth. Recovery is conventionally defined as all annealing processes occurring in deformed materials that occur without the migration of high angle grain boundaries [2], whose misorientation angles are typically defined as greater than 15°. Recovery is usually considered as a precursor of recrystallization, which involves migration of high angle grain boundaries. Recovery includes a number of different processes, which may also occur during plastic deformation as dynamic recovery.

The extensive early work on recovery has been reviewed previously, e.g. by Beck [3], Bever [4] and Perryman [5]. The early recovery studies were primarily focused on the release of stored energy and the changes in the physical and mechanical properties (see e.g. [6]), and the main interpretations of the recovery processes are mutual annihilation of dislocations with opposite Burgers vectors and rearrangement of dislocations into low angle subgrain boundaries [4,7]. It was later found by transmission electron microscopy (TEM) that subgrain growth is also an important recovery process, either by subgrain coalescence [8,9] or by subgrain boundary migration [10]. Recent reviews of recovery can be found in [1,11] and in the proceedings of the 36th Risø International Symposium on Materials Science [12].

The extensive work on high strain deformation of metals during the last three decades has shown that the heavily deformed microstructure is significantly different from that after a low strain [13–15].

Although the transition of microstructure with strain is gradual, we consider a true strain of 2 or above as a high strain (and below 2 as a low strain) in the present work to facilitate discussion. The high strain microstructure contains a high fraction of finely-spaced high angle boundaries and its recovery behaviour is significantly different from that after a low strain [16]. A unique recovery mechanism, namely Y-junction motion, was identified in these heavily deformed metals [16].

Aluminium is particularly prone to recovery, and more than half of the stored energy is typically released through recovery during annealing [17]. In the following, recent progress in the understanding of recovery of aluminium deformed to both low and high strains is presented.

2. Recovery mechanisms for aluminium deformed to low strains

2.1. Dislocation annihilation and rearrangement

During recovery of a deformed metal, dislocation annihilation may occur by glide, climb and cross slip of dislocations, reducing the stored energy. The kinetics of this recovery process is typically found to be close to logarithmic [18–20], with the apparent activation energy increasing as recovery proceeds.

There is a large fraction of dislocations in the deformed microstructure that are difficult to annihilate by the above mechanisms. During recovery (including dynamic recovery), these excess dislocations tend to form local low energy configurations such as dislocation walls and low angle boundaries, accommodating the lattice curvature imposed by the strain. In these configurations, dislocations mutually compensate their distortion fields and thus reduce the total elastic energy [21].

A simple case of forming a low angle boundary is by polygonisation, which was first demonstrated by Cahn [22]. The pioneering work shows that when a single crystal is plastically bent and annealed, excess dislocations of one kind distributed on the slip plane will rearrange into low angle boundaries perpendicular to the slip planes upon annealing, thereby reducing the total stored energy.

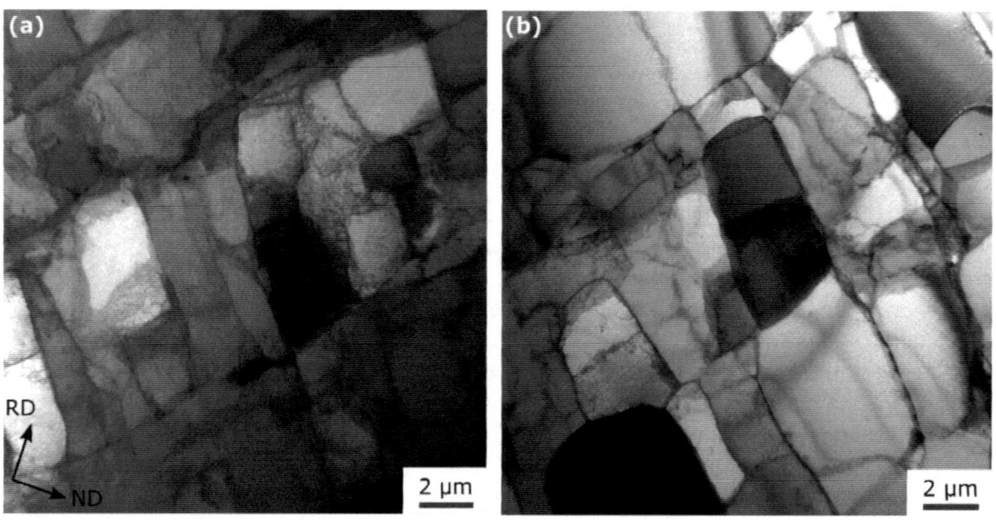

Figure 1. TEM micrographs showing microstructures in the longitudinal section of high purity aluminium (99.996 per cent purity) samples. (a) After 12 per cent cold rolling; (b) followed by annealing at 335 °C for 10 minutes.

In the general case of deformation of polycrystalline materials, dislocations of different Burgers vectors are generated and some of them remain in the structure. These dislocations react and arrange into three dimensional structures, e.g. a cell block structure with a high dislocation density in the cell walls and a low dislocation density in the cell interior [13,23] (see figure 1a). During annealing of

IOP Conf. Series: Materials Science and Engineering 1249 (2022) 012001 doi:10.1088/1757-899X/1249/1/012001

such a structure, recovery proceeds both in the cell walls and in the cell interior. Cell walls become sharper by removing dislocation pairs of opposite Burgers vectors through mutual annihilation and rearranging the excess dislocations into low angle boundaries. Such a structure (see figure 1b) with sharp low angle boundaries and almost dislocation free interior is typically called a subgrain structure.

2.2. *Subgrain growth*

The initial size of cells/subgrains depends on the materials and processing parameters, and is typically on the order of 1 μm. The remaining deformation stored energy is mainly stored in these cell/subgrain boundaries, and therefore there is a driving force for subgrain growth to further reduce the stored energy.

Migration of subgrain boundaries is considered as the dominant mechanism for subgrain growth [1,10,20,24]. The mobility of low angle subgrain boundaries is generally lower than that of high angle boundaries, and increases with increasing misorientation angle [1]. However, subgrain boundaries with very low misorientation angles (e.g. below 1°) may have a high mobility, especially under the influence of local stresses [25]. Moreover, those boundaries with very low misorientation angles may also disappear by continuously reducing their misorientation angles, leading to subgrain rotation and coalescence—an alternative mechanism of subgrain growth [26,27]. Such a mechanism may occur by extraction and emission of boundary dislocations to the subgrain interior followed by incorporation/annihilation at another subgrain boundary, or by absorption of boundary dislocations directly into an attached high angle grain boundary [26,28]. However, subgrain coalescence is generally considered to be less important than subgrain boundary migration.

The kinetics of subgrain growth was commonly analysed by analogy to grain growth, where a constant activation energy is assumed. Such an analysis neglects an important feature of recovery of a deformed microstructure, i.e. the increase of the activation energy during recovery. Consequently, many different growth exponents were reported in the literature.

2.3. *Characterization by synchrotron X-ray techniques*

Previous studies of recovery were typically either based on in-situ observations of thin foil samples by transmission electron microscopy (TEM) (e.g. [8,29]) or inferred from post-mortem observations on bulk samples. In recent years, significant progress in the development of synchrotron X-ray based experimental characterization techniques has provided new opportunities. The recovery in the bulk sample interior has been followed in-situ at synchrotron beamlines by analysing the changes in the diffraction spots, e.g. [30,31]. Moreover, the microstructural evolution may be directly characterized by two synchrotron based techniques, namely differential aperture X-ray microscopy (DAXM) [32] and dark field X-ray microscopy (DFXM) [33], both of which allow non-destructive 3D mapping of orientations with a sub-micrometre spatial resolution and an angular resolution of 0.01–0.02°.

Figure 2 shows results from a recent DAXM study of recovery in high purity aluminium (99.996 per cent purity) cold rolled to 12 per cent thickness reduction [34]. After annealing at 335 °C, the microstructure contained a large fraction of small subgrains (i.e. with equivalent spherical diameters less than 3 μm) as shown in figure 2a. Further annealing at 375 °C led to a significant drop of the volume fraction of small subgrains (figure 2b). The removal of the small subgrains during recovery is likely due to subgrain boundary migration, as the driving force from curvature is highest for shrinkage of small subgrains. There is also a slight increase in the size of big subgrains (figure 2c and 2d). Figure 2e and 2f show the biggest subgrain at the two recovery states, coloured according to the voxel-to-subgrain-average misorientation angle. This subgrain is composed of many small domains of slightly different orientations, and thus there is a quite large orientation spread inside this subgrain. After annealing at 375 °C, a few adjacent small domains joined this subgrain, increasing its size as well as its orientation spread. Such a subgrain growth pattern is consistent with subgrain coalescence, thereby providing new evidence of this recovery mechanism.

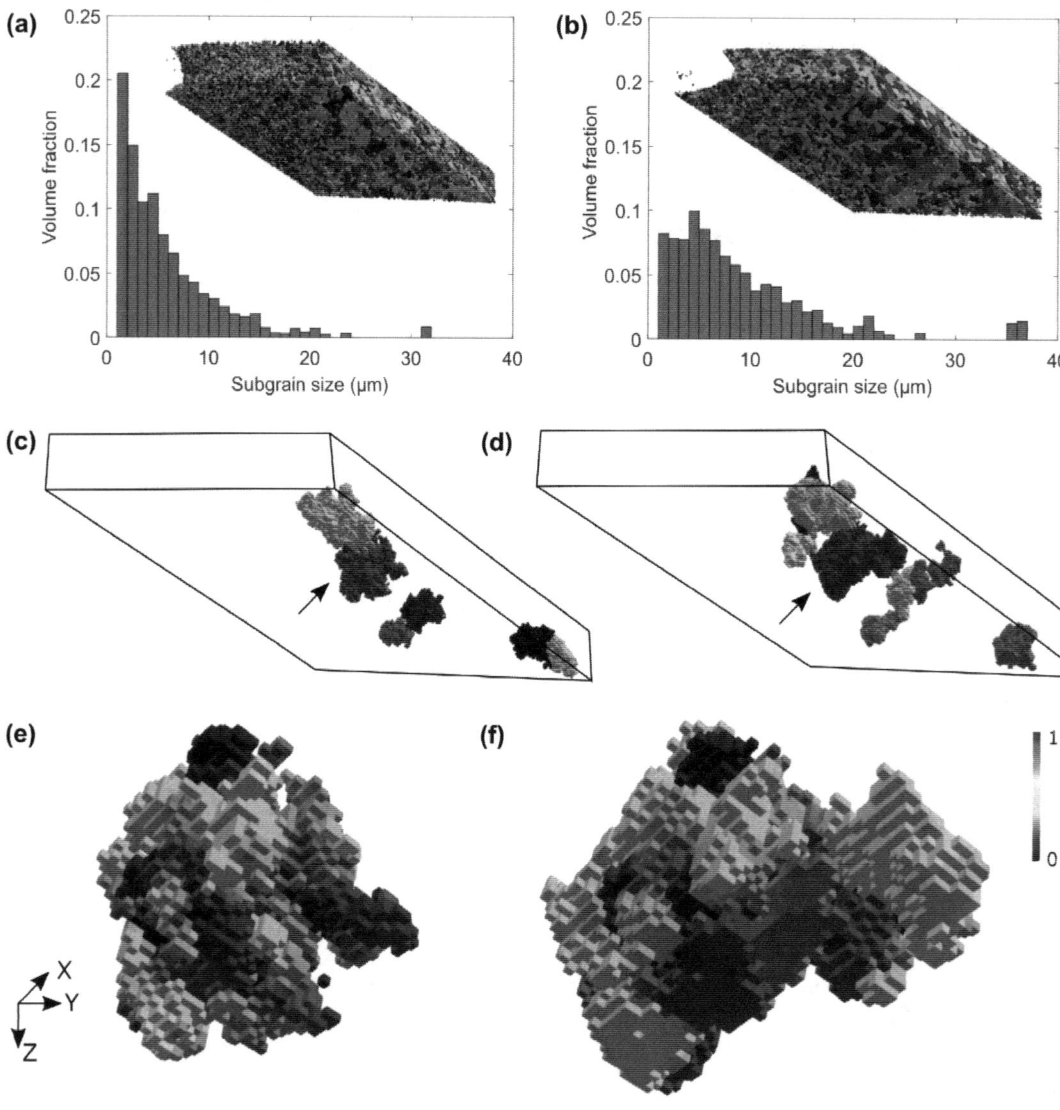

Figure 2. Subgrain evolution during recovery of 12 per cent cold rolled high purity aluminium (99.996 per cent purity). (a,b) Subgrain size (equivalent spherical diameter; 0.1° cut-off angle) distribution after annealing at 335 °C and 375 °C, respectively (randomly coloured subgrain maps are shown as insets); (c,d) The eight largest subgrains (randomly coloured) for the two recovery states; (e,f) The largest subgrain coloured based on the voxel-to-subgrain-average misorientation angle for the two recovery states (voxel size 1.5 μm) [34].

3. Recovery mechanisms for aluminium deformed to high strains

3.1. General observations

After deformation to a high strain, the deformation microstructure is significantly different from that in a lightly deformed metal. Taking cold rolling as an example, the microstructure after a high strain is characterized by finely-spaced lamellar boundaries, with interconnecting boundaries and loose dislocations in the region between extended lamellar boundaries [16]. Lamellar boundaries are typically of medium to high misorientation angles, while interconnecting boundaries are of low misorientation angles. An example of such a microstructure is shown in figure 3a for heavily cold rolled AA1050 aluminium [17]. Lamellar boundaries are not perfectly parallel or infinite. Instead, they form triple junctions, which have been named Y-junctions—each connecting three lamellar boundaries [16].

During recovery annealing of such microstructures, dislocation annihilation and microstructural coarsening occur [35]. As can be seen in the example shown in figure 3, there is an increase in the lamellar boundary spacing and a decrease in the dislocation density, but the lamellar morphology is largely maintained after the recovery annealing. Such a coarsening pattern is characteristic for heavily deformed microstructures of lamellar morphology, and such coarsening is particularly pronounced in aluminium [17].

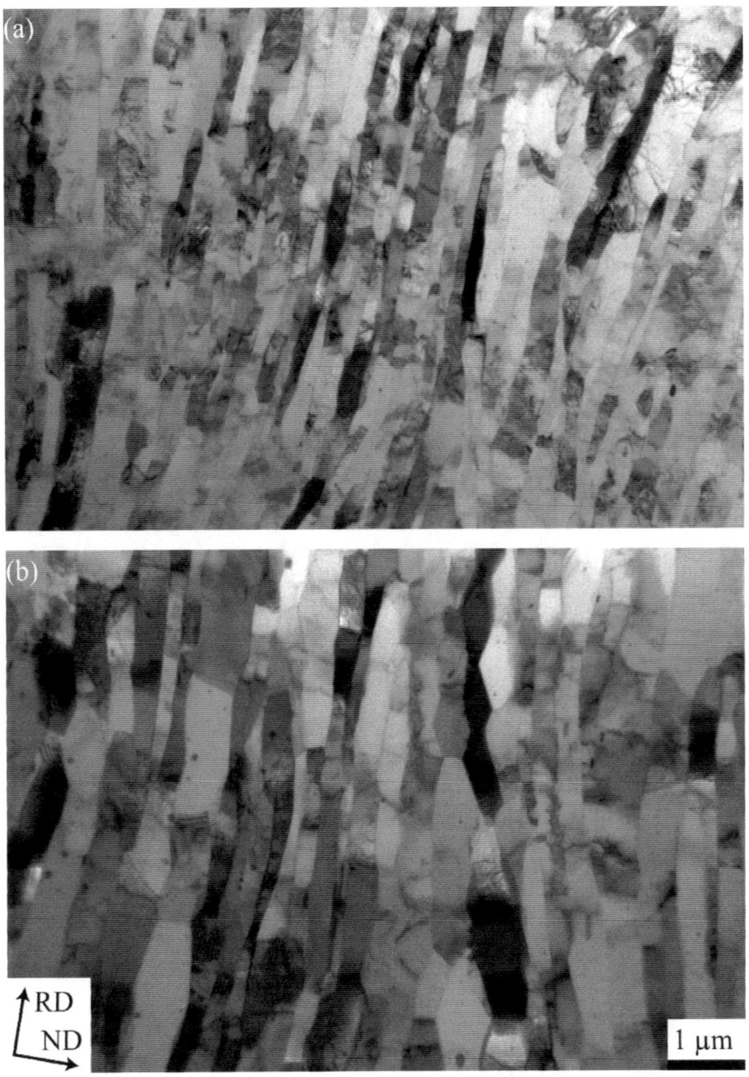

Figure 3. TEM micrographs showing microstructures in the longitudinal section of AA1050 aluminium samples. (a) Cold rolled to a thickness reduction of 99.6 per cent; (b) Followed by annealing at 180 °C for 1 h [17].

3.2. Uniform coarsening by Y-junction motion

Through direct in-situ and ex-situ TEM observations during recovery annealing of thin foil samples, as well as extensive studies by electron backscatter diffraction (EBSD) on bulk samples, an important recovery mechanism, namely Y-junction motion, has been identified in heavily deformed microstructures of lamellar morphology [16]. One example is shown in figure 4, where a Y-junction terminating the middle lamella of dark contrast migrated up during annealing, leading to a local increase of the lamellar boundary spacing but maintaining the lamellar morphology. This mechanism was found to be responsible for the uniform coarsening of deformed lamellar microstructures during recovery annealing [16,17]. During migration of the Y-junction shown in figure 4, an externally attached interconnecting boundary on the left-hand side interacted with the Y-junction, retarding its migration. Such a pinning effect was also found for attached lamellar boundaries, particles, and dislocations [29,36].

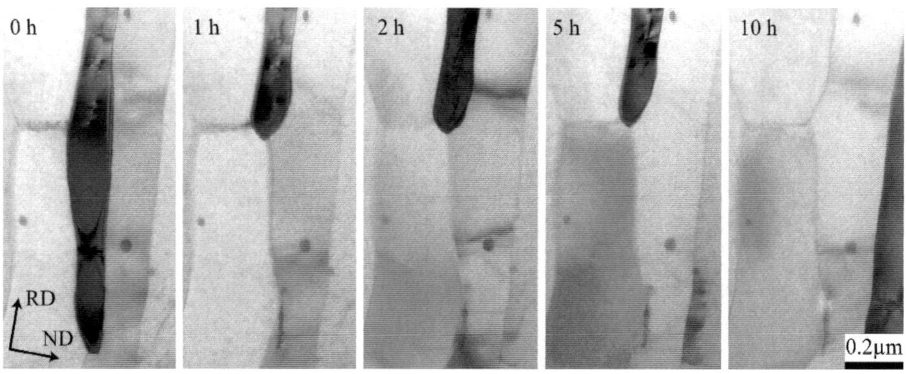

Figure 4. An example of Y-junction migration observed in the longitudinal section of AA1050 aluminium cold rolled to a thickness reduction of 99.6 per cent and annealed at 120 °C for different time intervals, as marked [16].

The geometry of a Y-junction is important in determining its stability, in particular the spacing D and the dihedral angle 2θ associated with the migrating lamella. Thin lamellae bounded by high angle boundaries have a high tendency to be removed by Y-junction migration due to a high driving force from the grain boundary surface tension, possibly also assisted by a high mobility. Consequently, thin lamellae of other texture components imbedded in rolling texture components are preferentially removed, leading to sharpening of the rolling texture when a dominant texture variant exists [37]. The preferential removal of thin lamellae also results in a uniform coarsening and a more symmetric distribution of the lamellar boundary spacing, i.e. the skewness of distribution decreases slightly [16]. However, Y-junctions may be pinned by attached lamellar and interconnecting boundaries, dislocations and particles, and their mobility may vary depending on their core structures and the mobility of the composing lamellar boundaries. Consequently, a large variation in the stability of Y-junctions were observed during recovery [16,29]. Meanwhile, interconnecting boundaries are found to be very stable in terms of migration during recovery annealing although their removal is inevitable when the lamellae that they reside in are removed by Y-junction motion.

4. Kinetics of recovery coarsening

The driving force for recovery coarsening comes from the stored energy in the deformed materials, mainly in the form of deformation induced boundaries. Following a classic recovery rate equation [19] and taking the stored energy in boundaries as the only source of driving force, the coarsening kinetics can be expressed as [38]

$$Ei(-\frac{k_2}{DT}) - Ei(-\frac{k_2}{D_0T}) = -k_1t, \tag{1}$$

$$k_1 = K_0 \exp(-\frac{Q_0}{RT}), \tag{2}$$

where Ei() is the exponential integral of the quantity inside the bracket, D is the subgrain size (or boundary spacing), D_0 is the initial subgrain size (or boundary spacing), t is the annealing time, T is the annealing temperature, R is the gas constant, and Q_0 is the apparent activation energy at the end of recovery. The three fitting parameters K_0, Q_0 and k_2 are associated with the operative recovery mechanisms. The apparent activation energy at any stage of coarsening (i.e. at boundary spacing D) can be expressed as

$$Q_{app} = Q_0 - \frac{k_2R}{D}. \tag{3}$$

Equation (1) describes the continuous coarsening of a deformation structure during isothermal annealing. The values of (k_2, k_1) can be fitted at each annealing temperature based on coarsening data, and Q_0 can then be determined by combining data of different annealing temperatures. The above coarsening model is considered to be universal for uniform coarsening of deformation microstructures [38,39], and it can be used to analyse both subgrain growth in lightly deformed aluminium and Y-junction migration in heavily deformed aluminium (figure 5). In both cases, the activation energy was found to increase approximately logarithmically with the annealing time, and at a given annealing time the activation energy increases with increasing annealing temperature. The estimated activation energies (Q_0) are consistent with the diffusion of Si and Fe in the Al-0.05Si alloy and AA1050 aluminium, respectively, suggesting that solute drag is an important rate controlling mechanism. Lower activation energies at early stages of recovery coarsening may be related to the deformed microstructure, which provides a strain field and short-circuit diffusion paths.

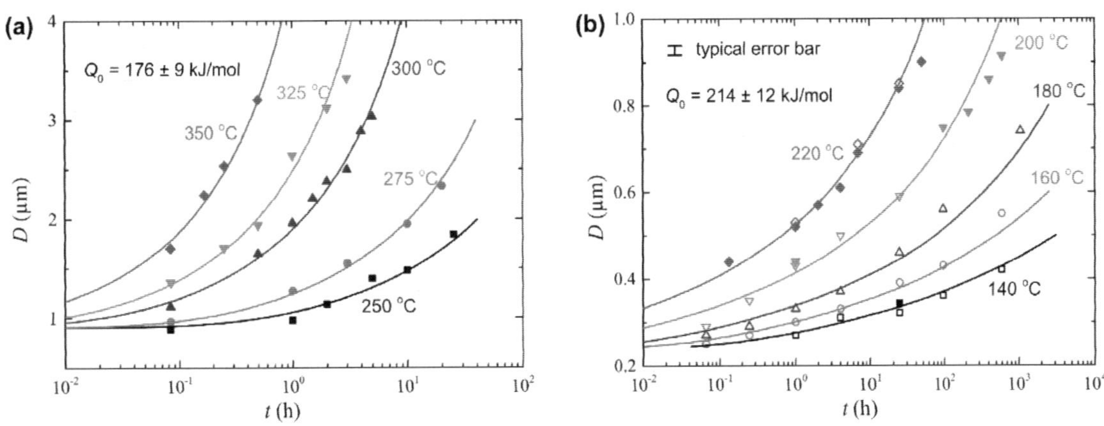

Figure 5. Recovery coarsening of deformation microstructures fitted by the unified coarsening model [38]. (a) Subgrain growth in Al-0.05Si single crystal channel die compressed by 70 per cent at room temperature [40]; (b) Y-junction motion in AA1050 aluminium cold rolled to a thickness reduction of 99.6 per cent [17].

5. Discussion

As the deformation microstructure changes significantly with increasing strain, the subsequent restoration processes also vary significantly depending on the deformation strain. In heavily deformed metals, uniform coarsening by Y-junction migration is followed by nucleation of recrystallization and growth of nuclei, and therefore Y-junction motion is considered as a recovery mechanism [16]—a high strain counterpart of subgrain growth. This recovery mechanism involves migration of deformation induced high angle boundaries and is therefore beyond the conventional definition of recovery, which is primarily based on lightly deformed microstructures. To be applicable to both low and high strain cases, the definition of recovery has to be extended to include all restoration processes in the non-recrystallized matrix. Our suggestion for such a definition is: *all restoration processes occurring* ***continuously*** *in deformed microstructures.* It follows that (i) recovery includes all types of uniform coarsening of deformed microstructures and (ii) recovery ends locally when the local deformed microstructure is replaced by recrystallized grains, which nucleate and grow in a **discontinuous** manner.

Both subgrain growth in metals deformed to low strains and Y-junction migration in metals deformed to high strains follow the same coarsening model, i.e. equation (1). However, the rate of coarsening is significantly different, and a higher recovery rate (more rapid decrease of hardness) is observed during isothermal annealing of aluminium deformed to a higher strain [41]. The high recovery rate at high strains results from the heavily deformed microstructure. As the strain increases, the boundary spacing decreases and the fraction of high angle boundaries with high energy and high mobility increases. As a result, migration of Y-junctions, which connect finely-spaced high angle lamellar boundaries, occurs much easier than growth of subgrains bounded by low angle boundaries. Thin lamellae and small subgrains are preferentially removed during recovery coarsening [16,34]. It was also observed in deformed commercial purity aluminium that bands containing boundaries of higher misorientation angles coarsen more rapidly than bands containing boundaries of lower misorientation angles [42]. The high angle boundaries can act as sinks of dislocations, and thus they may also enhance the rate of dislocation annihilation during recovery.

6. Summary

The microstructural evolution during recovery is reviewed in the present work, with a focus on recent progress in the understanding of recovery in aluminium. Subgrain growth by boundary migration and coalescence is an important recovery mechanism in aluminium at low strains, while uniform coarsening by Y-junction motion is the dominant recovery mechanism in aluminium at high strains. Both mechanisms follow the same coarsening model, and the coarsening rate increases with increasing strain. It is suggested that the conventional definition of recovery has to be extended to include new annealing phenomena in heavily deformed metals. A tentative suggestion for the definition of recovery is *all restoration processes occurring* ***continuously*** *in deformed microstructures.*

Acknowledgement

This project has received funding from the European Research Council (ERC) under the European Union's Horizon 2020 research and innovation programme (grant agreement No 788567). The author is grateful to Prof. D. Juul Jensen for valuable comments.

References

[1] Humphreys F J, Rollett A D and Rohrer G S 2017 *Recrystallization and related annealing phenomena.* (Amsterdam: Elsevier)
[2] Doherty R D, Hughes D A, Humphreys F J, Jonas J J, Juul Jensen D, Kassner M E, King W E, McNelley T R, McQueen H J and Rollett A D 1997 Current issues in recrystallization: a review *Mater. Sci. Eng. A* **238** 219–74
[3] Beck P A 1954 Annealing of Cold Worked Metals *Adv. Phys.* **3** 245–324
[4] Bever M B 1957 On the thermodynamics and kinetics of recovery *Creep and Recovery* ed R Maddin (Cleveland, OH: American Society for Metals) pp 14–51
[5] Perryman E C W 1957 Recovery of Mechanical Properties *Creep and Recovery* ed R Maddin (Cleveland, OH: American Society for Metals) pp 111–45

[6] Clareborough L, Hargreaves M and West G 1955 The Release of Energy during Annealing of Deformed Metals *Proc. R. Soc. London Ser. A-Mathematical Phys. Sci.* **232** 252–70
[7] Lytton J L, Westmaco K and Potter L C 1965 Relation between Flow Stress and Dislocation Structure during Recovery of High-Purity Aluminum *Trans. Metall. Soc. AIME* **233** 1757–65
[8] Hu H 1962 Direct Observations on Annealing of a Si-Fe Crystal in Electron Microscope *Trans. Metall. Soc. AIME* **224** 75–83
[9] Li J C M 1962 Possibility of Subgrain Rotation during Recrystallization *J. Appl. Phys.* **33** 2958–65
[10] Smith C J E and Dillamore I L 1970 Subgrain growth in high-purity iron *Met. Sci.* **4** 161–7
[11] Raabe D 2014 Recovery and Recrystallization: Phenomena, Physics, Models, Simulation *Physical Metallurgy* ed D E Laughlin and K Hono (Amsterdam: Elsevier) pp 2291–397
[12] Fæster S, Hansen N, Hong C, Huang X, Juul Jensen D, Mishin O V, Sun J, Yu T and Zhang Y B 2015 *Proceedings of the 36th Risø International Symposium on Materials Science: Effects of deformation induced structural variations on annealing mechanisms* vol 89 (Roskilde, Denmark: Technical University of Denmark)
[13] Bay B, Hansen N, Hughes D A and Kuhlmann-Wilsdorf D 1992 Overview no. 96 evolution of f.c.c. deformation structures in polyslip *Acta. Metall. Mater.* **40** 205–19
[14] Hansen N and Barlow C 2014 Plastic deformation of metals and alloys *Physical Metallurgy* ed D E Laughlin and K Hono (Amsterdam: Elsevier) pp 1681–764
[15] Liu Q, Huang X, Lloyd D J and Hansen N 2002 Microstructure and strength of commercial purity aluminium (AA 1200) cold-rolled to large strains *Acta Mater.* **50** 3789–802
[16] Yu T, Hansen N and Huang X 2011 Recovery by triple junction motion in aluminium deformed to ultrahigh strains *Proc. R. Soc. A* **467** 3039–65
[17] Yu T, Hansen N and Huang X 2013 Linking recovery and recrystallization through triple junction motion in aluminum cold rolled to a large strain *Acta Mater.* **61** 6577–86
[18] Kuhlmann D 1948 Zur theorie der nachwirkungserscheinungen *Z. Fur Phys.* **124** 468–81
[19] Borelius G, Berglund S and Sjoberg S 1952 Measurements on the evolution of heat during the recovery of cold-worked metals *Ark. Fys.* **6** 143–9
[20] Nes E 1995 Recovery revisited *Acta Metall. Mater.* **43** 2189–207
[21] Kuhlmann-Wilsdorf D 1995 Technological high strain deformation of 'wavy glide'metals and LEDS *Phys. status solidi* **149** 225–41
[22] Cahn R W 1949 Recrystallization of single crystals after plastic bending *J. Inst. Met.* **76** 121–43
[23] Jakobsen B, Poulsen H F, Lienert U, Almer J, Shastri S D, Sørensen H O, Gundlach C and Pantleon W 2006 Formation and subdivision of deformation structures during plastic deformation *Science* **312** 889–92
[24] Sandstrom R 1977 Subgrain Growth Occurring by Boundary Migration *Acta Metall.* **25** 905–11
[25] Bainbridge D W, Choh H L and H. Edwards E 1954 Recent observations on the motion of small angle dislocation boundaries *Acta Metall.* **2** 322–33
[26] Jones A R, Ralph B and Hansen N 1979 Subgrain Coalescence and the Nucleation of Recrystallization at Grain Boundaries in Aluminium *Proc. R. Soc. London Ser. A-Mathematical Phys. Eng. Sci.* **368** 345–57
[27] Doherty R D and Szpunar J A 1984 Kinetics of sub-grain coalescence—A reconsideration of the theory *Acta Metall.* **32** 1789–98
[28] Sandstrom R 1977 Recovery of Dislocations in Subgrains and Subgrain Coalescence *Acta Metall.* **25** 897–904
[29] Yu T, Hughes D A, Hansen N and Huang X 2015 In situ observation of triple junction motion during recovery of heavily deformed aluminum *Acta Mater.* **86** 269–78
[30] Gruber B, Grabner F, Falkinger G, Schökel A, Spieckermann F, Uggowitzer P J and Pogatscher S 2020 Room temperature recovery of cryogenically deformed aluminium alloys *Mater. Des.* **193** 108819

[31] Ahl S R, Simons H, Detlefs C, Juul Jensen D and Poulsen H F 2020 Subgrain dynamics during recovery of partly recrystallized aluminum *Acta Mater.* **185** 142–8
[32] Larson B C, Yang W, Ice G E, Budai J D and Tischler J Z 2002 Three-dimensional X-ray structural microscopy with submicrometre resolution *Nature* **415** 887–90
[33] Simons H, King A, Ludwig W, Detlefs C, Pantleon W, Schmidt S, Stöhr F, Snigireva I, Snigirev A and Poulsen H F 2015 Dark-field X-ray microscopy for multiscale structural characterization *Nat. Commun.* **6** 6098
[34] Yu T, Hong C S, Zhang Y B, Lindkvist A, Liu W J, Tischler J and Juul Jensen D 2022 Recovery of deformation microstructure in the bulk interior revealed by synchrotron X-ray micro-diffraction, **submitted**
[35] Yu T, Hansen N and Huang X 2012 Recovery mechanisms in nanostructured aluminium *Philos. Mag.* **92** 4056–74
[36] Shuai L, Huang T, Yu T, Wu G, Hansen N and Huang X 2021 Segregation and precipitation stabilizing an ultrafine lamellar-structured Al-0.3%Cu alloy *Acta Mater.* **206** 116595
[37] Mishin O V, Godfrey A, Juul Jensen D and Hansen N 2013 Recovery and recrystallization in commercial purity aluminum cold rolled to an ultrahigh strain *Acta Mater.* **61** 5354–64
[38] Yu T and Hansen N 2016 Coarsening kinetics of fine-scale microstructures in deformed materials *Acta Mater.* **120** 40–5
[39] Yu T and Hughes D A 2019 Strong pinning of triple junction migration for robust high strain nanostructures *Philos. Mag.* **99** 869–86
[40] Huang Y and Humphreys F J 2000 Subgrain growth and low angle boundary mobility in aluminium crystals of orientation {110}<001> *Acta Mater.* **48** 2017–30
[41] Yu T and Hansen N 2016 Recovery kinetics in commercial purity aluminum deformed to ultrahigh strain: model and experiment *Metall. Mater. Trans. A* **47** 4189–96
[42] Xing Q, Huang X and Hansen N 2006 Recovery of heavily cold-rolled aluminum: Effect of local texture *Metall. Mater. Trans. a-Physical Metall. Mater. Sci.* **37A** 1311–22

IOP Conf. Series: Materials Science and Engineering 1249 (2022) 012002 doi:10.1088/1757-899X/1249/1/012002

Microstructure and chemical characterization of Linear Friction Welded aerospace materials with focus on dissimilar joints

X Boyat, D Ballat-Durand, J Marteau, S Bouvier, M Risbet and J Favergeon

Université de technologie de Compiègne, CNRS, Roberval (Mechanics energy and electricity), Centre de recherche Royallieu - CS 60 319 - 60 203 Compiègne Cedex, France

E-mail : salima.bouvier@utc.fr

Abstract. This study is focused on a through characterization of the microstructure and chemical composition of dissimilar linear friction welded joints. First, Ti-6Al-2Sn-4Cr-2Mo (Ti6242) and Ti-5Al-2Sn-2Zr-4Mo-4Cr (Ti17) joints are described to identify the mechanisms behind the successful cohesion of the material surfaces. It was found that continuous dynamic recrystallization and cohesion took place through grain-boundary migration in a shared single phase β-domain obtained through high temperature and high strain conditions. Then, the possibility of assembling Ti6242 and Inconel 718 is examined. Chemical results combined with microstructure and crystal orientation analyses showed that the intermixing zone formed between the materials was particularly heterogeneous, with the development of an intermediate bimodal microstructure, mainly composed of Ti-β phase and Ti_2Ni intermetallic phase.

1. Introduction

Linear Friction Welding (LFW) is a solid-state assembling process. The weld is obtained by pressing a work-piece (forging part) into another one that describes a linear reciprocating motion (oscillating part). The control of the friction process is ensured through three independent parameters: the contact pressure, the oscillating amplitude and the frequency. With adapted process parameters, the local heat generation by friction at the interface and the consequent thermal softening cause the materials to plastically flow and lead to the coalescence of the rubbing parts into a thin metallurgical bond.

The work presented in this paper is part of a multipartite ANR project named OPTIMUM aiming at identifying the causal relationship of varied LFW conditions (i.e. process parameters and materials) on the microstructural changes involved in the weld and their resulting mechanical durability. Butt-weld test campaigns were carried out implying the joining of various aeronautical material combinations. First, two Ti-alloy grades, the near-α Ti6242 and the β-metastable Ti17, were linear friction welded in mono configurations. Second, the precipitation hardenable nickel superalloy Inconel 718 was linear friction welded through two initial metallurgical conditions: a soft solution annealed state and a hard precipitation strengthened state. Third, dissimilar welding of Ti6242-Ti17 and Ti6242-Inconel 718 were realized in view of understanding the bonding mechanisms at the interface. The ability to assemble dissimilar alloys is one of the most promising applications of LFW technology. Indeed, the latter is of high technological interest for replacing mechanical assemblies and to produce innovative multi-material and multi-functional parts. Literature review shows that assembling Ti-alloys can be done using dissimilar microstructures or different grades. Detailed descriptions of the central weld zone are often given but without identifying the phenomena at the origin of the bonds between materials. As an example, Guo *et al.* [1] gave a detailed description of the microstructure of a LFW joint between Ti-6Al-4V and Ti-6Al-2Sn-4Zr-6Mo and thoroughly characterized the texture of the central weld zone. However, they did not clarify the interfacial mechanisms responsible for solid state bonding. The underlying mechanisms leading to the bond formation in friction assembling processes and, more broadly, on mutually soluble solid-state assembling processes are still debated in the literature [2], [3]. In the present paper, based on extensive investigation of similar linear friction welded Ti-based and Ni-based components [4-8], dissimilar joints are studied in view of understanding the evolution of the interfacial properties and an attempt to clear up the interfacial mechanisms at a microstructural scale responsible for solid-state joining. Two configurations are studied: (a) two dissimilar, but mutually soluble, Ti-alloys: the near-α Ti-6Al-2Sn-4Zr-2Mo (Ti6242) and the β-metastable Ti-5Al-2Sn-2Zr-4Mo-4Cr (Ti17); (b) the near-α Ti-6Al-2Sn-4Zr-2Mo (Ti6242) and the Nickel superalloy grade Inconel 718. The second configuration hypothetically poses serious issues in terms of: mechanical compatibility considering a mutual bulk material plastic collapse during the friction welding procedure, sliding friction behavior and foremost, considering the complex chemistry of each of these alloys and their probable chemical reactivity at the weld interface. The first configuration was thoroughly discussed in a recent paper [9] and is recalled here in view of comparison with the second discussed configuration.

2. Experiment, materials and methods

A butt-welding test campaign is carried out using the commercial standard titanium alloy grades: Ti-6Al-2Sn-4Cr-2Mo (Ti6242) and Ti-5Al-2Sn-2Zr-4Mo-4Cr (Ti17), as well as Nickel superalloy grade Inconel 718 (IN718) in a solution annealed metallurgical state. Pairs of solid blocks of 15 x 80 x 70 mm^3 were cut by electrical-discharge machining from forged billets. Prior to the welding procedure, the raw assembling surfaces were polished with P120 SiC grinding paper. The extracted specimens were then linear friction welded in an open-air environment on the 15 x 80 mm^2 surface with the 80 mm edge parallel to the oscillation direction $\langle x \rangle$. It is also noted that the 80 mm edge is parallel with the billet axis. The Ti6242 block was used as the oscillating part whereas the Ti17 block (for the first dissimilar configuration) and the Inconel 718 block (for the second dissimilar configuration) were used as the forging part. The process parameters selected in this study were: a forging pressure of 90 MPa, a frequency of 50 Hz and a 2 mm amplitude in case of Ti6242-Ti17 and a forging pressure of 270 MPa, a frequency of 50 Hz and a 1 mm amplitude in case of Ti6242-Inconel 718. The microstructural characterization of the transition zone of the dissimilar joint is carried out on representative samples collected from cross-sectional cuts in the welded work-pieces, orthogonal to the friction direction of the welding process. The sample reference used to formulate the EBSD orientation data is defined with the "rolling direction" RD parallel with the friction direction $\langle x \rangle$. The "normal direction" ND is parallel with the forge direction $\langle z \rangle$ and the "transverse direction" TD is orthogonal to the (RD, ND) plane and corresponds to the $\langle y \rangle$ direction. Microstructural observations were performed using a tungsten Field Emission Gun Zeiss Σigma Scanning Electron Microscope (FEG-SEM). The Electron Back-Scattered Diffraction (EBSD) mapping was obtained using a Nordiff OptiPlex 7040 detector for Kikuchi pattern image capture. EBSD post-processing was performed using the Matlab open source toolbox MTEX v5.1.1. The local distribution of substitutional elements across the interfacial zone is investigated through energy-dispersive X-ray spectroscopy in TEM.

3. Results

3.1. Dissimilar Ti6242-Ti17 joint

3.1.1. Microstructure and texture evolution. The Ti6242 used in this study is characterized by an initial microstructure consisting mainly of coarse α nodules presenting an ellipsoidal shape with interspersed residual intergranular β matrix. On another note, the Ti17 exhibited an as-received

Widmanstätten $\alpha+\beta$ microstructure formed by the intragranular precipitation of fine entangled hardening α laths within large prior-β grains. Their respective β-transus temperatures specified by the manufacturer are approximately: 995°C for the Ti6242 and 890°C for the Ti17. Both microstructures are presented in Figure 1.

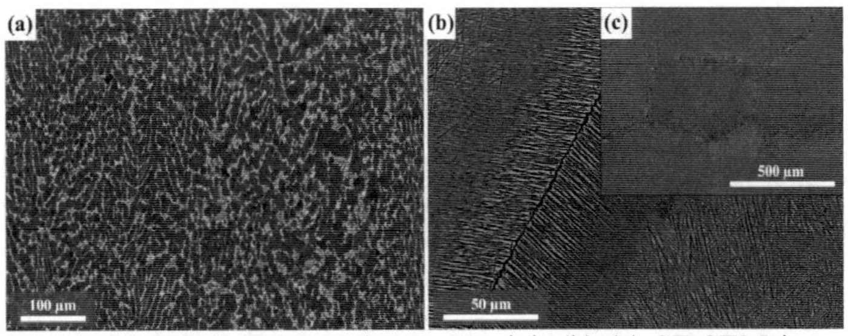

Figure 1. (a) SE-SEM image of the Ti6242 base material, (b)-(c) SE-SEM image of the Ti17 base material.

After welding, the Ti6242 and Ti17 alloys display strong differences in formed microstructures under rapid cooling from the Solution-Treated (ST) configurations (Figure 2.a) : the Ti6242 evolves towards an acicular HCP (Hexagonal Close Packed) α' martensitic laths microstructure (Figure 2.b) whereas the Ti17 evolves towards BCC (Body-Centered Cubic) β-metastable equiaxed grains (Figure 2.c). Such differences help to identify the welding interfacial zone as it leads to a sharp α'/ β_m phase contrast.

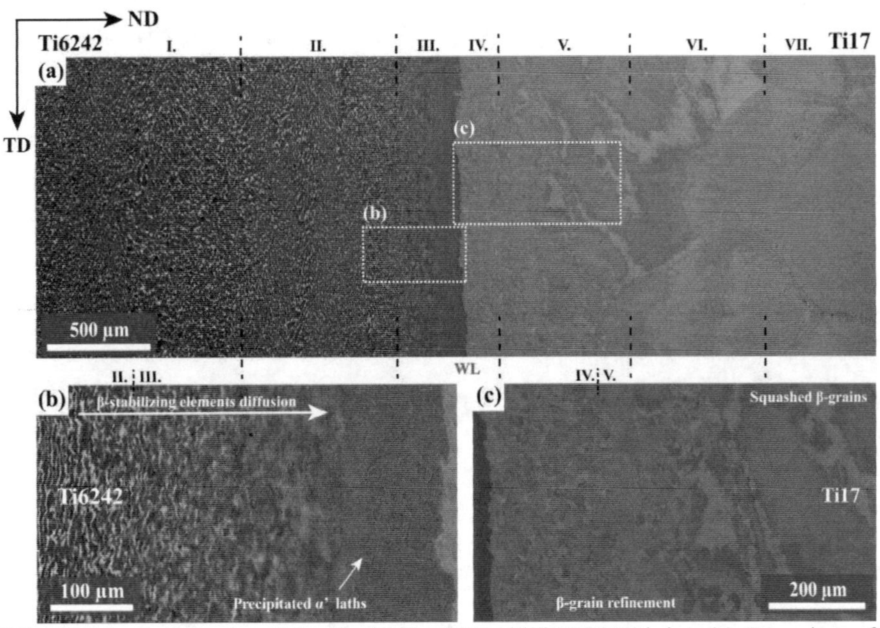

Figure 2. BSE-SEM micrographs of the dissimilar Ti6242/Ti17 LFW joint. (a) Overview of the weld interface; (b) zoom at the Ti6242 TMAZ; (c) zoom at the Ti17 TMAZ.

This is clearly shown in Figure 3, where SEM imaging carried out in the interfacial zone reveals the side-by-side coexistence of sharp needle-like α' structures with a β microstructure. Moreover, the microstructure analyses show that both phases appeared to be intermixed within apparently common β subgrains (see dashed area).

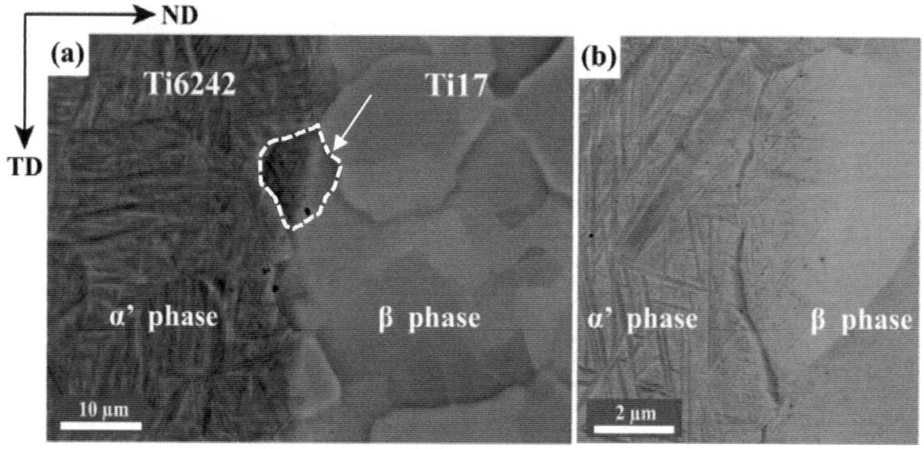

Figure 3. (a), (b) BSE-SEM image of the interfacial zone

In complement to imaging, orientation data measurements were carried out by EBSD technique to further investigate the microstructural changes across the joint. The resulting HCP and BCC orientation maps superimposed with the calculated grain boundaries are shown in Figure 4. Yet, since the crystal structures of α phase and α' martensitic phase are very similar, the EBSD indexing software was unable to distinguish the difference. Consequently, the α' phase would be indexed as α. A remarkable texture development is present in both Ti17 and Ti6242 indicating a significant crystal reorientation due to the deformation conditions during the process. This reorientation appeared to be rather limited to the WL in Ti17 but extended to 700 μm from the weld center line in Ti6242.

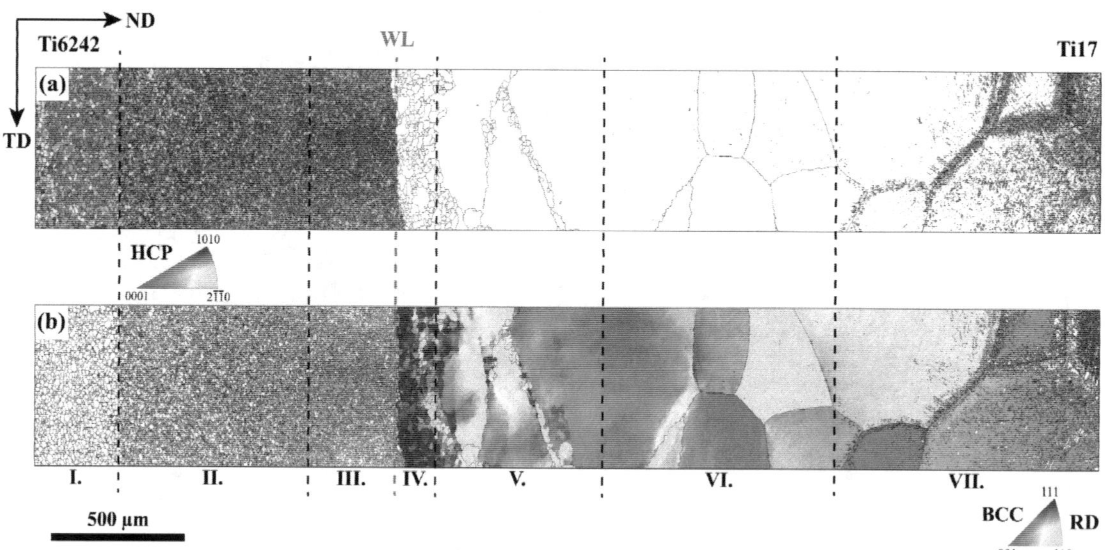

Figure 4. Inverse pole figures colored orientation maps obtained from EBSD analysis (0.5 μm step size) of the Ti6242/Ti17 LFW joint showing (a) the HCP and (b) BCC indexed data superimposed with the grain boundaries (5° misorientation angle threshold); the colors represent the crystallographic direction of the indexed lattice along RD direction.

The overviews of the microstructure in the joint (Figures 2 and 4) showed that both Ti17 and Ti6242 alloys present markers of an extensive α→β phase transformation that occurred during the process. In the Ti6242, these markers are present in Zones II and III. The former α nodules of the near WL area (Zone III) first undergo a β phase transformation during the hot deformation process. A significant microstructure refinement can be noted in this area that exhibits a remarkable fragmentation of the β matrix phase accompanied, in the Ti6242, by the intragranular precipitation of discernible acicular α' martensitic laths (as indicated in Figure 3). Neighboring the fully-transformed WL, an apparent Thermo-Mechanically Affected Zone (TMAZ) was revealed in the form of gradually plasticized grains and dissolved morphologies (i.e. smeared former β matrix in the Ti6242 and α precipitates dissolution in the Ti17). Seemingly unaffected microstructures were recovered at approximately 1 mm and 2.5 mm in the Ti6242 and Ti17 respectively. It is worth noting that for each material and in case of similar weld, similar microstructure morphologies, phase presence, and texture were reported in [4, 6-8], far from the WL area using analogous LFW conditions.

3.1.2. Chemical investigations at the transition zone. A detailed chemical analysis is conducted along with a cross-line profile shown in Figure 5.a. The resulting interdiffusion profiles across the interface present an interdiffusion band of approximately 4 μm in thickness and is visible in the transition zone. A basic one-dimension model is applied on the experimental data (Figure 5) with the aims of roughly estimating the order of magnitude of the interdiffusion phenomenon during the high-temperature processing time. One should also note that the centers of the error-function profiles do not actually correspond with the visible limit of the interface: a slight shift of the fitted curves towards the martensitic side is observed. The observed redistribution of the Ti6242 alloying elements is most likely caused by the progressive diffusion of the β stabilizing elements from the initially interspersed intergranular βrich matrix to the adjacent β transformed nodules. The gradual aspect of the diffusion phenomena is directly related to the local temperature distribution which lowers as moving away from the WL. It is worth noting that the intermixed α' and β microstructures observed in the WL zone are found within transitional subgrains belonging to the area of chemical composition gradient.

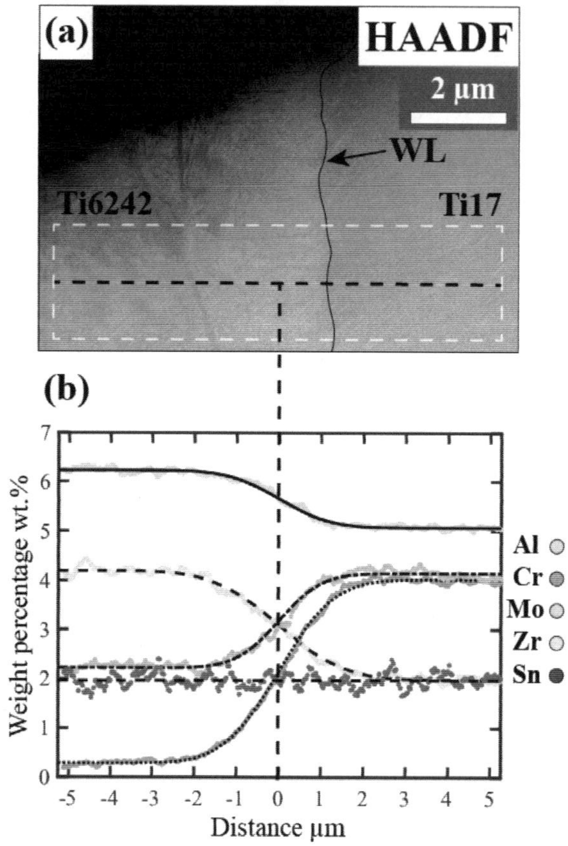

Figure 5. (a) HAADF image of the interfacial zone for EDX profile analysis; (b) line profile of element composition distribution across the Ti6242/Ti17 interface in wt.%.

3.1.3. Microstructural investigations by crystal orientation analysis at the transition zone. Additional crystal orientation investigations are carried out in the vicinity of the interfacial zone in order, first; to clarify a potential relationship between the in-contact β-metastable phase and the adjacent precipitated α' phase in the transitional zone and second; to reconstruct the microstructure history from the final state observations considering the thermo-mechanical processing conditions relating to the friction assembling process. EBSD analyses are first carried out in the vicinity of the interfacial zone. The resulting HCP and BCC inverse pole figure colored orientation maps are shown in Figure 6.a and 6.b respectively; superimposed grain boundaries are also provided. The change in crystallographic structure that operates under external constraints from the high-temperature β-parent lattice to the α' (or α) product crystal structure is known to theoretically follow coordinated rules minimizing the strain energy of phase transformation. Consequently, a self-developed subroutine is developed using the Matlab toolbox MTEX v5.1.1 and based on the work of Germain *et al.* [10, 11] which allows to reconstruct the parent β orientation prior to the $\beta_{lean} \rightarrow \alpha'$ among clusters of entangled HCP laths relying on Burgers Orientation Relationship (BOR). The β_{lean} phase is a homogenized phase in terms of chemical composition, with a lower content of β-stabilizers (e.g. Zr) and still has α-stabilizers (e.g. Al). The reconstructed β orientations are shown in Figure 6.c. The high-temperature β-reconstructed map confirms unequivocally the occurrence of a mutual β-phase mixture of the materials in the interfacial zone (existence of a β microstructural continuity between both alloys during the LFW). These peculiar interfacial grains revealed that the cohesive mechanisms between the rubbing surfaces occurred in the single-phase β domain under severe strain and high-temperature conditions.

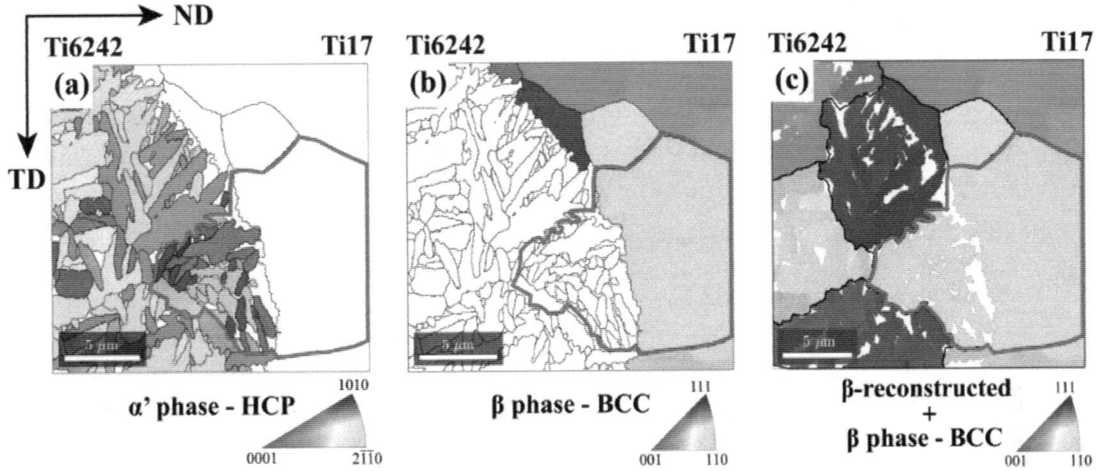

Figure 6. Inverse pole figure colored orientation maps from EBSD analysis at the Ti6242/Ti17 transition zone for (a) the HCP, (b) BCC indexed data and (c) the BCC reconstructed orientation.

3.1.4. Grain fragmentation. A classification of the grain boundary misorientations of the near-interfacial zone from EBSD investigations is combined with a Kernel Average Misorientation (KAM) cartography in Figure 7. A fragmented microstructure is visible in both materials characterized by a remarkably high fraction of low-angle grain boundaries. Such a microstructure would indicate that extensive grain fragmentation occurred accompanied by substantial crystallite rotations. These apparent crystallites constitute substructures with low internal misorientations formed within deformed large β grains.

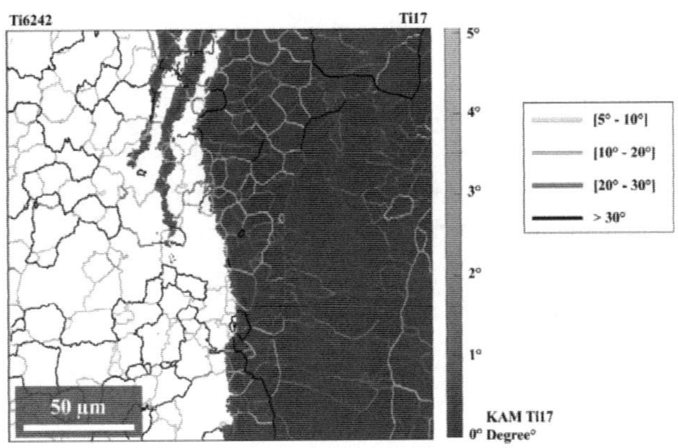

Figure 7. A classification of the grain boundary misorientations of the near-interfacial zone from EBSD investigations combined with a Kernel Average Misorientation (KAM) cartography.

The distributions of the misorientation angles in the reconstructed BCC Ti6242 microstructures (Figure 8.a) and the BCC Ti17 (Figure 8.b) corroborate the occurrence of significant grain fragmentation and subsequent texture development. Indeed, in both microstructures, the spatially correlated and uncorrelated misorientation distributions revealed the presence of a noteworthy statistical peak at the low angle boundaries; the former being significantly higher than the latter. Such a feature would indicate that a reorientation of the grains towards a specific direction occurred during LFW accompanied by gradual rotation of the crystallites from low-angle to high-angle grain boundaries. The mechanism forming such textured recrystallized-like microstructures is often referred to as a Continuous Dynamic Recrystallization (CDRX) process [12] and appeared to be more pronounced in the Ti6242 than in the Ti17.

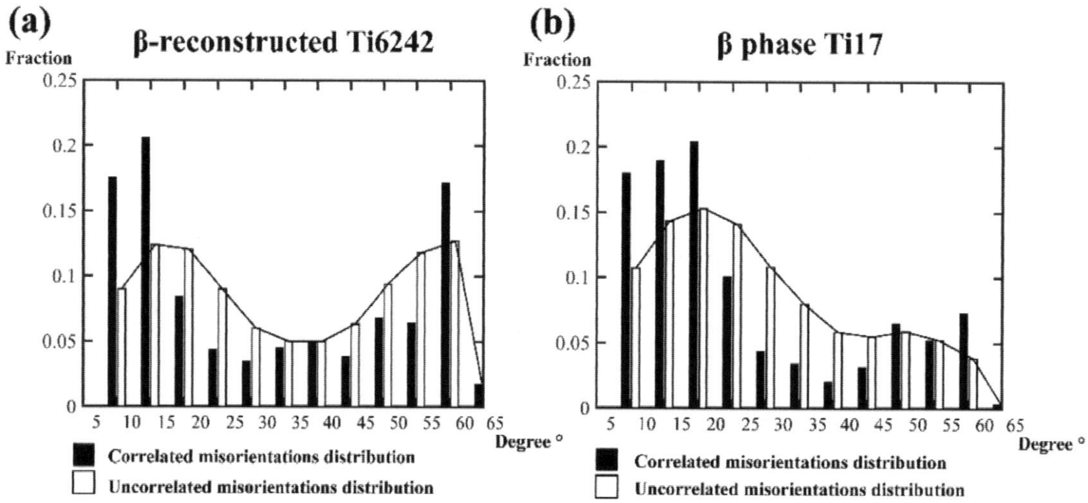

Figure 8. Spatially correlated and spatially uncorrelated misorientation angles distributions for (a) the Ti6242 β-reconstructed and (b) the Ti17 β microstructures.

3.1.5. Scenario for the interfacial bonding mechanisms. The overviews of the microstructure in the joint showed that both Ti17 and Ti6242 alloys present markers of an extensive $\alpha \rightarrow \beta$ phase transformation that occurred during the process. In Ti6242, the former α nodules of the near WL area (Zone III) first underwent a β phase transformation during the hot deformation process. The observed redistribution of the Ti6242 alloying elements was most likely caused by the progressive diffusion of the β stabilizing elements from the initially interspersed intergranular β_{rich} matrix to the adjacent β transformed nodules. The gradual aspect of the diffusion phenomena is directly related to the local temperature distribution which lowers as moving away from the WL. As a result, the rates of the transport phenomena in the WL zone were important enough to lead to substitutional elements partitioning close to a local β_{lean} composition on a 100 μm thick band of grain refined β microstructure. Hence, the β_{lean} composition results from the homogeneous partition of the alloy elements above the β-transus. This aspect agrees with the precipitation of the metastable phase α'. The latter results from the decomposition of a β_{lean} phase upon rapid cooling to room temperature. The cooling rates are high enough to hinder long-range diffusion phenomena. In the Ti17, the markers of $\alpha \rightarrow \beta$ transformation are present within a 2.5 mm thick band from the WL and extends to zone VII. The cooling rates from the β-domain were high enough to result in the stabilization of a single β-metastable phase at room temperature.

To identify the mechanisms resulting in bond formation, the β-grain boundaries of both materials were closely observed at the vicinity of the interfacial junction. In Figure 7, the features of β-grain fragmentation and rotation are clearly discernible in both materials with the presence of subgrain structures exhibiting low intragranular misorientation levels. These subgrains subdivide the prior-deformation β-parent grains in smaller crystallites. As illustrated in Figure 6, the interfacial zone exhibits an apparent continuity characterized by the presence of intermingled α'/β grains. The hypothetic parent crystallographic relationship between the α' phase and the β phase has been prospected via EBSD investigations in the junction zone (Figure 6). A perfect verification of the specific Burgers Orientation Relationship (BOR) was demonstrated in these peculiar intermingled structures between the precipitated α' laths of the Ti6242 and the corresponding β-metastable structure of the Ti17. Such findings reveal that both structures unequivocally shared the same β-parent crystallographic orientation above β-transus temperature conditions, and indicates that the cohesion mechanisms actually occurred between single-phase β structures.

As for grain fragmentation, during the LFW, the high-temperature conditions and the local mechanical loading due to friction and forging yield to an increase of density of dislocations and promote their mobility. Hence, the mutual interaction between dislocations yields to their patterning into dislocation wall structures (e.g. substructures with low internal misorientations in Figure 7). Such patterning, progressively subdivide the parent grains into smaller crystallites with low dislocation density with progressive gathering of the geometrically necessary dislocations into low angle boundaries and then transformation to high angle boundaries (Figure 7). Additionally, to this mechanism, a difference in local defect density and interfacial energy between the in-contact crystalline interfaces would promote a migration of the crystalline interfaces ultimately leading to the formation of a cohesive junction.

Once the reciprocating motion stopped, fast cooling caused both materials to experience either a $\beta_{lean}\rightarrow\alpha'$ or $\beta_{lean}\rightarrow\beta_{metastable}$ transformation in the interfacial zone depending on their local chemical composition. The limited process time and the subsequent hindered chemical homogenization at the transition zone lead to retaining the so-called intermingled α'/βm subgrains constituting the border between both Ti-alloys.

3.2. Dissimilar Ti6242-Inconel 718 joint

3.2.1. State of the art of joining Ti-Ni based alloys. In the past, several authors attempted to join nickel-based alloys to titanium alloys using different welding technologies: gas tungsten arc welding [13], explosive bonding [14], laser welding [15]. In most cases, unsuccessful welds were observed due to poor fusion interface, incomplete mixing of molten metals, cracking… Seretsky and Ryba [15] failed in joining pure titanium to pure nickel. The microstructure observation in the weld area reveals the presence of a heterogeneous chemical composition distribution inherited from poorly mixed liquids during welding. According to the authors, such non-equilibrium state can be explained by the short time for melting and solidification in laser welding. And they concluded that at short interaction times, there may be an apparent miscibility gap between certain compositions. Therefore, they introduced an intermetallic compound TiNi in an attempt to create a more gradual composition gradient from pure titanium to pure nickel in the weldment. However, unsuccessful weldments were obtained because of the cracks and all specimens fractured within the weld region. More recent works have shown that the brittle intermetallic compounds (IMCs) precipitation ($TiNi_3$ and Ti_2Ni) in the weldment are known to be the most detrimental for the quality of the weld in dissimilar Ti-Ni alloys joining, as recently reported by Ogura *et al.* [16]. Hence, in order to suppress the IMCs, some authors used interlayers made of different materials. Such interlayers act as both an IMC precipitation inhibitor and a diffusion barrier. In [17], the author introduced an Nb/Cu interlayer in order to join titanium rods of Ti64 and Ti6242 with Nickel alloy Inconel 718 by friction welding. Better understanding of species diffusion and interfacial microstructural evolution in case of Ti- and Nickel-based alloys assembling was proposed by Aleman *et al.* [18] in their study on the diffusion bonding of a titanium alloy Ti6242 to an Inconel 625. The authors highlighted a Kirkendall effect due to faster diffusion of the nickel in the titanium than the titanium in the superalloy. However, weld defects observation and joint strength have not been reported.

Since these earlier works, very few researches were published in the past three decades on dissimilar welding of Ni-Ti alloys as indicated in recent reviews [19-21]. There has recently been a resurgence of interest in welding Ni-Ti alloys due in particular, to the requirements for light weighting of high temperature structures in aerospace industry. Indeed, Ti- and Ni-based welded structures can potentially replace heavy Ni-based superalloy components and improve overall engine efficiency by reducing weight. In order to better understand the microstructural development in the weld pool, Chatterjee *et al.* [22, 23] investigated the weldability of pure Ti with pure Ni. The microstructure analyses in the weld area is discussed in the light of interplay of the transport processes in the weld with the thermodynamics of the Ti-Ni system. In case of sheet components, Chen *et al.* [24] used continuous fiber laser welding to successfully join Ti64 to Inconel 718. The authors indicated that the thickness of the intermetallic can be controlled by the process parameters. More recently, Topolski *et al.* [25] investigated the same system using explosive joining. Microstructure investigations showed that structure of the Ti and Ni components in the vicinity of the joint was refined, which strongly affected the local mechanical properties. The non-annealed interface in the welding area showed a 150 nm thick mixed Ti/Ni interlayer which evolved into three layers of intermetallics of a total 4 μm thickness after annealing. As a consequence, the annealed samples displayed lower mechanical properties. The extension of this work in case of titanium- and nickel-based alloys can be difficult because of the alloying elements and their effects on the transport phenomena and the thermodynamic equilibria. Ogura *et al.* [16] study the weldability of Ti-6Al-4V and Ni-based Inconel 718 alloy with a rotary friction technology. Sound joints are obtained using Au-18 mass% Ni interlayer. The authors show that the good mechanical properties of the joints are due to the formation of thin intermediate layer which forms by the mixture of brazing of the liquid interlayer and diffusion of the base metal to this interlayer. Such welding joints do contain alloying elements other than those of the base metals. This aspect could have other disadvantages with regard to certain physical properties (e.g. creep). Other interlayers (Ag-Cu, Ag-Cu-Pd) were investigated for the same system by Ogura *et al.* [26]. The joint strengths were improved compared to the Au-Ni interlayer owing to the near suppression of Ni-Ti intermetallic compound precipitation. Shehbaz *et al.* [27] studied another Ti-Ni system composed by a commercially pure titanium and Inconel 718 assembled by tungsten inert gaz welding and a vanadium interlayer. The latter remarkably suppressed Ti and Ni diffusion across it. However, still brittle response was observed near the V/Inconel 718 interface due to intermetallic compounds. In the different works mentioned above, a detailed microstructure examination of the weld area is not proposed. Investigating the local phenomena at the interface in joining titanium with nickel alloys remains relevant to pursue the understanding of the bonding mechanisms for Ti-Ni based alloys systems.

3.2.2 Macroscopic observation of the welded samples. An Inconel 718, in a solution annealed metallurgical state (called hereafter IN718-SA), was 'welded' with a Ti6242 block. The weld surface geometry is of 10 x 80 mm², both surfaces were grinded with a 120p standard grind paper prior welding. The associated visible light aspect of the dissimilar bond is given in Figure 9.

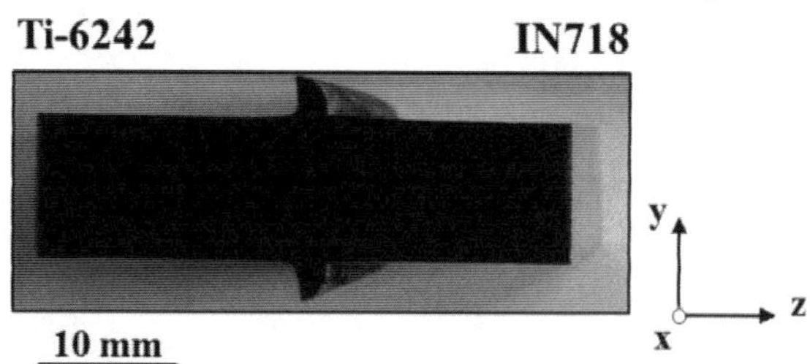

Figure 9. Photograph of the Ti-6242 / IN718-SA after transverse cross-sectional cutting procedure.

Only the Ti6242 alloy collapsed during the welding procedure, the IN718 part remained only little plastically affected by presenting a slight 'elephant foot' vertical squeezing due to the action of the

IOP Conf. Series: Materials Science and Engineering 1249 (2022) 012002 doi:10.1088/1757-899X/1249/1/012002

forging force. During the procedure, the material locally reached a hot-red aspect at the weld interface. Nevertheless, the thermo-mechanical conditions were not severe enough to induce subsequent plastification of the IN718 part (even considering its 'soft' initial metallurgical state).

3.2.3. Weld interface microstructure examination. The macrograph of the weld cross-section is shown in Figure 10. Weld cross-section showed good metallurgical bonding at Ti6242-IN718 interface. The Ti6242 (left) underwent important morphological modification, the white cloudy aspect within its microstructure indicate important β-stabilizer elements diffusion toward the α nodular microstructure due to high temperature condition. On another note, the IN718 (right) remained little affected by the weld procedure and only displayed a slight intragranular misorientation gradient testifying the occurrence of local plastic deformation. Nevertheless, the global shape of the IN718 grains remained quasi-unchanged. Between the parent materials, a peculiar intermediate microstructure developed presenting a quite heterogeneous aspect. The white / dark grey contrast seems to indicate bimodal microstructure.

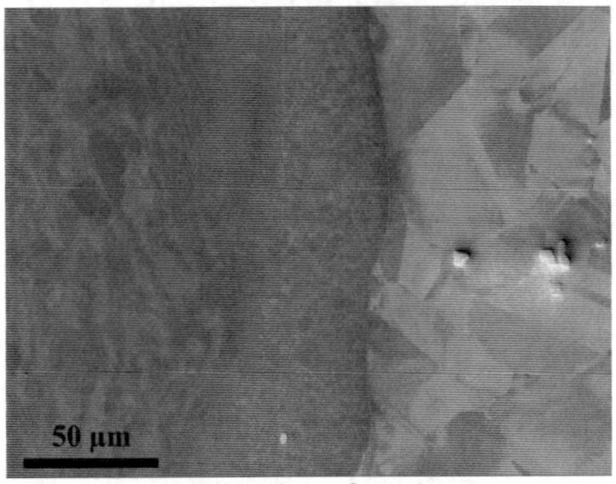

Figure 10. SEM-SE2 micrograph at the welding interface (transverse cross-section) of Ti6242 (left) and IN718 (right).

3.2.4. Phase identification and local chemical composition analysis. The global aspect of the intermediate phase presents a 50 µm total thickness with the remarkable occurrence of punctual irregularities in thickness and aspect along the bonding line (Figure 10). These irregularities are often accompanied by local excavation of the IN718 material (Figure 11 a, b), as if the austenitic phase of the IN718 had been locally dissolved within the intermediate phase. In addition to that, the intermediate layer locally takes the form of a fluid aspect flowing into the formed cavities (Figure 11). The intermediate layer of material has been further examined using EDX analyses. This, in an attempt to identify the chemical composition and the nature of the phase in presence at the weld interface. First, a cross-sectional profile was realized through the interfacial layer. The chemical results are combined with crystal orientation analyses and are presented in Figure 12 and Figure 13, respectively. Considering the potential reactivity between the alloys in contact, the occurrence of intermetallic compounds precipitation is highly suspected in the interfacial zone.

IOP Conf. Series: Materials Science and Engineering 1249 (2022) 012002 doi:10.1088/1757-899X/1249/1/012002

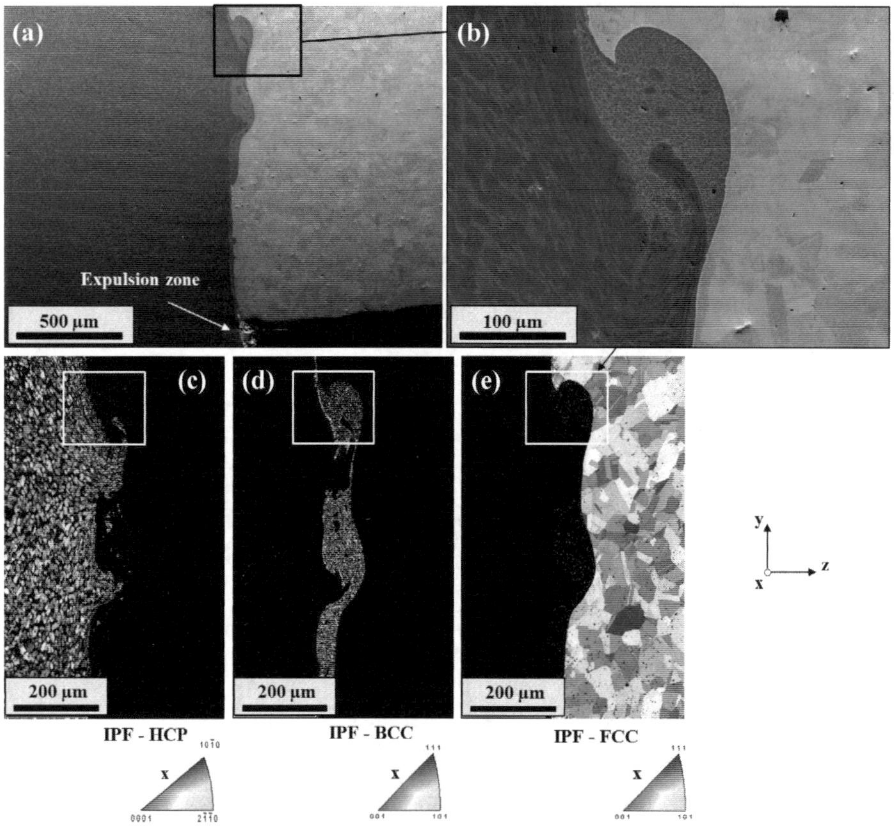

Figure 11. (a and b) SEM-SE2 micrograph at an interfacial irregularity, the associated crystal orientation signals are also provided on the :(c) HCP (Ti-α, phase), (d) BCC (Ti-β phase) and (e) FCC-γ (IN718) phases. The orientations are expressed by the direction along the friction direction x, normal direction of the maps.

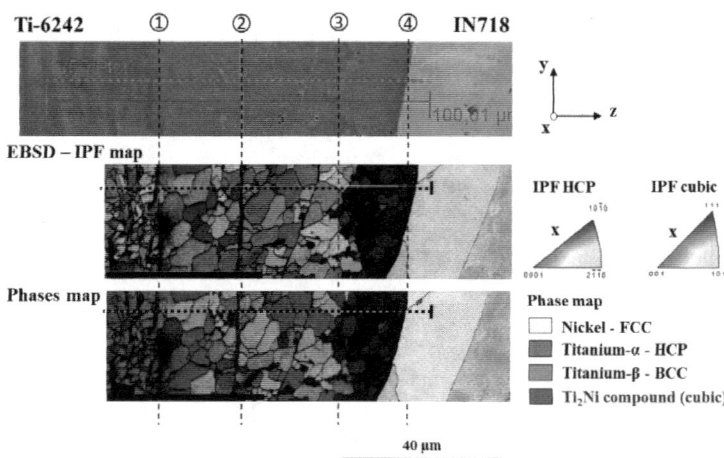

Figure 12. Inverse pole figure colored orientation maps combined with the phases map of the HCP (Ti-α, phase), BCC (Ti-β phase); FCC-γ (IN718) phases and Ti_2Ni (intermetallic phase) signal; CI index superimposed. The orientations are expressed by the direction along the friction direction x, normal direction of the maps.

In Figure 13, Position 1 marks the delineation between Ti6242 base material composition and progressive enrichment by IN718 elements. The stable chemical composition of structures observable between markers 3 and 4 indicate the formation of a Ni_2Ti intermetallic phase in the interfacial zone (probably responsible for cracking as indicated below). Such intermetallic compounds were also identified with the help of Kikuchi diffraction patterns. The latter notably matched with high confidence indexes with a cubic-Ni_2Ti crystal structure. Marker 2 also possesses the same local composition as the intermetallic Ni_2Ti phase. In addition to that, the subsequent element migration originating from the IN718 alloy composition locally enriched the Ti-alloy side in strong β-stabilizing elements. It resulted in the formation of a β phase at the weld interface. Thus, the interfacial microstructure is mainly composed of by a Ni_2Ti and the Ti-β two-phase compound. The local elements are then partitioned between a Ti-β (strongly) enriched phase and a Ti_2Ni intermetallic compound in a bimodal manner. The chemical analyses revealed quite stable local composition across the latter structures with well-defined composition once stabilized. The absence of chemical gradient indicates that a local thermodynamic equilibrium was reached in such structures, thus allowing to define them as phases.

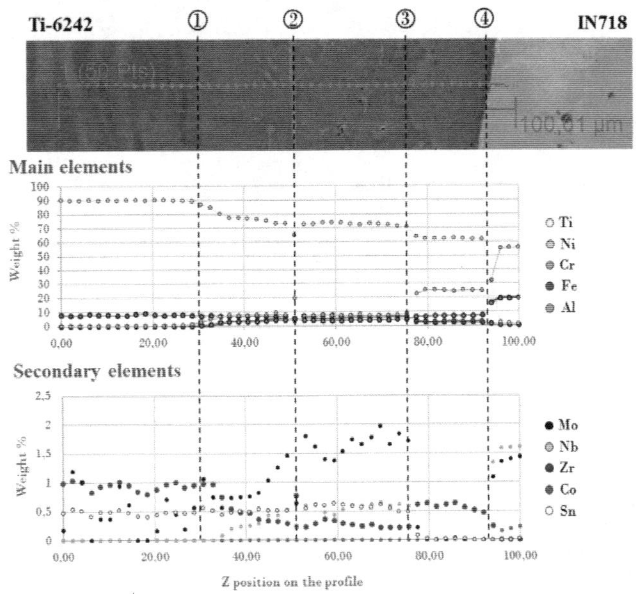

Figure 13. Composition analyses. The composition is given in weight %. 50 points of measure on a 100 µm length, 2 min acq/pts.

The intermixing zone is particularly heterogeneous and some peculiar microstructural features were also regularly observed within such intermediate microstructure. As an example, some isolated α-HCP nodules were identified in the interfacial junction with composition close to the Ti6242 base material composition in the center of the nodule (Figure 14). Such nodules were found in the middle of Ti-β enriched / Ti_2Ni regions. Subsequent chemical gradient is observed with the surrounding region, strongly enriched in IN718 additional elements (Figure 15). Such structures most likely remained in a far-from equilibrium state. Indeed, once the process stopped, the reaction kinetics most likely experienced a significant slowdown due to the critical cooling conditions, thus freezing the undergoing reaction. If the process had followed up, the important β-stabilizing effect brought by the IN718 elements would most likely favor the progressive stabilization of a room-temperature β metastable phase with the alternate and concomitant precipitation the Ti_2Ni intermetallic compound in a bimodal manner (depending on the local thermodynamic conditions).

3.2.5. Entrapped α-nodule analysis Closer observation carried out on the entrapped α-nodules in the welding interface area revealed internal structures with morphology close to HCP α' martensitic laths (Figure 16). The latter may have precipitated under critical cooling conditions once the friction welding process stopped. It also means that during the process, such nodule also underwent a α to β phase transformation during the thermomechanical processing. Then, a β→α' was triggered under cooling after the process stopped.

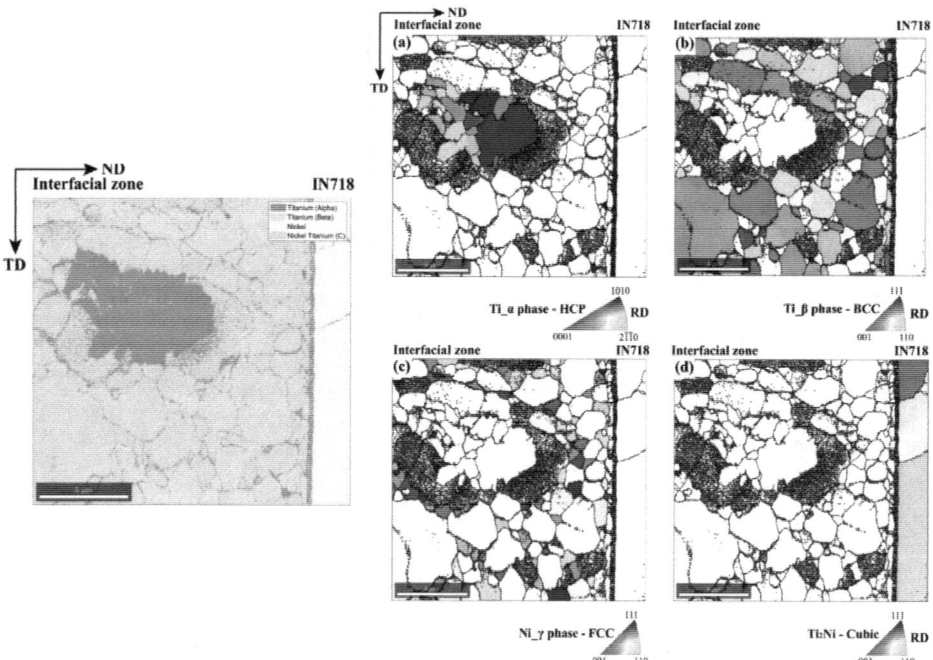

Figure 14. (Left) phase map at the interfacial zone; (right) Inverse pole figure colored orientation maps for each phase indexed, 60 nm step-size. The grain boundaries (if between same phase, 15° threshold) and phase boundaries are superimposed in black.

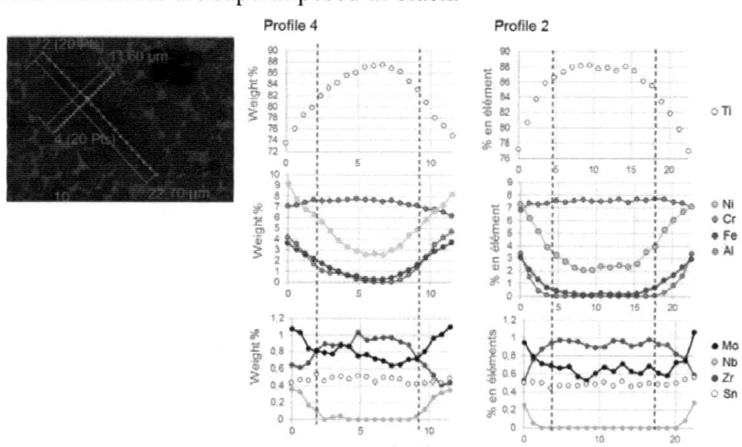

Figure 15. Island of HCP phase nodule in the middle of the intermetallic Ni_2Ti / Ti-β microstructure.

IOP Conf. Series: Materials Science and Engineering 1249 (2022) 012002 doi:10.1088/1757-899X/1249/1/012002

In addition to that, a marked edging phase was observed at the interface between the precipitated reactive phases and the IN718 base microstructure forming a whitish border in-between (Figure 16). Such features may correspond to the existence of another intermediate phase. The latter, however, could not be identified via Kikuchi diffraction pattern analyses. Nevertheless, the SEM-imaging grey contrast seems to indicate a fairly homogeneous chemical element partition that would suggest a phase precipitation occurrence.

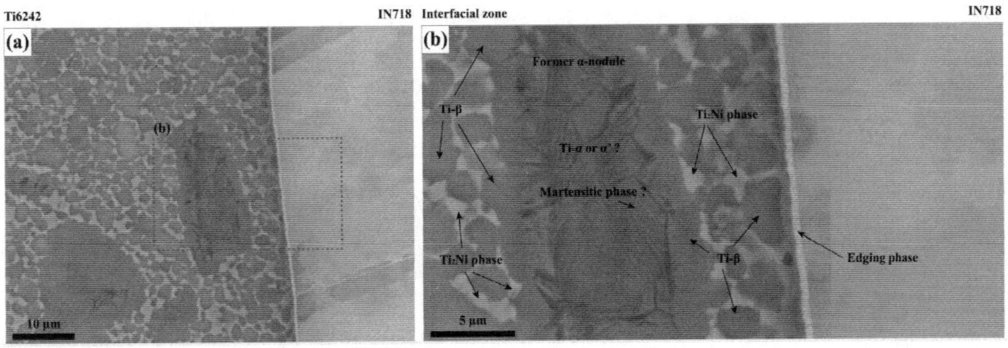

Figure 16. Detailed micrographs on a former α-nodule trapped in the intermediate Ni_2Ti / Ti-β microstructure.

The microhardness of the intermetallic phase reached high values, in the order of magnitude of 500 to 600 $HV_{0,3}$ and demonstrated a poor mechanical behavior (easy propensity to form cracks at the weld interface when indenting through hardness procedures). Cracks are notably observable near the weld expulsion zones (as illustrated in Figure 17).

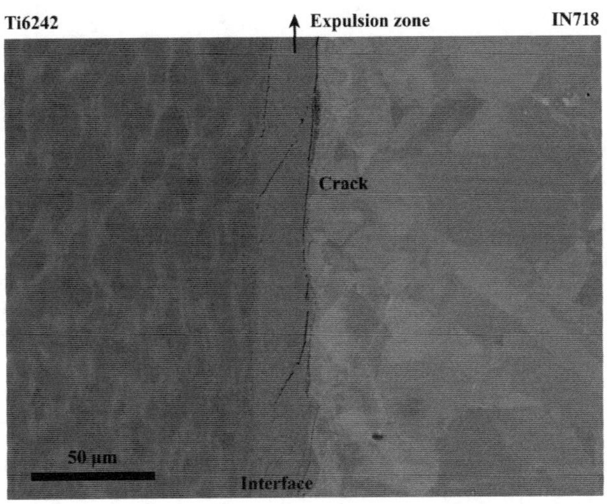

Figure 17. Island of HCP phase nodule in the middle of the intermetallic Ni2Ti / Ti-β microstructure.

3.2.6. Global description of the welding area. (1) The local intermixing at the weld interface led to the formation of a reactive interface between the friction-welded materials. Only the Ti6242 alloy collapsed during the process, the IN718 part remained slightly affected. (2) An intermediate microstructure develops; the latter presents a bimodal aspect and is mainly composed of Ti-β phase (rich in β stabilizing elements originating from IN718) and an intermetallic phase identified as a Ti_2Ni phase (cubic) between the contacting materials. (3) Such configuration does not appear feasible nor interesting from a practical and technical point of view. The unavoidable and spontaneous formation of an intermetallic structure at the weld interface was shown to be detrimental for the joint integrity.

4. Conclusion

This investigation was focused on a complete characterization of the microstructure and chemical composition of dissimilar linear friction welded joints. First, the LFW joint of two dissimilar titanium alloys (Ti6242 and Ti17) was examined. Thanks to a detailed description of the joint microstructure and chemical composition, it was found that, despite the observation of HCP-phase in Ti6242 and β-metastable phase in Ti17, some interfacial grains presented a common β-parent crystallographic orientation. This led to identify that the successful cohesion of the joint was obtained in the single-phase β-domain reached with the high-temperature and strains of the LFW process. This successful cohesion was the results of an active migration of the crystalline interfaces of the grains across the weld junction and was assisted by continuous dynamic recrystallization.

Then, the possibility of assembling Ti6242 and Inconel 718 was assessed. Successful welding of the materials was obtained but the microstructure investigation showed that the obtained intermixing zone was notably heterogeneous, with the development of a bimodal microstructure mainly composed of Ti-β phase and Ti_2Ni intermetallic phase. The formed intermetallic phase presented high hardness and had a propensity to lead to cracking. Therefore, despite the successful achievement of welding between Ti6242 and Inconel 718, the unavoidable formation of an intermetallic structure that was detrimental to the joint integrity led to conclude that such a material combination was not valuable from a technical point of view.

Acknowledgments

The authors acknowledge the financial support from the French National Research Agency (ANR) and FEDER through OPTIMUM ANR-DS0303 project. The authors are also grateful to ACB for providing LFW welded samples and Airbus for both their technical support.

References

[1] Guo Y, Chiu Y, Attallah M M, Li H, Bray S and Bowen P 2012 Characterization of Dissimilar Linear Friction Welds of α-β Titanium Alloys J. Mat. Eng. Performance 21 5 770-6
[2] Ward A A, Zhang Y and Cordero Z C 2018 Junction growth in ultrasonic spot welding and ultrasonic additive manufacturing Acta Mater. 158 393-406
[3] Sun L X, Li M Q and Li H 2018 Interface characteristics and recrystallization mechanism of dissimilar titanium bonds J. Mat. Sci. 53 7 5380-88
[4] Ballat-Durand D 2019 Microstructural characterization and optimization of linear friction welded titanium alloys joints through the influence of the process parameters and post-weld heat treatments PhD Thesis, Université de technologie de Compiegne, France
[5] Boyat X 2020 Contribution to the understanding of linear friction welding applied on aeronautical alloys PhD Thesis, Université de technologie de Compiegne, France
[6] Ballat-Durand D, Bouvier S, Risbet M and Pantleon W 2018 Multi-scale and multi-technic microstructure analysis of a linear friction weld of the metastable-β titanium alloy Ti-5Al-2Sn-2Zr-4Mo-4Cr (Ti17) towards a new Post-Weld Heat Treatment Mat. Charac. 144 661-70
[7] Ballat-Durand D, Bouvier S, Risbet M and Pantleon W 2019 Through analysis of the microstructure changes during linear friction welding of the near-α titanium alloy Ti-6Al-2Sn-4Zr-2Mo (Ti6242) towards microstructure optimization, Mat. Charac. 151 38-52
[8] Ballat-Durand D, Bouvier S and Risbet M 2019 Contributions of an innovative post-weld heat treatment to the micro-tensile behavior of two mono-material linear friction welded joints using: The β-metastable Ti–5Al–2Sn–2Zr–4Mo–4Cr (Ti17) and the near-α Ti–6Al–2Sn–4Zr–2Mo (Ti6242) Ti-alloys Mat. Sci. Eng. A 766 138334
[9] Boyat X, Ballat-Durand D, Marteau J, Bouvier S, Favergeon J, Orekhov A and Schryvers D 2019 Interfacial characteristics and cohesion mechanisms of linear friction welded dissimilar

titanium alloys: Ti–5Al–2Sn–2Zr–4Mo–4Cr (Ti17) and Ti–6Al–2Sn–4Zr–2Mo (Ti6242) Mat. Charac. 158 109942
[10] Germain L, Dey S R, Humbert M and Gey N 2007 Determination of parent orientation maps in advanced titanium-based alloys J. Microscopy 227 3 284-91
[11] Germain L, Gey N, Mercier R, Blaineau P and Humbert M 2012 An advanced approach to reconstructing parent orientation maps in the case of approximate orientation relations: Application to steels Acta Mater. 60 11 4551-62
[12] Gourdet S and Montheillet F 2003 A model of continuous dynamic recrystallization Acta Mater. 51 9 2685-99
[13] Gorin L G 1964 Welding titanium alloys to nickel-base alloys, Weld. Prod. 11 46-53
[14] Ells C E, Taylor G F and Mansey R C 1973 Joining Nickel to Titanium Titan. Sci. Technolo. 2527–34
[15] Seretsky J and Ryba E R 1976 Laser Welding of Dissimilar Metals: Titanium To Nickel, Welding Research Supp. 55 7 208s–11s
[16] Ogura T, Miyoshi K, Matsumura T, Imai T and Saida K 2019 Improvement of joint strength in dissimilar friction welding of Ti-6Al-4V alloy to type-718 nickel-based alloy using the Au–Ni interlayer Sci. Tech. Welding and Joining 24 4 327-33
[17] Kuo M 1993 Dissimilar Friction Welding of titanium alloys to Alloy 718 PhD thesis, The Ohio State University (USA)
[18] Aleman B, Gutierrez I and Urcola J 1995 Interface microstructures in the diffusion bonding of a titanium alloy Ti 6242 to an INCONEL 625 Metall. Mater. Trans. A 26 2 437–46
[19] Ramesh A P, Subramaniyan M and Eswaran P 2019 Review of friction welding of similar/dissimilar metals J. Physics: Conf. Series 1362 012032
[20] Quazi M M, Ishak M, Fazal M A, Arslan A, Rubaiee S, Qaban A, Aiman M H, Sultan T, Ali M M and Manladan S M 2020 Current research and development status of dissimilar materials laser welding of titanium and its alloys Optics and Laser Tech. 126 106090
[21] Sathishkumar G B, Sethuraman P, Chanakyan C, Sundaraselvan S, Joseph A Arockiam, Alagarsamy S V, Elmariung A, Meignanamoorthy M, Ravichandran M and Jayasathyakawin S 2021 Friction welding of similar and dissimilar materials: A review Mat. Today: Proc., in Press
[22] Chatterjee S, Abinandanan T and Chattopadhyay K 2006 Microstructure development during dissimilar welding: Case of laser welding of Ti with Ni involving intermetallic phase formation J. Mater. Sci. 41 3 643–52
[23] Chatterjee S, Abinandanan T and Chattopadhyay K 2008 Phase formation in Ti/Ni dissimilar welds Mater. Sci. Eng. A 490 Issue 1–27–15
[24] Chen H C, Pinkerton A J and Li L 2011 Fibre laser welding of dissimilar alloys of Ti-6Al-4V and Inconel 718 for aerospace applications Int. J. Adv. Manuf. Technol. 52 9–12 977–87
[25] Topolski K, Wiecinski P, Szulc Z, Galka A and Garbacz H 2014 Progress in the characterization of explosively joined Ti/Ni bimetals Mater. Des. 63 479–487
[26] Ogura T, Imai T and Saida K 2020 Joint Strength Improvement in Dissimilar Friction Welding of Titanium Alloy to Nickel Alloy with Interlayer Quarterly J. the Japan Welding Society 38 issue 2, 79s-83s
[27] Shehbaz T, Junaid M, Khan F N, Akbar U A and Hakeem A S 2022 Dissimilar tungsten inert gas welding between Inconel 718 and commercially pure titanium using a vanadium interlayer Proc.Institution of Mechanical Engineers, Part L: J. Mat: Design and Applications 236 issue 4715-29

IOP Conf. Series: Materials Science and Engineering 1249 (2022) 012003 doi:10.1088/1757-899X/1249/1/012003

Solid solution hardening effects on structure evolution and mechanical properties of nanostructured binary and high entropy alloys after high pressure torsion

T Keil, C Minnert, E Bruder and K Durst

Technical University of Darmstadt, Materials Science, Physical Metallurgy, Darmstadt 64287, Germany

E-mail: t.keil@phm.tu-darmstadt.de

Abstract. Two different alloy series (Cu-X, Ni-X) have been selected to investigate the effects of solutes on the saturation grain size, the thermal stability and mechanical properties after high pressure torsion. The results of the Cu-X series indicate that the saturation grain size does not correlate with the stacking fault energy but shows good agreement with solid solution hardening according to the Labusch model. This correlation does not only hold for binaries, but also for chemically complex high entropy alloys (Ni-X) in the form of $(CrMnFeCo)_xNi_{1-x}$, where the Varvenne model is used to describe solid solution hardening. The alloy series exhibit a grain size in the range of 50 – 425 nm after high pressure torsion and the solutes increase the strength as well as the thermal stability of the alloys after annealing. The nanostructured alloys exhibit an enhanced strain rate sensitivity exponent, as determined from nanoindentation strain rate jump and constant contact pressure creep testing, whereas an enhanced rate sensitivity is found at low strain rates. The relatively lower rate sensitivity of the alloys as well as their higher thermal stability indicate, that defect storage and annihilation is strongly influenced by a complex interaction of solutes, dislocations and grain boundaries.

1. Introduction

Severe plastic deformation methods like high pressure torsion (HPT) have been used for several decades to generate ultrafine-grained (UFG) or nanocrystalline (nc) bulk materials [1, 2]. Due to the quasi-constrained condition, a high strain can be applied to the material, leading to high dislocation densities of at least 10^{14} m^{-2} [2, 3]. During shear deformation, the dislocations are not randomly distributed, but rather form dislocation cell walls or sub-grain boundaries [4-7]. With increasing strain, the average grain boundary misorientation increases and reaches values >15° at high strain levels. While the grain refinement mechanisms are well established [8], the parameters influencing the saturation grain size d_s, i.e., the minimal achievable steady-state grain size at very large strain are not yet fully understood. The grain refinement can be described as a process of counterbalanced generation and annihilation (recovery, dynamic recrystallization) of dislocations during deformation [9-11]. From this perspective, it seems obvious that parameters such as the stacking fault energy (SFE) or the solid solution hardening (SSH) contribution play a decisive role and affect the grain refinement including the saturation grain size. It is known that the SFE strongly affects the grain refinement rate, especially at lower strains during HPT, but its effect on the saturation grain size is discussed controversially [1, 9, 12]. Investigations on a CuZn series showed that an increase in solute concentrations lowers the stacking fault energy substantially, which in turn results in a reduction in d_s [13]. However, a comparison of pure fcc elements with strongly

IOP Conf. Series: Materials Science and Engineering 1249 (2022) 012003 doi:10.1088/1757-899X/1249/1/012003

varying SFE shows similar d_s values after HPT deformation at the same homologous temperature [1]. Edalati et al. [9] empirically found a correlation between SSH contribution and d_s for various binaries and reasoned that the saturation grain size is mainly controlled by atomic size and modulus mismatch effects of solutes rather than by the SFE. The authors used the classical Labusch model [14, 15] to describe the solute - dislocation interaction and explained this correlation by stronger localized stress fields, which hinder the necessary dislocation motion and cause smaller d_s [9].

Bruder et al. [16] investigated a series of Cu-X alloys which, compared to each other, show either different SFE or SSH contributions. These results confirm the correlation given by Edalati et al. [9]. To further investigate the effects of solutes on the grain refinement, we go beyond the scope of the classical SSH models using a series of Ni-X alloys starting from pure Ni up to the equiatomic, five-component CrMnFeCoNi Cantor high entropy alloy (HEA) [17, 18]. This series is well suited to shed light on SSH effects, not only in dilute solid solutions but also in chemically complex HEAs, as the fcc crystal structure is present over the complete concentration interval [19-21]. For HEAs, the classical picture of solute and solvent species does not hold, hence, the Labusch model may not be applicable for this alloy class. Varvenne et al. [22-24] proposed a more recent approach for modeling SSH in HEAs, assuming the HEA as a homogeneous, monoatomic matrix with average properties, in which the individual constituents are embedded. In contrast to the Labusch model, which describes the interaction of dislocations and obstacles in dilute solid solutions based on an atomic size and modulus misfit, the Varvenne model describes the interaction between a dislocation and solutes only by an atomic size difference and neglects fluctuations in shear modulus.

This work summarizes and builds upon a series of publications by the present authors (Bruder et al. [16], Keil et al. [17, 18] and Minnert et al. [25, 26]) to further elucidate the effects of solutes, including the SFE and SSH, on the saturation grain size, the microstructural stability and the mechanical properties (hardness and strain rate sensitivity using advanced constant contact pressure (CCP) and strain rate jump (SRJ) testing) after HPT processing in binary and chemically complex alloys.

2. Experimental procedure

The Cu-X alloy series consists of binary Cu alloys with Al, Sn, or Zn as alloying elements. The concentration of the alloying elements is 5 wt.% (CuAl5, CuSn5 and CuZn5) or 30 wt.% (CuZn30). In addition, commercially pure Cu (99.95%) was tested for comparison. All alloys exhibit an alpha-Cu matrix with varying SFE and SSH contributions. The Ni-X series consists of three five-component alloys with the following compositions: $Cr_{2.5}Mn_{2.4}Fe_{2.6}Co_{2.9}Ni_{89.6}$ (nominal Ni92), $Cr_{10.5}Mn_{9.9}Fe_{10.3}Co_{10.6}Ni_{58.7}$ (Ni60) and $Cr_{20.6}Mn_{19.3}Fe_{19.9}$Co20.1$Ni_{20.1}$ (Ni20). Pure Ni with a purity level of 99.6% were used as reference materials with minor solute hardening effects. The alloy designation and exact compositions can be seen in table 1 and table 2.

Disc shaped samples were processed by HPT at room temperature using active water cooling to an equivalent plastic strain of $\varepsilon > 120$ at the outer radius of the sample to reach the saturation state. Subsequent isochronal heat treatments for 1 h were performed at different temperatures up to 500°C for the Cu-X series and up to 900°C for the Ni-X series. Flat sections of the samples were prepared by polishing down to colloidal silica. For the Cu-X series, Fe-(III)-nitrate solution was additionally used for the final polishing step. The viewing direction for microstructural analysis and the loading direction of the nanoindentation system are parallel to the loading direction of the HPT press. Nanoindentation testing was performed at room temperature (25 °C) on a G200 nanoindentation system (Keysight) using a Berkovich diamond tip (Synton-MDP) at a radial distance from the center above 0.5r. The hardness (H) was measured at a constant indentation strain rate (CSR) of $\dot{\varepsilon} = 0.05\,s^{-1}$ and the strain rate sensitivity was measured by means of indentation strain rate jump (SRJ) and constant contact pressure (CCP) testing. The applied strain rate for SRJ testing was varied every 500 nm indentation depth between 0.05 s^{-1}, 0.005 s^{-1} and 0.001 s^{-1}. For samples showing significant material pile-up during indentation, the contact area was corrected according to the procedure proposed by Kese et al. [27] (further information regarding the correction procedure is shown in Ref. [18] for the Ni-X series). In

IOP Conf. Series: Materials Science and Engineering 1249 (2022) 012003 doi:10.1088/1757-899X/1249/1/012003

addition, indentation creep experiments were carried out using the constant contact pressure approach [25, 26]. The initial penetration depth at the start of the creep segment was 500 nm, the creep time was set to 3000 s. The contact pressure, based on the contact stiffness signal S, was reduced in a single unloading event and subsequently held constant by controlling P/S^2, where P is the applied load.

The microstructural analysis was performed using a high resolution scanning electron microscope (TESCAN Mira 3) in backscattered electron (BSE) contrast mode. The grain size was measured by the line intercept method.

3. Results and Discussion

3.1 Saturation grain size

The saturation grain size after HPT deformation is derived for both alloy series from BSE images (Cu-X: figure 1, Ni-X: figure 2) and listed in table 1 (Cu-X) and table 2 (Ni-X) together with the relevant material properties like SFE and SSH. Within the respective alloy series, the pure elements show the largest saturation grain size. The obtained values for the saturation grain size are in good agreement with literature values of approximately 400 nm for Cu [11] and 170 nm for Ni [28], indicating that no slip occurred during deformation and the steady-state condition is reached. The addition of solutes to Cu and Ni results in decreasing grain sizes for both alloy systems (see table 1 & 2). The decreasing grain size with increasing Zn content is in good agreement with earlier investigations on CuZn alloys [9, 13, 29]. The evaluation of the saturation grain size of the Ni-X series clearly indicates the same trend as an increase in solute concentration results in smaller grain sizes from 142 nm for Ni100 down to 55 nm for Ni20.

Table 1. Alloy designation and composition in at.%, stacking fault energy [30-33], solid solution hardening contribution $\Delta\tau$ according to the Labusch model [14, 15], saturation grain size d_s and hardness H of the Cu-X series after HPT deformation.

Alloy designation and composition	SFE mJ/m^2	$\tau_{Labusch}$ MPa	d_s nm	H GPa
Cu (99.5 at.% Cu)	45	0	425 ± 115	1.32
CuAl5 (11.03 at.% Al)	28	24	133 ± 46	2.44
CuSn5 (2.74 at.% Sn)	40	51	62 ± 26	3.23
CuZn5 (4.87 at.% Zn)	43	15	304 ± 142	1.9
CuZn30 (29.4 at.% Zn)	14	40	100 ± 50	2.40

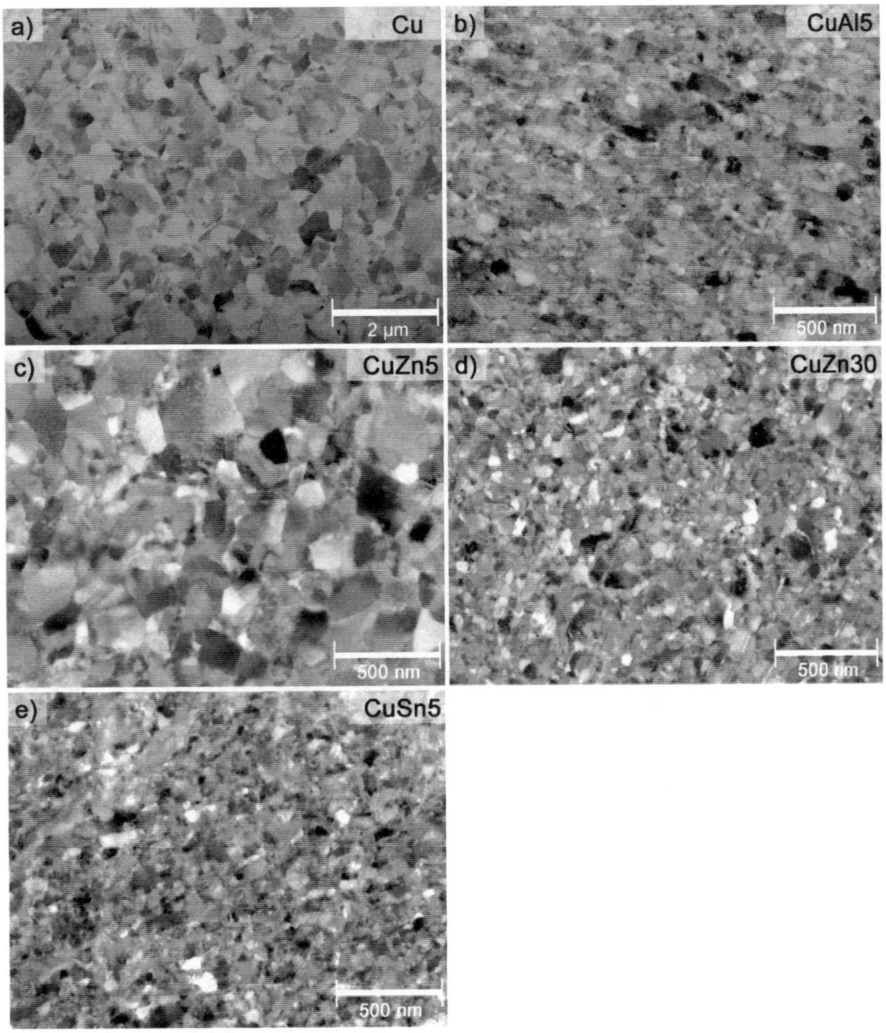

Figure 1. BSE images of the Cu-X alloys after HPT deformation: a) Cu, b) CuAl5, c) CuZn5, d) CuZn30 and e) CuSn5.

Table 2. Alloy designation and composition in at.%, solid solution hardening contribution Δτ according to the Labusch [14, 15] and the Varvenne model [22-24], saturation grain size d_s and hardness H of the Ni-X series after HPT deformation.

Alloy designation and composition	$\tau_{Labusch}$ MPa	$\tau_{Varvenne}$ MPa	d_s nm	H GPa
Ni100 (99.6 at.% Ni)	0	0	142 ± 57	3.93
Ni92 ($Cr_{2.5}Mn_{2.4}Fe_{2.6}$Co2.9$Ni_{89.6}$)	11	11	110 ± 45	4.30
Ni60 ($Cr_{10.5}Mn_{9.9}Fe_{10.3}Co_{10.6}Ni_{58.7}$)	33	40	62 ± 23	4.98
Ni20 ($Cr_{20.6}Mn_{19.3}Fe_{19.9}Co_{20.1}Ni_{20.1}$)	55	41	55 ± 17	4.82

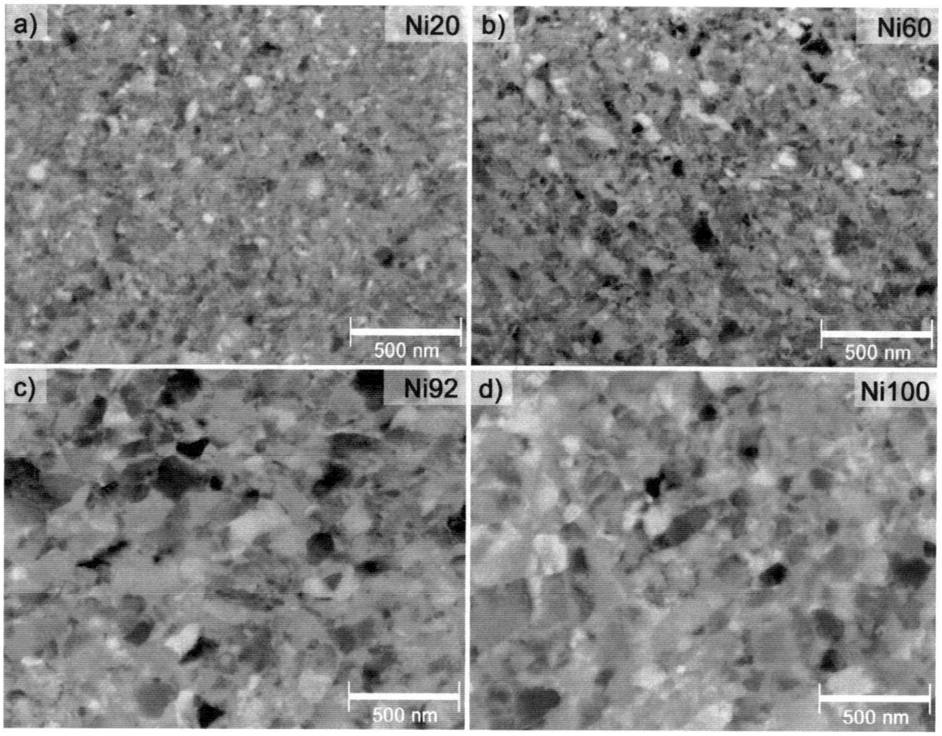

Figure 2. BSE images of the Ni-X alloys after HPT deformation: a) Ni20, b) Ni60, c) Ni92 and d) Ni100.

The SSH contributions for the binary alloys are calculated using the equations shown in Ref. [9] according to Labusch´s model [14, 15]. It shows that CuSn5 has an almost three times higher SSH compared to CuZn5, two times higher than CuAl5 and 27% higher than CuZn30 (table 1). The CuSn and CuZn alloys show either comparable SSH contribution (CuSn5, CuZn30) or similar SFE (CuSn5, CuZn5). This allows an independent analyzation of SFE and SSH on d_s as well as the mechanical properties. For the high entropy type Ni-X series, the SSH contributions are calculated via different SSH models and compared to the found saturation grain size to shed light on SSH effects in HEAs. The SSH contributions are calculated using either the Labusch model [14,15], assuming Ni as the matrix element, even though this is not strictly valid for the equiatomic Ni20 alloy, or using the more recent model from Varvenne et al. [22-24], which is specifically developed for HEAs. For the $\tau_{Varvenne}$ calculations, the notations from Ref. [23] are used.

A comparison of CuZn30 and CuSn5 reveals that d_s is governed by the SSH contribution but not by the SFE, as CuZn30 shows a much larger grain size in spite of having a three times smaller SFE. The proposed correlation between d_s and SSH contribution [9] does hold for the complete Cu-X series (figure 3), including different solute species (Al, Sn, Zn). On the other hand, there is no direct correlation between SFE and d_s. The results found for the Ni-X series indicate that the empirical correlation between $\tau_{Labusch}$ and d_s found by Edalati et al. [9] fails for highly concentrated solid solutions (figure 3 b). Ni20 and Ni60 exhibit a similar saturation grain size, although $\tau_{Labusch}$ is 63% higher for Ni20 compared to Ni60. Using the Varvenne model, similar SSH effects are obtained for Ni20 (41.36 MPa) and Ni60 (40.45 MPa), which is in good agreement with the observed d_s values. figure 3 b) clearly indicates that the saturation grain size correlates with τ not only in the dilute solid solutions but also in chemically complex compositions, given that the used SSH model provides proper values. The found correlation between SSH contribution and saturation grain size in binaries is based on the explanation of hindered dislocation motion, which is necessary for dislocation recovery or grain boundary migration [9]. The effect of alloying on the saturation grain size is not caused by the change in SFE, it is rather related to the change in grain boundary mobility [34]. In this respect, Keil et al. [18] showed that the solute content strongly effects the grain growth kinetics in the Ni-X system with higher thermal stability for higher solute concentrations. Very low activation volumes for grain growth have been found in the low temperature regime, which was attributed to the modification of grain boundaries during severe plastic deformation [18]. Such grain boundaries in the deformation-affected state show higher diffusivities and pronounced strain fields compared to relaxed grain boundaries [35]. Yet, only a fraction of the grain boundaries transforms during HPT into such a deformation affected state (≈ 20%), which leads to a composite structure of relaxed and deformation-affected grain boundaries [36]. This structure causes different properties along the boundary with locked regions of high resistance against displacement and thus high structural stability [36]. It is reasonable to expect that the addition of solutes further increases the complexity and inhomogeneity of the grain boundary structure, explaining the reduced grain boundary mobility and thus the smaller saturation grain size as well as the high thermal stability for alloys with a higher solute content. The microstructural stability can also be seen in the transient behavior during deformation at a change in strain rate, which is discussed in section 3.2.

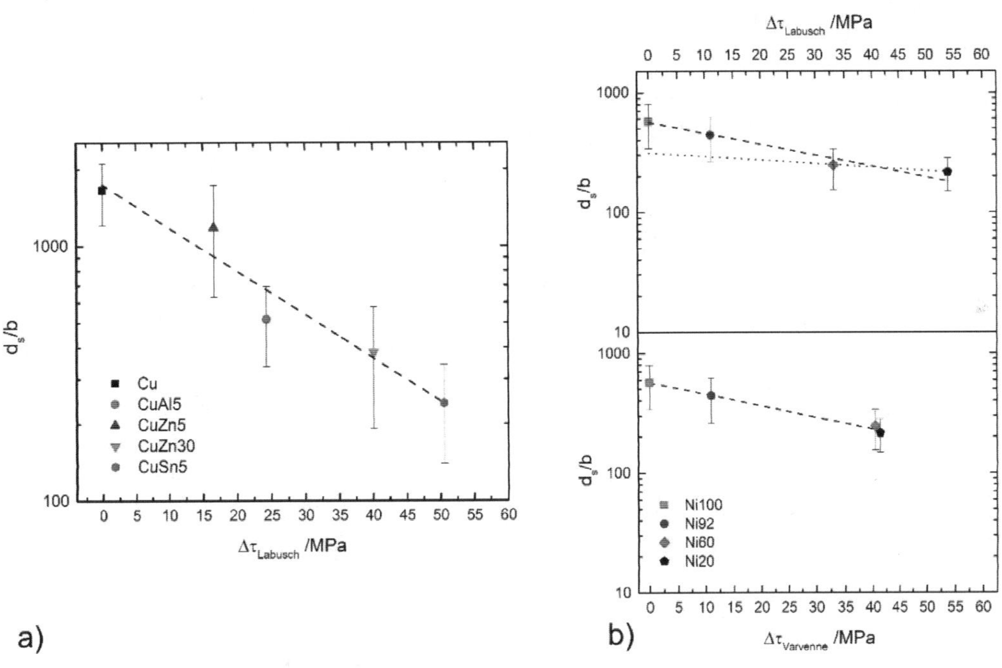

Figure 3. Normalized saturation grain size ds/b as a function of the SSH contribution Δτ, determined via the Labusch and the Varvenne model. The dashed lines are fitted to the complete data set, the dotted line represents only the extrapolation of Ni20 and Ni60.

3.2 Thermal stability and deformation mechanisms

3.2.1 Thermal stability. The thermal stability of the two alloy series is examined experimentally by the change in hardness after isochronal annealing. Figure 4 shows the effect of the annealing temperature on the nanoindentation hardness for Cu-X (a) and Ni-X (b). HPT deformation results in a significant increase in hardness compared to the coarse-grained states for both alloy series. The Cu-X series shows hardness values between 1.32 GPa for Cu and 3.23 GPa for CuSn5, whereas hardness values between 3.93 GPa for Ni100 and 4.98 GPa for Ni60 are found for the Ni-X series (figure 4). The hardness depends, on the one hand, on the SSH contribution and, on the other hand, on the grain size, which in turn is also affected by the SSH contribution. Within both series the trend of a higher hardness with increasing SSH contribution can be seen. Despite having a slightly larger d_s and smaller SSH, the CuAl5 alloy shows a similar or even higher hardness compared to CuZn30. The high hardness may be caused by secondary phases, even though the Cu-Al phase diagram clearly shows a single-phase field for CuAl5 and a supposedly single-phase microstructure is observable from BSE imaging at high magnification (figure 1 b).

The hardness evolution with annealing temperature shown in figure 4 unveils in general a higher microstructural stability for samples containing a certain amount of solutes, independent of the material system. It can be seen that three different stages upon annealing exist, first an almost constant hardness, second a significant hardness drop in a material specific temperature interval and third a slight decrease with further increase in annealing temperature. However, the stability range of the different compositions is largely different, with higher absolute transition temperatures for the Ni-X sample series. Figure 4 a) and b) shows the hardness as a function of absolute annealing temperature, whereas figure 4 c) and d) show the normalized hardness change due to annealing as a function of the homologous temperature (T/T_m).

The first stage of a nearly constant hardness is almost not observable for pure Cu, but extends to about 300°C for pure Ni. The hardness of pure Cu already starts to decrease at temperatures between 100 and 150 °C, whereas the other Cu-X compositions retain an almost constant hardness up to temperatures of 200 - 250 °C despite having higher driving forces (higher defect density) for grain growth. A significant effect on the microstructural stability by adding small amounts of solutes (8 at.%) cannot be seen in the present study for Ni-X, as the plateau in hardness covers the same temperature range as Ni100. Much higher solute concentrations (Ni20, Ni60) extend the hardness plateau to about 450 – 500 °C. Annealing at low temperatures may cause an increase in hardness, which can be seen for the Ni-X series as well as for CuAl5 and CuZn5. This phenomenon can be related to a hardening by annealing effect [37, 38], where dislocation sources and the density of free dislocations are reduced without significant grain growth. The transition regime is located at only 100 – 200 °C for pure Cu and at higher temperatures of approx. 250 – 500 °C for the Cu based samples containing a discrete amount of solutes. The transition regime is clearly observable for the Ni-X series, where it is located at temperatures between 300 – 800 °C for the dilute solid solutions and between 450 – 800 °C for the more complex Ni20 and Ni60 alloys. At temperatures above 500 °C, we do not expect significant changes in hardness for the Cu-X series. The transition from a strongly decreasing hardness to only slight changes with increasing temperature takes place at 800°C for the alloys with higher Ni content (Ni60, Ni92 and Ni100). The Ni20 HEA still shows a large drop in hardness up to 900°C. A further decrease at higher temperatures is also not expected in this case, as the hardness of the Cantor alloy after 1150°C annealing is in the same hardness range [39].

The grain growth kinetics in alloys are typically governed by the local elemental distribution e.g. the presence or absence of segregations (solute drag) or secondary phases (Zener drag). The Cu-X series remains single-phase upon annealing, indicating that the enhanced thermal stability of the alloys compared to pure Cu is related to the solute content, which provides dragging forces for grain boundary migration. A comparison of CuZn5 and CuZn30 indicates a higher microstructural stability for CuZn5, as CuZn5 shows a constant hardness up to 250°C where CuZn30 already starts to soften significantly.

Analyzing the results of CuZn30 in comparison to CuZn5 using the homologous temperature (figure 4c), a different picture emerges: Here, CuZn30 turns out to be more stable compared to CuZn5. This is related to the significantly reduced melting temperature of CuZn30 (T_m = 954 °C) compared to the other alloys, so that diffusional processes are enhanced at the same absolute T for CuZn30 in comparison to CuZn5. This could explain the different behavior of CuZn5 and CuZn30 at 250 °C, where CuZn5 shows an enhanced thermal stability compared to CuZn30 in spite of the lower SSH contribution.

In case of the Ni-X series, Ni20 as well as Ni60 show on the one hand an increased amount of solutes compared to Ni92 and Ni100, which may lead to a stronger solute drag. On the other hand, they exhibit decomposition tendencies at intermediate temperatures [18, 40], which results in additional pinning points by secondary particles and increased kinetic stabilization due to Zener pinning. In the CG states, the alloys with higher solute content show again a higher hardness, even though the different grain sizes have to be considered.

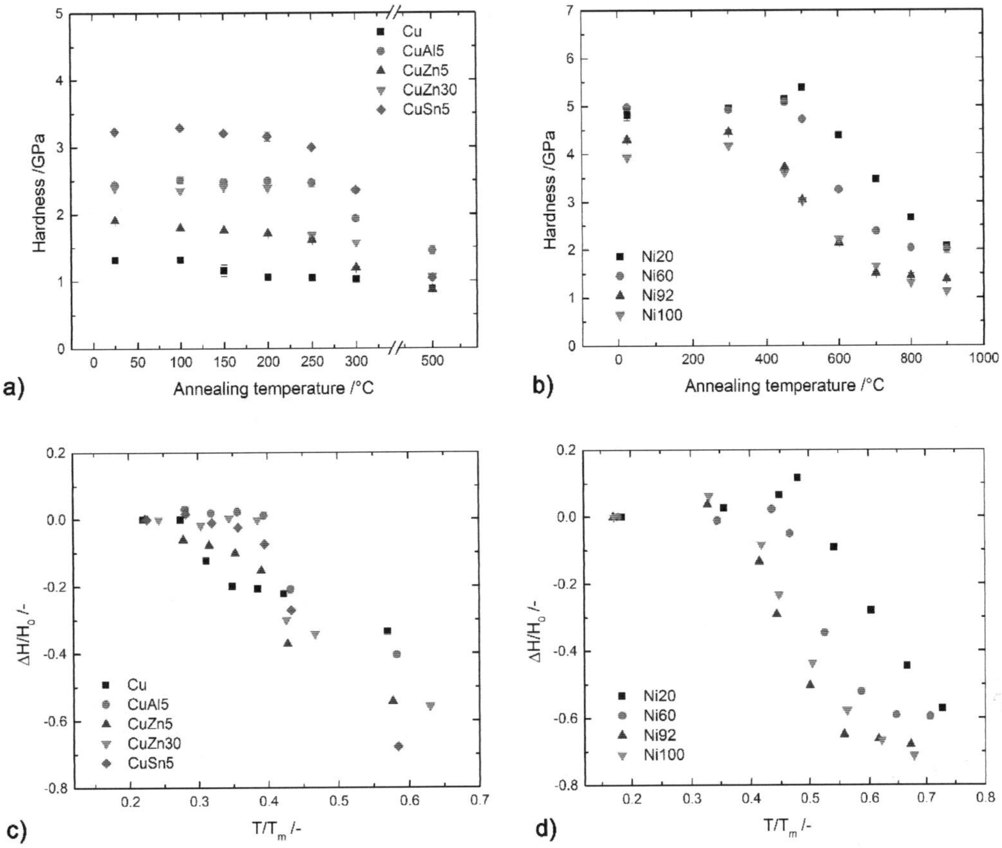

Figure 4. Hardness of the Cu-X (a) and Ni-X (b) series as a function of annealing temperature (1h) and normalized hardness change as function of the homologous temperature (T/T_m) for Cu-X (c) and Ni-X (d).

3.2.2. Deformation mechanisms. The strain rate sensitivity of the two alloy series is measured by means of nanoindentation strain rate jump and constant contact pressure tests according to Refs. [25, 26, 41, 42]. In figure 5, the hardness of Cu-X (a) and Ni-X (b) is plotted as a function of indentation depth and applied strain rate. The SRJ tests reveal an almost depth-independent hardness at indentation depths larger than 1000 nm but a pronounced strain rate sensitivity.

The observed SRS results in strain rate dependent hardness values, where an increase in $\dot{\varepsilon}$ is accompanied with higher H. The variation in strain rate occurs suddenly in an SRJ test, while the hardness exhibits a transient behavior until a steady state in deformation behavior is reached again. Analyzing these transients in more detail, significant differences between the individual solid solutions can be observed (see enlarged section of figure 5). Changing the strain rate from 0.05 to 0.001 s^{-1} results in softening of CuSn5, Ni20 and Ni60, followed by a short period of strain hardening. This behavior is most pronounced for the three alloys that show the highest SSH contributions (see table 1 & 2). This yield point phenomenon can still be seen for Ni92 and Ni100, but it is less pronounced. Cu, CuZn5, CuZn30 and CuAl5, on the other hand, exhibit only monotonic softening till a nearly constant hardness is reached with a short transient regime for CuAl5 and CuZn30 (medium SSH) and a long transient for Cu and CuZn5 (low SSH). Stress reduction tests [37, 38] suggest that dislocation-based plasticity is suppressed after a stress drop by temporary dislocation pinning of solutes. Consequently, thermally activated dislocation glide is hindered to such an extent that grain boundary-mediated mechanisms dominate over dislocation-based ones [37, 38]. Grain boundary-mediated mechanisms are of significant importance for dynamic recovery, which controls the strength by reducing the dislocation density while contributing to plastic strain [37, 38]. With further straining, mobile dislocations will be generated, which strengthens the materials till a steady state of dislocation annihilation and nucleation is reached. This explanation derived from stress reduction tests may also hold true for the strain hardening section after a strain rate jump as a reduction in strain rate is accompanied by a stress reduction and this yield phenomenon is only limited to alloys with a high SSH contribution. Moreover, alloys showing a very short transient coupled with a distinct change in hardness with changing strain rate seem to have a much more stable microstructure [33].

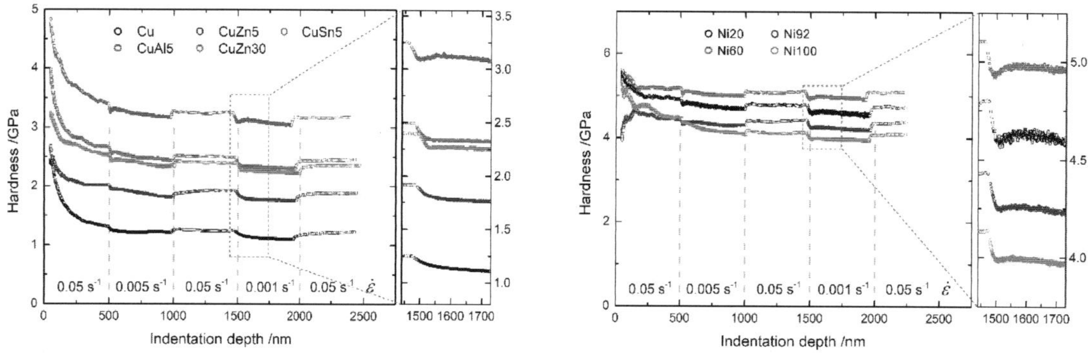

Figure 5. Indentation strain rate jump tests on HPT deformed Cu-X (left) and Ni-X (right).

Additional CCP indentation creep experiments are performed to access information on the deformation response and active deformation mechanisms at low strain rates (for further information regarding the CCP method, the authors refer to Refs. [25, 26]). The results of indentation creep tests performed at various contact pressures are shown in figure 6 a) for the Cu-X series and in b) for Ni20 and Ni100 as the two boundary compositions, in comparison with constant strain rate (CSR) and SRJ tests discussed previously. The data for all materials investigated show two different regimes, with a linear correlation between contact pressure and strain rate in each case. In the following, the two regimes are referred to as high stress and low stress regime. The transition between those two regimes takes place for all alloys at strain rates of about 10^{-5} - 10^{-6} s^{-1}. The slope of the linear fit, however, is significantly different for the two regimes indicating a change in the dominant mechanism carrying plastic deformation.

The material's response to a transient experiment such as an SRJ or CCP tests can be analyzed by the strain rate sensitivity exponent $m = \frac{d(lnH)}{d(ln\dot{\varepsilon})}$ and the corresponding activation volume $V = \frac{C \cdot \sqrt{3} \cdot k_B \cdot T}{m \cdot H}$ with C: constraint factor, k_B: Boltzmann constant, T: ambient temperature of 25 °C. In particular, V represents a signature for the dominant deformation mechanism. A higher SRS for nc or UFG metals can be explained by dislocation - grain boundary interactions, which are getting more pronounced for small grain sizes as the grain boundary density increases [41, 42, 44, 45]. Even though indentation SRJ testing is commonly used, it should be noted that the microstructure may not be stable within the plastic zone [33]. However, it should be noted, that different deformation mechanisms can be active simultaneously and that thus m and V only represent average values that provide information about the dominant deformation process.

The strain rate coefficients and activation volumes derived from the SRJ and CCP experiments are summarized in table 3. Both alloy series exhibit a SRS that is decreasing with grain size. This phenomenon is contrary to what is expected for fcc metals [41, 44]. Activation volumes of 19-33 b^3 (Cu-

IOP Conf. Series: Materials Science and Engineering 1249 (2022) 012003 doi:10.1088/1757-899X/1249/1/012003

X) or 17-20 b^3 (Ni-X) were determined in the high stress regime. In literature, activation volumes of 10-100 b^3 are discussed for dislocation-glide based plasticity (dislocation nucleation and/or dislocation cross-slip [25, 46-48]. Large stress reductions at the beginning of the CCP segment lead to a sharp drop in the activation volume. In the low stress regime, V values of 2 - 7 b^3 where derived for Cu-X and 2 - 3 b^3 for Ni20 and Ni100, indicating dislocation annihilation and grain boundary mediated processes become dominant [47, 48].

The creep behavior after a stress reduction is characterized by the net flow of dislocations. It has been found that large stress reductions can lead to a reversal in the direction of dislocation motion, causing annihilation at grain boundaries [37, 38, 49-51]. As a result, the dislocation density changes until equilibrium between dislocation nucleation and annihilation is obtained again at the new stress level. In a stress reduction test, this behavior is indicated by a negative strain rate sensitivity immediately after a sudden stress drop, but cannot be analyzed in a CCP test since the contact pressure changes incrementally, so only the long-term creep behavior (2500-3000 s) can be studied.

Table 3. SRS exponent m and activation volume V obtained by nanoindentation experiments for the HPT deformed Cu-X and Ni-X series at different stress levels.

	Strain rate sensitivity m		Activation volume V in b^3	
Alloy designation	High pressure (SRJ)	Low pressure (CCP)	High pressure (SRJ)	Low pressure (CCP)
Cu	0.024	0.501	33	2
CuAl5	0.017	0.164	25	3
CuZn5	0.037	0.318	19	3
CuZn30	0.020	0.068	20	7
CuSn5	0.012	0.111	23	3
Ni	0.014	0.112	20	3
Ni92	0.012	-	22	-
Ni60	0.013	-	19	-
Ni20	0.013	0.154	17	2

While the Cu-X series shows in the low stress regime quite similar activation volumes but strongly varying strain rate sensitivity exponents, m and V are almost similar for Ni20 and Ni100, despite having strongly varying solute content. This behavior further indicates an unstable microstructure for Cu and CuZn5, but also a comparably more stable microstructure for the Ni-X series as well as for the Cu-X samples showing medium to high SSH. A similar activation volume suggests that similar dislocation mechanisms are present in the alloys, however, grain coarsening processes also contribute to plastic deformation and may be the cause for higher m values. The Ni-X alloy series on the other hand exhibit similar m as well as V values within a stress regime, which points to a higher microstructural stability compared to the Cu-X series. It should be noted here that indentation testing was performed at room-temperature, which corresponds to a different homologous temperature for CuZn30 due to the lower melting point compared to the other Cu-X compositions. As we probe the time and temperature dependent material response during SRJ and CCP experiments, this has to be kept in mind. This fact may cause the higher activation volume for CuZn30 in the low pressure regime compared to the Cu-X series, indicating an easier dislocation motion at a given absolute temperature.

The change in hardness upon annealing as well as the differences in the transient and creep behavior at low strain rates indicate that SSH strongly affects the microstructural stability, i.e., the grain boundary mobility. A distinct hardness plateau to higher temperatures is observable for samples showing higher solute content, revealing lower grain boundary mobilities and thus higher structural stabilities. The indentation results regarding the transient behavior after a strain rate jump also indicate that samples with medium to high SSH contributions exhibit a higher microstructural stability. The reduced grain boundary mobility in turn explains the differences in the saturation of grain refinement for the different alloys and thus the different saturation grain sizes.

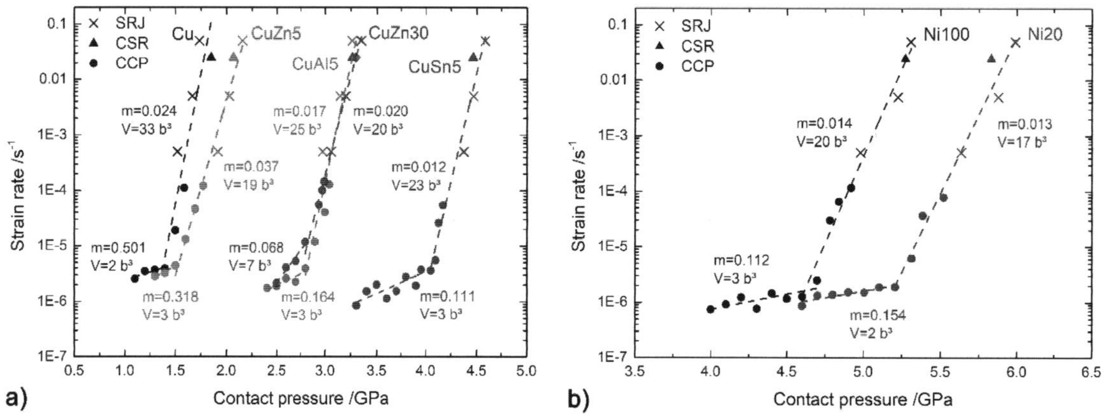

Figure 6. Stress dependent strain rate from nanoindentation constant contact pressure (CCP), constant strain rate (CSR) and strain rate jump (SRJ) testing of a) Cu-X and b) Ni20 and Ni100.

4. Conclusion/Summary

In summary, solid solution effects were studied for a Cu-X and a Ni-X alloy series with an emphasis on saturation grain size, thermal stability and corresponding deformation mechanisms. The Cu-X series allows a separation of alloying effects emerging from either SSH or SFE. The Ni-X sample series, on the other hand, targets a transition from dilute solid solutions to chemically complex HEAs in order to investigate solute effects in strongly varying environments.

The results indicate a strong correlation between SSH contribution and saturation grain size but not with the SFE in binary systems. Furthermore, this correlation holds also for chemically complex alloys using the Varvenne model of SSH for HEAs. The thermal stability of the different alloys after annealing was investigated using nanoindentation hardness testing and an enhanced stability was found for alloys showing an increased solute content (solute drag). For Ni20 and Ni60, Zener pinning sets in as an additional dragging force at intermediate temperatures due to the formation of secondary phases. Differences in the transient deformation behavior were observed after a strain rate jump. Alloys showing a high SSH contribution (CuSn5, Ni20, Ni60), exhibit a yield point phenomenon with a short strain hardening section, which is caused by dislocation - solute interactions. A reduced strain rate sensitivity with reduced grain size is observed for the Cu-X alloys, the Ni-X samples show very similar strain rate sensitivity exponents, despite having different grain sizes in the range from 55 nm (Ni20) to 142 nm (Ni100). The relatively high SRS of Cu and the CuZn alloys could be attributed to an unstable microstructure, whereas a low SRS in combination with reduced grain size refers to a higher microstructural stability. The observed activation volumes point towards dislocation-glide based plasticity (dislocation nucleation and/or dislocation cross-slip) being active in the as-deformed microstructures in the high stress range. The comparison of SRS data from different indentation methods reveals a transition in the dominant deformation mechanisms at strain rates of about 10^{-5} - 10^{-6} s^{-1}. Large stress reductions at the beginning of the CCP segment lead to a sharp drop in the activation volume. Activation volumes of 2-3 b^3 were determined by CCP tests in the low stress regime. This implies that at low strain rates, diffusion-controlled processes and dislocation annihilation are dominating the deformation behavior.

Acknowledgements

This work was funded by the German Research Foundation (DFG) under grant numbers DU 424/13-2 and DU 424/11-1. The authors thank Dr. Mathilde Laurent-Brocq for providing the Ni-X alloy systems and Dr. Achim Kuhn (Wieland-Werke AG, Germany) for providing the Cu solid solutions.

References

[1] Pippan R, Scheriau S, Taylor A, Hafok M, Hohenwarter A, and Bachmaier A 2010 Saturation of Fragmentation During Severe Plastic Deformation *Annu. Rev. Mater. Res.* **40** 319-43
[2] Zhilyaev A and Langdon T 2008 Using high-pressure torsion for metal processing: Fundamentals and applications *Prog. Mater. Sci.* **53** 893-979
[3] Valiev R Z, Estrin Y, Horita Z, Langdon T G, Zehetbauer M J, and Zhu Y T 2006 Producing bulk ultrafine-grained materials by severe plastic deformation *JOM* **58** 33-9
[4] Hansen N 2001 New discoveries in deformed metals *Metall. Mater. Trans. A* **32** 2917-35
[5] Hughes D A and Hansen N 1997 High angle boundaries formed by grain subdivision mechanisms *Acta Mater.* **45** 3871-86
[6] Humphreys F J and Hatherly M 2004 *Recrystallization and Related Annealing Phenomena* 2nd Ed Oxford Elsevier
[7] Hansen N and Kuhlmann-Wilsdorf D 1986 Low energy dislocation structures due to unidirectional deformation at low temperatures *Mater. Sci. Eng.* **81** 141- 61
[8] Edalati K et al. 2022 Nanomaterials by severe plastic deformation: review of historical developments and recent advances *Materials Research Letters* **10** 163-256
[9] Edalati K, Akama D, Nishio A, Lee S, Yonenaga Y, Cubero-Sesin J M and Horita Z 2014 Influence of dislocation–solute atom interactions and stacking fault energy on grain size of single-phase alloys after severe plastic deformation using high-pressure torsion *Acta Mater.* **69** 68-77
[10] Bachmaier A, Hafok M and Pippan R 2010 Rate Independent and Rate Dependent Structural Evolution during Severe Plastic Deformation *Mater. Trans.* **51** 8-13
[11] Hebesberger T, Stüwe H P, Vorhauer A, Wetscher F and Pippan R 2005 Structure of Cu deformed by high pressure torsion *Acta Mater.* **53** 393-402
[12] Mohamed F A and Dheda S S 2012 On the minimum grain size obtainable by high-pressure torsion *Mater. Sci.Eng. A* **558** 59-63
[13] Balogh L, Ungár T, Zhao Y, Zhu Y T, Horita Z, Xu C and Langdon T G 2008 Influence of stacking fault energy on microstructural characteristics of ultrafine-grain copper and copper–zinc alloys *Acta Mater.* **56** 809-20
[14] Labusch R 1970 A Statistical Theory of Solid Solution Hardening *Phys. Stat. Sol.* **41** 659-69
[15] Labusch R 1972 Statistische Theorien der Mischkristallhärtung *Acta Metall.* **20** 917-27
[16] Bruder E, Braun P, Rehman H u, Marceau R K W, Taylor A S, Pippan R and Durst K 2018 Influence of solute effects on the saturation grain size and rate sensitivity in Cu-X alloys *Scripta Mater.* **144** 5-8
[17] Keil T, Bruder E, Laurent-Brocq M and Durst K 2021 From diluted solid solutions to high entropy alloys: Saturation grain size and mechanical properties after high pressure torsion *Scripta Mater.* **192** 43-8
[18] Keil T, Taheriniya S, Bruder E, Wilde G and Durst K 2022 Effects of solutes on thermal stability, microstructure and mechanical properties in CrMnFeCoNi based alloys after high pressure torsion *Acta Mater.* **227** 117689

[19] Laurent-Brocq M, Perrière L, Pirès R, Prima F, Vermaut P and Champion Y 2017 From diluted solid solutions to high entropy alloys: On the evolution of properties with composition of multi-components alloys *Mater. Sci. Eng. A* **696** 228-35
[20] Bracq G, Laurent-Brocq M, Varvenne C, Perrière L, Curtin W A, Joubert J M and Guillot I 2019 Combining experiments and modelling to explore the solid solution strengthening of high and medium entropy alloys *Acta Mater.* **177** 266-79
[21] Keil T, Bruder E and Durst K 2019 Exploring the compositional parameter space of high-entropy alloys using a diffusion couple approach *Mater Des.* **176** 107816
[22] LaRosa C R, Shih M, Varvenne C and Ghazisaeidi M 2019 Solid solution strengthening theories of high-entropy alloys *Mater. Charact.* **151** 310-17
[23] Varvenne C and Curtin W A 2018 Predicting yield strengths of noble metal high entropy alloys *Scripta Mater.* **142** 92-5
[24] Varvenne C, Luque A and Curtin W A 2016 Theory of strengthening in fcc high entropy alloys *Acta Mater.* **118** p 164-76
[25] Minnert C and Durst K 2021 Nanoindentation creep testing: Advantages and limitations of the constant contact pressure method *J. Mater. Res.* **37** 567-79
[26] Prach O, Minnert C, Johanns K E and Durst K 2019 A new nanoindentation creep technique using constant contact pressure *J. Mater. Res.* **34** 2492-500
[27] Kese K O, Li Z C and Bergman B 2005 Method to account for true contact area in soda-lime glass during nanoindentation with the Berkovich tip *Mater. Sci. Eng. A* **404** 1-8
[28] Zhilyaev A P, Lee S, Nurislamova GV, Valiev R Z and Langdon T G 2001 Microhardness and microstructural evolution in pure nickel during high-pressure torsion *Scripta Mater.* **44** 2753-58
[29] Hafok M and Pippan R 2010 Influence of stacking fault energy and alloying on stage V hardening of HPT-deformed materials: Dedicated to Professor Dr. H.-P. Degischer on the occasion of his 65th birthday *International J. Mater. Res.* **101** 1097-04
[30] Freudenberger J, Kauffmann A, Klauß H, Marr T, Nenkov K, Subramanya Sarma V and Schultz L 2010 Studies on recrystallization of single-phase copper alloys by resistance measurements *Acta Mater.* **58** 2324-29
[31] Carter C B and Ray I L F 1977 On the stacking-fault energies of copper alloys *Philos. Mag.* **35** 189-200
[32] Gallagher PCJ 1970 The influence of alloying, temperature, and related effects on the stacking fault energy *Metall. Trans.* **1** 2429-61
[33] Maier-Kiener V, An X, Li L, Zhang Z, Pippan R and Durst K 2017 Influence of solid solution strengthening on the local mechanical properties of single crystal and ultrafine-grained binary Cu–AlX solid solutions *J.Mater. Res.* **32** 4583-91
[34] Renk O and Pippan R 2019 Saturation of Grain Refinement during Severe Plastic Deformation of Single Phase Materials: Reconsiderations, Current Status and Open Questions *Mater. Trans.* **60** 1270-82
[35] Wilde G, Ribbe J, Reglitz G, Wegner M, Rösner H, Estrin Y, Zehetbauer M, Setman D and Divinski S 2010 Plasticity and Grain Boundary Diffusion at Small Grain Sizes *Adv. Eng. Mater.* **12** 758-64
[36] Wilde G and Divinski S 2019 Grain Boundaries and Diffusion Phenomena in Severely Deformed Materials *Mater. Trans.* **60** 1302-15
[37] Sun Z, Van Petegem S, Cervellino A, Blum W and Van Swygenhoven H 2016 Grain size and alloying effects on dynamic recovery in nanocrystalline metals *Acta Mater.* **119** p 104-14
[38] Sun Z, Van Petegem S, Cervellino A, Durst K, Blum W and Van Swygenhoven H 2015 Dynamic recovery in nanocrystalline Ni *Acta Mater.* **91** 91-100
[39] Keil T, Utt D, Bruder E, Stukowski A, Albe K and Durst K 2021 Solid solution hardening in CrMnFeCoNi-based high entropy alloy systems studied by a combinatorial approach *J. Mater. Res.* **36** 2558-70

[40] Schuh B, Mendez-Martin F, Völker B, George E P, Clemens H, Pippan R and Hohenwarter A 2015 Mechanical properties, microstructure and thermal stability of a nanocrystalline CoCrFeMnNi high-entropy alloy after severe plastic deformation *Acta Mater.* **96** 258-68

[41] Maier V, Durst K, Mueller J, Backes B, Höppel H W and Göken M 2011 Nanoindentation strain-rate jump tests for determining the local strain-rate sensitivity in nanocrystalline Ni and ultrafine-grained Al *J. Mater. Res.* **26** 1421-30

[42] Durst K and Maier V 2015 Dynamic nanoindentation testing for studying thermally activated processes from single to nanocrystalline metals *Curr. Opin. Solid State Mater.Sci.* **19** 340-53

[43] Blum W, Li Y J and Durst K 2009 Stability of ultrafine-grained Cu to subgrain coarsening and recrystallization in annealing and deformation at elevated temperatures *Acta Mater.* **57** 5207-17

[44] Schwaiger R, Moser B, Dao M, Chollacoop N and Suresh S 2003 Some critical experiments on the strain-rate sensitivity of nanocrystalline nickel *Acta Mater.* **51** 5159-72

[45] Wei Q, Cheng S, Ramesh K T and Ma E 2004 Effect of nanocrystalline and ultrafine grain sizes on the strain rate sensitivity and activation volume: fcc versus bcc metals *Mater. Sci. Eng. A* **381** 71-9

[46] Wehrs J, Mohanty G, Guillonneau G, Taylor A A, Maeder X, Frey D, Philippe L, Mischler S, Wheeler J M and Michler J 2015 Comparison of In Situ Micromechanical Strain-Rate Sensitivity Measurement Techniques *JOM* **67** 1684-93

[47] Yang X S, Wang Y J, Wang G Y, Zhai H R, Dai L H and Zhang T Y 2016 Time, stress, and temperature-dependent deformation in nanostructured copper: Stress relaxation tests and simulations *Acta Mater.* **108** 252-63

[48] Monclús M A and Molina-Aldareguia J M 2018 *High Temperature Nanomechanical Testing* in Handbook of Mechanics of Materials 1-29

[49] Blum W, Eisenlohr P and Hu J 2016 Interpretation of unloading tests on nanocrystalline Cu in terms of two mechanisms of deformation *Mater. Sci. Eng. A* **665** 171-4

[50] Blum W, Eisenlohr P and Breutinger F 2002 Understanding creep—a review *Metall. Mater. Tran. A* **33** 291-303

[51] Milička K 1993 Constant structure creep in metals after stress reduction in steady state stage *Acta Metall.Mater.* **41** 1163-72

IOP Conf. Series: Materials Science and Engineering 1249 (2022) 012004 doi:10.1088/1757-899X/1249/1/012004

Microstructural Black Swans

B L Boyce[1]

[1]Sandia National Laboratories, Albuquerque, NM, 87185

E-mail:blboyce@sandia.gov

Abstract. Black swans are a metaphor for rare events with extreme consequences. In the domain of structural materials, black swans represent features in the microstructure that lead to catastrophic failure; as a result of their rarity, they are difficult to observe and often overlooked. These unusual weakest-link features are described variously as incipient, emergent, or anomalous. They give rise to localization, percolation, or avalanche events such as fracture, ductile rupture, dielectric breakdown, corrosion pit nucleation, and fatigue-crack initiation; as such, they are limiting cases in the concept of a representative volume. In this perspective, three examples are given of rare microstructural features and how they limit the mechanical reliability of structural metals. After taking stock of these examples, a future outlook considers the need for high-throughput testing and non-destructive characterization as well as detection algorithms and materials modelling strategies, including accelerated machine learning methods, that can capture anomalous events.

1. Introduction

In his 2007 book, statistician Nassim Taleb explored the notion of 'black swan' events: rare occurrences with significant consequences that are often overlooked as a result of our cognitive bias towards typical behavior [1]. For centuries, a swan with black features was inconceivable and the term was used as a metaphor to describe something that cannot exist. After the discovery of actual black swans in the late 17th century, the metaphor evolved to instead describe unexpected, rare events, often with significant consequences. The scientific method, in its requirement for a hypothesis formulation as a basis for investigation, along with the natural tendency for confirmation bias, often tends to minimize the study of such extreme outliers. In the field of materials science, unexpected material failures are caused by either overdriven external stimuli or rare weaknesses/imperfections in the material; the former referring to the driving-force for failure, the latter describing the material's resistance to failure.

Generally, all materials can be described by their composition and microstructure. The term 'microstructure' denotes a multiscale cornucopia of configurational variations: phase content, phase morphology, grain size, grain shape, crystallographic texture, inclusions, dislocation density, dislocation arrangements, residual stresses, grain boundary types, vacancy concentration, porosity, etc. (see Figure 1 in [2]). Each of these microstructural quantities have an associated statistical distribution and can be spatially heterogeneous. While some properties like density and melting point are controlled strongly by the composition and associated phase fractions with minimal sensitivity to microstructure parameters, other properties like ductile rupture and fatigue crack initiation are highly sensitive to local variations in the microstructure.

The effects of microstructural black swans have left an indelible mark on the engineering profession. To account for the possibility of undetected defects, the aerospace industry has adopted a damage-tolerant approach to turbine engine and airframe design since the 1970's [3, 4]. However, anticipating unexpected, rare microstructural features/defects can still be problematic. For example, in 1989, United Airlines Flight 232 from Denver to Philadelphia experienced an engine failure that led to extensive casualties. After thorough investigation, the failure was linked to the presence of an unexpected subsurface "hard-alpha" inclusion in the titanium alloy [5]. This discovery of a latent defect type has driven changes in the metallurgical processing of titanium alloys along with a rethinking of defect inspection.

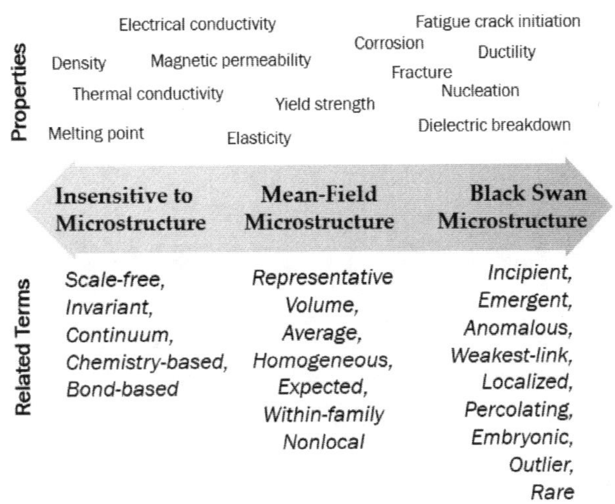

Figure 1. A qualitative mapping of different material properties on the spectrum that ranges from microstructurally-insensitive to highly sensitive to rare weakest-links.

In this perspective, three examples are described where black swan events have become apparent. These examples all pertain to structural metals: (1) high-cycle fatigue crack initiation in nanocrystalline metals, (2) weakest-link fracture processes in additively manufactured stainless steels, and (3) the incipient stage of deformation-induced void nucleation in ductile metals, and. Through this summary, we reflect on opportunities for high-throughput testing methods alongside high-throughput microstructural characterization methods and advanced data analytics including machine learning algorithms to identify microstructural black swans.

2. Examples of black swan microstructural events in structural metals

2.1. Fatigue-failures induced by abnormal grain growth

While thermal grain growth driven by a reduction in grain boundary area is a common occurrence, in this example, fatigue-loading of nanocrystalline metals is shown to induce only a few grains to grow, spurring the initiation of a fatigue crack and ultimately causing catastrophic failure. In *normal* grain growth the average grain size increases, yet the shape of the grain size distribution stays constant, as is typically characterized by a skewed distribution such as lognormal or gamma [6]. Analytic considerations suggested that the maximum grain size should be no greater than 2-3 times the average grain size [7] . Hillert described a far less common form of grain growth, *abnormal* grain growth, as having occurred when a minority of grains grow anomalously [8]; as a result of abnormal grain growth, the grain size distribution evolves and the upper limit on grain size is unbounded [7].

There has been a recent realization that abnormal grain growth could be induced by fatigue loading of nanocrystalline metals [9, 10]. While the initial grain size in these metals started <100 nm, after high-cycle fatigue loading, a few grains could grow to ~500 nm [10]. The fatigue cracks appeared to initiate in those regions of grain growth. The smaller nanocrystalline grains were considered too small to support the collective dislocation activity needed to form sufficient extrusions and initiate fatigue cracks [11].

The concept of fatigue-induced grain growth as a necessary precursor to crack initiation in nanocrystalline metals was challenging to confirm experimentally. Even in very small thin film tensile bars, with a gauge section of 5 μm x 1 mm x 5 mm, there would exist ~$2x10^{11}$ initial grains, yet only one large grain was needed to induce crack initiation. The rare abnormal grain growth event could be detected *post mortem* by identifying the site of crack initiation based on the fracture surface, and subsequently revealing the subsurface grain morphology via focussed ion beam dissection [11]. However, a novel experimental technique was required to detect such a rare event as it occurred, to track it's evolution, and to confirm that the grain growth preceded crack initiation, illustrated in Figure 2.

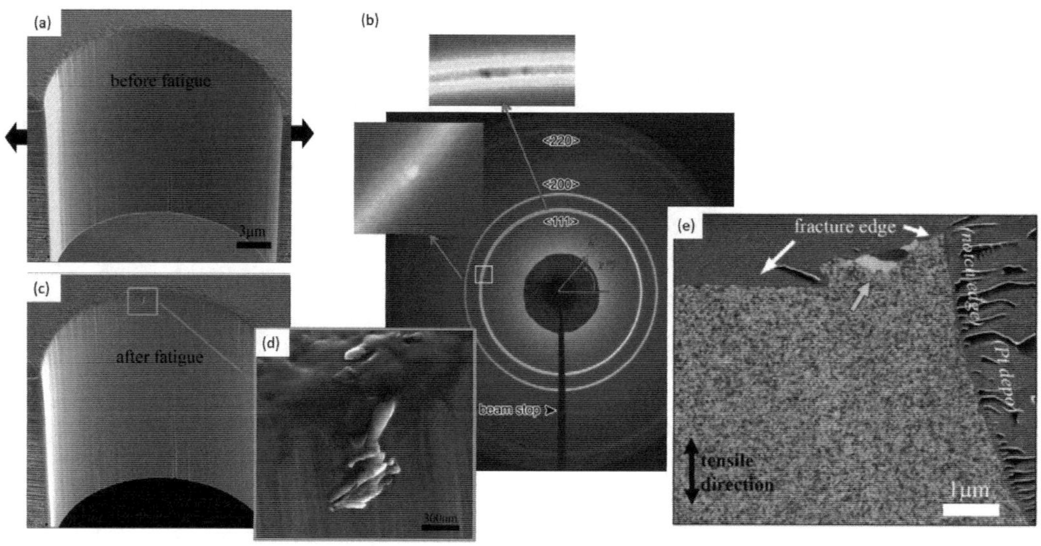

Figure 2. (a) A notched thin-film tensile fatigue specimen prior to fatigue loading. (b) Two separate examples of anomalous spikes in the Debye ring formed during transmission x-ray diffraction through a nanocrystalline film with anomalously large grains. (c) After the detection of anomalous diffraction spikes (red arrows), reinspection of the fatigue specimen revealed (d) localized surface extrusions and the formation of an incipient crack. (e) After fatigue failure, cross-sectional ion-channeling contrast imaging of the crack initiation zone confirmed the presence of a few large grains (green arrow) in a sea of nanocrystalline grains. For more details, see [14-16].

X-ray diffraction peak broadening analysis has been used for over a century to observe grain sizes in the sub-micrometre regime, using methods including Scherrer, Williamson-Hall, and Warren-Averbach [12, 13]. However, these techniques are not well suited to finding black swan crystallites. An alternative technique was developed that involved careful inspection of the Debye ring intensity profiles [14]. The Debye rings were created by monochromatic transmission x-ray diffraction through thin films of nanocrystalline metal. Smooth undulations in the intensity of the Debye rings corresponded to variations in the average crystallographic texture. But anomalous large individual grains caused isolated 'spikes' in the diffraction rings, as shown in Figure 2b. These local regions of intense diffraction were statistically anomalous with local signals rising >> 6 standard deviations over the Debye diffraction ring, caused by the anomalous volume of crystallite with a particular orientation. The new technique was first used to confirm isolated abnormal grain growth caused by fatigue loading, and subsequently used to observe the formation and evolution of the abnormal grains during in-situ high-cycle fatigue loading [15, 16]. These emergent grains, confirmed responsible for fatigue failure in nanocrystalline alloys, were exceedingly rare: occupying <0.00001% of the interrogation volume [14].

2.2. *Rare weakest-link failures in additively manufactured stainless steels*

While the previous example explored a weakest-link defect that emerged as a result of external stimuli, e.g., fatigue loading, this example explores a rare defect that was formed during material manufacturing. With the recent proliferation of additive manufacturing techniques for metals, there has been an urgent interest in understanding the reliability of these materials and accelerating their pathway to qualification [17, 18]. Traditional standardized test methods for measuring the mechanical reliability of metals are typically slow. While hardness testing and associated nanoindentation testing can be accelerated to provide high-throughput data collection [19, 20], they lack evidence of the tensile behaviour, especially in the post-necking and failure regime.

To address this shortcoming, a high-throughput tensile method was developed for additively manufactured metals. The initial development involved manual loading of each tensile specimen into a system that employed drop-in grips, non-contact strain measurement, and software automation [21]; while a subsequent version also employed automated sample loading and dimensional measurements [22]. The high-throughput methods allowed for 10-100 times improvement in sampling rate compared to conventional test methods, with essentially no loss in data fidelity.

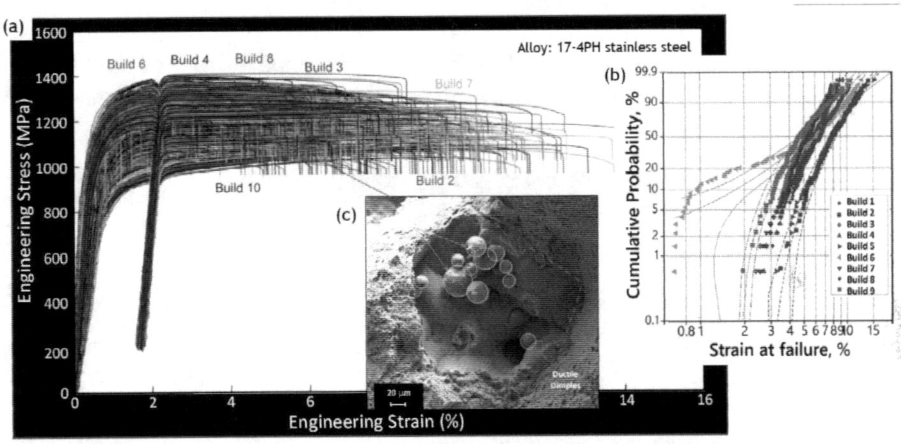

Figure 3. (a) Approximately 1000 tensile stress-strain curves collected under nominally identical conditions with distinct colors denoting the sequential build order; (b) stress-strain data was reduced into cumulative probability plots of properties such as ductility (strain at failure) as a function of build number. While most builds complied with a 3-parameter Weibull description, Build 6 exhibited anomalous bimodal response. (c) An example of a scanning electron microscope image of a lack-of-fusion pore that represents a typical defect in these materials. For more details, see [21].

In one such high-throughput dataset, shown in Figure 3, approximately 1000 tensile tests across several builds were collected under nominally identical conditions for an additively manufactured 17-4PH precipitation hardened stainless steel. While the tensile properties of most builds followed a 3-parameter Weibull distribution with only small variations in mean response, one build exhibited a significant fraction of anomalies. About 2% of the tensile tests experienced a different mode of failure that had never been observed before. The anomalous failure mode, deduced by subsequent fractographic examination, involved the coalescence of internal porosity into a pore network. Porosity is formed either by lack-of-fusion caused by insufficient volumetric energy density or keyhole collapse caused by excess volumetric energy density [23, 24]. When this porosity is sufficiently large, it becomes the ductility-limited microstructural feature. While all the specimens had some degree of internal lack-of-fusion porosity as shown in Figure 3c, the rare defect involved the link-up of several lack-of-fusion pores into a meta-cavity [21]. This meta-cavity was not only much larger, but it also connected to the free surface causing internal oxidation during heat-treatment. The formation of this rare meta-cavity was thought to be caused by the density and spatial correlation of lack-of-fusion pores reaching a percolation threshold [21].

2.3. *Incipient void nucleation in ductile metals*

The process of ductile tearing in metals often involves a sequence of void nucleation, void growth, and coalescence to final rupture [25]. Ultimately, a fracture surface of a ductile metal is littered with ductile dimples: after final failure, the process of nucleation-growth-and-coalescence does not appear to be rare. However, the very first sites of damage initiation start at the weakest links in the microstructure well before final rupture and kickstart the process that eventually causes catastrophic failure. Void nucleation is a not a rare event per say, but the embryonic conditions that lead to the initial onset of damage stem from very low probability events thought to be associated with triggering microstructural configurations. In metals with inclusions or hard phases, the nucleation process typically begins either via particle cracking or interfacial debonding [26]. However, in pure metals, less is pedagogically understood about the process. Early studies by Wilsdorf and others have indicated that the process initiates at deformation-induced dislocation substructures [27].

The voids that form during deformation initially occupy infinitesimally small region of the tensile bar, <1 ppb by volume, and are either a purely statistical occurrence, or more likely, they form at rare, triggering features in the microstructure. However, catching the incipient weakest link features that trigger void nucleation in metals is not trivial as it occurs subsurface and only after extensive

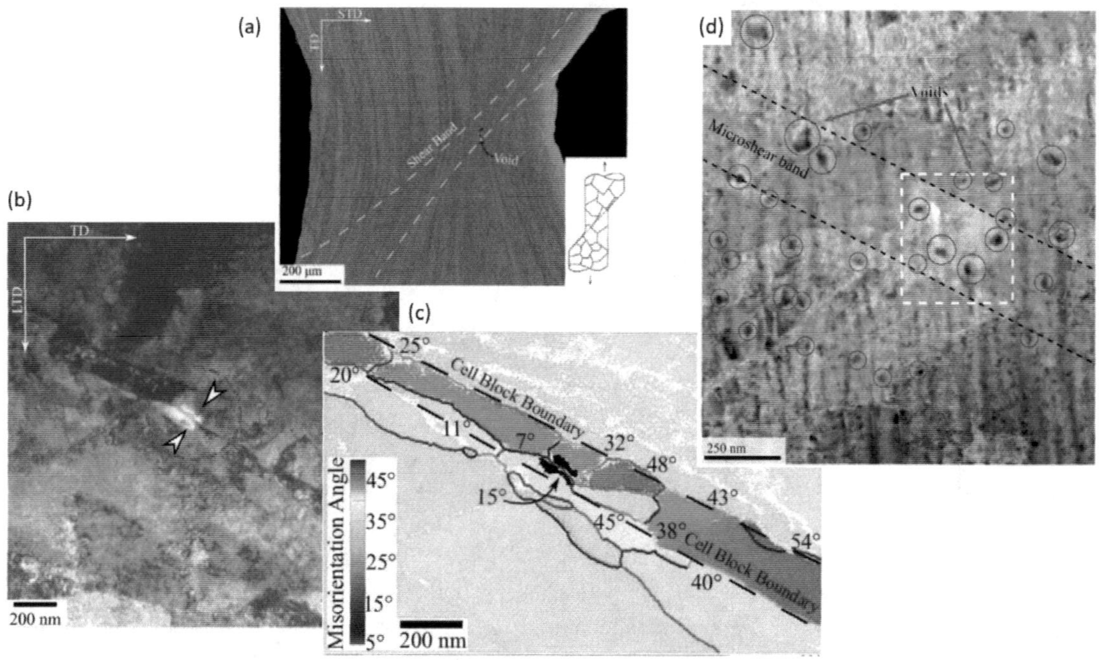

Figure 4. The observation of void nucleation events depends in part on the magnification / image resolution available. Here in high-purity nickel, a single void several micrometers in size is found inside a shear band after the onset of necking. (b) Here in high-purity tantalum at a higher magnification, incipient voids are found in the range of 100's of nanometers (c) formed at deformation-induced cell block boundaries. (d) And in high-purity copper, high-deformation regions possess incipient voids that are not rare, but rather prolific (red circles), apparently limited instead by their ability to grow. For more details, see [25, 30, 32].

deformation. While microtomography has emerged as a more widely used tool for such studies, it still only resolves micrometre scale voids [28, 29], which may not be at their incipient stage. An alternative method, albeit destructive in nature, is to interrupt the rupture process during a tensile test, and cross-section the partially deformed material to interrogate subsurface. As with microtomography, the interpretation of 'incipient' depends on the resolution with which voids can be resolved. Most images of deformation-induced void formation have focused on the microscale, as in Figure 4a. However, careful examination at the sub-micrometre scale using various forms of electron microscopy has revealed more comprehensive details of the unique and rare features in the microstructure where ductile voids form.

In tantalum, transmission electron microscopy coupled with transmission Kikuchi diffraction indicated that voids primarily initiate at and grow along high-angle dislocation cell-block boundaries [30]. However, there is not a perfect correlation between that microstructural feature and the void nucleation. That is to say, not every void formed precisely at a specific dislocation cell block boundary type; nor does every cell block boundary nucleate a void. The lack of definitive clarity could be because the surface observation does not reveal underlying 3-dimensional details. Or it could be a lack of resolution to identify the black swan feature that consistently drives nucleation. Or, perhaps most likely, it is the inability to observe the process dynamically in-situ, where for example, a dislocation channel may be actively pumping a heavy dose of crystallographic defects into a specific location.

In this case where experimental observation only gives partial details into the rare black swan microstructure, atomistic modelling has added further dynamical details. Through molecular dynamics simulation, a nanoscale void in the presence of a high density of glissile dislocations grew through a process of glide and cross-slip into the surface of the void [31]. The initial formation of a nanoscale void could occur through a process of vacancy condensation [32], itself a rare event given that the equilibrium concentration of vacancies is vanishingly small, one part in $\sim 10^{20}$. In ductile rupture, the microstructural 'black swan' event may require both the necessary strain-induced vacancy supersaturation [33] to nucleate the nanoscale void, and the high dislocation activity necessary to grow the void beyond the nanoscale. Recent observations in copper, Figure 4d, indeed found an unexpected plethora of nanoscale voids in the highly deformed region [32], suggesting that incipient nucleation was not rate-limiting but that the limiting condition also required a microstructural condition that drove early growth of the void.

3. Summary and Future outlook

The present perspective briefly reflected on three examples of black swan microstructural conditions that limit the mechanical performance of structural metals. There are numerous other examples in the literature within the domain of structural materials reliability. For example, Jha et al. found that the rarest and most extreme life-limiting features in a titanium alloy involved not just one microstructural feature, but a 'neighbourhood' of collaboratively-weak features [34]. Or, other researchers have revealed that the initial onset of plasticity occurs via dislocation avalanches that can be described through weakest link statistics [35]. Regardless of the examples chosen, these cases elucidate the benefits of robust statistical sampling of outlier events through high-throughput testing and characterization as well as stochastic modelling.

3.1. The role of high-throughput testing and characterization

In each of the three black swan examples provided herein employed distinct approaches to high-throughput characterization during or after exposure to stimuli that trigger weakest-link events. In the case of abnormal grain growth in nanocrystalline metals, a new approach to interpreting an x-ray diffraction pattern was required. Moreover, this technique for identifying statistically anomalous events required a dataset with rigorous statistical underpinnings; necessitating the exceptional counting statistics afforded by a synchrotron x-ray source. In the case of failure in additive manufacturing, a novel high-throughput tensile method was developed to enable >1000 tensile tests within the budget and time of an ordinary materials research program. And in the case of the ductile void nucleation problem, numerous high-resolution transmission Kikuchi diffraction patterns were feasible only due to the recent development of exceptionally high-speed CMOS-based electron detectors in electron microscopy [36]. Thus, in these examples, we see the benefits of: (1) high-intensity interrogation sources that enable energy-matter interactions with excellent counting statistics to detect anomalous events, (2) non-standardized test methods that reduce the cost and time of material preparation and testing over conventional methods, and (3) accelerated mapping techniques that provide larger surveys of materials and/or higher-resolution/multi-fidelity datasets.

A challenge remains that many commercially-available materials characterization instruments are not designed for the high-throughput statistical data collection needed to identify rare black swan events. Automated server upload, automated metadata tagging, accessible/robust API's, and automated sample loading are common limitations for many instruments. In addition, materials sample synthesis and specimen preparation techniques are not generally amenable to high-throughput. There are notable exceptions such as the RoboMET automated metallographic platform [37], used extensively for 3D materials characterization. Yet some of the most important characterization tools for structural metals, including transmission electron microscopy and electron backscatter diffraction, still typically require onerous sample preparation pathways.

3.2. The role of computational methods

Microstructural black swans may require new strategies to ICME (Integrated Computational Materials Engineering). Stochastic approaches to microstructural representation such as Ghosh's Statistically Equivalent Volume Elements [36] or Kalidindi's n-point spatial correlations [38] have only recently emerged as reduced-order representations for stochastic microstructural configurations and their adoption is not yet widespread. Moreover, high-throughput computational methods are allowing comprehensive exploration of the high-dimensional space of potential microstructural configurations, e.g. [39]. Accelerated computational methods such as phase field accelerated by long-short term memory [40] are also enabling a more rapid exploration of microstructural configurational space.

In addition to exploring stochastic configurational space, computational methods are being used in to identify anomalous features in experimentally-acquired datasets. In the nanocrystalline grain growth example provided in the previous section, the anomalous spikes in the diffraction pattern were identified automatically by an explicit search algorithm and a traditional statistical threshold approach based on the Z-score [14]. There are many different approaches for anomaly detection; relevant example references relating to anomaly detection in microstructure and materials manufacturing are provided here for several techniques: k-nearest neighbour [41], local outlier factor and isolation forests [42, 43], support vector machines [44], replicator neural networks [45], autoencoders [46], long short-term memory neural networks [47], Bayesian networks [48], hidden Markov models [49, 50], cluster-analysis methods [51, 52], and feature bagging [53]. While many of these techniques are unsupervised and can automatically detect anomalies, some methods like decision trees and k-nearest neighbours typically require supervised training by an expert and therefore foreknowledge of the anomaly. The different techniques also have varying degrees of computational expense; gpu-intensive machine learning approaches benefit from edge computing capabilities to perform real-time or near-real-time analysis for in-situ detection of material evolution and incipient events; over the past several years commercial vendors have begun to provide edge solutions for such applications.

A challenge that remains for computational approaches is the prediction of extreme outlier events. While high-throughput stochastic approaches can explore the epistemic space of known variables, those approaches may be limited by their assumptions on the shape of the distribution tails. The ubiquitous gaussian distribution is unbounded in the tails, a feature that is nonphysical for all quantities, necessitating the use of bounded distributions (e.g., Weibull). Alternatively, the Black Swan author Taleb advocates for the use of "fat-tail" distributions, noting that real-world "events are inherently fat tailed" [54, 55]. While fat tail distributions are still non-physical as they lack bounds and possess an ill-defined mean, the spirit of being cautious in the tails of distributions is laudable. Finally, the most extreme black swan events may involve microstructural features that are statistically out-of-family, stemming from a lurking second distribution, and unanticipated without experimental forensic evidence. Accounting for such black swan events will continue to challenge computational methods and alone forms a rational basis for the use of safety factors as a hedge in engineering design.

Acknowledgements
The work related to nanocrystalline alloys was supported by the United States Department of Energy (DOE), Office of Basic Energy Sciences (BES), Division of Materials Science and Engineering. The other two examples were funded by Laboratory Directed Research and Development. This work was performed, in part, at the Center for Integrated Nanotechnologies, an Office of Science User Facility operated for the U.S. Department of Energy. Sandia National Laboratories is a multi-mission laboratory managed and operated by National Technology and Engineering Solutions of Sandia, LLC., a wholly owned subsidiary of Honeywell International, Inc., for the U.S. Department of Energy's National Nuclear Security Administration under Contract No. DE-NA0003525. The views of the U.S. Department of Energy or the U.S. Government.

References

[1] Taleb NN 2007 *The black swan: The impact of the highly improbable* Random house
[2] Kalidindi SR, Buzzy M, Boyce BL and Dingreville R Digital Twins for Materials *Front. Mater.,* 48
[3] Broek D 2012 *The practical use of fracture mechanics* Springer Science & Business Media
[4] Newman Jr J 1998 The merging of fatigue and fracture mechanics concepts: a historical perspective *Prog. Aerosp. Sci.* **34** 5-6 347-90
[5] McEvily A, Ishihara S and Mutoh Y 2016 1989 DC-10 crash: A cold case mystery solved *Eng. Fract. Mech.* **157** 154-65
[6] Zöllner D and Streitenberger P 2010 Grain size distributions in normal grain growth *Practic. Metallog.* **47** 11 618-39
[7] Vaz MF and Fortes M 1988 Grain size distribution: The lognormal and the gamma distribution functions *Scripta Metall.* **22** 1 35-40
[8] Hillert M 1965 On the theory of normal and abnormal grain growth *Acta Metall.* **13** 3 227-38
[9] Vinogradov A and Hashimoto S 2003 Fatigue of severely deformed metals *Adv. Eng. Mater.* **5** 5 351-8
[10] Padilla H and Boyce B 2010 A review of fatigue behavior in nanocrystalline metals *Exper. Mech.* **50** 1 5-23
[11] Boyce BL and Padilla HA 2011 Anomalous fatigue behavior and fatigue-induced grain growth in nanocrystalline nickel alloys *Metall. Mater. Trans.* A **42** 7 1793-804
[12] Scardi P, Leoni M and Delhez R 2004 Line broadening analysis using integral breadth methods: a critical review *J. Appl. Crystall.* **37** 3 381-90
[13] Kunka C, Boyce BL, Foiles SM and Dingreville R 2019 Revealing inconsistencies in X-ray width methods for nanomaterials *Nanoscale* **11** 46 22456-66
[14] Boyce BL, Furnish TA, Padilla H, Van Campen D and Mehta A 2015 Detecting rare, abnormally large grains by x-ray diffraction *J. Mater. Sci.* **50**, 20, 6719-29
[15] Furnish TA, Mehta A, Van Campen D, Bufford DC, Hattar K and Boyce BL 2017 The onset and evolution of fatigue-induced abnormal grain growth in nanocrystalline Ni–Fe *J. Mater. Sci.* **52** 1 46-59
[16] Furnish TA, Bufford DC, Ren F, Mehta A, Hattar K and Boyce BL 2018 Evidence that abnormal grain growth precedes fatigue crack initiation in nanocrystalline Ni-Fe *Scripta Mater.* **143** 15-9

[17] Seifi M, Salem A, Beuth J, Harrysson O and Lewandowski JJ 2016 Overview of materials qualification needs for metal additive manufacturing *JOM* **68** 3 747-64
[18] Chen Z, Han C, Gao M, Kandukuri SY and Zhou K 2022 A review on qualification and certification for metal additive manufacturing *Virt. Phys. Prototyp.* **17** 2 382-405
[19] Parvinian S, Yabansu YC, Khosravani A, Garmestani H and Kalidindi SR 2020 High-throughput exploration of the process space in 18% Ni (350) maraging steels via spherical indentation stress–strain protocols and Gaussian process models *Integ. Mater. Manuf. Innov.* **9** 3 199-212
[20] Sau M *et al.* 2022 High-Throughput Nanoindentation Mapping of Additively Manufactured T91 Steel *JOM* **74** 4 1469-76
[21] Boyce BL *et al.* 2017 Extreme-value statistics reveal rare failure-critical defects in additive manufacturing *Adv. Eng. Mater.* **19** 8 1700102
[22] Heckman NM *et al.* 2020 Automated high-throughput tensile testing reveals stochastic process parameter sensitivity *Mater. Sci. Eng.* A **772** 138632
[23] Ferro P, Meneghello R, Savio G and Berto F 2020 A modified volumetric energy density–based approach for porosity assessment in additive manufacturing process design *The Int. J. Adv. Manuf. Tech.* **110** 7 1911-21
[24] Watring DS, Benzing JT, Hrabe N and Spear AD 2020 Effects of laser-energy density and build orientation on the structure–property relationships in as-built Inconel 718 manufactured by laser powder bed fusion *Addit. Manuf.* **36** 101425
[25] Noell PJ, Carroll JD and Boyce BL 2018 The mechanisms of ductile rupture *Acta Mater.* **161** 83-98
[26] Steglich D and Brocks W 1997 Micromechanical modelling of the behaviour of ductile materials including particles *Comput. Mater. Sci.* **9** 1-2 7-17
[27] Gardner RN, Pollock T and Wilsdorf H 1977 Crack initiation at dislocation cell boundaries in the ductile fracture of metals *Mater. Sci. Eng.* **29** 2 169-74
[28] Croom BP, Jin H, Noell PJ, Boyce BL and Li X 2019 Collaborative ductile rupture mechanisms of high-purity copper identified by in situ X-ray computed tomography *Acta Mater.* **181** 377-84
[29] King WE, Campbell GH, Haupt DL, Kinney JH, Riddle RA and Wien WL 1995 X-ray tomographic microscopy investigation of the ductile rupture of an aluminum foil bonded between sapphire blocks *Scripta Metall. Mater.* **33** 12
[30] Noell P, Carroll J, Hattar K, Clark B and Boyce B 2017 Do voids nucleate at grain boundaries during ductile rupture? *Acta Mater.* **137** 103-14
[31] Sills RB and Boyce B 2020 Void growth by dislocation adsorption *Mater. Res. Lett.* **8** 3 103-9
[32] Noell PJ, Sabisch JE, Medlin DL and Boyce BL 2020 Nanoscale conditions for ductile void nucleation in copper: Vacancy condensation and the growth-limited microstructural state *Acta Mater.* **184** 211-24
[33] Zehetbauer M, Steiner G, Schafler E, Korznikov AV and Korznikova E 2006 Deformation induced vacancies with severe plastic deformation: Measurements and modelling *Mater. Sci. Forum* (Trans Tech Publ) vol. 503, pp. 57-64
[34] Jha SK, Szczepanski CJ, John R and Larsen JM 2015 Deformation heterogeneities and their role in life-limiting fatigue failures in a two-phase titanium alloy *Acta Mater.* **82** 378-95
[35] Ispánovity PD *et al.* 2022 Dislocation avalanches are like earthquakes on the micron scale *Nat. Comm.* **13** 1 1-10
[36] Ghosh S and Groeber MA 2020 Developing virtual microstructures and statistically equivalent representative volume elements for polycrystalline materials *Handbook of Materials Modeling: Methods: Theory and Modeling* 1631-56
[37] Madison JD, Susan DF and Kilgo AC 2015 3D RoboMET Characterization Sandia National Lab.(SNL-NM), Albuquerque, NM (United States).

[38] Niezgoda SR, Yabansu YC and Kalidindi SR 2011 Understanding and visualizing microstructure and microstructure variance as a stochastic process *Acta Mater.* **59** 16 6387-400
[39] Homer ER 2019 High-throughput simulations for insight into grain boundary structure-property relationships and other complex microstructural phenomena *Comput. Mater. Sci.* **161** 244-54
[40] Montes de Oca Zapiain D, Stewart JA and Dingreville R 2021 Accelerating phase-field-based microstructure evolution predictions via surrogate models trained by machine learning methods *npj Comput. Mater.* **7** 1 1-11
[41] Sutrisno E, Fan Q, Das D and Pecht M 2012 Anomaly detection for insulated gate bipolar transistor (IGBT) under power cycling using principal component analysis and K-nearest neighbor algorithm *J. Wash. Acad. Sci.* 1-8
[42] Zhou L, Zhang T, Zhang Z, Lei Z and Zhu S 2021 A new online quality monitoring method of chain resistance upset butt welding based on Isolation Forest and Local Outlier Factor *J. Manuf. Process.* **68** 843-51
[43] Alsini R, Almakrab A, Ibrahim A and Ma X 2021 Improving the outlier detection method in concrete mix design by combining the isolation forest and local outlier factor *Constr. Build. Mater.* **270** 121396
[44] Garcia-Allende PB, Anabitarte F, Conde OM, Mirapeix J, Madruga FJ and Lopez-Higuera JM 2008 Support vector machines in hyperspectral imaging spectroscopy with application to material identification *Algorithms and Technologies for Multispectral, Hyperspectral, and Ultraspectral Imagery XIV* vol. 6966 pp. 557-67
[45] Dau HA, Ciesielski V and Song A 2014 Anomaly detection using replicator neural networks trained on examples of one class *Asia-Pacific Conference on Simulated Evolution and Learning* pp. 311-22
[46] Tsai D-M and Jen P-H 2021 Autoencoder-based anomaly detection for surface defect inspection *Adv. Eng. Inform.* **48**, 101272
[47] Becker P, Roth C, Roennau A and Dillmann R 2020 Acoustic anomaly detection in additive manufacturing with long short-term memory neural networks *2020 IEEE 7th International Conference on Industrial Engineering and Applications (ICIEA)* pp. 921-6
[48] Zhang A and Wang J 2020 A method for analyzing abnormality of automobile sunroof manufacturing process by using Bayesian method *2020 IEEE International Conference on Information Technology, Big Data and Artificial Intelligence (ICIBA)* **1** 233-7
[49] Kraljevski I, Duckhorn F, Tschöpe C and Wolff M 2021 Machine Learning for Anomaly Assessment in Sensor Networks for NDT *Aerospace IEEE Sens. J.* **21** 9 11000-8
[50] Görnitz N, Braun M and Kloft M 2015 Hidden markov anomaly detection *International conference on machine learning* 1833-42
[51] Dresvyanskiy D *et al.* 2019 Application of clustering methods to anomaly detection in fibrous media *IOP Conf. Ser.: Mater. Sci. Eng.* **537** 2 022001
[52] Zhou S, Zhou K, Wang J, Yang G and Wang S 2018 Application of cluster analysis to geochemical compositional data for identifying ore-related geochemical anomalies *Front. Earth Sci.* **12** 3 491-505
[53] Kharitonov A, Nahhas A, Pohl M and Turowski K 2022 Comparative analysis of machine learning models for anomaly detection in manufacturing *Proc. Comput. Sci.* **200** 1288-97
[54] Norman J, Bar-Yam Y and Taleb NN 2020 Systemic risk of pandemic via novel pathogens–coronavirus: a note *New England Complex Systems Institute* **26**
[55] Taleb NN, Bar-Yam Y and Cirillo P 2020 On single point forecasts for fat-tailed variables, *Int. J. Forecast.*

Thick interface size effect on dislocation transmission in nanolaminates

S Xu[1,*], J Y Cheng[2], N A Mara[2] and I J Beyerlein[1,3]

[1] Department of Mechanical Engineering, University of California, Santa Barbara, California 93106-5070, USA
[2] Department of Chemical Engineering and Materials Science, University of Minnesota, Minneapolis, MN 55455, USA
[3] Materials Department, University of California, Santa Barbara, California 93106-5050, USA

E-mail: shuozhixu@ucsb.edu

Abstract.
Recent experimental studies have reported that thick interfaces in nanolaminates can lead to greater strengths than conventionally sharp interfaces without sacrificing deformability. Using a multi-phase phase-field dislocation dynamics model, dislocation transmission across a compositionally graded, nanoscale thick interface is investigated. Thicker interfaces over a finite range are found to lead to greater resistance to transmission. The limit interface thickness at which the peak resistance is reached, and the strengthening capacity of the interface are greater when the dislocation is dissociated, as in a face-centered cubic lattice, than when it is compact, as in a body-centered cubic lattice. The composition transitions within the interface are treated with multiple sublayers, and it is found that the interface transmission barrier is as strong as its most resistance composition.

1. Introduction

Nanolaminates have drawn considerable attention due to their extraordinarily high strength, beyond the volume averaged strength based on their constituents [1, 2]. They consist of a stack of alternating layers of two or more dissimilar materials, wherein the individual layer thicknesses are nanoscale in dimension. Their superior strength, as well as many other notable properties, such as outstanding thermal stability [3], good room temperature ductility [4, 5], and excellent shock resistance [6, 7], have been attributed to their high density of interfaces. Interfaces in conventional face-centered cubic (FCC)/body-centered cubic (BCC) nanolaminates are sharp, composed of one atomic plane of Cu and one atomic plane of Nb [8]. In 2018, Chen et al. [9] synthesized Cu/Nb nanolaminates with interfaces that were on the contrary thick. As shown in figure 1, these thick Cu/Nb interfaces extend out of plane by at least a few nm and are chemically, crystallographically, and/or topologically distinct from those of the Cu or Nb crystals they join. Compression testing 40 nm/40 nm Cu/Nb nanolaminates revealed that when the interface thickness is about 10 nm, the yield stress is 50% higher and the flow stress 22% higher than when the interface is sharp, while retaining the same strain to failure of 14%. Another study revealed that even higher strengths and deformability were possible when both the interface and layer thicknesses equaled 10 nm [10]. Clearly, these thick or so-called 3D interfaces introduce newer and more variables to interface design than the commonly studied sharp or 2D interfaces.

In this work, we use a recently developed multi-phase phase field dislocation dynamics (PFDD) model [11] to investigate the role of 3D interface thickness h' and composition on the critical stress for dislocation transmission.

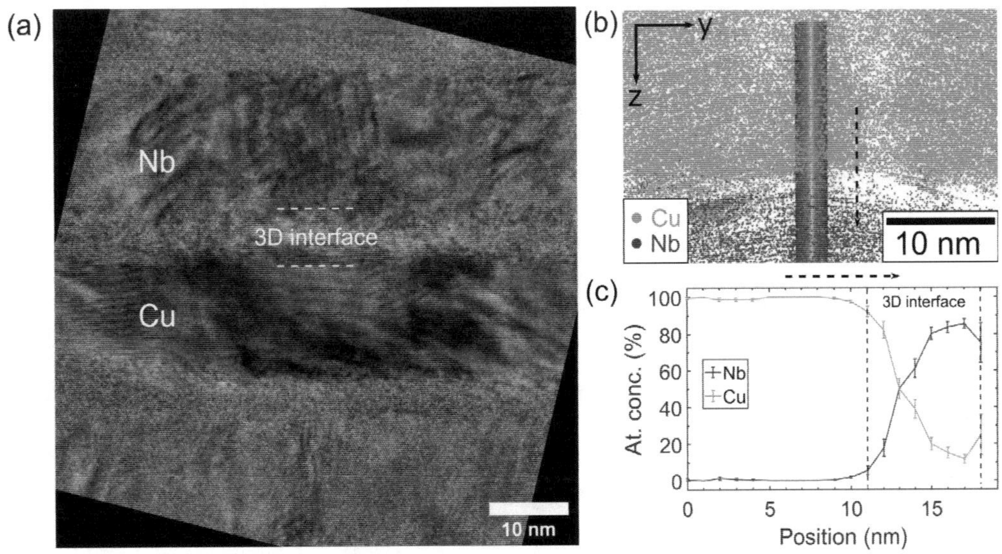

Figure 1. (a) A high-resolution transmission electron micrograph of a 3D Cu-Nb interface between a Cu layer and a Nb layer. More details were presented in Ref. [10]. (b) An atom probe tomography chemical map of the 3D interface. The purple cylinder represents a region of interest from which the line profile of chemical composition in (c) is extracted. The horizontal dashed arrow indicates the directionality in (b) and the two vertical dotted lines correspond to the two horizontal yellow lines in (a). More details were presented in Ref. [12].

2. Methods

2.1. PFDD formulation

In the formulation, each phase field order parameter is associated with an available slip system. Accordingly, the set of order parameters $\boldsymbol{\phi}$ consists of $n_{\rm op}$ order parameters, for which there are $N_{\rm sp}$ slip planes and $n_{\rm sp}$ slip directions per slip plane. Thus, $n_{\rm op} = N_{\rm sp} n_{\rm sp}$. The order parameter for a given slip system α, denoted by ϕ_α, is 0 for the unslipped state and 1 or more for the slipped state. The total system energy consists of energetic terms that are a function of the evolving order parameters [13]. For a multi-phase material, a second quantity, a virtual strain tensor $\boldsymbol{\epsilon}^{\rm v}$ independent of $\boldsymbol{\phi}$, is introduced and included in the solution [11].

The total energy density ψ at a given material point $\boldsymbol{x}$ contains the elastic $\psi_{\rm ela}$, lattice $\psi_{\rm lat}$, gradient $\psi_{\rm gra}$, and external $\psi_{\rm ext}$ energy densities [14],

$$\psi(\boldsymbol{x}) = \psi_{\rm ela}(\boldsymbol{x}) + \psi_{\rm lat}(\boldsymbol{x}) + \psi_{\rm gra}(\boldsymbol{x}) - \psi_{\rm ext}(\boldsymbol{x}) \tag{1}$$

where $\psi_{\rm lat}$ and $\psi_{\rm gra}$ only exist in materials that contain dislocations.

The first energy density, $\psi_{\rm ela}$, consists of two terms, the "equivalent" homogeneous elastic energy density, $\psi^{\rm eq}_{\rm ela}(\boldsymbol{x})$, and the "extra" elastic energy density, $\psi^{\rm ex}_{\rm ela}(\boldsymbol{x})$ [15], i.e.,

$$\psi_{\rm ela}(\boldsymbol{x}) = \psi^{\rm eq}_{\rm ela}(\boldsymbol{x}) + \psi^{\rm ex}_{\rm ela}(\boldsymbol{x}) \tag{2}$$

$$\psi_{\text{ela}}^{\text{eq}}(\boldsymbol{x}) = \frac{1}{2}[\boldsymbol{\epsilon}(\boldsymbol{x}) - \boldsymbol{\epsilon}^0(\boldsymbol{x})] \cdot \boldsymbol{C}[\boldsymbol{\epsilon}(\boldsymbol{x}) - \boldsymbol{\epsilon}^0(\boldsymbol{x})] \tag{3}$$

$$\psi_{\text{ela}}^{\text{ex}}(\boldsymbol{x}) = \frac{1}{2}\boldsymbol{\epsilon}^{\text{v}}(\boldsymbol{x}) \cdot \boldsymbol{M}^{[N]}(\boldsymbol{x})\boldsymbol{\epsilon}^{\text{v}}(\boldsymbol{x}) \tag{4}$$

where $\boldsymbol{C}$ is the elasticity tensor and $\boldsymbol{\epsilon}$ the strain tensor, given by $\boldsymbol{\epsilon} = \operatorname{sym}\boldsymbol{\beta}$, where $\boldsymbol{\beta} = \nabla \boldsymbol{u}$ is the distortion with $\boldsymbol{u}$ the displacement in material N, $1 < N \leq N_{\text{mat}}$. $\psi_{\text{ela}}^{\text{ex}}$ exists in all materials except material $N = 1$. The eigenstrain tensor $\boldsymbol{\epsilon}^0$ depends on the material, according to

$$\boldsymbol{\epsilon}^0(\boldsymbol{x}) = \begin{cases} \boldsymbol{\epsilon}^{\text{p}}(\boldsymbol{x}), & \boldsymbol{x} \in \text{material } 1 \\ \boldsymbol{\epsilon}^{\text{p}}(\boldsymbol{x}) + \boldsymbol{\epsilon}^{\text{v}}(\boldsymbol{x}), & \boldsymbol{x} \in \text{material } N \text{ where dislocations exist} \\ \boldsymbol{\epsilon}^{\text{v}}(\boldsymbol{x}), & \boldsymbol{x} \in \text{material } N \text{ where dislocations do not exist} \end{cases} \tag{5}$$

where $\boldsymbol{\epsilon}^{\text{p}}$ and $\boldsymbol{\epsilon}^{\text{v}}$ are the plastic strain tensor and virtual strain tensor, respectively. The tensor $\boldsymbol{\epsilon}^{\text{p}}$ is related to plastic distortion tensor $\boldsymbol{\beta}^{\text{p}}$ and order parameter $\boldsymbol{\phi}$ via

$$\boldsymbol{\epsilon}^{\text{p}} = \operatorname{sym}\boldsymbol{\beta}^{\text{p}} \tag{6}$$

$$\boldsymbol{\beta}^{\text{p}}(\boldsymbol{\phi}) = \sum_{\alpha=1}^{n_{\text{op}}} \frac{b_\alpha \phi_\alpha}{d_\alpha} \boldsymbol{s}_\alpha \otimes \boldsymbol{n}_\alpha \tag{7}$$

where $\boldsymbol{s}_\alpha$ is the slip direction unit vector, b_α is the slip vector magnitude, $\boldsymbol{n}_\alpha$ is the slip plane unit normal, and d_α is the interplanar distance for slip system α. In equation 4,

$$M_{ijkl}^{[N]}(\boldsymbol{x}) = -C_{ijmn}^{[1]} \left[\Delta C_{mnpq}^{[N]}(\boldsymbol{x})\right]^{-1} C_{pqkl}^{[1]} - C_{ijkl}^{[1]} \tag{8}$$

$$\Delta C_{ijkl}^{[N]}(\boldsymbol{x}) = C_{ijkl}^{[N]}(\boldsymbol{x}) - C_{ijkl}^{[1]} \tag{9}$$

The second energy density in equation 1, ψ_{lat}, is a function of $\boldsymbol{\phi}$ [16], i.e.,

$$\psi_{\text{lat}}(\boldsymbol{x}) = \frac{\gamma_{\text{gsf}}^{[N]}(\boldsymbol{\phi}(\boldsymbol{x}))}{l_{\text{gsf}}^{[N]}(\boldsymbol{x})} \tag{10}$$

where $\gamma_{\text{gsf}}^{[N]}$ is the generalized stacking fault energy (GSFE) and $l_{\text{gsf}}^{[N]}$ is the interplanar spacing for material N. For one system in the $\{110\}$ glide plane in the BCC lattice, $n_{\text{sp}} = 1$ and γ_{gsf} is represented by a GSFE curve along the $\langle 111 \rangle$ direction. For the FCC lattice, the three coplanar slip systems on the $\{111\}$ glide plane are used, and hence $n_{\text{sp}} = 3$ and γ_{gsf} is a GSFE surface.

The third energy density, ψ_{gra}, represents the energy associated with the change in $\boldsymbol{\phi}$ at the dislocation and hence is a function of the gradient $\nabla \phi_\alpha(\boldsymbol{x})$ [17]. It is related to the projection of the curl of the distortion tensor (Nye's tensor) in equation 7 onto each slip system as follows

$$\psi_{\text{gra}}(\boldsymbol{x}) = \sum_{\alpha,\beta=1}^{n_{\text{sp}}} \eta_{\alpha\beta}^{[N]} \nabla \phi_\alpha(\boldsymbol{x}) \cdot \boldsymbol{\Xi}_{\alpha\beta}(\boldsymbol{x}) \nabla \phi_\beta(\boldsymbol{x}) \tag{11}$$

where $\eta_{\alpha\beta}^{[N]}$ is a scaling factor for the gradient energy and has units of energy per unit length. It can be specified for a pair of order parameters $\alpha\beta$ in material N. The tensor $\boldsymbol{\Xi}_{\alpha\beta}$ depends on geometry and is:

$$\boldsymbol{\Xi}_{\alpha\beta} = \frac{\boldsymbol{b}_\alpha \cdot \boldsymbol{b}_\beta}{d_\alpha d_\beta}[(\boldsymbol{n}_\alpha \cdot \boldsymbol{n}_\beta)\boldsymbol{I} - \boldsymbol{n}_\beta \otimes \boldsymbol{n}_\alpha] \tag{12}$$

where $\boldsymbol{I}$ is the identity tensor. Here, for simplicity sake, we set $\eta_{\alpha\beta}^{[N]}$ the same for all $\alpha\beta$ pairs in the same material, and we drop the subscripts in what follows.

The last energy density in equation 1, the external energy density ψ_{ext}, is given by

$$\psi_{\text{ext}}(\boldsymbol{x}) = \boldsymbol{\sigma}_{\text{app}} \cdot \boldsymbol{\epsilon}^0(\boldsymbol{x}) \tag{13}$$

where $\boldsymbol{\sigma}_{\text{app}}$ is the applied stress tensor.

With the energy functional defined and order parameters initialized, the total system energy is minimized with respect to each order parameter ϕ_α and each virtual strain component ϵ_{ab} (where $a, b = 1, 2, 3$) via the time-dependent Ginzburg-Landau equations:

$$\dot{\phi}_\alpha(\boldsymbol{x}) = -m_0[\partial_{\phi_\alpha(\boldsymbol{x})}(\psi_{\text{ela}}(\boldsymbol{x}) + \psi_{\text{lat}}(\boldsymbol{x}) - \psi_{\text{ext}}(\boldsymbol{x})) - \nabla \cdot \partial_{\nabla\phi_\alpha(\boldsymbol{x})}\psi_{\text{gra}}(\boldsymbol{x})] \tag{14}$$

$$\dot{\epsilon}_{ab}^{\text{v}}(\boldsymbol{x}) = -m_0^{\text{v}}\partial_{\epsilon_{ab}^{\text{v}}(\boldsymbol{x})}(\psi_{\text{ela}}(\boldsymbol{x}) - \psi_{\text{ext}}(\boldsymbol{x})) \tag{15}$$

where the superposed dot denotes a time derivative. equation 15 is not applied to material 1. The Ginzburg-Landau coefficients, m_0 and m_0^{v}, are non-negative but not necessarily equal.

2.2. Simulation set-up

Figure 2 illustrates the 3D periodic simulation cell composed of an edge dislocation dipole bounded by a pair of 3D interfaces that lie perpendicular to the glide plane. To isolate the role of chemistry and interface thickness h', the two matrix crystals on either side of the interface belong to the same pure metal while the 3D interface consists of one or two Cu-Nb alloys with the same crystal structure as the matrix.

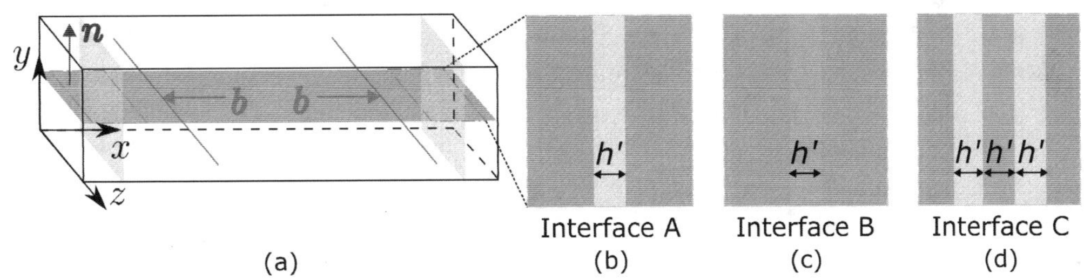

Figure 2. (a) Schematic of the simulation cell for an edge dislocation dipole interacting with two 3D interfaces in a Nb matrix or a Cu matrix. The grey parallelogram is the slip plane, while the interface plane normal is along the x axis. (b–d) In a Nb matrix, cyan and orange layers indicate BCC $Cu_{0.05}Nb_{0.95}$ (interface A) and BCC $Cu_{0.1}Nb_{0.9}$ (interface B), respectively; in a Cu matrix, cyan and orange layers correspond to FCC $Cu_{0.9}Nb_{0.1}$ (interface A) and FCC $Cu_{0.8}Nb_{0.2}$ (interface B), respectively.

For the BCC crystal, the glide plane is $\{110\}$ and the glide direction is $\langle 111 \rangle$. The corresponding lattice orientations of the cell are

$$(\boldsymbol{i}_x, \boldsymbol{i}_y, \boldsymbol{i}_z) = \left(\frac{\sqrt{3}}{3}[111], \frac{\sqrt{2}}{2}[\bar{1}10], \frac{\sqrt{6}}{6}[\bar{1}\bar{1}2]\right) \tag{16}$$

For the FCC crystal, the slip system is $\{111\}\langle 110 \rangle$ and the cell orientations are

$$(\boldsymbol{i}_x, \boldsymbol{i}_y, \boldsymbol{i}_z) = \left(\frac{\sqrt{2}}{2}[\bar{1}10], \frac{\sqrt{3}}{3}[111], \frac{\sqrt{6}}{6}[11\bar{2}]\right) \tag{17}$$

The lengths of the simulation cell, L_x, L_y, and L_z along the x, y, and z direction, respectively, are selected to minimize numerical artifacts. The number of grid points along the three directions, respectively, are 1024, 64, and 128. The two dislocations lie along the z axis on the mid-y plane, and are separated by $L_x/2$ along the x direction since this separation results in an equilibrium dipole position in its stress-free state. The two 3D interfaces are separated by $7L_x/8$ along the x direction. The distance between an interface and its adjacent dislocation is selected such that the two do not overlap even in the case of the thickest 3D interface studied ($\approx$ 15 nm).

The interface composition is selected such that the interface would form a coherent, cube-on-cube orientation relationship with the adjoining matrix material. In a prior work, with some additional calculations from this work, atomistic calculations were conducted to determine the structure, lattice parameters, elastic moduli, and GSFEs for the full spectrum of Cu-Nb compositions [15]. They identify that $Cu_{0.05}Nb_{0.95}$ and $Cu_{0.1}Nb_{0.9}$ are stable BCC alloys while $Cu_{0.8}Nb_{0.2}$ and $Cu_{0.9}Nb_{0.1}$ stable FCC alloys. The other compositions, with intermediate Cu and Nb compositions, are amorphous materials and not considered here for the interface phase.

A compositionally graded 3D interface is modeled as a series of three layers, wherein the two outer layers bound to the matrix have the same composition, while the inner layer has a different composition. This transition intends to mimic the gradual composition transitions within the 3D interface seen in figure 1 [12]. Figure 2(d) shows this "trilayer" representation of a graded 3D interface, in which each layer thickness is h' and the total thickness is thus $3h'$. For the BCC case, the matrix is Nb and the interface transitions from $Cu_{0.05}Nb_{0.95}$ then $Cu_{0.1}Nb_{0.9}$ and back to $Cu_{0.05}Nb_{0.95}$ and Nb. Likewise, for the FCC case, the matrix is Cu and the trilayer interface is first $Cu_{0.9}Nb_{0.1}$ then $Cu_{0.8}Nb_{0.2}$, and returns to $Cu_{0.9}Nb_{0.1}$. The interface thickness $3h'$ varies from 1.71 nm ($5.94b_{Nb}$) to 15.42 nm ($53.56b_{Nb}$) in a Nb matrix, and from 1.53 nm ($5.95b_{Cu}$) to 15.42 nm ($60.01b_{Cu}$) in a Cu matrix.

For comparison, separate simulations are performed considering 3D interfaces but with a single alloy composition, as depicted in figure 2(b,c). When the matrix is Nb, the monolayer interface is either BCC $Cu_{0.05}Nb_{0.95}$ alloy or BCC $Cu_{0.1}Nb_{0.9}$ alloy. The thickness h' varies from 0.57 nm ($1.98b_{Nb}$), approximately three times the inter-{111}-planar spacing, to 13.68 nm ($47.52b_{Nb}$). When the matrix is Cu, the interfaces considered are either $Cu_{0.9}Nb_{0.1}$ or $Cu_{0.8}Nb_{0.2}$. The h' ranges from twice the inter-{110}-planar spacing, 0.51 nm ($1.98b_{Cu}$) to 10.03 nm ($39.03b_{Cu}$).

Table 1. Lattice parameters a_0 (in Å), elastic constants C_{11}, C_{12}, C_{44} (in GPa), isotropic shear modulus in Voigt form $\mu = (3C_{44} + C_{11} - C_{12})/5$, and uniform gradient energy coefficients η (in μb^2) of the two pure metals and four random binary alloys used in this work.

	Lattice	a_0	C_{11}	C_{12}	C_{44}	μ	η
Nb	BCC	3.324 [18]	245 [19]	132 [19]	28.4 [19]	39.64	0
$Cu_{0.05}Nb_{0.95}$	BCC	3.287	236.76	139	35.66	40.95	0
$Cu_{0.1}Nb_{0.9}$	BCC	3.274 [15]	221.58 [15]	141.63 [15]	40.63 [15]	40.37	0
$Cu_{0.8}Nb_{0.2}$	FCC	3.759 [15]	195.16 [15]	128.44 [15]	79.32 [15]	60.94	0.19 [11]
$Cu_{0.9}Nb_{0.1}$	FCC	3.699 [15]	187.25 [15]	135.8 [15]	81.88 [15]	59.42	0.28 [11]
Cu	FCC	3.634 [20]	169 [19]	122 [19]	75.3 [19]	54.58	0.38 [11]

Table 1 and Table 2 present the input parameters for every material and, when available, their originating references. The uniform gradient energy coefficient η scales directly with the size of the partial dislocation core, excluding the intrinsic stacking fault (ISF) [21]. For undissociated dislocations, such as those in BCC metals, the gradient term is neglected (i.e., $\eta = 0$). For the dissociated dislocations in FCC materials, η is characterized by comparisons with the partial

Table 2. Unstable stacking fault energy (SFE) γ_{usf} and intrinsic SFE γ_{isf} of the six materials in Table 1, in mJ/m^2.

	Nb	$Cu_{0.05}Nb_{0.95}$	$Cu_{0.1}Nb_{0.9}$	$Cu_{0.8}Nb_{0.2}$	$Cu_{0.9}Nb_{0.1}$	Cu
γ_{usf}	676.78 [18]	710.13	703.46 [15]	151.18 [15]	155.19 [15]	160.52 [20]
γ_{isf}				38.17 [15]	40.67 [15]	41.83 [20]

core widths in the same material based on atomistic simulations [11]. The boundaries between different materials are assumed coherent since the lattice misfit between any adjoining matrix and interface phase is sufficiently small. Further, any residual Burgers vector left after dislocation transmission would also be small and due to the appreciable thickness of the 3D interfaces, would be distributed across the interface thickness.

A resolved shear stress is applied to the system to drive the dislocations to glide towards the 3D interfaces. For FCC materials, the edge dislocation under no stress dissociates into leading and trailing Shockley partials. Under the stress applied here, the magnitude of the shear stress resolved on each partial is the same and drives them in the same direction. The applied stress is incremented in order to identify σ_c, the critical stress for the dislocation to transmit the entire thick interface. Specifically, when the minimum applied stress for transmission σ_0 is found, we set $\sigma_c = \sigma_0 - \Delta\sigma_0/2$, where $\Delta\sigma_0 = 10^{-4}\mu$ is the smallest difference between the two stresses and μ is the isotropic shear modulus of the matrix. For each stress level, the iterations performed to solve equation 15 are terminated when the Euclidean norm of the difference in global vector of each order parameter and each virtual strain component between successive iterations is smaller than 10^{-5}. The Ginzburg-Landau coefficients m_0 and m_0^{v} are set to unity and the timestep used is $\Delta t = 0.02$.

3. Results

3.1. Monolayer interface

Figure 3 presents σ_c for both interface compositions as a function of h' in the BCC case.

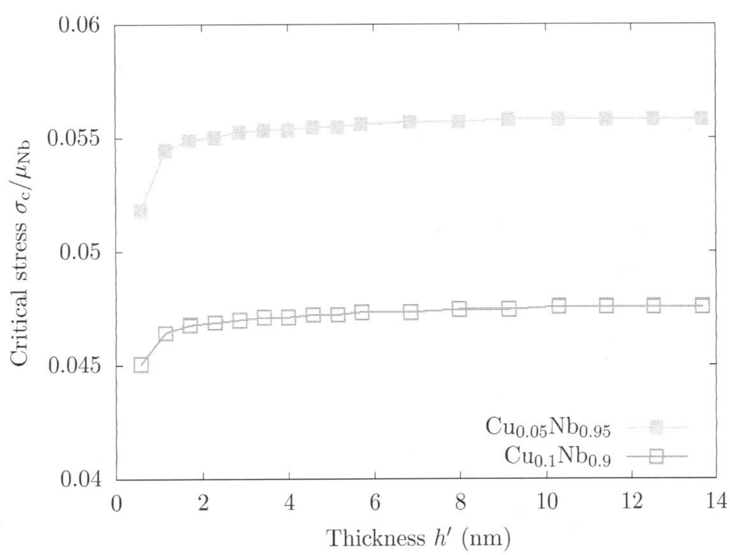

Figure 3. Critical stresses for transmission of an edge dislocation across thick interfaces in a Nb matrix, as a function of interface thickness.

Interfaces of any thickness substantially increase σ_c (30–60%) compared to the glide resistance in pure Nb, $0.033\mu_{Nb}$ (using the same model). The strengthening observed can be attributed to both higher μ and γ_{usf} of the interface. For the same h', the $Cu_{0.05}Nb_{0.95}$ interface composition, with the higher values of μ and γ_{usf}, provides the stronger barrier to transmission, than $Cu_{0.1}Nb_{0.9}$. The impact is large–with 4.8% higher γ_{usf} yielding on average 36% higher σ_c.

Importantly, thicker h' leads to higher σ_c. The positive size effect of h' on σ_c prevails, however, over a short range from 0.57 nm until $\approx$ 3 nm, beyond which there is no noticeable benefit to strengthening. The precise limit h' at which σ_c peaks is 9.12 nm for the $Cu_{0.05}Nb_{0.95}$ interface and 10.29 nm for the $Cu_{0.1}Nb_{0.9}$ one. Since σ_c is associated with a critical state of the dislocation, the locations of the dislocation in their sub-critical configurations are examined. Some configurations are shown in figure 4. They reveal that σ_c is needed to force transmission across the first boundary into the interface. For smaller h' where a positive size effect is seen, this first transmission event is aided by the image stresses from the interface boundary on the other side of the thick interface. For larger h', the aid diminishes and along with it, the size dependence.

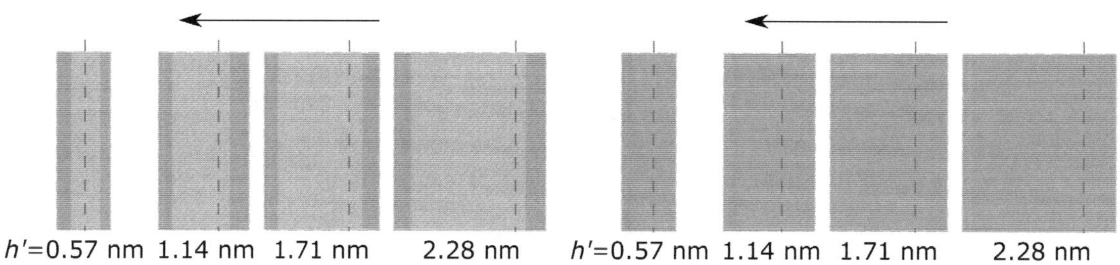

(a) $Cu_{0.05}Nb_{0.95}$ (b) $Cu_{0.1}Nb_{0.9}$

Figure 4. Position of the edge dislocation (vertical dashed line) with respect to the interface in a Nb matrix when the applied stress is slightly lower than the critical one. In each case, the dislocation moves from right to left, as indicated by the arrow. Partial views of the areas of interest are shown and do not represent the entire length of the simulation cell.

In the Cu matrix, introduction of a thick interface provides even greater increases in σ_c than in the Nb matrix. For comparison, the glide resistance for an edge dislocation in pure Cu using this model is $0.00035\mu_{Cu}$ [11]. As shown in figure 5, introduction of even the finest layer $\approx$ 0.6 nm increases σ_c at least 40 times. Further, unlike in the BCC case, the influence of 3D interface h' is strong and does not weaken until h' increases to around $\approx$ 4–5 nm. Between the two compositions, the $Cu_{0.9}Nb_{0.1}$ interface provides the greater transmission resistance than the $Cu_{0.8}Nb_{0.2}$ one, which would be expected since the former has the greater γ_{usf} and γ_{isf}.

The strengthening potential and strong size effect are counter-intuitive since transmitting and gliding across these interfaces from Cu would be expected to be easier than gliding in Cu alone since both interface compositions have a lower γ_{usf}, γ_{isf}, and their difference $\gamma_{usf} - \gamma_{isf}$ than pure Cu. It is also intriguing that the interfaces in the BCC counterpart did not provide strengthening to this extent although they have larger μ and γ_{usf} than the Nb matrix.

Analysis of sub-critical configurations in a Cu matrix suggests that the formation of the relatively low-energy ISF in the interface produces a strong pinning effect, that would not be realized in Nb. Figure 6 shows the positions of the leading (L) and trailing (T) partials in the sub-critical configurations, right before the pair (the full dislocation) completes transmission for a few h'. The L and T partials transmit across the first matrix/interface boundary and the interface thickness, maintaining approximately the same ISF width of $\approx$ 1.5–2 nm. This width is lower than the zero-stress ISF width in Cu, which is 2.53 nm ($= 9.85b_{Cu}$) and those in the alloys,

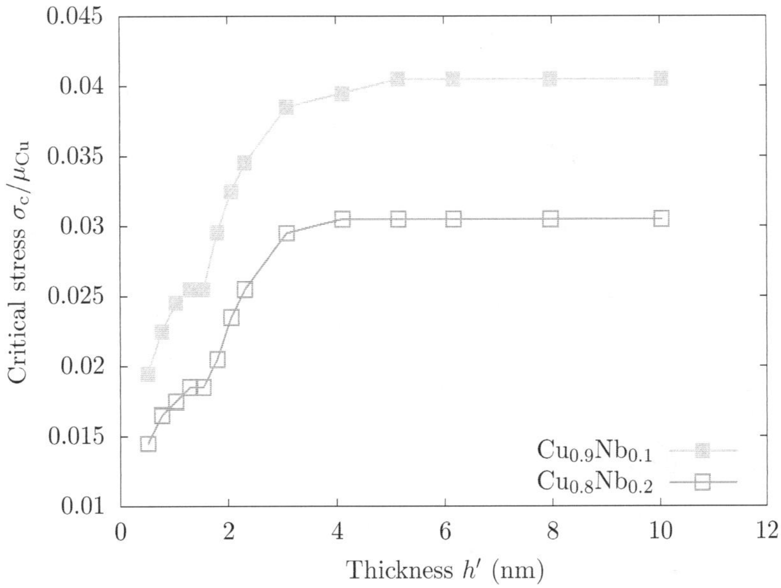

Figure 5. Critical stresses for transmission of an edge dislocation through two interface compositions in the Cu matrix, as a function of the interface thickness.

which are 2.59 nm and 2.63 nm, respectively [11]. For all thicknesses, the L partial transmits into the recipient Cu layer, while leaving the T partial in the interface and or the originating Cu layer. The σ_c is needed to push the T partial across the interface. As h' increases, the distance the L partial extends into Cu increases. The limit of h' is reached when the portion of the ISF width in the recipient Cu layer increases above that in the interface.

3.2. Trilayer interface

From the monolayer 3D interfaces, we turn our attention to the more compositionally complex trilayer interface, i.e., interface C in figure 2. We first consider the BCC case, in which the matrix is Nb and the interface transitions from $Cu_{0.1}Nb_{0.9}$ in the first outer layer adjacent to the Nb matrix to $Cu_{0.05}Nb_{0.95}$ in the central layer and back to $Cu_{0.1}Nb_{0.9}$ in the third layer. Figure 7 indicates that this gradient 3D interface of any thickness $3h'$ leads to an increase in σ_c above the Peierls stress in Nb, $0.03355\mu_{Nb}$ [15]. Importantly, these gradient interfaces have similar σ_c as the $Cu_{0.1}Nb_{0.9}$ monolayer interfaces for the same total thickness. Of the two compositions in the interface, $Cu_{0.05}Nb_{0.95}$ is the stronger one. An "inverse" weakest-link response is thus identified, wherein the barrier strength of the graded interface is as strong as its strongest composition.

Like the BCC monolayer interface, the size effect in $3h'$ on σ_c is positive for a short range until $\approx 3h' = 3.42$ nm when the size effect weakens and further increases in thickness produce little gain in barrier strength. For even greater $3h'$, a strict limit thickness is reached at which the peak σ_c is achieved. This limit $3h' = 10.26$ nm and is about three times the limit h' for either monolayer alloy interface. The locations of the dislocation at its critical point for a select set of $3h'$ are shown in figure 8a. In all cases, the critical stress is needed to force the dislocation to transmit across the first boundary into $Cu_{0.05}Nb_{0.95}$. Attaining this configuration is affected weakly by $3h'$. The limit $3h'$ is related to the distance beyond which the image stresses from the other boundaries become sufficiently minimized.

In a Cu matrix, the trilayer interface consists of two $Cu_{0.9}Nb_{0.1}$ layers adjacent to the matrix and one $Cu_{0.8}Nb_{0.2}$ layer in the middle. Like the monolayer interface, a strong size effect arises,

IOP Conf. Series: Materials Science and Engineering 1249 (2022) 012005 doi:10.1088/1757-899X/1249/1/012005

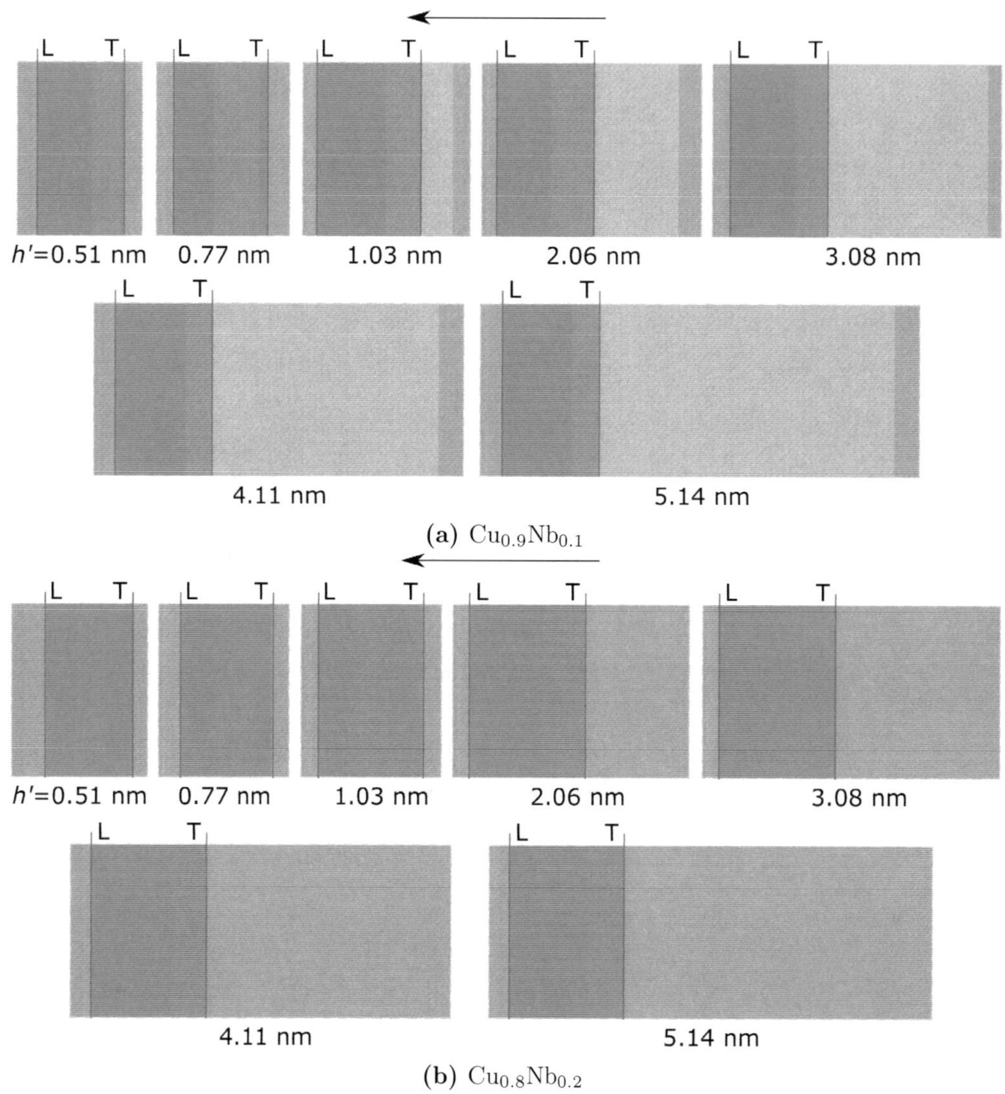

Figure 6. Position of the dissociated dislocation in its critical configuration in the Cu matrix when the applied stress is slightly lower than the critical one. Two Shockley partial dislocations are denoted by two vertical red lines, respectively, while the ISF by a semi-transparent red block between the two partials. "L" and "T" denote the leading and trailing partials, respectively.

as shown in figure 7, wherein σ_c nominally increases with $3h'$. The dependence on interface thickness is more pronounced than the gradient 3D interface in a Nb matrix. The σ_c peaks and then plateaus when $3h'$ reaches 9.24 nm, which is nearly twice the limit h' of the monolayer interfaces of the same compositions. The plateau value, $0.04\mu_{Cu}$, however, is close to that of the monolayer $Cu_{0.9}Nb_{0.1}$ interface, the stronger composition. Again, we find that the barrier strength of the graded interface of two compositions is as strong as its stronger composition.

Figure 8b presents the sub-critical configurations of the L and T partials for the broad range of $3h'$ over which σ_c varies. From $3h' \approx$ 1–4 nm, when h' is less than the ISF width of the dislocation, σ_c enables L to cross the first two layers and into the final $Cu_{0.9}Nb_{0.1}$ layer while T lies in the first $Cu_{0.9}Nb_{0.1}$ layer. Evidently σ_c is needed to force L and T to glide across the two

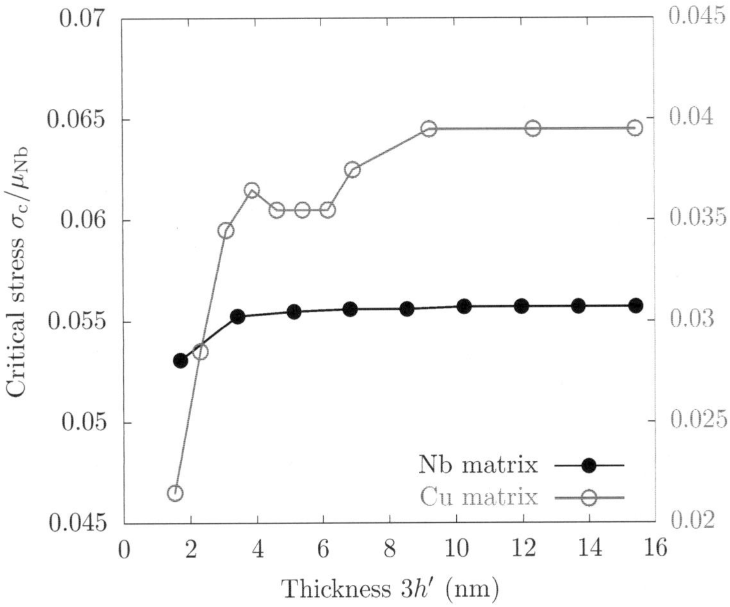

Figure 7. Critical stresses for transmission of an edge dislocation through the gradient interface in a Nb matrix or a Cu matrix, as a function of the total interface thickness.

stronger regions of the interface. The thicker h' these regions is, the greater the resistance. When $3h'$ increases above the ISF width from about 4 to 6 nm, however, σ_c still pushes L all the way into the last $Cu_{0.9}Nb_{0.1}$ layer but T lies in the weaker $Cu_{0.8}Nb_{0.2}$ layer. The σ_c decreases since achieving the critical configuration involves gliding through the weaker material. For greater $3h'$, σ_c reaches a limit value. The critical state at this limit corresponds to when the entire dislocation is on the verge of finally crossing the final interface/matrix boundary. The T partial has yet to transmit from the interface and back into the matrix (figure 8b at $3h' = 9.24$ nm).

4. Discussion

The results indicate a clear benefit in thickening the interface, where "larger is stronger" up to a limit thickness h'. These 3D interfaces can exhibit a non-negligible size effect on transmission resistance for thicknesses in the range of 2–10 nm. For some finely structured nanolaminates, these thicknesses are similar to or no smaller than the thickness of individual layers bounding the interfaces. With the added size effect from the interface h', altering both thicknesses independently to achieve optimal strength can be easily envisioned.

When treating a gradient interface, it is useful to consider the thickness of the composition within the interface that is the most resistant to glide. Calculations here forecast that rather than the weakest or even the averaged resistance across all compositions in the interface, it is the strongest portion of the interface and its volume fraction that dominates the resistance.

The dynamic, "composite" ISF width that transmits across the interface adds another important length scale. Although the resolved shear stresses on the L and T partials are the same, they often lie in different materials, differing in elastic moduli and lattice parameter. The results imply that greater differences in γ_{isf} between the matrix and 3D interface and lower γ_{isf} could possibly lead to even greater strengthening potential for 3D interfaces. Not only would the boosts in strengthening over the pure constituents be realized but also the range of h' could be greater than those seen in the composite system studied here.

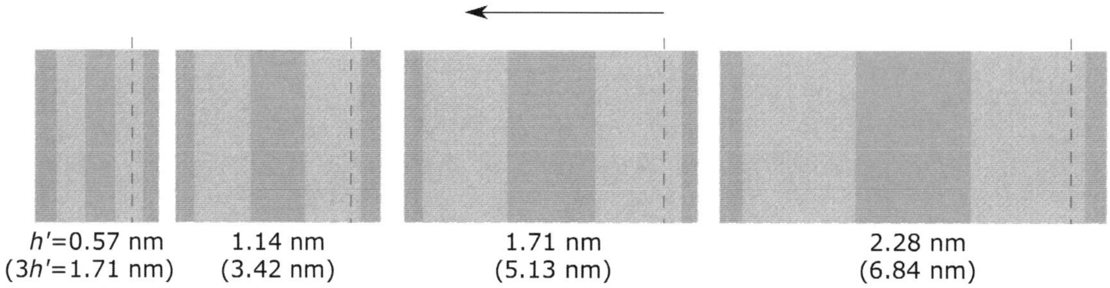

(a) Nb matrix

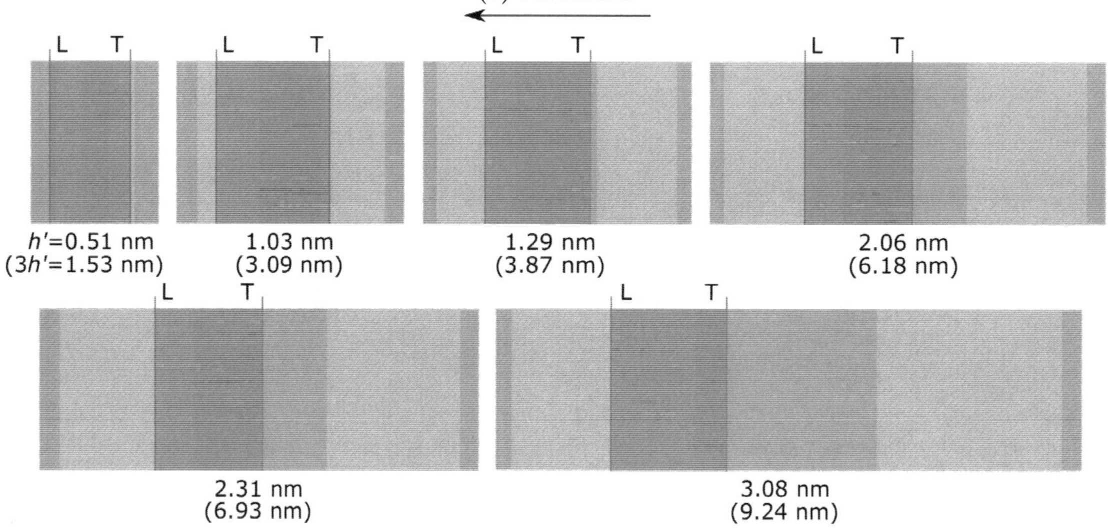

(b) Cu matrix

Figure 8. Position of the dissociated dislocation in their critical state, that is, when the applied stress is slightly lower than the critical one. Markers and parameters used are the same as those in figure 4 and figure 6.

5. Conclusions

In this paper, we utilize a recently developed multi-phase PFDD model [11] to study the transmission of an edge dislocation across compositionally gradient 3D interfaces of nanoscale thicknesses. The calculations investigate the role played by the 3D interface thickness h', composition, and dislocation core structure, either undissociated in BCC crystals or dissociated in FCC ones. One set-up consists of a fully BCC structure containing a thick 3D crystalline Cu-Nb alloy interface sandwiched between two Nb crystals of the same orientation. Similarly, the other one is composed of a completely FCC structure with two Cu crystals sharing one common thick 3D Cu-Nb alloy interface. Even for the finest h' ($\approx$ 0.5 nm), the interface increases the glide resistance, by 30–60% in the BCC structure and 40 times in the FCC one. For both crystal structures, a positive size effect in h' on transmission strength occurs up to a limit h' of $\approx$ 2–10 nm. We observe intrinsic stacking fault (ISF) strengthening, in which the dissociated core of the FCC dislocation leads to boosts in transmission strength over a broad range of nm thick interfaces. Comparisons with interfaces of uniform composition reveal that the transmission strength of gradient interfaces is controlled by its strongest composition. Thus, with the introduction of 3D interfaces, two important nm length scales arise, the thickness of the interface and the dynamic ISF width of the transmitting dislocation.

Acknowledgements
The authors gratefully acknowledge financial support from the Department of Energy, Office of Science, Basic Energy Sciences Program, USA DE-SC0020133. JYC is supported by DOE NNSA SSGF, USA under cooperative agreement number DE-NA0003960. Use was made of computational facilities purchased with funds from the National Science Foundation, USA (CNS-1725797) and administered by the Center for Scientific Computing (CSC), USA. The CSC is supported by the California NanoSystems Institute and the Materials Research Science and Engineering Center, USA (MRSEC; NSF DMR 1720256) at UC Santa Barbara.

References
[1] Mara N A and Beyerlein I J 2014 *J. Mater. Sci.* **49** 6497–516
[2] Mara N A and Beyerlein I J 2015 *Curr. Opin. Solid State Mater. Sci.* **19** 265–76
[3] Yang W, Beyerlein I J, Jin Q, Ge H, Xiong T, Yang L, Pang J, Zhou Y, Shao X, Zhang B, Zheng S and Ma X 2019 *Scr. Mater.* **166** 73–7
[4] Nizolek T, Beyerlein I J, Mara N A, Avallone J T and Pollock T M 2016 *Appl. Phys. Lett.* **108** 051903
[5] Pathak S, Velisavljevic N, Baldwin J K, Jain M, Zheng S, Mara N A and Beyerlein I J 2017 *Sci. Rep.* **7** 8264
[6] Han W Z, Cerreta E K, Mara N A, Beyerlein I J, Carpenter J S, Zheng S J, Trujillo C P, Dickerson P O and Misra A 2014 *Acta Mater.* **63** 150–61
[7] Zhang R F, Germann T C, Liu X Y, Wang J and Beyerlein I J 2014 *Acta Mater.* **79** 74–83
[8] Wang J, Zhang R F, Zhou C Z, Beyerlein I J and Misra A 2014 *Int. J. Plast.* **53** 40–55
[9] Chen Y, Li N, Hoagland R G, Liu X Y, Baldwin J K, Beyerlein I J, Cheng J Y and Mara N A 2020 *Acta Mater.* **199** 593–601
[10] Cheng J Y, Xu S, Chen Y, Li Z, Baldwin J K, Beyerlein I J and Mara N A 2022 *Nano Lett.* **22** 1897–904
[11] Xu S, Cheng J Y, Mara N A and Beyerlein I J 2022 *J. Mech. Phys. Solids* (under review)
[12] Li Z, Cheng J Y, Poplawsky J D, Xu S, Baldwin J, Beyerlein I J and Mara N A 2022 *Scr. Mater.* (under review)
[13] Xu S 2022 *Comput. Mater. Sci.* **210** 111419
[14] Xu S, Mianroodi J R, Hunter A, Svendsen B and Beyerlein I J 2020 *Int. J. Plast.* **129** 102689
[15] Xu S, Cheng J Y, Li Z, Mara N A and Beyerlein I J 2022 *Comput. Meth. Appl. Mech. Eng.* **389** 114426
[16] Xu S, Mianroodi J R, Hunter A, Beyerlein I J and Svendsen B 2019 *Philos. Mag.* **99** 1400–28
[17] Xu S, Smith L, Mianroodi J R, Hunter A, Svendsen B and Beyerlein I J 2019 *Modelling Simul. Mater. Sci. Eng.* **27** 074004
[18] Xu S, Su Y, Smith L T W and Beyerlein I J 2020 *J. Mech. Phys. Solids* **141** 104017
[19] Warlimont H and Martienssen W (eds) 2018 *Springer Handbook of Materials Data* 2nd ed (Springer International Publishing)
[20] Su Y, Xu S and Beyerlein I J 2019 *J. Appl. Phys.* **126** 105112
[21] Mianroodi J R, Hunter A, Beyerlein I J and Svendsen B 2016 *J. Mech. Phys. Solids* **95** 719–41

Local deformation mechanisms in metal systems with a tailored grain size distribution

A Godfrey[1], W Q Gao[1], G M Le[1], C L Zhang[1] and K N Zhu[2]

[1] Key Laboratory of Advanced Materials (MOE), School of Material Science and Engineering, Tsinghua University, Beijing 100084, P.R. China
[2] Jiangsu Provincial Department of Science and Technology, 39 Beijing East Road, Nanjing, Jiangsu, P .R. China

E-mail: awgodfrey@mail.tsinghua.edu.cn

Abstract. A large change in yield strength and ductility takes place in metals with a recrystallized average grain size in the near-micrometer grain size regime. This has been utilized to explore the influence of grain size heterogeneity on plastic deformation for recrystallized grains in aluminum prepared by spark plasma sintering, where tailoring of the grain size distribution and spatial arrangement can be achieved by selecting powder sizes and mixing conditions prior to sintering. The spatial variation of plastic deformation has been explored by digital image correlation (DIC), either at the sample scale or at the grain scale. For the latter colloidal SiO_2 markers also allow repeated collection of electron backscatter diffraction data over the same area during in-situ deformation, proving correlated information on local displacement gradients and crystal lattice rotations. For samples prepared from well-mixed powders only small deviations from a rule-of-mixtures in the yield strength are observed. A mesoscale pattern of strain distribution is observed from DIC investigations where this is related to, but not fully consistent with, the spatial arrangement of fine and coarse grains. In the case of macroscopically layered arrangements of coarse and fine grains, deformation is dominated by development of local strain concentrations, which take place preferentially in the fine grain regions.

1. Introduction

Much attention has been focused in recent years on the exploration of metals with heterogeneous microstructures, driven in many case by the objective of developing metals with enhanced combinations of strength and ductility [1-3]. A number of different approaches have been explored in this regard for single phase metals, including the use of laminated [4], harmonic [5], bimodal [6] and gradient [7,8] microstructures. The development of a heterogeneous and spatially varying grain structure forms the basis of many of these metals, where in a broad sense smaller grains provide high strengths, but larger grains contribute to enhanced plasticity to the material. The interaction between regions of differing grain size is, however, also important, as achieving good combination of strength and ductility requires both enhancing work-hardening and delaying the onset of strain localization, both of which are influenced by the spatial arrangement of grains of different size.

A related aspect is the observation that significant changes in mechanical properties take place in metals with a regular (i.e., approximately log-normal) grain structure, as the average grain size is decreased to the near-micrometer regime [9-12]. These changes include a transition from a continuous flow curve to one showing discontinuous yielding with an obvious yield point, together with an enhanced yield strength above that expected by extrapolation of the Hall-Petch slope to the near-micrometer regime. Both of these changes have been related to a form of dislocation source hardening, where there is an insufficient density of easily-operable dislocation sources in grains below a certain size to allow the onset of macroscopic slip [13], and have been observed for a number of different metals including Al [9], Cu [10], Fe [14], and stainless steel [15,16]. For aluminum, for example, a decrease of average grain size from around 5 μm to 1 μm results in a complete loss of uniform ductility (from ≈ 25%) with an ≈ 4-times increase in yield strength [9]. As such the development of grain size heterogeneity spanning this grain size regime presents a good opportunity to explore both the underlying deformation mechanisms in metals with grain size heterogeneity, and the relationship between local strain distribution and grain structure.

Achieving grain sizes in the near-micrometer regime in single phase metals presents certain experimental challenges, as grain sizes developed after traditional deformation + recrystallization treatments are typically limited to larger values. Two main routes have been explored for this purpose, namely through deformation to high plastic strain [17-19] followed by light annealing, and through the use of powder metallurgy methods [20,21]. Although an enhanced yield strength was observed in heavily deformed and then lightly annealed aluminum [13] the largest enhanced yield strengths have been observed in samples where the near-micrometer grains are in essentially a fully recrystallized state. It has been shown that this can be achieved by use of the spark plasma sintering (SPS) technique, as described in detail in the following section.

2. Spark plasma sintering

Spark plasma sintering was initially developed for fabrication of complex ceramics, but has now been demonstrated to be applicable to wide range of materials [22,23]. For metals this is essentially a powder metallurgy technique, involving application of heat and pressure to powder contained within a die. The general process requires applying a heating ramp under a minimal low pressure (for machine stability) until the target sintering temperature is reached, at which point the sintering pressure is increased and held for the desired sintering duration, followed by cooling. The process is usually carried out in an inert atmosphere or in vacuum. A distinguishing feature of the SPS process is that pulsed direct current used to provide Joule (resistance) heating to the powder compact. A consequence of this is a high heating rate that allows nearly full densification to be achieved by application of only moderate applied pressure and short sintering durations.

For Al the presence of a native oxide film on the surfaces of powder particles provides both a challenge and opportunities in terms of synthesis. As a challenge it has been found that a stepwise heating ramp is required to avoid issues with sample fracture, believed to be linked to egress of H_2O from the hydrated native oxide. During sintering this oxide film, typically of thickness 10 – 30 nm is broken up into fine nanoparticles resulting in a discontinuous network of particles in sintered samples that remain after sintering to decorate volumes corresponding to the original powder particle surfaces [9]. These particles provide a very strong Zener pinning effect, effectively limiting the grain size within each particle to the original particle size, even during annealing at temperatures very close to the melting point. As a result the grain size after SPS in such samples is limited by the initial powder size. In practice it has been found that for SPS of spherical Al powder particles of less than ≈ 5 μm size each powder particle results in a single grain after sintering, with larger powder particles resulting in the formation typically of a several grains. This in turn provides an opportunity using Al powders to control the grain size, and grain size distribution, at least in the grain size range of 5 μm and below.

Two other features of the grains produced by SPS of metal powders under optimal conditions should be noted. The first is that the resulting grains are essentially in a fully recrystallized state with a low dislocation density and only a small surface area per volume of low angle boundaries (typically generated inside larger initial particles), as has been revealed both by electron microscopy investigations of surfaces (via electron back-scatter diffraction, EBSD) or thin foil observations in the transmission electron microscope (TEM), and also by use of micro-diffraction using high-brightness X-ray synchrotron radiation to probe bulk material [24]. The other feature worthy of note is that the crystallographic texture of the grains in such SPS samples is close to random, as inherited from the random nature of the orientations in the powder in the die prior to sintering. These points are illustrated in figure 1 showing examples of samples of Al prepared by SPS from a powder with average particle size of ≈ 5 μm. In the interest of safety it should be added that the lower limit in grain size that can be achieved using this technique has not been explored by the present authors, as extremely fine Al powder particles present a formidable safety risk, and moreover SPS of ≈ 1 μm Al powder already yields sufficiently fine grains to achieve samples showing high yield stress without any uniform ductility [9].

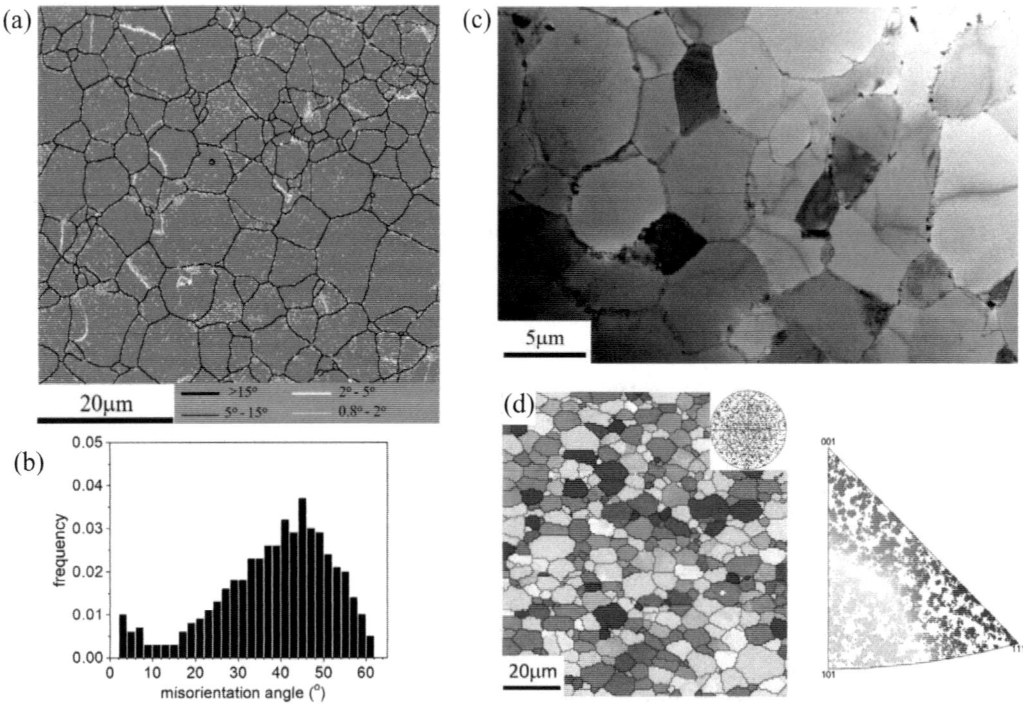

Figure 1. Key characteristics of Al samples with near-micrometer average grain size prepared by SPS from powder with initial average particle size of ≈ 5 μm: (a) EBSD boundary misorientation map and (b) boundary misorientation distribution showing the near-recrystallized state; (c) TEM micrograph showing oxide decoration along grain boundaries and the low dislocation density inside each grain; (d) EBSD map using inverse pole figure (IPF) coloring, together with {111} pole figure (inset), showing the near-random crystallographic texture.

3. Tailoring the grain size distribution in Al samples prepared by SPS

For Al, the one-to-one relationship between powder particle size and final grain size also allows the synthesis in a controlled manner of samples with a range of grain size distribution, by mixing together powder particles from batches with different average powder particle size. The extent of mixing can also be controlled to achieve either full mixing or some aggregation of the different powders, as shown in figure 2a for the case of a mixture in equal proportion of fine (≈ 1 μm average powder particle size) and coarse (≈ 5 μm average powder particle size) powders (nominal sizes as provided by the powder supplier). Here is seen that powder mixing by ultrasonication in ethanol leads to good mixing of the two powder sizes, while powder mixing using low energy ball milling results in significant clustering of each powder fraction. Examples of the grain structure after SPS of well-mixed Al powders with average powder particle sizes of ≈ 5 μm and ≈ 1 μm combined in different proportions are shown in figure 2b. Despite the appearance these samples are in fact only weakly bimodal, as the absolute difference in initial powder size is relatively small and as such there is significant overlap in the sizes of the grains resulting from the ≈ 5 μm and ≈ 1 μm powders. For this reason, as well as the fact that the grain sizes in this range straddle a significant transition in mechanical properties such a grain-size distribution has been termed 'transmodal' [25].

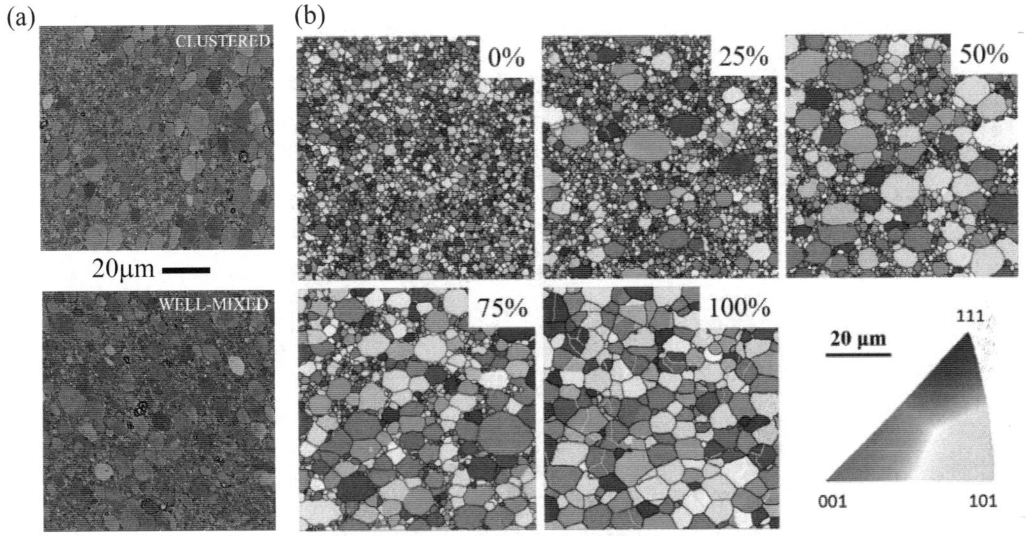

Figure 2. Grain structure in fully dense samples prepared by SPS of mixtures of fine (1 μm average powder particle size) and coarse (5 μm average powder particle size): (a) electron channeling contrast images of a 1:1 powder mixture mixed by low energy ball milling (clustered) or ultrasonication in ethanol (well-mixed); (b) EBSD maps (IPF coloring) according to sintering load direction for samples sintered from well-mixed powders with different fractions of coarse powder (as indicated in each image).

This can be seen clearly in figure 3, which shows example flow curves for a range of samples, each prepared using SPS from well mixed combinations of Al powders with initial particle sizes of ≈ 5 μm and ≈ 1 μm. Flow curves samples wholly prepared using ≈ 5 μm and ≈ 1 μm powders are also included, where it is seen that the sample prepared from the ≈ 5 μm powder exhibits continuous yielding, with a long range of work-hardening and good uniform ductility of ≈ 25%, whereas the sample prepared from the ≈ 1 μm powder exhibits a high strength but fails without exhibiting any uniform ductility. The samples prepared with different fractions of each powder show a range of intermediate mechanical properties. For samples prepared with up to 37% of the ≈ 1 μm powder, yielding takes place in a fully continuous manner with similar work-hardening after deformation beyond a few percent tensile strain. For the sample with 50% of the ≈ 1 μm powder the flow curve does not show a yield point, but an inflection is clearly visible. For samples with more than 50% of the ≈ 1 μm powder the flow curve shows either a yield point, or an absence of uniform elongation.

An interesting question to ask is whether the yield strength of the mixed powder samples varies linearly according to a rule of mixtures (ROM) calculation, or whether any additional strengthening is present, as has been claimed for some materials with heterogeneous microstructures [1-3]. The results of such an analysis are shown in table 1. One challenge here is in deciding how the yield strength should be defined in samples that do not yield in a continuous manner. For the analysis in table 1 the yield stress is defined as the conventional 0.2% proof stress for samples showing continuous yielding and as the upper yield point for the samples showing discontinuous yielding. No strong conclusion can be made on this point, although some suggestion of enhanced strengthening is seen for the samples with 37% and 50% of the ≈ 1 μm powder. The uniform ductility of these samples does not vary smoothly with powder mixture, and the data suggest the presence of some sample to sample variation. It is nevertheless clear that rapid loss of ductility once the fraction of the ≈ 1 μm powder exceeds 75%.

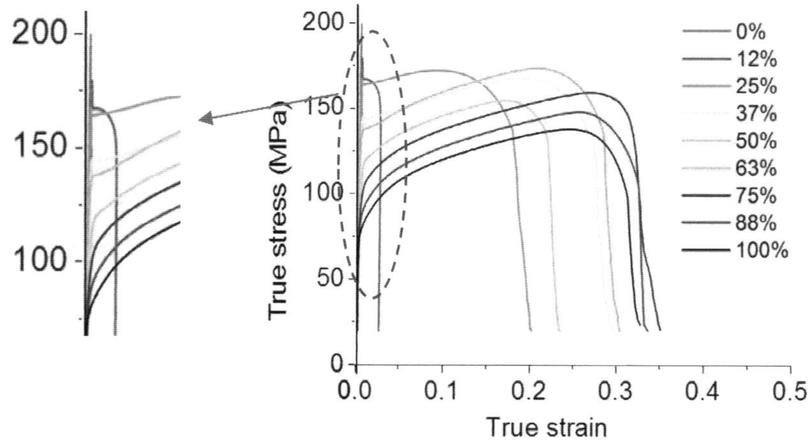

Figure 3. Tensile stress-strain curves for Al samples prepared by SPS from well-mixed powders of average particle size 5 μm and 1 μm in different proportions (the percentage indicates the fraction of the 5 μm powder). The near-yield region is magnified on the left-hand side to illustrate the transition from continuous to discontinuous yielding with increasing fraction of 1 μm powder.

Table 1: Comparison of experimental flow stress in well mixed (5 μm and 1 μm) powder samples with a rule-of-mixtures (ROM) calculation.

1 μm powder	5 μm powder	Yield stress (MPa)	ROM prediction (MPa)
0 %	100%	50	-
13%	87%	65	68
25%	75%	87	86
37%	63%	120	104
50%	50%	138	123
63%	37%	145[a]	141
75%	25%	167[a]	159
87%	13%	180[a]	177
100%	0%	195[a]	-

[a] Values where the yield stress is taken as the upper yield point (for all other cases this value is taken as the 0.2% proof stress).

4. Experimental investigation of deformation heterogeneity in mixed powder samples

4.1. Surface patterning to allow combined DIC and EBSD during in-situ deformation

In order to understand the influence on the mechanical properties of the variations in grain size distribution in the samples with a transmodal grain size distribution it is necessary to characterize the process of plastic deformation at a length scale that captures the behavior between and within individual grains. Two of the most commonly used experimental techniques for investigation of plastic deformation at this length scale are EBSD and digital image correlation (DIC) in the scanning electron microscope [26,27]. These techniques provide complimentary information regarding local plastic deformation, and as such it is of interest to collect both types of data over the same surface region of a sample, and moreover to be able to do this repeatedly during in-situ deformation experiments. The challenge here is that DIC measurements require surface markers to allow tracking of surface displacements. In some cases these markers can be microstructural features such as those visible after chemical etching [28]. This, however, is not compatible with collection of EBSD data on metallic samples where a highly polished (either mechanically or electrolytically) smooth flat surface is required.

The markers for DIC should be of a size considerably smaller than the length scale of interest in any particular investigation. For the samples considered in this work, with grain sizes in the near-micrometer regime and extending down to less than 1 μm, markers of very fine scale are therefore required. It has been shown that Au nanoparticles, either directly deposited or generated through a thin-film remodeling process [29] are well suited for this purpose, as such markers are both of a suitable size range and have high contrast. Unfortunately, however, the presence of such markers on a sample surface leads to a severe degradation of EBSD pattern quality, preventing their use in studies where it is desired to collect both types of data from the same region of a sample over a number of strain steps. As such the use of Au nanoparticle markers is limited to experiments of the type where EBSD microstructural data are collected before deformation, DIC data are collected over different strain steps during in-situ deformation, and then at the end of the experiment the markers are removed by a light re-polishing to allow EBSD characterization of final deformed microstructure.

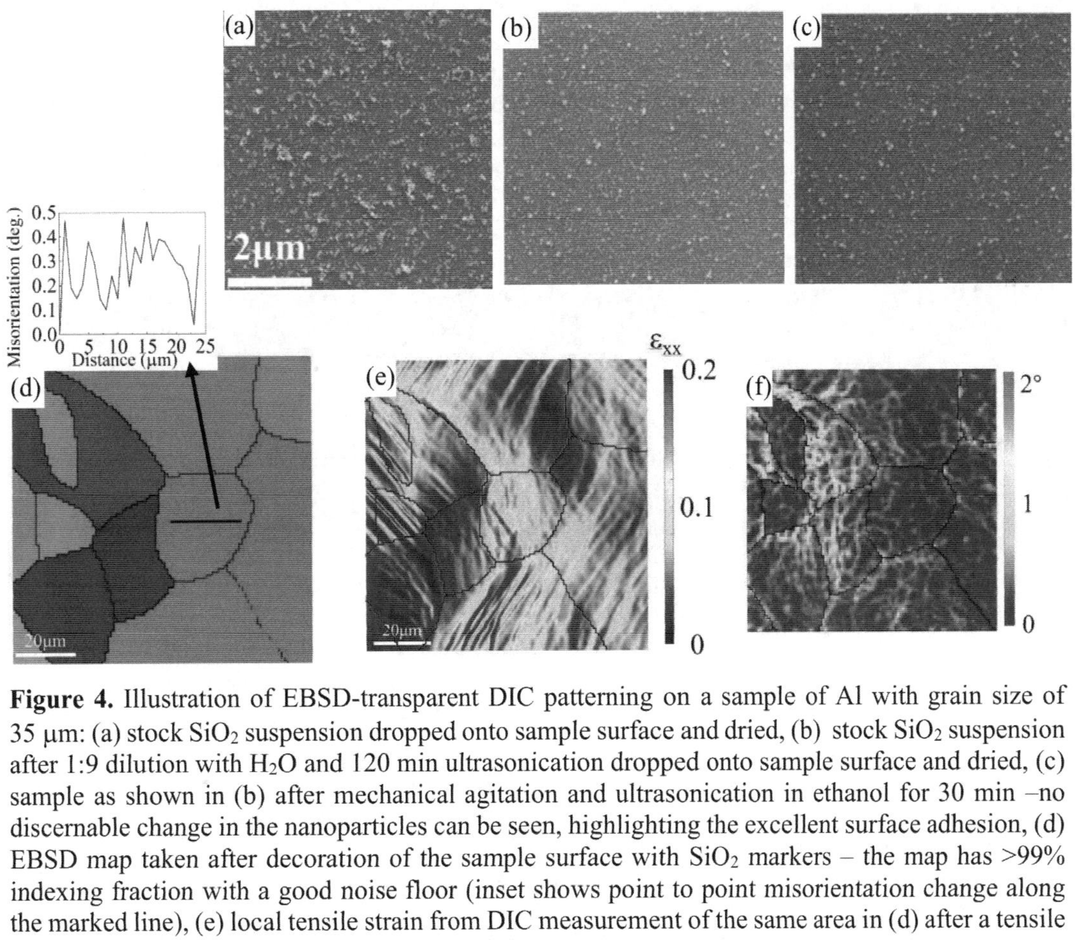

Figure 4. Illustration of EBSD-transparent DIC patterning on a sample of Al with grain size of 35 μm: (a) stock SiO_2 suspension dropped onto sample surface and dried, (b) stock SiO_2 suspension after 1:9 dilution with H_2O and 120 min ultrasonication dropped onto sample surface and dried, (c) sample as shown in (b) after mechanical agitation and ultrasonication in ethanol for 30 min –no discernable change in the nanoparticles can be seen, highlighting the excellent surface adhesion, (d) EBSD map taken after decoration of the sample surface with SiO_2 markers – the map has >99% indexing fraction with a good noise floor (inset shows point to point misorientation change along the marked line), (e) local tensile strain from DIC measurement of the same area in (d) after a tensile strain of 0.06, (f) kernel average misorientation map of the same region at a strain of 0.06.

As an alternative Yan et al. [30] proposed the use of colloidal SiO_2 nanoparticles as high-resolution DIC markers, introduced onto a sample surface by transfer from a platen through a variation of the drop-casting technique. These markers have less contrast than Au nanoparticles, and in general have a slightly larger average size, but have the advantage of being effectively transparent to the signal that give rise to EBSD patterns. It has recently been demonstrated that suitable coverage of markers can also be obtained pipetting a small amount of a colloidal SiO_2 suspension of controlled dilution onto a sample surface after long-duration ultrasonication (of up to 2 hours) to minimize any nanoparticle aggregation [31].

Examples of this patterning are shown in figures 4a,b, where the effect of dilution and ultrasonication is shown. The SiO_2 nanoparticles adhere strongly to the sample surface, such that once dried they are mechanically stable, as shown in figure 4c. Figures 4d-f show an example application of these DIC markers during in-situ tension of a sample of coarse grain size Al (average grain size of 35 μm) to a strain of 6%. The EBSD map in figure 4d was collected before deformation but after marker deposition, and has a very high indexed fraction of >99% (without any data cleaning) with a low orientation noise floor (see inset in figure 4d), clearly demonstrating that the markers do not influence the EBSD indexing.

The combination of high spatial resolution local strain and orientation data that can be achieved using the SiO_2 markers is shown in figure 4e,f, showing the tensile strain from DIC data and a kernel average misorientation map for the same region at the same tensile strain. It is interesting to note that there is very little correlation between the local scale features seen in the EBSD and DIC maps. Some differences in the two data sets can be attributed to the fact that they are subject to different limitations with regard to being based on 2D surface measurements. For the EBSD data out-of-plane lattice misorientations are not accessible without the use of serial-sectioning (not possible in such in-situ experiments as a destructive 3D measurement) and for SEM-based DIC data only the in-plane deformation gradients can be measured. More importantly though is the fact that local misorientation measures, such as the kernel average misorientation and other orientation gradient derived measures, do not in fact directly track plastic strain on the microstructural scale (though this may be possible in combination with lattice rotation calculations from a calibrated crystal plasticity model).

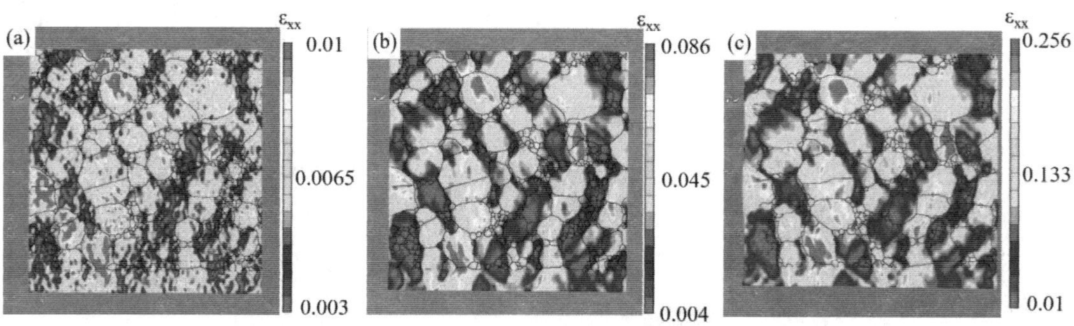

Figure 5. Evolution of local tensile strain of a well-mixed sample sintered from 63% of ≈ 5 μm Al powder and 37% of ≈ 1 μm Al powder at average tensile strains of (a) 0.006, (b) 0.04 and (c) 0.126.

4.2. Investigation of deformation heterogeneity in a well-mixed transmodal Al SPS sample

As an illustration of the experimental information available for investigation of the heterogeneity in plastic deformation in relation to the grain size distribution and spatial arrangement some results are presented here for a sample composed of 63% of ≈ 5 μm Al powder and 37% of ≈ 1 μm Al powder [25]. For this investigation a dog-bone shape tension sample was first cut from a spark plasma sintered disk. All surfaces were mechanically polished, then one flat surface was electropolished and finally coated with SiO_2 markers. The sample was then mounted in a tensile deformation stage inside a SEM and loaded to a final tensile strain of ε_{av} = 0.126. EBSD measurements were taken before deformation and at tensile strains of 0.006, 0.040, and 0.126 over a fixed region on the sample surface located near the center of the gauge length. DIC measurements were taken over the same region on the sample surface also at strains of 0.006, 0.040 and 0.126. For the DIC measurements images of size 4096 x 4096 pixels were collected and the DIC calculations performed using a window of 59 x 59 pixels with a step-size of 3 pixels. Complete details of the experimental procedure are given in [30]. It should be noted that the powders for this sample were taken from a different batch to those used to prepare the samples described in Section 3, resulting in some differences in mechanical properties (see figure 7d later in this work). These differences highlight the high sensitivity of mechanical properties to grain size in the near-micrometer regime, attributed here mainly due to a slightly smaller average grain size derived from the 1 μm powder fraction. For both sets of powders 37% was found to be fine (1 μm) powder that resulted in smooth continuous yielding and work-hardening up to the uniform elongation limit.

Figures 5a-c show the distribution of local tensile strain, as inferred from the DIC measurements, during loading of the sample. In these images the black lines show the grain structure, based on EBSD measurements taken at the same strain. Initial inspection of the data shows extensive heterogeneity in the local tensile strain, both between grains and within many of the grains. This is most apparent with regard to regions of lowest local tensile strain, where it is seen that many such regions correspond to locations consisting of fine grains. However, it is also evident that some parts of larger grains are also associated with regions of low strain. Regions of highest strain are in all but a few cases found inside the larger grains, sometimes in the grain interior and sometimes near grain boundaries or triple junctions, with localized strain concentration across grain boundaries observed in several places. Overall the strain distribution suggests a mesoscale patterning on a scale that is related to the spatial arrangement of the coarse and fine grains, though not precisely following the grain arrangements.

The micro-DIC show some level of strain transfer across grain boundaries in establishing this meso-scale patterning. This raises a question regarding the importance of the grain boundary strength, and therefore of the nanoscale oxide particle network, in determining the deformation behavior in these samples. In this regard it can be noted that the Hall-Petch slope (a measure of the grain boundary strength) for SPS Al with average grain size of less than 5 μm is around three times larger than for conventional coarse grain Al [9]. A similar difference is, however, also seen for near-micrometer grain size Cu [10], where grain boundaries are not decorated with oxides. As such that the enhanced grain boundary strength in these near-micrometer grain size samples cannot be directly attributed to the nanoscale oxides, though the high relatively high grain boundary strength may well influence the extent and arrangement of strain transfer across grain boundaries.

Table 2: Strain partitioning based on DIC measurements for a sample sintered from a well-mixed combination of 63% of ≈ 5 μm Al powder and 37% of ≈ 1 μm Al powder.

		$\varepsilon_{av} = 0.6\%$			$\varepsilon_{av} = 4.0\%$			$\varepsilon_{av} = 12.6\%$		
Grain-size class (μm)	$f = A_A$ ($= V_V$)	$\Sigma\varepsilon_{DIC}$ (x100)	$\Sigma\varepsilon_u$ (x100)	R	$\Sigma\varepsilon_{DIC}$ (x100)	$\Sigma\varepsilon_u$ (x100)	R	ε_{DIC} (x100)	ε_u (x100)	R
$d < 1$	0.036	0.0189	0.0226	0.84	0.0968	0.1540	0.63	0.1368	0.1868	0.73
$1 < d \leq 2$	0.129	0.0658	0.0800	0.82	0.2976	0.4527	0.66	0.8768	1.2027	0.73
$2 < d \leq 3$	0.092	0.0497	0.0570	0.87	0.2154	0.2983	0.72	0.7396	0.9743	0.76
$3 < d \leq 4$	0.043	0.0265	0.0265	1.00	0.1868	0.1784	1.05	0.5031	0.5149	0.98
$d > 4$	0.700	0.4618	0.4338	1.06	3.2111	2.9265	1.10	10.367	9.7414	1.06
$4 < d \leq 10$	0.376	0.2632	0.2331	1.13	1.9095	1.5723	1.21	6.0629	5.2335	1.16
$d > 10$	0.324	0.1975	0.2007	0.98	1.3035	1.3542	0.96	4.3043	4.5079	0.95

As a more quantitative analysis the average tensile strain for each grain in a tracked area containing approx. 320 grains has been determined at each strain step. The results are summarized in table 2 by dividing the data into a number of different size categories. In this table f is the area fraction in the tracked area of each grain class (stereologically equal to volume fraction for random sections of a non-oriented structure), ε_{av} is the average tensile strain over the entire tracked area, $\Sigma\varepsilon_{DIC}$ is the sum of tensile strain contribution from all map pixels within each grain-size class, and the value of $\Sigma\varepsilon_u$ is the expected sum of tensile strain for each grain class assuming a fully uniform strain distribution (determined for each grain-size class from the summed tensile strain over the investigated area multiplied by the area fraction of each grain size class). The value of $R = \Sigma\varepsilon_{DIC}/\Sigma\varepsilon_u$ is, therefore, a measure of how much more (or less) strain each grain-class takes compared to the case of fully homogenous deformation among all grains.

Three interesting observations can be made based on the data in table 2. Firstly, the fraction of strain in the finest grains is highest just after yielding ($\varepsilon_{av} = 0.6\%$), decreasing at $\varepsilon_{av} = 4\%$, and then increasing again at $\varepsilon_{av} = 12.6\%$ strain. The increase in fraction of strain in the finest grains at higher strains can be understood as with continued deformation dislocation storage takes place in larger grains, such that with further straining plastic deformation starts to be transferred more towards the finer grains. The more similar values of strain over all grain-size classes just after yielding (also seen from inspection of figure 5a) is more challenging to understand, but may be a reflection fact that at very low strain (near yielding) a fraction of strain comes from the movement of dislocations present in the as-sintered samples. In the finest grains these dislocations move to, and are annihilated at, grain boundaries without much interaction [32]. Secondly, grains above 10 μm take less average strain than grains in the range $4 < d \leq 10$ μm. The grains in each of these size classes cover a large area fraction and show a consistent trend so this cannot be discounted as an artifact. One thing that may be relevant here is that the largest grains are the most heterogeneous in their deformation, so that although the highest local strain values are found in these grains, they also contain regions with lower strain values. Thirdly, and most obvious in the data, is that in this sample significant redistribution of tensile strain to larger grains takes place for grains with size greater than 4 μm.

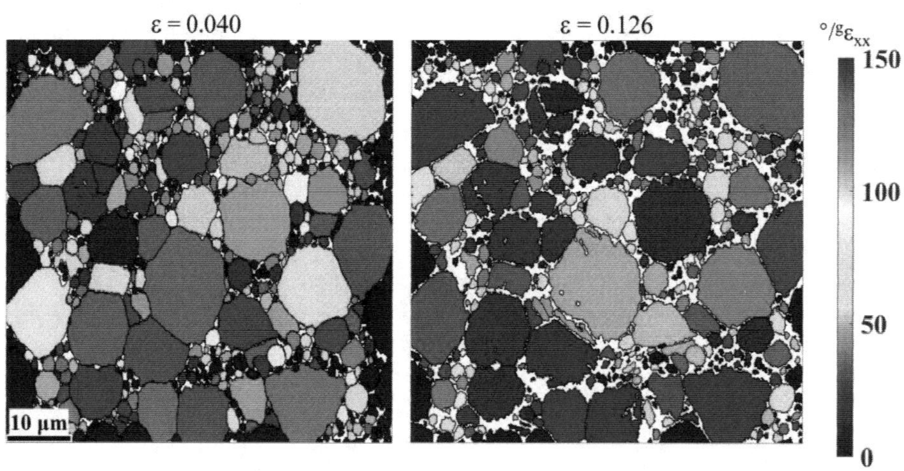

Figure 6. Strain normalized average grain rotation from combined in-situ DIC and EBSD data for tensile strains of 0.04 and 0.126 (defined as misorientation angle in degrees corresponding to change in average orientation of each grain divided by average local tensile strain for each grain).

As both EBSD and DIC data are available in this in-situ experiment, it is also possible to compare local tensile strains with the crystal lattice rotations developed during tensile deformation. This has been done here on a grain average basis by calculating the average rotation of each grain at each strain step (by comparison with the initial grain orientation) and combining this with the average DIC-determined tensile strain for each grain, to obtain a value for the average rotation per unit strain for each grain. Figure 6 shows the result of such calculations for a different region of the same sample at tensile strains of 0.04 and 0.126 (the analysis of this region of the sample is discussed in more detail in [30]). The data show a very wide variation in strain-normalized average grain rotation. Values for larger grains are mostly relatively low, as many large grains show some strain localization (contributing to a larger average strain value so a lower rotation per unit strain). More surprising is that many fine grains show relatively high values for strain-normalized average grain rotation, in some cases with these higher values in clusters suggesting some aggregate behavior, highlighting the fact that the plastic deformation in these heterogeneous grain structures follows a pattern that is not simply described by strain transfer from coarse grains to fine grains, despite the approximately 4-times difference of yield strength for grains of 1 μm and 5 μm size.

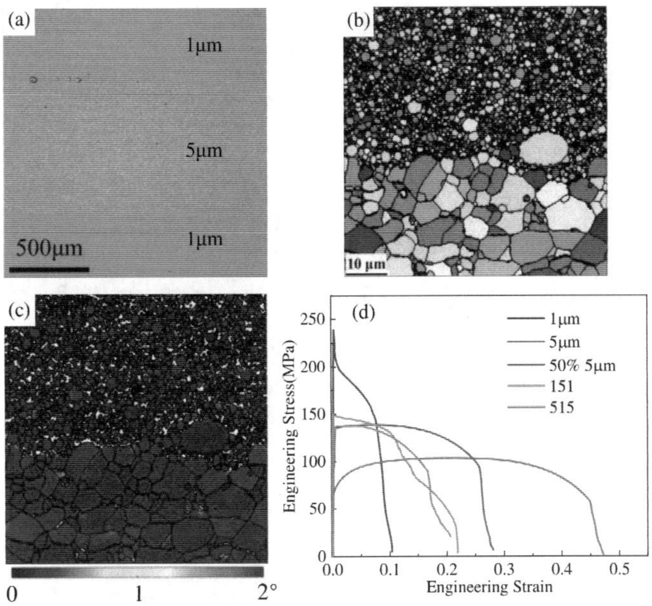

Figure 7. Macroscopically layered samples prepared using mixtures of 5 μm and 1 μm powder in three layers in equal volume fraction: (a) low magnification SEM channeling contrast image showing the near-complete cross section of a 151 sample, (b,c) EBSD maps using IPF coloring and kernel average misorientation coloring of an interface region, (d) example flow curves for the 151 and 515 samples, together with curves for samples sintered from just the 1 μm powder, just the 5 μm powder, and a well-mixed powder sample with 50% 5 μm powder fraction. During tension testing the loading axis is parallel to the macroscopic interface plane (horizontal in these images).

4.3. Macroscopically layered Al SPS samples

As an extreme case of mixing between the 1 μm and 5 μm powders the deformation of samples with macroscopic layers of powders has also been investigated, using here equal fractions of both powders (i.e., comparable to the 50% mixed powder sample). Samples were made with three layers as shown in figure 7a, with the two outer layers of equal thickness. Two types of sample were investigated, either with the 5 μm powder in the center layer and the 1 μm powder in the outer layer ("151"), or the opposite case with the 1 μm powder in the center layer ("515"). To make the samples powders were carefully added layer by layer to the powder die with a gentle pre-pressing applied after adding each layer to minimize intermixing of the layers. After adding all three layers the samples were sintered under the usual conditions for Al powders [9].

For the layered microstructures the internal interface quality is an important parameter in determining the deformation behavior. As shown for the 151 sample in figures 7a,b, the interfaces between the different layers are straight, distinct with very little intermixing of the two powder sizes. The 3 x 3 kernel average misorientation map in figure 7c shows very low values in both the coarse and fine powders, with no evidence of any local deformation at or near the interface region, highlighting the advantage of the SPS technique for preparation of samples with layered structure.

The strain-stress curves of the 151 and 515 samples, together with that for a 50% well-mixed powder sample, as well as those of the 1 μm and 5 μm sample, are shown in figure 7d. Note that these samples were made from a different batch of powder to the samples reported in figure 3, where for these samples the 1 μm powder shows a higher yield strength on account of a slightly smaller average powder size (and hence grain size in the sintered sample). The general flow behavior remains, however, similar, with the 1 μm powder sample showing a load drop after the peak stress and no uniform ductility, and the 5 μm powder showing a low yield strength with continuous yielding and a good uniform ductility. The

50% well-mixed powder sample shows an intermediate yield strength (at a value close, but slightly smaller than that expected from a rule-of-mixtures yield strength) with a sharp, though still continuous yield region, followed by low work-hardening to a maximum tensile stress at a strain of approx. 10%. The macroscopically layered 515 shows a similar flow curve to the well-mixed 50% sample in the near-yield region with samples, albeit with a reduced uniform elongation, but with extended post-uniform elongation. In contrast, the 151 sample shows a load drop after a higher peak stress compared to the 515 sample, with no uniform ductility but again an extended region of post-uniform elongation.

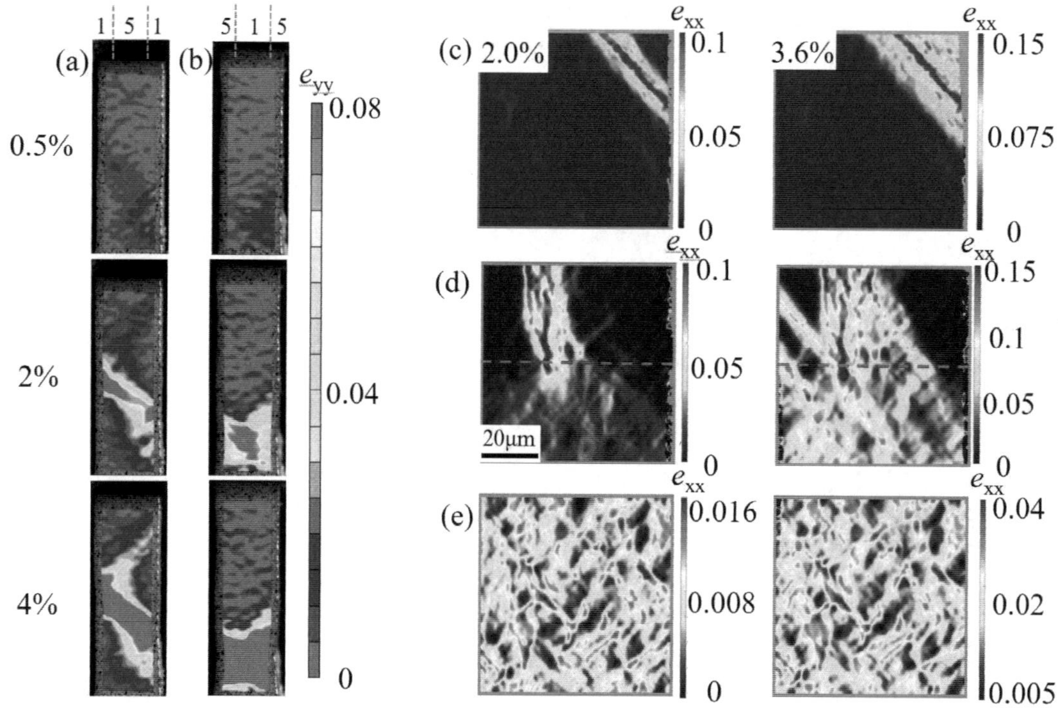

Figure 8. Investigation of local tensile strain distribution in macroscopically layered samples: (a, b) macro-DIC measurements on the 151 (a) and 515 (b) samples at tensile strains of 0.5%, 2% and 4% showing the different nature of strain localization in each sample; (c-e) micro-DIC measurements for the 515 sample at strain of 2.0% (left column) and 3.6% (right column) in the 1 μm grain size region (c), the interface region (d), and the 5 μm grain size region (e).

In order to further understand the deformation mechanisms of the macroscopically layered samples the strain distribution at different length scales was investigated using macro-DIC and micro-DIC measurements. The macro-DIC measurements were obtained on samples deformed using a laboratory testing frame on samples sprayed with a mixture of black and white paint. The micro-DIC measurements were carried out on samples of the same size deformed in a SEM using the colloidal SiO_2 nanoparticles as markers as described in section 2. The strain distribution over the sample gauge length for the 151 and 515 samples are shown in figures 8a and 8b, respectively at strains of 0.5%, 2% and 4%. Note that here the tensile loading direction is vertical on the page and the layer interfaces are edge-on. For both samples macroscopic strain concentration is already well developed at a strain of 2%, and already at 0.5% strain the beginning of this strain concentration can be seen. For the 151 sample this takes the form of a macroscopic shear band (with the development of a second shear band at 90° to the first also visible), whereas for the 515 sample the strain concentration is more broadly spread. Interestingly in both cases the regions of high strain concentration are associated with the volume containing the fine ≈ 1 μm grains indicating that strain concentration in the fine-grain region dominates macroscopic response.

The results of micro-DIC investigation of local tensile strain distribution in the 515 sample at strains of 2.0% and 3.6% are shown in figures 8c-e, showing the interface region (figure 8d), and locations far-away from the interface of the 1 μm grain size region (figure 8c) and the 5 μm grain size region (figure 8e). Note here that the sample orientation is rotated by 90° such that the tensile loading direction, and interface trace (marked by a dashed line in figure 8d) is horizontal. In the 1 μm grain size region (figure 8c) the strain is highly concentrated into a narrow shear band that broadens with increasing strain. In the near-interface region a shear band in the 1 μm grain size region results in transfer of strain across the interface to the adjacent 5 μm grain size region through the formation of a fan-shaped shear band affected zone (SBAZ), seen most clearly in the images at 2% strain. This SBAZ leads to a blunting of the shear band with increasing strain which slows down the rate of strain concentration in the 1 μm grain size region, thereby contributing to the more diffuse pattern seen in the macro-DIC observations. In the 5 μm grain size region (figure 8e) the strain distribution is more uniform, with no evidence of extended regions of strain concentration, and remaining similar at both the strains investigated. Note that the maximum local tensile strains plotted in figure 8e are considerably lower than those in figures 8c,d (hence the use of a different color scaling).

5. Summary

A significant transition in mechanical properties take places as the average grain size is decreased in the near-micrometer grain size regime, including large increases in flow stress and decrease in ductility. For Al, the use of SPS allows synthesis of samples with a variation both in grain size heterogeneity (in terms of the grain size distribution) covering this size regime, by taking advantage of the Zener pinning arising from nanoscale oxides associated with the native oxide layer on initial powder particles. As a powder-based method some control of the spatial arrangement of grains of different size class is also possible. Two extreme cases have been considered in this work, namely well-mixed combinations of fine (≈ 1 μm) and coarse (≈ 5 μm) grains, and a layered structure of these same grain sizes. Both of these highlight different aspects of the deformation mechanisms associated with improved combinations of strength and ductility with regard to the distribution of local deformation and its evolution during continued macroscopic plastic deformation. Investigation of such local deformation is aided by studies where both DIC and EBSD measurements can be made over the same area during in-situ deformation experiments, as these techniques both highlight different aspects of the local plastic deformation (and suffer from different aspects of being based on 2D-surface measurements). It should be noted also that studies in the transmission electron microscope remain the gold-standard for quantification of dislocation patterning and providing microstructural context for interpretation of EBSD data regarding local misorientations, where these reflect dislocation storage after deformation, whereas the DIC measurements reflect dislocation activity.

In the case of samples prepared from well-mixed powders of differing average size only small deviations from a rule-of-mixtures in the yield strength are observed, with differences in work-hardening only seen over a small range of a few percent after yielding, at least for samples with more than 50% by volume of coarse grains. For a well-mixed sample with 63% of coarse grains preferential deformation of the larger grains does take place, with evidence of a mesoscale strain distribution that is related to, but not completely determined by, the spatial arrangement of the coarse and fine grains. Strain partitioning decreases beyond a moderately low strain, as preferential dislocation storage in coarser grains leads to a decrease in relative strength compared to finer grains. Combined EBSD and DIC measurements show a more complex pattern of deformation with regard to the average grain rotation normalized by grain-average strain, with many smaller grains showing high values of this parameter. In the case of macroscopically three-layered arrangements of coarse and fine grains, the tensile deformation is dominated by the development of local strain concentrations, which take place preferentially in the fine grain regions. These strain concentrations only propagate weakly into coarse grain regions, suggesting that by increasing the number of macroscopic layers work-hardening (and uniform elongation) may be sustained over a larger strain.

Acknowledgements
Financial support from the National Natural Science Foundation of China (project numbers 51671113, 51471095 and 50971074) is gratefully acknowledged.

References
[1] Y. Zhu et al 2021 *Mater. Res. Lett.* **9** 1–31
[2] Yang M X, Yuan F P, Wu G L, Wei Y J, Huang X and Zhu Y T 2015 *Proc. Nat. Acad. Sci.* **112** 14501–5
[3] Ma E and Zhu T 2017 *Mater. Today* **20** 323–31
[4] Li J, Zhao M, Jin L, Wang F, Dong S and Dong J 2021 *J. Mater. Sci. Technol.* **71** 195–200
[5] Orlov D and Ameyama K 2020 *Mater. Sci. Technol.* **36** 517–26
[6] Vajpai S K, Ota M, Watanabe T, Maeda R, Sekiguchi T, Kusaka T and Ameyama K 2015 *Metall. Mater. Trans. A* **46** 903–14
[7] Fang T H, Li W L, Tao N R, Lu K 2011 *Science* **331** 1587–90
[8] Cheng Z, Zhou H F, Lu Q H, Gao H J and Lu L 2018 *Science* **362** eaau1925
[9] Le G M, Godfrey A, Hansen N 2013 *Mater. Des.* **49** 360–7
[10] Zhu K N, Godfrey A, Hansen N, Zhang X D 2017 *Mater. Des.* **117** 95–103
[11] Tsuji N, Ito Y, Saito Y and Minamino Y 2002 *Scripta Mater.* **47** 893–9
[12] Tian Y Z, Gao S, Zhao L J, Lu S, Pippan R, Zhang Z F and Tsuji N 2018 *Scripta Mater.* **142** 88–91
[13] Kamikawa N, Huang X, Tsuji N and Hansen N 2009 *Acta Mater.* **57** 4198–208
[14] Ruan Q, Yang M X, Liu W and Godfrey A 2019 *Mater. Sci. Eng. A* **744** 764–72
[15] Liu M W, Gong W, Zheng R X, Li J, Zhang Z, Gao S, Ma C L and Tsuji N 2022 *Acta Mater.* **226** 117629
[16] Flipon B, Keller C, de la Cruz L G, Hug E and Barbe F 2018 *Mater. Sci. Eng. A* **729** 249–56
[17] Xia S H, Vychigzhanina L V, Wang J T and Alexandrov I V 2008 *Mater. Sci. Forum* **584** 315–26.
[18] Orlov D, Todaka Y, Umemoto M and Tsuji N 2011 *Scripta Mater.* **64** 498–501
[19] Jin H and Lloyd D J 2004 *Scripta Mater.* **50** 1319–23
[20] Vajpai S K, Ota M, Watanabe T, Maeda R, Sekiguchi T, Kusaka T and Ameyama K 2015 *Metall. Mater. Trans. A* **46** 903–14
[21] Wang C W, Zhong W, Ping W W, Lin Z W, Wang R L, Dai J Q, Guo M, Xiong W, Zhao J-C and Hy L B 2021 *Adv. Sci.* **8** 2004229
[22] Omori M 2000 *Mater. Sci. Eng. A* **287** 183–8
[23] Weston N S, Thomas B and Jackson M 2019 *Mater. Sci. Technol.* **35** 1306–28
[24] Zhang C L, Godfrey A, Zhang Y, Wu G L, Xu R, Liu W and Juul Jensen D 2020 *Mater. Lett.* **269** 127653
[25] Gao W Q, Zhang C L, Yang M X, Zhang S Q, Juul Jensen D and Godfrey A 2021 *Mater. Sci. Engin. A* **828** 142010
[26] Pan B, Qian K M, Xie H M, Asundi A 2009 *Meas. Sci. Technol.* **20** 062001
[27] Khoo S-W, Karuppanan S, Tan C-S 2016 *Metrol. Meas. Syst.* **23** 461–80
[28] Weidner A and Biermann H 2021 *Adv. Eng. Mater.* **23** 2001409
[29] Harte A, Atkinson M, Preuss M and da Fonseca J Q 2020 *Acta Mater.* **195** 555–70
[30] Yan D S, Tasan C C and Raabe D 2015 *Acta Mater.* **96** 399–409
[31] Le G M, Godfrey A, Hansen N, Liu W, Winther G and Huang X 2013 *Acta Mater.* **61** 7072–86

IOP Conf. Series: Materials Science and Engineering 1249 (2022) 012007 doi:10.1088/1757-899X/1249/1/012007

Crystallographic Variability in Additive Manufacturing

Richard W. Fonda and David J. Rowenhorst

U.S. Naval Research Laboratory, Washington, DC, USA

E-mail: richard.fonda@nrl.navy.mil

Abstract. The crystallographic textures produced during additive manufacturing can be understood, predicted, and manipulated by varying the grain nucleation and growth processes. The resultant textures are primarily dictated by the melt pool geometry, which defines the local thermal gradient and thus the preferred crystal growth directions, as well as the scan strategy, which controls the propagation of grain orientations into subsequent layers. This texture can be diluted through heterogeneous nucleation of new grain orientations, which can occur through a variety of mechanisms. This ability to control the texture during additive manufacturing can enable the location-specific control of properties as a function of position in the build.

1. Introduction

Additive manufacturing (AM) is at the fundamental level a miniaturized version of conventional welding processes. Thus, much of the characteristic metallurgy and crystallography behavior in AM can be understood in terms of their more macroscopic welding counterparts, which have a more established history. Extending that more mature welding background to length scales where the weld-pool size is on the order of the grain size with potentially much larger thermal gradients provides a good foundation for describing the microstructural and crystallographic features observed in typical AM builds. It should be noted that this article mainly focuses on alloys that solidify with a single primary phase and do not undergo an allotropic phase transition on cooling. However, these foundations also apply to alloy systems that do exhibit multiple phase transformations, as will be briefly discussed.

The crystallographic texture that develops is primarily driven by the grain structures that develop during solidification. There are many factors that can affect the solidification of molten weld pools [1,2]. Since the weld pool is bounded by solid material, solidification usually occurs epitaxially from the pre-existing base metal crystalline structure without any nucleation barrier. In an isotropic system, solidification would proceed most quickly along the thermal gradient, which is perpendicular to weld pool wall. However, metals with a body-centered cubic (BCC) or face-centered cubic (FCC) crystal structure grow most favorably along the <001> directions. Thus, solidification typically occurs along the <001> direction that is most closely aligned with the local thermal gradient. This behavior is illustrated by the cellular solidification of AM 316L (Figure 1), which has cells oriented along <001> [1,3]. Even within the same grain, the solidification direction can switch to a different <001> variant if it is better aligned with the local thermal gradient (Figure 1, right). This behavior sets up a competitive growth between different grain orientations, with grains that have a <001> direction aligned closer to the thermal gradient dominating over more poorly-oriented grains. As illustrated in Figure 1, the type of solidification structures that form depend primarily on the amount of solute in the molten material and the compositional supercooling present in the weld pool. The compositional supercooling can be represented by the temperature gradient (G) divided by the solidification rate (R). Pure metals and alloys with low solute content have a planar solidification front; higher solute contents promote a cellular solidification, particularly at higher G/R ratios; and even higher solute contents, or lower G/R, produce a dendritic structure [2]. All three solidification structures are observed in AM builds.

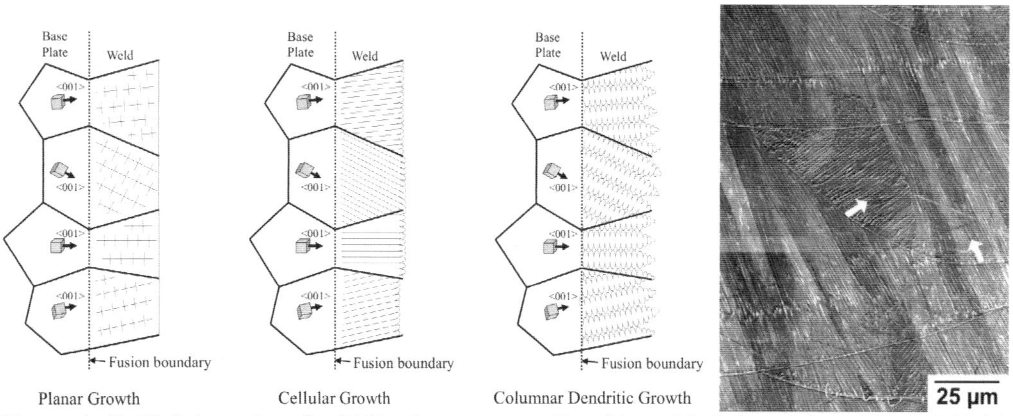

Figure 1. (Left) Schematics of solidification structures found in welds (adapted from [1]) and (right) cellular solidification within a single grain orientation across different thermal gradient directions.

2. Effect of Melt Pool Geometry

With the local thermal gradient dominating the selection of grain orientations during solidification, the three-dimensional melt pool geometry becomes the primary determinant of the resultant crystallographic texture. The extensive range of length scales and time frames possible with the various AM techniques [4,5] can produce a wide range of melt pool geometries, and can even vary within a single technique. Weld pools from the laser powder bed fusion AM process alone can range from a deep penetration keyhole geometry to a shallower conduction mode shape (Figure 2) as a function of laser power, scan speed, and equipment used [6]. And since the grains grow predominantly normal to the melt pool boundary, the resultant textures are also expected to vary as a function of that geometry.

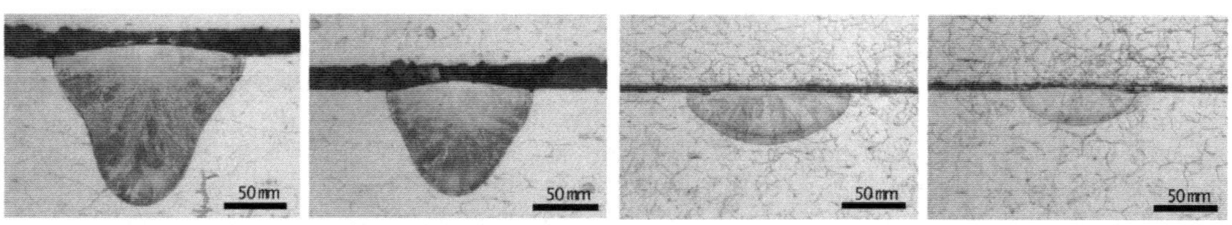

Figure 2. Weld pool cross sections of single-track laser powder bed fusion scans in Inconel 625 (adapted from [6]).

Shallower melt pool geometries (Figure 3, left) have a predominantly vertical thermal gradient within the melt pool. This can result from larger spot sizes [6] or faster scan speeds, such as those in electron beam AM. The epitaxial growth of grains from the underlying solid grain structure gives rise to a competitive growth process between the grains. Grains with a <001> direction aligned more closely with the vertical thermal gradient (and thus the vertical build direction) will grow at the expense of grains that are not as well oriented. This preferred grain orientation will increase in dominance to generate a columnar, vertical grain structure with <001> preferentially aligned along the build direction. The thermal gradient along the scan direction also influences the grain orientation, selecting orientation variants that not only have a <001> direction aligned with the build direction, but that also have another <001> direction aligned with the scan direction. This texture has been observed in both two-dimensional [7,8] and three-dimensional [9] analyses of such builds, and is shown at the left in Figure 4.

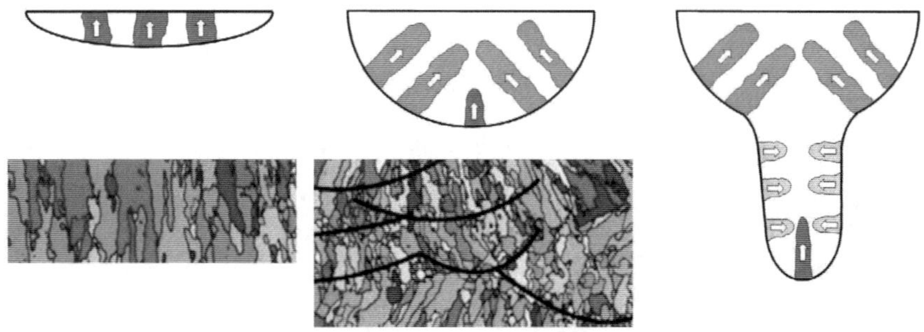

Figure 3. (Top) Melt pool geometries observed in additive manufacturing and the grain orientations that may form during solidification, and (bottom) kinetic Monte Carlo grain growth simulations of the corresponding grain structures (adapted from [7]).

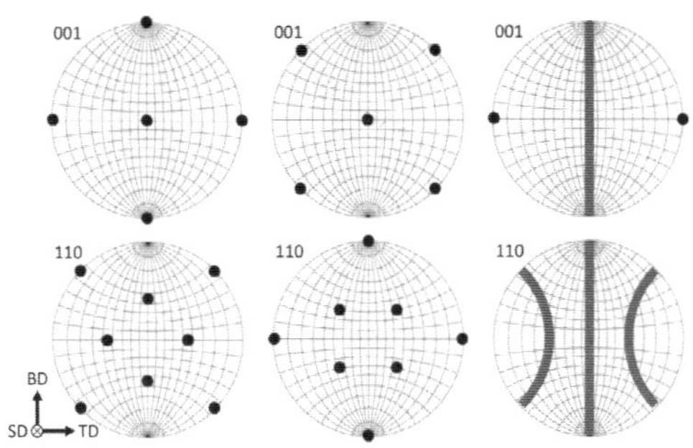

Figure 4. Schematic pole figures showing the crystallographic orientations typical of (left) shallow melt pools, (middle) deeper penetration melt pools, and (right) keyhole melt pools.

Higher energy density heat sources will penetrate deeper into the substrate, creating a melt pool with a more semicircular cross section (Figure 3, center and Figure 2, second from left). This equiaxed melt pool geometry is often observed in the laser powder bed fusion process. The deeper penetration aligns significant portions of the melt pool boundary approximately 45° from the vertical build direction, promoting <001> grain growth along those thermal gradient directions. At the same time, the tail of the melt pool tends to select variants of this orientation that also have a <001> direction aligned with the scan direction [10-12]. Competitive grain growth during solidification of subsequent layers that adheres to both the 45° orientation of <001> from the vertical build direction and the alignment of another <001> direction along the scan direction will produce an overall texture with <110> along the build direction [12], as shown in the middle of Figure 4.

The grain morphology and crystallographic texture can also vary as a function of changing deposition parameters (such as scan direction) from one layer to the next. For the simplest condition, where subsequent layers match the location and direction of the scans, the overall <110> || build direction and <001> || scan direction grain orientation is strengthened through the competitive grain growth process at each layer to generate a substantially single crystal orientation. Well-oriented grains will preferentially propagate into subsequent layers and more poorly oriented grains will be consumed. Then, even significant changes in the local thermal gradient, such as those accompanying the deposition of a new melt pool, will only change the solidification direction to a different <001> variant of the same crystal orientation because of the epitaxial growth constraint rather than nucleate a new grain orientation [13].

Changes in the scan direction from one layer to the next are common in laser powder bed fusion. The scan rotation causes grains that are ideally oriented for one layer to be not likely similarly favored for the next. Such changes maintain the preferred <001> grain growth at 45° to the build direction, but remove the constraint of <001> along the scan direction, resulting in a <110> fiber texture along the build direction. The propagation of grains from one layer to the next can produce complex three dimensional branching. Most grains are oriented with <110> close to the build direction, but the branches reflect the <001> growth at 45° to that direction. Figure 5 shows some examples of grain morphologies that can result from this branching and bending to maintain the <001> growth orientation.

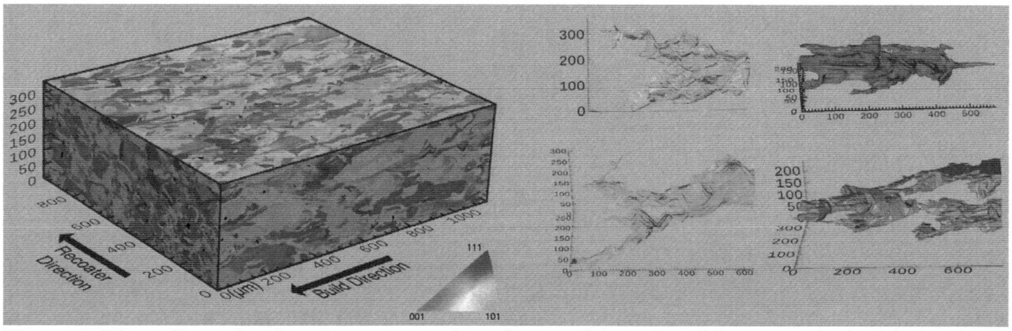

Figure 5. Three dimensional grain reconstructions from serial sectioned laser powder bed fusion 316L stainless steel, showing the complex branching morphology; IPF coloring is referenced to the build direction.

The melt pool geometry generated during laser powder bed fusion AM often falls between these two geometries. Particularly for builds with no scan rotation or displacement between layers, melt pool cross sections with an intermediate curvature can generate elongated vertical grains with <001> aligned to the build direction along the center of the melt pools (Figure 3, middle). These grain structures have been nicely characterized experimentally by Andreau et al. [11] and Sun et al. [12] and simulated by Pauza et al. [14].

Higher power density heat sources can produce a keyholing effect with deep penetration of the melt pool (Figure 3, right). This melt pool geometry has been studied and modeled by Pauza et al. [14]. At the center of the melt pools, vertical thin grains with <001> aligned with the build direction extend through the structure. The remainder of the build exhibits primarily horizontal grain growth within the keyhole region due to the transverse orientation of the thermal gradient in that region. However, there is no significant influence along the scan direction, resulting in a <001> fiber texture aligned with the transverse direction.

Employing compositions or parameters that retard the solidification rate will promote the formation of dendritic grains ahead of the solidification front. The equiaxed grains that form are most often found near the center or top of the melt pool. While these equiaxed grains are prevalent in conventional welding, the small sizes and fast cooling rates typically employed during additive manufacturing make these structures less common.

3. Control of Texture

The build parameters selected can be used to vary the crystallographic textures produced in a component, or even to choose different textures as a function of position within a component. One of the earliest and still best examples of this site-specific texture control is the use of three different sets of processing parameters by DeHoff and coworkers [15]. They varied the electron beam current and velocity conditions, guided by analytical models of heat transfer, to dynamically adjust the melt pool shape, size, temperature gradient, and interface velocity of electron beam AM to spell out the letters "DOE" (Figure 6). A <001> columnar solidification mode with epitaxial grain growth was used to generate the red background (red), use of a spot mode produced the highly misoriented but equiaxed grains of the letter outlines, and a dynamically varying spot mode was used to vary between the columnar and equiaxed solidification modes of the letter interiors.

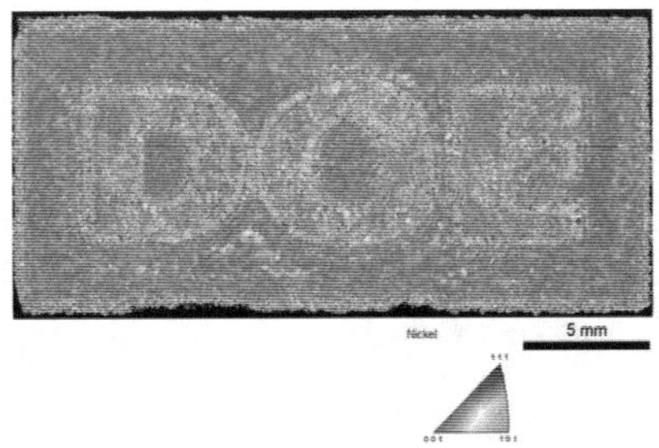

Figure 6. Inverse pole figure map displaying site-specific control of texture produced by variations in the processing parameters [15].

4. Heterogeneous Grain Nucleation

In addition to the epitaxial grain growth that predominates during AM solidification, it is also possible to promote heterogeneous grain nucleation. Heterogeneous nucleation generally gives rise to additional grain orientations without any preferred crystallographic relationship to the pre-existing solid substrate orientations or melt pool frame of reference. The effects of additional heterogeneous nucleation through a larger grain nucleation density and/or higher critical undercooling have been illustrated in cellular automata models by Herriott et al [16] and Teferra et al. [17] (see Figure 7). These results demonstrate both a refinement of the grain size, due to a greater density of small, equiaxed grains, and a randomization of the texture (indicated by more numerous color variations), arising from the introduction of new grain orientations, with increased heterogeneous grain nucleation rates.

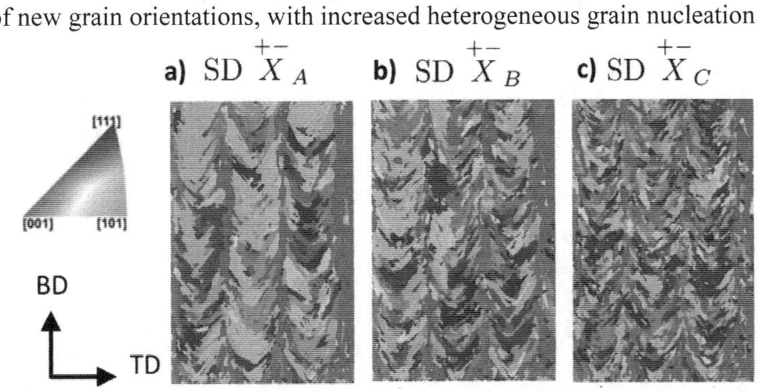

Figure 7. Cellular automata models showing the refinement of grain structure and randomization of crystallographic texture associated with increasing grain nucleation rates (from [17]).

There are multiple mechanisms for increasing the heterogeneous grain nucleation rate. In powder-based additive manufacturing techniques, one of the primary sources of heterogeneous grain nucleation sites is the powder feedstock itself. If the powder does not completely melt, the remaining grains within that unmelted powder can serve as epitaxial nucleation sites. Such nucleation from partially unmelted polycrystalline powder particles was demonstrated in a serial sectioning and reconstruction analysis by Rowenhorst and coworkers [18]. New grain orientations were shown to grow epitaxially from the fine grains on the surface of the unmelted solid powder particle and, while many of those grains were quickly consumed in the competitive grain growth process, some were oriented favorably for continued growth as the build progressed (Figure 8). They also showed that defects such as lack of fusion pores can also act as nucleation sites for new orientations. Similarly, unmelted powder particles adjacent to the melt-pool edge or solidified spatter impacting the top surface of the melt pool can also initiate growth of new grain orientations. Likewise, changes in the deposition parameters such as increases in the powder layer thickness or hatch spacing can decrease the grain size of the deposit by reducing the overlap between adjacent melt pools, allowing more fine grains to be retained from one layer to the next and hindering continued epitaxial growth of the large columnar grains [19].

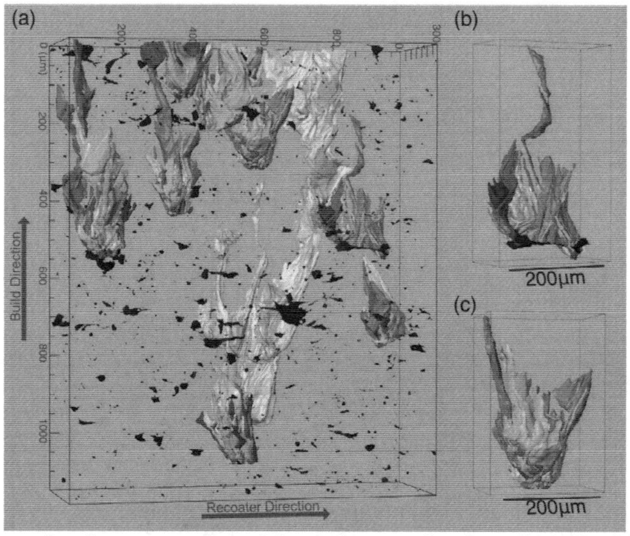

Figure 8. Three-dimensional reconstruction of the grains growing from (b) lack of fusion pores and (c) unmelted powder particles in a laser powder bed fusion 316L build [18].

Heterogeneous nucleation can also be promoted by introducing an inoculant that can serve as a nucleation site for the new grains. The effectiveness of this approach was demonstrated by Martin et al. [20], who showed that introduction of 1%Ta can replace the large, columnar grains of laser powder bed fusion deposited aluminum with equiaxed grains approximately two orders of magnitude smaller. This change also replaces the <110> || build direction fiber texture with a more random texture.

5. Allotropic Phase Transformations

Finally, it should be mentioned that many alloys, including most steels and titanium alloys, experience an allotropic phase transformation during cooling that will dramatically affect the final crystallographic texture of the build. Even if the initial solidification creates a strong texture, as discussed above, the subsequent transformation of that initial phase will generate multiple orientation variants of the lower temperature phase, refining the grain structure and obfuscating the overall texture.

6. Outlook/Future Directions

Additive manufacturing has tremendous potential, promising the ability to manufacture complex shapes with locally-optimized properties. However, the unique microstructures produced by AM, the associated crystallographic textures, and their complex relationships with both the processing parameters and the resultant properties are going to require additional research to fully realize the potential of this process.

While the capability to generate a few specific textures has been demonstrated, additional work remains to extend that capability to other textures and to understand better the effects of such changes. The melt pool shape and the scan path are the two primary parameters that will enable this. The 3D melt pool geometry defines the preferred grain orientations that will develop as the melt pool solidifies. The shape of the melt pool can be altered by varying the processing parameters as discussed above. Additional work is needed to explore the effects of possible melt pool shapes on the resultant texture. Recent research into changing the profile of the heat source or utilizing multiple heat sources may also enable the formation of new and/or refined texture orientations.

The propagation of grain orientations from one layer to the next is controlled by the scan orientation. Solidification is dominated by epitaxial growth of grains from pre-existing crystals in the underlying material, but it is a competitive process driven by the orientation of the local thermal gradient. Thus, if the substrate grain orientations are well-oriented with respect to the new local thermal gradient (and thus preferred growth direction), they will continue to propagate through the new layer. However, changes in the local thermal gradient direction due to changes in the scan orientation or other factors will inhibit continued growth of those grains, allowing new grain orientations to dominate as the new melt pool solidifies. The effect of the interlayer scan rotation on the textures that can be produced through this continually changing competition process remains to be fully examined.

Most importantly, it is critical to understand the effect of these texture orientations on the properties of the deposited material. What are the mechanical properties and corrosion behaviours of each texture? How does reducing the texture strength (and grain size) through enhanced grain nucleation change these properties? What is the impact of combining different textures? How are the microstructures, textures, and properties affected by heat treatments or HIP processing? It is important to understand the answers to these questions in order to intelligently control the process and achieve the desired properties.

There have been impressive demonstrations of our control of crystallographic variability in additive manufacturing, indicating an understanding of the basic principles driving the grain nucleation and growth processes that lead to specific crystallographic textures. These demonstrations reveal the potential of what can be achieved through the AM process. With the spatial control inherent in AM, this foretells the ability of AM to establish a specific texture and maintain it, or change it on demand, as a function of position within the build.

7. Summary

The crystallographic textures produced during additive manufacturing can be understood from the fundamental mechanisms of grain nucleation and growth. They originate from the epitaxial growth of grains from the pre-existing grain structure in the solid melt pool walls, and evolve through a competitive grain growth process that favors crystal orientations with a <001> direction more closely aligned to the local thermal gradient direction. This local thermal gradient is defined by the surface normal of the melt pool and can therefore be modified through changes in the melt pool geometry, which can be accomplished by varying the beam current, velocity, or scan conditions. This baseline texture can be modified by increasing the number of grain nucleation sites, whether from unmelted powder, second phase inoculants, or grain formation ahead of the solidification front. These heterogeneous grain nucleation sites will tend to reduce the average grain size and randomize the texture of the build.

References

[1] Kou S 1987 *Welding Metallurgy* (John Wiley & Sons) pp. 129-76.

[2] Linnert G E 1994 *Welding Metallurgy* (American Welding Society) pp. 787-802.

[3] Aroh J W, Rowenhorst D J, Feng C R, Rollett A D, and Fonda R W 2022 Stereographic Trace Analysis Reveals <100> Preferred Orientation of Cell Structures in Additively Manufactured 316L, manuscript in preparation.

[4] Lewandowski J J and Seifi M 2016 Metal Additive Manufacturing: A Review of Mechanical Properties *Annu. Rev. Mater. Res.* **46** 151-86.

[5] Cunningham R, Zhao C, Parab N, Kantzos C, Pauza J, Fezzaa K, Sun T, Rollett A D 2019 Keyhole threshold and morphology in laser melting revealed by ultrahigh-speed x-ray imaging *Science* **363** 849–52.

[6] Lane B, Heigel J, Ricker R, Zhirnov I, Khromschenko V, Weaver J, Phan T, Stoudt M, Mekhontsev S, and Levine L 2020 Measurements of melt pool geometry and cooling rates of individual laser traces on IN625 bare plates *Integrating Materials and Manufacturing Innovations* **9** 16-30.

[7] Sun Z, Tan X, Tor S B, and Chua C K 2018 Simultaneously enhanced strength and ductility for 3D-printed stainless steel 316L by selective laser melting *Asia Materials* **10** 127-36.

[8] Murray S P, Pusch K M, Polonsky A T, Torbet C J, Seward G G E, Zhou N, Forsik S A J, Nandwana P, Kirka M M, Dehoff R R, Slye W E, and Pollock T M 2020 A defect-resistant Co–Ni superalloy for 3D printing *Nature Communications* **11** 4975 1-11.

[9] Polonsky A T, Echlin M P, Lenthe W C, Dehoff R R, Kirka M M, and Pollock T M 2018 Defects and 3D Structural Inhomogeneity in Electron Beam Additively Manufactured Inconel 718 *Materials Characterization* **143** 171-81.

[10] Ma D, Stoica A D, Wang Z, and Beese A M 2017 Crystallographic Texture in an Additively Manufactured Nickel-Base Superalloy *Mat. Sci. Eng. A* **684** 47-53.

[11] Andreau O, Koutiri I, Peyre P, Penot J-D, Saintier N, Pessard E, De Terris T, Dupuy C, and Baudin T 2019 Texture control of 316L parts by modulation of the melt pool morphology in selective laser melting *J. Mater. Process Technol.* **264** 21–31.

[12] Sun S-H, Ishimoto T, Hagihara K, Tsutsumi Y, Hanawa T, and Nakano T 2019 Excellent mechanical and corrosion properties of austenitic stainless steel with a unique crystallographic lamellar microstructure via selective laser melting *Scripta Mater.* **159** 89-93.

[13] Polonsky A T, Lenthe W C, Echlin M P, Livescu V, Gray G T III, and Pollock T M 2020 Solidification-Driven Orientation Gradients in Additively Manufactured Stainless Steel *Acta Mater.* **183** 249-60.

[14] Pauza J G, Tayon W A, and Rollett A D 2021 Computer simulation of microstructure development in powder-bed additive manufacturing with crystallographic texture *Modelling. Simul. Mater. Sci. Eng.* **29** 055019 1-34.

[15] Dehoff R R, Kirka M M, Sames W J, Bilheux H, Tremsin A S, Lowe L E, and Babu S S 2015 Site specific control of crystallographic grain orientation through electron beam additive manufacturing *Mat. Sci. Tech.* **31**(8) 931-8.

[16] Herriott C, Li X, Kouraytem N, Tari V, Tan W, Anglin B, Rollett A D, and Spear A D 2019 A multi-scale, multi-physics modeling framework to predict spatial variation of properties in additive-manufactured metals *Modelling Simul. Mater. Sci. Eng.* **27** 025009 1-23.

[17] Teferra K and Rowenhorst D J 2021 Optimizing the cellular automata finite element model for additive manufacturing to simulate large microstructures *Acta Mater* **213** 116930 1-18.

[18] Rowenhorst D J, Nguyen L, Murphy-Leonard A D, Fonda R W 2020 Characterization of Microstructure in Additively Manufactured 316L using Automated Serial Sectioning *Current Opinion in Solid State and Materials Science* **24**(3) 100819 1-9.

[19] Pauza J and Rollett A 2021 Simulation Study of Hatch Spacing and Layer Thickness Effects on Microstructure in Laser Powder Bed Fusion Additive Manufacturing using a Texture-Aware Solidification Potts Model *J. Mat. Eng. Perf.* **30**(9) 7007-18.

[20] Martin J H, Yahata B, Mayer J, Mone R, Stonkevitch E, Miller J, O'Masta M R, Schaedler T, Hundley J, Callahan P, and Pollock T 2020 Grain Refinement Mechanisms in Additively Manufactured Nano-Functionalized Aluminum *Acta Mater.* **200** 1022-37.

Analysis of inter- and intragranular variations based on 3DXRD data and crystal plasticity

N Y Juul[1,4], J Oddershede[2], A Beaudoin[3] and G Winther[4]

[1]NOV Flexibles, Priorparken 480, DK-2605 Brøndby, Denmark
[2]Xnovo Technology Aps, Galoche Allé 15, 1st floor, DK-4600 Køge, Denmark
[3]Cornell High Energy Synchrotron Source (CHESS), 161 Synchrotron Drive, Ithaca, NY 14853, USA
[4]Civil and Mechanical Engineering, Technical University of Denmark, Produktionstorvet 425, DK-2800 Kgs. Lyngby, Denmark

Abstract. The synchrotron-based technique of 3DXRD/HEDM has proven a powerful tool to investigate individual grains in the bulk of a polycrystal during plastic deformation. Comparisons between such experimental data and crystal plasticity simulations typically show agreement with respect to overall qualitative trends. However, comparisons at the level of individual grains are in general less successful. Experimental observations and crystal plasticity simulations for a number of representative grains are analysed for similarities and differences in order to elucidate the origin of these. The comparison points at insufficient modelling of grain interactions, hardening and possibly also in the handling of ambiguities. Spatially resolved experimental data on the orientation spread is expected to provide the information needed to improve the predictive capacity at the level of individual grains.

1. Introduction

Crystallographic texture evolves as polycrystalline metals are mechanically processed to obtain the final shape of industrial products. The texture of, for example, rolled sheets control the mechanical anisotropy of the sheet during subsequent forming operations, typically leading to earing during deep drawing. Prediction and control of texture evolution is therefore a field of industrial interest, and over decades models of increasing complexity have been developed, which in general predict the overall texture well.

With the increasing availability of detailed three-dimensional (3D) grain maps experimentally characterised by synchrotron methods, e.g. three-dimensional X-ray microscopy (3DXRD) and high energy diffraction microscopy (HEDM) and the many variants of these, it has become possible to compare model predictions with experimental data directly on a grain-to-grain basis. Such comparisons indicate inadequacies of the models. Typically, the experimental data exhibit larger scatter than the model predictions [1]. This has, for example, been attributed to large effects of the initial stress state of each grain [2,3] as well as problems with the constitutive equations and material parameters employed in the models [4]. In particular, the relative strength of different slip system families in metals with hexagonal crystal lattices has been the subject of recent studies [5–7].

This study is a follow-up on a previous study of the stress state of individual grains in austenitic stainless steel elongated 1% [8]. In the present paper, the experimentally observed lattice rotations up to 5% elongation of the grains are compared to CPFEM predictions. A detailed analysis of selected grains is conducted in order to elucidate the successes and shortcomings of the model.

2. Experimental set-up and 3DXRD data analysis

The sample material is a fully recrystallized austenitic 316L stainless steel, with an average grain diameter of 90 μm. The 3DXRD experiment was conducted at Cornell High Energy Synchrotron Source (CHESS) at beamline F2. High-resolution spatial data on the illuminated volume was recorded using a near-field detector setup in order to obtain a 3D grain map, which was published in [9]. As the sample was deformed [10], the orientation change of each grain was recorded using a far-field detector. A schematic of the setup and details on the experimental conditions may be found in [8]. In total five consecutive layers along the tensile axis were scanned. Each layer had a height of 0.1 mm and the entire sample cross section of 0.7x0.7 mm was mapped out. The loading history of the sample was recorded using a digital image correlation (DIC) system [11] and a load cell. Diffraction data was obtained at 0.1, 1, 3 and 5% elongation to record the rotation path of each individual grain.

The diffraction data was reconstructed using the FABLE software package [12], where the initial indexing of the diffraction spots was performed with GrainSpotter [13] using the data from the undeformed sample state. For the undeformed state and the subsequent load steps, the centre of mass (CMS) orientations, positions, strain tensors and relative volumes of the individual grains were fitted by FitAllB [14]. For clarity and due to the cubic crystal symmetry, the crystallographic directions of the tensile axes for all grains are presented in the stereographic triangle spanned by [100], [110] and [111]. All slip system analyses also refer to this triangle, in which the primary slip system is $\overline{b2}$ (see table 1 for definition of the slip system notation). When analysing specific diffraction spots in section 4.3. , the spot indices obtained from GrainSpotter are employed, although slip systems analysis still refers to the grain orientation in the [100]-[110]-[111] triangle.

Table 1. Slip systems in Bishop-Hill notation.

Plane	(111)			(-1-11)			(11-1)			(1-11)		
Direction	[01-1]	[-101]	[1-10]	[0-1-1]	[101]	[-110]	[01-1]	[101]	[-1-10]	[0-1-1]	[-101]	[110]
Notation	a1	a2	a3	b1	b2	b3	c1	c2	c3	d1	d2	d3

The data reconstruction and parameter fitting were done one data set at a time. As five coherent layers were scanned for each strain step, grains spanning several layers need to be stitched together to gain a coherent description of the scanned volume. This is done by merging grain segments within set distance and misorientation tolerances. These tolerances increase with the level of deformation due to deformation-induced changes in the grain morphology and orientation. At 5% elongation grain segments were merged if their CMS positions were within 100 μm of each other and the CMS misorientation were within 2.7°. Due to increasing spot overlap, only 223 grains of the 357 grains indexed in the undeformed state could be tracked to 5% elongation.

3. Crystal plasticity modelling

The WARP3D crystal plasticity finite element code [15] was employed to simulate the behaviour of the grains during the tensile testing. The experimental 3D grain map in [9] was meshed into cubic 8-noded elements using Neper [16], each containing 4x4x4 voxels of dimensions 2.5x2.5x2.5 μm^3. The coarse meshing leads to loss of eight small grains and introduces slightly stepped grain boundaries.

Equations (1) and (2) describe the implementation of Voce's law of hardening in WARP3D:

$$\tau=\tau_{CRSS}+\tau_w \qquad (1)$$

IOP Conf. Series: Materials Science and Engineering 1249 (2022) 012008 doi:10.1088/1757-899X/1249/1/012008

$$\dot{\tau}_w = \theta_0 \left(1 - \tau_w/\tau_v\right)^m \sum_s \dot{\gamma}_s \tag{2}$$

In the equations, τ is the critical resolved shear stress (CRSS), τ_0 is the initial CRSS, τ_w is the change due to work-hardening [MPa], $\dot{\tau}_w$ is the work-hardening increment [MPa/s], θ_0 is the initial stage II / III hardening tangent [MPa], τ_v is the saturation CRSS-value [MPa], m is Voce's hardening exponent, and $\dot{\gamma}_s$ is the slip rate on slip system s [s^{-1}]. An initial guess for the constants was derived from fitting Voce's law to the experimental stress-strain curve. Subsequently, the constants were fine-tuned through several simulations, until the macroscopic work-hardening behaviour of the model agreed with the experimental curve.

4. Comparison of experimental and predicted rotations

The simulations were conducted on the entire grain map. To avoid potential effects of free sample surfaces, the present comparison of experiment and simulation, however, includes only 160 bulk grains.

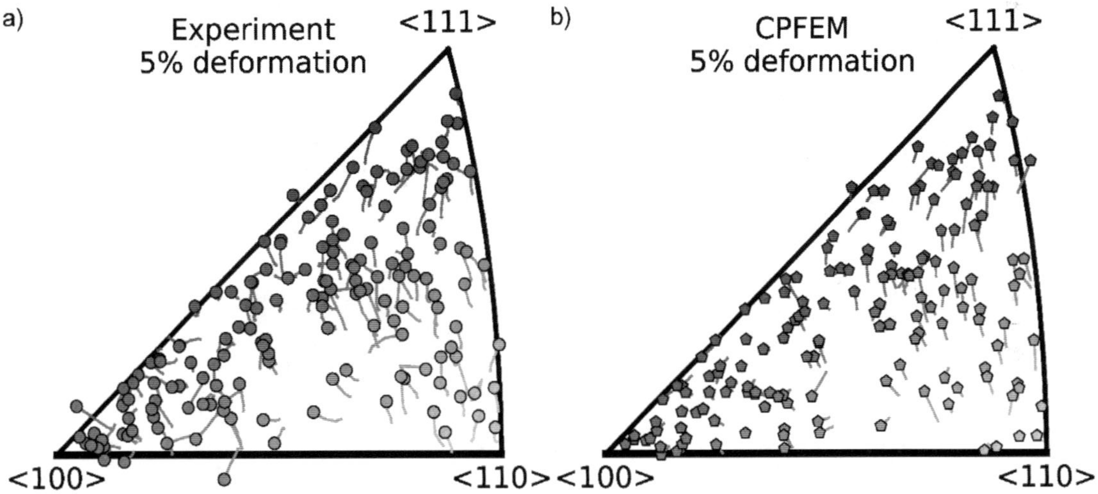

Figure 1. Rotation paths of the tensile axis in a) the experiment and b) the model prediction. The final orientation is marked by the symbol.

Figure 1 illustrates the rotation of the tensile axis of the individual grains as observed in the experiment and predicted by the model, presented in the stereographic triangle. It is clear that the deviations of the individual grains from the overall tendency are much larger in the experimental data than in the predictions. Furthermore, the measured lattice rotations are typically larger than the predictions as illustrated by the experimental mean value of 2.2° compared to a mean of 1.2° for the predictions. Only 21 out of the 160 grains rotate more in the model than in the experiment. However, the differences in rotation angle between experiment and model depends on the grain orientation. Experimental grains near [100] and [110] exhibit much larger rotations than predicted.

The rotation directions for the grains near [111] are well predicted and will not be considered further here. The directions for the [110] oriented grains are also fairly well predicted, although not as well as for [111]. By contrast, the predictions for the grains near [100] are poor; in fact most of the predicted rotation axes are more than 30° off.

IOP Conf. Series: Materials Science and Engineering 1249 (2022) 012008 doi:10.1088/1757-899X/1249/1/012008

4.1. The [100] region.

Most of the grains in the [100] region rotate more experimentally than predicted. However, three grains with very small (< 0.5°) experimental rotations were also observed. For comparison, the model predicts seven virtually non-rotating grains, but only one of these grains agrees with the experiment.

Grains near [100] are expected to have eight highly stressed slip systems, in agreement with the experimental stress state data at 1% elongation [8]. The inset of figure 2a illustrates the rotation directions induced by these systems. Within 15° of [100], 80% of the predicted mean slip activity of the 26 grains is on the $\overline{a2}$, $a3$, $\overline{b2}$ and $d3$ systems, i.e. on four of the eight systems. Their relative activities vary a lot between grains in agreement with the prediction of a wide range of rotation directions.

Four representative grains spanning the range of rotation directions are selected. To facilitate the subsequent analysis of the intragranular orientation spread, the selected grains have their rotation direction correctly predicted by the model. Grain D does not rotate its tensile axis experimentally (considering the experimental uncertainty of about 0.2°), grains B and C rotate more experimentally than predicted towards the [100]-[111] line, and grain A rotates more than predicted towards [100]. The rotation paths of these grains are illustrated in figure 2a. These rotations are in agreement with dominant activity on a combination of $\overline{b2}$, $a3$ and $\overline{a2}$. Rotation contributions from the fourth predicted system, $d3$, will counterbalance rotations from the other three systems, resulting in underestimation of the rotation angle just like the rotations from especially of $\overline{b2}$ and $\overline{a2}$, have large components that counterbalance each other.

In conclusion, the model overestimates the number of highly active systems, leading to too small predictions of rotation angles. However, the model qualitatively captures the observed spread in rotation directions, although not at the level of individual grains.

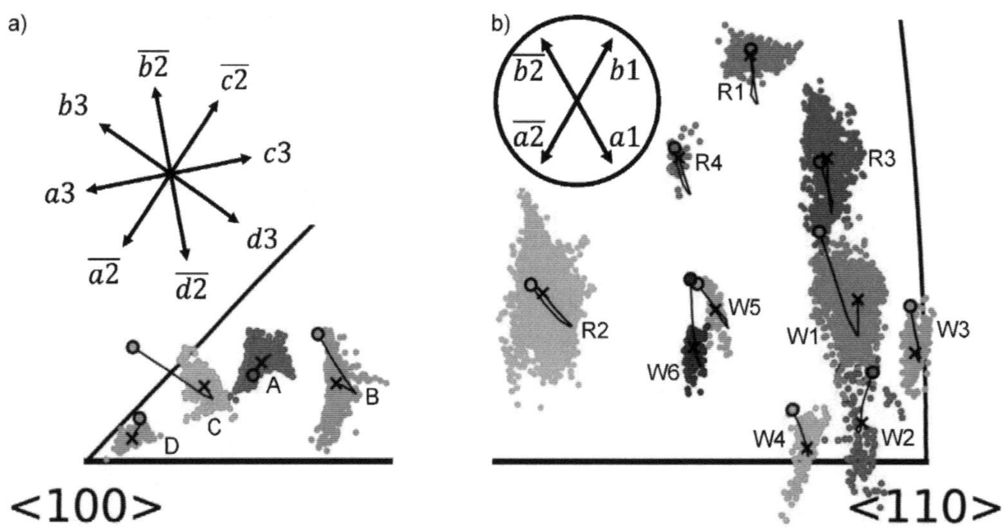

Figure 2. Stereographic triangle presenting the rotation of the tensile axis for the grains selected for detailed analysis (see text). Insets show the rotation directions induced by individual slip systems. The circles and crosses mark the measured and predicted orientations after 5% elongation, respectively. The lines show the rotation paths from the initial orientation. The coloured clouds correspond to the predicted orientation spread at 5% elongation. a) Four grains near <100>. b) Ten grains near <110>.

4.2. The [110] region.
In the region within 15° of [110], ten representative grains were studied in greater detail. The CMS orientations of all of these grains rotated away from the [110] corner and into the triangle as illustrated in figure 2b. The predicted orientations of four of the grains after 5% deformation were within about 0.5° of the experiment. For the other six grains the predicted rotations were in the wrong direction and/or smaller than in the experiment. For clarity, these two groups of grains will be termed right (R1-4) and wrong grains (W1-6) in the following.

When averaging over all the 23 grains within 15° of [110], 80% of the predicted slip activity is on the $\overline{a2}$, $b1$ and $\overline{b2}$ systems. The mean activity on $a1$ is small (6%). The inset in figure 2b shows the rotation directions induced by activity on each of these four systems. The figure also shows that all of the ten selected grains exhibit experimental rotations in agreement with dominant activity on the $\overline{b2}$ and/or $b1$ systems.

Analysis of the simulated elements making up the right grains and the wrong grains shows that although $\overline{b2}$ is the dominant system in most elements, more activity on $\overline{a2}$ than on $\overline{b2}$ is predicted for about 30% of the elements, more activity on $b1$ is predicted for 25% and more activity on $a1$ for 40 %. The corresponding numbers for the right grains are less than 5%, 10% and 1%. The grain-averaged slip patterns show a similar trend. The four right grains all have the primary system to be most active. By contrast, only two out of the six wrong grains have most activity on $\overline{b2}$, while $b1$ is most active in the others.

Summing up, the model overestimates the activities of systems $\overline{a2}$ and $b1$ while underestimating the activity of $\overline{b2}$ for the wrong grains. Overestimation of $a1$ and $\overline{a2}$ explains the small predicted rotation angles for the wrong grains because they induce rotations opposite to those of $\overline{b2}$ and $b1$.

4.3. Intragranular orientation spread
Due to the lack of near-field data for the deformed state, the intragranular orientation spread could not be experimentally quantified, as in e.g. [4,17–19]. Instead the predicted intragranular orientation spread is analysed in the stereographic triangle of figure 2 and compared to experiment by forward projection onto the far-field detector (Figures 3 and 4).

The FABLE polycrystal simulator, PolyXSim, was employed to provide forward projections of the simulated intragranular orientation spread as reflections onto a virtual far-field detector similar to the one used in the actual experiment. This is analogous to the virtual diffractometer previously applied in combination with crystal plasticity to study slip activity [20] and cyclic loading [20–22] . In the forward projections, the beam energy was assumed to be perfectly monochromatic, the experimentally determined spatial distortion of the detector was taken into account and a detector point spread of three pixels was added. Both the experimental and simulated diffraction spots are presented in the ω,η – coordinates, describing the rotation of the sample and the azimuthal direction on the detector, respectively (see [23] for details). This manner of comparison requires that the experimental average spot position is in reasonable agreement with the predicted one. This is the reason for selection of representative grains in the [100] direction for which the rotation directions were not completely wrongly predicted.

4.3.1 The [100] region. The predicted orientation spread of the four selected grains in the [100] region is illustrated in figure 2a. A detailed analysis of grain A is presented, and subsequently results from similar analyses of the three other grains is summarised. The predicted spread of this grain in figure 2a reveals spread in four directions from the initial orientation, corresponding to activities on one of $\overline{a2}$, $\overline{b2}$, $d3$ and $\overline{c2}$.

Figure 3 shows diffraction spots for grain A. In this case, {113} diffraction spots are chosen because these directions are close to the <112> rotation axes of the individual slip systems. Variations

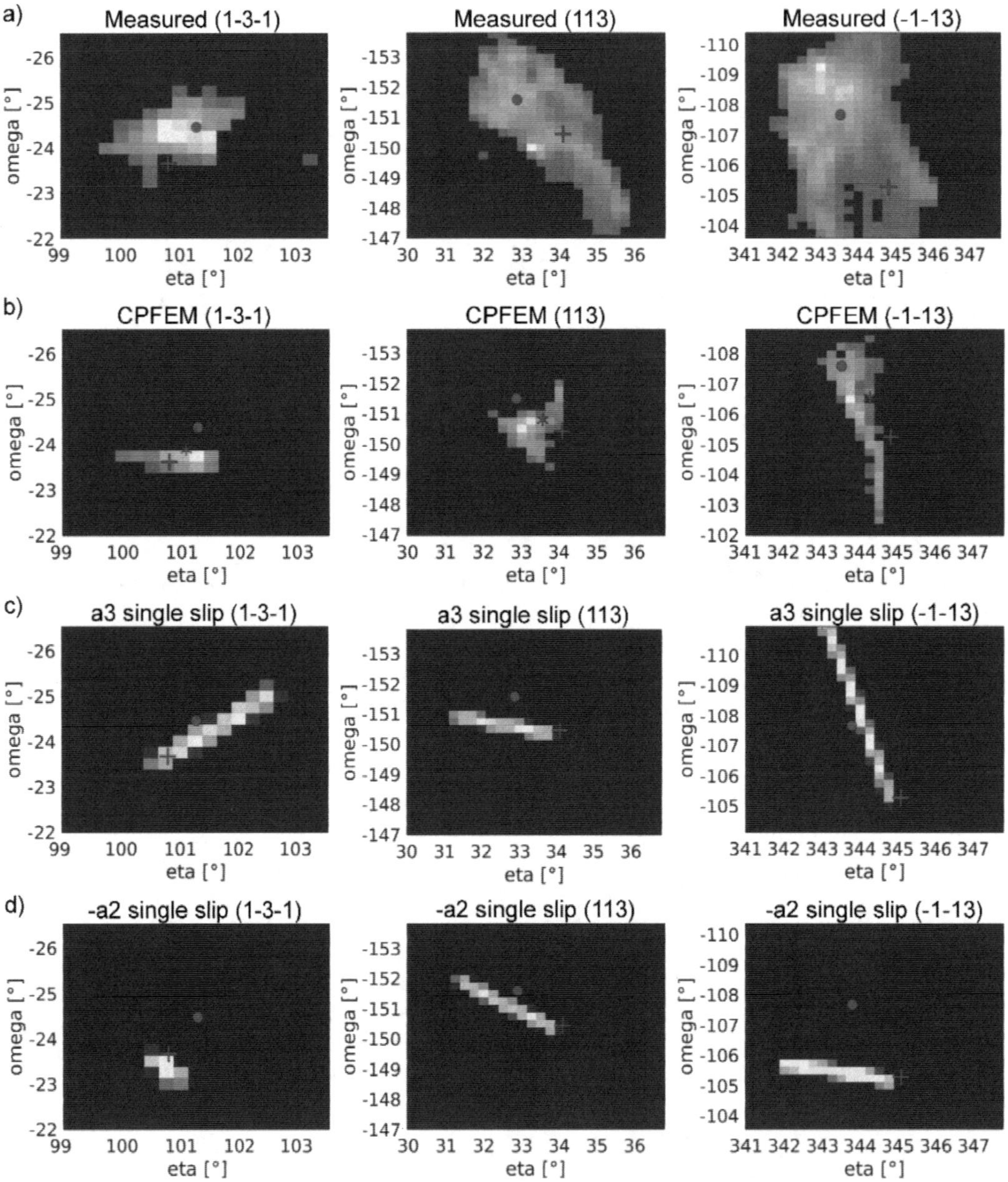

Figure 3. Diffraction spots from grain A at 5% elongation belonging to the {113} planes closest to the <112> rotation axes of the $\overline{a2}$, $a1$ and $c1$ slip systems. a) Measured spots. b) Simulated CPFEM spots. c) Spots from fits of single system activity on $a3$. d) Spots from fits of single system activity on $\overline{a2}$. The crosses mark the coordinates of the spot in the undeformed state and the circles and stars mark the CMS-coordinates at 5% elongation of the experiment and CPFEM model, respectively.

of the activity on a given system will not cause broadening of the spot corresponding to its rotation axes. However, slip variations on other systems result in spot broadening.

The three {113} spots in figure 3a represent the rotation axes of systems $\overline{a2}$, $a1$ and $c1$. The experimental (1-3-1) spot, which is the rotation axis for system $\overline{a2}$, reveals little spread. By contrast, the (113) and (-1-13) spots corresponding to expected inactive systems exhibit large spread with strong directionalities. Both have two intensity maxima, revealing that the grain splits up in two distinct parts.

Figure 3b illustrates the same reflections but the spots are simulated based on the orientation spread predicted by CPFEM. In this case, the (1-3-1) spot also exhibits little spread. The spread of the two other spots are underpredicted compared to the experiment. Nevertheless, the main directionality of both the (113) and (-1-13) spots along the rotation path of the CMS orientation are captured by the model. Also some splitting is predicted with an additional peak in the (-1-13) spot at low omega values, which is not found in the experimental spot.

The spots in figure 3c and d represent simulated rotation paths from single slip on systems $a3$ and $\overline{a2}$, respectively. Comparison of experimental spots with simulated ones originating from single slip activity and associated rotations is analogous to the procedure in [24]. In the present case, the simulations start from the initial orientation and continue up to an arbitrarily chosen shear of 0.1 on the active system. The directionalities of all three spots in figure 3c agree well with experiment and match one of the peaks from the grain splitting, the lower one in the centre image and the upper one in the right. This indicates that the grain leaves a tail of orientations, due to a gradient in the activity on the $a3$ system. The agreement of $a3$ variations with the spots simulated from the model prediction in figure 3b is also good.

The spots in figure 3d agree with the other part of the splitting grain. In addition, similar spot analysis not presented here reveals that parts of the grain have activity on $\overline{c2}$, leading to orientation spread in the opposite direction of the CMS rotation, as clearly demonstrated in the (113) spot in figure 3a. There is also some indication of this in the corresponding spot simulated from CPFEM results in figure 4b. Furthermore, spread in the activity of system $\overline{d2}$ is needed to explain the directionality of some of the spots. However, the model predicts substantial spread due to activity variations on $\overline{b2}$ and $d3$, which is not observed in the experimental spots.

Summing up, the orientation spread of the four investigated grains can be attributed to variations in the activities of the expected slip systems. For each of the selected grains the systems deduced from the orientation spread agree well with those corresponding to the CMS rotation. In general, the model appears to incorrectly predict the ratio between $\overline{a2}$ and $\overline{b2}$, underestimate the activity of $\overline{d2}$ and overestimate the activity of $d3$.

4.3.2 The [110] region. Figure 2b reveals the predicted orientation spread in the selected ten grains near [110]. In general, the spread is vertical in the stereographic triangle, in agreement with the predicted overall distribution of slip on systems $a1$, $\overline{a2}$, $b1$ and $\overline{b2}$.

Orthogonal {200} reflections are chosen to project the intragranular orientation spread from different directions. Figure 4 shows the diffraction spots from one of the wrong grains, W1. The experimental spots presented in figure 4a are split into two parts. By studying a series of spots from sample elongations of 1, 3 and 5% (not shown), it is clear that the splitting starts between 1 and 3% and that the rotation directions of the two parts are the same. Even the directionalities of the orientation spread in the two parts are the same, the smaller spot just rotates faster.

The simulated CPFEM spots are shown in figure 4b. As this is a poorly predicted grain, the absolute positions of the spots differ from the experimental ones. However, the width and height of the simulated spots are in overall good agreement with the experiment, with the (020) spot extending along η and the(002) spot along ω. The model therefore captures the magnitude and directionality of the orientation spread well, even though actual grain splitting is not predicted.

The mean slip activity leading to the predicted CMS orientation is on $b1$ (40%), $\overline{b2}$ (35%) and $\overline{a2}$ (10%). In order to investigate the origin of the orientation spread, another set of simulated spots are presented in figure 4c. These spots represent the orientation distribution arising from variations in the $b1/\overline{b2}$ ratio while keeping the activities of all other systems constant at the level predicted by the model. The $b1/\overline{b2}$ ratio takes values between 0 and 0.5. For comparison with figure 4a, the spots are artificially centred on the experimental CMS at 5% deformation. The directionalities of the spots in

IOP Conf. Series: Materials Science and Engineering 1249 (2022) 012008 doi:10.1088/1757-899X/1249/1/012008

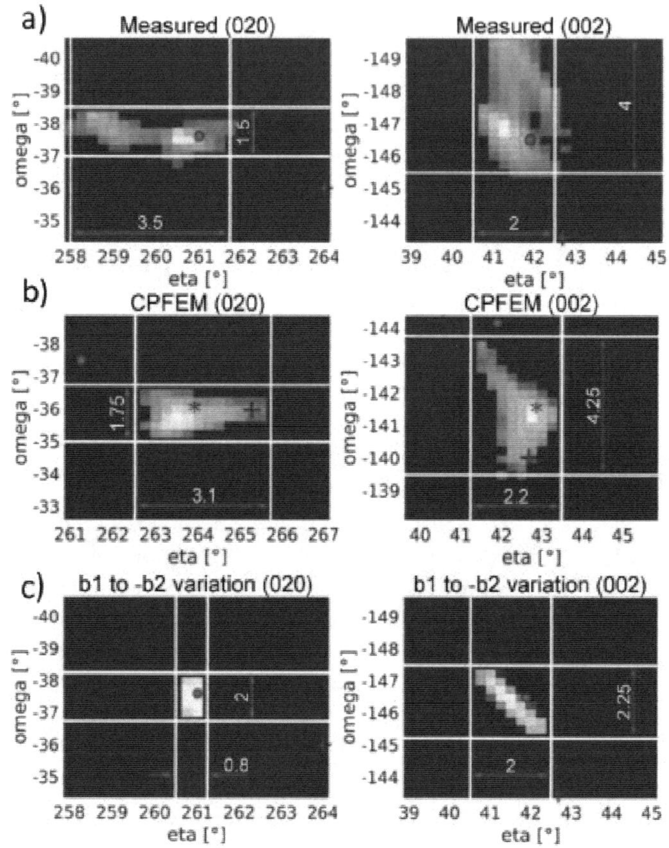

Figure 4. {200} diffraction spots at 5% elongation for grain W1. a) Measured spots. b) Simulated spots from CPFEM. c) Spots showing the orientation spread when shifting 50% of the $b1$ activity, in steps, to the primary system $\overline{b2}$. The crosses mark the CMS coordinates of the spots in the undeformed state, while the circles and stars mark the CMS coordinates of the measured and CPFEM spot at 5% elongation, respectivel,. The red circles in a) mark the CMS of the entire grain, which is present in more than one layer. This is the origin of the mismatch between the positions of the circle and the maximum intensity in the displayed spot. Note that the omega-eta interval in b) has been translated compared to a) and c) to center the simulated spot in image for easier comparison of its spread. The white lines serve as gauges to determine the extents of the spots.

figure 4a and c agree well, showing that a large part of the orientation spread is the result of relative variations in the activities of $\overline{b2}$ and $b1$ as also captured by the model. Relative variations between $\overline{b2}$ and $b1$ can, however, not explain the grain splitting. The orientation paths of the two parts of the grains have the same direction, which coincides with the direction caused by $\overline{b2}$. Grain splitting is therefore attributed to variations in $\overline{b2}$ activities, most likely due to variations in plastic strain.

Similar analyses were conducted also for the other nine grains near [110] selected for detailed studies. Themagnitudes and directions of the orientation spreads are well predicted by the CPFEM model for the grains examined. The orientation spread can be explained by relative variations in the $\overline{a2}$, $b1$ and $\overline{b2}$ systems predicted to dominate by the CPFEM model. Perhaps surprisingly, no general difference between right and wrong grains could be discerned.

5. Discussion

The origin of the discrepancies between the experimental and modelled rotations are discussed in the following.

5.1. Intergranular plastic strain differences

The CPFE-model predicts effective strain variations between the grains. Figure 5a plots the predicted mean plastic tensile strains vs. the predicted rotation angle for all the investigated grains. There is a tendency for large rotations in grains with large plastic strain, but as expected in a polycrystal this correlation is not strictly linear due to multislip conditions. It is, however, noteworthy that when plotting the predicted plastic strain vs. the experimentally observed rotation angle this correlation disappears, which may indicate a problem with the strain prediction in the individual grains.

The present study does not include any measurement of the plastic strain of individual grains. However, the plastic strain distribution between and within 23 grains has recently been measured in a tensile deformed Al-Cu alloy using Al_2Cu-particles as markers [25]. That study did not find any obvious correlations between the plastic strain and parameters relating to the individual grains but did find correlations between the shear and lateral contraction strains between neighbouring grains.

A detailed analysis of the predicted strain for grains near [110] reveals that except for a single grain, all the wrongly predicted grains have significantly smaller tensile strain (and also effective strain) than the correctly predicted ones, cf. figure 5b. This indicates that underestimation of the strain level may be part of the origin of the underestimated rotation angles in these grains.

Analysis of the strain predicted for the grains near [100] shows a much higher degree of uniaxial symmetry than for the grains of other orientations. This finding is in agreement with the deduced overprediction of the number of slip systems and may thus explain the consistent underprediction of the rotation angle.

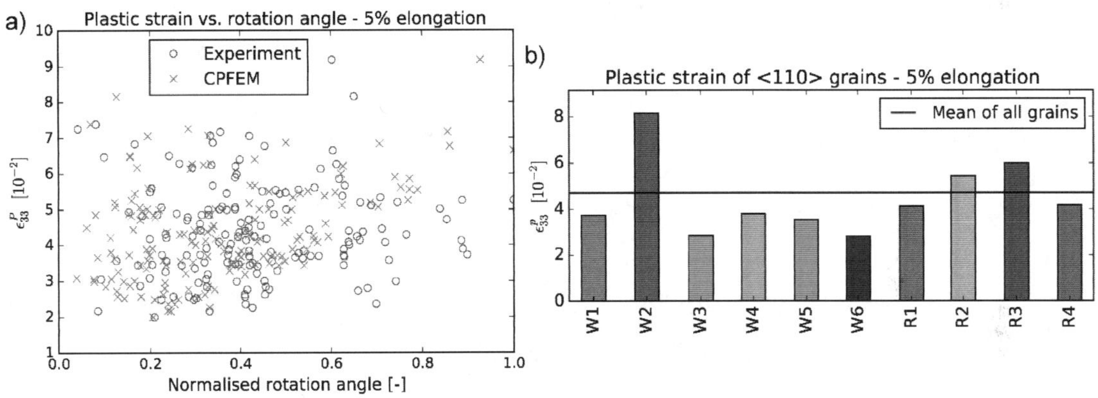

Figure 5. a) The modelled effective plastic strain plotted against the rotation angle. The circles represent measured rotation angles of the grains and the crosses the predicted rotation angles. b) the plastic tensile strain of the ten grains near [110] as described and seen in figure 2b.

5.2. Modelling method and parameters

Introduction of the initial residual stress of the individual grains have been seen to improve predictions in the elastic regime, but the effect when getting into the plastic regime was substantially smaller [3]. It is noted that the previous study [26] on the stress state of the same grains after 1% elongation revealed qualitative – although not quantitative in a grain-to-grain comparison – agreement between the experimental stress data and the CPFEM predictions. Residual stresses are therefore deemed unlikely as the major cause of the experimental rotation scatter.

A Voce type hardening law was employed in the modelling with isotropic hardening of all slip systems. The qualitative effects of anisotropic hardening may be assessed based on expected dislocation interactions.

5.3. Dislocation interactions and hardening

Four types of dislocation interactions have received particular attention in fcc metals. Hirth locks result from interaction of two dislocations with orthogonal Burgers vectors on different glide planes, resulting in a Burgers vector along <100>, i.e. not in a {111} glide plane. Lomer locks are reaction products with <110> Burgers vectors but with line directions that do not lie in a {111} plane. Glissile <110> junctions between dislocations on two different glide planes may also form. Collinear interactions occur between dislocations with the same Burgers vectors but on different planes. The relative strengths of these interactions have been evaluated by discrete dislocation dynamics simulations [27,28]. The collinear interaction is by far the strongest. The strength of Lomer and glissile interactions have been found to be approximately the same. The Hirth interaction is much weaker. Strong interactions are believed to contribute to strong latent hardening and eventually mutual exclusion of the slip systems on which the dislocations glide.

5.3.1 The [100] region. For grains near [100], the eight slip systems expected active and also predicted by the model consist of four pairs of collinear systems. The collinear interaction has been observed to shut down the activity of a system in each pair [29], leading to four remaining systems. Averaged over the 26 grains, the four systems, $\overline{a2}$, $a3$, $\overline{b2}$ and $d3$, were predicted most active by the model, as is also the case in each of the grains A-D selected for analysis of the orientation spread. Collinear systems are not among these four, and inclusion of collinear interactions in the modelling is therefore expected to have a minor effect. When also considering Lomer and glissile junctions, only two of the four predicted systems, namely $\overline{a2}/\overline{b2}$ or $a3/d3$, may be active in the same material volume. These are systems with large oppositely directed rotation components, i.e. the net result is a small rotation. The alternative is single slip conditions with large rotations, which will, however, induce large asymmetry in the plastic strain. This is why pure single slip is usually not expected in polycrystals. While both of these rotation behaviours are observed experimentally (viz. grains C and D), this analysis does not explain the large continuous spread in rotation directions. It is thus concluded that anisotropic hardening alone does not appear to explain the differences between the predicted and experimental rotation behaviour near [100].

Nevertheless, it is clear from the analysis of the experimental CMS and orientation spread data that most of the [100] grains exhibit slip on fewer systems than predicted by the CPFE model.

5.3.2 The [110] region. For the [110] oriented grains, the CMS rotation is grossly underestimated in the grains in the immediate vicinity of the ideal [110] orientation. As established above, the model predicts more activity on systems $\overline{a2}$, $b1$ and $a1$ in the wrongly predicted grains than in grains further from the ideal [110]. In addition, the experimental CMS rotations of data for the grains with wrong predictions suggest that the activities on $\overline{a2}$ and $a1$ are overestimated and that system $\overline{b2}$ is by far the most active.

Lomer locks, i.e. strong latent hardening, may form by interactions between $\overline{b2}$ and $a1$ as well as between $b1$ and $\overline{a2}$. Dominant double slip on $\overline{b2}$ and $b1$ may thus suppress $a1$ and $\overline{a2}$ due to latent hardening, resulting in larger rotation angles in better agreement with experiment. This mutual exclusion of slip systems is further supported by an analysis of deformation banding in a tensile-deformed aluminium single crystal of [110] orientation [30]. In that study, $b1$ and $\overline{b2}$ were found active within the same bands, whereas $\overline{a2}$ was active in other bands.

It should be noted that slip on each of the four slip systems considered here ($b1$, $\overline{b2}$, $a1$ and $\overline{a2}$) contributes almost in the same way to the tensile strain in the wrongly predicted grains as they are close to the ideal [110] orientation. Modulation of the distribution of slip activities among these systems will therefore not affect the tensile strain. The simulated difference in plastic strain between the [110] grains with wrong and right rotations therefore appears unrelated to latent hardening.

In conclusion, underestimation of the CMS rotation angle in the [110] corner of the stereographic triangle may be explained by details of anisotropic hardening not considered in the simulations, whereas the predicted strain differences may not.

6. Conclusions and Outlook

The comparison of the predicted rotations and orientation spread of individual grains for a tensile test of an austenitic stainless steel sample to 5% elongation showed a clear discrepancy between model and experimental results which further depends on the grain orientation.

In particular grains with the tensile direction near <100> or <110> show large discrepancies. These are the grain orientations which have the most solutions to the ambiguity problem of the standard Taylor model, i.e. multiple combinations of slip systems with the same strain but different rotations. This ambiguity leads to the wide range of rotations for the <100> region and the competition between slip systems on different planes in the <110> region. A major argument for the introduction of the rate-dependent crystal plasticity models is suppression of this problem, without explicitly making a choice between the various solutions. The present results indicate that this mathematical elimination of the problem may not be warranted by the underlying physics.

In some cases introduction of latent hardening justified by dislocation interactions is likely to introduce the selection of a physically correct solution. This appears to be the case for the grains near <110> but not for those near <100>.

It is further interesting that both for the grains near <100> and <110>, the predicted strain shows marked deviations from all the other grains. The plastic strain of the grains near <110> is smaller than average and the symmetry of the strain larger for those near <110>. Such effects may be real. For example, grains near <110> are known to prefer plane strain conditions over uniaxial strain, an effect leading to grain curling. The present simulations did, however, not detect such a situation. The predicted uniaxial symmetry of the grains near <100> is likely to be the origin of overprediction of the number of slip systems because multiple systems are needed to enforce the symmetry. While most likely related to the ambiguity, the many different rotation directions for these grains in both model and prediction are also expected to be the result of interactions with different environments of neighbouring grains. It must thus be concluded that the model probably does not grasp these correctly at the level of individual grains.

Summing up, improved algorithms to handle ambiguity, latent hardening and grain interaction mechanisms appear to be required to obtain more accurate predictions. In particular the grain interaction mechanisms will benefit if experimental data on spatially resolved orientation spread after deformation become available. Spot overlap in the diffraction patterns makes this difficult. However, scanning methods [17,30,31] may provide such data with the disadvantage of being significantly slower than the conventional 3DXRD/HEDM methods.

Acknowledgements
The Danish Independent Research Fund – Technology and Production is acknowledged for financial support through grant no. 0136-00194B.

References

[1] Winther G, Margulies L, Schmidt S and Poulsen HF 2004 *Acta Mater* **52** 2863
[2] Turner TJ, Shade PA, Bernier J V., Li SF, Schuren JC, Kenesei P, et al 2017 *Metall. Mater. Trans. A* **48** 627
[3] Tari V, Lebensohn RA, Pokharel R, Turner TJ, Shade PA, Bernier J V, et al 2018 *Acta Mater.* **154** 273
[4] Pokharel R, Lind J, Kanjarla AK, Lebensohn RA, Li SF, Kenesei P, et al 2014 *Annu. Rev. Condens. Matter Phys.* **5** 317
[5] Wang L, Zheng Z, Phukan H, Kenesei P, Park, Jun-SangLind J, Suter RM, et al 2017 *Acta Mater.* **132** 598
[6] Pagan DC, Shade PA, Barton NR, Park J-S, Kenesei P, Menasche DB, et al 2017 *Acta Mater.* **128** 406
[7] Pagan DC, Bernier J V., Dale D, Ko JYP, Turner TJ, Blank B, et al. 2018 *Scr. Mater.* **142** 96
[8] Juul NY, Oddershede J, Beaudoin A, Chatterjee K, Koker MKA, Dale D, et al 2017 *Acta Mater.* **141** 388
[9] Juul NY, Winther G, Dale D, Koker MKA, Shade P and Oddershede J 2016 *Scr. Mater.* **120** 1.

[10] Shade PA, Blank B, Schuren JC, Turner TJ, Kenesei P, Goetze K, et al 2015 *Rev. Sci. Instrum.* **86** 093902
[11] Guizar-Sicairos M, Thurman ST and Fienup JR. 2008 *Opt. Lett.* **33** 156
[12] Sørensen HO, Schmidt S, Wright JP, Vaughan GBM, Techert S, Garman EF, et al 2012 *Zeitschrift fur Krist.* **227** 63
[13] Schmidt S 2014 *J. Appl. Crystallogr.* **47** 276
[14] Oddershede J, Schmidt S, Poulsen HF, Sørensen HO, Wright J and Reimers W 2010 *J. Appl. Crystallogr.* **43** 539
[15] Messner M, Beaudoin A and Dodds R 2015 *Eng. Comput.* **32** 1526
[16] Quey R, Dawson PR and Barbe F 2011 *Comput. Methods Appl. Mech. Eng.* **200** 1729
[17] Pokharel R, Lind J, Li SF, Kenesei P, Lebensohn RA, Suter RM, et al 2015 *Int. J. Plast.* **67** 217
[18] Winther G, Wright JP, Schmidt S and Oddershede J 2017 *Int. J. Plast.* **88** 108
[19] Menasche DB, Lind J, Li SF, Kenesei P, Bingert JF, Lienert U, et al. 2016 *J. Appl. Phys.* **119** 154902
[20] Pagan DC, Miller MP 2014 *J. Appl. Crystallogr.* **47** 887
[21] Obstalecki M, Wong SL, Dawson PR and Miller MP 2014 *Acta Mater.* **75** 259
[22] Wong SL, Obstalecki M, Miller MP and Dawson PR 2015 *J. Mech. Phys. Solids* **79** 157
[23] Oddershede J, Wright JP, Beaudoin A and Winther G 2015 *Acta Mater.* **85** 301
[24] Pagan DC, Obstalecki M, Park JS and Miller MP 2018 *Acta Mater.* **147** 133
[25]. Toda H, Kamiko T, Tanabe Y, Kobayashi M, Leclere DJ, Uesugi K, et al 2016 *Acta Mater.* **107** 310
[26] Juul N, Oddershede J and Winther G 2020 *JOM* **72** 83
[27] Devincre B, Kubin L and Hoc T 2006 *Scr. Mater.* **54** 741
[28] Martínez E, Marian J, Arsenlis A, Victoria M and Perlado JM 2008 *J. Mech. Phys. Solids* **56** 869
[29] Madec R, Devincre B, Kubin L, Hoc T and Rodney D 2003 *Science* **301** 1879
[30] Wert JA, Huang X and Inoko F. 2003 *Proc. R. Soc. A Math. Phys. Eng. Sci.* **459** 85
[31] Hayashi Y, Setoyama D, Hirose Y, Yoshida T and Kimura H, 2019 *Science* **366** 1492
[32] Yu T, Du Y, Fan G, Xu R, Barabash R, Hansen N, Huang X and Zhang Y, 2021, *Acta Mater.* **202** 149

IOP Conf. Series: Materials Science and Engineering 1249 (2022) 012009 doi:10.1088/1757-899X/1249/1/012009

Topological aspects of mean-field crystallographically resolved models

L.A.I. Kestens[1,2], T. Nguyen-Minh[1], J. Ochoa Avendaño[2], H. Ghiabakloo[3] and A. Van Bael[3]

[1] Ghent University, Metals Science and Technology Group, Technologiepark 46, 9052 Ghent, Belgium
[2] Delft University of Technology, Materials Science and Engineering Dept, Mekelweg 2, 2628 CD Delft, Netherlands
[3] Katholieke Universiteit Leuven, Department of Materials Engineering (MTM), Kasteelpark Arenberg 44, 3001 Leuven, Belgium

E-mail: Leo.Kestens@ugent.be

Abstract. It is well-known that the crystallographic texture of a polycrystalline aggregate can be represented by the Orientation Distribution Function (ODF). A similar statistical approach can be extended to other microstructural state variables that are of relevance in the context of obtaining microstructurally based and quantitatively accurate structure-properties relations. In principle such statistical representations are of a non-topological nature, in contrast to an RVE (Representative Volume Element) description of the microstructure. However, by including additional variables to the statistical descriptor specific features of the topology may be taken into account. In this paper the example will be shown on how the plastic anisotropy simulation of a conventional deep drawing grade of Interstitial Free (IF) steel can be improved by considering the crystallographic misorientation of pairs of neighboring crystals, which represent the basic structural units of the 2-point mean field ALAMEL crystal plasticity model. In another example it will be shown how the recrystallization texture of the same deep drawing IF steel can be modelled with improved accuracy if the Strain Induced Boundary Mechanism (SIBM) is taken into account whereby a crystal orientation of low stored energy grows into a neighboring orientation of high stored energy.

1. Introduction

Metal sheet is one of the most widely used metal products that is employed in very broad sectors of industrial manufacturing such as automotive, house-hold appliances, aerospace, construction etc. It is well understood that there is a wide variety of mechanical sheet properties that are of a structure sensitive nature, i.e. they depend on microstructural state variables that are determined by the thermo-mechanical process through which the sheet is manufactured. It is generally accepted that the final microstructure of the finished sheet is formed throughout the sequence of solid-state transformations occurring during the manufacturing process. This may include phase transformations (e.g. for low-carbon steels and Ti-alloys), plastic deformation (at room temperature or elevated temperature) and ensuing thermally activated softening phenomena such as (static or dynamic) recrystallization and grain growth.

With growing understanding of the physical mechanisms underpinning the microstructural evolution processes increasingly accurate physical models were developed. This includes both microstructural evolution models (e.g. predicting plastic deformation textures) and constitutive models that simulate mechanical behavior on the basis of physical state-equations linking material properties to microstructural state variables (e.g. crystal plasticity models that predict plastic anisotropy on the basis of the orientation distribution function) [1]. Crystal Plasticity Finite Element (CPFE) models are among the most advanced and powerful tools to model deformation microstructures at the micrometer scale [2]. Recently, Traka et al [3] reported that a CPFE model in combination with a cellular automaton model can be employed to simulate thermally activated processes such as recovery and static recrystallization. Numerically CPFE models benefit from the Fast Fourier spectral solver method, which allows to speed up drastically numerical integration compared to conventional solvers [4, 5]. Such models are capable of simulation microstructural evolutions in a full-field mode taking into account the full spectrum of (sub-)structural heterogeneities.

In spite of its obvious merits the CPFE approach exhibits some intrinsic limitations, particularly when applied in 3D. One such limitations is the fact that a CPFE model requires a (3D) Representative Volume Element (RVE) as input. Such a RVE can be either numerically constructed, which involves a number of issues in its own right, or it can be derived from 3D microstructural images (e.g. 3D-EBSD scans) which are tedious and time consuming to collect. Moreover, even though the spectral solver method drastically reduces the computation time it still remains very considerable, particular if it is intended to combine a sufficiently large microstructural volume (in order to be statistically representative) with a sufficient spatial resolution (in order to capture the local topography in sufficient detail). It is not uncommon to require hours or even days of CPU time to model a medium to large rolling reduction on one single high resolution RVE with some dozens of initial grains. Therefore it is not trivial to apply the CPFE method in a multi-hierarchical approach to simulate macroscopic material behavior, in which the deformation history is different in all integration points of the FE simulation

Statistically based mean-field models offer a computationally less expensive alternative to full-field methods, but they lack the spatial resolution of the full-field models as they make abstraction of the local topological interactions of a polycrystalline aggregate. Instead, in a mean-field model the microstructure is represented by a set of relevant microstructural state variables and the (multi-variate) distribution function(s) of these state variables. A typical example is the Visco-Plastic-Self-Consistent (VPSC) model [6] which can predict texture evolution during plastic deformation by applying the macroscopic boundary conditions to a discrete set of crystal orientations and assuming that each crystal can be considered as an inclusion in the homogenized polycrystalline continuum. In spite of such drastic simplifications mean-field models can derive meaningful conclusions with regard to time-dependent evolution of selected microstructural state variables (e.g. crystal orientation) or for constitutive models in which these state-variables are the relevant microstructural descriptors (e.g. Reuss or Voigt homogenization approach for elastic properties of a polycrystalline aggregate). In a recent paper [7] Habraken et al. have shown a benchmark study where the cup height profile of a deep-drawn cup, simulated with an anisotropic yield criterion calibrated with a mean-field approach is of higher accuracy as compared to the one obtained with another yield criterion calibrated with a full-field approach (compare Figs. 106 and 111 in ref. [7]), and with the additional benefit of a drastically reduced computational cost.

The ALAMEL model by Van Houtte et al. [8] is a relaxed constraint mean-field crystal plasticity model that is capable of modelling how the macroscopically imposed boundary conditions are locally accommodated by a set of crystal orientations. In this model the crystallographic texture of the polycrystalline aggregate is represented by a set of N discrete orientations that pair-wise accommodate the imposed velocity gradient, whereby equal but opposed shear stresses are relaxed in a hypothetical grain boundary plane that separates the two crystal orientations of a pair. In the ALAMEL model orientations are paired arbitrarily. Rather than considering all possible $N(N-1)/2$ of two dissimilar orientations only a set of N/2 pairs is chosen from the set of N discrete orientations that represents the given texture (N is even). This implies that the ALAMEL model makes an implicit topological assumption, which takes into account, in a stochastic manner, the local neighborhood of a crystal orientation. In the present paper two examples will be considered in which the importance of local topology will be illustrated in a similar manner as applied in the ALAMEL model.

2. Modelling Lankford r-value profiles

The Lankford r-value (aka *normal* anisotropy) is the ratio of the width-to-thickness strain in a tensile test and it is used by the metal forming industry as a measure for the deep drawability of a sheet. The r-value profile, which displays the r-value as a function of the angle θ between the tensile axis and the rolling direction (RD) of the sheet, is a measure of the *planar* anisotropy of the sheet. It was shown by Sidor et al [9] that the ALAMEL code can be used to obtain improved simulations of r-value profiles as compared to the conventional Full Constraints Taylor (FCT) theory. In the ALAMEL code a set of N orientations is arbitrarily grouped in N/2 pairs, which introduces a stochastic element in the model. Pirani [10] has developed a code, which allows to order an array (g_1, .., g_i, g_j, … g_N) of N crystal orientations (N is even) such that the average misorientation $\overline{\Delta g}$ of two subsequent orientations in the array is as close as possible to a set target value ω_t; i.e. the code produces an array that minimizes the following object function:

$$\frac{1}{(N-1)}\left(\sum_{i=1}^{N-1}\left(|\omega_{i,(i+1)} - \omega_t|\right)\right) \quad (1)$$

Whereby $\omega_{i,(i+1)}$ is the misorientation between 2 successive orientations in the array. When $\omega_t = \omega_{min}$, ω_{max} it implies that the array is ordered such that the average misorientation is minimized or maximized, respectively.

Figure 1a shows the $\varphi_2 = 45°$ section of the ODF of a typical deep drawing grade Interstitial Free (IF) steel. It exhibits the characteristic γ-fiber (<111>//ND) texture, which provides excellent deep drawing properties to IF steels. This ODF was derived from the experimental EBSD scan (measured across the entire thickness of the sheet, cf. Figure 1b) by superposing Gaussian distribution functions with a spread of 5°. The orientation contrast map contained 1.87 10^6 pixels distributed over 11,600 grains (with a tolerance angle of 5°).

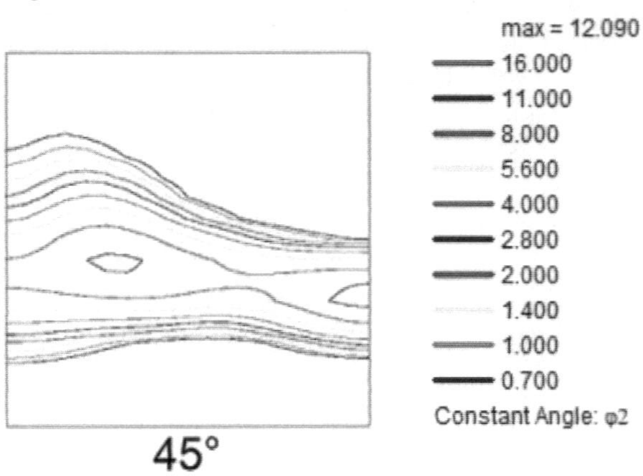

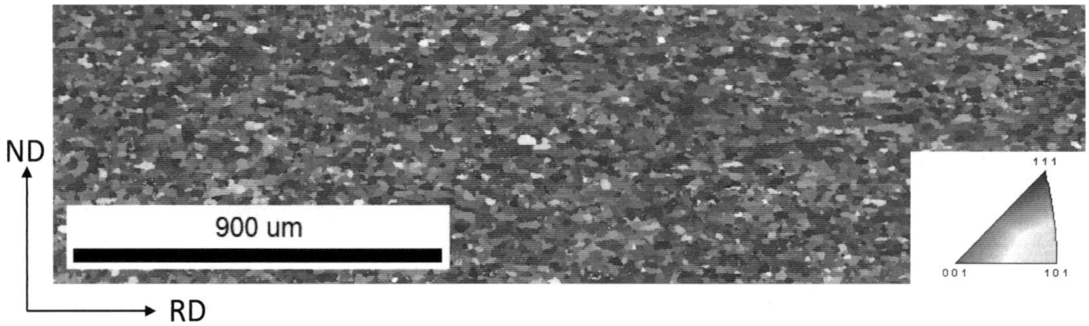

Figure 1. (a) φ_2 = 45° section of the ODF of a cold rolled and annealed IF steel; (b) Orientation map of the IF steel, color coded according to the ND Inverse Pole Figure key.

Figure 2 displays a number of simulated r-value profiles together with the experimentally measured r-values. The simulations were carried out by discretizing the continuous ODF function of Figure 1a in a set of 5,000 discrete orientations according to the procedure of Toth and Van Houtte [11]. The r-value profiles were calculated both with the conventional FCT theory and with the ALAMEL model considering {110}-{211}<111> slip systems and with identical critically resolved shear stresses on all slip systems (ignoring work hardening). For the ALAMEL model an input array of 5,000 orientations was ordered with to the above mentioned algorithm whereby $\omega_t = \omega_{min}$, 10°, 20°, 30°, 40°, 50°, ω_{max}. By evaluating the sum of squares (SS) of differences between the modelled and experimental r values (cf. Figure 2) it can be observed that the ALAMEL profiles always produce a better correspondence with the experimentally measured r-values as compared to the profile produced by the FCT model. The ALAMEL simulations with $\omega_t = \omega_{min}$, 10° produced the highest SS values as compared to the simulations with ω_t = 20, 40, ω_{max}. The profiles with $\omega_t = \omega_{min}$ and ω_t = 10° tend to converge to the FCT simulation results, which is no surprise because with smaller misorientations between paired crystals the pair tends to the single orientation behavior as considered by the FCT model.

Figure 2 also shows the result for $\omega_t = \omega_{exp}$, whereby ω_{exp} refers to the experimentally determined average misorientation derived from the measured EBSD scan shown in Figure 1b. To this purpose the EBSD map of Figure 1b was segmented in ~50,000 grain boundary (GB) segments of which the adjacent crystal orientation pairs were extracted together with the length of each segment. The ALAMEL simulations were carried out on 2,500 pairs of which the GB segments had the highest length (among the total of ~50,000 segments). Hence the ALAMEL simulation with $\omega_t = \omega_{exp}$ (ω_{exp} = 34.8° for the IF steel under consideration) is executed on a set of 2,500 neighboring pairs as observed in the correlated experimental mis-orientation texture. It can be seen that the $\omega_t = \omega_{exp}$ simulation result is nearly identical to ALAMEL simulations with ω_t = 20°, 30°, 40°. It is brought to memory that the latter simulations were carried out for a sequence of N/2 pairs without considering correlated neighborship. With the exception of ALAMEL simulations with $\omega_t = \omega_{min}$ and ω_t = 10° it can be concluded that for the present case under consideration the average misorientation is not significant, irrespective whether or not a sequence of truly correlated neighbors is considered.

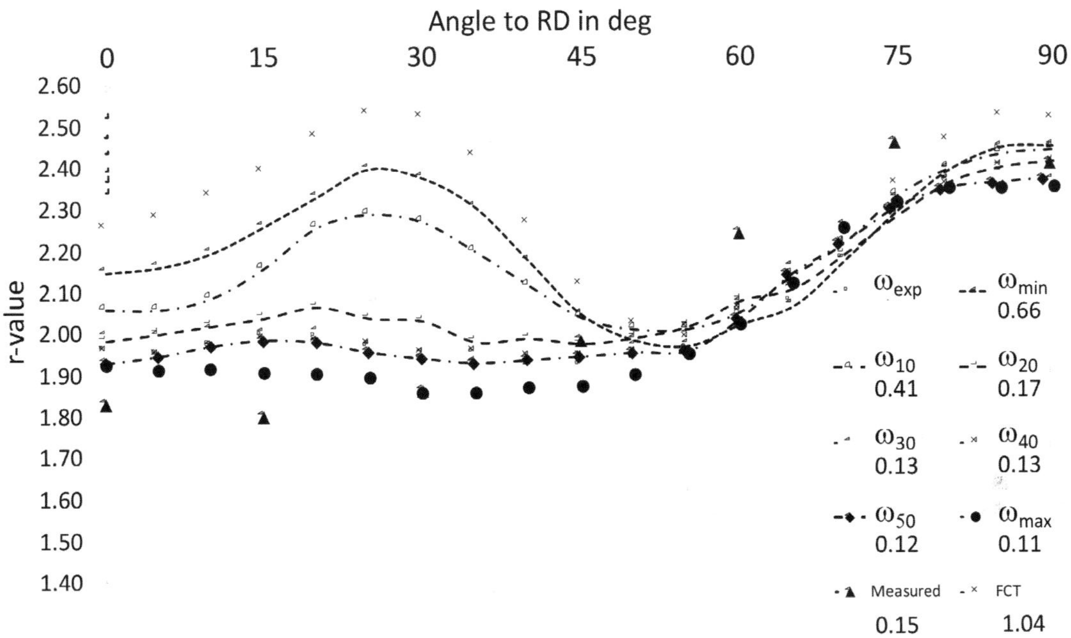

Figure 2. Lankford r-value profiles simulated with the FCT and ALAMEL models (with various values of ω_t) and the experimentally measured data. The legend also lists the Sum of Square differences of the simulation with respect to the experimental values.

3. Modelling the recrystallization texture

Kestens and Jonas [12] have published a mean-field recrystallization texture model whereby the recrystallization texture is regarded as the result of an orientation selective nucleation and growth process. In this model it is assumed that the orientation dependence of the stored energy of plastic deformation can be reliably quantified by the Taylor Factor (TF) [13] which is a dimensionless measure for the orientation (and velocity gradient) dependent plastic power dissipation. The model by Kestens and Jonas [12] allows to predict with reasonable accuracy the γ-fiber recrystallization texture (<111>//ND) of deep drawing grade IF steels of the type shown in Fig. 1a. Afterwards the model was also successfully applied to simulate the particle stimulated nucleation texture during recrystallization of an Al 6016 alloy [9] and the non-conventional recrystallization texture of a non-oriented electrical steel that was processed by asymmetrical rolling [14, 15]. In the model of Kestens and Jonas [12] the deformation texture is discretized in a set of N discrete orientations g_i. To each of these orientations a nucleation probability is attributed according to the following expression, which defines the nucleation operator $P_N(g_i)$:

$$P_N(g_i) = 1 \qquad \text{if } M(g_i) \geq M_0 \qquad \text{(Nucleation Law 1)}$$
$$P_N(g_i) = 0 \qquad \text{if } M(g_i) < M_0 \qquad (2)$$

whereby $M(g_i)$ is the TF of orientation g_i and M_0 is the critical threshold value for nucleation. Nucleation Law 1 simply assumes that high stored energy orientations (i.e. grains with a TF exceeding a critical value) are selected for nucleation. It is well-documented that high-stored energy orientations are selected for nucleation after cold rolling in low-carbon steels [16, 17].

The nucleation texture $f_N(g_i)$ is obtained by applying the nucleation operator $P_N(g_i)$ on the deformation texture $f_D(g_i)$:

$$f_N(g_i) = cP_N(g_i)f_D(g_i) \qquad (3)$$

whereby the constant c normalizes the nucleation texture to 1 (i.e. $\sum f_N(g_i) = 1$); $f_D(g_i)$ and $f_N(g_i)$ represent the volume fraction of orientation g_i in the deformation and nucleation texture, respectively. Nucleation Law (1) prevents {001} orientations to be present among the recrystallization nuclei as it is well known that the {001} orientations have the lowest TF for plane strain compression (rolling) [18, 19]. However, it is well documented that {001} orientations appear in the early nucleation texture [20]. Figure 3a shows the modelled α/γ deformation texture of a cold rolled IF steel after a rolling reduction of 80%. This texture was obtained by applying the ALAMEL model on a set of 5,000 initial orientations representing the experimentally measured hot band texture (not shown here). After an annealing treatment of 30min at 630°C the sample exhibited a recrystallized volume fraction of ~15%, of which the EBSD measured texture is shown in Figure 3b. It can be seen that the latter texture, which will be considered here as the early nucleation texture, exhibits a small but significant maximum on a {001} component. In general, the nucleation of low TF orientations is associated with Strain Induced Boundary Migration (SIBM), which enhances crystallite volumes of low stored energy (i.e. with low TF) to bulge out and grow into regions of high stored energy (i.e. with high TF) [19].

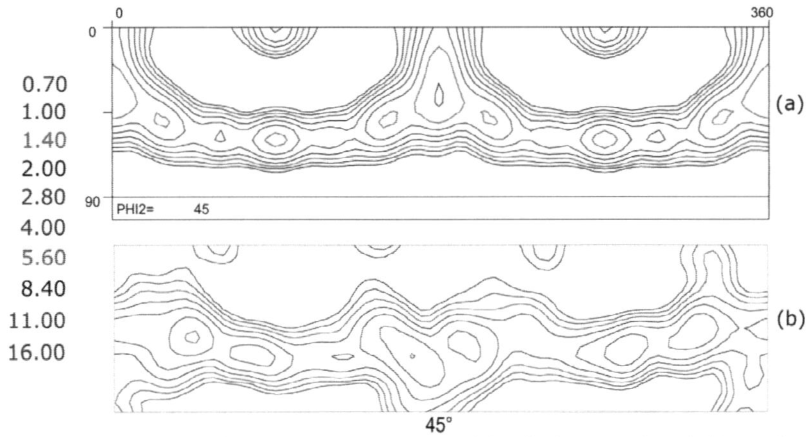

Figure 3. (a) Modelled deformation texture of a 80% cold rolled IF steel and (b) nucleation texture of recrystallized grains after annealing at 630°C for 30min.

In the model by Kestens and Jonas [12] the TF was calculated with the Full Constraint Taylor (FCT) theory according to the linear optimization method by Van Houtte [21]. In the ALAMEL model crystal orientations are considered pair-wise and thus, the TF of a grain is dependent of the adjacent grain to which it is paired. The implicit presence of orientation topology in the ALAMEL model allows to model the implication of the SIBM mechanism by assuming a nucleation condition of the type:

$$P_N(g_i) = 1 \quad \text{if } |\Delta M(g_i, g_j)| \geq |\Delta M_1| \quad \text{(Nucleation Law 2)}$$
$$P_N(g_i) = 0 \quad \text{if } |\Delta M(g_i, g_j)| < |\Delta M_1| \quad (4)$$

Hereby g_i is the orientation with the lowest TF of an arbitrary pair (g_i, g_j) (i.e. $M(g_i) < M(g_j)$); $\Delta M(g_i, g_j) = M(g_i) - M(g_j)$ and $|\Delta M_1|$ is the critical difference between low and high TF to trigger SIBM nucleation.

Figure 4a shows the nucleation texture that was derived by applying Nucleation Law 1 (Eq. 2) on the deformation texture $f_D(g)$ shown in Figure 3a. In the present model Taylor factors were normalized between 0 and 1 (TFn = (TF-min(TF))/(max(TF)-min(TF)) As expected only high Taylor factor γ-fiber grains appear in the nucleation texture as this nucleation law excludes all low TF orientations from nucleation. In Figure 4b the nucleation texture is shown that is obtained by applying Nucleation Law 2 (Eq. 4), which simulates the effect of SIBM. It can be seen that in this case a nucleation texture is derived which exhibits both α- and γ-fiber grains, albeit with significantly lower intensity than observed in the nucleation texture of Figure 4a. It is obvious that Nucleation Law 2 grossly over-estimates the nucleation of α-fiber grains, which are barely present in the experimentally observed nucleation texture. By combining both nucleation laws of Eqs. 1 and 3 with a weight ratio of 5:4, the nucleation texture of Figure 4c is obtained. It can be seen that by this combined law it is possible (i) to include a minor {001}<110> component in the nucleation texture and (ii) to slightly modify the skeleton line (curvature) of the γ-fiber so as to obtain a better correspondence with the experimental result.

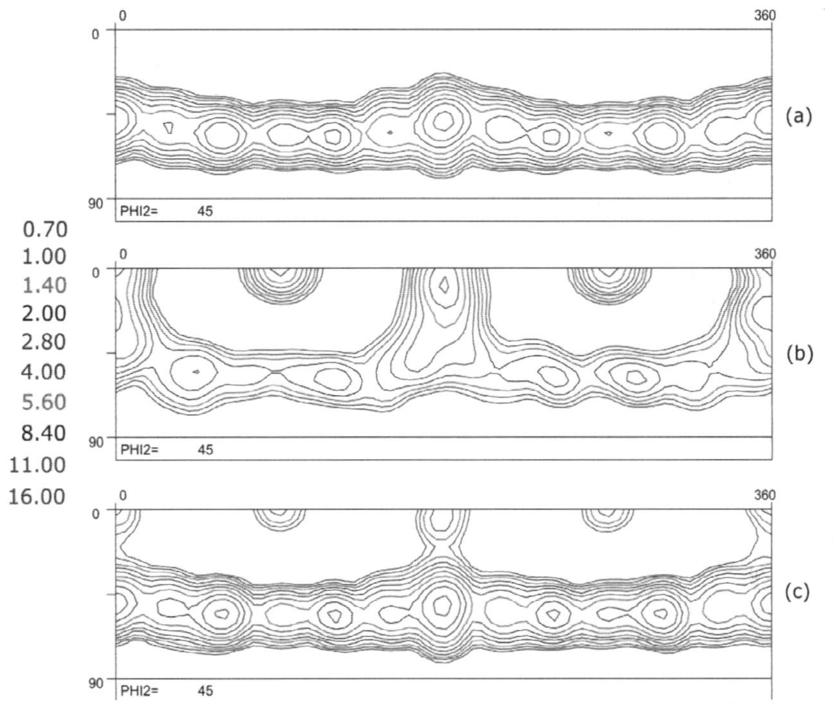

Figure 4. Nucleation texture obtained by applying (a) Nucleation Law (1) and (b) Nucleation Law (2) on the deformation texture of Figure 1a. (c) Nucleation texture obtained by combination both nucleation laws (1) and (2). Simulations for $M_0 = 0.75$ and $|\Delta M_1| = 0.8$.

4. Discussion

It was clearly illustrated that the r-value profiles simulated with the ALAMEL model produce a better correspondence with the experimental data (as compared to the FCT model) on condition that the average misorientation ω_t between subsequent orientations in the array is larger than the minimum value or 10°. Hence, for the accuracy of the model and for the present case under consideration it is not required that a coherent set of physically neighboring orientations is considered. An arbitrary set of orientation pairs, randomly sampled from the texture, does produce a similar accuracy on condition that the average misorientation is beyond the limit of a typical misorientation corresponding to a low-angle grain boundary.

The algorithm by which the ordered arrays of discrete orientations were produced is of a stochastic nature., i.e. running the algorithm repeatedly for identical ω_t values will produce different sequences. The r-value profiles of Figure 2 for different ω_t values represent the average of 3 simulation runs for each ω_t value. The scatter in r-value for a given ω_t value is very small, though (of the order of 0.02-0.07). Hence, it was chosen not to display scatter bars for the data of Figure 2.

Another element of topological information that plays a role in the ALAMEL model is the morphology of grains. In the ALAMEL model it is assumed that shear stresses are relaxed in the GB plane separating the 2 crystals of a pair. Hence, the model considers a distribution of GB plane normal vectors $\bar{n}$, essentially reflecting the grain shape morphology. In the present work a spatially uniform distribution of GB normals is considered, which is equivalent to assuming that the average shape of a grain is equi-axed. The model could be further refined by extracting an approximation of the GB normals distribution from the experimental EBSD scans measured on 3 perpendicular sections.

The recrystallization model presented in Section 3 is only a very brief summary of the model that will be presented in the PhD thesis of Jhon Ochoa [22]. The effect of model parameters such as M_0, ΔM_1 and weight factors of nucleation laws need to be assessed in detail. The strength of a mean-field approach is that it is computationally effective and for specific microstructural state variables or state properties the mean-field modelling results are of comparable accuracy as obtained with full field models. One of the limitations of a mean-field model is its limited scope as it only can derive conclusions for the microstructural descriptors that are considered in the model. It implies that phenomena that are difficult to be characterized by statistically representative state variables cannot be the subject of the simulation process. E.g. the presence of local shear bands as source of recrystallization nuclei is challenging to be included in a mean-field model. A similar observation applies to secondary recrystallization phenomena whereby only a handful of grains consume the entire initial structure. If phenomena are of such local character that a statistical representation becomes obsolete then a full-field approach is required.

5. Conclusion

In the present paper the role of microstructural topology was considered in mean-field crystallographically resolved models. In the proposed analysis the topological features of the microstructure were described according to a method applied in the ALAMEL model by Van Houtte et al. [5], whereby a texture is represented by an even number of N orientations that are grouped in pairs. By ordering these grains in a linear sequence and selecting the pair groups on the basis of the set average pair misorientation, topological aspects can be added to the model.

A state model is considered that simulates the Lankford r-value profile by applying the ALAMEL model on a set of N (even) discrete orientations representing the texture of an interstitial free deep drawing steel. It was shown that the ALAMEL predictions of the r-value profile can be improved by considering the appropriate average misorientation in the array of N/2 orientation pairs that represent the texture. The condition for obtaining a reasonable accuracy of the simulation was that the average misorientation of the paired orientations is $\gtrsim 15°$, corresponding to the high-angle misorientation threshold. Considering a set of pairs consisting of physical neighbors did not improve the accuracy of the model in the present case of an IF steel with characteristic deep drawing texture.

A recrystallization texture model was briefly outlined which intends to model the recrystallization texture of a cold rolled and annealed IF steel. The model describes oriented nucleation by combining the high stored energy nucleation law (which enhances nucleation of γ-fiber grains) with another nucleation law that simulates Strain Induced Boundary Migration (SIBM). The latter nucleation law considers ALAMEL orientation pairs, whereby the orientation with the lower Taylor Factor (the low stored energy grain) is allowed to grow into the orientation with the higher Taylor Factor (the high stored energy grain) on condition that the difference in Taylor Factor exceed a critical minimum. SIBM enhances nucleation of {001} and α-fiber grains.

Acknowledgments

A. Van Bael acknowledges financial support from the FWO (K801421N).

References

[1] Van Houtte P, Kanjarla A K, Van Bael A, Seefeldt M. and Delannay L 2006 Multiscale modelling of the plastic anisotropy and deformation texture of polycrystalline materials Eur J Mech A solids 25 634-48

[2] Roters F,Diehl M., Shanthraj P, Eisenlohr P, Reuber C, Wong S L, Maiti T, Ebrahimi, A, Hochrainer T, Fabritius H O, Nikolov S, Friák M, Fujita N, Grilli N, Janssens K G F, Jia N, Kok P J J, Ma D, Meier F, Werner E, Stricker M, Weygand D and Raabe D, 2019 DAMASK – The Düsseldorf Advanced Material Simulation Kit for modeling multi-physics crystal plasticity, thermal, and damage phenomena from the single crystal up to the component scale. Comput Mater Sci 158 420-78

[3] Traka K, Sedighiani K, Bos C, Galan Lopez J, Angenendt K, Raabe D and Sietsma J 2021 Topological aspects responsible for recrystallization evolution in an IF-steel sheet – Investigation with cellular-automaton simulations. Comput Mate Sci 198

[4] Sedighiani K, Diehl M, Traka K, Roters F, Sietsma J and Raabe D 2020 An efficient and robust approach to determine material parameters of crystal plasticity constitutive laws from macro-scale stress-strain curves Int J Plast 134

[5] Sedighiani K, Shah V, Traka K, Diehl M, Roters F, Sietsma J and Raabe D 2021 Large-deformation crystal plasticity simulation of microstructure and microtexture evolution through adaptive remeshing. Int J Plast 146

[6] Lebensohn, R A and Tomé C N 1993 A self-consistent anisotropic approach for the simulation of plastic deformation and texture development of polycrystals: Application to zirconium alloys. Acta Metallurgica Et Materialia 41 2611-24

[7] Habraken A M, et al. 2022 Analysis of ESAFORM 2021 cup drawing benchmark of an Al alloy, critical factors for accuracy and efficiency of FE simulations. Int J Mater Form in press

[8] Van Houtte P, Li S, Seefeldt M and Delannay L 2005 Deformation texture prediction: From the Taylor model to the advanced Lamel model. InterJ Plast21 589-624

[9] Sidor J J, Petrov R H and Kestens L A I 2011 Modeling the crystallographic texture changes in aluminum alloys during recrystallization Acta Mater 59 5735-48

[10] Pirani A 2021 Investigation of crystallographic disorientation topology in simulation of rolling textures with the ALAMEL model Master thesis, Delft University of Technology, Delft, http://resolver.tudelft.nl/uuid:faf40cc7-3e1c-4cf9-90f0-1a9ac879ef4f.

[11] Tóth L S and Van Houtte, P 1992 Discretization Techniques for Orientation Distribution Functions. Textures and Microstruct 19[12] Kestens L and Jonas J J 1996 Modeling texture change during the static recrystallization of interstitial free steels. MMTA 27 155-64

[13] Kestens L A I and Pirgazi H 2016 Texture formation in metal alloys with cubic crystal structures. Mater Sci Technol 32 1303-15

[14] Nguyen-Minh T, Sidor J, Petrov R and Kestens L A I 2012Texture evolution during asymmetrical warm rolling and subsequent annealing of electrical steel. In Proceedings of the 16th Int Confe on Textures of Materials, ICOTOM 16, (Materials Science Forum) 702-703 758-61

[15] Nguyen-Minh T 2021 Crystallographic texture control in a non-oriented electrical steel by plastic deformation and recrystallization. PhD Thesis, Delft University of Technology, Delft

[16] Ray R K, Jonas J J and Hook R E 1994 Cold rolling and annealing textures in low carbon and extra low carbon steels Int Mater Rev 39. 129-72

[17] Hutchinson B 1999 Deformation microstructures and textures in steels. Philos Trans Royal Soc A Philos T R Soc A 357 1471-85

[18] Dillamore I L, Morris P L, Smith C J E, Hutchinson W B and Raynor G V 1972 Transition bands and recrystallization in metals. Proc of the Royal Society of London. A. Mathematical and Physical Sciences 329 405-20

[19] Hutchinson WB 1989 Recrystallisation textures in iron resulting from nucleation at grain boundaries. Acta Metallur 37 1047-56

[20] Takajo S, Vogel S C and Hayakawa Y 2019 Strongly developed texture components near rotated cube orientation during recrystallization in a severely cold rolled low carbon steel. ISIJ International 59 541-50

[21] Van Houtte P A 1988 Comprehensive Mathematical Formulation of an Extended Taylor–

Bishop–Hill Model Featuring Relaxed Constraints, the Renouard–Wintenberger Theory and a Strain Rate Sensitivity Model. Textures and Microstruc 8
[22] Ochoa Avendaño J 2023 Continuum Modelling of Recrystallization Textures in Deep Drawing Grade IF Steels. PhD Thesis (in progress) Delft University of Technology, Delft

Grain growth in thin film under strong temperature gradients

Dana Zöllner[1,2] and Wolfgang Pantleon[2]

[1] Graduate Academy, Otto von Guericke University Magdeburg, Universitätsplatz 2, 39106 Magdeburg, Germany
[2] Department of Civil and Mechanical Engineering, Technical University of Denmark DTU, Produktionstorvet 425, 2800 Kgs. Lyngby, Denmark

E-mail: dana.zoellner@ovgu.de

Abstract. Grain growth in thin films has attracted lots of attention due to the numerous applications of polycrystalline films, on one hand, and modified properties due to changes in the microstructure caused by thermal effects, on the other hand. While the phenomenon of grain growth is well understood in general, effects of e.g. temperature gradients leave questions open still today. In the present study, we investigate the influence of strong temperature gradients on grain growth in thin films. To that aim, a modified three-dimensional Potts model algorithm is employed, where the annealing temperature depends on the position within the sample leading to spatial heterogeneities in the grain boundary network. As a consequence of different mobilities, a drag effect occurs on the boundary network evolution that has serious consequences for the microstructures evolving during grain growth.

1. Introduction

Grain boundaries in polycrystalline materials are inherent structural features usually on the micrometer size scale. They are of particular interest as almost all mechanical properties, such as hardness or strength, depend on the distance between grain boundaries, i.e. the grain size [1, 2]. Changes in the boundary network lead to a change in properties, which may render materials unsuitable for certain applications. Understanding how such changes in the boundary network progress during deformation, recrystallization or grain growth is of utmost importance to predict the materials behavior and materials life spans during operation at elevated temperatures. At first glance, grain growth appears to be a rather simple process: at elevated temperatures, grain boundaries move in such a manner that larger grains grow at the expense of smaller neighbors. This leads to a reduction of the total grain boundary interfacial area and therewith of the Gibbs free energy of the microstructure.

On a more detailed level, this coarsening process is very complex as there are a number of influential factors. One of them are open surfaces, which gain importance in thin films. It has been observed in relatively thick [3] as well as very thin [4] metal layers that when the average grain size reaches the size of the layer thickness, grain growth slows down and comes to a halt [5,6]. Because of the rather large number of applications of thin films, e.g., as protective coatings or in semiconductor technology, understanding this is of importance as an increasing number of overview publications on the topic of polycrystalline grain microstructures and grain growth shows [7-10]. Quite a large number of these investigations analyze the problem in two dimensions (2D). Even though this is in accordance with traditional metallography, i.e. a two-dimensional characterization of materials microstructures by optical microscopy, and a number of important findings have been achieved, inherent problems of such 2D investigations are well known. Any two-dimensional section through an object in three dimensions (3D) may yield only a poor idea about true size or shape of the object. This becomes even more important for polycrystalline materials in the presence of heterogeneities within the sample.

In thin films, coarsening during grain growth may become even more complicated when temperature gradients are considered. While temperature gradients may be very small and therewith negligible in some applications, in others — such as in nuclear reactors [11] or in the heat affected zone during welding [12] — strong spatial temperature gradients have to be taken into account. In a recent paper, Zöllner [13] modeled the impact of a strong temperature gradient on grain growth in films using a three-dimensional Potts model. She found that the growth of grains at the hot open surface, where the boundary mobility is higher, is restricted by the need to maintain contiguity with smaller grains in colder regions with lower grain boundary mobility, whereas the growth of grains at a cold open surface is enhanced due to required contiguity with larger grains in warmer regions. In compliance with the temperature gradient, a visible gradient in the microstructure is observed.

In the present investigation, we model a strong temperature gradient, which influences the three-dimensional grain growth via the grain boundary mobility. Additionally, we construct an artificial microstructural gradient from the three-dimensional microstructure evolution during isothermal annealing to distinguish effects of locally different temperatures from that of a temperature gradient.

2. Theoretical Background

Changes in a grain boundary network develop thermally activated and gradually in time by grain boundary migration. The higher the temperature T, the faster the progress occurs, provided, that the driving force from a reduction in stored energy is the same. The temperature dependence of the grain boundary mobility m follows at realistic temperatures an Arrhenius dependence [14]

$$m(T) = m_0\, e^{-\frac{Q^*}{k_B T}}, \qquad (1)$$

determined by a quasi temperature-independent pre-exponential factor m_0, the Boltzmann constant k_B and the activation energy Q^* for thermally activated boundary migration. For achieving the same progress at two different constant temperatures T and T', different corresponding annealing times t and t' are required, which are related to each other according to

$$m(T)t = m(T')t'. \qquad (2)$$

This relation can be utilized for predicting the progress of grain growth at a certain temperature T based on the progress at another temperature T' taking Eq. (1) into account.

3. Simulations

For about 40 years, computer algorithms have enabled researchers to simulate recrystallization, grain growth, and similar phenomena such as Ostwald ripening in polycrystalline materials. In particular, mesoscopic simulation methods work well on a microstructural level and allow a selective tailoring of materials parameters. Different models were developed over the years, such as the Monte Carlo Potts model and the phase-field method (cf., [15-19]). In particular, the Potts model has a relative simple algorithm that can be modified to tailor and model complex problems [13,20-22].

In the Potts model all grain boundaries are sharp interfaces with zero width. Only two materials parameters characterize the boundaries, the boundary mobility m and the boundary energy γ. Commonly, both depend on the misorientation between two adjacent grains, but also on the annealing temperature T. Where the boundary mobility depends strongly on temperature following the Arrhenius relation of Eq. (1), the boundary energy has only a minor influence within the investigated temperature range and, therefore, is considered constant in the current investigation.

The Potts model is utilized here for simulation of grain growth in a strong linear temperature gradient through a thin film (see figure 1) for up to 1000 Monte Carlo time steps (MCS). The simulation box (see scheme in figure 1a) contains $N = 250 \times 250 \times 75$ voxels on a 3D cubic lattice.

The length of each voxel along the coordinate axis is called Monte Carlo Unit (MCU). Specifically, the layer thickness, i.e. the size in z-direction, has been selected such as to simulate a statistically relevant number of grains in the gradient direction and to circumvent simulation effects of extremely thin layers ($z \gg 1$ MCU). A linear relation between temperature and position along the z-direction in the material assuming steady-state thermal conduction as in [13] is applied (figure 1b). It is assumed that the top surface at $z = 1$ MCU is kept at a constant temperature $T_{top} = 300$ °C whereas the bottom at $z = 75$ MCU is kept at room temperature $T_{bottom} = 25$ °C. The activation energy of grain boundary motion is selected such that the mobility, which is highest at the top surface $z = 1$ MCU, decreases throughout the layer to $m_{bottom} = m_{top}/10$ at the bottom in an exponential manner (figure 1c).

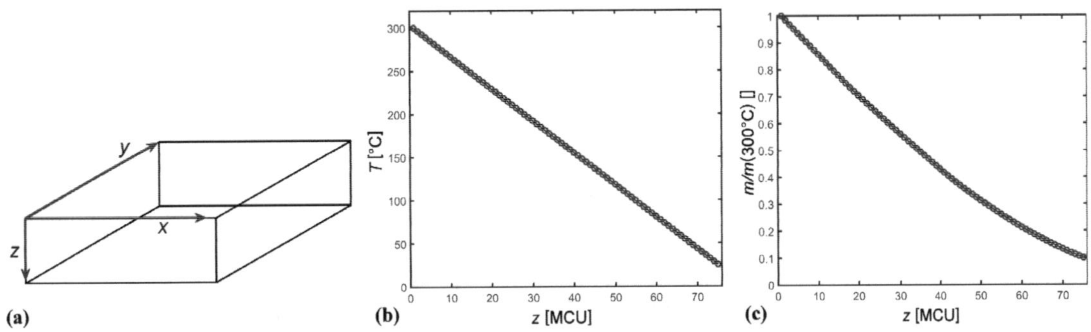

Figure 1. (a) Simulation box of size 250 × 250 × 75 MCU^3 with coordinate axes. (b) Imposed temperature gradient as function of film depth. (c) Resulting depth dependence of grain boundary mobility from Eq. (1).

For any given voxel within the lattice, a neighborhood of the size 3 × 3 × 3, i.e. the point of interest and its 26 nearest neighbors, is considered for local interactions. For the margins of the simulation box, specific boundary conditions are presumed: periodic boundary conditions for the surfaces parallel to the temperature gradient and free boundary conditions at the top and bottom surface (parallel to the xy-plane perpendicular to the temperature gradient). An initial microstructure is generated and projected on the simulation box with 25000 grains following a normal grain size distribution.

Grain growth under the temperature gradient is simulated up to 1000 MCS. The grain structure resulting after 800 MCS is illustrated in figure 2e. Obviously, due to the temperature gradient, the progress of grain growth is different in each of the layers (at different depth). The coarsening has proceeded most at the highest temperatures towards the top and least at the lowest temperatures at the bottom (at least as long as surface effects do not play a role).

In an alternative approach to achieve the microstructure resulting from the linear temperature gradient, an artificial microstructural gradient is constructed from a separate also three-dimensional, but isothermal simulation of grain growth at a constant temperature $T' = T_{top} = 300$ °C. In order to mimic the depth-dependent differences in evolution under a temperature gradient, layers from corresponding different isothermal annealing times $t'(z)$ are assembled. Each layer in depth z in the temperature gradient $T(z)$ is represented by a layer of the isothermal simulation at the same height after an effective isothermal annealing time

$$t'(z) = \frac{m(T(z))}{m(T')} t = e^{-\frac{Q^*}{k_B T(z)} + \frac{Q^*}{k_B T'}} t, \quad (3)$$

which follows directly from Eqs. (1) and (2). For this effective annealing time, the progress in microstructural evolution in the layer in the isothermal simulation is the same as for the simulation under the temperature gradient after $t = 800$ MCS. This artificially reconstructed microstructural gradient takes into account the different thermal activation in each layer depending on its depths, but not the same spatial coupling as occurring in the grain growth simulation under the temperature gradient. Nevertheless, this approach has not only the advantage that the underlying isothermal simulations are generally faster than real temperature gradient models, but also it enables the reuse of older simulation data and, hence, calculating microstructures with different temperature profiles from one and the same isothermal simulation.

4. Results

In a three-dimensional grain structure, where the grain size is comparably small in contrast to the film thickness, the grain structure within the film evolves mostly just as within a bulk sample. The average grain size — in terms of the average radius R — follows the standard parabolic growth during normal grain growth independent of dimensionality [23,24]. Analogously, the average apparent grain area $\langle A \rangle$ measured from a 2D section increases linearly with time t. Of course, this is only true for normal grain growth, where all boundaries have the same properties. Under this condition, the parabolic growth law is also obeyed in thin films, when the grain size is indeed smaller than the film thickness. In general, even for a mostly columnar grain microstructure the parabolic growth law holds as well with a smaller growth rate at the surfaces [25,26].

In the current case, the evolution of the simulated grain microstructure under isothermal conditions is shown in figure 2 (a to c) for three different annealing times. Grain coarsening is clearly visible and no preferred direction of growth is detected. The latter is even true for the simulation of grain growth in the linear temperature gradient (figure 2e). The grains still appear equiaxed without preferred growth direction, but a strong microstructural gradient develops throughout the thickness with large grains close to the top surface and much smaller grains at the bottom.

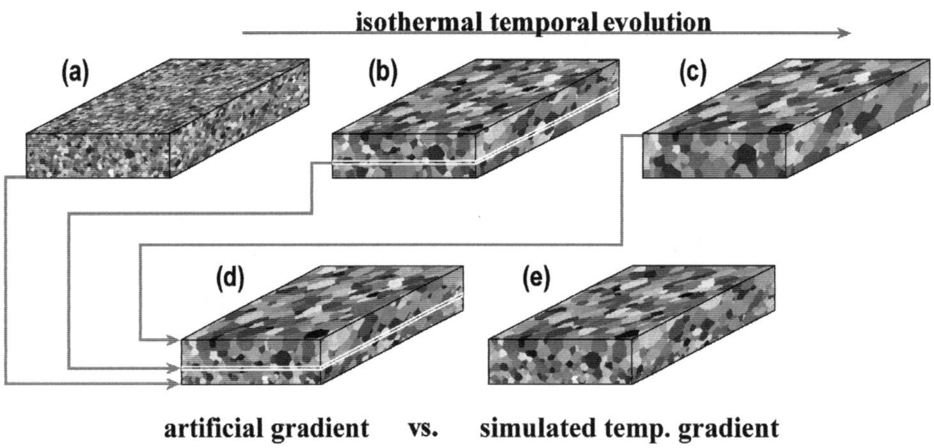

Figure 2. Simulated grain growth in thin films: the top row shows the 3D microstructural evolution under isothermal conditions. The grain structures represent annealing at 300 °C for (a) 78 MCS, (b) 402 MCS and (c) 800 MCS. The bottom row shows the (d) constructed artificial microstructural gradient in comparison to (e) the microstructural gradient evolving under the temperature gradient simulated for 800 MCS.

Applying the above idea of creating a microstructural gradient artificially, 75 layers each in its own height are taken from 75 different annealing times given by Eq. (3) of the isothermal simulation to construct the 3D microstructure such that the result is supposed to describe a microstructure having evolved under the temperature gradient after simulation for t = 800 MCS. The respective image is shown in figure 2d. As the different sections of the artificial microstructural gradient are taken from different annealing times of the isothermal simulation, here also a clear microstructural gradient becomes visible. The top of the artificial microstructure has a clearly larger average grain size as it is taken from a later annealing time ($t' = 800$ MCS) compared to the bottom of the gradient sample gathered from ($t' = 78$ MCS). At a first glance, the artificially created microstructural gradient has a very similar appearance compared to the simulated microstructure under the temperature gradient.

For a more detailed analysis, the simulated microstructures have been characterized regarding the evolution of the average grain size. To that aim, the average grain areas $\langle A \rangle$ were followed over time for different sections (z = {1, 5, 10, … 75} MCU). The results are shown in the top row of figure 3. The case of isothermal annealing at 300 °C is illustrated in figure 3a. The data for the two surfaces (z = {1, 75} MCU) are colored in red, the next nearest analyzed sections (z = {5, 70} MCU) in magenta, and all other inner sections from the bulk in black. The presence of a surface effect on top and bottom is apparent. While all inner sections yield similar results (in particular up to 500 MCS) with some statistical fluctuations, both open surfaces evolve faster. This is a result of the fact that the sample terminates at these surfaces (which is not so for the faces of the simulation box along the temperature gradient where periodic boundary conditions maintain continuity). Grain boundaries align themselves perpendicular to the open surfaces and no grains from regions outside the film can grow into these. As an additional result, the sections close to the surface experience a slight drag effect: Their average area increases faster compared to those further inside; but not as fast as that at the surfaces. Consequently, the resulting average grain area plotted as a function of film thickness z for t = 400 MCS in figure 3c shows a statistically homogeneous microstructure in the film center with an average grain area $\langle A \rangle$ = 142 MCU^2. Towards both surfaces, regions up to about 15 layers show enhanced coarsening.

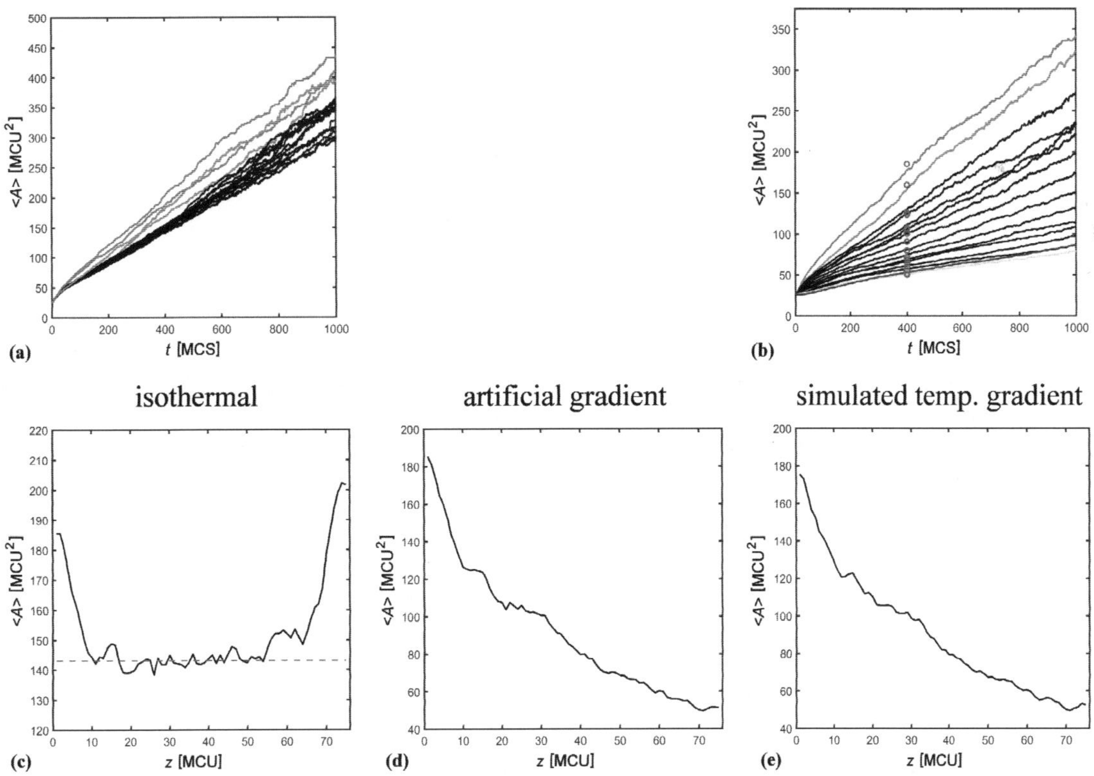

Figure 3. Average grain size: In the upper row, the temporal development of the average grain area is shown for (a) isothermal coarsening and (b) coarsening in a linear temperature gradient, both for different sections along the z-direction through the films. In the lower row, the average areas after 400 MCS are given as a function of the layer depth z within the films for all three analyzed cases: (c) the isothermal simulation, (d) the artificially constructed microstructure, and (e) the simulation with linear temperature gradient.

For the simulation with the linear temperature gradient as in figure 1b, the evolution of the mean average grain area in the different layers (figure 3b) shows that there is indeed a microstructural gradient resulting from the underlying mobility gradient. The hot surface associated with the highest mobility evolves fastest (plotted in red). The nearest analyzed section is given in magenta and has a slightly retarded growth. All inner sections ($z = \{10, 15, \dots 65\}$ MCU) follow a systematic, where the lower the temperature, the smaller the growth rate. This is even true for $z = 70$ MCU shown in cyan. The cold surface (plotted in blue) shows again faster coarsening, however, faster with respect to the slow coarsening of the nearest sections (plotted in cyan). The resulting dependence of the average grain area on the depth z within the film is given in figure 3e for $t = 400$ MCS.

Finally, the obtained average sizes are compared to that of the artificial microstructural gradient. Here, a special feature arises: As the microstructure of the artificial gradient is calculated from different time instants of an isothermal simulation, one cannot follow its temporal evolution easily. Nevertheless, one can obtain grain microstructures for a specific time t. In particular, for the artificial microstructural gradient constructed for $t = 400$ MCS, the average grain sizes from the sections $z = \{1, 5, 10, 15, \dots 75\}$ MCU each taken at their respective t' are added to figure 3b (green circles) comparing the artificial microstructural gradient directly to the microstructural gradient simulated for the temperature gradient. The small deviations visible are fully within the range of statistical fluctuations. The good correspondence between both microstructures becomes even more apparent when comparing the depth-dependent average grain areas $\langle A \rangle(z)$ for both gradients as shown in figure 3d and 3e.

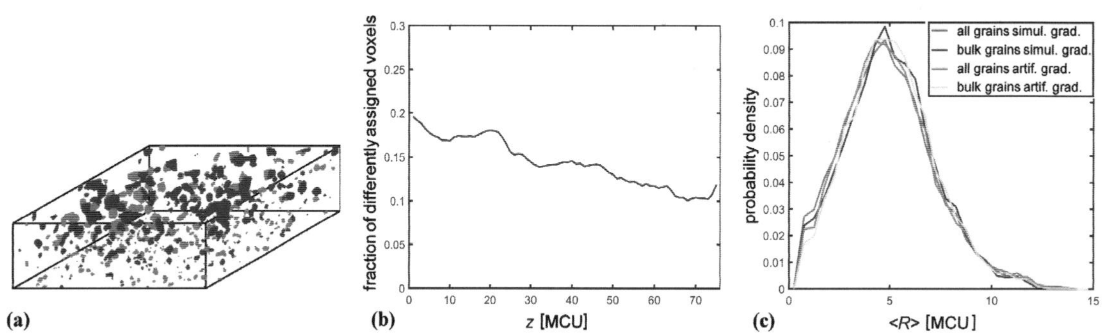

Figure 4. Comparison between simulated and artificial microstructural gradient: (a) The voxels of non-matching grains are highlighted; grains in the microstructure gradient simulated under the temperature gradient, not present in the artificially constructed gradient are shown in blue. Red grains are present solely in the artificial microstructural gradient. (b) Fraction of differently assigned voxels in the volume. (c) Distribution of equivalent spherical radii from real and artificial gradient independently for all grains and bulk grains only.

Further comparison between the microstructure caused by the temperature gradient and the artificial microstructural gradient created to resemble the different progress in growth requires additional quantitative analysis. Figure 4 presents the results of a point-to-point comparison between both microstructures after 400 MCS. From the 25000 grains in the initial configuration, only 5502 survived in the simulated gradient and 5566 in the artificial one. The non-matching grains are illustrated in figure 4a in blue, if present in the proper simulation with the temperature gradient, but not in the artificially created one, and red in the opposite case. Apparently, there are a non-negligible number of grains in both simulations that evolve differently on the individual grain level, even when the general results agree well on the microstructural level according to figure 3. As shown in figure 4b, the average fraction of differently assigned voxels in the entire volume is 15% with largest deviations at the hot surface. Despite these differently assigned points, the averaged grain size defined as equal spherical radius derived from the volumes of all grains differs only slightly: 9.93 MCU (10.17 MCU) for the simulated gradient vs. 9.83 MCU (10.04 MCU) for the artificially created one. The numbers in brackets refer to bulk grains only. The corresponding distributions for the equivalent spherical radii are matching closely (cf. figure 4c). Kolmogoróv-Smirnov testing allows quantifying similarities between the obtained distributions: Where the distributions from the artificial gradient microstructure and the simulated one do not differ significantly, the hypothesis that the distribution of all grains and bulk grains only originate from the same distribution is rejected at a significance level of 5%. The artificially created gradient assembling layers corresponding to different annealing times of an isothermal simulation resembles closely the real microstructural gradient from simulating a temperature gradient. Consequently, the microstructure established in a temperature gradient is a sole consequence of the locally different mobilities and no effects from growing in a temperature or mobility gradient are discerned. In both microstructural gradients, the grains do not exhibit any preferred growth direction and maintain equiaxiality.

Still the question remains, how does the microstructural gradient develop? The first effect that comes to mind is the fact that at higher temperatures grain boundaries move faster due to their higher mobility. Hence, one is inclined to presume that a growing grain may grow faster into the hotter than into the colder region. One may even expect a certain skewness of the grain shape developing.

This is shown exemplarily for the simulation of three grains undergoing recrystallization in a temperature gradient in figure 5. Here recrystallization is used as it allows an easy tracking of the boundaries of the growing grains. Three seeds are put into a three-dimensional box of size 120 × 120 × 350 MCU^3 on a straight line along the temperature gradient. In figure 5 the temporal development of the section through the center of the three grains is shown. All three seeds grow into small grains. If one focusses on the central green grain, one realizes that within the first 10 MCS of annealing, the seed grows and the center of mass z_c does not change significantly along the z-direction, that is the direction of the temperature gradient. When the grain has reached a substantial size, its center moves to lower z, i.e. upwards and therewith into the hotter region. This indicates clearly that recrystallizing grains grow indeed faster into the hotter region than into the colder, which is in agreement with the above expectations. However, when the simulation continues, at about 25 MCS the growing grain starts to impinge on the upper grain. At this point, the center stops moving upwards. As the grain can not continue to grow further into the hot region, the center starts even to move downwards. After 45 MCS, the shift downward exceeds the maximum shift upwards.

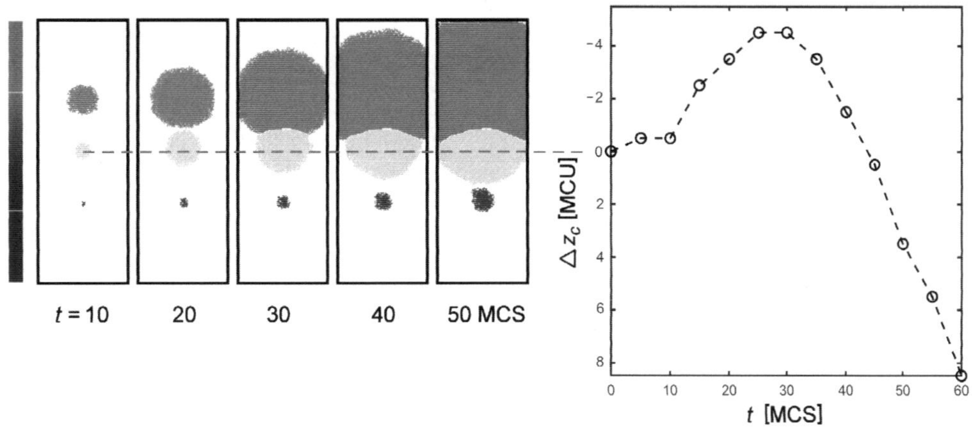

Figure 5. Three-dimensional simulation of recrystallization originating form three seeds under a linear temperature gradient along z. (a) On the left, the temporal development of a section is shown. The top of the samples is hot, the bottom cold. (b) On the right, the shift of the center of mass of the green grain along z-direction resp. along the temperature gradient is given.

While this is easy to understand for such a simplified case, for a more complex 3D distribution of many seeds, the individual behavior of the center of mass of a grain during recrystallization will depend not only on its location and therewith on temperature and local temperature gradient, but also on the neighboring seeds. This is also the case, for the above simulated case of grain growth in a temperature gradient. In average, the grains develop neither a pronounced skewness along the direction of the temperature gradient, nor a systematic shift in their center of mass position. Their coarsening depends rather, on one hand, on the local temperature and local grain boundary mobility, respectively. On the other hand, due the complex boundary network, effects of the grain sizes and their number of edges are non-negligible for the evolution and an equiaxed microstructure is maintained.

5. Summary and Discussion

While in general the phenomenon of grain growth in polycrystalline materials is well understood, effects of temperature gradients leave questions open still today. Therefore, in the present study, we investigated the influence of strong temperature gradients on grain growth in thin films. To that aim, we modified a three-dimensional Potts model algorithm such that the annealing temperature varies within the thickness of the sample producing spatial heterogeneities in grain boundary mobility.

We have shown that in contrast to isothermal grain growth, under a temperature gradient the grain microstructure evolves from a spatially homogenously distributed structure with equiaxed grains to a distinct microstructure with visibly larger grains at the hot surface compared to the colder surface. In addition, the grain boundaries at the open surfaces align perpendicular to the surface, while the average areas at the surfaces evolve slightly faster compared to the inner regions close by. In the bulk of the volume, the grains remain equiaxed despite the temperature gradient.

Furthermore, we constructed an artificial model for the evolving microstructure gradient, where the specific microstructure of each layers originates from the time evolution of isothermal grain growth. While this does not predict the actual evolution of a certain grain, the temporal evolution of the average grain size as well as the resulting grain size distributions agree quite well.

Acknowledgement

DZ gratefully acknowledges the support from the Otto Mønsted foundation for an Otto Mønsted visiting professorship at DTU.

References

[1] Henkel D and Pense AW 2002 *Structure and Properties of Engineering Materials*. McGraw-Hill.
[2] Callister WD and Rethwisch DG 2014 *Materials Science and Engineering* Wiley.
[3] Beck PA, Holzworth ML and Sperry P 1949 *Trans. AIME* **180** 163.
[4] Palmer JE, Thompson CV and Smith HI 1987 *J. Appl. Phys.* **62** 2492.
[5] Thompson CV 1990 *Ann. Rev. Mater. Sci.* **20** 245.
[6] Thompson CV 2000 *Ann. Rev. Mater. Sci.* **30** 159.
[7] Karoutsos V, Toudas M, Delimitis A, Grammatikopoulos S and Poulopoulos P 2012 *Thin Solid Films* **520** 4074.
[8] Liu Z-J and Shen YG 2004 *Acta Mater.* **52** 729.
[9] Hallberg H and Olsson PAT 2016 *J. Mechan. Phys. Sol.* **90** 160.
[10] Barmak K, Eggeling E, Kinderlehrer D, Sharp R, Ta'asan S, Rollett AD and Coffey KR 2013 *Progr. Mater. Sci.* **58** 987.
[11] Olander DR 1976 *Fundamental aspects of nuclear reactor fuel elements*. Springfield (VA): Technical Information Center, Energy Research and Development Administration.

[12] Zhang S, Ma Y, Huang S, Youssef SS, Qi M, Wang H, Qiu J, Lei J and Yang R 2019 *J. Mater. Sci. Tech.* **35** 1681.
[13] Zöllner D 2022 *Model. Sim. Mater. Sci. Eng.* **30** 025010.
[14] Turnbull D 1951 *Trans. AIME* **191** 661.
[15] Steinbach I, Pezzolla F, Nestler B, Seeßelberg M, Prieler R, Schmitz GJ, Rezende JLL 1996 *Phys.* **D 94** 135.
[16] Weygand D, Brechet Y, Lepinoux J and Gust W 1999 *Phil. Mag.* **B 79** 703.
[17] Barrales-Mora L, Gottstein G and Shvindlerman LS 2008 *Acta Mater.* **56** 5915.
[18] Anderson MP, Srolovitz DJ, Grest GS and Sahni PS 1984 *Acta Metall.* **32** 784.
[19] Zöllner D and Streitenberger P 2004 *Mater. Sci. Forum* **467-470** 1129.
[20] Zöllner D 2016 *Comput. Mater. Sci.* **118** 325.
[21] Adam K, Zöllner D and Field DP 2018 *Model. Sim. Mater. Sci. Eng.* **26** 035011.
[22] Zöllner D and Zlotnikov I 2019 *Adv. Theo. Sim.* **2** 1900064.
[23] Burke JE and Turnbull D 1952 *Progr. Metal Phys.* **3** 220.
[24] Humphreys FJ and Hatherly M 1995 *Recrystallization Related Annealing Phenomena* Pergamon Press Oxford.
[25] Zöllner D 2016 *Comput. Mater. Sci.* **125** 51.
[26] Zöllner D 2021 *Comput. Mater. Sci.* **200** 110803.

IOP Conf. Series: Materials Science and Engineering 1249 (2022) 012011 doi:10.1088/1757-899X/1249/1/012011

Tensile deformation-induced twinning in an austenitic low density steel over a wide range of temperature

Hui Wang[1,2], Jiangxiong Liang[1], Zhiyue Shi[1], Cunyu Wang[1], Xiaodan Zhang[3] and Wenquan Cao[1]

[1]Central Iron and Steel Research Institute (CISRI), Beijing 100081, China
[2]School of Materials Science and Engineering, Chongqing University, Chongqing 400044, China
[3]Section of Manufacturing Engineering, Department of Mechanical Engineering, Technical University of Denmark, 2800 Kgs, Lyngby, Denmark

E-mail: caowenquan@nercast.com

Abstract. A Fe30Mn11Al1CNbV low-density steel was processed by solid solution treatment and deformed at different temperatures from -196°C to +200°C by tensile testing. It was found that the deformation microstructure is mainly dominated by thin dislocation slip bands during tensile deformation, and deformation twinning was found near grain boundaries at a strain of 0.5, regardless of deformation temperature. The stacking fault energy is calculated by consideration of both chemical composition and the deformation temperature, which is high and normally should not initiate deformation twinning. However, deformation twinning is occasionally observed in the deformation microstructure at room temperature, and becomes to be one of the dominating features at liquid nitrogen temperature. Based on the analysis of work hardening and microstructure evolution, it is proposed that the deformation behaviour of the austenitic low density steel is controlled by the continuous thinning of dislocation slip bands at temperatures of 25°C and 200°C, but by a synergistic behaviour of slip bands and deformation twinning at -196°C.

1. Introduction

Low-density steels based on the FeMnAlC system have attracted much attention in recent years due to their relatively low density, and to an excellent combination of strength and ductility [1-3]. For example, the density of such steels can be tuned down by nearly 20% to as low as 6.5g cm^{-3} by the combined addition of Al and C into a high manganese steel [4,5], and at the same time their yield strength can be improved to be higher than 600MPa, combined with a ultimate tensile strength of around 1000 MPa and a total elongation of larger than 50% [4,6]. Such a combination of both low density and good mechanical properties is in high demand by the vehicle industry to fulfill requirements of enhanced safety, weight-reduction and easy-fabrication, in response to present concerns over resource limitations and environmental pollution [7].

However, in contrast to traditional twin-induced plasticity (TWIP) steels, low-density FeMnAlC steels are designed to contain a considerable amount of light elements, with a typical composition of 6–12% Al, 20–30% Mn and 0.80–1.50% carbon [1,2,4]. The addition of Al to a high Mn austenitic steel results in a high stacking fault energy, where planar dislocation glide is reported to be the main deformation mechanism at room temperature [1,8,9]. However, the twinning phenomenon was found in our previous study of the low density Fe-30Mn-11Al-1.2C steel during rolling at room temperature, indicating a transition in deformation mechanism from slip bands to deformation twinning at a critical thickness reduction of 50%, which was explained by the deformation inhomogeneity [10]. As it is well known, the deformation transition from dislocation glide, to TWIP and to transformation-induced plasticity (TRIP) has mainly been attributed to the stacking fault energy, which is directly related to the deformation temperature. However, no deformation temperature effects on the deformation microstructure of low density steels have been reported in the literature.

In this study, a micro-alloyed low density steel of composition Fe30Mn11Al1C0.1Nb0.1V was used to study the effects of deformation temperature on the microstructure and mechanical properties by tensile testing at different temperatures of -196°C, 25°C (room temperature, RT) and 200°C, with a focus on the stacking fault energy and the underlying deformation mechanism.

2. Experiment

A low-density steel with nominal chemical composition of Fe30Mn11Al1C0.1Nb0.1V was fabricated by melting in a high frequency vacuum induction furnace with a capacity of 50 kg, followed by casting into rod ingots with a diameter of 150 mm. The ingots were homogenized at 1200 °C for 5h and then were hot rolled into rods with a diameter of 16 mm, followed by air-cooling to room temperature. The rods were then solution-treated at 1050 °C for 2 hours followed by water quenching. Tension tests were carried out on standard round dog-bone specimens, with a gauge length of 25 mm, a diameter of 5 mm, with the tensile axis taken parallel to the rod axis. The tensile data of the solution-treated specimens were recorded from tensile tests carried out under a strain rate of 1.1×10^{-3} s^{-1} at different temperatures of -196°C, +25°C and 200°C using an MTS tensile machine.

The microstructure of the low-density steel was examined before and after solution processing (solution treatment at 1050°C for 2 h and water quenching) using an electron back scattered diffraction (EBSD) system attached to a field emission gun scanning electron microscope (SEM, FEI Quanta 650, Thermo Fisher Scientific, Hillsboro, OR, USA), operated at 20 kV. A detailed investigation of the deformation microstructure in samples deformed at different temperatures to fracture was carried out on samples using a transmission electron microscope (TEM, Tecnai G2 F20 S-TWIN, Thermo Fisher Scientific, Hillsboro, OR, USA) at an operating voltage of 200 kV. Specimens used for EBSD observation were first mechanically polished and then electro-polished using a 7% Nital solution at room temperature. Grain boundary maps were obtained from the EBSD data using the Channel 5 post-processing software. Thin foils for TEM analysis were twin-jet electro-polished at -20 °C using a mixture of 10% perchloric acid and 90% ethanol at 20V.

3. Results and discussion

Figure 1(a) shows the microstructure of the Fe30Mn11Al1C0.1Nb0.1V steel in the hot rolled state (a) while figure 1(b) shows the microstructure in the solution treated state. The microstructure after hot rolling consists predominately of the FCC austenite phase with a small fraction (< 5%) of the BCC ferrite phase. As shown in figure 1(a), a high fraction of ∑3 twin boundaries corresponding to about 43% out of all the high angle boundaries with misorientation angles > 15°, is observed. The average grain size of the austenite phase is about 5.6 μm (taking also into account twin boundaries as high angle boundaries), which is larger than that of the ferrite phase, where the average grain size is around 4.3 μm. As shown in figure 1(b), after solution processing, the grain size is 10.2 μm and the fraction of ∑3 twin boundaries is about 60% with the BCC phase fraction decreased to 1.7%, but the average BCC grain size increased to 4.8 μm. This change of the microstructure through the solution treatment implies that grain growth of the austenite phase takes place during the solution treatment. However the grain growth mechanism is still to be further explored.

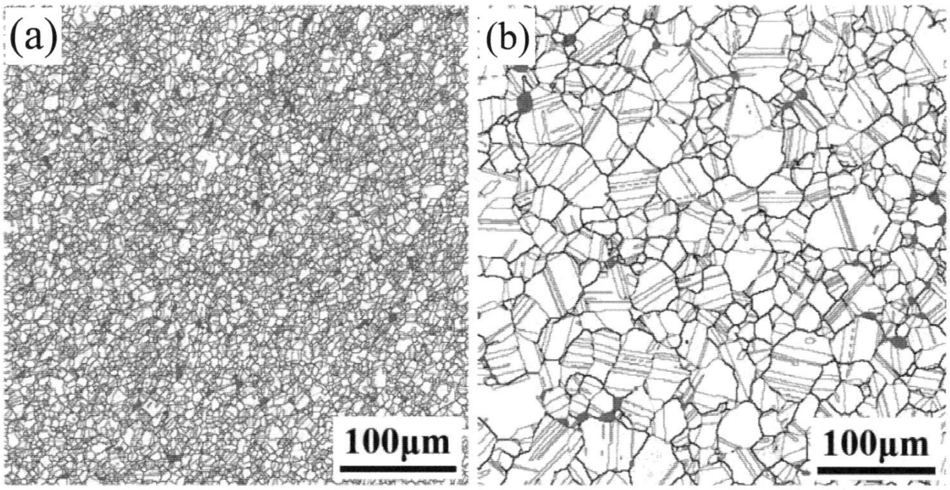

Figure 1. EBSD microstructures of the experimental steel with phase maps showing FCC (white) and BCC (green), superposed by twin and non-twin boundaries. (a) Hot-deformed state and (b) solution-processed at 1050°C for 2 h followed by water quenching.

Figure 2(a) shows the engineering tensile stress-strain curves of the low density steel deformed at -196 °C， +25 °C and 200 °C after solution treatment at 1050°C for 2 h and water quenching. It can be seen that both yield strength and tensile strength monotonically increase from 462 MPa and 654 MPa at 200 °C, to 692 MPa and 915 MPa at room temperature, and to 1270 MPa and 1630 MPa at -196 °C. At the same time, the ductility is about 63 % at 200 °C, 70 % at room temperature and 50 % at -196 °C, which means that a significantly increase of the yield strength is realized at liquid nitrogen temperature while the ductility is not significantly reduced with the decreased deformation temperature, as shown in figure 2(a).

Figure 2(b) shows the true stress-strain curves calculated from figure 2(a), which are used to examine the work hardening rate of the low density steel to aid understanding of its high ductility during tensile deformation. The variation of work hardening rate depends strongly on the deformation temperature, as shown in figure 2(c) and figure 2(d). The work hardening rate does not continuously decrease with increasing deformation strain, instead showing a behavior that is very similar to that observed in steels controlled by TWIP or TRIP effects [11,12]. The work hardening rates at 200 °C and room temperature are similar, while the work hardening rate at 200 °C fluctuates slightly before a true tensile strain of 0.23 and fluctuates strongly after this strain. However, up to a true tensile strain of 0.35 the work hardening rate at -196 °C is much higher than the work hardening rate at the other two higher temperatures.

Figure 3 shows the deformation microstructure at the tensile fracture strain at 200°C and RT. Dislocation slip bands dominate the deformation microstructure in both cases. This result is in a good agreement with the finding that the deformation behavior of low density steels is controlled by dislocation planar slip, and therefore dislocation slip bands, which bring a large strength contribution to the flow stress [13,14]. An interesting point here is the fast refinement of the dislocation slip bands, as shown in figure 3, where the dislocation slip band is about 50 nm, indicating a very strong suppression of the recovery of the dislocation structure. Further research on microstructural evolution during tensile deformation will be focused on to understand this fast slip band refinement in the near future. Quite surprisingly, only a very little amount of deformation twinning is observed in the fractured specimen deformed at room temperature. This means that a critical strain or critical stress is the key for the onset of deformation twinning, rather than the stacking fault energy as has been proposed to be the controlling factor for deformation twinning [1,9].

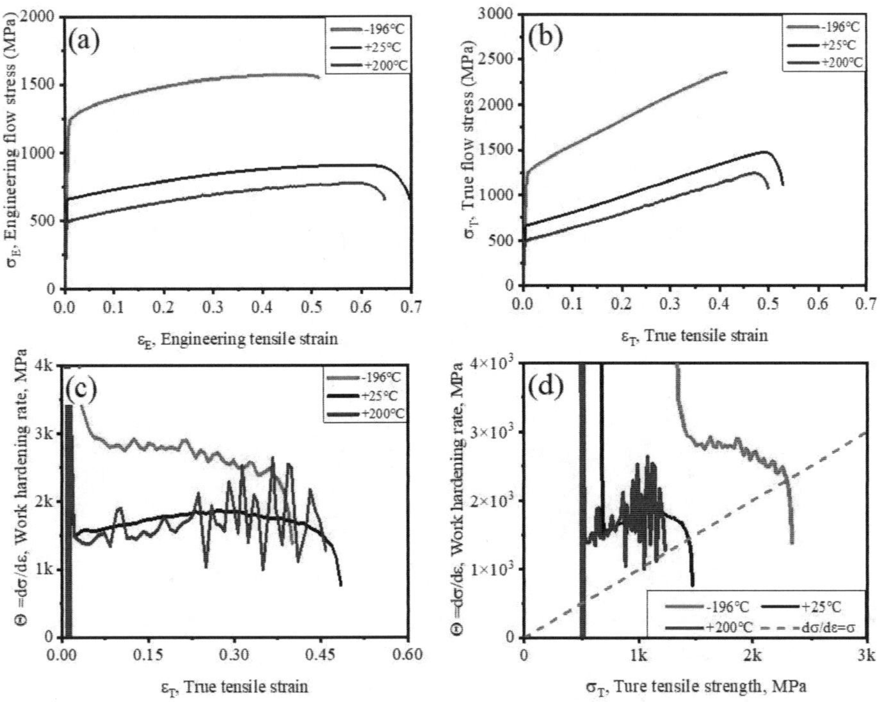

Figure 2. Stress strain curves and work hardening behavior of the low-density steel deformed at different temperatures: (a) engineering stress-strain curves; (b) true stress-strain curves; (c) work hardening rate versus true tensile strain and (d) work hardening rate versus true tensile strength.

Figure 3 shows the deformation microstructure at the tensile fracture strain at 200°C and RT. Dislocation slip bands dominate the deformation microstructure in both cases. This result is in a good agreement with the finding that the deformation behavior of low density steels is controlled by dislocation planar slip, and therefore dislocation slip bands, which bring a large strength contribution to the flow stress [13,14]. An interesting point here is the fast refinement of the dislocation slip bands, as shown in figure 3, where the dislocation slip band is about 50 nm, indicating a very strong suppression of the recovery of the dislocation structure. Further research on microstructural evolution during tensile deformation will be focused on to understand this fast slip band refinement in the near future. Quite surprisingly, only a very little amount of deformation twinning is observed in the fractured specimen deformed at room temperature. This means that a critical strain or critical stress is the key for the onset of deformation twinning, rather than the stacking fault energy as has been proposed to be the controlling factor for deformation twinning [1,9].

In order to investigate the deformation mechanism and the high work hardening rate at -196 °C (liquid nitrogen temperature, LNT) the microstructure after tensile deformation to different strains have been examined and are shown in figure 4. It can be seen that after 5% and 10% deformation, the microstructure is dominated by dislocation slip bands, as shown in figure 4 (a) and (b). However, deformation twinning takes place in the region near grain boundaries after 20% tensile deformation at LNT, as shown in figure 4(c) and (d), which is very similar to that found in the specimen after fracture at room temperature (Fig. 3(d)). In contrast to the dislocation slip band microstructure, extensive deformation twinning is observed after 30% and 50% tensile deformation at LNT, as shown in figure 4(e) and (f). This means that the deformation behaviors transitions from slip band formation at low strains to deformation twinning development at medium and high strains when the low density steel is deformed at LNT. It should be pointed out that the area fraction of twinning after 50% tensile deformation at LNT is about 35%, but this twinning mainly found in regions close to high angle grain boundaries.

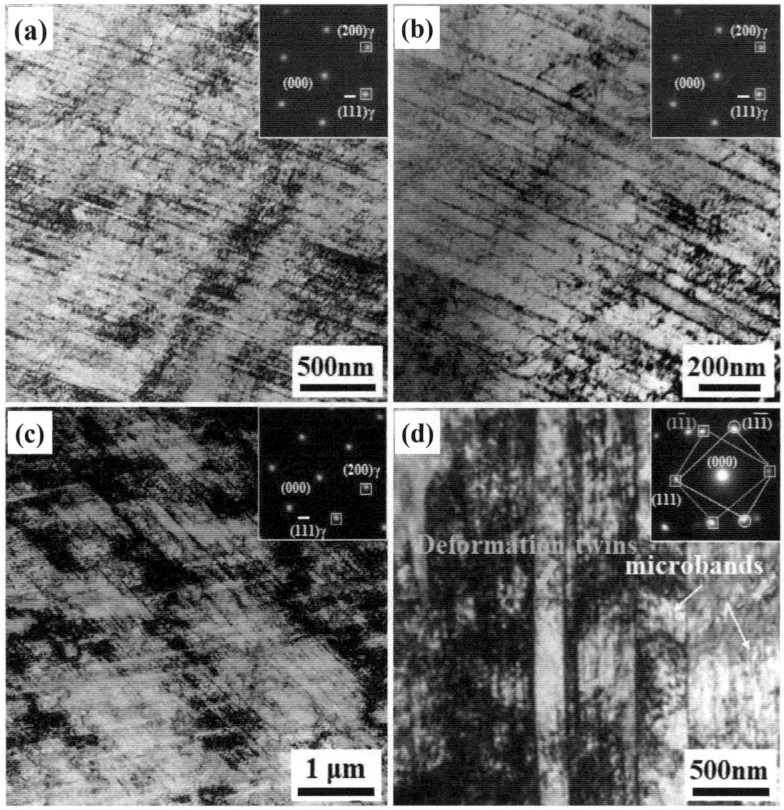

Figure 3. Microstructures of the designed steel after deformation at different temperatures: (a) and (b) 200°C, (c) and (d) RT, showing a planar slip dominated deformation mechanism, evidenced by the diffraction patterns shown in the inset images.

In order to understand the deformation behavior, the stacking fault energy has been calculated as a function of deformation temperature [15,16]. The results show that the stacking fault energy significantly decreases with decreasing deformation temperature, as shown in figure 5(a). The stacking fault energy is calculated to be 107 mJ m^{-2} at 200°C, 75 mJ m^{-2} at room temperature, and 54 mJ m^{-2} at liquid nitrogen temperature, which means that the stacking fault energy is outside of the range 20 – 40 mJ m^{-2} that is considered suitable for the formation of deformation twinning [17,18]. The critical stress is estimated from the tensile tests based on the assumption that the true stress is the critical stress at the deformation strain where deformation twinning is observed by TEM, as shown in figure 4. Based on this assumption, the critical deformation twinning stress (σ_{CT}) at LNT temperature is estimated to be 1834 MPa at a true strain of 0.18 from figure 2(b), because deformation twinning is observed at this true strain, as shown in figure 4(c) and (d). It could also be argued that deformation twins may be found just after 10% elongation (corresponding to true stress of 1630 MPa), even though the deformation microstructure at this strain is not examined in this study. Thus it is reasonable to estimate that the critical stress for the onset of twinning is in the range of 1630 – 1834MPa, which gives an average value of 1732 ± 102MPa. The critical stress at room temperature is about 1472 MPa at true strain 0.5 from figure 2(b), because deformation twinning is only observed at this true strain as shown in figure 3(d). However at 200 °C, no-deformation twins are found during tensile deformation as shown in figure 3(a) and (b), thus the ultimate strength of 1250 MPa is simply taken as the minimum critical twinning stress at this temperature. This may, however, underestimate the critical stress for twinning nucleation at 200 °C. Theoretically, according to the nucleation theory of deformation twins, the critical stress for twin nucleation can be calculated according equation (1) [13], which gives the critical stress of the deformation twinning as about 2142 MPa at LNT, 2980 MPa at room temperature and 4252 MPa at 200°C.

$$\sigma_{\mathrm{CTC}} = \mathrm{M}\frac{2\gamma}{b}, \quad (1)$$

where γ is the stacking fault energy, b =1.54 is the burgers vector, M is the Taylor factor of 3.06 and σ_{CTC} is the critical stress for the deformation twin nucleation calculated by equation (1).

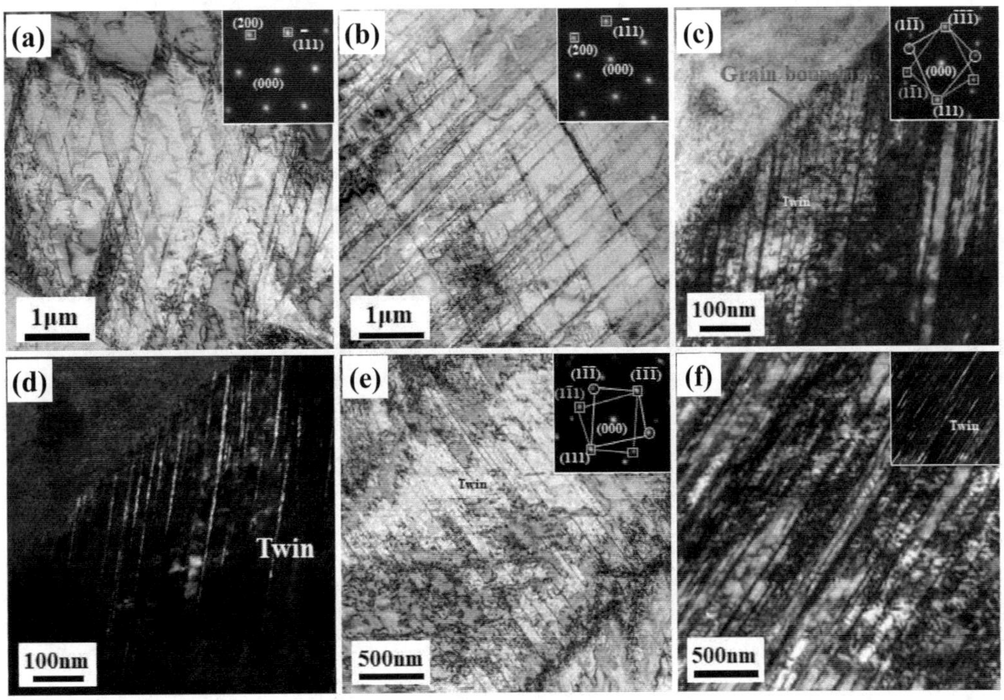

Figure 4. TEM images of the microstructure after LNT tensile deformation to different strains: (a) 5%, (b) 10%, (c) and (d) 20%, (e) 30%, and (f) 50%, in which the inserted diffraction patterns are used to reveal the deformation mechanism.

It can be seen that the measured critical stress values are not only significantly lower than the critical stress values predicted by equation (1), but also show a different evolution trend with deformation temperature. This differing trend could be attributed to fact that the critical stress for twinning as estimated from the stress strain curves is not the effective critical stress but a composite stress, including the friction stress, grain boundary strengthening and dislocation strengthening. For simplicity, the effective critical stress σ_{CTE} for deformation twinning can be calculated subtracting the yield stress $\sigma_{0.2}$ from the measured critical stress σ_{CTM}, which gives an effective critical stress of about 462 MPa at LNT, 780 MPa at room temperature and 788 MPa at 200°C, where these values show an increasing trend in critical stress with increasing deformation temperature. However, this effective critical twinning stress σ_{CTE} is only about 1/4 of the calculated values σ_{CTC} based on equation (1). It is postulated that stress concentrations contributed to this difference between the experimental results estimated from the stress strain curves and the theoretical values calculated from equation (1).

Based on above analysis, it can be concluded that the nucleation of deformation twinning is not only determined by the stacking fault energy, but also by the yield strength and by local stress concentrations. Due to the fact that there is a strong contribution from the friction stress to the yield strength with decreasing temperature, the factors affecting deformation twinning can be summarized as stacking fault energy, the friction stress, and stress concentrations, all of which can be utilized to understand the difference between the experimental results and the calculated data.

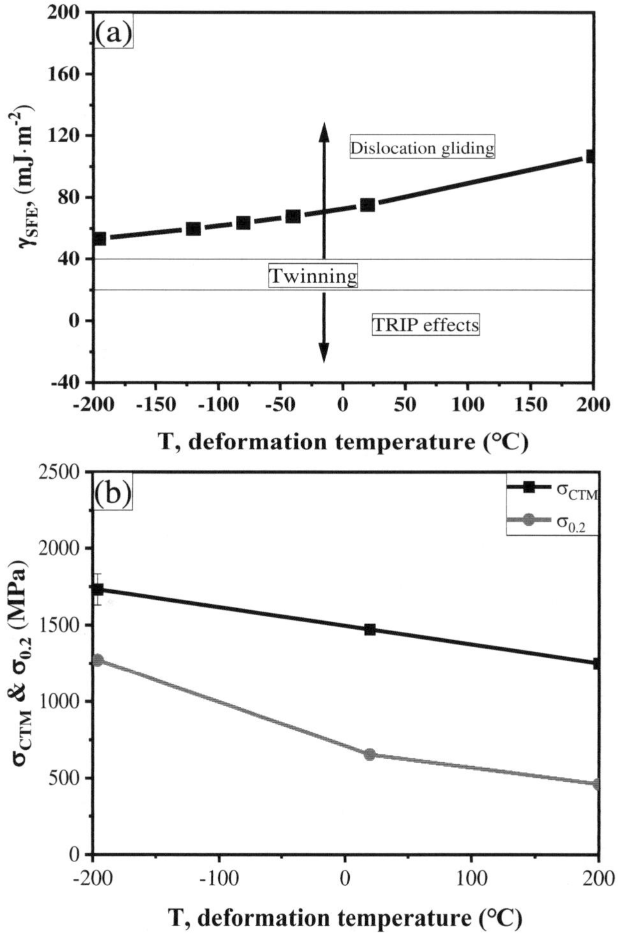

Figure 5. Stacking fault energy and critical twinning stress for the low density steel: (a) calculated stacking fault energy as a function of temperature, and (b) measured critical stress for twinning and yield strength versus deformation temperature.

4. Summary

In this study the influence of deformation temperature on the deformation of a low-density steel of composition Fe30Mn11Al1C0.1Nb0.1V has been investigated by tensile deformation at different temperatures of -196°C, 25°C and 200°C. It is found that the deformation microstructure is dominated by the formation and thinning of dislocation slip bands during tensile deformation at deformation temperatures of 25°C and 200°C. However, extensive deformation twinning is observed in the deformation microstructure after 20% tensile deformation at liquid nitrogen temperature. A transition in deformation mechanisms from dislocation slip to deformation twinning with decreasing temperature and increasing deformation strain are found. Both the high friction stress and the presence of stress concentrations are found to be responsible for the difference between the estimated critical deformation twinning stress and theoretically calculated results. Further research on rapid slip band refinement and the strongly suppressed recovery in low density steel should be carried out, as the factors are essential for the understanding of its strengthening and work hardening behavior.

Acknowledgements
The authors gratefully acknowledge the support from the National Natural Science Foundation of China (Grant No. 51871062). X. Zhang acknowledges the support from the European Research Council under the European Union Horizon 2020 research and innovation program (grant agreement No 788567-M4D).

References
[1] Chen S, Rana R, Haldar A and Ray R K 2017 *Prog. Mater. Sci.* **89** 345–91
[2] Gutierrez-Urrutia I 2021 *ISIJ Int.* **61** 16–25
[3] Frommeyer G, Brüx U and Neumann P 2003 *ISIJ Int.* **43** 438–46
[4] Kim H, Suh D W and Kim N J 2013 *Sci. Technol. Adv. Mater.* **14** 014205
[5] Frommeyer G, Drewes E J and Engl B 2000 *Rev. Met. Paris.* **97** 1245–53
[6] Yoo J D, Hwang S W and Park K T 2009 *Metall. Mater. Trans. A Phys. Metall. Mater. Sci.* **40** 1520–3
[7] Zhilyaev A P and Langdon T G 2008 *Prog. Mater. Sci.* **53** 893–979
[8] Jo M C, Choi J H, Lee H, Zargaran A, Ryu J H, Sohn S S, Kim N J and Lee S 2019 *Mater. Sci. Eng. A* **740-741** 16–27
[9] Song W, Ingendahl T and Bleck W 2014 *Acta Metall. Sin. (English Lett.* **27** 546–556
[10] Ren P, Chen X P, Mei L, Nie Y Y, Cao W Q and Liu Q 2020 *Mater. Sci. Eng. A* **775** 138984
[11] Gutierrez-Urrutia I and Raabe D 2011 *Acta Mater.* **59** 6449–62
[12] Choudhary B K 2014 *Metall. Mater. Trans. A Phys. Metall. Mater. Sci.* **45** 302–16
[13] Yoo J D and Park K T 2008 *Mater. Sci. Eng. A* **496** 417–24
[14] Welsch E, Ponge D, Hafez Haghighat S M, Sandlöbes S, Choi P, Herbig M, Zaefferer S and Raabe D 2016 *Acta Mater.* **116** 188–99
[15] Olson G B and Cohen M 1976 *Metall. Trans. A* **7** 1897–904
[16] Saeed-Akbari A, Imlau J, Prahl U and Bleck W 2009 *Metall. Mater. Trans. A Phys. Metall. Mater. Sci.* **40** 3076–90
[17] Pierce D T, Jiménez J A, Bentley J, Raabe D, Oskay C and Wittig J E 2014 *Acta Mater.* **68** 238–53
[18] Pierce D T, Jiménez J A, Bentley J, Raabe D and Wittig J E 2015 *Acta Mater.* **100** 178–90

IOP Conf. Series: Materials Science and Engineering 1249 (2022) 012012 doi:10.1088/1757-899X/1249/1/012012

Mechanics of accelerated strain hardening in harmonic-structure materials

Joséphine Chatellier[1], Elis Sjögren-Levin[1], Kei Ameyama[2] and Dmytro Orlov[1*]

[1]Department of Mechanical Engineering Sciences, LTH, Lund University, P.O. Box 118, Lund 22100, Sweden;
[2]Department of Mechanical Engineering, College of Science and Engineering, Ritsumeikan University, 1-1-1 Noji-higashi, Kusatsu, Shiga 525-8577, Japan;

*E-mail: dmytro.orlov@material.lth.se; tel +46 46 - 222 9095

Abstract. Harmonic-structure (HS) design is one of the most efficient in the family of architected gradient-structure materials recently attracting increasing attention in global material science community due to leading structural performance characteristics. This work studies *in situ* the mechanics of plastic deformation in HS materials on the example of commercially pure nickel (Ni) during uniaxial tensile loading. It reveals that strain partitioning between ultrafine and coarse grain fractions in HS Ni commences already at early stages of plastic deformation. Slow accumulation of strain in the ultrafine-grain regions leads to the slow consumption of ductility resource, while accelerated accumulation of strain in the coarse-grain regions leads to an accelerated strain hardening. These effects combined with evolving variations of strain tensor components have synergetic effect leading to the unique plastic behaviour and excellent structural performance of HS materials.

1. Introduction

Extensive studies over the past decade have convincingly demonstrated that materials having gradient structures architected down to nanoscale offer excellent and often unique opportunities for the optimisation of structural properties including strength-ductility balance [1-4]. To date, such materials have the best potential of approaching the structural efficiency of natural counterparts that also have heterogeneous structures evolved over time in dead [5] and live [6-8] materials.

Harmonic-structure (HS) is one of the most efficient design concepts in the family of materials with micro-architected gradient structures. It is shown schematically in figure 1a. From an idea generated slightly over a decade ago in the group of Ameyama at Ritsumeikan University, supported by a series of early studies and clearly formulated in [9], HS has become a firmly established approach proven for a variety of pure metals and alloys by now. The achievements of several research groups working around the globe with HS materials have been briefly assessed in [10] and comprehensively reviewed in [11] recently. Harmonic-structure materials have bimodal grain size distribution ordered such that one (ultra-)fine grain (UFG) fraction forms honeycomb-style periodic 3D skeleton with embedded islands of another coarse-grain (CG) fraction. Such a 3D topology makes HS materials heterogeneous on micro- but homogeneous on macro-scales leading to a unique set of properties with rather low variation of performance characteristics [12].

Although substantial contributions towards understanding the rheology of HS materials have been made recently [11], plastic deformation mechanisms operating during their mechanical loading remain largely an open question. On one hand, they should be governed by physical mechanisms typical for metallic materials, as comprehensively studied in the community of Risø symposium attendees and organisers, see examples in [13-16]. On the other hand, HS materials typically demonstrate somewhat unusual combination of simultaneously elevated strength and ductility characteristics along with accelerated strain hardening. These most likely stem from the unique for HS topological arrangement of UFG / CG fractions and demand further exploration. An experimental approach for such studies based on synchrotron X-ray diffraction is introduced in [17].

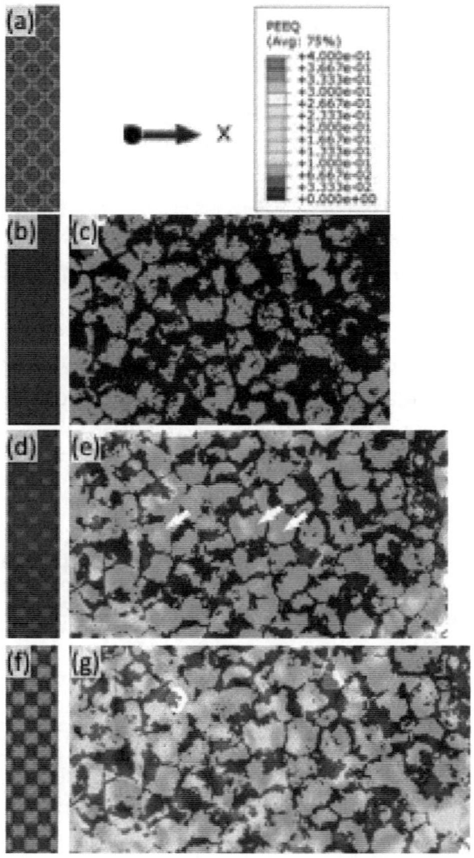

Figure 1. Evolution of local equivalent strain distribution in HS Ni with 40% UFG at various levels of global strain in uniaxial tension: e=0.002 (b,c), e=0.02 (d,e), and e=0.04 (f,g). The direction of sample extension x is shown by black arrow in (a) along with the model of HS material (where green and grey colours only indicate the locations of CG and UFG fractions, respectively) used in finite-element simulations [18] and a legend indicating the colour-scale of strain levels. The distribution of local equivalent strain is shown in (b,d,f) based on finite-element simulations, and in (c,e,g) on digital image correlation analysis with underlaying microstructure at corresponding global strain levels.

In this paper, we explore the mechanics of HS materials using digital image correlation (DIC) analysis. The latter is based on laboratory-scale *in situ* monitoring of pure nickel with harmonic structure during monotonic tensile loading. The study reveals significant strain partitioning from the early stages of plastic deformation, which is consistent down to quantitative level with our phenomenological and finite-element (FE) models as well as reports from other scientists. Following analysis brings further insights in the rheology of HS materials.

IOP Conf. Series: Materials Science and Engineering 1249 (2022) 012012 doi:10.1088/1757-899X/1249/1/012012

2. Experimental materials and methods

The following procedures were used for the fabrication of samples, specimen preparation and testing.

2.1. Material fabrication

As-received nickel powder of commercial purity (99.2%) with a mean particle size of 150 μm was mixed with SUJ2 steel balls at a mass ratio of 1.8:1, placed in a SKD11 vial and mechanically milled (MM) in a planetary ball mill Fritch P-5 at ambient temperature for 100h with a rotation speed of 150 rpm. Afterwards, the powders were consolidated by spark-plasma sintering (SPS) in vacuum at 1073 K and applied pressure 100 MPa for 30 min. In our previous studies [19, 20], it was reported that such a procedure leads to the formation of harmonic structure with a nominal UFG fraction of 40%.

2.2. Specimen preparation and characterisation of microstructure

Specimens for mechanical testing and microstructure characterisation were cut from the samples by electric-discharge machining. Afterwards, they were ground on SiC paper and then polished in diamond suspensions and colloidal silica on a last step. Finally, they were etched in sulphuric-nitric acids-based solution for revealing material microstructures in optical microscopy.

The resulting contrast was sufficient for digital image correlation analysis in tensile testing using an optical stereo-microscope Olympus SZX16 equipped with polarized light and digital camera.

2.3. Mechanical testing with DIC analysis

Due to the limited dimensions of fabricated samples, tensile tests were carried out on miniaturised dog-bone shaped specimens with nominal gage dimensions (length x width x thickness) 5.0x1.5x1.0 mm^3. The initial strain rate was 3.6×10^{-3} s^{-1}. At least three specimens were tested for each material condition.

In-house built miniaturised tensile rig was used for the majority of tensile testing, and Instron ElectroPulse-10000 machine with custom-built jaw faces to accommodate the miniaturised specimens and a special horizontal set up of the stereomicroscope for *in situ* test monitoring was used for high-accuracy testing. The digital camera in the microscope was directly connected to the control PC of the instrument allowing time synchronisation. DIC analysis was carried out using GOM Correlate software suite.

3. Results and analysis

The analysis of microstructure evolution in this study is intentionally limited to magnifications attainable in optical microscopy. Such magnifications are convenient for direct comparison with the earlier results of FE simulations [18] and are particularly interesting for revealing meso-scale phenomena in plastic deformation of HS materials. Most of the analysis focuses on correlation between 'global' strain at the scale of entire specimen and 'local' strain fields on a scale within a representative volume element (RVE), which consists of a CG island encompassed by UFG shell as illustrated by inserts in figures 2,3. The latter is further resolved into 'neck' and 'junction' corresponding to narrowest and widest spacings across UFG skeleton regions between CG islands, respectively, as illustrated by inserts in figure 4.

Figure 1 reveals the evolution of microstructure in HS nickel sample at three representative strain levels: (i) developed stage of globally elastic deformation at equivalent strain level e=0.002 immediately before global elastic-plastic transition, figure 1c; (ii) characteristic global strain level e=0.02 when strain partitioning becomes significant enough for visual identification in DIC maps, figure 1e; and (iii) characteristic global strain level e=0.04 when strain partitioning in a DIC map, figure 1g, can still be conveniently correlated with local strain distribution calculated in FE simulations, figure 1f.

The examination of figure 1c reveals homogeneous strain field, which is consistent with the results of FE simulations in figure 1b. This confirms rather homogeneous co-deformation of UFG and CG fractions over the entire range of global elastic strains. The increase of global strain level leads to the development of local strain gradients in the microstructure with higher strain levels being in CG area centres, as indicated by white arrows in figure 1e for example. At the same time, the UFG skeleton appears to preserve integrity while distorting proportional to the change of specimen shape. Such a behaviour suggests strain partitioning between the fractions and agrees with the findings of Park et al in [21]. With further increase of global strain level, the gradients evolve to become more frequent and higher in amplitude, figure 1g. Such a distribution of strain is consistent with the results of FE simulations presented at respective global strain levels in figure 1d,f, although some deviations from the pattern can be found due to non-ideal real-life HS. It is interesting to note that the strain distributions are similar at both qualitative and quantitative levels, albeit the gradients are sharper in FE simulations. The later can be explained by very high spatial resolution in FE model, while relatively large subset in DIC analysis leads to the smoothing of strain gradients due averaging of strain levels over many pixels.

Quantitative analysis of DIC data until the global strain levels of e=0.14 is presented in figure 2. At this stage, the distortion of specimen surface in plastic flow becomes so high that we cannot distinguish between CG and UFG areas anymore, and at some stage beyond that the correlation algorithm fails due to excessive surface reflectivity in our experimental setup. The local strain values presented by green and grey curves in figure 2 were calculated as an average of at least ten correlation areas within CG and forty within UFG fractions that are shown schematically in the insert.

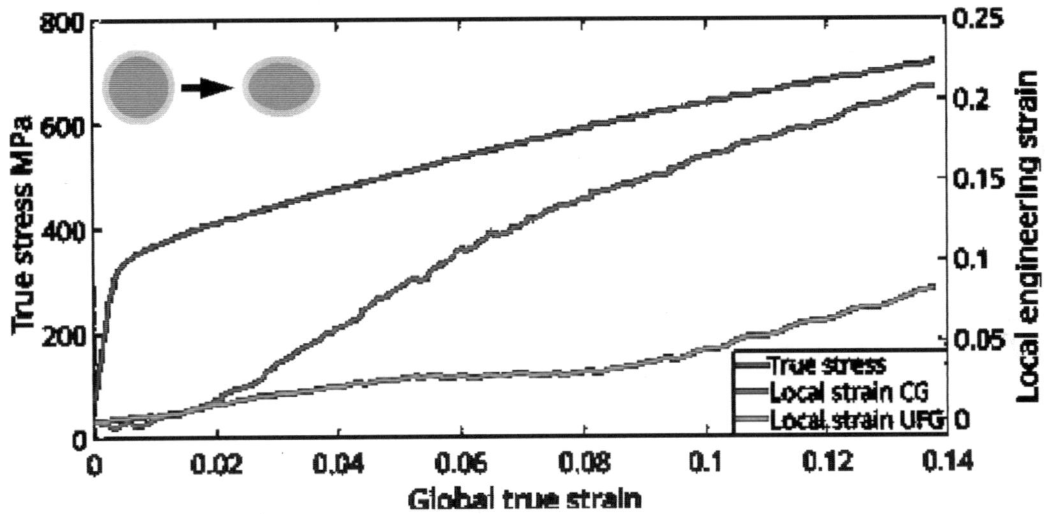

Figure 2. Experimental results from uniaxial tensile testing of HS Ni revealing the dependence of true stress (left axis) and local engineering strain in the direction of loading ε_x separately for CG (core) and UFG (shell) regions on global true strain.

It can be seen that during the linear part of the stress-strain curve, both grain fractions accumulate local strain homogeneously at the same rate. Close to the global elastic-plastic transition, strain partitioning starts taking place, but the trend is somewhat erratic making conclusive analysis difficult at this stage. This must be attributed to the limitations of accuracy in our measurements at this global strain level. The accumulated strain level also appears somewhat elevated on the diagram but actually varies around zero level in numerical values. The accumulation of local strain in UFG fraction is virtually monotonic and linear until e=0.05, then becomes stagnant until e=0.08, and increasing monotonically again until the end of measured global strain range. The CG fraction reveals non-monotonic accumulation of average local strain until e=0.015 with large variation in values (not presented in the diagram for figure clarity). Afterwards, it increases monotonically at a very high rate until $e\approx0.06$ and at a substantially lower rate yet higher than in the UFG fraction subsequently.

The trends of local strain accumulation in CG / UFG fractions determined experimentally in this study is generally consistent with a theoretical 2D model of HS material deformation developed in [22]. According to this model, the mode of deformation in UFG skeleton is significantly different from that in the CG islands. At the relatively small levels of strain under consideration in present study, deformation should proceed in agreement with the Le Chatelier's principle (or the principle of similitude

[23]). Namely, the geometric distortion of a single structural unit in our specimen (one CG island fully surrounded by UFG shell) is proportional to the deformation of entire specimen. The illustration of that is presented graphically in the inset of figure 3. In such a case, deformation in the CG island should proceed in plain-strain mode, while in the UFG skeleton it should be elongation according to the increasing perimeter of the CG region circumference in RVE. The kinetics of local strain accumulation depending on global tensile deformation of a specimen according the model from [22] is presented in figure 3 for the range of global strain in our present study. General trends do appear similar to those in figure 2, while the kinetics of local strain accumulation determined experimentally is still somewhat different from the theory. As mentioned above, experimental trends have substantial variations in the rate of strain accumulation of both grain fractions as well as slightly lower values of accumulated strain in CG while higher in UFG. These can be explained by two somewhat connected factors. First is the lower-than-desired resolution of DIC data leading to somewhat exaggerated values of local strain in UFG fraction. Second is the actual size (thickness) of UFG areas, which is presumed to be constant in the model, but should actually expand during plastic deformation due to strain hardening. Although this effect may significantly affect the kinetics of local strain accumulation, its detailed study is outside of scope in the present work. It is reserved for a follow-up publication.

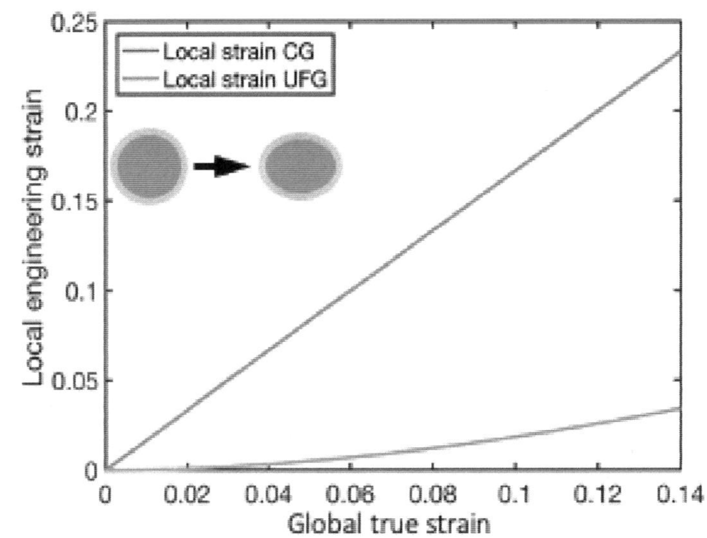

Figure 3. Analytical estimate of engineering strain evolution in CG core (green) and UFG shell (grey) regions assuming purely geometrical transformation of a circle into an ellipse during uniaxial tension according to a model developed in [22]. The insert illustrates transformation of a circle at a maximum global strain level indicated on the diagram.

Such a divergence in the evolution of accumulated strain levels between CG and UFG fractions again confirms significant strain partitioning from the very early stages of plastic deformation of HS materials. Significantly faster kinetics of accumulation along with higher total values of local strain in the CG fraction leads to its significantly faster strain hardening. Experimental evidence of this effect in terms of homogenisation in nano-hardness levels between grain fractions was reported in [24]. At the same time, significantly slower accumulation of strain in the UFG fraction allows delaying its fracture to much later stages of global specimen deformation in spite of typically limited ductility of UFG and nanocrystalline materials. Enhanced ductility of HS materials compared to their UFG counterparts is well documented [11, 12, 20]. These two synergetic effects can be leading, if not dominant, factors in the excellent structural performance of HS materials reported earlier [10, 11].

Other factors related to the interface between UFG/CG fractions are further contributors into the unique behaviour of harmonic structures. The dynamics of dislocation structure evolution in HS titanium was studied in [11] using transmission electron microscopy. The authors reported the formation of large shear bands in coarse grains within CG fraction along with strong pileups and accumulation of dislocations close to the UFG interfaces. These results agree well with our explanation of minor discrepancies in the accumulation of strain in the UFG fraction between the experiment and the model. Our own *in situ* studies of stress partitioning in HS nickel using synchrotron X-ray diffraction will be reported soon in an upcoming publication.

Looking deeper into the accumulation of strain within the UFG skeleton, two principal directions of extreme difference can be distinguished. First one is a 'neck by design', which is the narrowest region of UFG fraction formed between two CG islands. Second one is a 'junction' connecting three CG islands in 2D sections (or four CG islands in 3D) in real-life HS, which is the widest region of UFG fraction. At the same time, let us point out that if close-packed structure is assumed in 3D, at a distance approximately equal to half-diameter of the CG island in the out-of-plane direction, the 'neck' becomes the 'junction' and *vice versa*. It is interesting to check if the kinetics of local strain accumulation is different between them. Respective data are presented in figure 4, where dependencies of both local engineering strain in the direction of loading ε_x and local equivalent strain on global equivalent strain are shown.

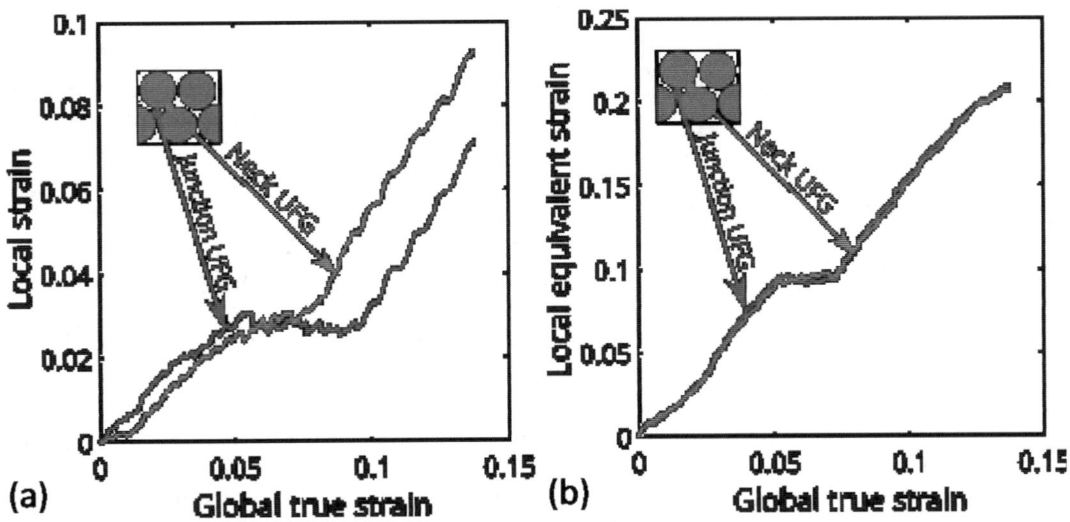

Figure 4. DIC analysis of experimental data from uniaxial tensile testing of HS Ni revealing the dependence on global true strain of (a) local engineering strain in the direction of loading ε_x and (b) local equivalent strain e^l for the UFG skeleton separately for 'neck' and 'junction' regions.

In the case of only looking into the local engineering strain component of strain tensor ε_x, figure 4a, the kinetics of evolution resembles averaged trend presented in figure 2. However, the similarity between the neck UFG and the junction UFG regions is limited to the rates of local strain accumulation before and after stagnation. The junction UFG regions show somewhat higher values before the stagnation and significantly lower after that. The extent of stagnation itself in the junction UFG regions is double that in the neck. These differences may indicate that the large-volume junction regions experience certain level of plastic deformation at the early stages of loading, but then stagnate and continue to accumulate strain much slower than the necks. At the same time, it is very interesting to notice that the dependencies of local equivalent strain accumulation on global equivalent strain in both the regions are virtually the same, figure 4b. Such results indicate significant contribution of strain components other than ε_x in the accumulation of local equivalent strain e^l, which will be further investigated in following studies.

4. Conclusions

In this work, the mechanics of plastic deformation in harmonic-structure materials was studied *in situ* during uniaxial tensile testing using digital image correlation analysis of optical images. It has been shown that strain partitioning between UFG skeleton and CG islands commences already at early stages of plastic deformation. Slow accumulation of strain in the skeleton leads to the slow consumption of ductility resource in the UFG regions, while accelerated accumulation of strain in the islands leads to an accelerated strain hardening in the CG regions. These are further enhanced by the significant contributions of non-linear strain tensor components into the total accumulated equivalent strain. These effects have synergetic effect leading to the unique plastic behaviour and excellent structural performance of harmonic-structure materials.

Acknowledgements

The authors acknowledge Swedish Research Council (Vetenskapsrådet) for the financial support of this study through a research grant # 2016-03811 and the Swedish Foundation for International Cooperation in Research and Higher Education (STINT) for the support of international collaboration through grants # SJ2017-7432 and MG2019-8450.

References

[1] Editorial 2021 Architectures down to nano Nature Materials 20 1451
[2] Zhu Y, Ameyama K, Anderson P M, Beyerlein I J, Gao H, Kim H S, Lavernia E, Mathaudhu S, Mughrabi H, Ritchie R O, Tsuji N, Zhang X and Wu X 2021 Heterostructured materials: superior properties from hetero-zone interaction Mat Res Lett 9 1-31
[3] Ma E and Zhu T 2017 Towards strength–ductility synergy through the design of heterogeneous nanostructures in metals Mat Today 20 323-31
[4] Zhou M 2013 Exceptional Properties by Design Science 339 1161-2
[5] Beygelzimer Y, Kulagin R, Fratzl P and Estrin Y 2020 The Earth's Lithosphere Inspires Materials Design Adv Mat 2005473
[6] Mandelbrot B B 1983 The Fractal Geometry of Nature (New York City: Henry Holt and Co)
[7] Wegst U G K and Ashby M F 2004 The mechanical efficiency of natural materials Phil Mag 84 2167-86
[8] Meyers M A, McKittrick J and Chen P-Y 2013 Structural Biological Materials: Critical Mechanics-Materials Connections Science 339 773-9
[9] Orlov D, Fujiwara H and Ameyama K 2013 Obtaining Copper with Harmonic Structure for the Optimal Balance of Structure-Performance Relationship Materials Transactions 54 1549-53
[10] Orlov D and Ameyama K 2020 Critical Assesment 37: Harmonic-structure materials - idea, status and perspectives Mat Sci and Tech 36 517-26
[11] Ameyama K, Cazes F, Couque H, Dirras G, Kikuchi S, Li J, Mompiou F, Mondal K, Orlov D, Sharma B, Tingaud D and Vajpai S K 2022 Harmonic structure, a promising microstructure design Mat Res Lett 10 440-71
[12] Zhang Z, Orlov D, Vajpai S K, Tong B and Ameyama K 2015 Importance of Bimodal Structure Topology in the Control of Mechanical Properties of a Stainless Steel Adv Eng Mat 17 791-5
[13] Hansen N and Juul Jensen D 1999 Development of microstructure in FCC metals during cold work Philosophical Transactions of the Royal Society A: Math, Phys and Eng Sci 357 1447-69
[14] Pantleon W and Hansen N 2001 Dislocation boundaries--the distribution function of disorientation angles Acta Mat 49 1479-93
[15] Hansen N, Huang X and Huges D A 2004 Deformation Microstructures. In: 2004 TMS Annual Meeting and Exhibition, ed T G L Y.T. Zhu, R.Z. Valiev, S.L. Semiatin, D.H. Shin, T.C. Lowe (Charlotte, North Carolina, USA: TMS (The Minerals, Metals & Materials Society)) pp 67-74
[16] Hughes D A and Hansen N 2014 Exploring the Limit of Dislocation Based Plasticity in Nanostructured Metals Phys Rev Lett 112 135504

[17] Sjögren-Levin E, Pantleon W, Ahadi A, Hegedüs Z, Lienert U, Tsuji N, Ameyama K and Orlov D 2022 Separation of XRD peak profiles in single-phase metals with bimodal grain structure to analyze stress partitioning IOP Conf Ser: Mat Sci and Eng In Press
[18] Shokry A, Ahadi A, Ståhle P and Orlov D 2021 Improvement of structural efficiency in metals by the control of topological arrangements in ultrafine and coarse grains Sci Rep 11 17445
[19] Sawangrat C, Kato S, Orlov D and Ameyama K 2014 Harmonic-structured copper: performance and proof of fabrication concept based on severe plastic deformation of powders J of Mat Sci 49 6579-85
[20] Zhang Z, Vajpai S K, Orlov D and Ameyama K 2014 Improvement of mechanical properties in SUS304L steel through the control of bimodal microstructure characteristics Mat Sci and Eng A 598 106-13
[21] Park H K, Ameyama K, Yoo J, Hwang H and Kim H S 2018 Additional hardening in harmonic structured materials by strain partitioning and back stress Mat Res Lett 6 261-7
[22] Orlov D, Kulagin R and Beygelzimer Y 2020 Strain partitioning and back-stress evaluation in harmonic-structure materials Mat Lett 275 128126
[23] Rayleigh 1915 The Principle of Similitude Nature 95 66-8
[24] Orlov D, Zhou J, Hall S, Ota-Kawabata M and Ameyama K 2019 Advantages of architectured harmonic structure in structural performance IOP Conf Ser: Mat Sci and Eng 580 265-72

IOP Conf. Series: Materials Science and Engineering 1249 (2022) 012013 doi:10.1088/1757-899X/1249/1/012013

Grain boundary evolution in nanograined metals: from relaxation to Schwarzation

X Li and K Lu

Shenyang National Laboratory for Materials Science, Institute of Metal Research, Chinese Academy of Sciences, 72 Wenhua Road, Shenyang 110016, China

E-mail: xyli@imr.ac.cn

Abstract. High density of grain boundaries (GBs) in nanograined metals may become unstable and evolve in various ways such as migration, sliding, or relaxation under thermal or mechanical stimuli. In this report, we will present our recent studies on structural evolution of GBs in nanograined Cu during intensive plastic deformation. As the grain size decreased below 200 nm, GB migration become obvious assisted by dislocation activities. Below 70 nm, accompanied by a transition in dominant deformation mechanism from full dislocations to partials, GB migration decays gradually, while the GB structures relax into lower energy states through their interactions with partials. The population of low-energy GBs increases with a decreasing grain size. Below 10 nm, further refinement of grains results in a violent structural evolution of the GB network into a 3-dimensional periodic minimal surface structure, Schwarz-D, through a phase-transformation-like process, which is termed as Schwarzation.

1. Introduction

Grain boundary (GB) plays an important role in the deformation behaviors and mechanical properties of metals. Usually, in coarse grained and submicron grained metals, GBs are strengtheners by impeding dislocation slip with an ability to absorb and emit dislocations. The more population of GBs, i.e., the smaller the grain size, the higher the strength. However, as grain size decreases to the nanometer scale, high density GBs become unstable, especially in pure metals, resulting in grain coarsening and softening [1]. The poor stability is one of the major bottlenecks of nanograined metals, limiting their processing and applications.

Various GB activities have been reported in nanograined metals, such as GB migration [2], GB sliding [3], GB rotation [4], and so on. The commonly observed GB motion is migration accompanied by concurrent grain coarsening as the nanograined metals are heated or stressed. The thermally-induced GB migration is similar as that subgrain growth and grain growth after recrystallization in the deformed coarse-grained metals, except that it occurs at much lower temperatures than the ordinary recrystallization. Different from that in the coarse-grained metals, GB migration in nanograined metals can also be triggered by mechanical loading. GBs tend to move in a direction perpendicular to the tangent plane of the boundary by applying stress or strain move in response to mechanical loading with concomitant coarsening of nanograins, as evidenced by experiments and by atomistic simulations in a number of nanograined metals and alloys under various loading conditions such as tension [5], compression [6], and indentation [7], even at cryogenic temperatures. The mechanically induced GB migration leads to apparent softening and deterioration of the tensile ductility of the nanograined metals.

Besides migration, other GB motions may also operate in nanograined metals to accommodate plastic deformation, such as GB sliding, generating a relative displacement of neighboring grains and GB rotation, changing the misorientation of neighboring grains [3].

From another point of view, dislocation motions in nanograined metals are also alerted from that in their coarse-grained counterparts. According to atomistic simulations of Yamakov et. al., the splitting distance of dislocation increases with a decreasing grain size, which corresponds to an increase of applied stress on the grains [8]. The deformation mechanism will transfer from full dislocation slip to partial dislocations when grain size is below a critical size that is equal to the dislocation splitting distance, which depends on the resolved shear stress and stacking faults energy. Swygenhoven et al. further pointed out that the dislocation splitting distance is affected by both the stable and unstable stacking faults energy [9]. Experimentally, a large number of twins were observed in deformed nanograined Al [10] and Ni [11], while coarse-grained Al and Ni with relatively high stacking fault energies always deformed by full dislocations without twinning.

The GB behaviors and the deformation mechanism alteration are regarded as fundamental features distinguishing the nanograined metals from their coarse-grained and ultrafine-grained counterparts.

Recently, we observed autonomous GB modification and relaxation [12] in Cu and other face centered cubic (FCC) nanograined metals below a critical grain size. Different from other GB activities mentioned above, GB relaxation leads to a structure modification of GBs to lower energy states and enhances the thermal stability and mechanical stability of nanograined metals significantly [12,13]. Continuously reducing the grain size to extremely fine range (< 10 nm), along with the GB relaxation, the nanograined structures transform into a metastable state where GBs network evolves into a triple periodic minimal surface (TPMS) topology [14-16]. We referred the metastable structure to as "Schwarz crystal" after the mathematician who discovered the TPMS in 1865, and the transformation as "Schwarzation" correspondingly.

In this paper, the deformation behavior in Cu below 200 nm and the plastic deformation induced GB activities in Cu are reviewed with a decreasing grain size, from GB migration to GB relaxation, and finally to the Schwarzation process.

2. GB migration (< 200 nm)

By means of the surface mechanical grinding treatment, a gradient nanograined surface layer was prepared on a coarse-grained pure Cu rod, which was employed to investigate the grain size dependence and the orientation dependence of the mechanically-induced GB migration in the nanograined structures. The average grain size is about 40 nm in the topmost surface. It increases gradually with depth from the treated surface to about 100 nm at 100 μm depth. A tension strain of about 30% was applied on the gradient nanograined sample. It is found that under tension deformation, as grain size decreases from submicron, GB migration appears at about 200 nm, accompanied with apparent grain coarsening. Below 200 nm, GB migration intensifies with a decreasing grain size to about 70 nm. The grains with an initial size of 75 nm coarsen to about 135 nm in size after tension [17].

Although a few mechanisms have been proposed for the mechanically induced GB migration, why GB migration becomes active in grains smaller than 200 nm and decays below 70 nm remains ambiguous. It is assumed that in nanograined metals, GB migration prefers to operate to accommodate plastic strain only when dislocation motions are suppressed and less effective in carrying plastic strain in nanograins below a critical size. However, the traditional coupling model and the dislocation model of GB migration in coarse-grained metals suggest that GBs migrate through obvious atomic shuffling in the boundary region and coordinate motion of dislocations. During GB migration, the normal motion of a GB is accompanied by a relative in-plane translation, which is deduced theoretically to occur by generation and motion of GB dislocations. The relationship between GB migration and dislocations needs to be clarified.

Our investigation on the orientation dependence of GB migration shows that GB migration is more significant in the nanograins with greater Schmid factors where dislocations are easier to slip, indicating that dislocation motion plays a key role in mechanically induced GB migration. Compared with the as-prepared samples, more grains with higher Schmid factors present in the as-tensioned samples, while the number of grains with lower Schmid factors decreases obviously [17]. That is, during GB migration, dislocations emitted from GBs are prone to slip into grains with higher Schmid factors, causing the grains to grow simply under the action of stress.

The preferred dislocation motion in specific grains cause the variation of the texture of sample with the <111> component reduce from 23.4% in the as-prepared samples to 8.53% in the as-tensioned, while the <001>texture components increase from 4.51% to 15%. The interaction of those dislocations with GBs during GB migration also leads to changes of GB characteristics. After GB migration, the fraction of high angle GBs decreases and the fraction of low angle GBs between 2° and 15° increased from 8.5% in the as-prepared sample to 16.6% in the as-tensioned. It implies that GB migration can relax the GB structure to some extent and reduce the total excess energy.

3. GB relaxation (< 70 nm)

In the gradient nanograined sample, when the grain size is smaller than 70 nm, GB migration and consequent grain coarsening decrease at smaller grain sizes (Figure 1). The average grain size in the topmost layer increases slightly from 44 nm to 48 nm after tension with a uniform elongation of 31%, in contrast to the apparent coarsening in the subsurface layer. Many through-grain twins and stacking faults were detected in the deformed grains with initial sizes below 70 nm and the statistic twin density increases tremendously after tension [13].

It is indicated that dislocations and GB migration in larger grains below 70 nm is substituted by formation of twins and stacking faults for accommodating strains. The transition of deformation mechanism can be inferred from the classical dislocation theory. In terms of Orowan relation, full dislocation multiplication requires that Frank-Read-type sources have a minimum grain size (D^*) at the yield strength (σ_y), i.e., averagely full dislocation bowing out is inhibited in grains smaller than D^*

$$D^* = \frac{\mu b}{\sigma_y(D^*)/m} \quad (1)$$

where μ is the shear modulus, b is the Burgers vector of dislocation, σ_y is the stress applied on the grains, $1/m$ is the average Schmid factor for polycrystals (1/3). Based on the measured yield strength of the nanograined metals, the calculated critical size D^* in Cu is 70 nm. Therefore, in Cu below 70 nm, full dislocation and GB migration are suppressed while partials are activated forming twins or stacking faults.

During the plastic deformation process, formation of through-grain stacking faults or twins by emitting partial dislocations from and terminating at GBs can induce GB relaxation to lower energy states via dissociation processes. An obvious drop in GB energy is noticed as grain sizes decease below about 100 nm, the measured specific GB excess energy for grain size of ~50 nm is about 0.23-0.27 J/m^2, only a half of conventional GB energy for Cu of 0.52±0.03 J/m^2.

Consequently, not only the mechanical stability but also the thermal stability of the nanograined Cu is enhanced significantly. As grain size is below 70 nm, contrary to the "smaller less stable" trend commonly observed, the apparent grain coarsening temperature (T_{GC}) increases with a decreasing grain size. Grains with sizes of 70 nm exhibit the lowest thermal stability with a T_{GC} of 393 K, about 0.28 T_m (T_m is the equilibrium melting point of Cu). In contrast, T_{GC} of 50-nm grains with relaxed GBs is about 623 K (0.46 T_m), even higher than the conventional recrystallization temperature of deformed coarse-grained pure Cu (0.4 T_m) [12].

GB relaxation is observed not only in Cu, but also in other FCC metals with higher stacking fault energies (such as Ni and Al) and metals with lower stacking fault energies (Ag). The GB relaxation found in the nanograined metals indicated that below a critical grain size, the dislocation motion changes and interacting with dislocations will also change the GB state simultaneously [18,19]. The extensively-existing GB structure modification and GB relaxation in nanograined metals prepared by plastic deformation can be utilized to develop stable nanostructured metals and alloys for high temperature applications.

4. Schwarzation (<10 nm)

Further plastic deformation induces continuous grain refinement accompanied by progressive GBs relaxation. So, what is the ultimate state of the nanograined structure with continuing relaxation of GBs is a problem worth exploring. From the conventional wisdom, one may expect a transformation from polycrystalline structures into an amorphous state when grains are extremely refined to a limit, say, few nanometers. Indeed, amorphization has been observed in Ni-P, Ni-W and other metallic alloys with extreme refinement of grains. However, this inference is based an assumption that GBs remain disordered in structure and their population increases constantly with a decrease of grain size. However, with GB relaxation, GB structures change, leading to an elevated thermal stability. The thermal stability of grains with size of 50 nm in Cu (coarsening at 623 K) is much higher than the structure stability of amorphous Cu alloys (crystallization temperature of about 450-500 K). It is hard to imagine that further GB relaxation and stabilization will lead to amorphization. Analysis from the GB relaxation process, there are hints for the ultimate state of GBs: more and more low energy GBs, a certain amount of twin boundaries and a very high thermal stability.

We cut the top surface layer with fine nanograins from the as-prepared gradient nanograined samples and deformed it further by using high pressure torsion at cryogenic temperature. The Cu sample after torsion consists of extremely fine grains with an average size of about 5 nm. No amorphous phase was detected. A manifold structure is observed in which irregularly shaped aggregates connected to each other. The aggregates are made up of individual grains of a few nanometers in size and truncated octahedron in shape (Kelvin polyhedron). A large amount of Σ3 twin boundaries (~25%) and other low-Σ CSL boundaries (~20%) were detected, underlying the typical structures of relaxed GBs.

We then performed a MD simulation by constructing an extended Kelvin supercell with 16 truncated-octahedral-shaped grains of equal size and twin boundaries embedded in it. After relaxation and annealing, or deformation, the GBs in the extended Kelvin polycrystal developed into a form that topologically resembles a TPMS - Schwarz D-surface through extensively partial motions and GB reaction and migration events [14]. The transformation from Kelvin polyhedron to Schwarz D is thermodynamically driven by further minimizing the total boundary area, which means that the formation of the TPMS structure is an energetically more-favorable GB relaxation process under thermal and mechanical stimuli [15].

Cu with Schwarz-D crystal structure exhibits a superior thermal stability and mechanical stability. No apparent grain coarsening occurs up to temperatures near the melting point (1348 K). The nanohardness is as high as 4.7 ± 0.2 GPa, equivalent to a yield strength of 1.57 GPa (~72% of the ideal shear strength of Cu at 0 K). The pronounced stability originates from the D minimal surface topology of the GB network. The mean curvature equals to zero everywhere in the surface, indicating that there is no drive force for grain boundary migration according to the grain boundary migration velocity equation (2) deduced from Laplace-Young equation [20].

$$V = M_0 e^{(-\frac{Q_m}{RT})} \cdot H \cdot \gamma \tag{2}$$

where M_0 is the GB mobility, Q_m is the activate energy for GB migration, R is the gas constant, T is the temperature, H is the curvature of GB, which is zero in minimal surface and γ is the GB energy. In addition, there is no triple junction or qudra-pole, which is highly mobile in minimal-surface-GBs.

The inherent stability observed in the Schwarz-D structured Cu with extremely fine grains is consistent with the stability trend of nanograined structures with GB relaxation in the grain size range of tens of nanometers. It means that the Schwarzation can be regarded as an extension of the GB relaxation. Thus, we may suggest that the nanograined metals in which GB relaxation operates would transform into the Schwarz crystal structure eventually with an extreme grain refinement.

Besides high thermal stability and high strength, the unique topologic minimal surface type GB network may bring other fundamental changes into the behavior of metals under mechanical or thermal stimuli. We prepared supersaturated Al-15 wt. % Mg alloy with the Schwarz crystal structure by high pressure torsion and found that several diffusion-controlled processes including precipitation of intermetallic phase, grain coarsening, and even melting are all suppressed in the alloy [16]. After annealing at different temperatures in the precipitation-phase zone below 700 K, the single-phase zone at 700-736 K or even the melting zone above 736 K, no precipitation and apparent grain coarsening were detected. Partial melting was found as annealed above 805 K, 69 K higher than the solidus temperature of the alloy. The calculated diffusion rate is only about 1.5×10^{-20} m^2s^{-1}, several orders of magnitude lower than that in the coarse-grained and nanograined sample. It is revealed that the Schwarz crystal structure is effective in suppressing atomic diffusion in the alloy. Other behaviors and properties related to dislocations, GBs and diffusion in Schwarz crystal metals are expected to be distinct from those in conventional coarse-grained and nanograined ones.

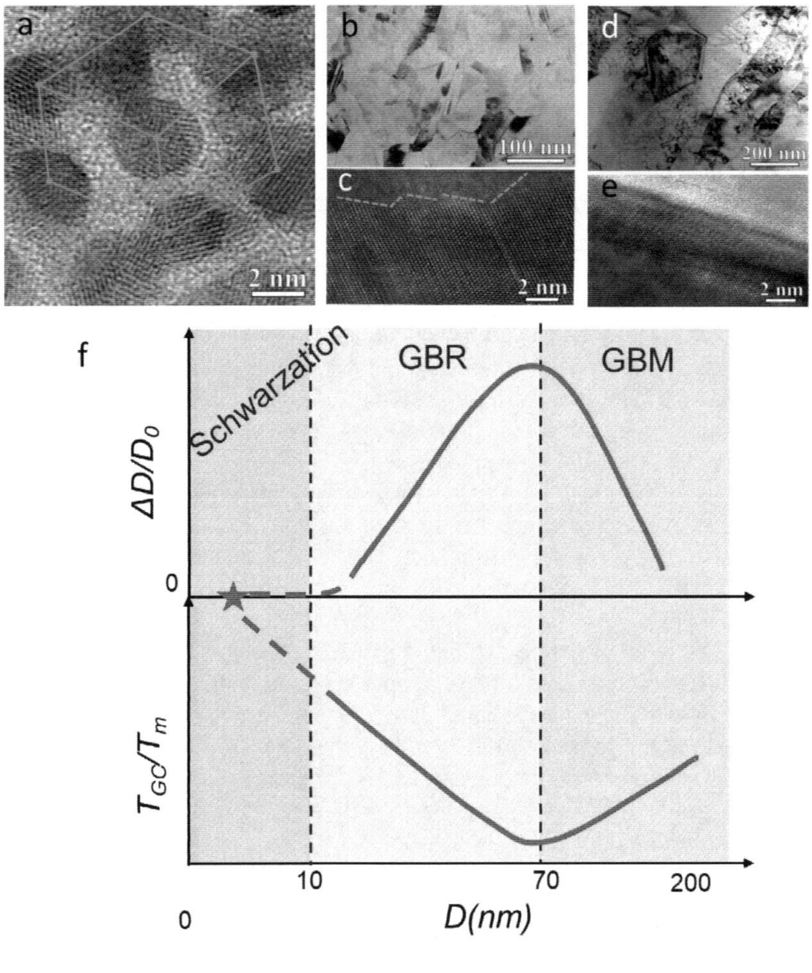

Figure 1. Variation of grain boundary activities with grain sizes in nanograined Cu samples and the corresponding stabilities. (a) Schwarzation: Schwarz crystal with a grain size of about 5 nm; (b) and (c) GB relaxation in 40 nm grains with through-grain stacking faults and twins and a relaxed grain boundary showing facet characteristics; (d) end (e) GB migration in 150 nm nanograined Cu with dislocations and a normal grain boundary; (f) the mechanical and thermal stabilities of nanograins in three ranges with different grain boundary activities, where $\Delta D/D_0$ represents the relative grain size change under same strain, inverse to mechanical stability and T_{GC}/T_m represents the thermal stability by the relative apparent grain coarsening temperature.

5. Summary

In this paper, our recent progress in the plastic deformation behaviors, especially the activities of grain boundaries in Cu below 200 nm has been reviewed. When the grains are smaller than 200 nm, grain boundary migration becomes pronounced in the nanograins with active dislocation motions. Below 70 nm, grain boundary migration is suppressed and grain boundaries relax autonomously to low-energy states due to their interactions with the activated partial dislocations during deformation, resulting a significant enhancement of stability of nanograins. Further refining the grains to a few nanometers with straining, a new metastable structure Schwarz crystal is formed where the grain boundaries evolve into a three-dimensional periodic minimal surface constrained by twin boundary networks. The Schwarz crystal structure in Cu exhibits a high stability close to the equilibrium melting point and a high strength close to the theoretical value.

Acknowledgments

We acknowledge the financial support of the Ministry of Science & Technology of China (Grant Nos. 2017YFA0700700 and 2017YFA0204401).

References

[1] Lu K 2016 *Nat. Rev. Mater.* **1**, 16019
[2] Zhang K, Weertman J R and Eastman J A 2004 *Appl. Phys. Lett.* **85** 5197
[3] Swygenhoven H. Van and Derlet P M 2001 *Phys. Rev. B* **64**, 224105
[4] Wang L, et. al 2014 *Nat. Commun.* **5** 4402
[5] Rupert T J, Gianola D S, Gan Y and Hemker K J 2009 *Science* **326** 1686
[6] Brandstetter S, Zhang K, Escuadro A, Weertman J R and Swygenhoven H Van 2008 *Scr. Mater.* **58** 61
[7] Jin M, Minor A M, Stach E A and Morris J W 2004 *Acta Mater.* **52** 5381
[8] Yamakov V, Wolf D, Phillpot S R, Mukherjee A K and Gleiter H 2003 *Nat. Mater.* **3** 43
[9] Swygenhoven H Van, Derlet P M and Frøseth A G 2004 *Nat. Mater.* **3** 399
[10] Liao X, Zhou F, Lavernia E J, He D and Zhu Y 2003 *Appl. Phys. Lett.* **83** 5062
[11] Wu X and Zhu Y 2008 *Phys. Rev. Lett.* **101** 025503
[12] Zhou X, Li X and Lu K 2018 *Science* **360** 526
[13] Zhou X, Li X and Lu K 2019 *Phys. Rev. Lett.* **122** 126101
[14] Li X, Jin Z, Zhou X and Lu K 2020 *Science* **370** 831
[15] Jin Z, Li X and Lu K 2021 *Phys. Rev. Lett.* **127** 136101
[16] Xu W, Zhang B, Li X and Lu K 2021 *Science* **373** 683
[17] Hou J, Li X and Lu K 2021 *J Mater Sci Technol* **68** 30
[18] Li X and Lu K 2021 *Acc. of Mater. Res.* **2**, 108
[19] Zhou X, Li X and Lu K 2020 *Scr. Mater.* **187** 345
[20] Peng H, Gong M, Chen Y and Liu F 2017 *Int. Mater. Rev.* **62** 303

IOP Conf. Series: Materials Science and Engineering 1249 (2022) 012014 doi:10.1088/1757-899X/1249/1/012014

Revisiting the phase transformations involving Cu_6Sn_5 intermetallic: resolving local domain structures induced by ordering

S Martin[1], A Winkelmann[2] and A Leineweber[1]

[1] Institute of Materials Science, TU Bergakademie Freiberg, Germany
[2] Academic Centre for Materials and Nanotechnology, AGH University of Science and Technology in Kraków, Poland

E-mail: stefan.martin@iww.tu-freiberg.de

Abstract. Cu_6Sn_5 is one of the most important intermetallics present in solder joints based on Cu and Sn. Due to still uncertain reasons for failure of these joints, detailed microstructural research can help to figure out the underlying mechanisms. For Cu- and as well for Sn-rich environments of the Cu_6Sn_5 the microstructural occurrence of the phase transformations are investigated. During the formation of stable η' from the disordered high temperature η phase the formation of metastable modulated incommensurately ordered η'' takes place. For the copper rich samples long-range ordering and twin-related domain formation within η'' was observed. Although conventional Hough-based EBSD analysis was not able to identify the correct domain orientation, Kikuchi pattern matching using simulated patterns was able to resolve the correct microstructure. As ordered intermetallics tend to be brittle in nature, the current microstructure research provides a basis to investigate unresolved problems in the initiation of failure.

1. Introduction

The soldering process is based on the fact that alloys exhibiting a low melting temperature are heated in order to connect electrical conductors, e.g. copper wires, with the solidifying solder material during their re-solidification. With its low melting point of approx. 232 °C, tin is an ideal metal for soldering, which forms various intermetallic phases with transition metals showing higher melting temperatures involving peritectic and peritectoid reactions, to increase the strength of the interconnect. In combination with Cu the Cu_6Sn_5 "intermetallic" is, besides Cu_3Sn, one of the most important intermetallic compounds occurring in the microstructure upon soldering. Note that by using the terms Cu_6Sn_5 or Cu_6Sn_5 *intermetallic* we summarize all *phases* having approximately that composition and closely related crystal structures. Polymorphic phase transformations between such Cu_6Sn_5-related phases [1] and associated stresses due to changes of the specific volume of the associated phases throughout production or during life time of the joints are often assumed to be the reason for failure [2]. The involved phase transformations, are driven by order-disorder transitions.

During the formation of stable Cu_6Sn_5 metastable states involving modulated incommensurately ordered η'' and commensurately modulated ordering in η' occur [3]. Especially the order formation of the η'' phase from the disordered high temperature state has not been investigated in detail and was just recently being elucidated [4]. This ordering phase transformation involves reduction of symmetry

(*6/mmm* → *2/m*) and is prone to show a domain formation to accommodate the transformation stresses. In the present paper it is possible to make these domains visible by the adequate processing of EBSD Kikuchi pattern by pattern matching, as conventional Hough-based EBSD is not able to resolve the subtle differences from different domain orientations reliably. In concert with the description of the microstructure changes as precipitation of Cu_3Sn and interdiffusion processes during annealing of previously quenched high temperature states, a comprehensive image is drawn to understand the mechanisms during the phase formation of the Cu_6Sn_5 intermetallic. Thereby, the present paper interconnects the results which have been published in Refs. [3–5]. To give an overview on the phase transformations which are involving the Cu_6Sn_5 intermetallic a separate chapter, following next, is dedicated to highlight recent findings.

2. Peculiarities of the phase transformation in Cu_6Sn_5

The widely accepted binary Cu-Sn phase diagram [6] is shown in figure 1(a) and depicts the Cu_6Sn_5 compound as an disordered high temperature phase η and an ordered low temperature phase η′ with a composition dependent transition temperature in between 189 °C to 186 °C. The range of homogeneity of the Cu_6Sn_5 intermetallic has recently been reassessed [3], as the Sn-content was underestimated as shown in figure 1(b).

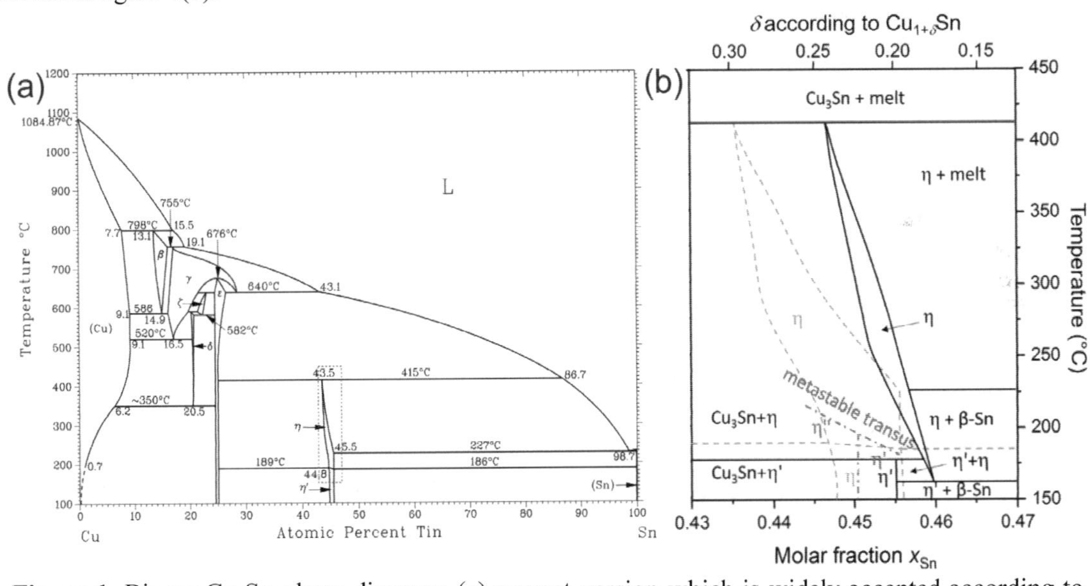

Figure 1. Binary Cu-Sn phase diagram: (a) current version which is widely accepted according to Refs. [6, 7]; (b) modified excerpt from (a) (see red dashed box) with incorporation of the metastable transformation temperatures referring to the reassessed diagram (solid lines) (according to [8, 9]).

The disordered high temperature phase shows a NiAs/Ni_2Sn-type structure with *P6₃/mmc* space-group symmetry, based on a NiAs-type substructure Cu(1)Sn as well as additional Cu(2) on trigonal-bipyramidal interstices. Those interstices are occupied with δ Cu(2) atoms and $1-\delta$ vacancies. This results in an effective formula of $Cu(1)Cu(2)_\delta Sn$ or $Cu_{1+\delta}Sn$ (compare figure 1(b)). The η″ phase exhibits a monoclinic superstructure [10], due to the ordering of δ Cu(2) atoms vs. $1-\delta$ vacancies. To describe the crystal structure a *C*-centered unit cell is employed with a 10 times higher volume compared with the hexagonal subcell of the disordered η phase with an ideal superstructure at $\delta = 0.2$, yielding Cu_6Sn_5. Whereas the thermodynamic stable η phase is assumed as a line compound [9], metastable states appear to develop with a higher copper content during low temperature annealing experiments [11] (see figure 1(b)). Some previously reported long-range ordered structure models of Cu-rich Cu_6Sn_5 (e.g. η^8

and η^{4+1}) , were interpreted to be approximant descriptions of the incommensurate η'' phase with continuously varying structure parameters [3, 8].

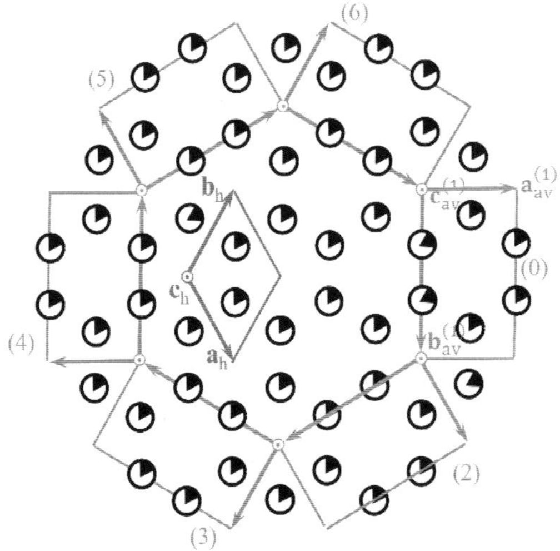

Figure 2. Crystallographic relation between the disordered hexagonal high temperature η phase (red coordinate system - subscript h) and the six resulting orientations of η'' described by the average structure (light blue) model. Only the Cu(2) sites are shown (with their partial occupation corresponding to *δ*) forming a hcp-like arrangement.

As described in more detail in Ref. [4], due to loss of rotational symmetry in particular the 6-fold rotational symmetry of the high-temperature η, there should coexist 6 different orientational variants of the η'' phase (c.f. figure 2). Only a slight unit-cell distortion and the weak diffraction effects due to the incommensurate ordering of the Cu(2) atoms vs. vacancies will result for the individual variants. Hence, it is expected that all kinds of diffraction patterns, including EBSD Kikuchi pattern will show severe pseudosymmetry, which have to be overcome by suitable evaluation methods to make the domain structure visible.

Model structure	Space group	Lattice parameters	Comment
η - disordered	*P*6₃/*mmc*	$a = 4.22$ Å, $c = 5.11$ Å	Conventional description of disordered η phase
η'' - average structure	*C*2/*c*	$a_{av} = 4.22$ Å, $b_{av} = 7.32$ Å, $c_{av} = 5.11$Å, $\beta_{av} = 90.23°$	Crystallographic description of the incommensurate η'' structure, which considers the average distortions of structure due to ordering, but not the actual displacive and occupational order, see also[8].
η^{4+1} approximant structure		$a'_m = 81.137$ Å, $b'_m = 7.314$ Å, $c'_m = 9.848$ Å, $\beta'_m = 93.145°$	*Approximant* structure of the modulated description of η'' [8]; structure corresponding to transformed unit cell of η^{4+1} superstructure formulated in Ref. [12].

Table 1. Crystal structure models used to describe the specific phases of the Cu_6Sn_5 intermetallic for the evaluation of the XRD and EBSD data.

3. Experimental Details

To investigate the homogeneity range of the η/η′ phase, the microstructure evolution during ordering and the formation of the stable phases, alloys of $Cu_{57}Sn_{43}$ and $Cu_{52}Sn_{48}$ target composition were prepared mainly by melting and homogenization of pure elements at 800 °C for 24 h and cooled by water quenching in sealed fused silica tubes described in detail in Ref. [8]. The same tube was then heated to 380 °C for 120 h followed by rapid quenching in ice water, to obtain defined homogenised states and equilibrated microstructures of η + ε phase and η + β-Sn phase. The homogenized alloys were disintegrated with a mortar into smaller pieces, which were then resealed for additional heat treatments in a salt bath (240 °C) or in an oil bath at lower temperatures.

To investigate the phase composition, to measure the lattice parameters, and to determine the degree of ordering of the Cu_6Sn_5 intermetallic, X-ray diffraction (XRD) analysis was conducted from powder that was obtained by crushing the heat-treated specimens with an agate mortar. A Bruker D8 ADVANCE diffractometer was used equipped with a Co tube and a quartz-crystal Johansson monochromator in the primary beam supplying monochromatic Co-$K\alpha_1$ radiation and a LYNXEYE position-sensitive detector. The Rietveld technique applied in TOPAS software was used to evaluate the data, using independent pseudo-Voigt functions for each phase and taking the specimen displacement error into account. For evaluation of XRD patterns from the η phase, the corresponding disordered hexagonal structure model was used. This was also done, if very broad satellite reflections occurred, e.g. as an early stage of η″ formation, due to insufficient quenching. If the presence of the η″ phase was indicated by well-established satellite reflections, a disordered structure model with a monoclinic *average* unit cell was used for evaluation of the XRD patterns (see table 1), thus ignoring the relatively weak satellite reflections due to the incommensurate ordering in the course of the refinements. Note that quantitative evaluations of XRD patterns from the η″ phase under consideration of the satellite reflections has been reported in previous works [3, 8].

Alloy pieces were cold-embedded in epoxy resin in order to avoid further heating of the samples for SEM analysis. Final polishing was done using colloidal silica (40 nm) to obtain sufficient surfaces for back-scattered electrons imaging and electron backscatter diffraction (EBSD) measurements. A JEOL JSM-7800F field emission gun-SEM was employed, equipped with EDAX energy-dispersive X-ray analysis (EDX) and EBSD systems. Primary electrons were accelerated using 20 kV, resulting a probe current of approximately 7 nA.

Backscattered electron (BSE) contrast imaging was used to obtain microstructural information regarding the local phase distribution, grain size and shape of the η″ domains. EBSD measurements with a 2 × 2 pixel binning of the Hikari Super camera were conducted. For post processing and pattern matching all Kikuchi patterns were stored. The structure model of β-Sn with $I4_1/amd$ space group symmetry and a = 5.83 Å and c = 3.18 Å was used, whereas the ε-Cu_3Sn phase was identified with a disordered hexagonally closed packed structure with $P6_3/mmc$ space group symmetry and a = 2.76 Å and c = 4.32 Å, thus avoiding pseudosymmetry issues due to its orthorhombic superstructure.

To evaluate the domain structure of the η″ phase, pattern matching was used, as conventional EBSD was not able to distinguish the individual orientations [5]. A two-step approach was used to resolve the pseudosymmetry issue. A spherical Kikuchi pattern master reference was simulated for the η^{4+1} approximant structure, to account for the description of the ordering in η″ applying a Bloch wave based method [13]. In the first step the orientation was determined roughly with a resolution of approximately 2° for binned patterns with 80 × 80 pixels. The orientation was refined in a second step, taking into account the orientation from the first step as starting parameter under full resolution 480 × 480 pixel of the pattern. As the pseudosymmetric solutions are eventually not yet identified correctly, the diffraction effects due to the incommensurate Cu(2) ordering were considered. That was achieved by using the "conventional" 3-dimensional periodic η^{4+1} approximant structure model, a variant of what has been reported previously [11]. Simulated Kikuchi patterns were always considered for all 6 possible pseudosymmetric orientations, differing by multiples of 60° rotation around the $[001]_{av}$ direction (c.f. figure 2). Therefore, for every data point of an EBSD map six orientations are matched. The one with the highest normalized cross correlation coefficient (NCCC) was treated as final solution. The details are described in Ref. [4].

4. Results and discussion

To investigate the microstructure changes and to observe phase transformations involving metastable η'' during the formation of the stable η/η′ phase of the Cu_6Sn_5 intermetallic in detail, annealings at distinct temperatures several 10 K above the η′ transformation temperature were conducted. The microstructure of the as-melted samples (800 °C for 24 h) in figure 3(a) shows mainly the presence of the ε-Cu_3Sn, the η phase and remaining β-Sn for both compositions ($Cu_{57}Sn_{43}$ and $Cu_{52}Sn_{48}$). According to the phase diagram (figure 1) the γ phase could have formed as the first crystalline phase to a small amount. But due to the efficient quenching no γ was detectable in the as-melted microstructure. The ε-Cu_3Sn forms instead during quenching to a large amount followed by the peritectic reaction ε-Cu_3Sn + Sn(l) → η. Quenching terminates this phase transformation before thermodynamic equilibrium was achieved. To get closer to equilibrium the samples were homogenized at 380 °C for 120 h followed by quenching and the resulting microstructures are shown in figure 3(b,d). These states are characteristic representatives of the disordered high temperature η phase. The $Cu_{57}Sn_{43}$ sample (figure 3(b)) shows grains of equiaxed η as majority phase and elongated ε-Cu_3Sn, whose shape arises from the primary dendritic solidification of ε-Cu_3Sn. The $Cu_{52}Sn_{48}$ sample (figure 3(d)) instead shows large globular η crystallites and β-Sn in between, which was liquid at the treatment temperature. Due to the interaction with liquid Sn, larger crystallites are forming compared with the Cu-rich samples. Within the solidified remaining melt, small dendrites of η can be found (not shown in the image), following the eutectic reaction upon solidification of the liquid. During annealing at lower temperatures, the η phase should get enriched by tin [3]. For the two chosen compositions this is achieved by distinct processes. Within the copper-rich samples ($Cu_{57}Sn_{43}$) ε-Cu_3Sn is precipitated at grain and phase boundaries (figure 3(c)), and, therefore, η is enriched with Sn. For the Sn-rich samples ($Cu_{52}Sn_{48}$) Sn is diffusing into η, as shown in figure 3(e), where an EDS line scan is superimposed onto an η grain showing the inwards diffusion of Sn after 30 h annealing time at 240 °C [5].

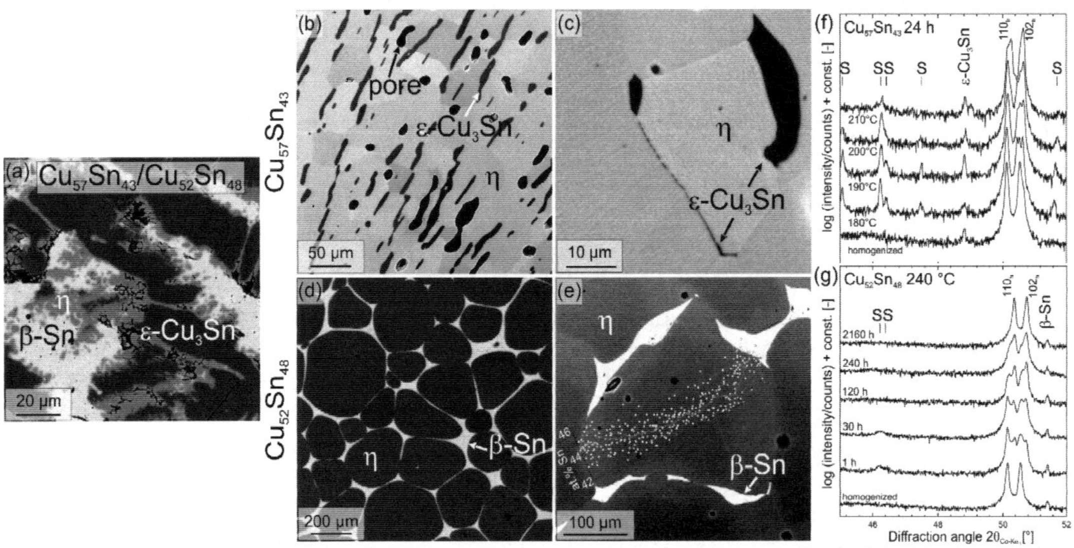

Figure 3. SEM micrographs of different heat-treated states:(a) as-melted (800°C for 24 h + quenching); (b) $Cu_{57}Sn_{43}$ homogenized (380 °C for 120 h + quenching); (c) $Cu_{57}Sn_{43}$ homogenized + 240 °C for 1h + quenching; (d) $Cu_{52}Sn_{48}$ homogenized (380 °C for 120h + quenching), (e) $Cu_{52}Sn_{48}$ homogenized + 240 °C for 30h + quenching and superimposed EDS line scan data of the local Sn content of an η-grain embedded in β-Sn; XRD pattern from the annealed samples of the: (f) Cu-rich $Cu_{57}Sn_{43}$ 24 h for isochronous treatments at various temperatures showing η″ order and ε-Cu_3Sn formation and (g) Sn-rich $Cu_{52}Sn_{48}$ for isothermal treatments at various times showing a Sn-enrichment of the η phase. Data was reproduced from Ref.[5].

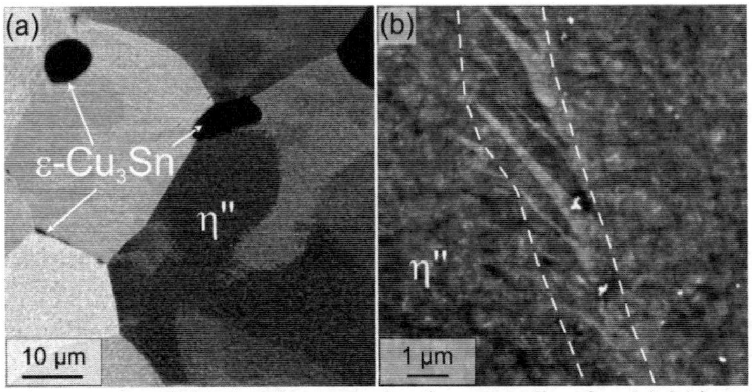

Figure 4. SEM micrographs of the $Cu_{57}Sn_{43}$ sample after annealing for: (a) 24 h at 180 °C showing the pronounced domain formation; (b) 24 h at 150 °C depicting the discontinuous domain interface (delineated by dashed lines) between two domains.

The process of diffusion in the $Cu_{52}Sn_{48}$ is also remarkably shown by the interpretation of the XRD data (figure 3(g)) for isothermal heat treatments at 240 °C. The increasing shift of the diffraction maxima at increasing holding times are related to the diffusion of Sn from the β-Sn grains into η. The observed increasing intensity (max. intensity) of the shifted 110_h and 102_h maxima can be interpreted as distinct saturation of the Sn diffusion for small and large crystallites (see distinct grain sizes in figure 3(d)) of the η phase. These could be treated as an individual Sn-poor and Sn-rich phase [5]. Smaller η grains have reached their equilibrium Sn-content faster and they represent intensity which was shifted at shorter times and vice versa for the larger grains. Small intensities dedicated to an η″ superstructure after 1 h and 30 h annealing time indicate an insufficient quenching and a higher tendency towards ordering for Cu-rich compositions. For the Cu rich samples ($Cu_{57}Sn_{43}$) the same treatment (240 °C for various hours) would result in continuous shift of the diffraction peaks (not shown here) to higher angels and an increase of maxima related to Cu_3Sn phase [5]. An isochronous heat treatment (24 h at various temperatures) is shown here instead, elucidating the process of order-formation (figure 3(g)). Besides the expected increase in intensity of the Cu_3Sn maxima, the pronounced superstructure reflections indicate the formation of the η″ phase due to ordering. The splitting of 102_h reflection into three peaks with approximately the same intensity (indices $11\overline{2}_{av}$, 022_{av}, 112_{av}) are characteristic features of the ordering process accompanied with lowering the symmetry.

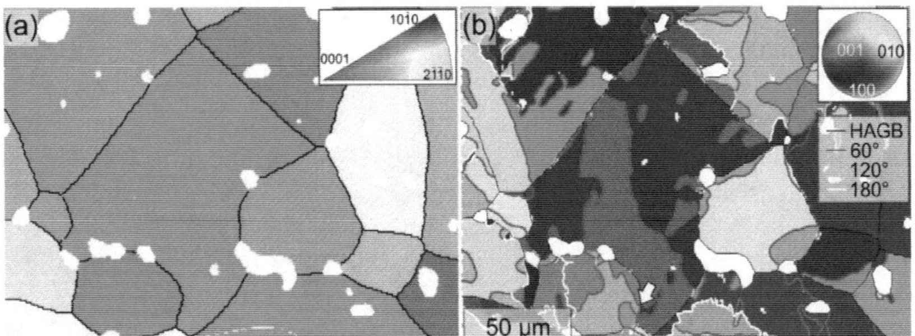

Figure 5. EBSD characterisation of the η″ microstructure: (a) orientation map of a η″ domain structure after 180 °C for 24 h using the disordered high temperature model η for Hough based indexing, reveals the initial η high-angle grain boundaries (HAGB) highlighted by black lines; (b) orientation map of an η″ domain structure resolving the correct domain structure by employing the pattern matching algorithm for the η^{4+1} model as closest approximant for the incommensurate η″ structure. Images are adapted and modified from [4].

The potential domain formation during the ordering process marking the η→η″ transformation is observed in the BSE images shown in figure 4. In figure 4(a) intensity variations within the former η grains indicate the different domains, which are otherwise often associated with the formation of subgrains. The fuzzy appearance of the domain boundaries in figure 4(a) is underlined by investigations at higher magnification in figure 4(b) showing a faceting at the interface. Trace analysis of these ledges revealed that the facets are related to strain free interfaces between the different domain orientations [4].

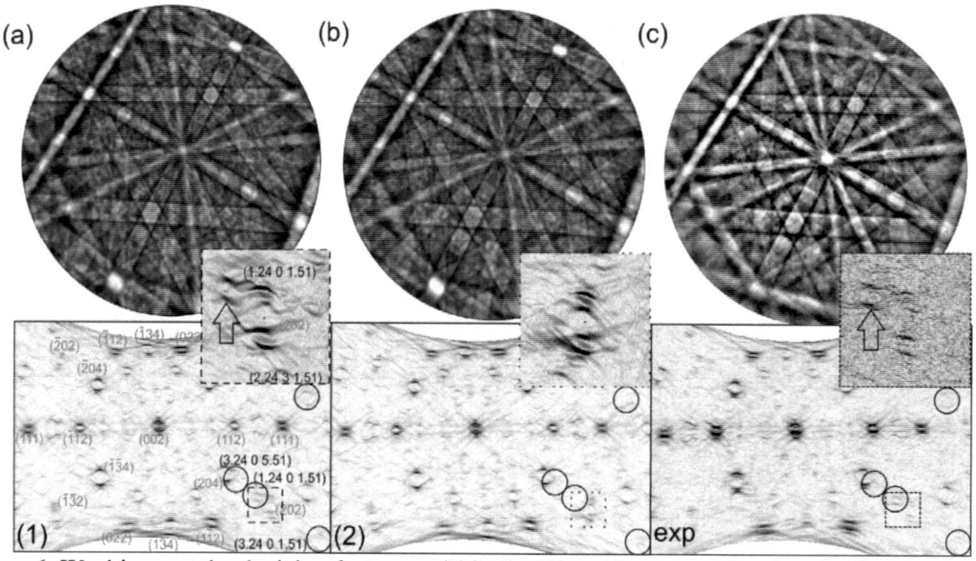

Figure 6. Working out the decisive features within the Kikuchi pattern allowing for the correct domain identification: (a) simulated pattern for the 0° variant; (b) simulated pattern for the (60°) variant; (c) corresponding experimental pattern; corresponding Funk transform of the pattern (a)-(c) with insets from the signatures from the most prominent superstructure bands in the vicinity of the signature of the (202) band. (a)-(c) reproduced from [4].

To determine the orientation of these η″ domains EBSD analysis was conducted in several ways. If indexing was done assuming the undistorted η phase, it is shown, that BSE intensity variations observed in figure 4(a) for example are not visible in the orientation spread and, therefore, cannot be related to subgrain boundaries within the former η grains separated by high angle boundaries of (figure 5(a)). Indexing using the average structure (table 1), as model structure for the description of the η″ phase, did not yield any reliable orientation determination. The detected orientation changes did not coincide with the intensity distribution of the BSE images (not shown here). The distortion of the unit cell apart from parent η structure is too small to discriminate the different variants. Further, by using the structural model of the average structure (table 1), the superstructure reflections are not considered in the reflector list. Nevertheless, the intensity of the superstructure bands would be too weak within the EBSD patterns to be detected by the conventional Hough based algorithm, as discussed later. Consequently, pattern matching employing the method described in the experimental section was employed, to work out the subtle differences. By the use of the η^{4+1} structural model (c.f. table 1) for the η″ phase, the intensity of the weak bands from ordering is included in the simulated pattern and can be compared with the measured intensities. As result a clear distinction of the individual domains is shown in figure 5(b), where the initial parent η grains are populated with up to six variants with uniform orientation. Also imaged BSE intensities were reproduced accordingly [4]. 60°, 120° and 180° misorientations around the $[001]_h/[001]_{av}$ can be found as misorientations between neighboring domains (figure 5(b)), representing twin relation for that system, adopting the relations from figure 2.

Although the results seem straight forward, a verification whether the correct orientations were identified is lacking. Therefore, the Kikuchi pattern were screened for the presence of bands due to order formation. In figure 6(a,b) two potential solutions for fitting the experimental pattern (figure 6(c)) are presented. The inspection by eye does not resolve clear differences. Consequently, the Funk transform of the pattern [14] was analyzed for presence of differences or similarities between the experimental and the 6 simulated pseudosymmetric solutions. In case of the presented orientation in figure 6(a), the correct Kikuchi pattern shows the same maxima at the same places in Funk space (see black circles and compare the enlarged view of the inset for the $(1.24\ 0\ 1.51)_{\eta4+1}$ band near the 202 band) as they are observed in the experimental one. The simulated pattern in figure 6(b) represents the simulated orientation that is rotated 60° around $[001]_h/[001]_{av}$, where the intensities from the bands from ordering are located at different areas of the Funk transform. For all inspected solutions of the mapping shown in figure 5(b), these positions were matching with the correct solution from the pattern matching approach. Although the decisive parameter ($0 < NCCC < 1$) for the evaluation of the correct solutions did only vary about 0.03, for total values in the order of 0.7, this difference was sufficient to achieve the correct solutions.

5. Conclusion

From the presented results by annealing of the disordered high temperature η phase at lower temperatures, several microstructural processes and reactions during stable and metastable phase transformations of the Cu_6Sn_5 intermetallic are elucidated. In a Sn-rich surrounding η enriches gradually by the diffusion of Sn. Within the Cu-rich samples, the Sn enrichment of η is facilitated by the formation of Cu_3Sn at the grain boundaries and at pre-existing Cu_3Sn. Further ordering was observed yielding a distortion of the hexagonal η crystal structure, which is described as η''. This distortion enables the proposed and observed domain formation. By the use of pattern matching the correct identification of the domain orientation was enabled, which was not successful using the conventional Hough based methods. Having developed the possibility of the correct domain identification, potential failure analysis in Cu-Sn solder joints can be revisited for possible domain/domain or domain/grain interactions.

Acknowledgement
A. Winkelmann acknowledges the support from Polish National Science Centre (NCN) grant number 2020/37/B/ST5/03669. For sample production and metallographic preparation M. Loeffler, B. Wahl and D. Hübgen are acknowledged.

References

[1] Ghosh G and Asta M 2005 J. Mater. Res. 20 3102–17
[2] Mu D K, McDonald S D, Read J, Huang H and Nogita K 2016 Curr. Op. Sol. Stat. Mater. Sci. 20 55–76
[3] Leineweber A, Wieser C and Hügel W 2020 Scr. Mater. 183 66–70
[4] Martin S, Winkelmann A and Leineweber A 2022 Acta Mater. 229 117828
[5] Leineweber A, Löffler M and Martin S 2021 J. Electr. Mater. 50 5898–914
[6] Saunders N and Miodownik A P 1990 Bull. Alloy Phase Diagr. 11 278–87
[7] Massalski T B and Okamoto H (eds) 1990 Binary alloy phase diagrams (Materials Park, Ohio: ASM International)
[8] Leineweber A, Wieser C and Hügel W 2020 Z. Kristallogr. 235 445–57
[9] Wieser C, Hügel W, Walnsch A and Leineweber A 2020 J. Electr. Mater. 49 245–56
[10] Larsson A-K, Stenberg L and Lidin S 1994 Acta Crystallogr. B 50 636–43
[11] Leineweber A 2011 Int. J. Mater. Res. 102 861–73
[12] Wu Y Q, Barry J C, Yamamoto T, Gu Q F, McDonald S D, Matsumura S, Huang H and Nogita K 2012 Acta Mater. 60 6581–91
[13] Winkelmann A, Trager-Cowan C, Sweeney F, Day A P and Parbrook P 2007 Ultramicroscopy 107 414–21
[14] Nolze G, Tokarski T, Rychłowski Ł, Cios G and Winkelmann A 2021 J. Appl. Crystallogr. 54 1012–22

A phase-field approach to study BCC precipitate evolution in FCC $Al_xCrFeNi$ medium entropy alloy including elastic effects

X Zuo and N Moelans

KU Leuven, Department of Materials Engineering, Kasteelpark Arenberg 44, box 2450, 3001 Leuven, Belgium

E-mail: xiaojing.zuo@kuleuven.be; nele.moelans@kuleuven.be

Abstract. A phase-field approach is presented to simulate the evolution of a BCC precipitate in an FCC matrix for $Al_xCrFeNi$ based medium entropy systems. The CALPHAD composition dependence of the Gibbs energies and diffusion mobilities for this quaternary alloy are introduced in the model using ThermoCalc with TCHEA2 and MOBNI4 databases. Heterogeneous elastic properties are assumed for the FCC and BCC phase computed for the quaternary system. The effect of the elastic stresses due to lattice mismatch on precipitate shape, evolution and matrix composition is presented for a single BCC precipitate in and FCC matrix. We find that the elastic effects are considerable.

1. Introduction

Research on high-entropy alloys (HEAs) and multicomponent alloys (MCAs) has become very popular today. High-entropy alloys have shown to have, amongst others, unique mechanical properties [1,2]. Moreover, the high number of elements gives wide flexibility in tailoring the microstructure towards specific properties and even combinations of properties. For example, Yang et al. [3] investigate the effect of Al on the microstructure and mechanical properties of $Al_xCoCrFeNi$ alloys. Their results show that with the increase of the amount of Al, the microstructure changes from a single-phase FCC structure to a two-phase FCC+BCC structure, and, for even higher Al-content to a two-phase BCC(A2+B2) structure. The hardness of the alloy increases with Al-content, while ductility decreases. Furthermore, to reduce the cost due to the high cobalt content in the $Al_xCoCrFeNi$ based alloys, the microstructure and properties of Co-free $Al_xCrFeNi$ multicomponent alloys have also been studied [4,5]. It is shown that Co-free $Al_xCrFeNi$ alloys can also give an excellent combination of strength and ductility, as well as improved wear and corrosion resistance.

While experimental studies have thus successfully proposed new and more advanced alloys, further optimization of these alloys based on experiments only, tends to be costly and time-consuming, due to the large composition space. The further development and application of computational tools towards multicomponent alloys is therefore receiving an increasing amount of attention. Simulations may allow us to study the microstructure evolution behavior as a function of various parameters in a more efficient way, before selecting and conducting experiments. However, there are still difficulties with multicomponent alloy simulations. For example, it is not straightforward to obtain an accurate representation of the thermodynamic and elastic properties of the different phases in a phase-field model applied to the phase transformation behaviors in HEAs.

The influence of the misfit strain occurring at the interface changes the morphology of the phases and precipitates. This is well studied for binary systems. However, less work is reported on the investigation of elastic effects for multicomponent alloys. To the authors' knowledge, only Li et al. [6] studied coherent precipitation in BCC/B2 microstructures in AlNiCoFeCr HEA through phase-field modeling including elastic effects. However, a simplified second order composition dependence was used to represent the thermodynamic free energies of the phases and a same constant diffusion mobility for two phases was assumed while the elastic constants were selected based on literature data for binary alloys.

The aim of this paper is to present initial results using a phase-field approach investigating microstructure evolution in the presence of elastic effects due to lattice misfit including CALPHAD data on the composition dependence of the thermodynamic and diffusion mobilities. Zuo et al. [7,8] recently introduced a phase-field model for multicomponent precipitate growth and coarsening taking into account in a highly efficient way the CALPHAD Gibbs energy and diffusion mobility data for the quaternary alloy. In the current paper, elastic effects are added to this model description and initial simulations considering a single BCC precipitate in an FCC matrix are presented showing the effects of the elastic mismatch on precipitate shape, precipitate volume fraction and composition in the matrix as a function of time. Especially for this last effect, it is important to describe the realistic composition dependence of the Gibbs energies of the phases in the model. In the next section, the model and parameter choice is described. In the third section, the simulation results are presented and discussed.

2. Phase-field model and parameter values

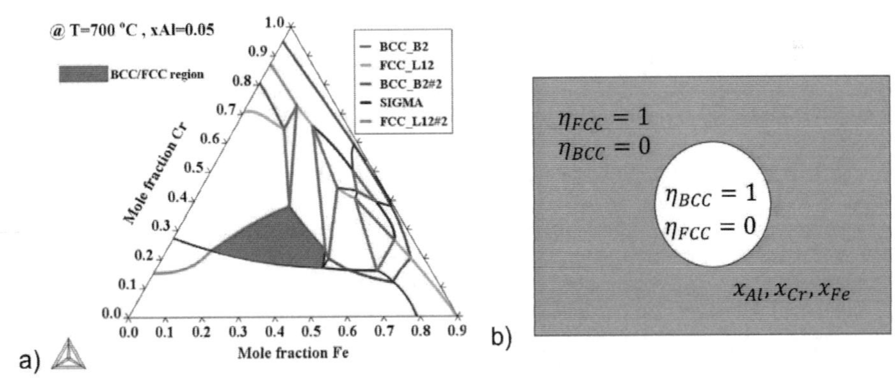

Figure 1. a) AlCrFeNi phase diagram section at T=700°C and x_{Al}=0.05 as computed using Thermo-Calc with the TCHEA2 database. The two-phase region BCC+FCC considered in the simulation is indicated in red. b) Schematic representation of the system considered in the simulation.

2.1. Model

In the simulations, a BCC precipitate in an FCC matrix is considered, assuming a quaternary AlCrFeNi alloy. The two-phase region is indicated in the red area of the phase diagram section in Figure 1 a). Figure 1 b) gives a schematic representation of the simulation system considering the evolution of a BCC precipitate in an FCC matrix. Two non-conserved order parameters are needed to represent the FCC and BCC phases, namely $\eta_{BCC}(x,y,t)$ and $\eta_{FCC}(x,y,t)$ (strictly speaking, we could have used a single order parameter distinguishing between two phases; however, to allow for extension towards three-phase and polycrystalline microstructures, a multi-phase field formulation is adopted). Within the BCC phase, η_{BCC}=1 and η_{FCC}=0, while in the FCC phase, η_{BCC}=0 and η_{FCC}=1. In the phase-field approach interfaces are assumed to have a finite width. At the interface between the two phases, the order parameters show a gradual transition between the values 0 and 1 over a narrow region. The local composition in the system is represented by the local mole fractions of Al, Cr, and Fe, using three conserved mole fraction fields $x_{Al}(x,y,t)$, $x_{Cr}(x,y,t)$ and $x_{Fe}(x,y,t)$. The mole fraction of Ni is considered as the dependent variable and is obtained as $x_{Ni}=1-x_{Al}-x_{Cr}-x_{Fe}$. A 2-dimensional system is considered, with x and y the coordinates along the x- and y-direction and t, time.

In a phase-field model, the evolution equations of the order parameters and mole fraction fields are derived from a free energy functional. The total free energy F of the system consists of contributions from interface energy, bulk chemical energy and elastic energy, and is formulated as a function of the order parameters representing the different phases η_{BCC} and η_{FCC} and the mole fractions as:

$$F = \int_V \frac{k}{2}[(\nabla\eta_{FCC})^2 + (\nabla\eta_{BCC})^2] + mf_0(\eta_{BCC},\eta_{FCC}) +$$
$$h_{FCC}(\eta_{BCC},\eta_{FCC})f^{FCC}(x_{Al}^{FCC},x_{Cr}^{FCC},x_{Fe}^{FCC}) + h_{BCC}(\eta_{BCC},\eta_{FCC})f^{BCC}(x_{Al}^{FCC},x_{Cr}^{FCC},x_{Fe}^{FCC}) + f^{el}dV.$$

The first two terms represent the contributions to the interface energy, with f_0 a fourth-order Landau polynomial of the order parameters

$$f_0 = \left(\frac{1}{4}\eta_{FCC}^4 - \frac{1}{2}\eta_{FCC}^2\right) + \left(\frac{1}{4}\eta_{BCC}^4 - \frac{1}{2}\eta_{BCC}^2\right) + 1.5\eta_{FCC}^2\eta_{BCC}^2.$$

The model parameters m and k relate to the interfacial energy δ and diffuse interface width l^{int} as [9]

$$k = \frac{3}{4}l^{int}\delta \text{ and } m = \frac{6\delta}{l^{int}}.$$

The next two terms represent the bulk chemical contribution with $f^{FCC} = \frac{G_m^{FCC}}{V_m}$ and $f^{BCC} = \frac{G_m^{BCC}}{V_m}$, the free energy densities as a function of composition, G_m^{BCC} as G_m^{FCC} Gibbs energy functions of the FCC and BCC phase and $V_m = 10^{-5}\ \mathrm{m}^3/\mathrm{mol}$ a reference molar volume (difference in molar volumes of the different phases are considered through the eigenstrains in the elastic contribution). The Gibbs energy functions are obtained fitting higher order polynomial functions through a large set of Gibbs energy data calculated for the composition domain of interest using Thermo-Calc using the TCHEA2 database, as discussed in detail in previous papers [7,8].

The interpolation functions h_{FCC} and h_{BCC} are formulated as [10]

$$h_{FCC}(\eta_{BCC},\eta_{FCC}) = \frac{\eta_{FCC}^2}{\eta_{FCC}^2+\eta_{BCC}^2} \text{ and } h_{BCC}(\eta_{BCC},\eta_{FCC}) = \frac{\eta_{BCC}^2}{\eta_{FCC}^2+\eta_{BCC}^2}.$$

These interpolation functions play a key role at the diffuse interfaces, where they interpolate various material properties between their values in the two phases.

The last term f^{el} represents the elastic contribution to the total free energy, formulated for a multi-phase system and following a micro-elasticity approach [11],

$$f^{el} = \frac{1}{2}\epsilon_{ij}^{el}C_{ijkl}\epsilon_{kl}^{el},$$

where the elastic strain is defined as the difference between the total strain and the eigenstrain, $\epsilon_{kl}^{el} = \epsilon_{kl}^{tot} - \epsilon_{kl}^{*}$. The total strain, ϵ_{kl}^{tot}, is formulated as the sum of the homogeneous strain, $\overline{\epsilon_{kl}}$, and local heterogeneous strain, $\delta\epsilon_{kl}$, $\epsilon_{kl}^{tot} = \overline{\epsilon_{kl}} + \delta\epsilon_{kl}$, with the heterogeneous strain related to the displacement as $\delta\epsilon_{kl} = \frac{1}{2}\left[\frac{\partial u_k(\vec{r})}{\partial r_l} + \frac{\partial u_l(\vec{r})}{\partial r_k}\right]$ and $\overline{\epsilon_{kl}}$, such that $\int_V \delta\epsilon_{kl}\, d^3r = 0$. A mixture model is used for the stiffness tensor, $C_{ijkl} = h_{FCC}(\eta_{BCC},\eta_{FCC})C_{ijkl}^{FCC} + h_{BCC}(\eta_{BCC},\eta_{FCC})C_{ijkl}^{BCC}$, and for the eigenstrain, $\epsilon_{kl}^{*} = h_{FCC}(\eta_{BCC},\eta_{FCC})\epsilon_{kl}^{*,FCC} + h_{BCC}(\eta_{BCC},\eta_{FCC})\epsilon_{kl}^{*,BCC}$. The displacement fields are obtained by imposing mechanical equilibrium, $\frac{\partial\sigma_{ij}}{\partial r_j} = 0$, with $\sigma_{ij} = C_{ijkl}\epsilon_{kl}^{el}$, assuming small strains within the linear elasticity limit.

For the bulk chemical energy contribution, the Kim-Kim-Suzuki [12] approach (KKS) is applied to assure local equilibrium over the diffuse interfaces. Phase compositions $x_i^{FCC/BCC}$ namely, x_{Al}^{FCC}, x_{Cr}^{FCC}, x_{Fe}^{FCC}, x_{Al}^{BCC}, x_{Cr}^{BCC} and x_{Fe}^{BCC}, are introduced. They are functions of the local composition fields,

x_{Al}, x_{Cr}, x_{Fe} and the order parameters η_{BCC} and η_{FC} and are determined at each position, such that for each of the components, the diffusion potential is equal in the coexisting phases, assuming local equilibrium, giving the following set of equations:

$$\frac{\partial G_m^{FCC}}{\partial x_{Al}^{FCC}} = \frac{\partial G_m^{BCC}}{\partial x_{Al}^{BCC}} = \tilde{\mu}_{Al}, \quad \frac{\partial G_m^{FCC}}{\partial x_{Cr}^{FCC}} = \frac{\partial G_m^{BCC}}{\partial x_{Cr}^{BCC}} = \tilde{\mu}_{Cr}, \quad \frac{\partial G_m^{FCC}}{\partial x_{Fe}^{FCC}} = \frac{\partial G_m^{BCC}}{\partial x_{Fe}^{BCC}} = \tilde{\mu}_{Fe},$$

where G_m^{FCC} and G_m^{BCC} are the Gibbs energies of the FCC and BCC phases respectively and $\tilde{\mu}_i$ represents the diffusion potential of component i (i=Al, Cr, Fe). Moreover, the phase compositions $x_i^{FCC/BCC}$ should relate to the overall mole fractions x_i , as

$$x_{Al} = h_{FCC}(\eta_{BCC}, \eta_{FCC}) x_{Al}^{FCC} + h_{BCC}(\eta_{BCC}, \eta_{FCC}) x_{Al}^{BCC}$$
$$x_{Cr} = h_{FCC}(\eta_{BCC}, \eta_{FCC}) x_{Cr}^{FCC} + h_{BCC}(\eta_{BCC}, \eta_{FCC}) x_{Cr}^{BCC}$$
$$x_{Fe} = h_{FCC}(\eta_{BCC}, \eta_{FCC}) x_{Fe}^{FCC} + h_{BCC}(\eta_{BCC}, \eta_{FCC}) x_{Fe}^{BCC}.$$

The evolution equations for the mole fractions, describing the diffusion of the different components, are derived starting from the following equations, which ensure mass conservation:

$$\frac{1}{V_m}\frac{\partial x_i}{\partial t} = \sum_{j=1}^{3} \nabla M_{ij} \nabla \tilde{\mu}_j$$

with x_i the mole fraction of component i, $\tilde{\mu}_j$ the diffusion potential of component j and M_{ij} the diffusion mobilities formulated as an interpolation between the values of the diffusion mobilities of the FCC and BCC phases as

$$M_{ij} = h_{FCC} M_{ij}^{FCC} + h_{BCC} M_{ij}^{BCC}$$

with M_{ij}^{FCC} and M_{ij}^{BCC} the diffusion mobilities of the FCC and BCC phase, respectively. For each phase (FCC/BCC), the interdiffusion mobilities $M_{ij}^{FCC/BCC}$ are taken composition dependent using the expression from [13]

$$M_{ij}^{FCC/BCC} = \sum_{k=1}^{C} (\delta_{ik} - x_i) \sum_{l=1}^{C} (\delta_{jl} - x_j)\, L_{kl}^{FCC/BCC},$$

with $L_{kl}^{FCC/BCC} = \frac{x_l}{V_m} \Omega_l^{FCC/BCC}$ for $l = k$ and $L_{kl} = 0$ for $l \neq k$, $\Omega_l^{FCC/BCC}$ the atomic mobilities of the 4 elements (Al, Cr, Fe and Ni) in the FCC and BCC phase, which are implemented in the phase-field simulation code. In this work, the atomic mobilities $\Omega_l^{FCC/BCC}$ are assumed to be independent of composition, but different in each phase. Their values were calculated using Thermo-Calc software with MOBHEA2 database and assuming the equilibrium composition of each phase. It was verified that the atomic mobilities are indeed almost constant over the considered concentration region and that the composition dependence of the diffusion in this quaternary alloys is well represented [7,8].

For the non-conserved order parameters, the time evolution is calculated applying an Allen-Cahn type of equations,

$$\frac{\partial \eta}{\partial t} = -L_\eta \frac{\delta F}{\delta \eta},$$

with F the total free energy. For each term the derivative with respect to the order parameter has to be made, giving

$$\frac{\partial \eta_{FCC}}{\partial t} = -L_\eta \left[m(\eta_{FCC}^3 - \eta_{FCC} + 3\eta_{FCC}\eta_{BCC}^2) - k_i \nabla^2 \eta_{FCC} + \frac{\partial h_{FCC} f^{FCC}}{\partial \eta_{FCC}} + \frac{\partial h_{BCC} f^{BCC}}{\partial \eta_{FCC}} + \frac{\partial f^{el}}{\partial \eta_{FCC}} \right]$$
$$\frac{\partial \eta_{BCC}}{\partial t} = -L_\eta \left[m(\eta_{BCC}^3 - \eta_{BCC} + 3\eta_{BCC}\eta_{FCC}^2) - k_i \nabla^2 \eta_{BCC} + \frac{\partial h_{BCC} f^{BCC}}{\partial \eta_{BCC}} + \frac{\partial h_{FCC} f^{FCC}}{\partial \eta_{BCC}} + \frac{\partial f^{el}}{\partial \eta_{BCC}} \right]$$

To obtain diffusion controlled-interface migration in a phase-field model, the kinetic coefficient $\boldsymbol{L}_{\eta}$, should be taken as $\boldsymbol{L}_{\eta} = \frac{4m}{3k\zeta}$, where

$$\zeta = \sum_{k=1}^{C-1} (x_{FCC,eq,k} - x_{BCC,eq,k}) \sum_{m=1}^{C-1} (M_{km}^{FCC/BCC} V_m)^{-1} (x_{FCC,eq,k} - x_{BCC,eq,k}) \text{ [10].}$$

2.2. Simulated system and model parameters

Simulations are performed at 700°C. Table 1 lists the overall composition of the system, which is in the two-phase BCC+FCC region (see Figure 1). The equilibrium compositions of the two phases are also listed. The simulation is initialized with the equilibrium composition in the two phases.

Table 1. Equilibrium phases, equilibrium compositions (mole fraction), and equilibrium volume fraction calculated with Thermo-Calc using TCHEA2 database and used to initialize the system.

T (°C)	Phases	Cr	Al	Fe	Ni	Volume Fraction
700	Overall	0.3166	0.05	0.3166	0.3167	
	BCC(EQ)	0.83419	0.00075	0.15739	0.00767	0.23660
	FCC(EQ)	0.15991	0.06491	0.36480	0.41038	0.76340

Gibbs energy data are computed using Thermo-Calc in combination with the TCHEA2 database for high-entropy alloys. The atomic mobilities are computed using the database MOBNI4 and assuming the equilibrium compositions of the phases as given in Table 1, giving $\Omega_{Cr}^{FCC} = 5.3388 \cdot 10^{-23}$, $\Omega_{Al}^{FCC} = 1.93 \cdot 10^{-22}$, $\Omega_{Fe}^{FCC} = 4.54 \cdot 10^{-23}$, $\Omega_{Ni}^{FCC} = 8.57 \cdot 10^{-23}$; $\Omega_{Cr}^{BCC} = 5.56 \cdot 10^{-2}$, $\Omega_{Al}^{BCC} = 2.3 \cdot 10^{-2}$, $\Omega_{Fe}^{BCC} = 2 \cdot 10^{-25}$, $\Omega_{Ni}^{BCC} = 2.13 \cdot 10^{-25}$.

The lattice constants and stiffness tensors are taken as $a^{BCC} = 3.0\,\mathring{A}$, $C_{11}^{BCC} = 632$ GPa, $C_{12}^{BCC} = 283$ GPa, $C_{44}^{BCC} = 632$ GPa, $a^{FCC} = 3.6\,\mathring{A}$, $C_{11}^{FCC} = 240$ GPa, $C_{12}^{FCC} = 117$ GPa, $C_{44}^{FCC} = 185$ GPa. These values were obtained from first principles calculations using VASP (Vienna ab initio simulation package). Special quasirandom structures (SQS) [14] were generated in the Alloy-Theoretic Automated Toolkit (ATAT) to represent the complex multicomponent atomic structure for each phase assuming its equilibrium composition as given in table 1. A detailed description of the calculation method will be given in a future publication, as it is quite lengthy, while the focus of this paper is on the phase-field simulations. The FCC phase is taken as the reference phase for computing the eigenstrain, $\epsilon_{kl}^{FCC,*} = 0$, while the eigenstrain of the BCC phase is computed based on the lattice constants of the two phases as $\epsilon_{kl}^{BCC,*} = \frac{|a_{BCC} - a_{FCC}/\sqrt[3]{2}|}{a_{BCC}}$.

For the remaining simulation parameters, we considered: grid spacing $\Delta x = 1 \cdot 10^{-9}$ m, interface energy $\delta = 1$ J·m^{-2}, system size along each side 10^{-7} m, approximate width of the diffuse interface $l^{int} = 3 \cdot 10^{-9}$ m and kinetic coefficient in the Allen-Cahn equations $L_{\eta} = 1.8374 \cdot 10^{-10}$ m^{3}J^{-1}s^{-1}.

3. Results and discussion

The evolution of a single BCC precipitate in an FCC matrix is simulated for the same initial conditions with and without the consideration of elastic stress. The aim is to analyse the effect of elastic misfit stresses on precipitate shape and stability and on the matrix and precipitate composition for an individual BCC precipitate.

Simulation images representing the BCC precipitate shape and size evolution with and without elastic effect are shown in Figure 2 (a) and (b). The initial precipitate volume and phase concentrations were the same for both cases and are those obtained from the Thermo-Calc equilibrium calculation. Results show that, without elastic effects, the precipitate remains circular and hardly changes in time. In contrast, when elastic effects are considered, the precipitate shape evolves into a cubic form, reflecting the anisotropic stiffness tensors, and the size of the precipitate is largely reduced.

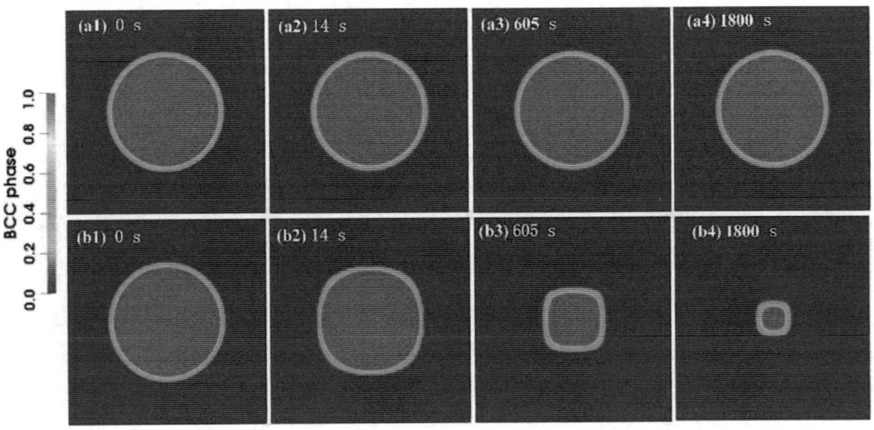

Figure 2. Simulation snapshots of the microstructure evolution of BCC precipitate (a) without and (b) with elastic effects.

The corresponding mole fraction fields for the two cases at 1800 s of simulation time are shown in Figure 3. In the system without elastic effect, the mole fractions in the phases stay close to the equilibrium values. When elastic effects are introduced, there are clear effects on the mole fractions in the two phases.

The stress components σ_{11}, σ_{22} and σ_{12} and the Von mises stress obtained in the simulation with elastic effects at different time steps are shown in Figure 4. For the considered system, the magnitude of the normal stress components and the von Mises stress are highest inside the precipitate. The shear stress reaches its highest value along the diagonal direction. The elastic stress decreases as the precipitate is shrinking.

Chen et al. [15] measured the yield and fracture strength of $Al_xCrFeNi$ alloys for different Al contents. According to their results, we expect a yield strength between 0.7 and 1 GPa (while the fracture strength is around 3-3.5 GPa). In the simulation, the von Mises stresses exceed these values. Further research is needed to find out the reason for this deviation. It is possible that the ab initio calculated eigenstrain or stiffness coefficients are somewhat overestimated, or, possibly, the precipitate-matrix interface is semi-coherent (instead of fully coherent in the current model) making it possible to relax the elastic stresses to a certain extent.

4. Summary and outlook

A phase-field study considering the effect of elastic coherency stresses on a single BCC precipitate evolution in an FCC matrix for $Al_xCrFeNi$ medium entropy alloys is presented. The CALPHAD composition dependence of the Gibbs energies and diffusion mobilities is accurately taken into account in the model. The eigenstrain and stiffness coefficients of the two phases were obtained from ab initio calculations for the quaternary system assuming the equilibrium compositions of the phases. The simulations show that for the computed eigenstrain and elastic constants, the BCC precipitate obtains a cubic shape and continues to shrink during the simulation, indicating that the stability of the BCC phase is drastically reduced due to the coherency strains. Since the von Mises stress measured in the simulations exceeds the estimated yield stress at several positions in the system, it is concluded that plasticity plays a role and that including plasticity into a future model may improve the accuracy.

In the future, this model can be applied considering larger system sizes, a large number of BCC precipitates, and/or several FCC grains. Since lattice parameters may vary with composition, computation of the eigenstrains and stiffness tensors for different compositions and using them in the proposed phase-field method may bring interesting insights in how precipitate shape, size and distribution can be manipulated in high and medium entropy alloys by modifying composition.

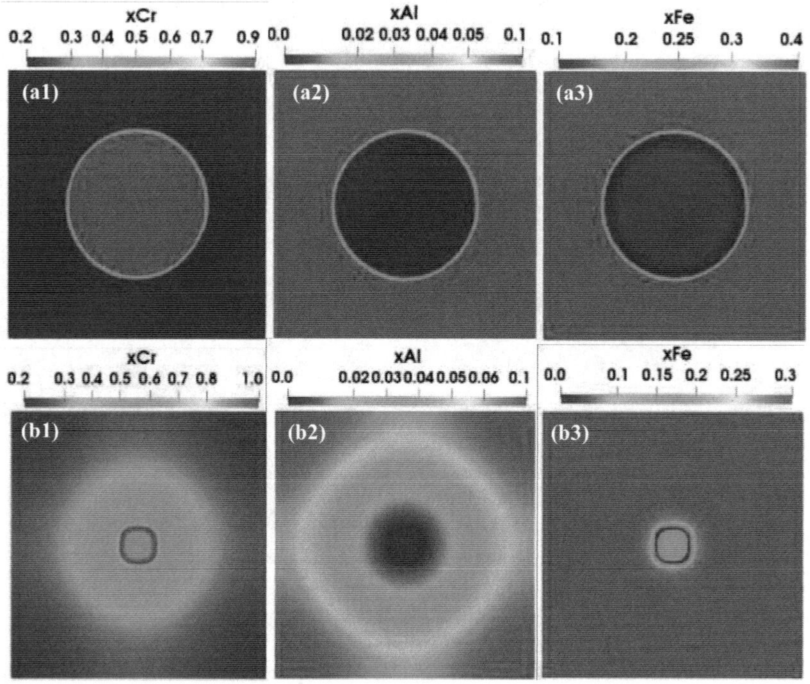

Figure 3. Mole fraction fields at 1800 seconds of simulated evolution time. (a1) x_{Cr}, (a2) x_{Al}, (a3) x_{Fe} the mole fraction fields of respectively Cr, Al and Fe when no elastic stresses are considered. (b1) x_{Cr}, (b2) x_{Al}, (b3) x_{Fe} the mole fraction fields of respectively Cr, Al and Fe when elastic stresses are considered.

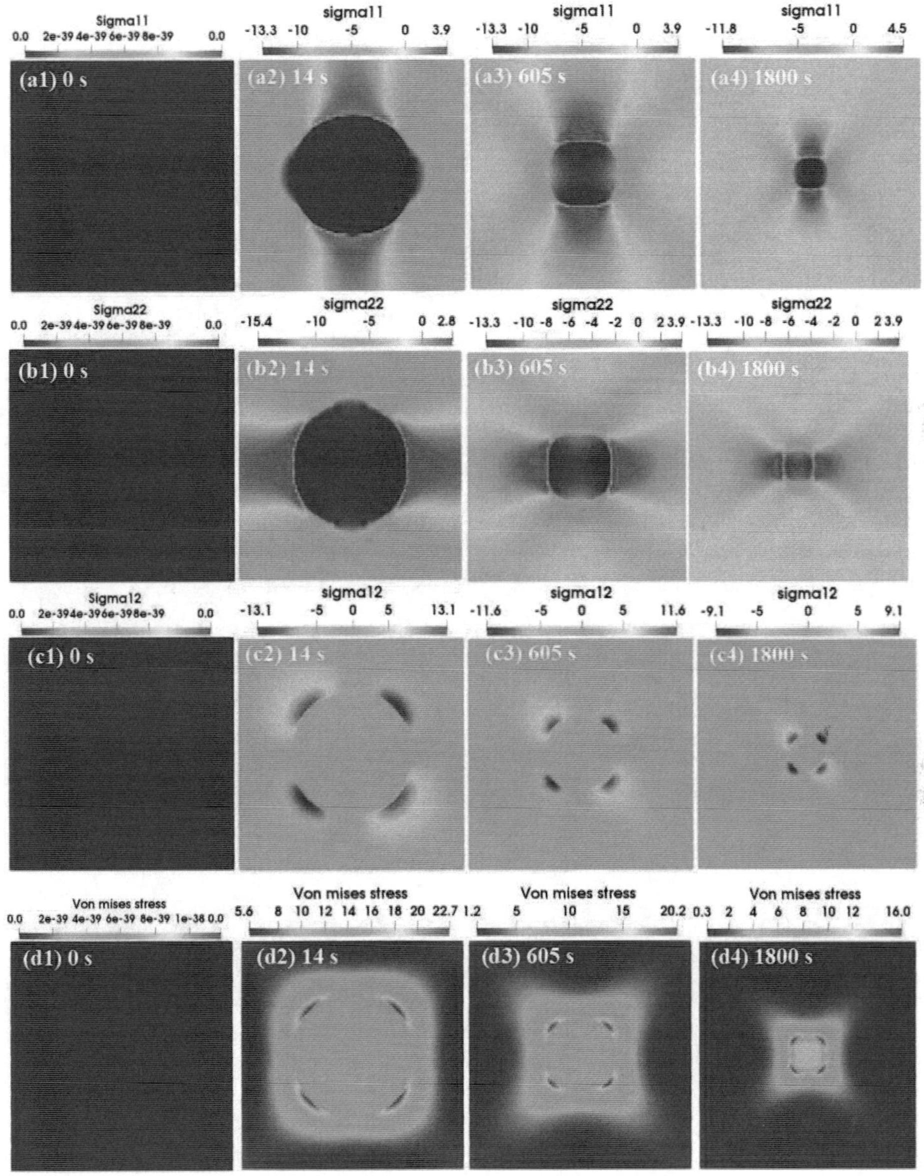

Figure 4. Simulation snapshots of the different stress components a) σ_{11}, b) σ_{22}, c) σ_{12} and d) von Mises stress, expressed in GPa.

References

[1] Yeh J W, Chen S K, Lin S J, et al. 2004 *Adv. Eng. Mater.* **6** 299
[2] Cantor B, Chang I T H, Knight P and Vincent A J B 2004 *Mater. Sci. Eng.* A **375-377** 213
[3] Yang T, Xia S, Liu S, et al 2015 *Mater. Sci. Eng.* A **648** 15
[4] Shang C Y and Wang Y 2017 *MSF* **898** 628
[5] Wu H, Xie J, Yang H-Y, et al. 2022 *J. Mater. Eng. and perform,*

https://doi.org/10.1007/s11665-021-06563-w
[6] Li J L, Li Z, Wang Q, Dong C and Liaw P K 2020 *Acta Mater.* **197** 10
[7] Zuo X J, Miotti Bettanini A, Hilhorst A, et al 2021 *J. Phase Equilib. Diffus.* **42** 794
[8] Zuo X J, Coutinho Y, Chatterjee S and Moelans N 2022 *Mater. Theory* **6** 12
[9] Moelans N, Blanpain B and Wollants P 2008 *Phys. Rev.* B **78** 024113
[10] Moelans N 2011 *Acta Mater.* **59** 1077
[11] Durga A, Wollants P and Moelans N 2020 *Acta Mater.* **188** 241
[12] Kim S G, Kim W T and Suzuki T, 1999 *Phys. Rev.* E 60 **7**186
[13] Andersson J and Agren J 1992 *J. Applied Phys.* **72** 1350
[14] van de Walle A, Tiway P, de Jong M, et al. 2013 *Calphad* **42** 13
[15] Chen X, Wi J Q, Sui Y W et al. 2017, *Mater. Sci. Eng.* A **681** 25

IOP Conf. Series: Materials Science and Engineering 1249 (2022) 012016 doi:10.1088/1757-899X/1249/1/012016

Origin of strain localization at twin boundary in Inconel 718 superalloy fabricated by laser powder bed fusion

X C Li[1], Y N Wu[1], R Yang[1,2] and Z B Zhang[1]

[1] Center for Adaptive System Engineering, ShanghaiTech University, Shanghai 201210, China
[2] Institute of Metal Research, Chinese Academy of Science, Shenyang 110016, China

E-mail: zhangzhb1@shanghaitech.edu.cn

Abstract. Microscale strain evolution and mechanistic origin of strain localization of Inconel 718 superalloy fabricated by laser powder bed fusion was investigated. Strain maps obtained by in-situ high resolution digital image correlation (HRDIC) demonstrate that localized dislocation activities and severe strain localization occur along the coherent twin boundaries (TBs). Such localized dislocation plasticity along TBs exhibits no dependency on the elastic anisotropy and the length of TBs, and the determining factor of the occurrence of localization is the maximum Schmid factor of the TBs. By employing correlative characterizations of TBs, γ'' precipitate and dislocation slip bands, it is found out that the precipitation of V-shaped γ'' along TBs incurs the formation of precipitates denuded zones, which takes the responsibility for localized plasticity and severe strain localization at TBs.

1. Introduction

Inconel 718 Ni-based superalloy (hereafter abbreviated as IN718) is a typical chromium-rich precipitation-hardened alloy with a high strength and high corrosion resistance at high temperatures, so it is widely used in turbine engines, nuclear and petrochemical plants [1]. Traditional routes to manufacture IN718 components require excessive machining and tooling, which is still a challenge in particular for components with complex shapes [2]. Laser-based additive manufacturing, such as laser powder bed fusion (LPBF), has the advantages of integrated manufacturing with high freedom in the component's shape [3]. Most of Ni-based superalloys with precipitates strengthening, however, have a low formability using additive manufacturing due to their poor weldability. Fortunately, IN718 is demonstrated to be one of the most suitable commercial alloys for additive manufacturing because of its low content in aluminum and titanium [4].

Although crack-free IN718 components can be facilely manufactured by LPBF, its practical applications are still hampered by its special microstructure and distinct mechanical behavior compared to its cast and wrought counterpart, which originate from the very high cooling rate, steep thermal gradient and complex thermal cycles during LPBF [4, 5]. The as-built IN718 by LPBF is usually characterized by a strong texture along building direction, submicron scale cellular structure and micro-segregation of Nb, Ti, Cr along cell walls [6-8]. Elemental segregations lead to the formation of brittle Laves phase and δ phase, which consumes the γ'' forming elements and reduces the ductility [7-9]. Therefore, subsequent solution treatment is required to homogenize the chemistry, dissolve the brittle phases and diminish the texture before aging treatment. Although LPBF IN718 can develop a microstructure similar to that produced by cast and wrought after homogenization and aging treatment, its fatigue properties are still found to be inferior to the wrought IN718 alloy [4].

For polycrystalline Ni-based superalloys, strain localization at grain boundaries, in particular at coherent twin boundaries (TBs), is frequently reported as one of the dominant reasons for crack initiation [10-13]. This phenomenon and related mechanism have been extensively explored in Ni-based superalloys strengthened by γ' precipitate, but it rarely known in γ''-strengthened superalloys. As a γ''-strengthened Ni-based superalloys, IN718 alloy readily develops a high fraction of TBs during casting and annealing due to its low stacking fault energy [14]. Although the as-built IN718 by LPBF does not contain many TBs because of the extremely fast cooling rate during solidification, considerable TBs may generate during recrystallisation when IN718 alloy is treated above 1000 °C for homogenization [15, 16]. Therefore, it is of high necessity to examine whether TBs in such LPBF fabricated IN718 alloy promote strain localization and uncover the underlying mechanism accounting for it. These understandings are supposed to be crucial to optimize the mechanical properties by tailoring the grain boundary distribution via tuning the parameters during LPBF and subsequent annealing processes.

In this study, in-situ high resolution digital image correlation (HRDIC) was carried out to investigate the strain evolution and source of strain localization during uniaxial loading of LPBF-fabricated IN718 alloy after annealing. Severe strain localization was detected along TBs and the underlying reason was revealed via correlative microscopy characterizations.

2. Experimental

Commercial Inconel 718 alloy powders was supplied by the SLM Solutions Group AG. Laser powder bed fusion (LPBF) was conducted on a SLM Solution 280 Twins machine with two 400 W fiber lasers. The main parameters for the LPBF process are: laser power 300 W, scanning speed 1300 mm/s, layer thickness 0.04 mm, hatching distance 0.09 mm, stripe scanning strategy with 67° rotation. Vertical blocks with dimensions of 12×12×120 mm^3 were built within an argon atmosphere. Solution treatment at 1080 °C for 3 hours and a subsequent two-step aging at 720 °C for 8 hours and 620 °C for 8 hours were conducted on the as-built samples. Scanning electron microscope (SEM) was used to characterize the microstructure and electron backscatter diffraction (EBSD) was used to obtain the orientation maps and texture of the as-built and annealed IN718 alloy. Both EBSD and SEM were carried out on a Tescan Mira4 FEG SEM equipped with the Symmetry EBSD detector and with Aztec analysis system (Oxford Instruments).

In-situ high resolution digital image correlation (HRDIC) was performed using a Kammrath&Weiss GmbH mini-tester in the Tescan Mira4 SEM. HRDIC experiment is consist of sample preparation, SEM imaging at different strain steps and data analysis. Flat dog-bone sample (gauge geometry of 25 × 3 × 1 mm) for HRDIC was ground and polished with a quality for EBSD. Nanoscale gold speckles were prepared on the sample via gold sputtering and gold remodeling. More details about the HRDIC technique and analysis can be found in previously reports by the group from the University of Manchester [17, 18]. SEM images (11×8 frames) covering an area of 500×375 μm^2 (50 × 50 μm^2, 2048 × 2048 pixel for each frame) were acquired before and after tensile deformation to different strains, and these SEM images were processed using La Vision DaVis software and strain field data was then generated. Cross correlation of the images prior to and after deformation was done with a final interrogation window size of 39 × 39 pixels and a step size of 10 pixels. Effective shear strain maps were obtained using the DaVis software with the protocol proposed by the Manchester group [17, 18]. The DefDap python script was used to link the strain map and EBSD data [19].

3. Results and discussion

Figure 1 shows the microstructure of the IN718 alloy in the as-built condition (figure 1a-c) and after subsequent homogenization at 1080 °C for 3 hours and aging treatments (figure 1d-f). Columnar structure along the building direction (BD) is observed in the as-built IN718 alloy (figure 1a). A strong cube texture developed (figure 1b) during LPBF with <100> crystallographic directions parallel to BD and transverse directions (TD). This is resulted from the epitaxial growth of <100>-oriented columnar cells across the building layers along the thermal gradient. Backscattered electron (BSE) image in figure 1c demonstrates that the as-built IN718 is consist of sub-micron cells, and elemental segregation at cell walls is evident, which is consistent with other reports [6,7]. After annealing treatments, the columnar structure is replaced by nearly equiaxed microstructure, as seen in the orientation maps (figure 1a and figure 1d), which indicates the occurrence of recrystallization during solution treatment. The {100} pole figures (figure 1b and figure 1e) show that recrystallization completely removes the initial cube texture, and the cellular structure and chemical micro-segregation are also eliminated (figure 1f). Moreover, the characters of grain boundaries dramatically change during recrystallization (figure 1b and figure 1e): the high fractions of low angle grain boundaries disappear, whereas grain boundaries with misorientation angle near 60° become dominant (65.6%) in the sample after annealing. These boundaries are identified as Σ3 coherent twin boundaries (TBs).

In-situ HRDIC was then performed to record the strain maps and visualize the strain evolution on sub-grain scale during uniaxial loading. Effective shear strain maps of the sample after tensioning to four different plastic strains, namely 0.30%, 0.42%, 0.63%, and 0.82% (ε_{xx} obtained from the strain maps), are presented in figure 2a-d. Orientation data of this region before tensioning was obtained by EBSD, so grain boundaries are overlaid on the strain map. The strain maps indicate that dislocation slip in IN718 alloy exhibits high planarity and high localization, which leads to the formation of discrete and straight dislocation slip bands (DSBs). Such slip characters are known to be resulted from its low stacking fault energy and the dispersed γ'' precipitates in IN718 [20, 21]. With the increasing of external strain, more DSBs generate, and the existing slip bands develop higher shear strain and become longer if it was not constrained by grain boundaries yet. Some straight DSBs develop much higher strain than their adjacent regions, and by correlating with the grain boundary map it is found out that these DSBs with severe strain localization are mostly along TBs. Strain intensity profiles across several TBs of four selected regions (marked as I, II, III and IV in figure 2a-e) are shown in figure 2f. Due to the fact that the maximum applied plastic strain is only 0.82%, effective shear strains of the areas without visible DSBs are very low (less than 0.2%) in most cases. The shear strain along TBs are up to two orders of magnitude higher than that of adjacent regions without DSBs, for instance in the case of region I, and the degree of strain localization increases when the external strain increases from 0.30% to 0.82%. Some low strain peaks appear in the line profile when the strain reaches 0.82%, which are associated with the DSBs in the grain interior, but these strains are considerably lower than that along TBs.

Obviously, the strain localization along TBs is a result of localized dislocation activities at TBs. In order to uncover the origin of such localized dislocation plasticity along TBs, we first performed statistical analyses on the relationship between the formation of DSBs and characters of TBs. Three characters of TBs are supposed to be taken into considerations, namely the lengths of TBs, the maximum Schmid factors associated with each TB, and the elastic anisotropy across TBs. The maximum Schmid factor of a TB is calculated in the following way: the TB is along one {111} plane, which is the dislocation slip plane, and for each {111} plane there are three possible <110> Burgers vectors; the Burgers vector with the highest Schmid factors is assumed to be activated, and the value is the maximum Schmid factor associated with this TB. The elastic anisotropy of a TB is the difference in Young's modulus across the TB when the crystal is uniaxially loaded along specific directions. Stinville and his coworkers studied the role of TBs in strain localizations in some γ'-strengthened Ni-based superalloys and concluded that the elastic anisotropy across TBs (i.e. difference) plays a significant role in the strain localization and cracking at TBs [11, 12]. We tensioned a sample to a strain of ~1% and analyzed whether the development of DSBs along TBs has any dependency on the TBs characters. A representative region for this analysis and statistic results are shown in figure 3. It demonstrates that the initiation and non-initiation of DSBs along TBs are independent of the elastic anisotropy across TBs and the length of TBs. The only factor that affects the tendency of DSBs formation along TBs is the maximum Schmid factor of the TBs. Therefore if a TB is oriented with a high Schmid factor with respect to the loading direction, localized dislocation slip and severe strain localization will occur along the TB, irrespective of the elastic anisotropy and the length of the TB. This is in contrary to the findings in γ'

strengthened Ni-based superalloys by Stinville et al [11, 12], which indicates the role of precipitates, i.e. γ' or γ'', might be significant.

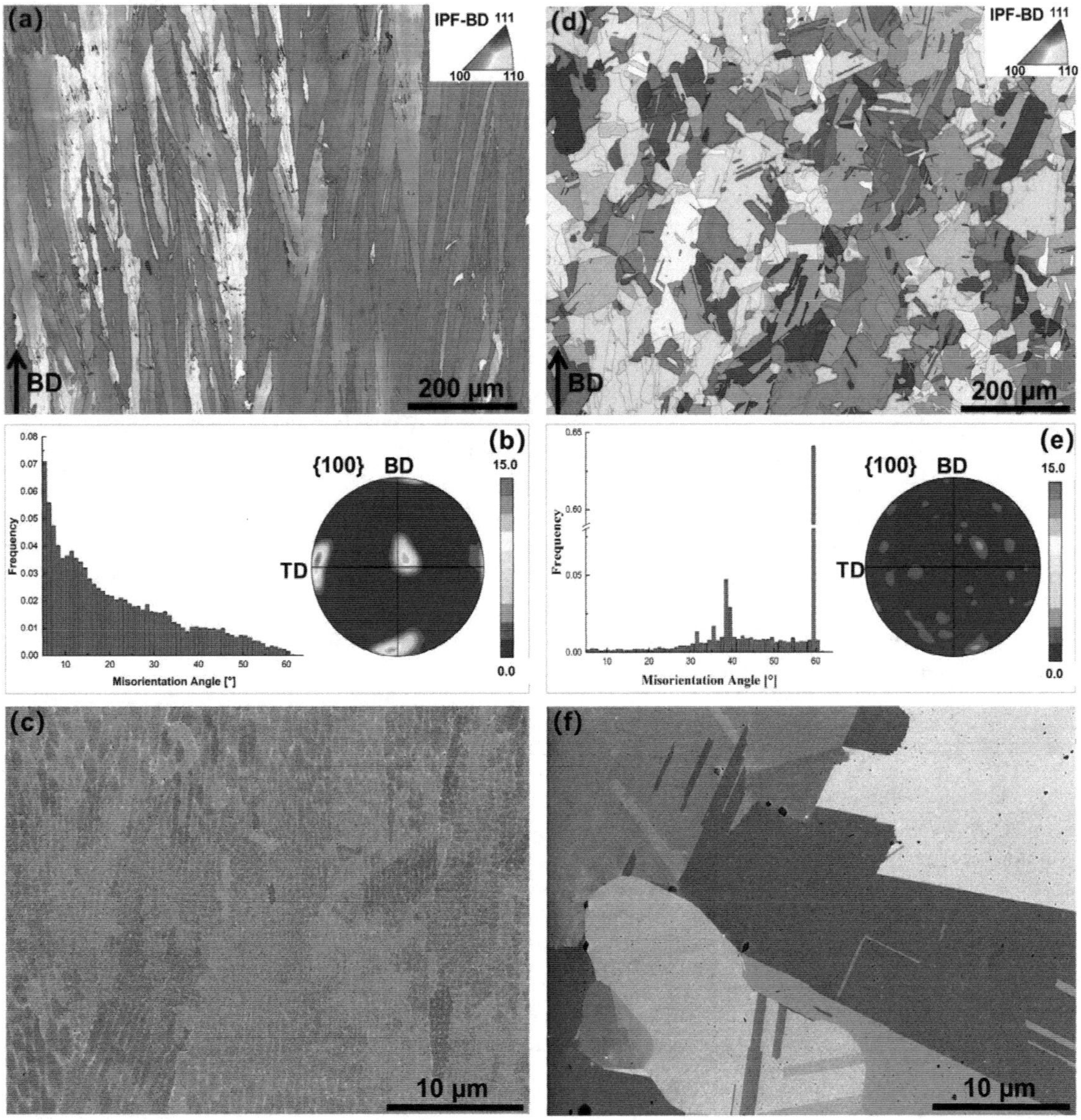

Figure 1. Microstructure of Inconel 718 alloy (a)-(c) fabricated by LPBF and (d)-(f) after solution and aging treatment: (a) and (d) inverse pole figure orientation maps obtained by EBSD; (b) and (e) histograms of misorientation distribution and {100} pole figures; (c) and (f) backscattered electron images.

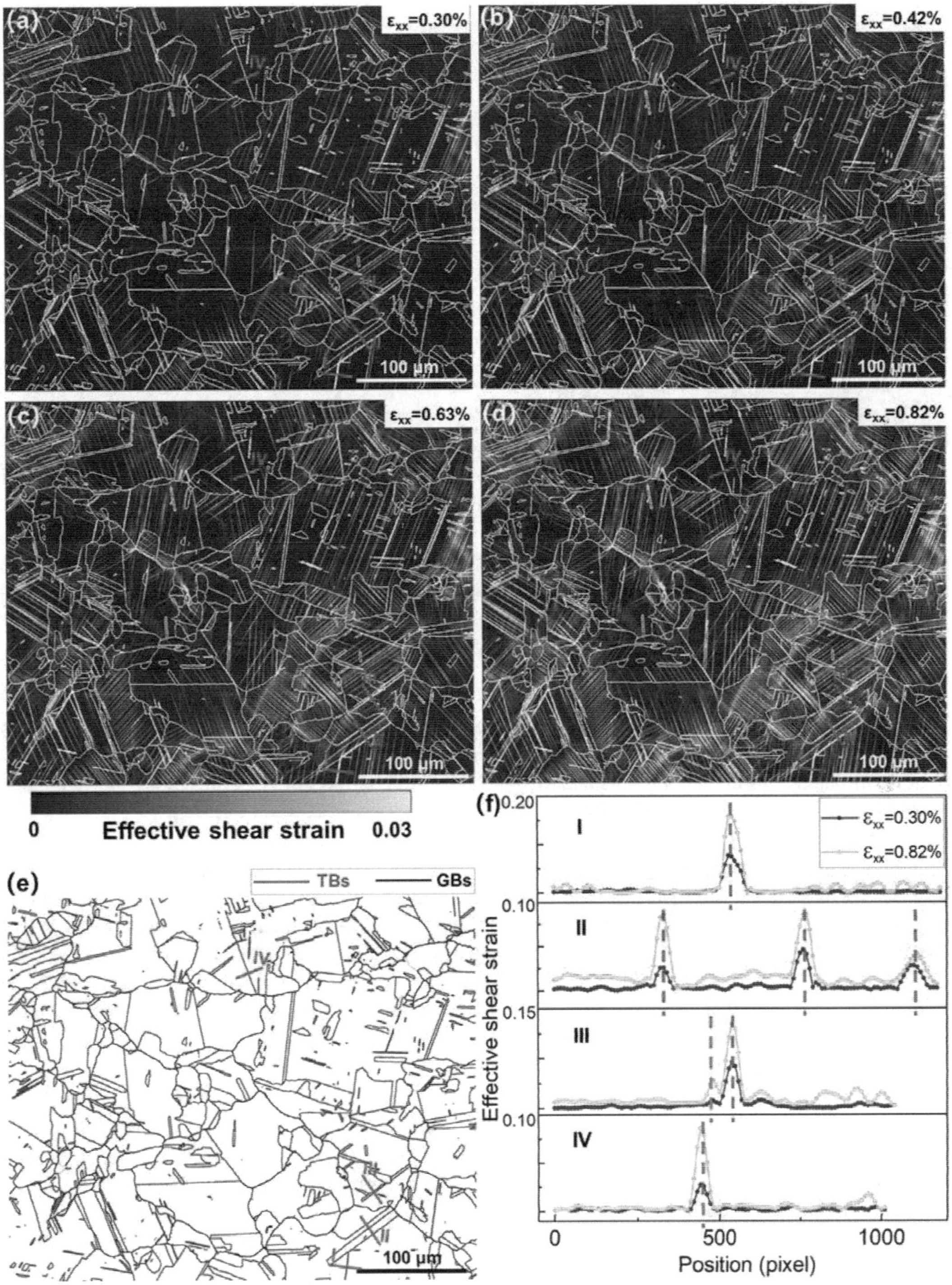

Figure 2. Strain evolution of Inconel 718 alloy after solution and aging treatment during tensioning: (a)-(d) effective shear strain maps of a region of interest in the sample after tensioning to plastic strains of 0.30%, 0.42%, 0.63%, 0.82%, respectively; (e) the corresponding grain boundary map obtained by EBSD, where blue lines are twin boundaries (TBs) and black lines are general grain boundaries (GBs); (f) shear strain profiles across several TBs indicated in by red lines in (a)-(e)

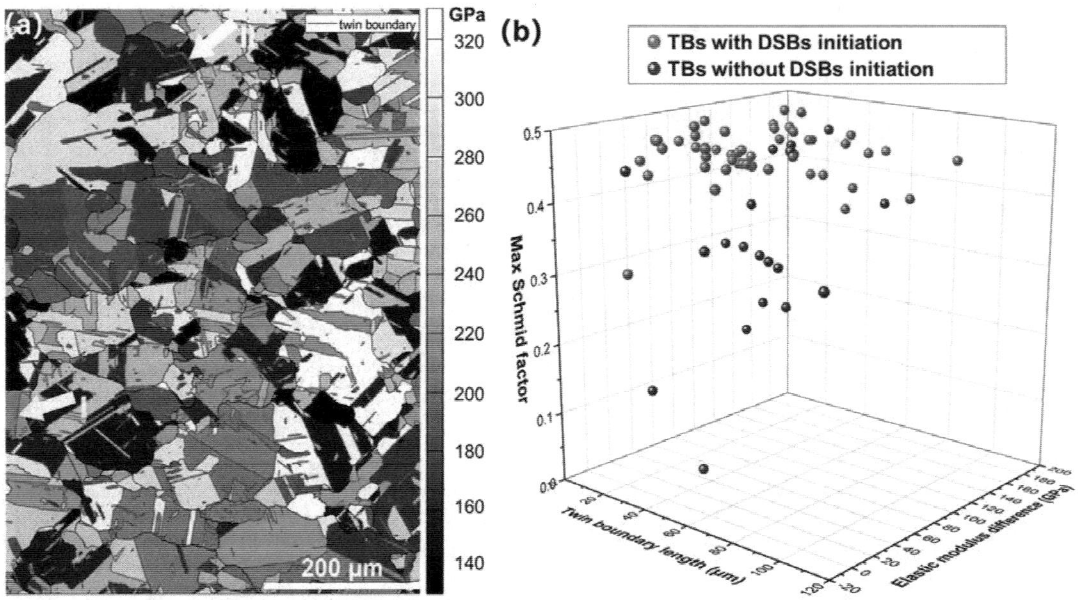

Figure 3. Analyses on the development of slip traces of a region in a sample after tensioning to a strain of ~1%: relationships between the characteristics of TBs and the development of slip traces : (a) Young's modulus map obtained from the orientation data and elastic constants of Alloy 718; (b) Statistics of 49 TBs which have developed slip traces (red symbols) and 19 TBs which have not developed slip traces (blue symbols).

High resolution backscattered electron (BSE) image (figure 4a) shows that V-shaped γ'' with one-to-one correspondence formed along TBs in IN718 alloy, which is similar to our previous finding in another γ''-strengthened superalloy (Incoloy 945X) [22]. The formation of such symmetric γ'' along TBs is attributed to the extremely low free energy in forming γ''-TB-γ'' structure compared to forming other morphologies, which originates from the interplay between the crystallographic structure of γ'' and TB of Ni matrix [22]. We examined the TBs which have developed DSBs using high resolution SEM to reveal the role of the special γ'' in the initiation of DSBs and strain localization. Two examples are shown in figure 4b-d, and they are from the regions marked in figure 3a. We select these two TBs as representatives to demonstrate the negligible role of elastic anisotropy in the initiation of DSBs, because the elastic anisotropy of these TBs are trivial (less than 10 GPa) but DSBs formed along them. The BSE images show that the DSBs are not exactly along TBs but along a plane defined by the cusps of the V-shaped γ'' along the TB. The precipitation of V-shaped γ'' along TBs leads to the formation of two planes that are parallel to TBs, i.e. they are along {111} crystallographic planes, and these planes are denuded in precipitation strengthening, as seen in the BSE images, so they serve as the weak channels for dislocation activation and slip. In addition, the two TBs in figure 4c and 4d both have high Schmid factors for {111}<110> slip (0.44 and 0.45, respectively), so they are readily the easy path for dislocation activities. In addition, Zhang et al [22] found out that dislocations are activated at a much lower stress level at these weakened planes compared to that in grain interiors of Incoloy 945X, which results in severe strain localization. Similar phenomenon was also observed in the in-situ HRDIC results in figure 2. Therefore, the formation of V-shaped γ'' along TBs is the root cause for the localized dislocation plasticity and strain localization at TBs in IN718 alloy.

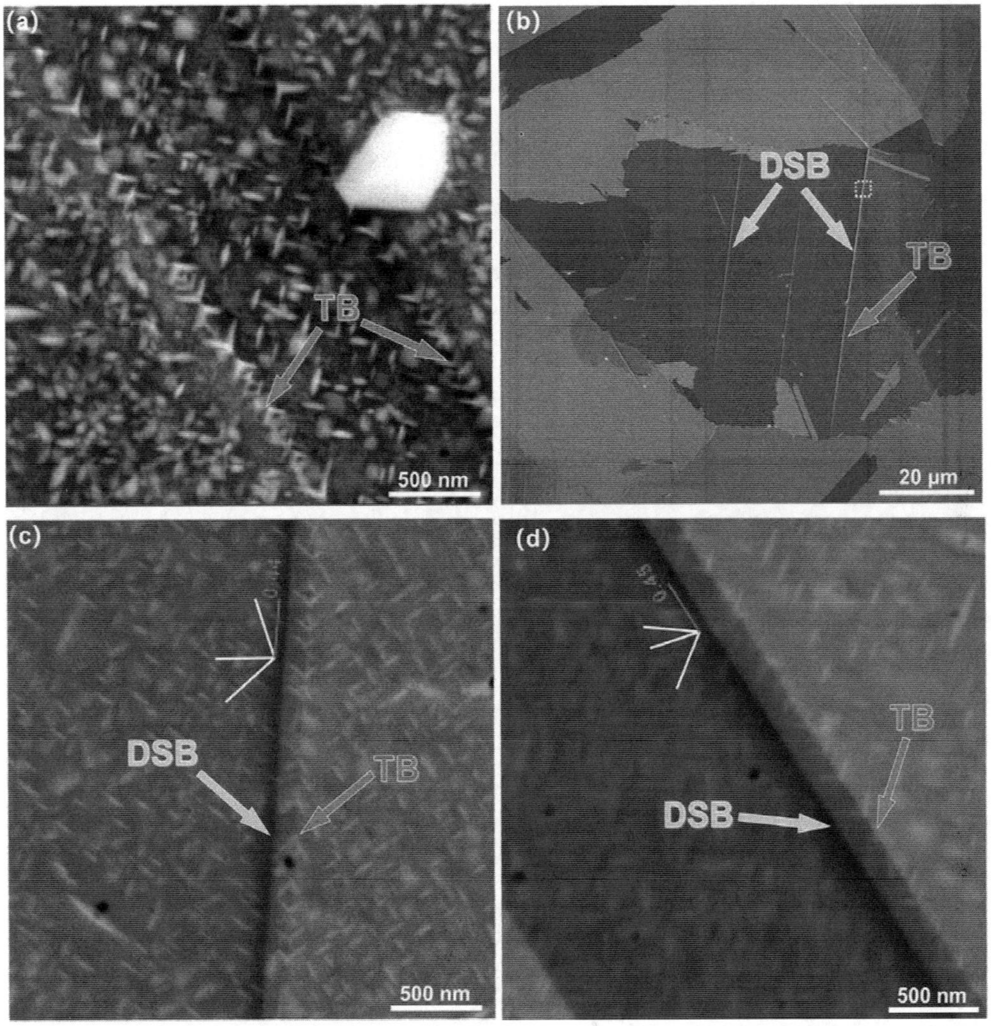

Figure 4. Precipitates and dislocation slip bands (DSBs) at twin boundaries: (a) V-shaped γ″ along TBs in the aged sample; (b)-(d) DSBs at TBs in the sample after tensioning to a strain of ~1%; (c) is an enlarged image of the framed region in (b), and the TB in (c) is the one noted as I in figure 3a; the TB in (d) is the one noted as II in figure 3a. {111} plane traces are indicated in (c) and (d), and the ones parallel to the TB and DSB are presented in red lines with Schmid factors.

4. Conclusion

In summary, substantial TBs generate when the LPBF-fabricated IN718 alloy is subjected to homogenization treatment and aging. Using in-situ HRDIC technique, it is found out that severe strain localization occurs at TBs. The localized dislocation plasticity at TBs has no dependency on the elastic anisotropy and the length of TBs, and the maximum Schmid factor of the TBs is the only determining factor for the occurrence of strain localization at the TBs. Correlative characterizations of the TBs, γ″ precipitates and dislocation slip bands at TBs demonstrate that the precipitation of V-shaped γ″ along TBs results in the formation of precipitates denuded zones, which takes the responsibility for localized dislocation plasticity and severe strain localization at TBs. These findings indicate that TBs in LPBF-fabricated IN718 alloy are the vulnerable sites for fatigue crack initiation, and tailoring the character of grain boundary and precipitation behavior by tuning the processing parameters during LPBF and subsequent annealing treatment is potentially viable to enhance the fatigue performance of IN718 alloy fabricated by LPBF.

Acknowledgement
Financial supports from the National Natural Science Foundation of China (Grant No. 52001212) and ShanghaiTech University Startup fund are gratefully acknowledged.

References

[1] Reed R C 2006 *The Superalloys: Fundamentals and Applications* (Cambridge: Cambridge University Press)
[2] Dudzinski D, Devillez A, Moufki A, Larrouquere D, Zerrouki V and Vigneau J 2004 *Int. J. Mach. Tools Manuf.* **44** 439-56
[3] DebRoy T, Wei H L, Zuback J S, Mukherjee T, Elmer J W, Milewski J O, Beese A M, Wilson-Heid A, De A and Zhang W 2018 *Prog. Mater. Sci* **92** 112-224
[4] Hosseini E and Popovich V A 2019*Addit Manuf* **30** 100877
[5] Herzog D, Seyda V, Wycisk E and Emmelmann C 2016 *Acta Mater.* **117** 371-92
[6] Pröbstle M, Neumeier S, Hopfenmüller J, Freund L P, Niendorf T, Schwarze D and Göken M 2016 *Mater. Sci. Eng. A* **674** 299-307
[7] Deng D, Peng R L, Brodin H and Moverare J 2018 *Mater. Sci. Eng. A* **713** 294-306
[8] Li X, Shi J J, Wang C H, Cao G H, Russell A M, Zhou Z J, Li C P and Chen G F 2018 *J. Alloys Compd.* **764** 639-49
[9] Zhang D, Niu W, Cao X and Liu Z 2015 *Materials Science and Engineering: A* **644** 32-40
[10] Stein C A et al 2014 *Curr. Opin. Solid St. M.* **18** 244-52
[11] Stinville J C, Lenthe W C, Miao J and Pollock T M 2016 *Acta Mater.* **103** 461-73
[12] Stinville J C, Lenthe W C, Echlin M P, Callahan P G, Texier D and Pollock T M 2017*Int. J. Fracture* **208** 221-40
[13] Seita M, Hanson J P, Gradecak S and Demkowicz M J 2015 *Nat. Commun.* **6**
[14] Zhang Z, Obasi G, Morana R and Preuss M 2016 *Acta Mater.* **113** 272-83
[15] Holland S, Wang X, Fang X Y, Guo Y B, Yan F and Li L 2018 *Mater. Sci. Eng. A* **725** 406-18
[16] Zhao Y, Li K, Gargani M and Xiong W 2020 *Addit Manuf* **36** 101404
[17] Di Gioacchino F and Quinta da Fonseca J 2015 *Inter. J. Plasticity* **74** 92-109
[18] Harte A, Atkinson M, Preuss M and Quinta da Fonseca J 2020 *Acta Mater.* **195** 555-70
[19] Atkinson M D, Harte A, Crowther P and Quinta da Fonseca J. 2020*10.5281/zenodo.3688097*
[20] Sundararaman M, Mukhopadhyay P and Banerjee S 1988 *Acta Metall.* **36** 847-64
[21] Charpagne M A, Hestroffer J M, Polonsky A T, Echlin M P, Texier D, Valle V, Beyerlein I J, Pollock T M and Stinville J C 2021 *Acta Mater.* **215** 117037
[22] Zhang Z, Yang Z, Lu S, Harte A, Morana R and Preuss M 2020 *Nat. Commun.* **11** 4890

A new rejuvenation heat treatment of crept Ni-based superalloy single crystals

Sicong Lin[1], Hao Shen[1], Guangni Zhou[1], Weifeng He[2], Kai Chen[1*]

[1] Center for Advancing Materials Performance from the Nanoscale (CAMP-Nano), State Key Laboratory for Mechanical Behaviour of Materials, Xi'an Jiaotong University, Xi'an, Shaanxi 710049, P. R. China
[2] State Key Laboratory for Manufacturing Systems Engineering, Xi'an Jiaotong University, Xi'an, Shaanxi 710049, P. R. China

E-mail: kchenlbl@gmail.com

Abstract. Ni-based superalloys are widely used in harsh environments due to their exceptional high-temperature mechanical properties, and their microstructures are inevitably subjected to degradation, including accumulation of strain/stress, generation of high density of dislocations, and rafting of γ'-particles, under creep damage. To prolong their service life, rejuvenation heat treatment is proposed to restore their microstructure. In this study, a Ni-based superalloy single crystal specimen with 0.15% creep strain is taken as a surrogate of partially served components. The microstructure, including the dislocation density and distribution as well as the size and morphology of γ'-particles, is investigated. Then two comparative heat treatment protocols are applied to the crept superalloy single crystal. Because super-solvus homogenization treatment triggers recrystallization, a recovery pre-annealing step at sub-solvus temperature is designed to relieve the stored deformation energy. After homogenization and aging heat treatment, the single crystalline structure is preserved, and in the meanwhile the γ'-particles are restored to uniform cuboidal morphology and the dislocations are fully eliminated. Our findings provide a route to rejuvenate crept superalloy single crystals.

1. Introduction

Ni-based superalloy single crystals are being widely used as blades and vanes in gas turbines and aero-engines due to their excellent combination of high-temperature strength, toughness, corrosion, and oxidation resistance [1–3]. Under the harsh service environment of complex stresses at elevated temperatures, creep damage occurs over time. The major strengthening γ'-phase particles lose their designed cuboidal shape and become coarsened and rafted [4–6], leading to significant degradation of mechanical properties such as creep resistance. As a result, the demands are pressing to develop efficient heat treatment techniques to restore the microstructure and extend the service lives of the superalloy single crystal components.

At present, two methods have been proposed to extend the lives of crept superalloys, which are hot isostatic pressing [7,8] and rejuvenation heat treatment [9–11]. Hot isostatic pressing shows great advantages in closing up the creep holes, while rejuvenation heat treatment, which employs much less complex equipment and auxiliaries, is more suitable to superalloy single crystal components after a certain period of service without many creep holes generated. However, for the rejuvenation heat treatment, scientific and technical problems remain unsolved. The essence is to recover the creep induced plastic deformation and avoid the disastrous recrystallization, which may lower the creep, fatigue and oxidation resistance of superalloys by one order of magnitude [12]. By applying the industrial standard heat treatment protocols to the crept superalloy single crystals, although the stored deformation energy can be relieved completely, the single crystals are under the risk of recrystallization during the super-solvus homogenization annealing even if the crept deformation strain is only 0.5% [10]. Alternatively, Rettberg et al. carried out the rejuvenation heat treatment containing a sub-solvus homogenization annealing and then 'conventional' aging annealing that is exactly the same as the industrial standard protocol in term of temperature and time [9]. By this means, the remaining γ'-particles during the sub-solvus annealing suppressed recrystallization successfully; but in the meanwhile, stored deformation energy associated with dislocations cannot be relieved completely also because of the impediments by the residual γ'-particles. Thus after a few cycles of creep-rejuvenation tests, with the accumulated deformation energy, recrystallization sets in rapidly, resulting in fractures along the high-angle grain boundaries.

In this study, two rejuvenation heat treatment protocols are employed and their effects are comparatively studied on the microstructure of a creep deformed second generation Ni-based superalloy single crystal. With industrial standard heat treatment protocol, recrystallization occurs in the crept specimen with 0.2% deformation strain. By applying a recovery annealing prior to the super-solvus homogenization heat treatment, recrystallization is avoided, and the final product after full heat treatment is dislocation free and composed of regular-shaped cuboidal γ'-particles that are identical with those before creep test. Our results provide a strategy to solve the paradox between the complete relief of the stored deformation energy and the growth of recrystallized grains.

2. Experimental

2.1. Materials and creep test

A second generation CMSX-4 Ni-based superalloy single crystal with a nominal composition of Ni-5.6Al-1.0Ti-6.5Cr-9.6Co-2.09Ta-6.4W-0.6Mo-6.5Ta-3Re-0.1Hf (wt%) was used in this study. The superalloy was composed of dendrites with their primary dendrite width of hundreds of microns. In each dendrite, face-centered cubic γ-phase matrix formed a continuous framework, filled with cuboidal $L1_2$ γ'-phase strengthening precipitates with 300 - 350 nm side length and high volume fraction (60 - 70%). The orientation along the longitudinal axis of the superalloy single crystal bars used for present work was within 15° deviated from the [001] crystallographic direction.

The single crystal bars were firstly cut into 76 mm in length by electrical discharge machining, and were further subjected to standard heat treatment, including a two-step super-solvus homogenization heat treatment at 1290 °C for 1 h and then 1300 °C for 6 h, followed by a two-step aging treatment at 1140 °C for 4 h and 870 °C for 16 h [13]. The single crystal was air cooled after each step of the heat treatment. Afterwards the superalloy single crystal was machined into creep specimens with the gauge length of 23 mm and diameter of 5 mm, as shown in Fig. 1a. The creep tests were carried out in air for 20 hours at constant temperature of 1050 °C and tensile stress of 160 MPa and repeated in two identical specimens. After the specimens were furnace cooled, the plastic strains measured in the specimens were about 0.15%.

2.2. Rejuvenation and microstructure characterization

The crept specimen was firstly sectioned transversely into 4 mm in thickness, and then cut along the longitudinal direction into two halves for the comparative study of the rejuvenation heat treatment effects with different protocols. To simplify the microstructure characterization, a coordinate system was established. The tensile stress loading direction, parallel to the [001] crystallographic direction, was defined to be y-axis. The longitudinal plane normal, denoted to be the z-axis, was along the [100] direction. The first rejuvenation heat treatment (Protocol 1) included two parts, which were exactly the same as the standard heat treatment (Fig. 1b). The second rejuvenation heat treatment (Protocol 2), in contrast, introduced a recovery annealing at 1100 °C for 2 h prior to Protocol 1 (Fig. 1c).

Prior and after the rejuvenation heat treatment, the microstructure of the superalloy single crystals were characterized on both the transversal and longitudinal planes. The specimens were mechanically ground and finely polished with 3.5 μm diamond paste. After electro-etching in aqueous phosphoric acid (25 vol%) at 5 V for 5 s, the samples were investigated using scanning electron microscopy (SEM) under secondary electron imaging mode. Electron backscattered diffraction (EBSD) was used to map the grain orientations with the step size of 10 μm. The EBSD samples were prepared by electrochemical polishing in 10 vol% perchloric acid alcohol at the voltage of 25 V for about 20 s. For analysis of the dislocation densities, transmission electron microscope (TEM) was employed.

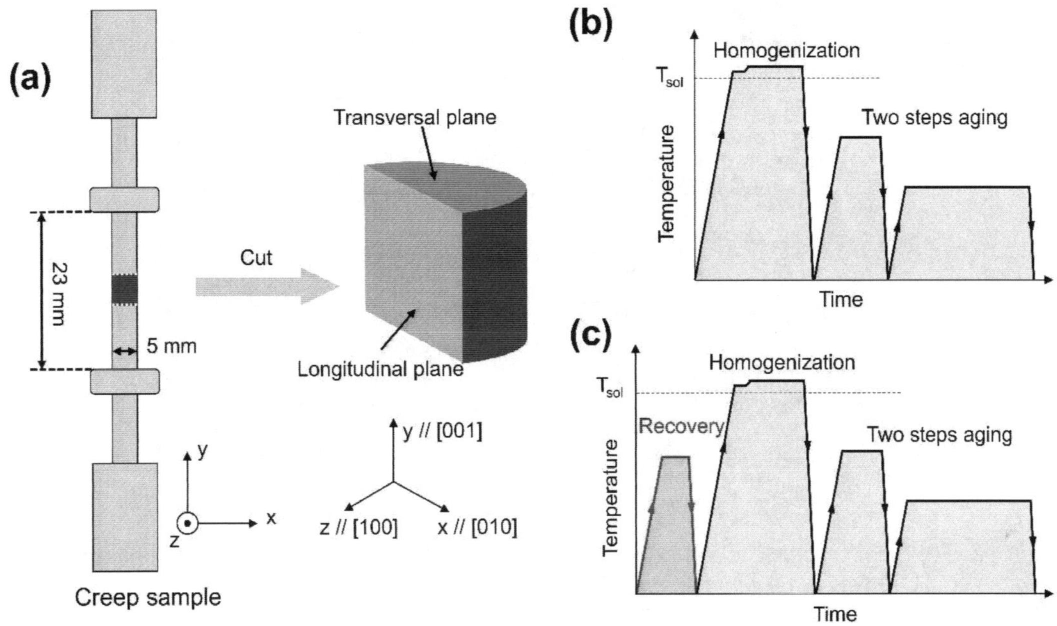

Figure 1. Schematic of the creep test specimen (a) and the comparison of rejuvenation heat treatment effects with two different protocols (b - c).

3. Results

3.1. Microstructure of the crept superalloy

The morphology of the γ' strengthening precipitates and the distribution of dislocations of the crept superalloy single crystals are investigated on both the transversal and longitudinal planes and displayed in Fig. 2. On the transversal plane (Fig. 2a), coarsened γ' particles are observed, with rough surfaces. During the creep deformation, numerous dislocations cut into the γ' precipitates, and each of them creates a kink on the surface once they leave the γ' precipitates. Most of the dislocation line traces are 45° inclined with respect to the γ' precipitate edges, while some are roughly parallel, indicating that different slip systems are activated during the creep test. In the observed region of the TEM micrograph (Fig. 2b), dislocations are aligned parallel, while networks are not maturely developed yet, probably due to the short period of creep test and lack of diffusion time.

The rafting morphology of the γ' precipitate is more clearly visualized on the longitudinal plane in Fig. 2c. Because of the negative lattice misfit between γ and γ', the coarsening direction is perpendicular to the tensile stress [4]. Along the coarsening direction, the γ' precipitates grow up to dozens of microns in length, with a thickness of only half micron or so. The width of the γ channel becomes rather non-uniform after creep test, ranging from tens to one hundred nanometres, which is resulted from the atomic diffusion assisted by the inhomogeneously distributed dislocations. From the TEM image in Fig. 2d, most of the dislocations are observed lying in the γ channels or at the γ/γ′ interfaces. Only a small fraction of the dislocations, like the one highlighted by the yellow arrow, are found staying in the γ′ particles.

To quantify the dislocation density, a line-intercept method is applied to the TEM micrographs on the transversal and longitudinal planes [14]. Generally, the dislocation densities measured in both directions are consistent, approximately 2×10^{14} m^{-2}. ′

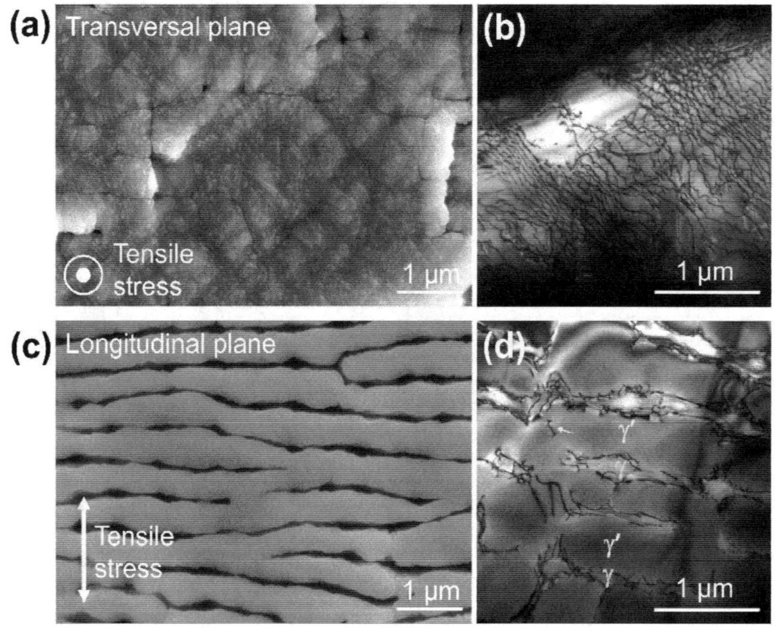

Figure 2. The SEM (a, c) and TEM (b, d) micrographs of the transversal and longitudinal planes of the 0.15% crept superalloy single crystal, showing the rafted γ′ microstructure and non-uniform dislocation distribution.

3.2. Recrystallization induced by standard heat treatment

When applying Protocol 1, which is adopted from the industrial standard heat treatment and without any specifically designed recovery annealing, to the crept CMSX-4 Ni-based superalloy single crystals, recrystallization sets in from the corners and surfaces and the single crystal structure is destroyed (Fig. 3a). So far no preferred orientations are detected in the recrystallized grains, and thus the misorientation angle/axis pair between the recrystallized grains and the original grain is relatively arbitrary, except that annealing twins are widely observed among the newly formed recrystallized grains. After heat treatment, the kernel average misorientation (KAM) ranges within 1° in the original grain, except the low angle grain boundaries between the adjacent dendrites (Fig. 3b). The geometrically necessary dislocation (GND) density, computed from the misorientation angle between each pair of adjacent scanning pixels [15], is found to be on the order of 10^{13} m^{-2} (Fig. 3c), which is close to the resolution of the EBSD method. Furthermore, the distribution map of the grain orientation spread (GOS), which is defined as the average of the misorientation angles to the grain mean orientation, is plotted in Fig. 3d. From the GOS map, it is clearly seen that the crystallographic orientation in the recrystallized grains is more uniform than that in the original grain, confirming that during the process of recrystallization, the microscopic defects, such as the GNDs, are completely eliminated as the high angle grain boundaries sweep through. In summary, although the dislocation density is lower down significantly after the heat treatment obeying Protocol 1, recrystallization renders the microstructure of the Ni-based superalloy into polycrystalline. Consequently, standard heat treatment is not acceptable for superalloy single crystals with creep induced plastic strain as tiny as 0.15%.

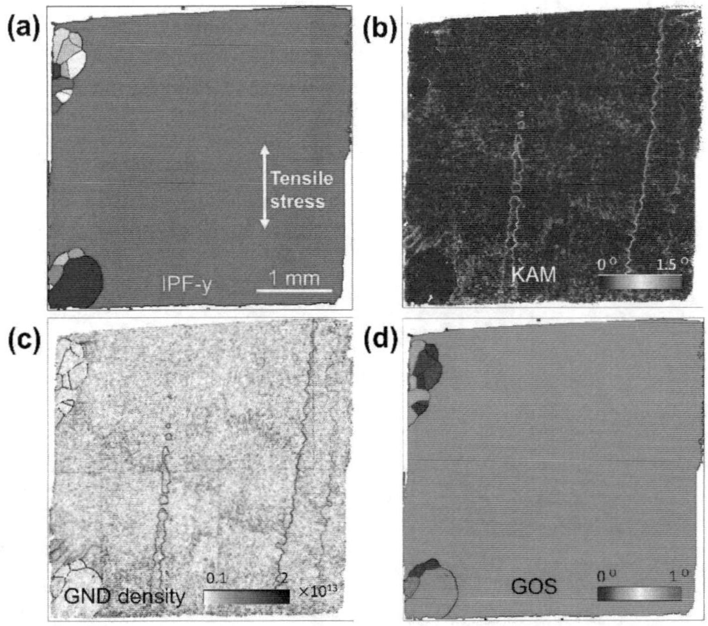

Figure 3. EBSD characterization of the crept superalloy after standard heat treatment. (a) The inverse pole figure along y-axis (IPF-y), (b) KAM, (c) GND density, and (d) GOS maps are plotted.

3.3. Recovery induced by rejuvenation heat treatment

With the help of recovery annealing at 1100 °C for 2 h prior to the standard heat treatment, recrystallization is not triggered and thus single crystalline microstructure is preserved. The EBSD IPF-y map displays a uniform crystallographic orientation distribution without any high angle grain boundaries (Fig. 4a). In the KAM map, only low angle grain boundaries with less than 1° misorientation between dendrites that have formed during the solidification are visible (Fig. 4b). The misorientations within each dendrite are even smaller, corresponding to an average dislocation density of lower than 10^{13} m^{-2} (Fig. 4c). The GOS of the scanned region is similar to that of the original grain after heat treatment Protocol 1, but higher than the recrystallized grain (Fig. 4d).

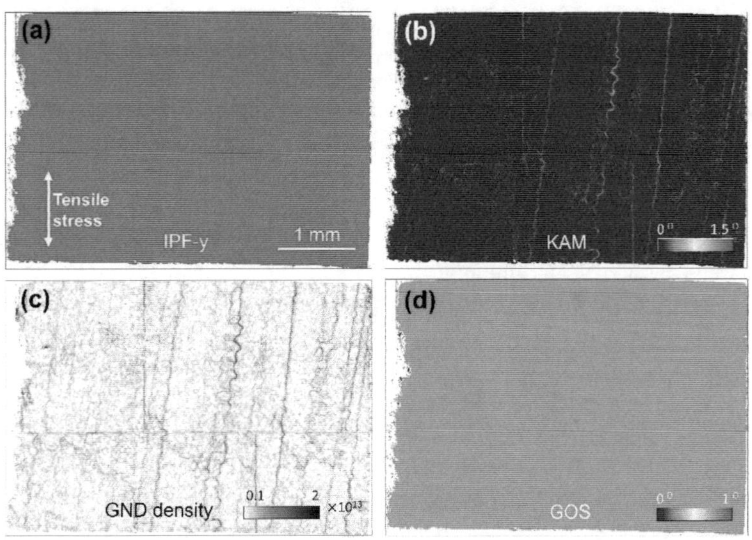

Figure 4. EBSD characterization of the crept superalloy after heat treatment with Protocol 2.

After the rejuvenation heat treatment, the γ'-precipitates restored to the regular cuboidal morphology, with an average side length of about 350 nm (Fig. 5a). In these γ'-precipitates, no dislocation cutting traces are visible because the specimen has been annealed at a temperature that is above the solvus point for sufficiently long period time to allow all the 'old' deformed and coarsened γ'-precipitates to fully dissolve into the γ-matrix. In the following air cooling process, γ'-particles re-precipitate, and then grow bigger in size and more regular in shape during the final two-step aging. From the TEM bright-mode image demonstrated in Fig. 5b, no dislocations are detectable in the viewed area at all. After checking many TEM specimens, dislocations are only occasionally observed.

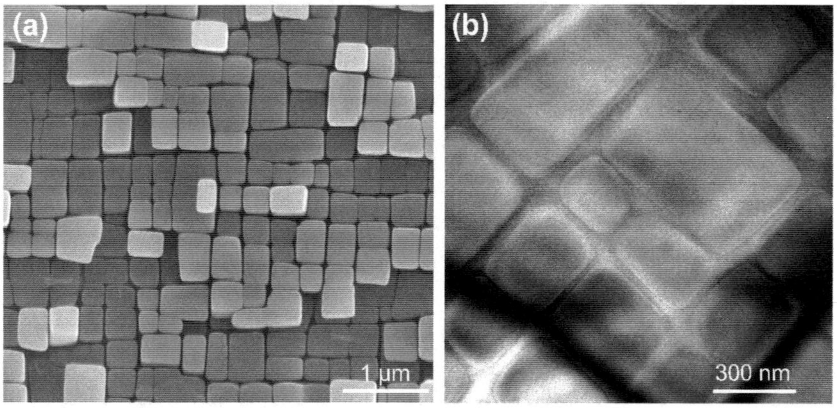

Figure 5. Microstructure investigation of the crept superalloy after rejuvenation heat treatment adopting Protocol 2 using (a) SEM and (b) TEM.

4. Discussion

Ni-based superalloy single crystals are known to be prone to recrystallizations. In previous reports, uni-axial compressive and tensile deformations at room and elevated temperatures could trigger recrystallization readily [16,17]. The higher the annealing temperature was, the lower the critical plastic strains to trigger recrystallizations were. For super-solvus annealing treatments, 2% plastic compressive strain could cause the recrystallized grains to grow at the superalloy surfaces. With higher plastic strains such as ~4% imposed at room temperature, even sub-solvus would induce recrystallization [18]. About why recrystallization is readily triggered in superalloy single crystals with such low plastic strain, there is lack of in-depth investigation in the published literature, except a rather rough explanation from the angle of stacking fault energy. Based on TEM measurements [19], simulations [20] and computations [21], the stacking fault energy of Ni-based superalloy single crystals is approximately 20 mJ/m^2, leading to the dissociation of dislocations into immobile partial ones. The resulted partial dislocations are hard to climb or cross-slip, which are necessary for dislocations to move, react, and annihilate to relieve the stored deformation energy [22].

The magnitude of the critical plastic strain to induce recrystallization is dependent of not only deformation temperature, but also the mode. For example, in this study, the creep strain is only 0.15%, about one order of magnitude lower than the reported uni-axial compression and tension, while recrystallization is still inevitable when the crept superalloy is subjected to super-solvus heat annealing without deformation energy relief treatment. This finding is similar to the observations reported in previous study [10]. What is novel in this study is that we developed a pre-annealing to relieve the stored deformation energy efficiently. Our heat treatment design is completely different from the strategy reported previously [9], in which the homogenization annealing was carried out at the temperatures much lower than the solvus point to avoid recrystallization. It is expected that with such sub-solvus heat treatment, the strain/stress are not fully relieved, and some dislocations are still blocked by the undissolved γ'-particles. These leftover dislocations are known to behave as atomic diffusion paths, and the elastic strain associated with them influence the direction of the γ'-particle coarsening [23].

Consequently, after aging annealing, the size and shape of the γ′-particles will not be fully restored to the state prior to creep test. Combining the residual strain/stress, microscopic defects, a suboptimal γ′-precipitates, the high temperature mechanical properties of the superalloy single crystals can hardly reach their designed level.

Our new heat treatment protocol provides a novel strategy to solve the recrystallization issue. Similar to what has been discovered by us in 3D-printed Ni-based superalloys [24], by the recovery pre-annealing, the stored deformation energy is effectively relieved, securing that no recrystallization is activated even when heating the superalloys higher than the solvus point. Moreover, the recovery treatment takes much shorter time than the standard heat treatment protocol, and thus is completely affordable. However, it is still an open question worth further systematic study regarding the critical plastic strain introduced by creep test.

5. Conclusions

In this study, a new rejuvenation method is developed to restore the microstructure of the creep deformed Ni-based superalloy single crystals. It has been demonstrated that the strengthening phase γ′-particles are readily rafted even with a tiny deformation strain after creep test, along with high density of dislocations distributed in the matrix γ-channels. Direct super-solvus homogenization annealing of the crept superalloy activates recrystallization. The newly formed crystal grains grow up to hundreds of microns, rendering the unacceptable polycrystalline structure. Thus a recovery heat treatment is designed prior to the super-solvus homogenization heat treatment to relieve the residual stresses, annihilate dislocations, and eliminate the driving force for recrystallization. Then the following homogenization and aging heat treatment result in regular cuboidal γ′-particles, achieving the restoration of the microstructures to the states prior to the creep tests. Our findings pave a path to the effective rejuvenation of crept superalloy single crystals.

Acknowledgements

This work is supported by the National Natural Science Foundation of China (Grant No. 51927801) and 111 Project 2.0 (BP0618008).

References

[1] Pollock T M 2016 Alloy design for aircraft engines *Nat. Mater.* **15** 809.
[2] Pollock T M and Tin S 2006 Nickel-based superalloys for advanced turbine engines: chemistry, microstructure, and properties *J. Propuls. Power.* **22** 361–374.
[3] Fullagar K P L, Broomfield R W, Hulands M, Harris K, Erickson G L and Sikkenga S L 1996 Aero engine test experience with CMSX-4® alloy single crystal turbine blades *J. Eng. Gas Turbines Power.* **118** 380-388.
[4] Nabarro F R N 1996 Rafting in superalloys *Metall. Mater. Trans. A Phys. Metall. Mater. Sci.* **27** 513–530.
[5] Matan N, Cox D C, Rae C M F and Reed R C 1999 On the kinetics of rafting in CMSX-4 superalloy single crystals *Acta Mater.* **47** 2031–2045.
[6] Reed R C, Cox D C and Rae C M F 2007 Kinetics of rafting in a single crystal superalloy: Effects of residual microsegregation *Mater. Sci. Technol.* **23** 893–902.
[7] Ruttert B, *et al,* 2018 Rejuvenation of single-crystal Ni-base superalloy turbine blades: unlimited service life? *Metall. Mater. Trans. A Phys. Metall. Mater. Sci.* **49** 4262–4273.
[8] Ruttert B, *et al,* 2017 Rejuvenation of creep resistance of a Ni-base single-crystal superalloy by hot isostatic pressing *Mater. Des.* **134** 418–425.
[9] Rettberg L H, Callahan P G, Goodlet B R and Pollock T M 2021 Rejuvenation of directionally solidified and single-crystal Nickel-base superalloys *Metall. Mater. Trans. A Phys. Metall. Mater. Sci.* **52** 1609–1631.

[10] Tan K J, Wang X G, Liang J J, Meng J, Zhou Y Z and Sun X F 2021 Effects of rejuvenation heat treatment on microstructure and creep property of a Ni-based single crystal superalloy *J. Mater. Sci. Technol.* **60** 206–215.
[11] James A 2001 Review of rejuvenation process for nickel base superalloys *Mater. Sci. Technol.* **17** 481–486.
[12] Koff B L 2004 Gas turbine technology evolution: A designer's perspective *J. Propuls. Power.* **20** 577–595.
[13] Parsa A B, *et al,* 2015 Advanced scale bridging microstructure analysis of single crystal Ni-base Superalloys *Adv. Eng. Mater.* **17** 216-230.
[14] Norfleet D M, Dimiduk D M, Polasik S J, Uchic M D and Mills M J 2008 Dislocation structures and their relationship to strength in deformed nickel microcrystals *Acta. Mater.* **56** 2988–3001.
[15] Pantleon W 2008 Resolving the geometrically necessary dislocation content by conventional electron backscattering diffraction *Scr. Mater.* **58** 994-997.
[16] Cox D C, Roebuck B, Rae C M F and Reed R C 2003 Recrystallisation of single crystal superalloy CMSX-4 *Mater. Sci. Technol.* **19** 440–446.
[17] Mathur H N, Panwisawas C, Jones C N, Reed R C and Rae C M F 2017 Nucleation of recrystallisation in castings of single crystal Ni-based superalloys *Acta Mater.* **129** 112–123.
[18] Bürgel R, Portella P D and Preuhs J 2000 Recrystallization in single crystal of nickel base superalloy. in *superalloys 2000* (eds. Pollock, T. M. & Kissinger, R. D.) 229–238.
[19] Pettinari F, *et al,* 2002 Stacking fault energy in short-range ordered γ-phases of Ni-based superalloys *Mater. Sci. Eng. A* **325** 511–519.
[20] Benyoucef M, Décamps B, Coujou A and Clément N 1995 Stacking-fault energy at room temperature of the γ matrix of the MC2 Ni-based superalloy *Philos. Mag. A* **71** 907–923.
[21] Tian S, Zhu X J, Wu J, Yu H C, Shu D L and Qian B J 2016 Influence of temperature on stacking fault energy and creep mechanism of a single crystal Nickel-based superalloy *J. Mater. Sci. Technol.* **32** 790–798.
[22] Humphreys F J and Hatherly M 2004 Chapter 6 Recovery after deformation, in *Recrystallization and related annealing phenomena* 169–213.
[23] Kontis P, Li Z M, Collins D M, Cormier J Raabe D and Gault B 2018 The effect of chromium and cobalt segregation at dislocations on nickel-based superalloys *Scr. Mater.* **145** 76–80.
[24] Chen K, *et al,* 2020 Rafting-enabled recovery avoids recrystallization in 3D-printing-repaired single-crystal superalloys *Adv. Mater.* **32** 1907164.

Network science approach to predict the yield strength of metals containing soft particles and pores

T Andriollo

Department of Mechanical and Production Engineering, Aarhus University, DK
E-mail: titoan@mpe.au.dk

Abstract. Local fluctuations in particle size and spatial distributions can substantially affect the yield strength of metals containing soft particles and/or porosities. The phenomenon can be studied via computational homogenization techniques, which, however, can be highly computationally demanding when large representative volume elements are needed. A more efficient approach to model the plastic deformation that controls the yield strength of metals containing pores has recently been proposed by the author and co-workers. The key idea is to treat the material as a network rather than as a continuum. In fact, plastic deformation in porous metals occurs via the formation of shear bands connecting the pores, and it can thus be modelled as a time-evolving network where the nodes represent the pores and the links represent the shear bands. So far, the applicability of this new approach has only been tested against 2D synthetic microstructures. The present work takes it a step further by applying the approach to real spheroidal graphite iron. Based on strain data collected in-situ via digital volume correlation, it is shown that the new network approach can explain the spatial variations of the plastic deformation that arise from local variations of the particle distribution throughout a tensile specimen.

1. Introduction

Many metallic materials contain soft particles and porosities at the microstructural scale. These can either be introduced on purpose in order to achieve specific properties, like in composites and porous orthopedic implants [1], or be an undesired outcome of the manufacturing process, like porosity in 3D printed metallic parts [2]. As the impact of soft particles and porosities – henceforth simply called particles – on the material yield strength can be substantial, the availability of predictive tools that are both accurate and efficient is paramount.

Most predictive tools that are available today assume the material to be a continuum and make use of concepts and methods proper to continuum mechanics. Analytical approaches are fast but consider almost exclusively the effect of the particle volume fraction [3]. As the particle size, shape and spatial distributions can also have a significant impact on the material strength [4], direct numerical simulation techniques based on numerical models of the whole material microstructure have become popular in recent years [5]. However, the maximum number of particles that can be dealt with in a 3D finite element model is now in the order of a few hundred [6], which is still very small compared to the number of µm-sized particles forming the microstructure of a real metallic part. A workaround is to limit the analysis to a representative volume element (RVE) of the microstructure and to estimate its macroscopic strength via computational homogenization techniques [7]. Yet, as the RVE must be sufficiently large to be statistically representative, this approach can still be computationally very expensive [8]. The reason is that a boundary value problem involving non-linear differential equations must be solved over the entire RVE domain, a task that is normally carried out by means of finite element or fast Fourier transform methods [7,9]. Another problem is that the definition of a suitable RVE is impossible when the particle distribution is not regular enough at intermediate scales [8].

The absence of a technique that is both accurate enough to account for microstructural variability in terms of particle size, shape and spatial distributions as well as fast enough to enable strength predictions based on a very large number of particles has two adverse consequences. First, it precludes a synergistic coupling with modern, data-intensive characterization techniques tools like e.g. X-ray tomography, which can now provide full 3D quantifications of the distribution of μm-sized particles over several mm^3 [10]. Second, it hinders the development of new functionally graded materials with tailored, non-uniform spatial distribution of the particles throughout the microstructure [11,12], as it prevents a full understanding of the mechanisms governing the plastic deformation and therefore the yield strength in the presence of many and unevenly distributed particles.

To overcome the limitations of the current predictive tools, an innovative approach was recently proposed by the author and co-workers, which merges continuum micro-mechanics with concepts from network science [13]. In the following, the method is first described and subsequently applied to explain the local variations of the plastic deformation field measured with digital volume correlation during uniaxial tensile loading of spheroidal graphite iron (SGI).

2. Network approach

The novel idea introduced in ref. [13] is to study the plastic deformation at the μm-scale of a metal containing particles using network science instead of continuum approaches. It is in fact well known that plastic deformation in metals involves the formation of plastic shear bands [14], which in the presence of soft particles or voids spread by linking the latter in a progressive fashion [15,16]. The phenomenon is illustrated in figure 1, which shows the evolution of the plastic deformation field during virtual tensile loading of a 2D finite element model of a metal microstructure containing a random distribution of pores. It can be noted that plastic deformation starts around the pores due to local stress concentration (figure 1 (b)). However, the onset of the plastic deformation occurs at a level of macroscopic strain $\bar{\varepsilon}_{pdo}$ at which the material macroscopic response is still seemingly linear (see figure 1 (a)). Non-linearity at the macro-scale becomes apparent only when plastic shear bands connecting the pores start to form (figure 1 (d)). New shear bands form and further develop into a complex network as the load increases progressively up to the macroscopic 0.2% offset yield stress $\bar{\sigma}_y$ (figure 1 (e) and (f)).

The features of the deformation process highlighted in figure 1 suggest that the macroscopic yielding is not controlled by the onset of plastic deformation around the pores, but rather by the subsequent formation and evolution of the shear bands. This is confirmed by the absence of correlation between $\bar{\sigma}_y$ and $\bar{\varepsilon}_{pdo}$ evidenced by the statistical box plots of figure 2, which are based on data for $\bar{\sigma}_y$ and $\bar{\varepsilon}_{pdo}$ predicted by 100 finite element models, each featuring a different random spatial distribution of the pores. That is, earlier onset of plastic deformation at the microstructural level does not necessarily mean lower yield strength and vice versa.

In order to model the evolution of the plastic shear bands that, as just pointed out, control the macroscopic yield stress of metals containing soft particles, the concept of *shear band network (SBN)* was presented in ref. [13]. The SBN is defined as a time-dependent, unweighted and undirected spatial network [17] in which every *node* is associated with a particle and every *link* is associated with a pair of nodes for which the corresponding particles are connected by at least one shear band. The concept is exemplified in figure 3, where the simulated plastic deformation field in a magnified region of the finite element model is shown next to the corresponding SBN. The following conclusions were obtained by analyzing a large amount of SBN data generated via finite element modelling of 2D random microstructures [13]:

- Despite the SBN is not known a priori, it turns out that the SBN is a subset of the Delaunay network defined by the particle centers. It follows that the evolution of the SBN can be considered as a progressive activation of the links of the Delaunay network, which can be constructed from knowledge of the microstructural topology, e.g. from a micrograph.

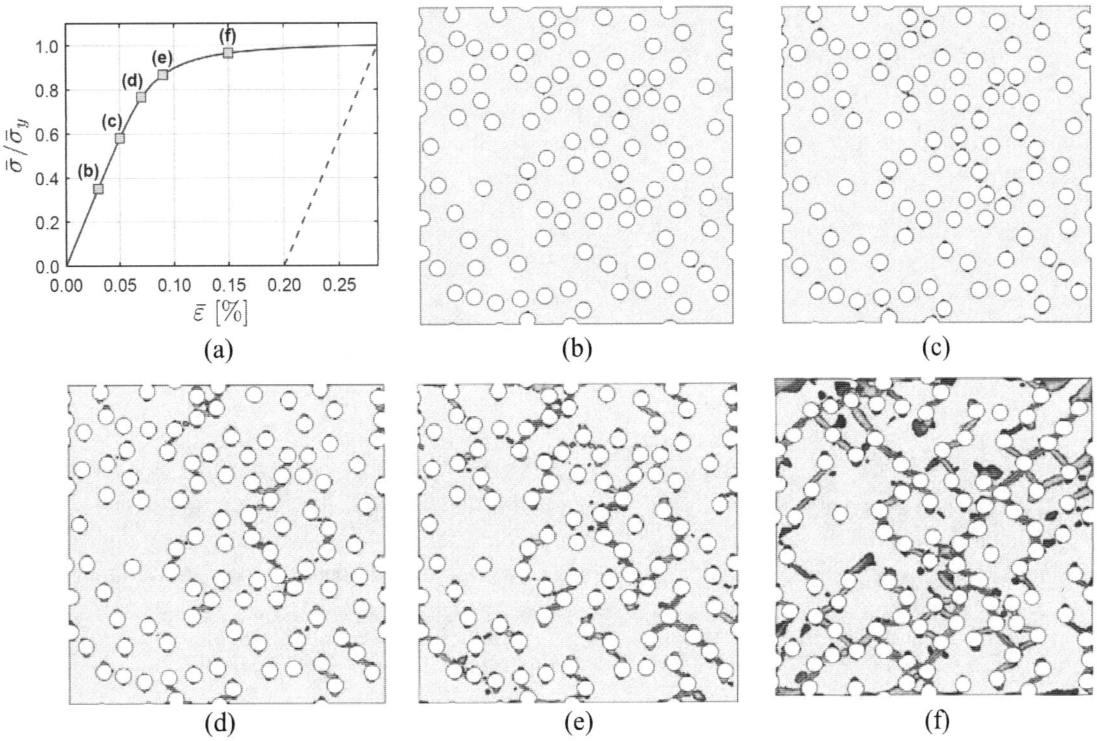

Figure 1. (a) Simulated macroscopic stress-strain curve and (b-f) evolution of the plastic deformation field during virtual tensile testing of a 600x600 μm^2 finite element model of the microstructure of a metal containing a random distribution of circular pores. The model is equivalent to that described in detail in ref. [13]. The level of macroscopic strain $\bar{\varepsilon}$ in the contours (b-f) is indicated in (a). The color in (b-f) indicates the equivalent plastic strain magnitude: blue=low strain, red=high strain. Pores and areas undergoing elastic deformation only are shown in white and light gray, respectively.

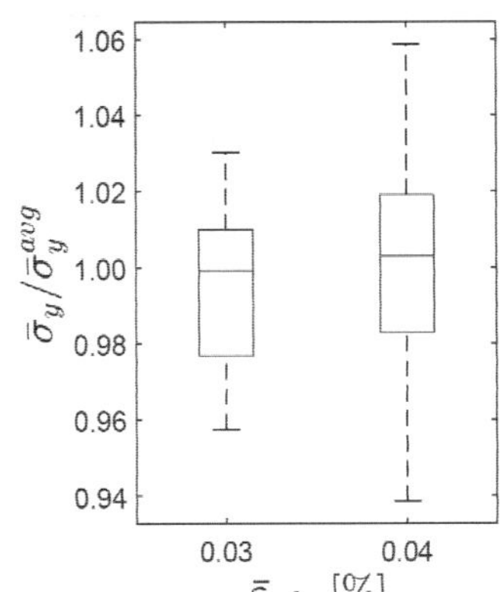

Figure 2. Statistical box plots based on macroscopic values of the yield stress and strain at plastic deformation onset predicted by 100 finite element models. The models are equivalent to that in ref. [13] except for the spatial distribution of the pores, which is generated randomly in each of them.

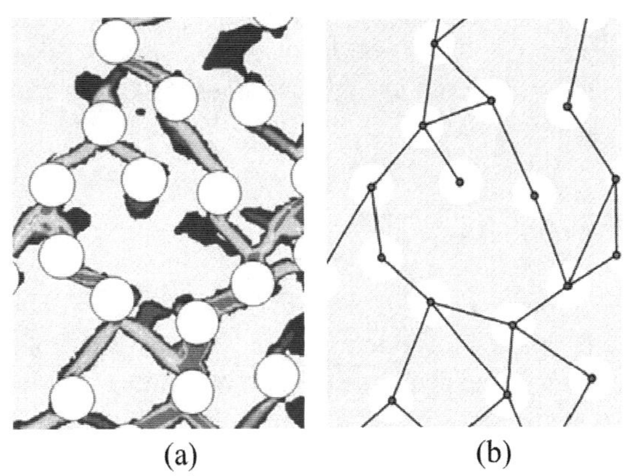

Figure 3. (a) Close-up of plastic shear bands predicted in a region of the finite element model (see figure 1 for color interpretation) and (b) corresponding shear band network. Black solid lines represent the network links and red circles represent the network nodes.

- During the early deformation stages, the number of Delaunay links of length ℓ and orientation θ w.r.t. the tensile loading direction that activate per unit area follows the distribution function

$$f_{sb}(\ell,\theta) \propto \frac{1}{\ell}\chi(\theta) \tag{1}$$

where

$$\chi(\theta) = \begin{cases} \alpha\dfrac{2-\xi(\theta)}{\xi(\theta)^2-\xi(\theta)+1} & \text{if } \theta > \theta^{cr} \\ 0 & \text{if } \theta \le \theta^{cr} \end{cases} \tag{2}$$

In the last expression, $\theta^{cr} \approx 35°$, the normalization factor α is equal to 0.0108 and the function $\xi(\theta)$ is implicitly defined by the equation

$$\cos(2\theta) = \frac{1+\xi}{3(\xi-1)} \tag{3}$$

It is important to remark that, in Eq. (1), the length ℓ of a link is defined as the center-to-center distance between the corresponding voids, minus the sum of the voids'equivalent radii.

- The macroscopic yield strength correlates well with a new indicator I_s expressing the material susceptibility to plastic yielding, which is computed as follows. First, based on the particle distribution in the microstructure, the set of potential shear bands are identified with the set of links of the Delaunay network. Next, the Delaunay links are classified into bins, based on their length ℓ and orientation θ. Finally, I_s is computed as

$$I_S = \frac{1}{n_L^{tot}}\frac{\sum_{i,j} f_{sb}(\ell^i,\theta^j)n_L(i,j)}{\sum_{i,j} f_{sb}(\ell^i,\theta^j)} \tag{4}$$

where ℓ^i and θ^j denote the mean values of ℓ and θ in the bin (i,j), $n_L(i,j)$ is the number of Delaunay links in the bin (i,j), n_L^{tot} is the total number of Delaunay links and $f_{sb}(\ell^i,\theta^j)$ is computed via Eqs. (1)–(3).

In essence, the work [13] shows that by exploiting the concept of SBN, a yield strength indicator can be derived that is inexpensive, parameter-free and can account for the combined effect of the particle *spatial distribution* and *mean volume fraction*. However, such an approach was only validated against data from virtual tensile testing of 2D artificial microstructures. The aim of the present work is to take a first step to verify the applicability of the method to real microstructures and real testing conditions.

3. Methodology

3.1. Experimental

A ferritic SGI was chosen as test material, due to its composite microstructure consisting of soft, spherical graphite particles embedded in a ductile matrix of ferrite. A dog bone shaped specimen with gauge length 10 mm and 1x2 mm rectangular cross-section was machined from a cast block. The specimen was mounted on a displacement-controlled tensile stage equipped with a load cell inside a Zeiss Xradia 520 Versa Computed Tomography system and loaded incrementally up to a strain of ≈ 7%. At each load increment, X-ray tomography scans of the central volume of the specimen (see figure 4 (a)) were acquired in-situ using energies up to 150 keV, resulting in a voxel size of 1.359 μm. More details of the experimental setup and of the specific material used can be found in ref. [6].

The global digital volume correlation (G-DVC) algorithm implemented in the software Avizo® was used to compute the strain field at the last load increment over the internal Y-Z section of the specimen shown in figure 4 (b). A mesh composed of tetrahedral elements with side-length 60 μm was

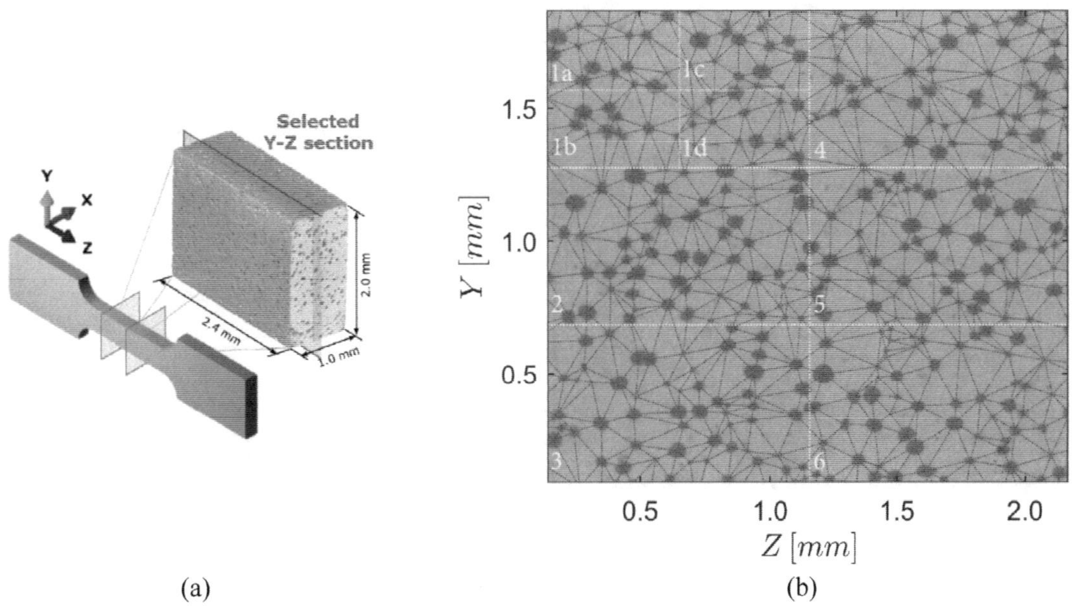

Figure 4. (a) Geometry of the tensile specimen with indication of the volume that was scanned with X-ray tomography. (b) Slice of the X-ray tomogram corresponding to the longitudinal Y-Z section of the specimen that is highlighted in (a). The dashed yellow lines indicate the division into regions and subregions, whereas the red segments and dots represent the links and nodes of the Delaunay network associated with the graphite particles.

adopted, which provides a strain resolution – determined from two distinct X-ray tomograms of the undeformed specimen – that is 5 to 10 times smaller than the measured strain values. The von Mises equivalent strain from G-DVC was then averaged over the six rectangular regions delimited by dashed yellow lines in figure 4 (b) – henceforth denoted as "Region 1", .., "Region 6" – in order to reveal variations of the mean plastic deformation associated with local variations of the particle size and spatial distributions. This strain averaging process was repeated over the 24 subregions obtained by dividing each region into 4 parts A, B, C, D as shown in the top-left corner of figure 4 (b).

3.2. Image analysis
The slice of the X-ray tomogram shown in figure 4 (b) was segmented using a standard thresholding technique in order to separate the graphite phase from the surrounding ferrite. Next, the individual particles were identified by means of a watershed procedure [18] and their equivalent radii and centroids were calculated. This information was used to compute the Delaunay network, which is shown in red in figure 4 (b). The indicator I_S was then calculated for Region 1 – 6 based on Eq. (4), considering, as already mentioned, the length ℓ of each Delaunay link equal to the center-to-center particle distance minus the sum of the particle equivalent radii [13]. Finally, the computation of I_S was repeated for each of the 24 subregions mentioned above, to investigate the sensitivity of the results to the size of the area over which I_S is calculated.

4. Results and discussion
The values of the mean equivalent strain measured with G-DVC over each region are listed in Table 1. Variations in the order of ± 5% can be noticed, which become as large as ± 10% when the strain averaging is carried out at the level of the subregions. This indicates a strong impact of the microstructure on how plastic deformation takes place locally during tensile loading.

Table 1. G-DVC measured von Mises equivalent strain averaged over the regions and subregions defined by the dashed yellow lines in figure 4 (b). Values are in %.

	Region 1	Region 2	Region 3	Region 4	Region 5	Region 6
Subregion A	7.49	6.39	7.27	6.76	7.38	7.47
Subregion B	6.75	6.71	6.94	6.64	7.28	7.38
Subregion C	7.40	7.04	6.73	7.45	7.58	7.67
Subregion D	7.60	7.01	7.15	7.45	7.91	7.36
Mean	7.31	6.88	7.02	7.07	7.54	7.47

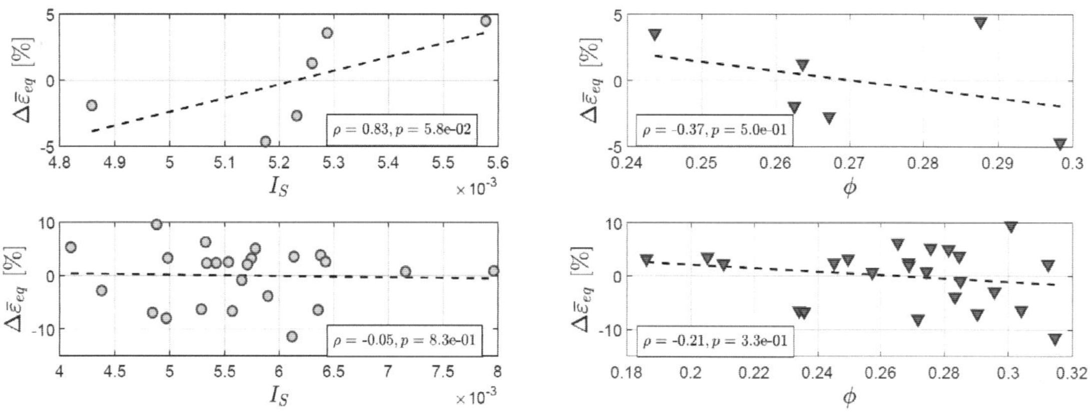

(a) (b)

Figure 5. Correlation between the spatial variation of the mean equivalent strain, $\Delta\bar{\varepsilon}_{eq}$, and (a) the strength indicator I_s and (b) the disorder parameter ϕ. In both (a) and (b), the top plot is based on data computed at the level of each region, whereas the bottom plot is based on data computed at the level of each subregion. Also, in each plot the dashed black line represents a linear fit of the data, whereas ρ and p indicate the value of the Spearman's rank coefficient and its p-value, respectively.

The top plot of figure 5 (a) compares the variation of the mean strain across Region 1–6 to the corresponding values of I_s. The values of the Spearman's rank coefficient and of its p-value (reported in the figure) confirm that there is significant positive correlation between the two quantities, meaning that the larger I_s and the larger the mean strain. This is in agreement with the fact that I_s was introduced to quantify the material susceptibility to plastic yielding [13], and therefore areas with higher values of I_s are expected to yield earlier and develop larger plastic deformation upon loading. On the other hand, some non-negligible scatter in the plotted data exists. The fluctuation in the graphite volume fraction – which varies from a minimum of 12.3% to a maximum of 16.6% across Region 1–6 – should not play a role in this respect, because, as already mentioned, the yield strength indicator I_s is expected to account for the combined effect of the particle spatial distribution and volume fraction. Therefore, the observed scatter can raise the suspicion of deficiencies either in the definition of I_s or, more in general, in the network approach described in Section 2. However, it must be pointed out that the scatter can also be associated with other reasons. In fact, it should be kept in mind that the indicator I_s was developed based on 2D considerations [13]. Therefore, it can be applied in the present analysis under the assumption that the material that extends from the considered Y-Z section to the lateral faces of the specimen (see figure 4) affects Region 1–6 to the same degree. Another implicit assumption is that the strain variability is only caused by fluctuations in the graphite particle size and spatial distributions. That is, the ferrite matrix surrounding the graphite behaves as a homogeneous isotropic medium with uniform properties. In reality, depending on the solidification history, local property variations can exist in the matrix of SGI, which can have a pronounced effect on the local plastic deformation [19,20]. The extent to which these two assumptions are violated is hard to estimate, but it can be expected that the extent increases with decreasing size of the region over which the strain is averaged, as the impact of local phenomena is amplified. The absence of correlation between I_S and the mean strain when the data for the 24 smaller subregions is considered (see bottom plot of figure 5 (a)) suggests that this might indeed be the case. On the other hand, such finding may also be linked to the inability of I_S to give a meaningful estimate of the yielding susceptibility when computed over an area that is smaller than a certain threshold. In the light of this, a definitive conclusion regarding the applicability of the indicator I_S to real materials cannot be reached just based on the present data.

The current study confirms, however, that the plastic deformation of metals containing soft particles or voids cannot be predicted by means of methods based on the concept of elastic stress concentration. In fact, figure 5 (b) shows that there is no correlation between the variation of the mean strain and the corresponding values of the disorder index ϕ. This last quantity represents the magnitude of the local porosity (or particle volume fraction) fluctuations relative to the macroscopic value, and it is known to capture the impact of variations in the pore (or particle) distributions on the strength of brittle elastic materials [21]. In the present case, a large value of ϕ should be associated with reduced local strength and thus higher plastic strain, which is nevertheless the opposite of what the data in figure 5 (b) seems to suggest.

5. Conclusions and outlook

A novel network approach was recently proposed by the author and co-workers to model the plastic deformation that controls the yield strength of metals containing soft particles and/or pores. In this work, the capability of the new approach to explain the variations of the plastic deformation arising from fluctuations in the distributions of the graphite particle in a tensile specimen of ferritic SGI was tested. A correlation was detected between the yielding indicator I_S and the equivalent strain computed locally throughout the microstructure. However, a high scatter was observed, along with a marked decrease of the correlation quality upon reduction of the area used to compute the values of the strain and of the indicator I_S. Further studies are needed to clarify these aspects and shed further light on the applicability of this new promising approach. In this respect, it is worth emphasizing that the indicator I_S is based on simple physical arguments that considers only the length and orientation distributions of the Delaunay links as parameters affecting the yield strength. The great opportunity offered by the concept of SBN introduced in [13] is that it paves the way for developing improved indicators by taking advantage of the powerful tools available in the network science community – a path that has never been exploited so far to predict the mechanical properties of metals containing soft particles and/or pores.

Acknowledgements

The X-ray data used in this article was collected within the framework of a previous work, ref. [6], thanks to the invaluable help of Dr. Søren Fæster and Dr. Yubin Zhang from the Technical University of Denmark.

References

[1] Aboudi J, Arnold S and Bednarcyk B 2012 *Micromechanics of Composite Materials: a Generalized Multiscale Analysis Approach* (Butterworth-Heinemann)

[2] Ben D D, Ma Y R, Yang H J, Meng L X, Shao X H, Liu H Q, Wang S G, Duan Q Q and Zhang Z F 2020 Heterogeneous microstructure and voids dependence of tensile deformation in a selective laser melted AlSi10Mg alloy *Mater. Sci. Eng. A* **798** 140109

[3] Rousselier G 2021 Porous plasticity revisited: Macroscopic and multiscale modeling *Int. J. Plast.* **136** 102881

[4] Sieradzki K and Rong Li 1986 Fracture behavior of a solid with random porosity *Phys. Rev. Lett.* **56** 2509–12
[5] Bishop J E, Emery J M, Battaile C C, Littlewood D J and Baines A J 2016 Direct numerical simulations in solid mechanics for quantifying the macroscale effects of microstructure and material model-form error *JOM - J. Miner. Met. Mater. Soc.* **68** 1427–45
[6] Andriollo T, Zhang Y B, Fæster S, Thorborg J and Hattel J 2019 Impact of micro-scale residual stress on in-situ tensile testing of ductile cast iron: Digital volume correlation vs. model with fully resolved microstructure vs. periodic unit cell *J. Mech. Phys. Solids* **125** 714–35
[7] Matouš K, Geers M G D, Kouznetsova V G and Gillman A 2017 A review of predictive nonlinear theories for multiscale modeling of heterogeneous materials *J. Comput. Phys.* **330** 192–220
[8] Bargmann S, Klusemann B, Markmann J, Schnabel J E, Schneider K, Soyarslan C and Wilmers J 2018 Generation of 3D representative volume elements for heterogeneous materials: A review *Prog. Mater. Sci.* **96** 322–84
[9] Segurado J, Lebensohn R A and LLorca J 2018 Computational homogenization of polycrystals *Advances in Crystals and Elastic Metamaterials, Part 1 Advances in Applied Mechanics* vol 51, ed M I Hussein (Elsevier) pp 1–114
[10] Villarraga-Gómez H, Herazo E L and Smith S T 2019 X-ray computed tomography: from medical imaging to dimensional metrology *Precis. Eng.* **60** 544–69
[11] Sola A, Bellucci D and Cannillo V 2016 Functionally graded materials for orthopedic applications – an update on design and manufacturing *Biotechnol. Adv.* **34** 504–31
[12] Wu X and Zhu Y 2017 Heterogeneous materials: a new class of materials with unprecedented mechanical properties *Mater. Res. Lett.* **5** 527–32
[13] Andriollo T, Kouznetsova V and Alessandretti L A network approach for the prediction of the irreversible deformation of heterogeneous solids *Phys. Rev. Lett.* Under review
[14] Ashby M, Shercliff H and Cebon D 2010 *Materials: engineering, science, processing and design* (Oxford, UK: Butterworth-Heinemann)
[15] Buljac A, Helfen L, Hild F and Morgeneyer T F 2018 Effect of void arrangement on ductile damage mechanisms in nodular graphite cast iron: In situ 3D measurements *Eng. Fract. Mech.* **192** 242–61
[16] Bilger N, Auslender F, Bornert M, Michel J C, Moulinec H, Suquet P and Zaoui A 2005 Effect of a nonuniform distribution of voids on the plastic response of voided materials: A computational and statistical analysis *Int. J. Solids Struct.* **42** 517–38
[17] Barthelemy M 2018 *Morphogenesis of spatial networks* (Cham, Switzerland: Springer International Publishing)
[18] Jain A 1989 *Fundamentals of Digital Image Processing* (Prentice-Hall)
[19] Xu C L, Andriollo T, Zhang Y B, Hernando J C, Hattel J and Tiedje N 2020 Micromechanical impact of solidification regions in ductile iron revealed via a 3D strain partitioning analysis method *Scr. Mater.* **178**
[20] Fernandino D O and Boeri R E 2019 In-situ microscopic analysis of ferritic ductile iron during tensile loading: Relation between matrix heterogeneities and damage mechanisms *Fatigue Fract. Eng. Mater. Struct.* 1–12
[21] Laubie H, Radjai F, Pellenq R and Ulm F J 2017 Stress Transmission and Failure in Disordered Porous Media *Phys. Rev. Lett.* **119** 1–6

IOP Conf. Series: Materials Science and Engineering 1249 (2022) 012019 doi:10.1088/1757-899X/1249/1/012019

Simulated migration behavior of metastable Σ3 (11 8 5) incoherent twin grain boundaries

Eric R Homer[1], Akarsh Verma[1], Darcey Britton[1], Oliver K. Johnson[1], and Gregory B. Thompson[2]

[1]Dept. of Mechanical Engineering, Brigham Young University, Provo, UT 84602, USA.
[2]Dept. of Metallurgical and Materials Engineering, University of Alabama, Tuscaloosa, AL 35401, USA.

E-mail: eric.homer@byu.edu

Abstract. Molecular dynamics simulations are used to examine the migration behavior of one incoherent twin, a Σ3 (11 8 5) / (8 11 5) grain boundary. The boundary is known to exhibit non-Arrhenius boundary migration that slows as temperature increases. This behavior is examined in 165 metastable structures of the same boundary and in a large simulation cell where smaller length scales can have less of an effect. The metastable boundaries show diverse migration behaviors from non-Arrhenius to Arrhenius, though the majority of them exhibit non-Arrhenius behavior. The large simulation cell sizes show no dependence on system size, eliminating concerns about the effect of facet nucleation on the migration in periodic simulation cells. However, facet structures play an important role in the migration of the metastable boundaries. Boundaries with larger facets typically migrate faster than those with smaller facets and with defects in the boundary structure.

1. Introduction

Migration of incoherent twin grain boundaries (GBs) has attracted attention in recent years due to their interesting behaviors. For example, in a survey of 388 nickel GBs, Olmsted et al. noted that a large number of incoherent twin GBs exhibited "non-activated" motion that contrasts traditional thermally activated expectations of GB mobility [1]. It is noted by Olmsted et al. that the mobility is inversely proportional to temperature, decreasing as temperature increases. This behavior has been confirmed in numerous simulations [2–6]. Experiments of nanoindentation at cryogenic temperatures suggest that twins play an important role in the observed coarsening behaviors [7].

In recent years more attention has been paid to the fact that GBs can exhibit a multiplicity of atomic structures even when the macroscopic, crystallographic character is fixed. For example, several works have examined metastable configurations rather than focusing only on the minimum energy GB structure [8–12]. Furthermore, Race et al. have demonstrated that size of the simulation cell can play an important role in the observation of GB migration since some GBs require nucleation of new facets and the GB plane size affects the kinetics of that nucleation [13]. To gain insight into these two factors, our simulations examine metastable GB structures and simulations in large cells.

In this work, we examine the temperature dependent migration behavior of one incoherent twin, a Σ3 (11 8 5) / (8 11 5). This particular GB has been studied previously to determine the mechanisms for migration [3], which are consistent with mechanisms for other incoherent twin boundaries [14]. In this same GB, Chesser and Holm later showed that the interatomic potential and the number and height of the facet junctions play a critical role in whether non-Arrhenius migration characteristics are exhibited [4]. Furthermore, a correlation was found betweeen generalized interface fault energy (GIFE), defined in [15], with lower GIFE exhibiting non-Arrhenius behavior and higher GIFE exhibiting Arrhenius behavior.

More recently, some of the authors revived a classical model of grain boundary migration that, in its unsimplified form, is able to capture a natural transition from Arrhenius to non-Arrhenius migration, which occurs when barrier heights of migration transition from large to small [6]. The model is derived from two thermally activated processes of forward and backward jumps that facilitate or detract from the net migration of the boundary. In this model the GB velocity, v, is related to the intrinsic barrier height of migration, Q, and driving force for migration, p, by the following relationship:

$$v = v_o \exp\left(-\frac{Q}{k_B T}\right) 2\sinh\left(\frac{p}{2k_B T}\right) \tag{1}$$

where v_o is a velocity prefactor and k_B is Boltzmann's constant. It is worth noting that in the limit of small driving force p relative to temperature T (i.e. $p \ll k_B T$), the sinh function is approximately linear and can be simplified using a first-order Taylor series expansion (linear term only). In this limit, something very similar to the traditional relationship between velocity and driving force is recovered, $v = Mp$, where the two values are related by the mobility, M. Nevertheless, as defined in Eq. 1, the velocity can be predicted to exhibit both Arrhenius and non-Arrhenius behaviors, depending upon the magnitude of the intrinsic barrier height, Q, and the magnitude of the driving force, p [6].

Some of the results in this work were presented in [6] as evidence to support the form of the classical model, as defined in Eq. 1. In the work presented here, we examine these results in additional detail to better understand how the metastable structure of a GB can lead to diverse behaviors. We also examine new simulations with large simulation cells to determine how the migration behavior is affected by this size effect.

2. Methods

In this work we examine a single incoherent twin using atomistic simulations. It is a $\Sigma 3$ coincidence site lattice (CSL) GB with a 60° rotation about the [1 1 1] axis. The boundary planes for the GB of interest are defined from both sides of the GB as (11 8 5) / (8 11 5).

To understand how metastability and system size affect the migration of a boundary known to exhibit non-Arrhenius migration, we examine two different sets of simulations. The first set of simulations is aimed at examining how different metastable GB structures affect the overall migration. The second set of simulations is aimed at examining a larger simulation cell size. Both sets of simulations examine nickel with the embedded atom method (EAM) potential developed by Foiles and Hoyt [16]. Finally, we employ methods to examine the variability of the migration velocities and use this to determine uncertainty of measured migration characteristics.

2.1. Simulations to examine metastability

The first set of simulations, examining metastability, have a simulation cell that is 41.8 Å by 30.3 Å in the GB plane and more than 85 Å in both directions normal to the boundary plane. The simulation cells are periodic in the boundary plane, but not in the direction normal to the boundary plane, thus containing one GB. Each simulation cell has ~20,500 atoms.

We followed standard construction methods to create numerous possible starting configurations before using conjugate gradient minimization to determine the minimum energy structure of each starting configuration [17–19]. Variables influencing the starting configurations

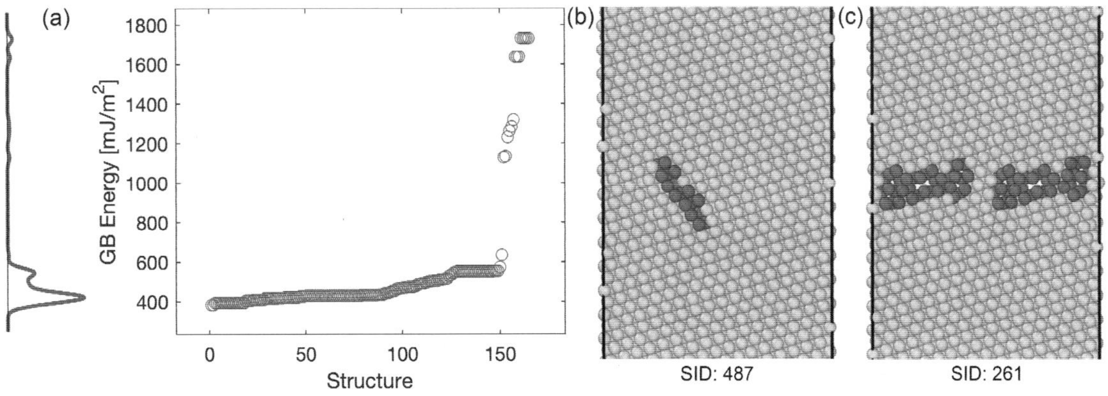

Figure 1. (a) Monotonic ordering and distribution of grain boundary interface energy for all 165 $\Sigma3\langle 11\,8\,5\rangle$ metastable grain boundary structures. (b-c) Atomic structure plots of the (b) minimum and (c) maximum GB energy structures, respectively.

of the atoms include relative placement of the two grains, location of the boundary plane, allowed proximity of atoms, and deletion of atoms from specific crystals when allowed proximity is exceeded, all of which are discussed in more detail in [18]. This resulted in 504 possible starting points from which 165 unique GB structures were obtained; a distribution of GB energies of these 165 structures can be seen in figure 1a. These structures are labeled throughout the manuscript with the structure ID (SID) assigned when the GBs were created.

Snapshots of the minimum and maximum energy GB structures are shown in figures 1b and 1c. The minimum energy structure has a single, large coherent twin facet separated by a step that is comprised of 6 $\{1\,1\,1\}$ planes. The maximum energy structure has two much smaller coherent twin facets separated by much larger defects.

We define each of the 165 structures as a metastable configuration since each has unique configurations of atom positions. It is possible that some of these structures that are unique at 0 K may adopt similar structures at finite temperatures. Determining uniqueness among metastable configurations is a topic of active research and is beyond the scope of the present work [8, 10, 11]. This sampling of metastable structures is not meant to be representative of an ensemble; we simply note that these 165 metastable boundaries represent a range of structures this GB could adopt.

Instead of simulating the migration of just the minimum energy grain boundary structure, as is most commonly done, in the first set of simulations we examine the migration of all 165 metastable structures. Molecular dynamics is used to simulate the grain boundary migration using LAMMPS [20]. Standard energy conserving orientational (ECO) force methods [21, 22] are used to induce migration with a driving force of 0.01 eV/atom. The ECO driving force uses an orientation parameter to determine which atoms belong to which grain and then adds energy to or subtracts energy from atoms in the two grains across the boundary. The difference in energy induces atoms in the high energy grain to join the lattice of the low energy grain, thereby reducing the energy in the system and causing migration of the boundary. Each boundary is subjected to the ECO driving force until the boundary sweeps entirely through one crystal or for a maximum time of 400 ps. If the boundary does not travel more than 10 Å in the 400 ps, it is designated as immobile, with a velocity of zero.

Since molecular dynamics requires a random seed for the initial velocity distribution of the atoms, different simulation trajectories are obtained with each unique random seed of a simulation run. To understand how this variation in trajectories affects the simulations, we

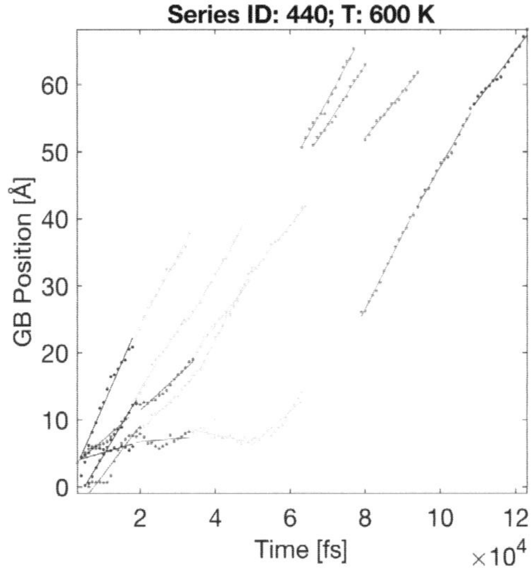

Figure 2. Temporal subdivision velocity fits to GB position as a function of time for 5 different replicates of one metastable grain boundary structure. Figure reproduced from [6].

examined 5 replicates of each grain boundary migration simulation.

Each simulation cell uses an NPT thermostat-barostat and is equilibrated for 50 ps to ensure adequate convergence of temperature to a desired value of 200, 400, 600, 800, 1000, 1200, or 1400 K, at a desired pressure of 0 MPa. Most temperature and pressure fluctuations about the target temperature and zero pressure are ± 10 K and ± 100 MPa, respectively. In total, 5775 simulations were run (165 unique grain boundaries $\times$ 5 replicates $\times$ 7 temperatures).

To account for the fact that not all boundary position vs. time data are linear over the entire range, we employ a temporal subdivision method to fit the velocity over short times [23, 24]. This enables the measurement of velocity variation even under conditions of stick-slip migration. Other methods to quantify boundary migration uncertainty also exist [25]. An example of this temporal subdivision applied to the 5 replicates of a single GB is given in figure 2. Here, the boundary position is plotted as a function of time and the velocity is fit over 15 ps windows. The subdivisions are disjoint and result in a distribution of measured velocities. One can combine the measured velocities for all the replicate runs to obtain a distribution of velocities. This velocity distribution for a given structure at a given temperature accounts for variations in velocity from the various replicates, as well as variation in velocity over time during a given simulation obtained by the temporal subdivision methods used here.

2.2. Simulations to examine system size

The second set of simulations, examining system size, have a simulation cell that is much larger at 500 Å by 517 Å in the GB plane and 408 Å in the direction normal to the GB. The methods to determine dimensions of simulation box are from [26]. This simulation cell size is in the regime required for multiple-island nucleation [13].

In this large simulation work, cells are periodic in all three dimensions and thus have two GBs. The starting configuration for this GB is obtained by deleting the overlapping nickel atoms. No effort is made to find minimum energy configurations. Each simulation cell has more than 9.6 million atoms. Boundary migration is induced using the ECO driving force as noted above using a driving force of 0.01 eV/atom. Migration is measured in the same manner as described for the metastable GB simulations.

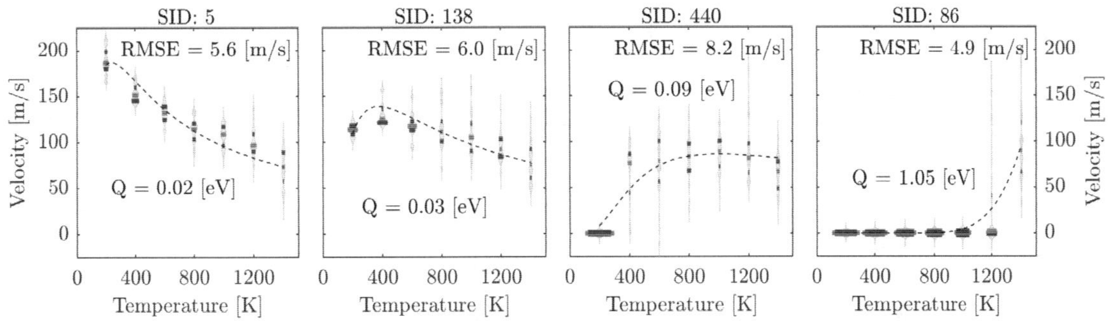

Figure 3. Velocity distributions and corresponding fit for migration of 4 different metastable structures. The red and blue lines mark the median and 1st/3rd quartiles of the distribution, respectively. Figure adapted from [6].

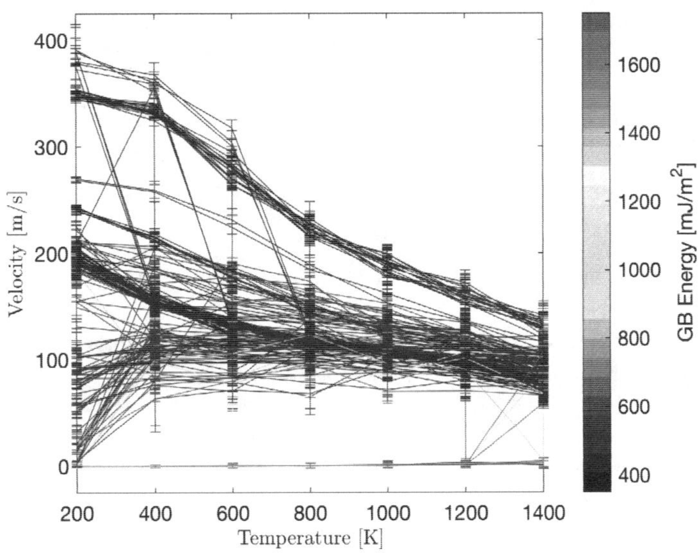

Figure 4. Velocity distributions for all 165 metastable structures as a function of temperature, colored by GB energy. Figure reproduced from [6].

3. Results

3.1. Simulations to examine metastability

We first examine the simulations of all 165 metastable structures. Four representative velocity distributions over the full range of temperatures simulated are shown in figure 3. Here one can see the velocity distributions at each simulation temperature, accounting for variation from different replicate simulations as well as the temporal subdivision. As can be seen, the migration behavior of these 4 structures is notably varied, from entirely non-Arrhenius, with inverse temperature dependence, to entirely Arrhenius, with immobile boundaries at low temperatures, but whose velocity increases exponentially at the highest two temperatures.

To illustrate the velocity trends among all 165 metastable structures, figure 4 plots the median velocity of the velocity distributions with error bars that show the first and third quartiles of the measured velocity distributions at each temperature.

Bayesian regression is used to fit Eq. 1 to the velocity distributions over the full range of temperatures, as illustrated with the dashed line in figure 3. The values and uncertainty of the

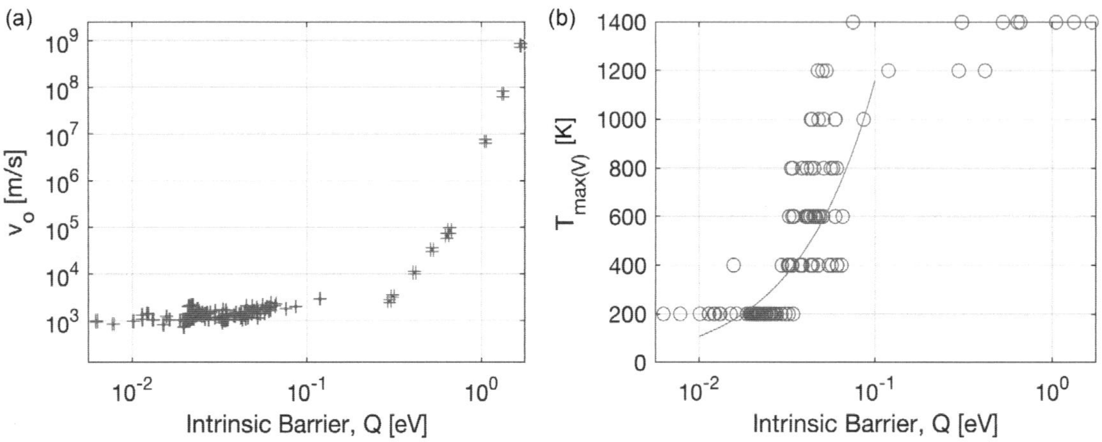

Figure 5. (a) Distribution of fit parameters. Note that one standard deviation of the uncertainty on the fits is of the size of the datapoints in the plot. (b) Plot of temperature at which each metastable structure has its maximum velocity as a function of the fitted intrinsic barrier height Q. Red line is a plot of Eq. 2. Figures reproduced from [6].

fit variables, Q and v_o, are plotted in figure 5a. The goodness of fit for all 165 structures is given by the weighted root-mean-square error (RMSE), the average and maximum of which are 6.8 and 26 m/s, respectively. The uncertainty of the fit variables is illustrated by the error bars on each datapoint in figure 5a.

Figure 5b plots the temperature at which the velocity of each simulation is maximum $T_{max(v)}$ as a function of the fitted intrinsic barrier height Q, which is obtained from the fit of the data to Eq. 1. It can be seen that as the intrinsic barrier height Q drops, the temperature at which the velocity is a maximum moves to lower temperatures. This is overlaid with an analytical solution, in red, which can be determined by solving for the temperature, T, when Eq. 1 is at its maximum, which is given by

$$T_{v_{max}} = \frac{p}{2k_B \tanh^{-1}\left(\frac{p}{2Q}\right)}. \tag{2}$$

The analytical solution captures the general trend in the data, which is one indicator that Eq. 1 describes the trends in all 165 metastable structures. Note that $T_{max(v)}$, as obtained from the data, only takes values corresponding to the simulation temperatures, which occur at 200 K intervals.

3.2. Simulations to examine system size

The velocities of the two GBs in the large simulation cell are shown as a function of temperature in figure 6a. Note that the velocity values are in the top and bottom range of those illustrated in figure 4. Note also that the intrinsic barrier height values, Q, are also in the range of those expected for the temperature at which the velocity is a maximum, as illustrated in figure 5b.

It is noteworthy that the two interfaces with the same overall boundary plane migrate at such different velocities. However, the two GBs have different faceting structures, as illustrated in figure 6b where the atoms are colored by the centrosymmetry parameter [27]. The bottom GB, which migrates more quickly, has large twin facets separated by large near $\{1\,1\,0\}$ facets, which migrate in an almost planar fashion. In contrast, the top GB has smaller twin facets that are separated by alternating near $\{1\,1\,2\}$ facets. The near $\{1\,1\,2\}$ facets are illustrated more

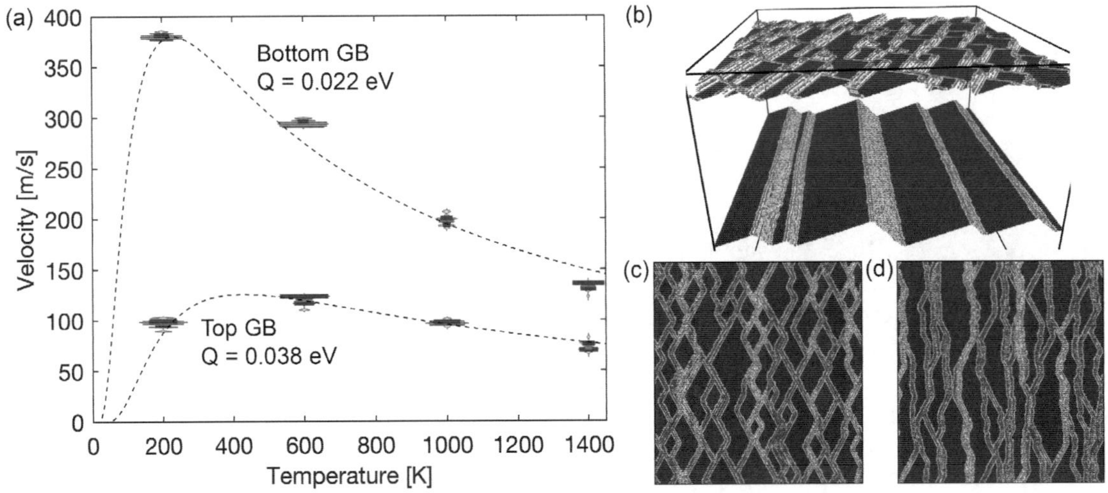

Figure 6. (a) Velocity as a function of temperature for the large system size simulation cell containing two GBs. Intrinsic barrier heights, Q, are fit using Eq. 1. (b) Structure of the large system size simulation cell at 200 K showing the different structure in the top and bottom GBs. (c) Top down view showing the alternating $\{1\,1\,2\}$ facets in the top GB at the beginning of the simulation that evolve to be more like near $\{1\,1\,0\}$ facets in (d) at the end of the simulation. The atoms in (b-d) are colored according to their centrosymmetry parameter value.

Table 1. List of the number of facets for the top and bottom GBs in the large simulation cell as a function of temperature.

	Number of Facets			
	200 K	600 K	1000 K	1400 K
Top GB	14-16	10-11	12-14	15
Bottom GB	4-5	6	5-7	8

clearly with a top down view in figure 6c at the beginning of the simulation. Interestingly, the near $\{1\,1\,2\}$ facets are largely immobile but the near $\{1\,1\,0\}$ facets emerge from the $\{1\,1\,2\}$ facet junctions and the boundary migrates by the motion of the near $\{1\,1\,0\}$ facets that merge. This is visible in a top down view of the top GB at the end of the simulation, as illustrated in figure 6d.

Table 1 lists the range of the number of facets, counted by looking at the simulation from the edge, that exist in each cell at the different temperatures. The faster, bottom GB has about half as many facets as the slower, top GB.

4. Discussion

First, we note that that the diverse migration behaviors illustrated in figures 3 and 4 span the full range from non-Arrhenius, with inverse temperature dependence, to Arrhenius, where velocity increases with Arrhenius temperature dependence. The majority of the metastable GBs exhibit non-Arrhenius behavior, confirming this type of behavior for the $\Sigma3\,(11\,8\,5)$ GB. We also note that the classical velocity equation, Eq. 1, is able to account for the full range of behaviors by fitting the prefactor, v_o, and the intrinsic barrier height, Q, of the process. In short, at small values of Q, the temperature dependence in the sinh term of Eq. 1 dominates and results in the non-Arrhenius inverse temperature dependence. At larger values of Q, the exponential term of Eq. 1 dominates and results in the expected Arrhenius temperature dependence. This classical model is described with more detail in [6]. In what follows, we discuss what is learned about the structure of these GBs that leads to the diverse behaviors observed.

We begin with the large system size simulations where the two GBs, with different structures, exhibit different velocity profiles, both of which are non-Arrhenius. As noted above, the system size of these large simulations is within the range specified by Race et al. for multiple-island nucleation [13]. Hadian et al. also showed that stepped, or faceted, GBs have lower nucleation barriers allowing them to migrate faster than flat GBs where nucleation would play a bigger role [28]. This is critical because small simulation cells can lead to structures where a given facet might persist and allow the boundary to migrate quickly, whereas in a large simulation cell, that facet might have to nucleate to allow that boundary to migrate quickly. The nucleation time can be the limiting factor in the migration of some boundaries, leading to faster velocities in small simulation cells and slower velocities in larger simulation cells. The velocities of the two GBs in the large simulation cells is well within the range of the velocities exhibited by the smaller metastable GBs (c.f. figures 4 and 6). Furthermore, the faceted boundaries in this work have the whole facet migrate together, not nucleating single steps of the facet as shown in other works [29]. Thus, at least in the cases examined here, nucleation of new facets does not appear to be a limiting factor to enable the migration of these GBs when using simulation cells that are periodic.

However, if one were to use non-periodic simulation cells, once each facet migrated to the edge, the boundary would be a coherent twin and have a new boundary plane orientation. At this point, any further migration would require nucleation and growth of some sort. Furthermore, at these large, non-periodic scales, one would likely need to analyze boundary migration in the context of a microstructural network where different triple junctions could influence the types of disconnections that would nucleate and migrate, as has been examined in [30].

In the large simulation cell sizes it is clear that the two GBs that migrate at different velocities and have different intrinsic barrier heights also have very different facet structures. Humberson and Holm have shown that system size affects the type of facets that can emerge in the GB studied here [3], which has an effect on the observed migration behaviors as evidenced by the data in figure 6. As noted in the results, the bottom GB migrates more quickly and has larger facets that are simpler in structure. In contrast, the top GB structure has twin facets separated by alternating near $\{1\,1\,2\}$ facets, which is a faceting structure that is discussed in more detail in [3]. However, as has been reported in [14], a boundary with a $\{1\,1\,2\}$ normal doesn't migrate as quickly as a boundary with a $\{1\,1\,0\}$ normal. Thus, it is not surprising that the boundary moves under the motion of near $\{1\,1\,0\}$ facets, as shown in figure 6d.

It is also worth noting that all the facets between the coherent twins in figure 6 have a height with a multiple of 3 $\{1\,1\,1\}$ planes. This fits with observations of simulated migration of incoherent twin facets where each set of 3 $\{1\,1\,1\}$ planes of atoms exhibits unique and specific atomic motions along Shockley partial directions [3, 14]. This is also supported by reports of incoherent twin boundary migration examined in terms of different combinations of disconnections that allow the boundary to migrate with or without strain [30]; these disconnection combinations involve specific numbers of $\{1\,1\,1\}$ planes.

While this analysis provides insight into the factors that impact the velocity of the top and bottom GBs in the large simulation cell, it does not clearly determine why the bottom GB is faster than the top GB. As noted above, the two both migrate from the motion of near $\{1\,1\,0\}$ facets. The top GB has nearly twice as many facets as the bottom GB (c.f. Table 1). Is there nucleation involved in the migration of the top GB by the emerging $\{1\,1\,0\}$ facets or accommodation of the structure in the $\{1\,1\,2\}$ planes as they are consumed? Is there some interaction between facets that slows GB migration, and which impacts the top GB more since it has more facets? While we are unable to determine which, if any, of these is the exact cause of the slower migration, we can gain additional insight by examining the structure and other

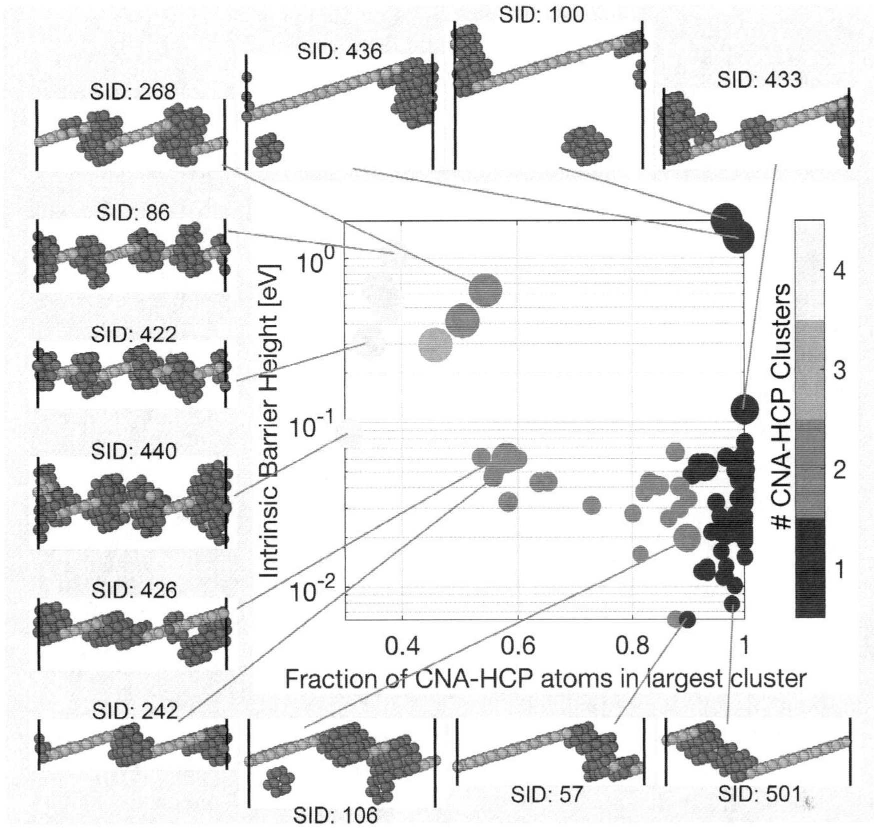

Figure 7. Plot of the intrinsic barrier height,Q, as a function of the fraction of CNA-HCP atoms in the largest coherent twin facet. The markers are colored by the number of CNA-HCP atom clusters with more than 10 atoms, and the size of each data point scales with the GB energy, where large points have higher GB energy and vice versa. This plot is surrounded by snapshots of various metastable GBs, identified by their SID. Atoms identified as HCP from the CNA analysis are cyan and atoms that are not FCC, HCP, or BCC are red.

aspects of the metastable GBs.

Figure 7 illustrates a number of the metastable GB structures during migration at 200 K. In these depictions, common neighbor analysis (CNA) is used to determine which atoms belong to the GB by eliminating all atoms with an FCC environment. This leaves behind atoms with an HCP environment, which is exhibited by atoms in a coherent twin ($\{1\,1\,1\}$) plane, and which are colored in cyan. The red atoms are those that don't have an FCC, BCC, or HCP neighborhood. It is clear that in many metastable GBs, a single twin facet is present in the structure. However, figure 7 also illustrates a variety of other structures where multiple twin facets are present. It is worth noting that in all cases when 4 clusters are present, the twin planes are not separated by a multiple of 3 $\{1\,1\,1\}$ planes. The structures labeled as SID 268 and 422 even have edge dislocations in the red atom clusters.

Figure 7 also plots the fitted intrinsic barrier height, Q, as a function of the fraction of CNA-HCP atoms in the largest cluster; the marker size correlates with the 0 K GB energy while the marker colors indicate the number of CNA-HCP atom clusters with more than 10 atoms in the cluster. This plot illustrates a weak negative correlation between the two variables, with a correlation coefficient of -0.36. However, there are clearly outliers. SIDs 100 and 436 both have

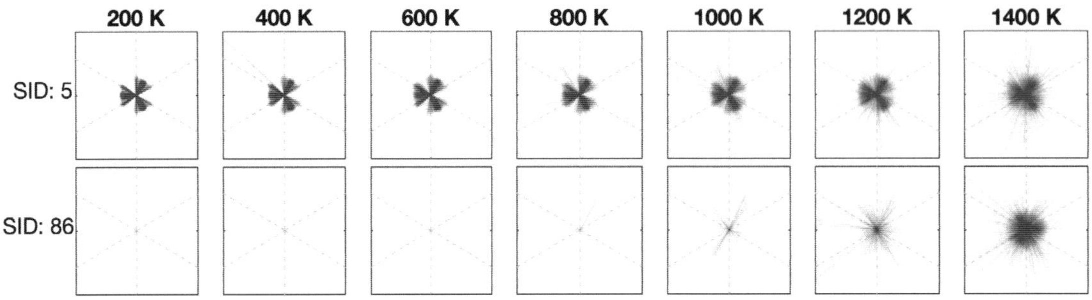

Figure 8. Comparison of slip vector distributions for two different structures of the $\Sigma3\langle 11\,8\,5\rangle$ GB over a range of temperatures. These motions are plotted in the $\{1\,1\,1\}$ plane and the dashed lines mark the $\langle 1\,1\,\bar{2}\rangle$ directions in this plane.

a single facet but have high intrinsic barrier height values, Q; note however that these both have vacancy cluster defects that have been passed through or left behind during the migration. In the case of SID 100, in most simulations the high energy GB is unable to migrate with its high energy structure in the incoherent facet. But in some cases, and as illustrated in figure 7, the boundary manages to restructure itself and migrate away and leave the defect. Since most of the simulations of this boundary did not observe this restructuring and migration away from the defect except at higher temperatures, this boundary behaved in an Arrhenius manner and was fit with a high intrinsic barrier height, Q. SID 106 also has higher GB energy and two point defects but they are smaller and in all cases this boundary migrated easily, meaning that it must have been able to quickly restructure and leave behind those point defects.

It is also worth noting that in the case of the metastable structures with more facets, the inherent atomic motions change. As noted above, previous observations of incoherent twin facet migration have observed that each set of 3 $\{1\,1\,1\}$ planes exhibits specific motions along Shockley partial directions [14]. The same type of analysis is shown in figure 8 for low and high Q value metastable GBs, SIDs 5 and 86, respectively. Figure 8 shows slip vector values, calculated using methods from [31], that enable relative atom motions to be identified as a boundary sweeps through a given region. Note that the low Q value GB, SID 5, has well-defined atomic motion directions that become more diffuse with increasing temperature. This is the thermal disordering at higher temperatures that frustrates the simple atomic motions and is the presumed source of the non-Arrhenius temperature dependence. On the other hand, the high Q value GB, SID 86, is not mobile at low temperatures. However, when it is mobile at high temperatures, the atom motion directions are not the same as those for the low Q value GB. In other words, the nature of the atomic migration has changed and appears to be more random, though they are still almost entirely limited to the $\{1\,1\,1\}$ planes in which they reside.

Together, the metastable GB simulations and the large system size simulations provide considerable insight into the effects of different boundary structures and are supported by other published works. In particular, Chesser and Holm examined migration in the same type of GB as those examined here ($\Sigma3\,(11\,8\,5)$) in a variety of FCC metals. Some metals minimized into structures with multiple facets whereas other metals minimized into a single facet. They showed that facet coalescence is a necessary but insufficient condition for non-Arrhenius boundary migration [4]. From a different perspective, this work supports the importance of having facets of specific structure, but suggests that variations in the structure, including defects in the GB, can interfere with the ability of a boundary to migrate. This points back to the importance of examining the role of the disconnections available to each GB [32]. In particular, the distributions of disconnections available to a given GB may be important [33]. Since disconnections are based on the displacement shift complete (DSC) lattice of the misorientation, it is not clear at this point how one can account for variability in the GB structure. Different disconnections have different energies associated with their formation and migration [33]. Will different disconnections have a greater ability to accommodate the migration of different metastable GB structures? The answer to this question will require an in-depth analysis of the mechanisms of migration in the context of disconnections.

The atomic motions of most of the metastable GBs examined in this work match those shown by SID 5 in figure 8, suggesting they likely migrate using the same combination of disconnections. Variability due to thermal noise is not likely a cause for the large variation because each individual simulation was run with 5 replicates and, as can be seen in figure 3, most of the variation is tightly distributed. The variability in their behaviors is most likely attributed to small differences in the GB structure that may be minor but sufficient to cause the diverse behaviors exhibited in figure 4. The presence of extended defects can have a considerable impact on boundary migration, fundamentally altering the nucleation process to allow boundaries to migrate [34]. Thus, while some defects may slow down a GB, other defects may lower the barrier to nucleation and migration of disconnections.

5. Conclusions

In this work we have examined the migration behavior of an incoherent twin GB in nickel. In particular we have examined how metastable structures lead to a range of diverse GB migration behaviors. From this examination, the following conclusions can be made

- The $\Sigma3\,(11\,8\,5)$ GB exhibits a range of velocity temperature dependence from non-Arrhenius to Arrhenius, which can be quantified by the intrinsic barrier height fit to the observations. However, the majority of the metastable GBs exhibited non-Arrhenius behavior with inverse temperature-dependence in support of previous results.
- The system size can affect the facet structure of GBs that are created, which has an affect on the velocity of a GB. However, the system size does not appear to play a critical role in periodic simulation cells where these GBs have persistent facets that facilitate the migration thereby negating the need for facet nucleation. A different analysis would be required for non-periodic simulation cells where facet persistence would not be guaranteed.
- The variability in the migration behaviors appears to be mostly affected by the faceting structure, where large facets that enable Shockley partial triplets on 3 separate $\{1\,1\,1\}$ planes migrate quickly and smaller facets that don't allow these atomic motions migrate slower.
- Certain GBs, even with large facets, appear to possess structures that do not readily allow the Shockley partial triplets and thus lead to immobility or much slower migration; however, it is possible for the GB to leave the defect behind and then migrate at a faster pace.

In summary, the ability of a GB to take on a multiplicity of structures means that the challenge of predicting GB migration is more complicated. Fortunately, the atomic motions among the metastable configurations of a GB are consistent, at least in the case of the GB examined here. Furthermore, the consistency of GB migration in experiments suggests that one can expect certain behaviors even if it is unlikely that all atoms along a GB structure are exactly the same. While it remains an open challenge to determine how to incorporate the variability of metastable GB structures into continuum models of GB migration, there is evidence that this can be captured in some statistical manner.

Acknowledgements
This work was supported by the U.S. Department of Energy (DOE), Office of Science, Basic Energy Sciences (BES) under Award #DE-SC0016441.

References

[1] Olmsted D L, Holm E A and Foiles S M 2009 *Acta Mater.* **57** 3704–13
[2] Homer E R, Holm E A, Foiles S M and Olmsted D L 2014 *JOM* **66** 114–20
[3] Humberson J and Holm E A 2017 *Scripta Mater.* **130** 1–6
[4] Chesser I and Holm E 2018 *Scripta Mater.* **157** 19–23
[5] Humberson J, Chesser I and Holm E A 2019 *Acta Mater.* **175** 55–65
[6] Homer E R, Johnson O K, Britton D, Patterson J E, Sevy E T and Thompson G B 2022 *npj Computational Materials* **8** 157
[7] Brons J G, Hardwick J A, Padilla II H A, Hattar K, Thompson G B and Boyce B L 2014 *Mater. Sci. Eng* A **592** 182–8
[8] Frolov T, Olmsted D L, Asta M and Mishin Y 2013 *Nat. Comm.* **4** 1899
[9] Cantwell P R, Tang M, Dillon S J, Luo J, Rohrer G S and Harmer M P 2014 *Acta Mater.* **62** 1–48
[10] Han J, Vitek V and Srolovitz D J 2016 *Acta Mater.* **104** 259–73
[11] Hickman J and Mishin Y 2017 *Phys. Rev. Mater.* **1** 010601
[12] Zhu Q, Samanta A, Li B, Rudd R E and Frolov T 2018 *Nat. Comm.* **9** 467
[13] Race C P, Pezold J v and Neugebauer J 2014 *Phys. Rev.* B **89** 214110
[14] Priedeman J L, Olmsted D L and Homer E R 2017 *Acta Mater.* **131** 553–63
[15] Barrett C D, El Kadiri H and Moser R 2017 *Int. J. Solids Struct.* **110-111** 106–12
[16] Foiles S M and Hoyt J J 2006 *Acta Mater.* **54** 3351–7
[17] Olmsted D L, Foiles S M and Holm E A 2009 *Acta Mater.* **57** 3694–703
[18] Homer E R 2015 *36th Risø International Symposium on Materials Science* (Roskilde: IOP Publishing) p 012006
[19] Homer E R, Hart G L W, Owens C B, Hensley D M, Spendlove J C and Serafin L H 2022 *Acta Mater.* **234** 118006
[20] Plimpton S J 1995 *J Comput. Phys.* **117** 1–19
[21] Ulomek F, O'Brien C J, Foiles S M and Mohles V 2015 *Model. Simul. Mater. Sc.* **23** 025007
[22] Schratt A A and Mohles V 2020 *Comp. Mater. Sci.* **182** 109774
[23] Zhou X and Foiles S M 2017 *Uncertainty Quantification and Model Calibration* ed Hessling J P (Rijeka: IntechOpen)
[24] Johnson O K, Homer E R, Fullwood D T, Page D E, Varela K F and Baird S G 2021 *Acta Mater.* **215** 116967
[25] Race C P 2014 *Mol. Simulat.* **41** 1069–73
[26] Tschopp M A and McDowell D L 2007 *Philos. Mag.* **87** 3147–73
[27] Kelchner C L, Plimpton S J and Hamilton J C 1998 *Phys. Rev.* B **58** 11085 – 8
[28] Hadian R, Grabowski B, Race C P and Neugebauer J 2016 *Phys. Rev.* B **94** 165413
[29] Hadian R, Grabowski B, Finnis M W and Neugebauer J 2018 *Phys. Rev. Mater.* **2** 043601
[30] Xu M, Chen K, Cao F, Estrada L V, Kaufman T M, Ye F, Hahn H, Han J, Srolovitz D J and Pan X 2022 *arXiv* (*Preprint* 2201.04190)
[31] Zimmerman J A, Kelchner C L, Klein P A, Hamilton J C and Foiles S M 2001 *Phys. Rev. Lett.* **87** 165507
[32] Han J, Thomas S L and Srolovitz D J 2018 *Prog. Mater. Sci.* **98** 386 – 476
[33] Chen K, Han J and Srolovitz D J 2020 *Acta Mater.* **194** 412–21
[34] Race C P, Hadian R, Pezold J v, Grabowski B and Neugebauer J 2015 *Phys. Rev.* B **92** 174115

IOP Conf. Series: Materials Science and Engineering 1249 (2022) 012020 doi:10.1088/1757-899X/1249/1/012020

Characterization of hierarchical lath martensite microstructure in low carbon steels using ultra-high voltage TEM and SEM-EBSD analysis

M Sugiyama[1], M Takei[2], S Sekida[1] and N Maruyama[1]

[1]Nippon Steel Research Alliance Laboratories, Graduate School of Engineering, Osaka University, 2-1 Yamadaoka, Suita, Osaka 565-0871, JAPAN
[2]Division of Materials and Manufacturing Science, Graduate School of Engineering, Osaka University, 2-1 Yamadaoka, Suita, Osaka 565-0871, JAPAN

E-mail: sugiyama@mat.eng.osaka-u.ac.jp

Abstract. Lath martensite is a heterogenous microstructure in low carbon steels and plays an important part in microstructure for the high strength steels. The complex martensitic structure has been clarified as the hierarchical block and packet structures using SEM observations and electron backscatter diffraction (EBSD) analysis. As the result of variant analysis, it is found that the block consists of sub-blocks and the block size and distribution seem to be new key factors to control the mechanical properties. Each block is further subdivided by laths, which are the finest structural units in lath martensite. The morphology of laths and their variants in low carbon steels has been further investigated using the ultra-high voltage TEM (UHVEM) with 3MeV in accelerated voltage. Based on a comparison of UHVEM and SEM observations for the same field of view, it is found that the laths have a further hierarchical organization with twinned morphology and the fragmented structure within the laths. Two plane crystal analysis of the twin in a lath has been carried out using SEM-EBSD analysis after Xe-FIB large area fabrication process, resulting in the identification of the twin morphology such as the {112} twin plane perpendicular to the {110} habit plane and along the <111> direction. Although several types of twinned morphology such as the long and short length are observed in the lath microstructure under TEM observations, the twinning structure observed in various forms is explained by a simple model in the present study.

1. Introduction

In the field of automobile bodies, the technical issue has been placed on reducing both collision safety and weight to improve driving fuel economy. Furthermore, the recent acceleration of electrification to reduce greenhouse gas emissions requires further weight reduction to lower the weight burden of the battery. Therefore, in the next generation of ultralightweight automobile bodies, research and development will be conducted to freely utilize ultra-high-strength steel plates of 1.0 GPa class or higher [1]. Optimal alloy design and material microstructure design using computational science is also advancing [2], and fundamental knowledge such as the understanding of the existing distribution of alloying elements and the quantification of material microstructure from micro to nano scale is becoming more important.

The current mainstream material structure of high strength steel plate is composed of low-carbon lath martensite, represented by the Fe-C steel [3]. Although the lath martensite in the Fe-C system appears below 0.6 mass%C [4], the area where research on ultra-high strength steel sheets is mainly in the low carbon range of 0.1~0.2 mass%C. To improve hardenability, 1~2 mass%Mn is often added.

IOP Conf. Series: Materials Science and Engineering 1249 (2022) 012020 doi:10.1088/1757-899X/1249/1/012020

During rapid cooling from the high-temperature austenite phase, martensitic transformation occurs in a wide temperature range from about 723K to 473K, depending strongly on the carbon content. The martensitic transformation occurs at the Ms temperature without any atomic diffusion, and the low-carbon lath martensite itself has a hierarchical microstructure called packets or blocks, and sub-blocks. The morphology and crystallography of lath martensite in Fe-C alloys have been extensively investigated using SEM and EBSD analysis by Morito et al.[5], and the variant analysis method for the lath and blocks including the same orientation lath units has been developed based on the K-S orientation relationship between the FCC austenite to the BCC(BCT) martensite.

In the low carbon martensite transformation, the carbon atom can still diffuse sufficiently in the Ms temperature range. After the martensitic transformation, the carbon atoms move to stable positions as dislocations, various interfaces and boundaries, and even clustering or formation of carbides in the lath martensite phase. Since it is difficult to track the behavior of the carbon, it is important to clarify the morphology of the lath unit, including the several kinds of lattice defects in the matrix. Based on the success of variant analysis for the hierarchical lath martensite structure, the characteristic features of sub-blocks in low carbon steels are gradually made clear, for example, the existence of a strict rule for the variant selection in the sub-blocks [6], and the tendency of preferential formation of variants pairs in the relation to the shape strain minimization [7].

The internal defect structure of lath martensite has been extensively investigated by TEM observations [8-9]. The substructure of lath martensite produced by water or oil quenching consists of high densities of tangled dislocations, reflecting lattice invariant deformation and volume accommodation effects during athermal transformation at relatively high Ms temperatures. With increasing in the carbon content, it has been considered that the internal defect structure changes from dislocations to the twinned structure [10]. Since the Ms of Fe-C martensite is still higher than room temperature even in Fe-0.6%C, the internal microstructure of lath martensite has been investigated in the Fe-20%Ni-5%Mn alloys which transformed to lath martensite isothermally below room temperature [11]. On the other hand, it is reported by the present authors that the twins are introduced into the lath martensite obtained even at high temperature by the normal quenching rate [12].

Since martensitic transformation is also a non-diffusion transformation between solid phases, there is a phenomenological assumption that the boundary surface between the matrix and martensite must be an invariant surface [13], which has been macroscopically verified in the past [14]. However, this invariant surface cannot be created only by lattice deformation from FCC to BCC (BCT), so in addition to lattice deformation, slip or twin deformation must be introduced as a second deformation in martensitic transformation. In Fe-Ni alloys, it is known that the transformation point (Ms) decreases with increasing Ni concentration and the lattice-invariant strain changes continuously from dislocation to twinning [15]. However, in the Fe-C system, there are various reports, and the correlation between a lower transformation point and twinning has not necessarily been obtained. Recently, it is reported by Ping et al. that the lath martensite seems to have the twins as the dominant microstructural feature in the low-carbon Fe-0.1%C alloys quenched at temperatures only from 1323K to 1473K [16]. Under the TEM observations, several types of twinned morphologies such as short twins normally inclined to the lath boundary, and long twins having more straight or flat boundary planes are reported.

Recent crystallographic research progress on lath martensite is remarkable, and crystallographic analysis using SEM-EBSD and three-dimensional morphological analysis using FIB-SEM have advanced the crystallographic understanding of the packets, blocks, and sub-blocks in the transformation from FCC to BCC (BCT) [17,18]. On the other hand, different views on the microstructure of the lath unit have emerged, including recent research results, and the number of unresolved interpretations for the lath formation is conversely increasing.

In this study, as a means of connecting these observation scales, we focused on the ultrahigh-voltage TEM of 2000 kV or higher, which can observe a region about five times thicker than the general-purpose 200 kV TEM. The purpose of the study is to make clear the three dimensional lath morphology and properties of twins introduced to the lath, and to clarify the crystallographic findings of the hierarchical lath martensite microstructure in low carbon steels.

2. Experimental procedure

An ingot of 50kg in weight was melted in a laboratory vacuum furnace. After hot forming, a homogenizing heat treatment was carried out at 1373K for 1 hr, and the slab was roughly hot-rolled from 30 mm to 3.4 mm in thickness. After removing the surface decarbonized layer, the plate was rolled at room temperature to 1.2 mm in thickness. In order to obtain the full martensite phase, a small piece of plate is sealed in a vacuum in a quartz tube, and annealed at 1373K for 3 minutes, then quenched to iced water. The sample composition is chemically analyzed to be Fe-0.07C-1.0Mn(mass%).

A sample surface was mechanically polished and prepared by electropolishing using a solution of 95 vol.% acetic acid and 5 vol.% perchloric acid for SEM observations. The SEM used is a Hitachi S-5000 equipped with an EBSD detector for crystal orientation analysis. The Φ3mm thin foil samples for TEM observations were electropolished by the TenuPol-5 using the same electrolyte. The microstructure was observed with a conventional 200 kV TEM, and 3MV ultra-high voltage TEM (Hitachi H-3000). The photograph of UHVEM is shown in figure 1[19]. The electron cathode of UHVEM is a LaB_6 single crystal. By increasing the acceleration voltage, the penetrability of the electron beam through the iron sample is improved, and it is possible to observe a region about five times thicker than the specimen observed by the conventional 200kV TEM. A large area focused ion beam (FIB) fabrication was utilized by the Xe plasma ion FIB made by TESCAN (AMBER X), in order to carry out the two-plane crystallographic analysis for the twin introduced in the lath martensite.

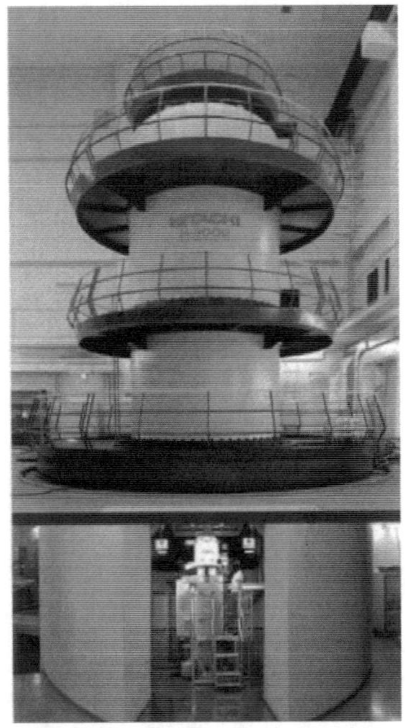

Figure 1. Photograph of UHVEM.

3. Results

The microstructure of the sample quenched from 1373K is the hierarchical morphology of laths, blocks, and packets, in which the minimum lath width is about 200nm, and the lath extends from several micrometers to several tens of micrometers. Lattice defects inside lath martensite are mostly dislocations, but there are also regions where a large amount of twins are observed in some places. Under the thick area observations using UHVEM, it is confirmed that mainly two types of twining morphology. One is a long twin domain along the lath boundary as shown in figure 2(a), and the other is a short twin contrast introduced with a certain angle to the lath interface, as seen by the black contrast in figure 2(b). The short twins are also observed in the middle part of a lath when the lath contrast is observed in the plane morphology. It must be noted here that the description of the length of the twins is simply a notation based on how they appear as contrast in the TEM image.

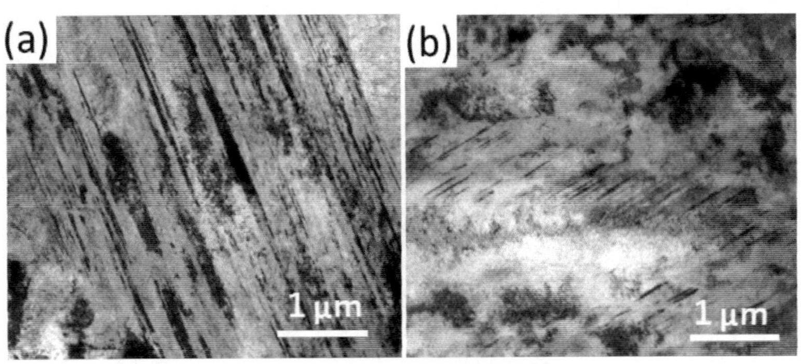

Figure 2. TEM images of twins in a lath showing different contrast regions such as the long (a) and the short striations (b).

By the dark field imaging and corresponding electron diffraction analysis, the nature of long twins along the lath interface is examined, as seen in figure 3. Figure 3(a) is a bright field image and figure 3(b) is the corresponding electron diffraction pattern and figure 3(c) is the dark field image taken from 1-10 diffraction spot. The incident electron beam is parallel to [110] direction, and (-1 12) twinning spots appear on the diffraction pattern. Under the incident beam condition, the interface of twin planes is satisfied with the edge-on condition, and the thickness of twinned regions is estimated to be several tens of nanometers. The length direction of twins is parallel to <111> direction under the present observation condition. The twinning microstructure is observed more often by UHVEM in a thick area than by the conventional 200kV TEM in a thin region.

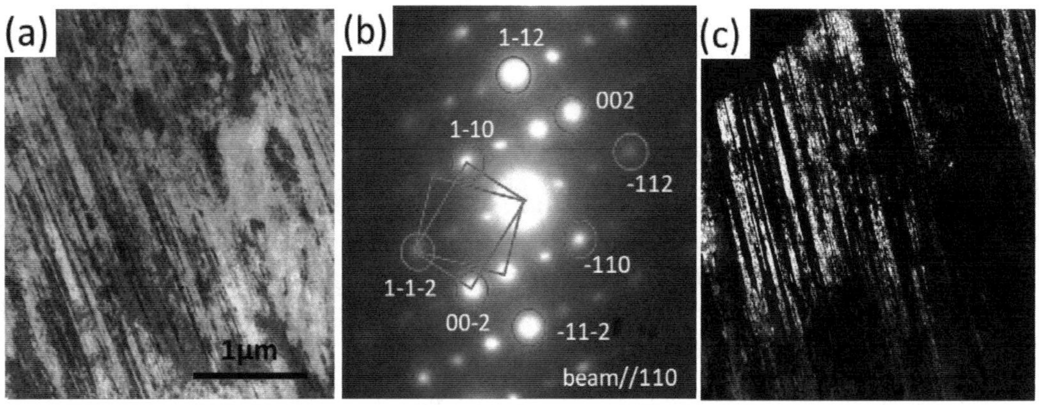

Figure 3. Diffraction analysis of long twins in the lath, (a) bright field image, (b) electron diffraction pattern showing two orientations with twin relationship, the matrix (110) parallel to the twin (1-1-2), (c) dark field image taken from 1-10 reflection.

Since the internal microstructure of thick areas can be observed with an UHVEM, the same sample can be used to check the internal microstructure of a large area with an SEM in the same field of view. As seen in figure 4(a) and 4(b), the same microstructure and contrast are obtained between them, in which a red rectangular indicates the same lath martensite block. The contrast of the UHVEM image in figure 4(b) is reversed in comparison with that of SEM, because the TEM image is formed by transmitted electrons. In both micrographs, lots of fragmentation and plate like contrast as indicated by the white circle are seen in a lath. Since a one-to-one correspondence between SEM and TEM images can be easily obtained by utilizing UHVEM, crystal orientation analysis can be carried out using the SEM-EBSD method in order to make clear the crystal orientation relationships of the regions represented as

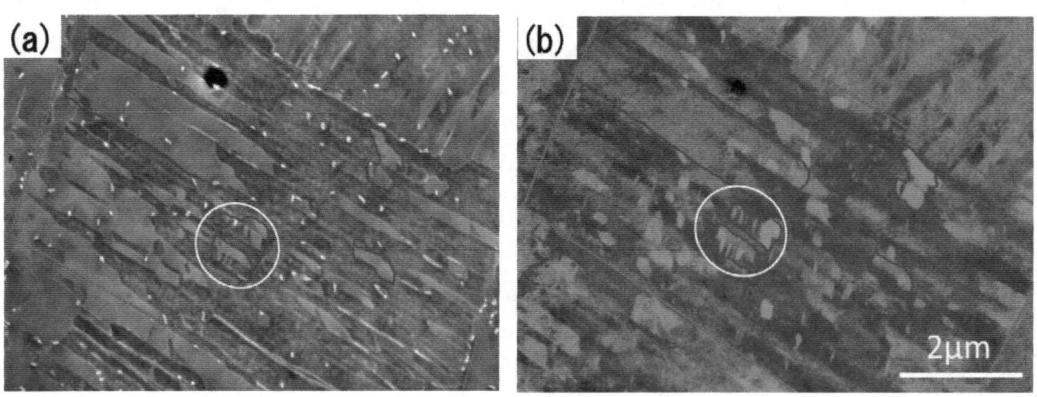

Figure 4. One-to-one correspondence of lath block images taken by SEM (a) and UHVEM (b).

the fragmentation and line contrast. The fragmentation contrast is found to be different variant lath, as discussed later.

Figure 5(a) is a large area SEM micrograph showing the lath martensite, and the corresponding inverse pole figure map measured by EBSD is indicated in figure 5(b). Figure 5(c) is the enlarged SEM image of the area enclosed by the rectangle in figure 5(b), and the fragmentations are also seen. The upright area of figure 5(c) is further enlarged as seen in figure 5(d). There is clear line contrast within the lath unit in figure 5(d), which is the twin boundary contrast identified by the comparison of the UHVEM observation.

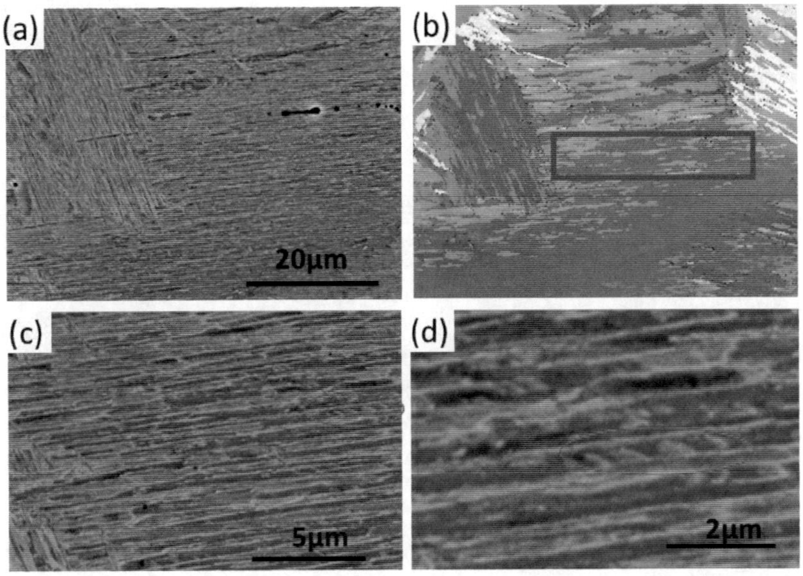

Figure 5. SEM observations and EBSD analysis. Same areas taken by low mag (a) and enlarged images of (c) and (d). The area (d) is from an enlarged image of rectangle zone in the IPF map (b).

The twinned area indicated by figure 5(d) has been used for the following two-planes crystallographic analysis using the Xe plasma ion FIB system. Figure 6(a) is the SEM image showing the overall shape after processing of the Xe plasma ion FIB fabrication. After the fabrication of the area from the direction of parallel and perpendicular to the lath unit, indicated in figure 6 (a), it is confirmed that the thin plate lath units are stacked diagonally as indicated by figure 6(b) and (c). Figure 6(c) is a cross sectional view of the portion indicated by a red circle in figure 6(a). The morphology of the laths is found to be more like a thin plate than a thin needle. Figure 6(d) is the schematic diagram showing the size dimension of lath morphology reported in literature [20], which will be

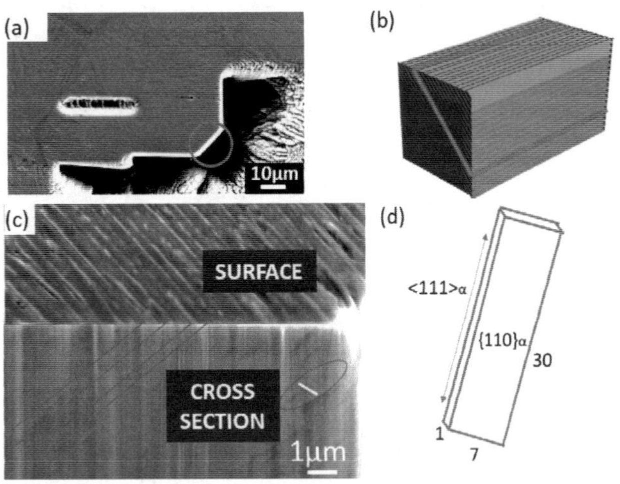

Figure 6. Two-plane analysis for the rectangle area in figure 5(b). (a) is low mag image after Xe-plasma ion FIB fabrication, and (c) is the cross-sectional SEM image. The morphology of laths is schematically indicated in (b) and (d).

reconfirmed by the present research. Although the flat surfaces of the laths near the habit plane are as large as several tens of μm, large-area processing using Xe plasma ion FIB allowed us to examine the morphology of the entire lath from many angles.

4. Discussion

This study shows that a one-to-one microstructure correspondence between TEM and SEM images can be easily achieved using UHVEM, and twin contrast in a lath is also identified by the SEM image. As the result of two-plane analysis indicated in figure 6, the lath morphology is close to the thin plate one. Since the habit plane is close to the {110} and the length direction is along to <111>, a (112) plane can be taken to be the remaining vertical one. Based on these conditions, the lath variants and twining plane observed in figure 5(d) have been examined.

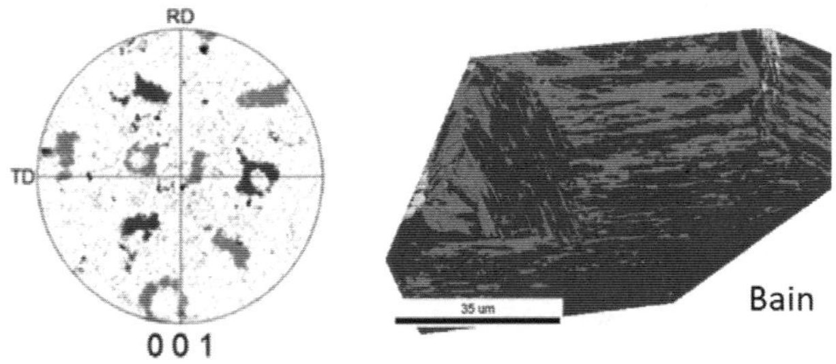

Figure 7. Bain map of lath martensite for figure 5(b).

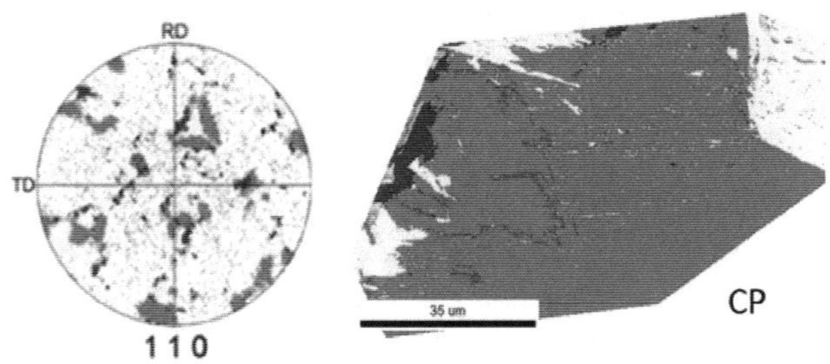

Figure 8. Close-packed plane map of lath martensite for figure 5(b).

When the BCC martensite is transformed from the FCC austenite, three $\langle 001\rangle_F$ axes are equally selected as the Bain orientation relationship, that is represented by $\{111\}_F//\{011\}_B$. The Bain relationship of each lath unit is indicated by the three colours map, as seen in figure 7. It is found that the two lath units indicated by blue and pink colours in previous figure 5(b) have a different Bain axis relationship. As the next, since there are four equivalent closed packed planes of $\{111\}_F$ in the austenite phase, they are indicated by four colours in figure 8, which correspond the packet region. It is found that the two lath units indicated by blue and pink colours in figure 5(b) have been transformed from the same closed packed plane in austenite. Finally, the boundary where the orientation difference in the <111> direction is rotated to 60° in BCC crystals is shown by the red line in figure 9. It is found that the red lines correspond to the interface between the blue and pink colour region in figure 5(b). Judging from all of three data shown in figures 7,8,9, the blue and pink colour regions consist of V1-V2 variant pairs determined by the Morito's notation [5]. As shown in figure 5, the V1 and V2 are corresponding to the fragmentation contrast as seen in figure 4. Next, the twin morphology observed in the rectangular region of figure 5(b) has been examined. The perpendicular direction of the twinned region showing the blue colour in the IPF map is close to <111>, but the laths are stacked in a tilted orientation in this region, as shown in figure 6(b). Thus, the real length direction of the lath unit with the <111> direction is different one inclined from perpendicular to the paper surface.

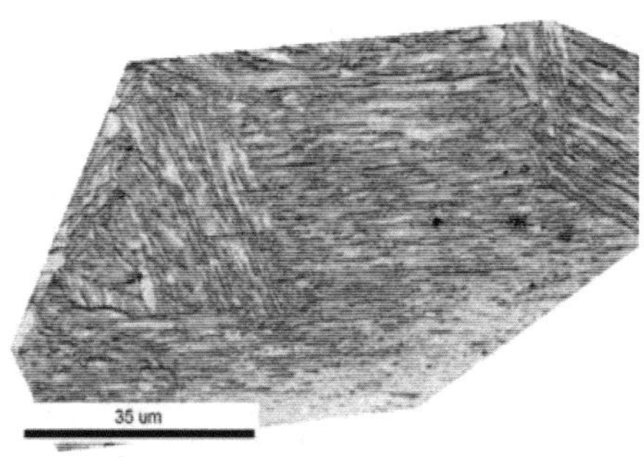

Figure 9. Boundary map satisfying with 60° rotation in <111> direction.

Since the three-dimensional morphology of lath is found, the orientation information measured by EBSD is converted to the lath crystal coordinate system with a {110} habit plane, extending in the <111> direction, and the remaining vertical plane being {112}. Based on the coordinate transformation, the crystal orientation of the observation plane normal of the V1 lath is obtained to be <8-9-6> direction in the lath crystal coordinate system. By the orientation analysis under this condition, the line contrast of figure 5(d) is a coincidence to the trace of the vertical plane being (112) plane, as indicated in figure 10(a). By the similar procedure, the line contrast observed at the cross section area of figure 6(c) is also identified to be the trace of (112) plane in the lath unit. Both of the analysis, the line contrast of lath units observed in SEM and UHVEM is identified as the same type of (112) twin plane. That is, the twinned plane in the lath is perpendicular to the habit plane and elongation to the <111> direction. It must be noticed here that the (112) twin may be introduced to the compression direction of <001> during the FCC to BCC martensite transformation, as the result of the double shear mechanism.

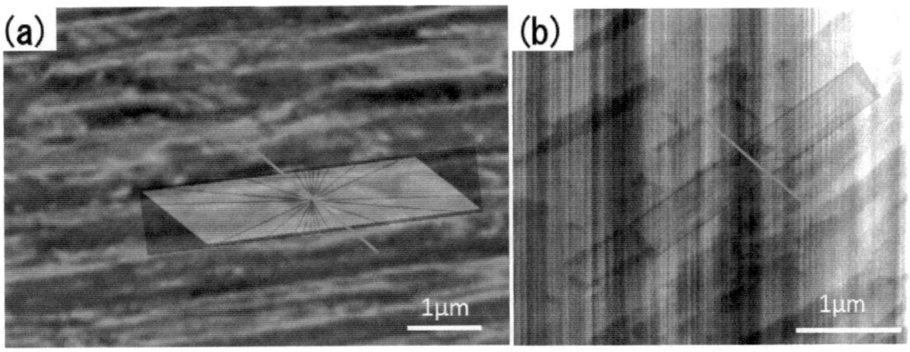

Figure 10. Schematic diagram showing the (112) plane trace among 12 trace directions of {121} and the relationship to the twin plane traces of figure 6 in the surface (a) and cross-section (b).

Based on the present analysis, the twinning plane is uniquely determined in the lath crystal coordinate system. The orientation relationship is also satisfied with the long contrast twins as indicated by figure 3. Furthermore, the same procedure has been done for the short contrast twins, as indicated in figure 11. The trace direction of the short twins observed from the lath interface is also explained by the proposed twin models in the lath.

Although the number of twins in the lath being examined for the orientation is still limited, at the present, its validity will be verified by examining the morphology of twins observed in the TEM from various crystal orientations.

Figure 11. Trace analysis of short contrast twins.

5. Conclusions

By utilizing the UHVEM and SEM-EBSD analysis, the microstructural morphology in the lath martensite and corresponding twining structure has been investigated, and the following findings on the hierarchical structure are obtained in Fe-0.07mass%C-1.0mass%Mn alloy.

(1) Since the high voltage TEM allows observation of thicker regions of the sample, a one-to-one correspondence of the microstructure between SEM and TEM is possible, and twinning introduced into the lath in low carbon martensite is also identified by the SEM observations.

(2) Utilization of Xe-FIB enabled cross-sectional machining of large areas, the two-plane analysis of three-dimensional lath morphology and corresponding twin plane characteristics has been successfully performed.

(3) From the two-plane analysis, the {112} twinning plane of lath martensite is uniquely determined with the characteristic plane perpendicular to the {110} habit plane and extending in the <111> longitudinal direction.
(4) Based on the present simple twin plane model, at the present, it is explained as the same nature of different twin contrast such as a long or a short one under the TEM observations.

Acknowledgment
We would like to express our sincere thanks to Dr. Kodama and Dr. Okawa of TOYO corporation for their cooperation with the use of the Xe plasma ion FIB processing equipment.
A part of this work was supported by "Advanced Characterization Nanotechnology Platform, Nanotechnology Platform Program of the Ministry of Education, Culture, Sports, Science and Technology (MEXT), Japan" at the Research Center for Ultra-High Voltage Electron Microscopy (Nanotechnology Open Facilities) in Osaka University. We sincerely appreciate for the cooperation of Prof. Ichikawa in the use of UHVEM.

References
[1] Okitsu Y, Naito T, Takaki N, Sugiura T and Tsuji N 2010 *SAE Technical Paper* 01-0438.
[2] Uenishi A 2019 *Nippon Steel Technical Report,* **122** 7-12
[3] Krauss G 1999 *Mat. Sci. Eng.* A. **273-275** 40-57
[4] Hutchinson B, Hagstrom J, Karlsson O, Lindell D, Tornberg M, Lindberg F and Thuvander M 2011 *Acta Mater.* **59** 5845-58
[5] Morito S, Tanaka H, Konishi R, Furuhara T and Maki T 2003 *Acta Mater.* **51** 1789-99
[6] Kitahara H, Ueji R, Tsuji N and Minamino Y 2006 *Acta Metall.* **54** 1279-88
[7] Morito S, Huang X, Furuhara T, Maki T and Hansen N 2006 *Acta Metall.* **54** 5323-31
[8] Sandvik B P J and Wayman C M 1983 *Metallography* **16** 199-227
[9] Morsdorf L, Tasan C C, Ponge D and Raabe D 2015 *Acta Mater.* **95** 366-77
[10] Wayman C M 1990 *Mater. Sci. Forum* **56-58** 1-32
[11] Wakasa K and Wayman C M 1981 *Acta Metall.* **29** 991-1011
[12] Sugiyama M, Sawa G, Hata K and Maruyama N 2019 *Materials Sci. and Eng.* **580** 012045
[13] Wechsler M S, Lieberman D S and Read T A 1953 *J Metals* **5** 1503
[14] Bowles J B and Mackenzie J K 1954 *Acta Met.* **2** 129
[15] Patterson R L and Wayman C M 1966 *Acta Met.* **14** 347
[16] Ping D H, Guo S Q, Imura M. Liu X, Ohmura T, Ohmuma M, Lu X, Abe T and Onodera H 2018 *Scientific Reports* **8** 14264
[17] Stormvinter A, Miyamoto G, Furuhara T, Hedstrom P, Borgenstam A 2012 *Acta Materialia* **60** 7265-74
[18] Kinney C C, Pytlewski K R, Khachaturyan A G and Morris Jr. J W 2014 *Acta Mater.* **69** 372-85
[19] Takaoka A, Ura K, Mori H, Katsuta T, Matsui I and Hayashi S 1997 *J. Electron Microsc.* **46** 447-56
[20] Wakasa K and Wayman C M 1981 *Acta Metall.* **29** 973-90

IOP Conf. Series: Materials Science and Engineering 1249 (2022) 012021 doi:10.1088/1757-899X/1249/1/012021

The Neper/FEPX Project: Free / Open-source Polycrystal Generation, Deformation Simulation, and Post-processing

Romain Quey [1] **and Matthew Kasemer** [2]

[1] Mines Saint-Étienne, Univ Lyon, CNRS, UMR 5307 LGF, F – 42023 Saint-Etienne, France
[2] Department of Mechanical Engineering, University of Alabama, Tuscaloosa, AL 35487 United States

E-mail: `romain.quey@mines-stetienne.fr`

Abstract. In the past couple of years, efforts have lead toward the convergence of two established software packages: Neper, developed primarily at CNRS and Mines Saint-Étienne, and FEPX, developed historically at the Deformation Process laboratory at Cornell University and (since 2020) at the Advanced Computational Materials Engineering Laboratory at the University of Alabama. The objective was to create an extensive and homogeneous ensemble for polycrystal plasticity studies that includes polycrystal generation and meshing, parallel finite-element crystal-plasticity simulation, post-processing and visualization. The two programs were upgraded to common standards, both in terms of usage and resources. Collectively, the project is under active development and distributed under free / open-source license (see both: `https://neper.info`, `https://fepx.info`).

1. Introduction

The Neper project was started in 2003 and first released as free / open-source software in July 2009, and has been developed by Romain Quey, primarily at CNRS and Mines Saint-Étienne. The original intent of the effort was to generate random polycrystalline microstructures, as Voronoi tessellations, and to mesh them for further use in finite-element crystal-plasticity simulations [1], including at large strains [2]. The proposed set of methods can be efficiently applied to virtually any "CAD-type" polycrystal models (i.e. defined in a vectorial way, using vertices, edges, (planar) faces and polyhedra). Further efforts were made on the representation of more complex microstructures, such as those found in steels or titanium alloys [3], and on the generation of polycrystals of specified grain properties [4], be it from experiments, e.g. to replicate typical grain size and shape statistics, or from fundamental arguments, e.g. to isolate the influence of specific factors, in purely numerical studies [3]. For many years, Neper's development was mainly application-driven, but the program was sufficiently general to be of interest to others (see an exhaustive list of articles at `https://neper.info/applications.html`).

The FEPX project was started in the late 1990's in Paul Dawson's group, the Deformation Processes Laboratory (DPLab) at Cornell University, and since 2020 has been maintained and developed in Matthew Kasemer's group, the Advanced Computational Materials Engineering Laboratory (ACME Lab) at the University of Alabama. The original intent of the effort was to develop a code to support the simulation of polycrystalline deformation by explicitly incorporating both the anisotropic elastic and plastic response. Early iterations of the framework focused on development for simulation of sheet forming operations [5] and texture evolution [6]. Eventually, the framework included the formulation of polycrystal models considering local homogenized response [7, 8] and robust parallelization for large-scale computation [9]. Further development saw the application of the framework to full-field simulations (for which grains are typically discretized into tens to thousands of elements) using regular-shaped grains [10], and finally using generalized, random-shaped grains via Voronoi tessellations in the first convergence with the Neper project [2].

After a decade of scientific collaboration and cross-usage of the two programs between the two research groups, and in a continuing effort to make FEPX publicly available, it was decided in April 2020 to make Neper and FEPX two "companion programs", to form an extensive and homogeneous ensemble dedicated to the analysis of polycrystal deformation, which lead to first common releases in July 2020. The original goal was to ease the concomitant use of the two programs, via the design of a shared file format, and to upgrade the programs to common standards in terms of usage (user interface, program inputs and outputs, etc.) and resources (website, documentation, source code management, user support, etc.). Since then, the Neper/FEPX project has continued to develop in all its aspects, and has started to integrate capabilities of other, "satellite" projects developed by the two groups and specifically dedicated to result post-processing, such as Orilib and Hermes (CNRS / Mines Saint-Étienne), two general programs for (EBSD-type) data analysis, and ODFPF (DPLab / ACME Lab), a program for orientation distribution function and pole figure related computations. The following provides an overview of the Neper/FEPX project, with a focus on recent advances.

2. Description

Neper and FEPX are programs designed to be used in succession. Neper itself is made of several modules ("-T" for tessellation, "-M" for meshing, "-S" for simulation and post-processing, and "-V" for visualization). Neper is written in the C language and multithreaded using OpenMP, and is typically used on a personal computer or workstation (i.e., a single computational node, $\mathcal{O}(10)$ of processing units). It uses several external libraries, for various tasks ranging for distance computation to meshing and image generation [11, 12, 13, 14, 15, 16, 17, 18, 19]. FEPX is written in the Fortran language and parallelized using Open MPI, and is typically used on computer clusters (i.e., multiple computational nodes, $\mathcal{O}(10-1000)$ of processing units). A typical Neper/FEPX workflow is illustrated in figure 1 and detailed in the following.

2.1. Tessellation Generation (Neper -T)

Polycrystals are most often represented as *tessellations* defined under a vectorial format, i.e., using vertices, edges, faces and polyhedra (or "cells"), and cells are generally convex. Such a "CAD-type" geometrical representation allows for the modelling of a wide variety of polycrystalline microstructures while still being prone to meshing by standard methods. The tessellations include *single-scale tessellations*, which are able to represent simple polycrystals, and *multi-scale tessellations*, which are more adapted to represent polycrystalline microstructures for which subdivisions (i.e., intra-cell or intra-grain features) associated with phase transformations occur during the material processing, as is the case in most steels, titanium alloys, etc.

2.1.1. Single-scale Tessellations. Any convex-cell tessellation (or, more specifically, *normal* tessellation [20]) can be generated, independently of the sizes, shapes and arrangements of its cells. This is made possible using Laguerre tessellations, which can intrinsically generate any such tessellations. The cells of a Laguerre tessellation, C_i, are defined from a set of seeds, S_i, of

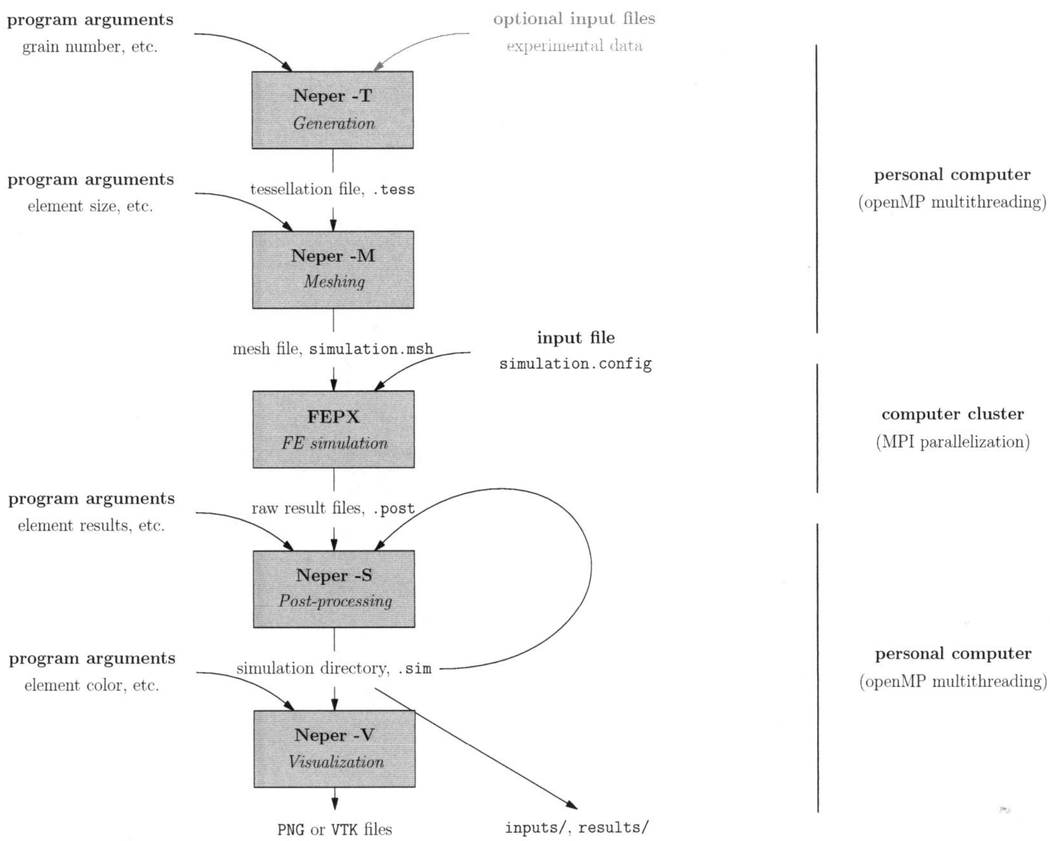

Figure 1. A typical Neper/FEPX workflow, including the successive steps, and input and output files. The different Neper modules and FEPX are run in succession, take inputs as arguments (for Neper) or external files (optionally for Neper, exclusively for FEPX), and exchange result files and directories.

given positions, x_i, and weights, w_i:

$$C_i = \left\{ P(x) \in D \mid d(P, S_i)^2 - w_i < d(P, S_j)^2 - w_j \ \forall j \neq i \right\}, \tag{1}$$

where $d(\bullet, \bullet)$ is the Euclidean distance. The seed positions and weights are determined by a constraint-free optimization. This is different from more standard approaches, for which seed attributes are pre-determined, e.g. from a grain size distribution and a dense sphere packing algorithm or from individual grain properties [4]. Optimization is carried out so as to obtain specified cell properties (grain size or shape distributions, morphological texture, etc.), or grain positions, sizes or shapes, or even a rasterized image of the polycrystal, as may be provided by electron backscatter diffraction or synchrotron X-ray diffraction experiments [21, 4]. Examples of single-scale tessellations are provided in figure 2a-c.

2.1.2. Multi-scale Tessellations. Several scales are obtained by generating tessellations in the individual cells of the previous, upper-scale tessellation(s), mimicking the actual grain subdivisions occurring during material processing. Arbitrarily complex microstructures can be generated, where the cell properties of the tessellation are specified at each scale, as in the case of a single-scale tessellation. For instance, subcells can be used to represent bainitic packets in steel or, equivalently, colonies in titanium alloys. Lamellar structures can also be generated. Different phases can be applied to the different cell subdivisions. Orientation relationships between successive scales can also be taken into account [3]. An example of a multi-scale tessellation is provided in figure 2d.

2.2. Meshing (Neper -M)

Tessellations can be discretized via unstructured meshing into tetrahedral elements, which provides a uniform element size distribution and maximizes the element shape quality factors. Meshing is usually preceded by a tessellation regularization procedure that removes the smallest features of the tessellation [2], which is essential to obtain good-quality meshes (both in terms of element size and shape factor), facilitates the numerical convergence of the simulation and allows for large strain simulations (up to about 40%). To reach larger strains, re-meshing can be applied, by which a new mesh is constructed over a deformed mesh and the state variables are mapped to the new mesh [2]. Remeshing does not significantly affect the simulation results and, using all methods combined, polycrystals have been deformed to strains of the order of 1.2–1.4 for aluminium in plane strain compression [2, 22, 23], and larger strains can be attained. Example polycrystal meshes are provided in figure 2e,f.

2.3. Crystal-Plasticity Simulation (FEPX)

Broadly, a polycrystal is defined via the mesh and crystal orientations (defined at elements), while the crystal behavior is defined via its elastic response (Hooke's law) and plastic response (slip systems, slip kinetics, and hardening), and the associated modeling parameters. Loading is applied as a function of time by specifying velocities or forces as boundary conditions.

2.3.1. Crystal Response. A truncated description of the base models employed in FEPX is given below (see `https://fepx.info` for a description of all available models). The stress, σ, is related to the elastic strain, ε, via Hooke's law:

$$\sigma = \mathcal{C}\varepsilon, \tag{2}$$

where $\mathcal{C}$ is the stiffness tensor. Cubic, hexagonal and tetragonal symmetries can be considered.

The kinematics of slip are described by a power law:

$$\dot{\gamma}^\alpha = \dot{\gamma}_0 \left(\frac{|\tau^\alpha|}{g^\alpha}\right)^{1/m} \operatorname{sgn}(\tau^\alpha), \tag{3}$$

where $\dot{\gamma}_0$ is the fixed-rate strain rate scaling coefficient, τ^α is the resolved shear stress on a slip system, α, g^α is the current critical resolved shear stress, and m is the rate sensitivity exponent.

For an isotropic hardening assumption, slip system strength evolution (hardening) is modeled by:

$$\dot{g^\alpha} = h_0 \left(\frac{g_{s0} - g^\alpha}{g_{s0} - g_0}\right)^n \dot{\gamma}, \tag{4}$$

where h_0 is the fixed-state hardening rate scaling coefficient, g_{s0} is the initial slip system saturation strength, g_0 is the initial slip system strength, and n is the non-linear Voce hardening exponent. In the above equation, $\dot{\gamma}$ is calculated as

$$\dot{\gamma} = \sum_\alpha |\dot{\gamma}^\alpha| . \tag{5}$$

FEPX has the ability to handle other crystal behaviors, such as cyclic hardening, saturation strength evolution, latent hardening, etc., further described in FEPX's documentation.

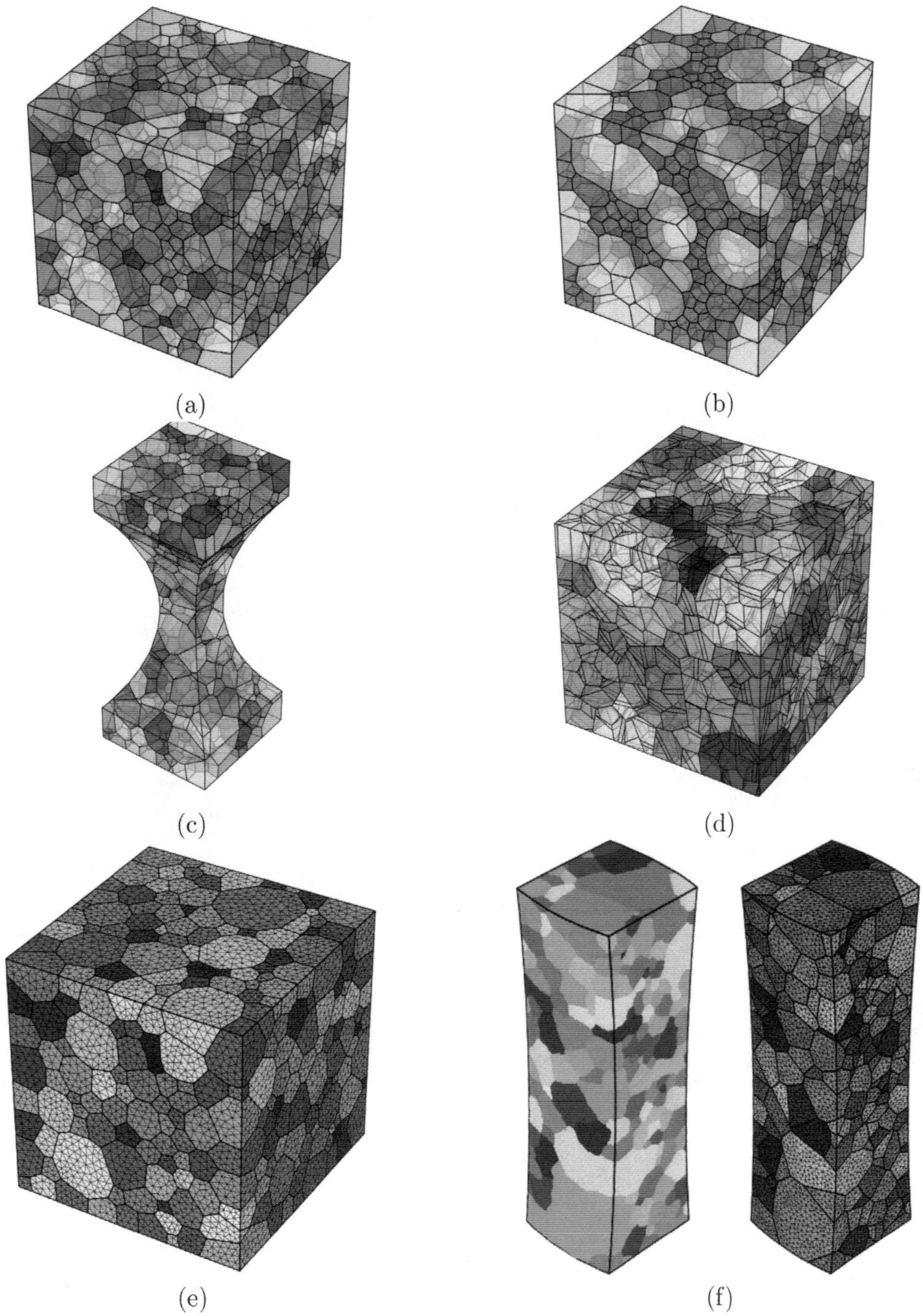

Figure 2. Typical microstructures and meshes generated by Neper. (a) Typical grain-growth polycrystal, (b) 2-phase polycrystal, (d) non-convex polycrystalline specimen, (d) 3-scale polycrystal representative of a (bainitic) steel microstructure, (e) meshed grain-growth polycrystal, and (f) meshed DCT polycrystal.

IOP Conf. Series: Materials Science and Engineering 1249 (2022) 012021 doi:10.1088/1757-899X/1249/1/012021

A complete description of the base constitutive model and the finite element method implementation can be found in Refs. [7, 8] and are also documented in a theory manual, which describes FEPX's pre-2020 state [24].

2.3.2. Deformation History. A simulation is defined as a set of steps through time, in which (generally) either a user-specified strain or load state is targeted at the end of each time step. FEPX considers either uniaxial deformation conditions or principal-triaxial deformation conditions. Simulations may also be performed considering either constant loading rates or constant strain rates.

2.3.3. Boundary Conditions. Boundary conditions must be applied to control the deformation response. FEPX allows for generalized boundary conditions, though comes pre-loaded with a variety of conditions suited to the base deformation histories (uniaxial and principal-triaxial) described above. Generally, boundary conditions are defined at a set of nodes and are specified as the prescribed velocities in three orthogonal directions at each node (aligned with the sample coordinate system).

2.3.4. Running Conditions. Simulation runtime is dependent on both the size and difficulty of the simulation, and the size of computer/cluster available to perform the simulations.

Concerning the former, many aspects will influence simulation runtime, including the difficulty/non-linearity of the problem, the computational expense of the models employed, and other factors. However, simulation runtime will be primarily affected by the size of the mesh and the number of load steps. Generally speaking, the larger the mesh (i.e., the larger the number of elements/nodes), or the higher number of load steps, the longer the simulation will take to complete on a given number of computational units.

Concerning the latter, simulations are typically performed in parallel on multiple processing cores—either on typical computers with multi-core processors ($\mathcal{O}(10)$ processing units), computational workstations with multiple multi-core processors ($\mathcal{O}(10-100)$ processing units), or computational clusters with potentially many computational nodes each with multiple multi-core processors ($\mathcal{O}(10-1000)$ total processing units).

Approximate runtimes for a simulation with 10 equally-spaced load steps to 5% strain utilizing isotropic hardening, utilizing a computational workstation with 24 processing units are summarized in Table 1. Note that these values do not imply scaling behavior, nor are they meant to demonstrate optimal running conditions, but instead are meant to give a rough estimate of how long simulations of a certain size will take to complete given a number of processing units, and one would expect more efficient computation of larger simulations performed on more processing units.

Number of Nodes	Approximate Runtime (min)
10,000	1
30,000	5
100,000	60
400,000	300

Table 1. Estimates of representative simulation runtimes

IOP Conf. Series: Materials Science and Engineering 1249 (2022) 012021 doi:10.1088/1757-899X/1249/1/012021

2.4. Post-processing (Neper -S)

Post-processing involves the archiving of a simulation (inputs and results) and further processing of the results. Typically, this is done using a custom file and directory structure, and custom scripts and program written in a general language. Neper aims to simplify and standardize this process via the definition of a *simulation directory* and the ability to modify or complete it using simple procedures, with a focus on operations specific to polycrystal computations.

A *simulation directory* has a simple and adaptable hierarchical file and directory structure, as shown in figure 3, which includes the simulation inputs and results, at successive deformation steps. A *simulation directory* allows for post-processing operations such as direct computation of mesh properties (grain centers and volumes, etc.), new simulations results from the mesh or existing simulation results (standard computations such as that of the equivalent stress from the full stress tensor, or arbitrary computations via generalized mathematical expressions), grain-averaged or sample-averaged results, results over pre-defined sets or elements, etc. The new results are in turn included into the *simulation directory*. Practically, the new results become attributes of the entities to which they relate, which allows for the reverse approach consisting of defining mesh regions from the simulation results, as is useful for fine analyses or to identify microstructure–property relationships.

A *simulation directory* can also organize experimental data as input (optionally results), such as 2D or 3D orientation maps provided by EBSD or synchrotron X-ray diffraction techniques. This makes it possible to use the same tools and procedures to process data obtained by different methods, which is particularly interesting in the context of experiment-simulation comparisons [25].

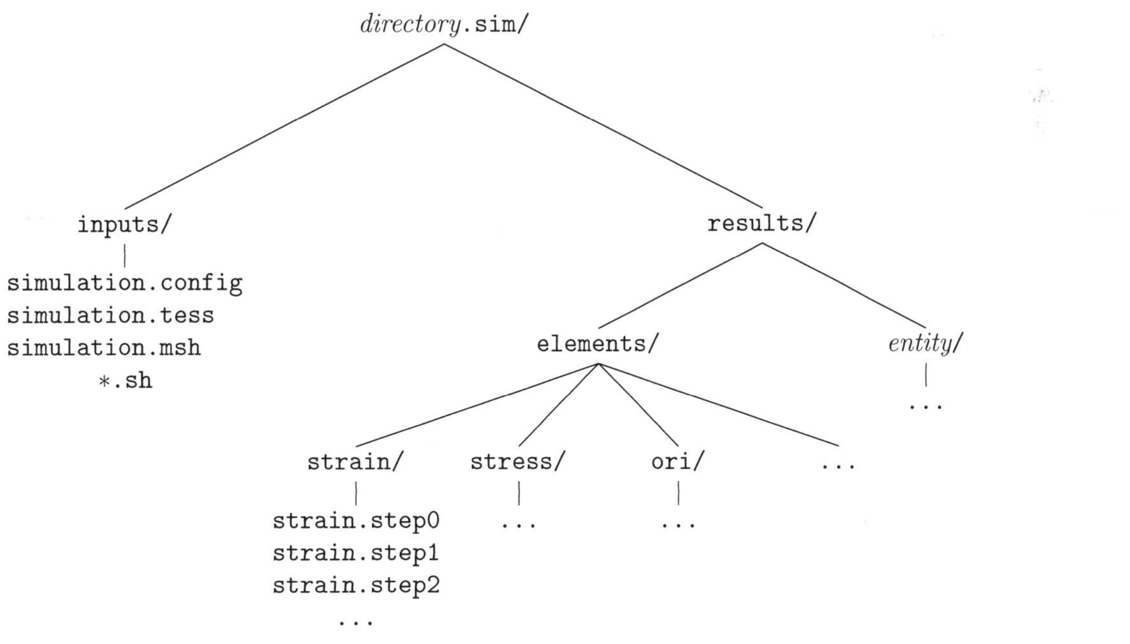

Figure 3. Typical structure of a *simulation directory*. A *simulation directory* contains two self-defined subdirectories, `inputs` and `results`. The `results` directory itself is divided into one directory for each entity, which can be either the built-in `elements`, `elsets` (for grains), `nodes` and `mesh` (for the full sample), or custom entities, and each entity directory is itself divided into result directories, themselves containing the actual data files.

2.5. Visualization (Neper -V)

Visualization mainly operates in real, physical space, to show a mesh and its simulation results, but can also be done as pole figures or in orientation space, to analyse the results. Images are generated non-interactively, at the PNG or PDF formats, or at the VTK format for interactive visualization using, e.g., Paraview. 3D scenes (in real space and orientation space) are rendered as rasterized images, by ray-tracing [18], while 2D scenes (in pole figure space) are rendered as vectorial images [19]. Example visualizations are provided in figure 4. Results can be visualized at successive simulation steps or at different view angles, and animations can be created accordingly, as MP4 files. In real space, a displacement field can typically be applied at nodes, and any result field (stress, strain, orientation, etc.) can be visualized at elements, as shown in figure 4a. Different data sets (such as those obtained at successive steps of a simulation) can be superimposed onto the same image to produce advanced plots, as shown for a pole figure in figure 4b. Finally, an example of plot in Rodrigues orientation space is provided in figure 4c.

3. Conclusions

The Neper/FEPX project aims to built an extensive and homogeneous ensemble for polycrystal plasticity studies. Both Neper and FEPX are products of over 40 collective years of active development and use, for various applications. Recent efforts have upgraded both programs to common standards, which lead to simplified usage, increased robustness and improved resources (websites, documentations, source code management, user support, etc.). The two programs can be used efficiently to simulate polycrystal deformation in a fully-automated way. Current efforts are focused on the improvement and extension of post-processing, mostly by integrating otherwise-existing capabilities, and on strengthening further synergies between the two programs.

References

[1] Barbe F, Quey R, Musienko A and Cailletaud G 2009 *Mech. Res. Com.* **36** 762–768
[2] Quey R, Dawson P and Barbe F 2011 *Comput. Methods Appl. Mech. Eng.* **200** 1729–45
[3] Kasemer M, Quey R and Dawson P 2017 *J. Mech. Phys. Solids* **103** 179–98
[4] Quey R and Renversade L 2018 *Comp. Methods Appl. Mech. Eng.* **330** 308–33
[5] Mathur K 1995 *Simulation of Materials Processing: Theory, Methods and Applications - NUMIFORM 95*
[6] Dawson P and Beaudoin A 1997 *JOM* **49** 34–41
[7] Marin E and Dawson P 1998 *Comput. Methods Appl. Mech. Eng.* **165** 1–21
[8] Marin E and Dawson P 1998 *Comput. Methods Appl. Mech. Eng.* **165** 23–41
[9] Mika D and Dawson P 1999 *Acta Mater.* **47** 1355–69
[10] Ritz H and Dawson P R 2009 *Model. Simul. Mater. Sci. Eng.* **17** 1–21
[11] GNU Scientific Library URL `http://www.gnu.org/software/gsl/`
[12] muparser - fast math parser library URL `https://github.com/beltoforion/muparser`
[13] nanoflann: a C++ header-only fork of FLANN, a library for nearest neighbor (NN) with kd-trees URL `https://github.com/jlblancoc/nanoflann`
[14] Montanari M and Petrinic N 2018 *SoftwareX* **7** 352–5
[15] Geuzaine C and Remacle J F 2009 *International Journal for Numerical Methods in Engineering* **79** 1309–1331
[16] The NLopt nonlinear-optimization package URL `http://ab-initio.mit.edu/wiki/index.php/NLopt`
[17] Scotch URL `https://www.labri.fr/perso/pelegrin/scotch`
[18] *http://www.povray.org*
[19] Asymptote: the Vector Graphics Language URL `http://asymptote.sourceforge.net`
[20] Lautensack C 2008 *J. Appl. Stat.* **35** 985–995
[21] Kasemer M, Echlin M, Stinville J, Pollock T and Dawson P 2017 *Acta Mater.* **136** 288–302
[22] Quey R, Driver J and Dawson P 2015 *J. Mech. Phys. Solids* **84** 506–27
[23] Quey R, Dawson P and Driver J 2012 *J. Mech. Phys. Solids* **60** 509–24
[24] Dawson P and Boyce D 2015 *ArXiv* arXiv:1504.03296 [cond–mat.mtrl–sci]
[25] Renversade L and Quey R 2019 *9 IOP Conf. Ser.: Mater. Sci. Eng.* **580** 012022
[26] Li R, Quey R, Zhang Y, Kobayashi M, Oddershede J and Juul Jensen D to appear *IOP Conf. Ser.: Mater. Sci. Eng.*

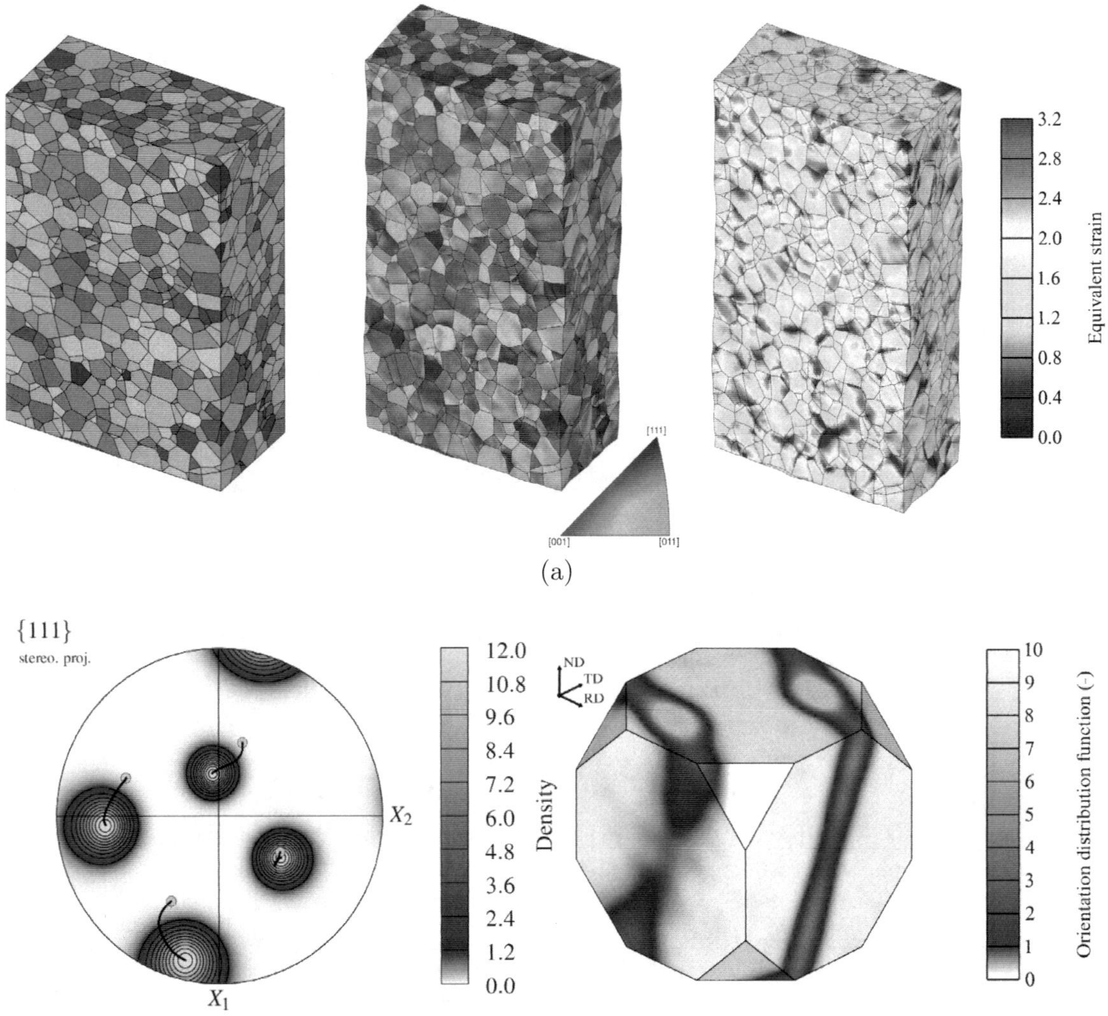

Figure 4. Typical simulation results obtained using Neper and FEPX. (a) Al–4%Cu polycrystal deformed to 16% in uniaxial tension (left: undeformed polycrystal, middle: deformed polycrystal colored by orientation, right: deformed polycrystal colored by equivalent strain) [26], (b) grain lattice rotation on a pole figure, (c) rolling orientation distribution function over Rodrigues orientation space.

Recrystallization kinetics of cold rolled Gum Metal

A Lindkvist, T Holm Johansen, L Kørup Pedersen, Y Zhang and D Juul Jensen

Department of Civil and Mechanical Engineering, Technical University of Denmark, DK-2800 Kgs Lyngby, Denmark

E-mail: adli@mek.dtu.dk

Abstract. A wire of a β-Titanium alloy known as Gum Metal was cold rolled to a reduction of 80% and heat treated for times in the range 3-120 minutes at a fixed temperature of 740 °C to yield a set of partially recrystallized microstructures. The recrystallization course was evaluated by light optical microscopy and Electron Backscatter Diffraction, based on which the recrystallization kinetics was analysed using the Johnson-Mehl-Avrami-Kolmogorov (JMAK) model. This analysis showed that the kinetics of this material does not follow the traditional JMAK behaviour but has two distinct stages with very different Avrami exponents.

1. Introduction

Gum Metal is a β-titanium alloy which has been shown to have excellent mechanical properties such as high strength at both high and low temperatures [1,2]. It is especially promising for biomedical applications due to the natural bone-bonding properties of titanium as well as the extremely low elastic modulus and the absence of toxic elements [3,4]. The alloy also benefits from high ductility (sustaining strains around 20% in tension before failure [1]) as well as Invar and Elinvar-like properties (i.e. nearly no density or modulus change with temperature) along with the high strength. While the large ductility is useful, it is often more desirable to exchange some of it for even higher strength levels. The thermomechanical processing steps used to tune the mechanical properties are similar to other, more traditional alloys. The first step will be a cold deformation to increase the strength via work hardening (and to form the part into the desired shape), followed by one or several heat treatment steps to recover all or part of the ductility via recrystallization and recovery. Interestingly, Gum Metal has shown unexpectedly high strength in *partially* recrystallized states, which is at least partly explained by the preferential formation of nano-precipitates of the ω phase in the recrystallized grains [5].

In order to design and tune a recrystallization heat treatment, the kinetics of the process must be quantified. While this has been examined previously by several authors, the results seem inconsistent even for similar procedures, and the details of the kinetics are unclear [5,6]. In this work the kinetics will be analyzed by the Johnson-Mehl-Avrami-Kolmogorov (JMAK) model [7–9]. This model describes the recrystallized volume fraction V_V as a function of time during isothermal annealing as:

$$V_V = 1 - \exp(-kt^n) \quad (1)$$

Where t is the annealing time, n is the Avrami exponent, and k is a constant related to the nucleation- and growth rate. In an idealized case, where the nuclei are assumed to be randomly distributed and the growth rate constant in all three dimensions, the Avrami exponent is between 3 and 4 (depending on how nucleation is occurring). The JMAK model also presumes that the material is homogeneous, which is rarely the case for recrystallization, as the deformation microstructure is not uniform in most cases.

In this work, the recrystallization kinetics of Gum metal will be examined by characterizing the microstructure throughout a recrystallization treatment. This will be done by a combination of optical microscopy and Electron Backscatter Diffraction (EBSD) for samples that have been partially recrystallized to different extents.

2. Experimental

2.1. Material

A commercial Gum Metal wire with a diameter of 1 mm and a composition of Ti-35.7Nb-1.9Ta-2.8Zr-0.4O (wt.%) was annealed in vacuum at 800 °C for 18 hours to ensure a fully recrystallized structure. The wire was then flat rolled to a reduction of 80% at maximum height in several passes. It is worth noting that since the wire has a circular cross section, only the center of the wire experiences the nominal reduction. In addition, the narrowness of the wire means that plane strain deformation is not fulfilled. In order to account for this, the samples were cut perpendicularly to the rolling direction (RD), the microstructural variation along the normal direction (ND) and more notably the transverse direction (TD) can then be readily seen. In order to minimize the effect that this variation has on the results, all measurements were taken as close to the center of the sample as possible.

2.2. Heat treatments

Annealing at 740 °C, 770 °C, and 800 °C were initially used to find the best temperature for this study. Since all temperatures produced a similar amount of secondary phases the lowest temperature of 740 °C was chosen to keep the recrystallization rate relatively low. Samples were heat treated at this temperature for different time periods in a range of 3-120 minutes. The samples were air cooled to minimize α'' formation at rapid cooling rates and excessive ω formation at slow cooling rates.

2.3. Microstructure characterization

The samples were polished and etched to reveal grain boundaries which were examined using a Leica Model DM2500M light optical microscope. The images were analyzed and the degree of recrystallization was quantified using the line counting method, where a set of lines were drawn across the micrographs and the line fraction that crossed recrystallized grains was measured.

The samples were also examined using EBSD. The EBSD scan was performed with a step size of 1 μm. The EBSD maps were processed using the Matlab software package MTEX. The recrystallized grains were identified using a set of criteria, specifically the grain orientation spread (GOS) as well as an orientation difference threshold for adjacent pixels to characterize a grain boundary. The maximum GOS for a grain to be considered recrystallized was set to 1.5°, the grain boundary threshold was set to 2° for samples heat treated for up to 15 minutes and 1.5° for those longer than 15 minutes. There was also a minimum grain size for the recrystallized grains, which was set to 9 pixels for the shorter heat treatments and 16 pixels for the longer heat treatments. These changes were made for the more annealed samples to avoid parts of the recovered matrix to be considered recrystallized.

3. Results and discussion

3.1. Quantification of recrystallization

In figure 1, optical micrographs and orientation maps are shown for samples heat treated for different amounts of time. From these maps it appears that the nucleation does not seem to be significantly clustered early in recrystallization (figure 1i), but some clusters are observed after 6 minutes (figure 1j), with a slight tendency to form bands. After 15 minutes, most of the recrystallized grains are impinged upon other recrystallized grains (figure 1k). Some darker bands are seen at the sample surface in the optical micrographs (figure 1a-d), especially after longer annealing times, which is believed to be a secondary phase. While some recrystallized grains do appear even in these regions, they are unlikely to be representative of the recrystallization behavior of the base material. In the analysis of recrystallization kinetics, those areas are excluded, they are therefore cropped out of the maps in figure 1i-l.

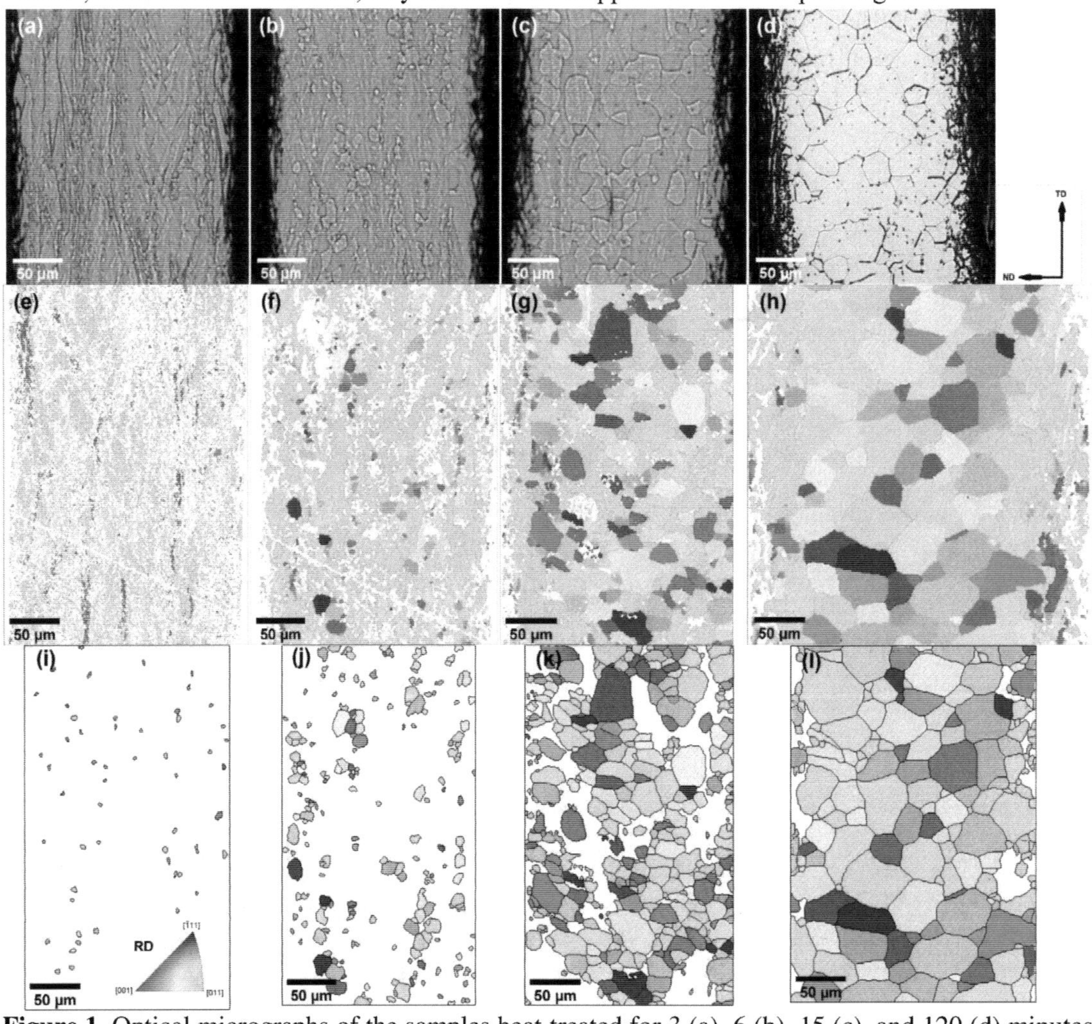

Figure 1. Optical micrographs of the samples heat treated for 3 (a), 6 (b), 15 (c), and 120 (d) minutes. Orientation maps of the same samples are seen in (e-h). (i-l) show the orientation maps showing only the recrystallized grains in the same samples. Coloring is based on the included IPF key.

The recrystallized volume fractions were obtained from both EBSD maps and from manual line counting based on light optical images. The two sets of results are plotted in figure 2a. It is seen that the values obtained from the optical micrographs are consistently lower than those obtained from the EBSD analysis. Reasons include significant uncertainty in identifying recrystallized grains from the optical micrographs as well as the fact that in the analysis of the optical data, the whole sample width was included, i.e. also the regions near the wire edge which has undergone less plastic deformation. Both sets of data were however fitted to equation 1. As it is clear from In the Avrami plot (see figure 2b), the fitting of the data points with a single straight line is very poor for both sets of measurements, but significantly better with two separate lines. The fitted values for the Avrami exponent are reported in figure 2b.

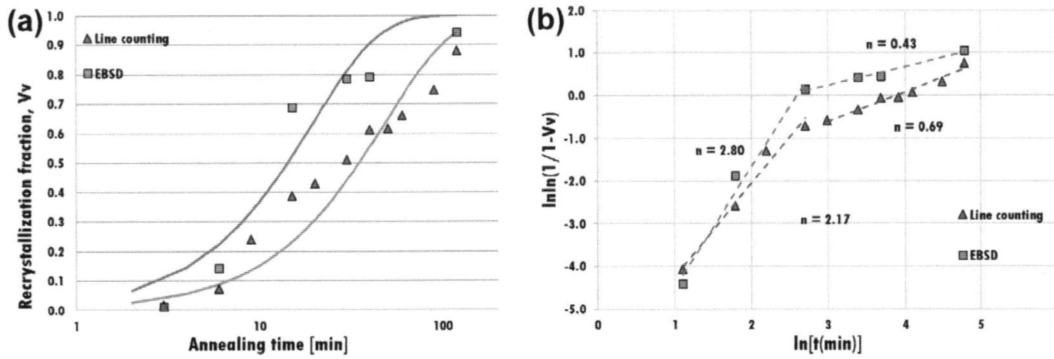

Figure 2. a) Data points with a fitted sigmoidal JMAK line for EBSD (green squares) and for line counting (blue triangles). b) Avrami plot. The dashed lines correspond to fits based on short and long heat treatments.

If only the early measurements are considered, the Avrami n-value is 2.8 based on EBSD measurements. The latter part of the recrystallization process has an Avrami exponent of merely 0.43. To further understand the recrystallization kinetics, the growth and nucleation rates need to be examined.

3.2. Investigation of nucleation and growth rates

To provide more information about what is happening at the different stages of the recrystallization process, it is useful to look at the number of recrystallized grains and the average size of the grains, see figure 3. The number of recrystallized grains per unit volume is estimated from the EBSD data as $N_V = N/(A \times D)$ where N is the total number of recrystallized grains in the map, A is the area of the map, and D is the mean grain size. The equivalent circular diameter was used for the grain size. It is clear that nucleation takes place early in the recrystallization process. It is worth noting that some nuclei are likely to be too small to be detected and we are therefore likely to underestimate the number of nuclei in the sample heat treated for only 3 minutes. It is thus suggested that site-saturation is a good approximation for the nucleation. This together with the measured Avrami exponent close to 3 at the early stage of recrystallization, suggests that the growth is nearly three-dimensional at this stage. The grains are observed to grow rapidly to a size of approximately 15 μm after 15 min annealing. At the later stages, the average grain size only increases slightly, which is clearly an effect of impingement. The low Avrami exponent in the later stage, however, further suggests that the growth rate or the growth dimensionality is reduced. The present analysis does not allow a proper evaluation of these two possibilities.

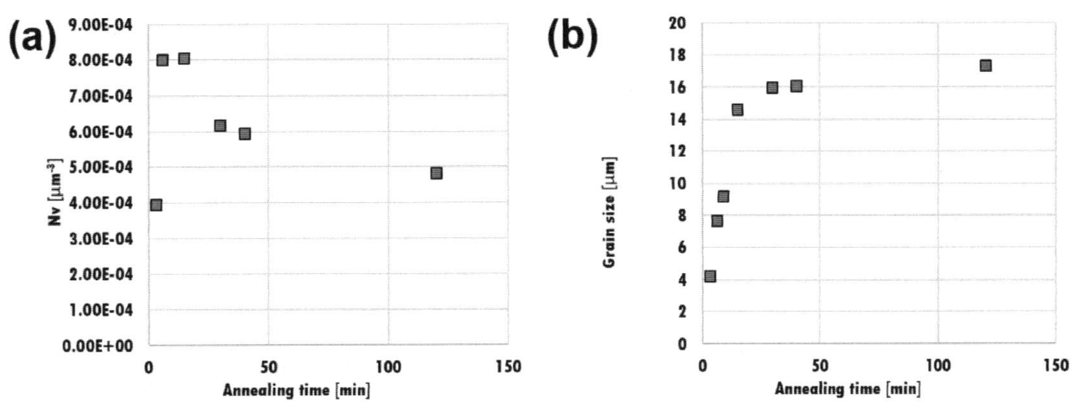

Figure 3. a) The number of recrystallized grains per unit volume. b) The average equivalent circular diameter of the recrystallized grains.

A similar low Avrami exponent has been observed previously in e.g. Ni, Cu, and Mg [10–12]. Possible reasons for the low Avrami exponent include:

1. A possible reduction in growth dimensionality. This is however not expected to be a dominating effect in the present sample, because the clustering does not seem to be very directional
2. The high energy regions in the deformed matrix are already consumed by recrystallized grains during the earlier stages of the process, and the driving force is thus reduced continuously as the recrystallization proceeds.
3. Recovery of the deformed matrix will also reduce the driving force for recrystallization, which continuously decreases the driving force and therefore also the growth rate.
4. The grain boundary misorientation relationship between the recrystallized grain and the surrounding matrix may change during recrystallization. This would naturally manifest in a gradual slowing of the growth rate as the boundaries with a more favorable misorientation to the matrix will grow rapidly in the early stages, but successively impinge as the process goes on. Slower and slower moving boundaries are then responsible for the growth rate at the later stages.
5. Nanoprecipitation of ω phase will slow down boundary migration via Zener pinning. This type of precipitation has been shown to occur for similar heat treatments and would become more and more severe as annealing continues [5].

Future work including measurements of the stored energy within the deformed matrix as well as quantification of the nanoprecipitation at various stages of recrystallization, may conclude which of these is the primary driver of the observed effect.

4. Conclusion

The microstructural evolution of Gum Metal rolled to a reduction of 80% in thickness has been investigated for isothermal recrystallization. The JMAK model was used to obtain information about the kinetics of the process. The results can be summarized as:

- The Avrami plot show two distinct different kinetic intervals with Avrami exponent of 2.1-2.8 in the early stages (3-15 min) and 0.48-0.69 in the later stages (30-120 min) of recrystallization.
- The recrystallization rate of Gum Metal is rapid in the early stages facilitated by site saturation nucleation and fast 3D growth of the recrystallizing grains. During later stages, it is suggested that the growth rate decreases very significantly, likely caused by factors including nanoprecipitation, recovery, and heterogeneities in the deformed microstructure.

References

[1] Saito T et al. 2003 Multifunctional alloys obtained via a dislocation-free plastic deformation mechanism. *Science*. **300(5618)** 464–7

[2] Kuramoto S, Furuta T, Hwang J, Nishino K and Saito T 2006 Elastic properties of Gum Metal. *Mater. Sci. Eng. A*. **442(1-2 SPEC. ISS.)** 454–7

[3] Gordin DM, Ion R, Vasilescu C, Drob SI, Cimpean A and Gloriant T 2014 Potentiality of the "gum Metal" titanium-based alloy for biomedical applications. *Mater. Sci. Eng. C*. **44** 362–70

[4] Tanaka M et al. 2014 Bone bonding ability of a chemically and thermally treated low elastic modulus Ti alloy: Gum metal. *J. Mater. Sci. Mater. Med.* **25(3)** 635–43

[5] Zhang JL, Tasan CC, Lai MJ, Yan D and Raabe D 2017 Partial recrystallization of gum metal to achieve enhanced strength and ductility. *Acta Mater.* **135** 400–10

[6] Guo WY, Xing H, Sun J, Li XL, Wu JS and Chen R 2008 Evolution of microstructure and texture during recrystallization of the cold-swaged Ti-Nb-Ta-Zr-O alloy. *Metall. Mater. Trans. A Phys. Metall. Mater. Sci. A*. **39(3)** 672–8

[7] Kolmogorov AN 1937 On the statistical theory of the crystallization of metals. *Bull. Acad. Sci. USSR, Math. Ser.* **1**

[8] Avrami M 1939 Kinetics of phase change. I: General theory. *J. Chem. Phys.* **7(12)** 1103–12
[9] Johnson WA and Mehl RF 1939 Reaction kinetics in processes of nucleation and growth. *Trans. Am. Inst. Mining, Metall. Pet. Eng.* **135(1089)** 416–42
[10] Chao HY, Sun HF, Chen WZ and Wang ED 2011 Static recrystallization kinetics of a heavily cold drawn AZ31 magnesium alloy under annealing treatment. *Mater. Charact.* **62(3)** 312–20
[11] Mishin O V., Zhang YB and Godfrey A 2017 The influence of multiscale heterogeneity on recrystallization in nickel processed by accumulative roll bonding. *J. Mater. Sci.* **52(5)** 2730–45
[12] Lin F, Zhang Y, Pantleon W and Juul Jensen D 2018 Importance of Non-uniform Boundary Migration for Recrystallization Kinetics. *Metall. Mater. Trans. A Phys. Metall. Mater. Sci.* **49(11)** 5246–58

IOP Conf. Series: Materials Science and Engineering 1249 (2022) 012023 doi:10.1088/1757-899X/1249/1/012023

Microstructural evolutions induced by an electrical breakdown in a binary Cu-25Cr alloy

L Varoto[1,2], M Chosson[2], J-J Blandin[1], A Papillon[2], S Roure[2] and G Martin[1]

[1]Univ. Grenoble Alpes, CNRS, Grenoble INP, SIMaP, F-38000 Grenoble, France
[2]Schneider Electric Industries, F-38000 Grenoble, France

E-mail: lucas.varoto@grenoble-inp.fr

Abstract. Cu-Cr based alloys with a high Cr content are widely used as electrical contact materials in vacuum interrupters due to their good combination of electrical, thermal, and mechanical properties. During a current interruption process in a vacuum, the establishment of an electrical arc results in a severe local thermal input onto the contact material surface. Few studies have closely investigated the microstructural evolution during this phenomenon as well as the Cu-Cr alloy properties evolution. This study reports the microstructural, electrical conductivity, and hardness evolution of a Cu-25Cr (25 wt%Cr) solid state sintered alloy as electrical contact before and after successive current interruptions. Electron scanning and optical microscopy show that this microstructural evolution is a result of a complex phenomenon. Local and heterogeneous thermal conditions take place resulting in significant microstructural evolution. The resulting microstructure is related to out-of-equilibrium solidification conditions. Electrical conductivity is decreased by half in the whole thermally affected zone by the electrical arc and hardness is multiplied by three to four times.

1. Introduction

Cu-Cr alloys with Cr content from 5 to 50 %wt are mainly used as electrical contacts for vacuum interrupters for medium voltage applications. These properties are the result of a good electrical conductor, copper, and a semi-refractory metal, chromium, that can withstand the thermal constraints imposed by such applications [1]. Different manufacturing techniques have been used to produce these binaries Cu-Cr alloys, such as vacuum induction melting (VIM) [2-3], solid state sintering [4-6], and more recently vacuum arc remelting (VAR) as well as powder bed fusion additive manufacturing techniques [7-10]. It has been shown that the properties of these alloys are strongly dependent on their composition and microstructure, especially when it comes to the size and morphology of the Cr-rich phase. Some studies show that the refinement of the Cr-rich phase as well as its morphology play a key role to improve the properties of Cu-Cr alloys as electrical contacts [3-5]. However, there is no consensus for now in the literature regarding microstructure optimization: size, morphology, and spatial distribution of the Cr-rich phase. The optimal proportion of the Cr-rich phase is yet to be established for Cu-Cr based alloys as electrical contacts [1].

During the vacuum interrupter's switching operation, when the electrical contacts separate from each other to interrupt the electrical current, a vacuum electrical arc is generated across the contacts as illustrated schematically in figure 1a-b. This process establishes an intense and severe thermal and electrical input onto the surface and consequently alters the microstructure of the contact. The generated electrical arc will move around the surface for several milliseconds (ms) until the alternative current drops to zero [1]. During this phenomenon, the Cu-Cr electrical contact is melted and rapidly solidified resulting in significant microstructural evolutions. Figure 1c indicates the region of interest to investigate microstructural evolutions induced by the electrical arc.

It has been reported that in this case and in other applications or manufacturing techniques with high cooling rates that Cu-Cr undergoes out-of-equilibrium solidification [6,7]. Under such out-of-equilibrium conditions, the Cu-Cr system shows a large liquid metastable miscibility gap (MG) [14-21]. Therefore, once the solidification path enters the MG, the final microstructure will be a result of a liquid phase separation process. This has also been seen and reported in the Cu-Fe and Cu-Co systems, see e.g. [22-25]. These findings suggest that significant microstructure evolutions and thus properties changes can be expected during the Cu-Cr alloy lifecycle as electrical contacts. In the present work, the microstructure evolution post-electrical arc is investigated. The microstructure and electrical and mechanical performances of a Cu-25Cr binary alloy (25wt% Cr) processed by solid state sintering in the as-fabricated state as well as after several electrical breakdowns are characterized. The mechanisms responsible for such microstructural evolutions are discussed.

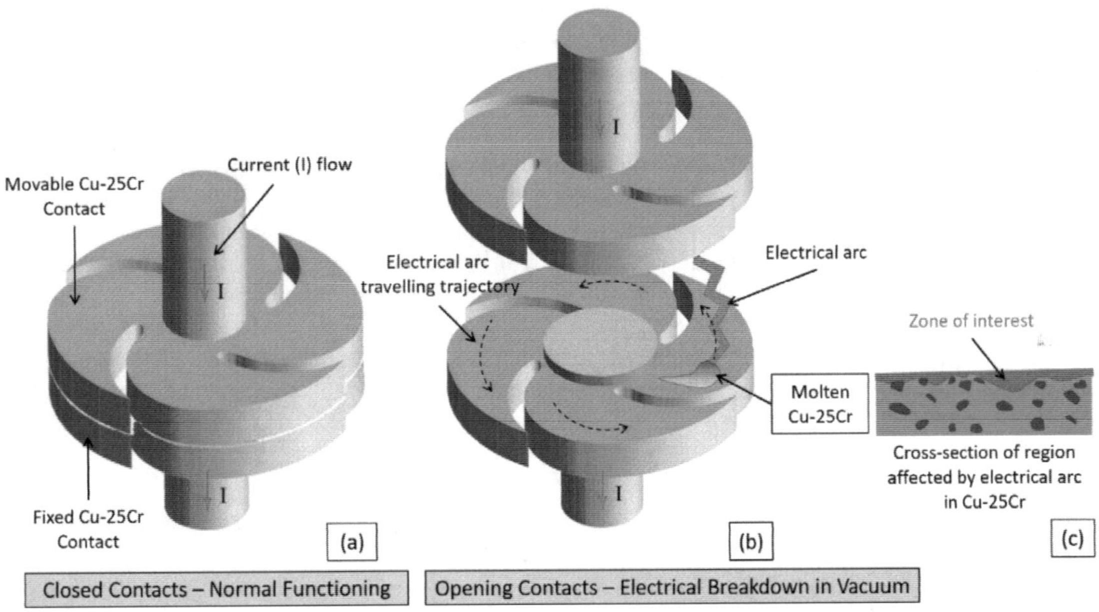

Figure 1. Schematic illustration of (a) closed electrical contacts under standard operating conditions. (b) opening contacts with the establishment of a vacuum electrical arc traveling onto the Cu-25Cr surfaces (electrical breakdown). (c) Representation of the microstructure in cross-section located in the region of interest, i.e. region affected by the electrical arc.

2. Materials and Methods

Cu-25Cr (25 %wt in Cr) samples were fabricated by solid state sintering. Pure Cu powder and pure Cr powder are mixed in a 3:1 mass ratio, respectively. A net-shape uniaxial pressing results in a green body that is then sintered in a secondary vacuum near copper's melting point to maximize solid state diffusion. Such samples were used as electrical contact materials in this study. Irregular Cr inclusions have a particle distribution from 50 to 80 μm in equivalent diameter. The samples analyzed were cut parallel to the pressing direction allowing the microstructure to be observed in cross-sections. Samples were mechanically polished and finished using a vibratory polisher with a solution containing 0.03 μm colloidal silica particles.

The electrical tests were conducted to reproduce the severe electrical and thermal constraints developed during the electrical contact lifecycle. Thus, vacuum interrupters were assembled with Cu-

25Cr sintered alloy as electrical contacts. The short circuit current conditions were reproduced by using capacitors bench discharges. An oscillating circuit is added to the system to generate the transitional recovery voltage. The experimental methodology can be summarized as follows: from a nominal short circuit current, a series of successive current interruptions is conducted. If the current interruption is a success, the next interruption is then conducted with an increase of 10% in the nominal short circuit current. If not, the next interruption is done with a 10% decrease in the nominal short circuit current. This series of successive interruptions proceeds until the total energy (accumulative) reaches over 90% of the contact electrical endurance capacity (number of times capable of current interruption). Then, the vacuum interrupters were disassembled and the Cu-25Cr contacts were recovered for microstructure analyses and properties evaluation.

The microstructure evolution of the Cu-25Cr alloy was analyzed using a Zeiss Ultra-55-FEG scanning electron microscope equipped with an energy dispersion spectrometer (EDS) by Bruker (XFlash Detector 5030) and a backscattered electron detector (BSD). An accelerating voltage of 15 kV and a high current mode were chosen for SEM-BSD and EDS analysis. Care was taken to select a region of interest with a long enough distance between Cr-rich inclusion in the plane of observation. At least five EDS measurements were conducted for each region of interest analyzed. Optical microscopy was also used to observe the whole affected zone of the microstructures after an electrical breakdown. Samples were etched with a 5ml:5ml:1g water:ethanol:iron nitrate solution to reveal the grain boundaries in the Cu-rich phase.

The electrical conductivity as well as the penetration depth, δ_f, were measured by Foucault's current (Eddy current) method in a FischerScope MMS PC2 Sigmascope equipped with a 3 mm probe at 480 kHz. The hardness (HV/5) of the as-fabricated (after solid state sintering) and micro-hardness (HV/0.005) of the post-electrical breakdown microstructure were done using a Presi HZ50-4 hardness tester. Under such testing conditions, the micro indentations deform a region sufficiently large to reflect the hardness of the microstructure in the region affected by the electrical arc. Even though different loads were employed to measure the hardness in both cases, it was considered enough to establish comparisons with the as fabricated. The mean and standard deviation from twenty measurement data for hardness and electrical conductivity were used. X-ray tomography was performed on cylindrical samples with a 10 mm height and nearly 0.5 mm in diameter for the as-fabricated alloy. An EasyTom XL was used at 100 kV, with a 0.5 mm aluminum filter, a voxel size of 0.8 μm, and 2 seconds acquisition per image with a total of 1408 projections.

3. Results

3.1. Microstructure before and after electrical breakdowns

An overview of the microstructural characteristics of the Cu-25Cr alloy processed by solid state sintering is given in figure 2a. The microstructure consists of two phases: a Cu-rich phase (matrix) and a coarse Cr-inclusions phase with an average size of 50-80 μm. From image analysis and the assistance of X-ray computed tomography (XCT), the volume fraction of coarse Cr-inclusions in the microstructure is around 33 vol%. In addition, there is a second population of irregular small Cr precipitates with a typical size of the order of 1 to 5 μm distributed in the Cu matrix, see Figure 2b. During the sintering heat treatment, the Cr diffusion in the Cu matrix occurs from the Cr/Cu interface. As the solid solubility of Cr in Cu is very limited, with a maximum of 0.7 wt%Cr at the eutectic temperature (1077 °C), the precipitation of the Cr-rich phase in the Cu matrix during the cooling stages post-sintering takes place. Residual porosity of about 4 %vol was measured based on the 3D image from the XCT scan. The pore population can be classified into two categories: spherical pores located in the Cu matrix, and Cu/Cr interfacial porosity, with a ratio of 1:3 respectively. The pores are often filled with matter, mostly copper, during surface preparation before microstructure observations. Therefore, the porosity cannot be properly revealed by conventional surface preparation. Figure 3 shows a 2D cross-section parallel to the pressing direction extracted from the volume determined from the XCT scan where pores appear in black due to the absorption contrast.

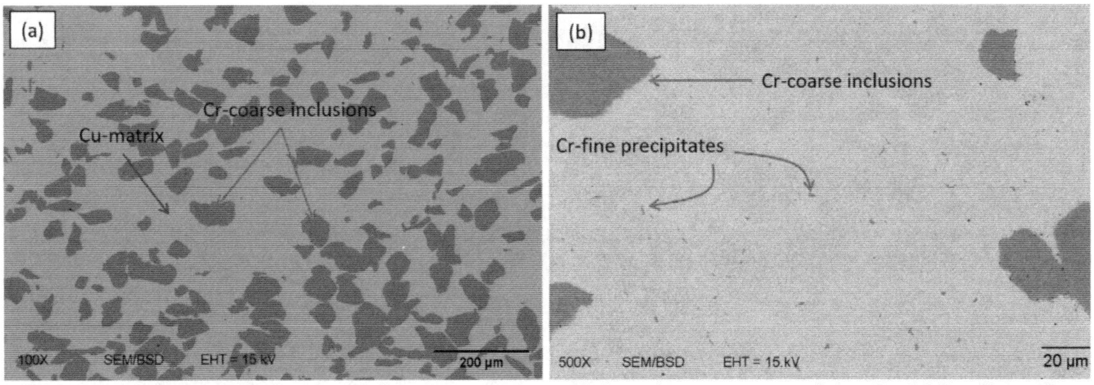

Figure 2. (a) SEM-BSE image giving an overview of the microstructure of the Cu-25Cr alloy after solid state sintering. (b) Enlarge view illustrating the finer Cr-precipitates distributed in the Cu-matrix.

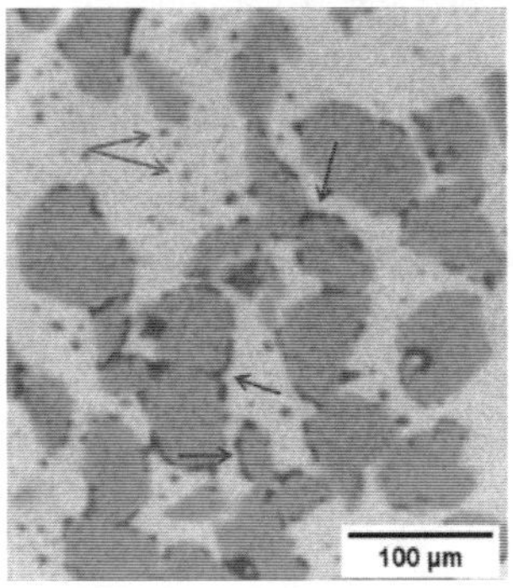

Figure 3. A 2D cross-section extracted from an XCT scan acquired in a Cu-25Cr solid state sintered sample parallel to the pressing direction. The porosity appears in black, the Cu matrix in light gray, and the Cr-rich phase in dark gray.

Figure 4a shows an SEM-BSD image revealing the microstructure post-electrical arc. There is a drastic change in Cr-phase size and morphology due to the intense thermal input of a localized vacuum electrical arc. In this case, the region affected by the thermal input consists of two zones: an upper layer (*Zone 1*) and a second zone just underneath (*Zone 2*). From microscopy analysis of several different regions, this two-zone configuration seems to be the general case all around the Cu-25Cr electrical contact microstructure after successive electrical breakdowns. Figure 4b shows these distinct zones and that there is a transition between them as well as different microstructural characteristics for each one. In *Zone 1,* Cr-inclusions show a spherical morphology with a typical size between 200 nm and 1 µm. A single morphology of the Cr-rich phase is observed in *Zone 1*. In *Zone 2,* just underneath, the Cr-phase shows different morphologies, and size variations were also detected. Very fine spherical Cr-rich inclusions, from about 150 nm to 300 nm were found in the lower region of *Zone 2*, see figure 4c evolving to micro-sized Cr-dendrites. Figures 4c and d illustrate this morphological change of the Cr-rich phase in *Zone 2*. In addition, some small spherical Cr-rich phase is observed just underneath *Zone2*, see figure 4c. Partially melted and un-melted coarse Cr-inclusions are also present in *Zone* 2. The average depth where microstructure changes are detected (i.e. *Zone 1* + *Zone 2*) reaches from 80 to 120 µm. When revealing the Cu grain boundaries using chemical etching (Figure 5a-b) the depth of the region affected by the electrical arc can be seen. Columnar Cu grains are observed as part of *Zone 2* and they can reach nearly 200 µm in length, see Figure 5.

EDS analysis indicates an increase of Cr content in the Cu matrix from about an order of magnitude: 0.4 wt% Cr in the Cu-matrix not affected by the electrical arc (after solid state sintering) to around 4 wt% Cr in *Zone 1*. These EDS results help to roughly estimate the content of Cr in Cu after passing the electrical arc onto the surface. However, it must be highlighted that more resolved characterization methods such as atom probe tomography would be required to properly quantify the evolution of the chemical composition of the Cu matrix in the region affected by the electrical arc.

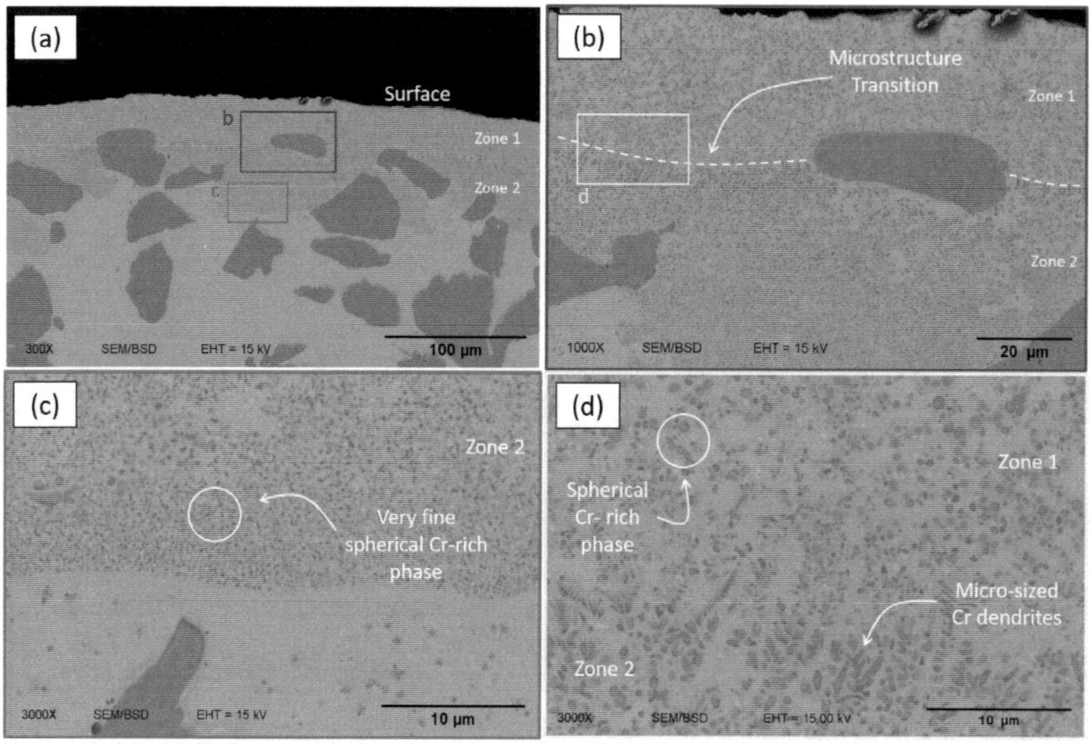

Figure 4. SEM-BSE images (a) showing an overview of the drastically changed microstructure after an electrical breakdown. (b) two distinct zones: Zone 1 and Zone 2, and their associated microstructure. (c) the small spherical Cr-precipitates from *Zone 2,* and a transitional microstructure showing the transition from the Cu columnar grains to a dense spherical Cr-rich phase microstructure. (d) an enlarged view of the transition zone showed in (b) where the micro-sized Cr-dendrites from *Zone 2* can be easily distinguished from the small spherical Cr-rich phase observed in *Zone 1.*

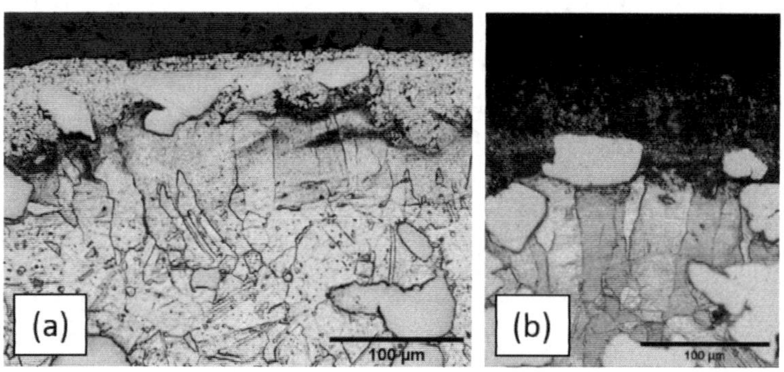

Figure 5. Optical microscopy image after chemical etching of two different regions from two different Cu-25Cr sintered electrical contacts ((a) and (b), respectively) under the same electrical test conditions. Columnar Cu grains can be seen and are helpful to reveal the whole region affected by the vacuum electrical arc.

3.2. Hardness and electrical conductivity measurements

The as-fabricated and post-electrical arc microstructure of the same Cu-25Cr alloy show different mechanical and electrical properties resulting from the microstructural changes illustrated previously. Micro-hardness (HV/0.005) measurements performed in the different zones affected by the electrical arc show a significant hardness increase. While the as fabricated Cu-25Cr solid state sintered microstructure has an average hardness of 60 ± 3 HV/5, *Zone 1* and *Zone 2* hardness increase to 182 ± 6 and 208 ± 7 HV/0.005, respectively. Therefore, *Zone 2* presents the highest hardness of the different microstructures observed. As-fabricated Cu-25Cr sintered samples have an average of 31 ± 2 MS/m electrical conductivity. After successive electrical tests, measurements throughout Cu-25Cr sintered contacts surface resulted in an average of 16.4 ± 2 MS/m. With a penetration depth of roughly 180 ± 13 µm, one can consider that in these measurement conditions, the electrical conductivity represents that of the region affected by the electrical arc. Thus, there is a significant increase in resistivity in this region in comparison with the as-fabricated microstructure. These results are summarized in figure 6 where the corresponding properties throughout the region affected by the electrical arc are given.

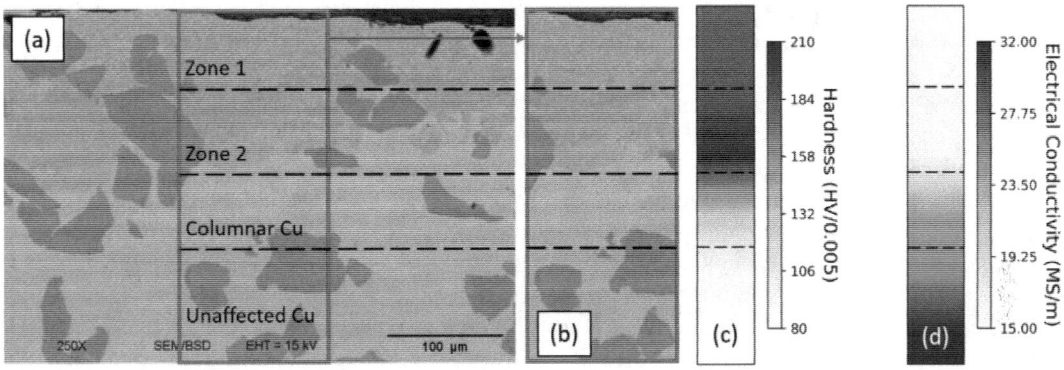

Figure 6. (a) SEM-BSE image of the microstructure resulting from the electrical arc. The micro-hardness and electrical conductivity have been measured in the region of interest highlighted in red in (b). Color maps show respectively (c) the hardness and (d) the electrical conductivity variation in (b).

4. Discussion

4.1. Melting conditions and associated microstructure evolutions

To date, few studies have closely studied these microstructural changes as well as their consequences on the mechanical and electrical properties [6, 8, 15]. As illustrated in figures 4 and 5, the resulting microstructure shows a two-zones configuration where a clear microstructural evolution can be seen. These microstructures are the result of successive electrical breakdowns, thus, several melting and solidification events will occur by the end of the Cu-25Cr alloy lifecycle. It is known that under these experimental conditions, the vacuum electrical arc moves around onto the Cu-25Cr alloy contact material [1,27] as illustrated in figure 1b. This establishes a dynamic arc traveling and therefore a dynamic and complex melting and solidification conditions that can impact the microstructure.

It has been reported that for a single vacuum electrical breakdown in Cu-25Cr solid state sintered alloy as contact materials the total heat affected zone is very similar to that found in *Zone 2* [8]. However, such gradient in Cr-morphology has not been yet reported as well as an upper layer as *Zone 1*. The microstructural observations suggest that the *Zone 1* + *Zone 2* configuration is a result of a highly complex and heterogeneous temperature and thermal distribution during successive electrical breakdowns in a vacuum. In addition, it has been reported that extremely high cooling rates from solidification are reached, from 10^4 to 10^6 K/s [6, 8].

One can say that as no Cr-phase evolution is seen in the extent of the Cu columnar grains until a clear highly fine spherical Cr-phase distribution (figure 4c), the temperature did not reach the Cr phase melting conditions, but only the Cu matrix one. When approaching the surface, a clear Cr evolution can be seen in the microstructure. Temperature conditions were such that both phases, namely the Cu-rich and Cr-rich were melted. The Cr morphological and size evolution, from small spherical Cr-rich phase to Cr micro-sized dendrites in *Zone 2* suggest that there are also local variations of the solidification conditions from the bottom to the top.

Since there is a dynamic melting and solidification process during the electrical arc traveling, liquid Cu-Cr on the surface gets either ejected or flows with the vacuum arc traveling at extremely fast speeds in a short time frame (some ms) [1]. Such convective flows in the molten bed are thought to be responsible for the transport of matter to adjacent regions on the surface and, once solidified, accumulate locally. The latter phenomenon is believed to contribute to the microstructural variability observed in the different zones. This supports the complexity of the thermal input throughout successive electrical breakdowns. Thus, further investigations are needed to better understand the heat distribution after the vacuum electrical arc phenomenon.

Un-melted and partially melted Cr-rich coarse inclusions can also be found in *Zone 2*. Figure 4a and figure 5 show the presence of such Cr coarse inclusions. In this case, the relatively large size of Cr inclusions makes the melting of the microstructure more difficult during the vacuum arc given the time scale. The Cu/Cr interfacial porosity can also play an important role locally in the heat exchange between the vacuum arc and the microstructure. As there is a significant fraction of porosity at the interface between Cu and Cr inclusions, the heat transfer from Cu to Cr is reduced in comparison with the true solid-to-solid interface, or the molten Cu that eventually wets the surface of Cr inclusions.

The Cr-rich phase shows two different morphologies in the region affected by the electrical arc: (i) fine spherical Cr-rich phase morphology, and (ii) micro-sized Cr with dendritic morphology. Both morphologies are solidification structures resulting from variations in solidification conditions. It has been reported that the Cu-Cr system has a large metastable liquid-liquid miscibility gap. Thus, in out-of-equilibrium solidification conditions, such as at extremely high cooling rates and high undercooling, the solidification of the Cu-Cr system goes through a liquid phase separation (LPS) process [14-17]. In this process, when the liquid Cu-Cr enters the miscibility gap it separates into a Cu-rich liquid and a Cr-rich liquid. This liquid phase separation creates, for an instance of time, Cr-rich liquid spheroids immersed into the Cu-rich liquid. As solidification evolves, the Cr-rich liquid spheroids tend to agglomerate and grow until complete solidification. This process results in the Cr-morphology (i) found post vacuum electrical arc and has been reported and discussed in different cases where Cu-Cr alloys undergo out-of-equilibrium solidification [8-12,14-17]. Cr-morphology (ii) suggests typical dendritic nucleation and growth as primary Cr solidifies. These two Cr phase morphologies also support the difference in local solidification conditions from the beginning of the Cu columnar grains to the surface.

4.2. Electrical conductivity and hardness evolution

Under the thermal conditions created by the electrical arc, a supersaturated solid solution of Cr in the Cu-matrix can be expected. The temperature and cooling rate gradient discussed support the trend deduced from EDS analyses: higher cooling rates help retain more Cr in the Cu. In the as-fabricated Cu matrix, the Cr content was roughly measured to be about 0.4 wt%Cr vs. 1 wt%Cr in the columnar Cu grains in *Zone 2* and reached about 4 wt%Cr in the Cu-matrix in *Zone 1*. This trend, though being deduced from EDS, agrees with the literature [6, 8]. The significant decrease in electrical conductivity is attributed to the increase of Cr in solid solution in the Cu-matrix. The increase of Cr retained in the Cu-matrix solid solution increases the resistivity of the Cu lattice since Cr atoms act as scattering centers. It has been reported that the base resistivity of pure copper (1.71 $\mu\Omega\cdot$cm) increases by 4.12 $\mu\Omega\cdot$cm/at% Cr [26]. Therefore, considering that the penetration depth average represents the overall affected zone, one can say that from the un-changed matrix to the surface there is a loss of half of the electrical conductivity.

Besides, there is a significant increase in hardness compared to the hardness of the as-fabricated Cu-25Cr microstructure. As reported, there is a gradient in hardness from the columnar Cu grains to a maximum hardness of 210 HV/0.005 in *Zone 2* and 180 HV/0.005 in *Zone 1* while the unaffected microstructure of the as-fabricated alloy has an average hardness of 60 HV/5. Since the solid solution strengthening of Cr in Cu has been reported to have a very low contribution, these hardness values seem to be correlated to the Cr-rich phase size where the maximum hardness is reached in the region having the finest spherical Cr-phase inclusions in *Zone 2* [28].

This microstructure evolution during successive current interruption tests can be thus seen as the signature of the passing of the electrical arc. However, even with a local decrease in electrical conductivity, the electrical contact is still functional during each successive current interruption under vacuum conditions but with a drastically changed microstructure. Further investigations would be required to understand more deeply this microstructure evolution and used to further optimize the microstructure of the electrical contacts.

5. Conclusion

The complex microstructural evolution after successive vacuum electrical breakdown is characterized as well as the electrical conductivity and hardness. After successive vacuum electrical breakdowns, two zones characterize the total affected zone by a vacuum electrical arc. First, an upper one (*Zone 1*), close to the surface, which is the result of liquid flow transported by the vacuum electrical arc movement that solidifies. The fine spherical Cr-rich phase, from 500 nm to 1 μm, is dominant in this microstructure (*Zone 1*). And a second region (*Zone 2*) where a very fine spherical Cr-rich phase, from 150 nm to 300 nm evolving to micro-sized Cr-dendrites can be observed. This two-zones configuration can penetrate down to 200 μm into the bulk. These microstructures result from complex and out-of-equilibrium solidification conditions, established mainly by high cooling rates during vacuum electrical arc traveling. Under these conditions, the small spherical Cr-rich phase is the result of the liquid phase separation process. Electrical conductivity decreases significantly from 31 ± 2 MS/m in the as-fabricated state to 16.4 ± 2 MS/m in the highly transformed microstructure due to the increase of Cr content in solid solution in Cu. Hardness increases from 60 ± 3 HV/5 for the as-fabricated Cu-25Cr microstructure to a maximum of 208 ± 7 HV/0.005 in the region affected by the electrical arc.

References

[1] Slade P G 2008 *The Vacuum Interrupter: Theory, Design, and Application* (Boca Raton: CR Press Taylor & Francis Group)
[2] Zhang C Y, Wang Y P, Yang Z M, Guo Y and Ding B J 2004 *J. Alloys Compd.* **366** 289–92
[3] Rieder W F, Schussk M, Glatzle W and Kny E 1989 *IEEE Trans CPMT* **12** 273-83
[4] Wang Y P and Ding D J 1999 *IEEE Trans. CPMT* **22** 467-72
[5] R Muller 1988 *Siemens Forsch. Entwicklungsber* **17** 105-11
[6] Wei X, Wang J, Yang Z, Sun Z, Yu D, Song X, Ding B and Yang S 2011 *J. Alloys Compd.* **509** 7116-20
[7] Papillon A, Missiaen J, Chaix J, Roure S and Schellekens H 2016 *Int. J. Refract. Met. Hard. Mater.* **65** 9-13
[8] Hauf U, Kauffmann A, Kauffmann-Weiss S, Feilbach A, Boening M, Mueller F E H, Hinrichsen V and Heilmaier M 2017 *Metals* **7** 478
[9] Ding B J, Yang Z M and Wang X T 1996 *IEEE Trans. CPMT* **19** 76–81.
[10] Uchida S, Kimura T, Nakamoto T, Ozaki T, Miki T, Takemura M, Oka Y and Tsubota R 2019 *Mater. Des.* **175** 107815
[11] Szemkus S, Kempf B, Jahn S, Wiehl G, Heringhaus F and Rettenmayr M 2018 *J. of Mater. Processing Tech.* **252** 612-17
[12] Momeni S, Guschlbauer R, Osmanlic F, Korner C 2018 *Mater. Lett.* **223** 250-52
[13] Zhao F, Yang Z M and Ding B J 2000 *Trans. Nonferrous Met. Soc. China* **10** 73–5
[14] Zhou Z M, Gao J, Li F, Zhang Y K, Wang Y P and Kolbe M 2009 *J. Mater. Sci.* **44** 3793-99
[15] Zhang L, Yu G, He X, Li S, Ning W and Wang X 2019 *J. Alloys Compd.* **772** 209-17
[16] Si S-H, Zhang H, He Y-Z, Li M-X and Guo S 2015 *Metals* **5** 2119-27
[17] Jacob K T, Priya S and Waseda Y 2000 *Z. Metallkd.* **91** 594-600
[18] Zhou Z, Zhou T, Chai L, Tu J, Wang Y, Huang W, Xiao H, Xiao Z 2015 *Mater. Res.* **18** 34-39
[19] Zhou Z M, Gao J, Li F, Wang T P and Kolbe M 2011 *J. Mater. Sci.* **46** 7039-45
[20] Sun Z, Zhang C, Zhu Y, Yang Z, Ding B, and Song X 2003 *J. Alloys Compd.* **361** 165-168
[21] Zhou Z M, Wang Y P, Gao J and Kolbe M 2005 *Mater. Sc. Eng.* A **398** 318-22
[22] Munitz A and Abbaschian R 1996 *Metall. Mater. Trans.* A **27** 4049-59
[23] Curiotto, Pryds N, Johnson E and Battezzati L 2006 *Metall. Mater. Trans.* A **37**.
[24] Zhang J, Cui X and Wang 2018 *J. Cast Mater. Res.* **31** 87-92
[25] Munitz A 1987 *Metall. Mater. Trans.* B **18** 565-75
[26] Tenwick M J and Davies H 1988 *Mater. Sci. Eng.* **98** 543-46.
[27] Wang L, Wang H, Wang C, Wang K, Lin R and Jia S 2022 *IEEE Trans. Plasma Sci.* **Early access**
[28] Liu P, Kang B X, Cao X G, Huang J L and Gu H C 2000 *J. Mater. Sci.* **35** 1691-94

IOP Conf. Series: Materials Science and Engineering 1249 (2022) 012024 doi:10.1088/1757-899X/1249/1/012024

The heterogeneous microstructure in an Al-Zn-Mg-Cu-Sc-Zr alloy produced by selective laser melting

Y Y Zhu [1,2], W Pan [3], Y Y Li [1,2], T L Huang [1,4,*], Y Z Zhang [3] and X Huang [1,2]

[1] *International Joint Laboratory for Light Alloys (MOE), College of Materials Science and Engineering, Chongqing University, Chongqing 400044, China*
[2] *Shenyang National Laboratory for Materials Sciences, Chongqing University, Chongqing 400044, China*
[3] *National Engineering & Technology Research Center for Non-Ferrous Metals Composites, GRINM Group Corporation Limited, Beijing 101407, China*
[4] *Electron Microscopy Center of Chongqing University, Chongqing University, Chongqing 400044, China*

Email: huangtl@cqu.edu.cn

Abstract. Fully dense and crack-free Al-Zn-Mg alloys with Sc and Zr additions were successfully fabricated by selective laser melting. The microstructures of as-built alloys with different Sc and Zr additions were characterized by scanning electron microscopy and transmission electron microscopy. The microstructure was composed of coarse columnar grains without Sc and Zr additions. In contrast, a fine heterogeneous microstructure, consisting of equiaxed grains at the bottom of the melt pool and columnar grains in the remaining volumes, was formed by adding 0.6 wt.% Sc and 0.26 wt.% Zr. In addition, a high density of precipitates was identified as η' and *I*-phases that were distributed at grain boundaries and inside grains.

1. Introduction

As an additive manufacturing technology, selective laser melting (SLM), combined with computer-aided design, has unique advantages in efficiently fabricating complex metallic objects [1-5]. SLM technology has been widely studied in the manufacture of aluminum alloys, and high-quality components of some aluminum alloys, such as Al-Si and AlSi10Mg [6-8], have been successfully printed. However, SLM manufacturing of heat treatable high-strength aluminum alloys, such as 2xxx, 6xxx, and 7xxx series Al alloys, has still been a challenge due to their poor liquidus fluidity [9-11].

Recently, some studies showed that high quality samples of Al-Zn-Mg-Cu (7xxx series) alloys could be successfully fabricated by adding the elements Sc and Zr [12-14]. However, the microstructure of the SLMed Al-Zn-Mg-Cu with Sc and Zr additions has not been well characterized, and the effect of Sc and Zr additions on the microstructure heterogeneity has not been well understood. In this study, the microstructures of SLMed Al-Zn-Mg alloys with different Sc and Zr additions were characterized by electron microscopy, and the effects of Sc and Zr additions are discussed.

2. Materials and experimental procedures

In this study, Al-Zn-Mg (-Cu-Sc-Zr) alloys with different contents of Sc and Zr were fabricated by SLM. The alloy chemical compositions, measured by inductively coupled emission plasma spectrometry, are shown in table 1.

The microstructures were characterized by means of scanning electron microscopy (SEM) and transmission electron microscopy (TEM). SEM samples were prepared by electropolishing in a solution of 10% perchloric acid ($HClO_4$) and 90% ethanol (C_2H_5OH) and observed with a JEOL JSM-7800F scanning electron microscope. TEM characterization was conducted in a JEOL JEM-2100 electron microscope operated at 200 kV. Thin foils for TEM were prepared by twin-jet electropolishing at -20 °C in a solution of 30% nitric acid (HNO_3) and 70% methanol (CH_3OH). The grain size was measured using the Nano measure software based on SEM images by use of line intercept measurements, with these measurements taken perpendicular to the build direction (BD) for the columnar grains.

Table 1. Chemical compositions of original powders and as-SLMed alloys, wt.%.

		Zn	Mg	Cu	Sc	Zr	Al
Alloy 1	powder	6.2	2	0	0	0	Bal.
	as-SLMed	4.52	1.71	0	0	0	Bal.
Alloy 2	powder	6.2	2	0	0.2	0.12	Bal.
	as-SLMed	4.74	1.70	0	0.23	0.14	Bal.
Alloy 3	powder	6.2	2	0	0.4	0.24	Bal.
	as-SLMed	4.45	1.64	0	0.39	0.21	Bal.
Alloy 4	powder	6.8	2	0.35	0.60	0.36	Bal.
	as-SLMed	4.47	1.22	0.36	0.60	0.26	Bal.

3. Results and discussion

Figure 1 and figure 2 show the microstructures of as-built alloys with different content of Sc and Zr. The microstructure can be classified into two types. Figure 1 shows the first type, which can be recognized as columnar grains. Figure 1(a) and (b) show the microstructure of alloy 1 with no addition of Sc. The irregular columnar grains were of large sizes in both the width and length directions. The coarse columnar grains grow epitaxially through the melt pool boundaries (MPB, marked by yellow dash lines). White dash lines in figure 1(b) indicate the columnar grain boundaries. In addition, cracks and pores can easily be found in the microstructure. Most of the cracks propagate along the boundaries between columnar grains. These cracks were considered to be hot cracks, which were formed during the solidification of SLM [15]. It should also be noticed that near the bottoms of the melt pools, the densities of precipitates are high. The microstructure of alloy 2 (0.23 Sc+0.14 Zr) was depicted in figure 1(c) and (d). The morphology is very similar to that shown in figure 1(a) and (b). Large-scale cracks and pores have not been eliminated. However, the grain size is slightly smaller, and several smaller-sized grains (indicated by black arrows in figure 1(d)) can be seen at the melt pool boundary. Densities of the precipitates at the bottoms of the melt pools are apparently higher than that in figure 1(a) and (b).

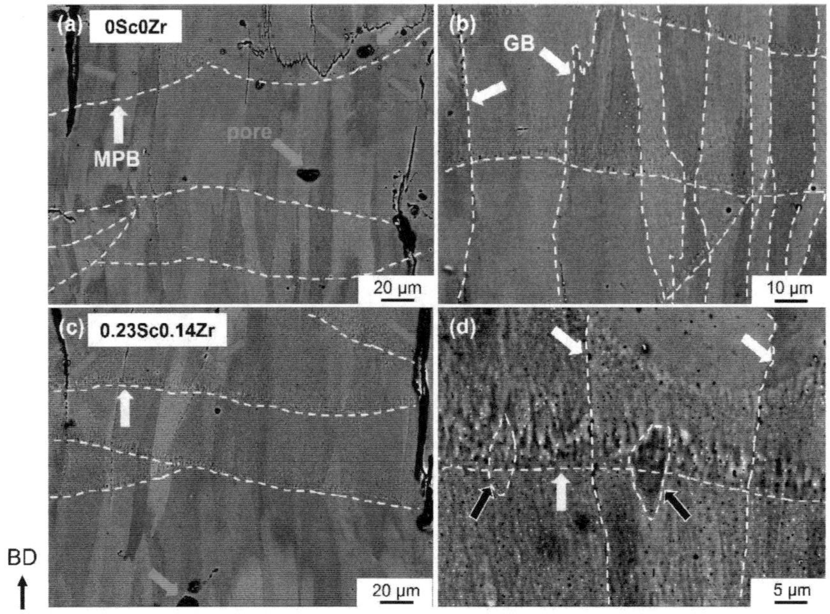

Figure 1. SEM-BSE images showing the coarse columnar grains traversed through melt pools of as-built alloys from the cross-section parallel to the build direction (BD). (a, b) alloy 1; (c, d) alloy 2. Melt pool boundaries, MPBs, are marked by yellow dashed lines; the grain boundaries, GBs, of columnar grains are marked by white dashed lines.

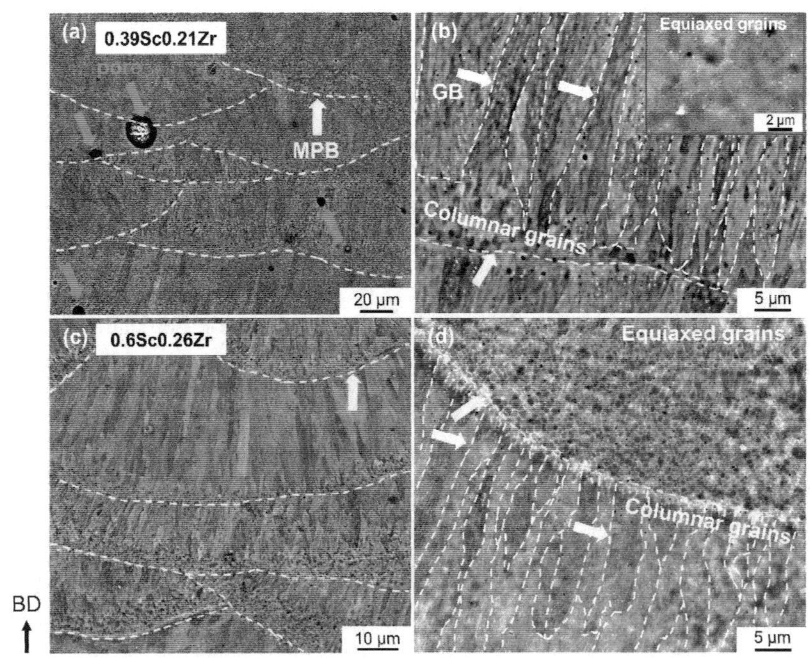

Figure 2. SEM-BSE images showing the fine heterogeneous microstructure of as-built alloys from the cross-section parallel to the build direction (BD). (a, b) alloy 3; (c, d) alloy 4. Melt pool boundaries, MPBs, are marked by yellow dashed lines; grain boundaries (GBs) of columnar grains, are marked by white dashed lines.

The second type of microstructure, which was observed in the samples with higher Sc and Zr additions, can be recognized as a fine heterogeneous microstructure. Figure 2(a) and (b) show the microstructure for alloy 3 with 0.39 wt.% Sc and 0.21 wt.% Zr. Compared with alloy 1 and alloy 2, the microstructure is composed of grains with two different morphologies, namely columnar and equiaxed. The columnar grains show similar morphology to that seen in figure 1, but have much smaller sizes. More importantly, these columnar grains are confined to the inside of one melt pool, indicating that they did not grow epitaxially during manufacturing. The equiaxed grains are formed at the bottom of melt pools. Figure 2(b) illustrates more details of the columnar grains and equiaxed grains, which were decorated with a large number of second phase particles (appearing with brighter contrast) at grain boundaries. Overall, large scale cracks cannot be observed, except for near some large pores. Figure 2(c) and (d) show the microstructure of alloy 4 with 0.6 wt.% Sc and 0.26 wt.% Zr addition. As seen in figure 2(c), a typical heterogeneous microstructure was developed, composed of fine equiaxed grains (at the bottom of melt pools) and columnar grains. Compared to figure 2(a), the areas of fine equiaxed grains are more extensive. In addition, no cracks and pores are present. Note that some black dots seen in the image were caused by loss of particles in the process of electropolishing. Figure 2(d) depicts more details of the distribution of precipitates associated with equiaxed grains and columnar grains. The many second phase particles are mainly distributed at the grain boundaries. Furthermore, many particles are distributed along the length direction of the columnar grains. Also be noticed is that the densities of the second phase particles are much higher in figure 2(d) than in figure 1(b) and (d).

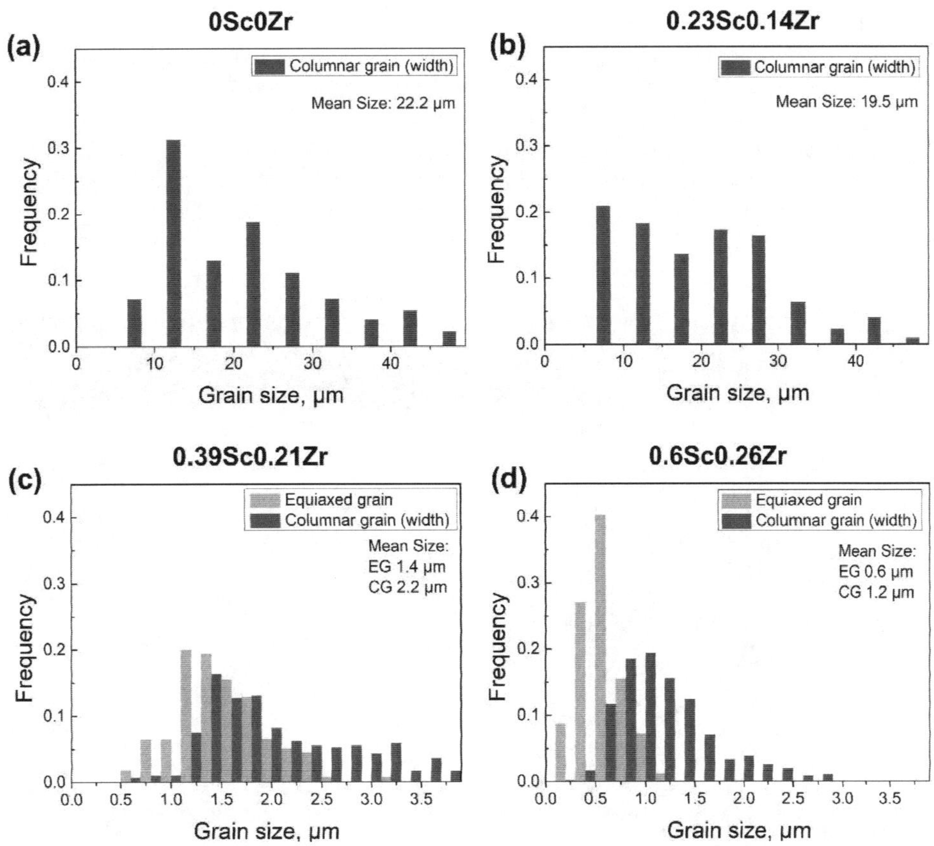

Figure 3. Histograms showing grain size distributions for (a) alloy 1; (b) alloy 2; (c) alloy 3; (4) alloy 4.

Figure 3 illustrates the evolutionary trends of grain size for as-built alloys with increasing amount of of Sc and Zr additions. In alloy 1 and alloy 2, there are only coarse columnar grains without equiaxed grains. The widths of columnar grains in alloy 1 and alloy 2 are large, and their mean values are 22.2 μm and 19.5 μm, respectively. The addition of 0.23 wt.% Sc and 0.14 wt.% Zr (alloy 2) contributes to reducing the width of columnar grains, but the effect is limited. In alloy 3 and alloy 4, heterogeneous microstructures composed of equiaxed grains and columnar grains are formed, figure 3(c) and (d). The mean size of the equiaxed grains in alloy 3 is 1.4 μm, while it is 0.5 μm in alloy 4. Similarly, the mean widths of columnar grains in alloy 3 and alloy 4 are 2.2 μm and 1.2μm, respectively. Clearly, the grain size is reduced significantly increasing amount of Sc and Zr addition. The depth of equiaxed grains region and the depth of melt pool were measured from the SEM-ECC images and their ratios are shown in table 2. It is evident that the formation of a heterogeneous microstructure is enhanced with increasing addition of Sc and Zr in SLMed Al-Zn-Mg alloys and that the equiaxed grains occupy almost 20% of the microstructure in alloy 4, which has the highest addition of Sc and Zr in the present study.

Table 2. Ratio of the depth of equiaxed grains region to the depth of melt pool.

	Alloy 1	Alloy 2	Alloy 3	Alloy 4
Ratio (equiaxed grain/melt pool)	0	~ 0	1:20	1:5

Figure 4 illustrates TEM observations of the precipitation behavior in alloy 4. Fine η' phase particles are precipitated at grain boundaries and along the dislocations within columnar grains. Some coarse particles are aligned along the length direction of columnar grains, which correspond to the particles seen in the ECC images in figure 1 and figure 2. These coarse particles are the *I*-phase, often reported for the SLM-produced 7xxx series Al alloy [16, 17]. The *I*-phase is rich in Mg and Zn and has an icosahedral quasicrystal structure. It is believed that during SLM processing, the high density of second phase particles suppresses grain growth and enhances the formation of the fine heterogeneous microstructure.

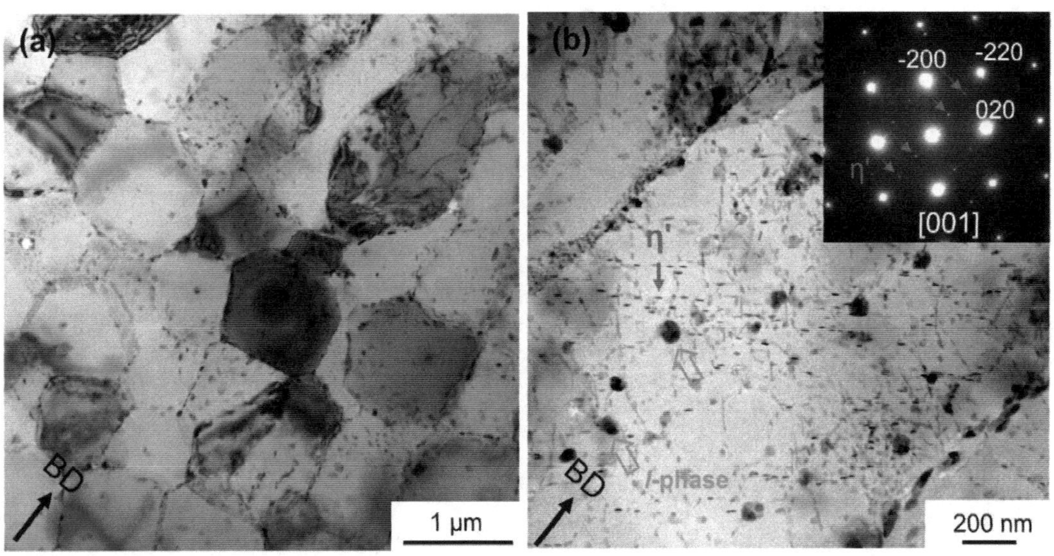

Figure 4. TEM images showing the precipitation behaviour in the as-printed alloy 4 observed from the cross-section parallel to the build direction (BD). (a) Equiaxed grains; (b) Columnar grains.

The present results demonstrate that the microstructure in SLMed Al-Zn-Mg alloys transforms from a coarse columnar grain structure without Sc and Zr addition to a heterogeneous microstructure composed of fine columnar grains and equiaxed grains with an addition of 0.6 wt.% Sc and 0.26 wt.% Zr. With the formation of fine heterogeneous microstructure, cracks and pores are also reduced significantly and even eliminated. The elimination of these defects is considered to relate to the formation of heterogeneous microstructures, which effectively disperses stresses generated during the thermal cycling process [18]. The formation of fine heterogeneous microstructures is expected to improve the mechanical properties of the alloys fabricated by SLM.

4. Conclusions

The microstructures of as-built Al-Zn-Mg alloys with different Sc and Zr additions have been characterized using SEM and TEM. The main observations are summarized as follows:

- The microstructure in SLMed Al-Zn-Mg alloys transforms from a coarse columnar grain structure without Sc and Zr addition to a heterogeneous microstructure composed of fine columnar grains and equiaxed grains with an addition of 0.6 wt.% Sc and 0.26 wt.% Zr;
- The formation of fine heterogeneous microstructures is associated with a significantly reduced number of cracks and pores, leading to remarkable improvement in the SLMed sample quality of Al-Zn-Mg alloys.

Acknowledgments

The authors wish to thank the financial support of the National Key Research and Development Program of China (2021YFB3702101) and the National Natural Science Foundation of China (Grant Nos. 51971043 and 52130107).

References

[1] DebRoy T, Wei H L and Zuback J S 2018 *Prog. Mater. Sci.* **92** 112-224
[2] Herzog D, Seyda V and Wycisk E 2016 *Acta Mater.* **117** 371-92
[3] Liu Z, Zhao D and Wang P 2022 *J. Mater. Sci. Technol.* **100** 224-36
[4] Galy C, Le G E and Lacoste E 2018 *Addit. Manuf.* **22** 165-75
[5] Frazier W E 2014 *Journal of Mater. Eng. Perform.* **23** 1917-28
[6] Thijs L, Kempen K and 2013 *Acta Mater.* **61** 1809-19
[7] Roth C C, Dejean T T and Mohr D 2021 *Addit. Manuf.* **43** 101998
[8] Suryawanshi J, Prashanth K G and Scudino S 2016 *Acta Mater.* **115** 285-94
[9] Enz J, Riekehr S and Ventzke V 2016 *J. Mater. Process. Technol.* **237** 155-62
[10] Enz J, Riekehr S and Ventzke V 2016 *Metall. Mater. Trans. A* **47** 2830-41
[11] Brüggemann J P, Risse L and Kullmer G 2018 *Procedia Struct. Integr.* **13** 311-6
[12] Martin J H, Yahata B D and Hundley J M 2017 *Nature* **549** 365-9
[13] Zhou L, Pan H and Hyer H 2019 *Scr. Mater.* **158** 24-8
[14] Zhu Z, Ng F L and Seet H L 2022 *Mater. Today* **52** 90-101
[15] Tan Q, Fan Z and Tang X 2021 *Mater. Sci. Eng.* A. **821** 141638
[16] Gharbi O, Kairy K S and De P R 2019 *npj Mater. Deg.* **3** 40
[17] Kairy S K, Gharbi O and Nicklaus J 2018 *Metall. and Mater. Trans. A* **50** 529-33
[18] Mehta A, Zhou L and Huynh T 2021 *Addit. Manuf.* **41** 101966

IOP Conf. Series: Materials Science and Engineering 1249 (2022) 012025 doi:10.1088/1757-899X/1249/1/012025

Revisiting plastic strain ratio determination in aluminium using crystal plasticity

D Lindell[1]

[1] Swedish institute for mining metallurgy and materials (Swerim), Isafjordsgatan 28, 164 40 Kista, Sweden

Abstract. Determination of plastic strain ratio (*r*, also known as the Lankford coefficient) from texture data is one of the most widely used applications of crystal plasticity methods. Early attempts using Taylor theory and x-ray texture data already provided practically useful information of e.g. drawability of sheet metals. Developments in recent decades within both electron backscatter diffraction and computational mechanics have significantly improved the accuracy of these calculations. Predictions can now be made that compares relatively well with experiments making *r*-value predictions from texture data a routine method within material and process development. The current work reviews recent studies on plastic strain ratio determination by computational means for aluminium sheet metal, with the aim to assess the expected accuracy using modern crystal plasticity methods. Furthermore, we investigate the major sources of error in these calculations by comparison between previous studies and with new experiments and calculations. In particular, we observe and investigate a maximum in the absolute error occurring typically at 45° to the rolling direction. The sources of this error are discussed in terms of both model conditions and the material heterogeneities giving rise to anisotropy.

1. Introduction

Accurate numerical methods capable of bridging microscopic and macroscopic length scales are of outermost importance in the design and development of better materials. Crystal plasticity models capable of capturing micromechanical response of polycrystals have seen an impressive development in the last decades. For example, full field crystal plasticity at reasonable low computational cost using spectral solver [1-2] and fast mean field models incorporating special schemes make the simulations more realistic without expense on cost. Developments within the mean field models include, but are not limited to grain interaction e.g. [3-6], intragranular misorientations and fragmentation [7-8] and full elastic-viscoplastic descriptions including back stresses [9-11]. The crystal plasticity methods, and the mean field codes in particular, are not solely of academic interest. Sheet metal forming is an area where crystal plasticity methods potentially can be very useful within applied material and process development. The same type of models can be used to predict both the development of the microstructure during processing and to predict forming behaviour of the final product. The most easily assessable formability criteria from crystal plasticity models are the plastic strain ratio r, also known as the Lankford coefficient. Early attempts using Taylor model could in fact estimate the *r*-value [12] reasonably well simply by selecting the *r*-value corresponding to the lowest Taylor factor from a series of simulations. State of the art models provide large flexibility allowing predictions of *r*-value even for complex materials deforming by multiple slip and twinning systems, such as Mg [13]. In view of literature, the accuracy in *r*-value determination is rather good when using modern mean field crystal plasticity approaches and that is also our experience. However, that is not always the case, during the course of our work we have found the aluminium alloy 3003 in H14 condition to be particularly challenging with large mismatch between calculated *r* using crystal plasticity and experiments, in particular for 45° loading relative the rolling direction. Engler [14] has reported similar phenomena for AA7021 in various tempers. This uncertainty is not fully satisfactory considering the large potential for these methods in applied research. The aim of the current work was to investigate the main sources of error in *r*-value determination of aluminium sheet metal using crystal plasticity methods.

2. Methodology

The current work is based on literature sources and investigations of aluminium alloy 3003 sheets in O and strain hardened H14 tempers. The 1.4 mm sheet in O temper has a fully recrystallised microstructure and rather weak texture but with influence of cube texture. The 1.0 mm sheet in the strain hardened H14 temper originates from the same heat as the material in O-temper, and has a more typically fcc cold rolling texture. Texture analysis was performed using electron backscatter diffraction (EBSD) on the cross section, total area covered was ~30 mm^2 (full thickness times ~20 mm) and the number of unique grains were 3000 – 8000 depending on sample. 1000 orientations from the EBSD datasets were used in the crystal plasticity simulations. These were extracted from the EBSD dataset by choosing a step size corresponding resulting to 1000 orientations, i.e. maximising the distance between the points. Synthetic datasets were also made with use of the MTEX toolbox [15]. *r*-values were modelled using the viscoplastic self-consistent code (VPSC) [16]. The simulations were performed using only the octahedral {111}<110> slip system. A large parameter set has been evaluated, and for the simulations given here the affine linearisation scheme has been used. For H14 a grain aspect ratio of 10:2:1 (RD:TD:ND, RD = rolling direction, TD = transverse direction, ND = normal direction) was employed whereas for O temper equiaxed grains were used. Reason for using non-equiaxed grain morphology in case of H14 is because it is experimentally justified and gives better agreement with experiment. Engler and Aegerter [17] used a similar type of aspect ratios for *r*-value calculations of AA5005-H16. The simulations of *r* given in the results sections corresponds to the initial *r*-value, but is also valid for somewhat larger strains as well as the variation with strain is not very large in this material.

Tensile testing was performed for the same specimens using 50 mm gauge length and transverse extensometer. The *r*-value was determined at different angles relative the rolling direction, $\alpha = 0$, 22.5, 45, 67.5 and 90°. Each direction was tested three times and the values given corresponds to the average. The scatter between different points is within 0.1. The uniform elongation is only a few percent for H14, the *r*-value is therefore determined between 1 - 2% whereas for the O temper the *r*-value is determined as the average in the range 2 - 20%. For O-temper the contribution from elastic strain to the total strain is negligible but for H14-temper it gives a small offset in the order of 0 - 0.05 for H14 in the direction of $r = 1$, when assuming isotropic elasticity. The quantity δ is used throughout the results section to describe the absolute difference in *r*-value between experiment and simulation.

3. Results and discussion

Figure 1 and figure 2 show calculated *r*-values compared to experiments which illustrates the phenomena of uncertainty pointed out in the introductory section. For the O temper, the *r*-value prediction is in very good agreement with experiment. The variation of *r* with angle is not large but the trend is captured, only exception being for TD where the trend is not fully captured but the absolute difference between experiment and prediction δ is still <0.1. However, for H14 temper given in figure 2 the situation is completely different. While the experiments show a rather monotonic increase from $r = 0.3$ at $\alpha = 0°$, to $r = 0.65$ at $\alpha = 67.5°$, the simulations show a steeper increase up to $r = 1.1$ at $\alpha = 45°$ and then decreases. Hence, at $\alpha = 45°$ the absolute difference between experiment and prediction δ is ~0.5.

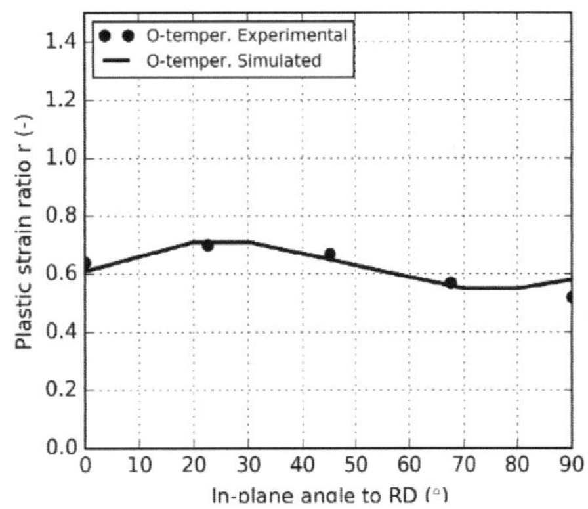

Figure 1. r-value calculated for AA3003 in O-temper and comparison with experiment.

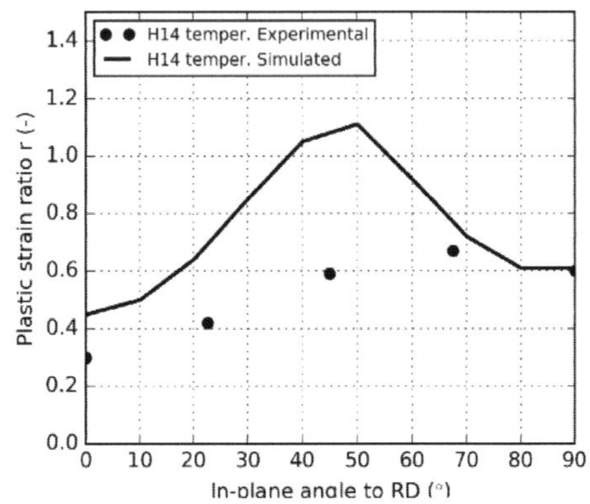

Figure 2. r-value calculated for AA3003 in H14 temper and comparison with experiment.

3.1 Alloys and tempers
Engler and Aegerter [17] have used VPSC to simulate *r*-values of AA5005 in O, H16 and H24 tempers. Agreement with experiment is generally very good for 13 of the 15 datapoints with $\delta \sim 0.1$. However rather large deviations were observed for TD in case of O and H16 tempers. In these two cases $\delta \sim 0.5$. In both cases the model seems to overestimate the *r*-value.

In [18] Engler has studied AA5182 in H19, H26 and H28 tempers. The materials have fcc rolling texture and therefore similar trend with α as given in figure 2. For H19 and H28 tempers the model overestimates *r* with 0.2 - 0.3 for $\alpha = 45°$ but very good agreement is found for H26. Notably, the prediction for TD is also poor with $\delta \sim 0.5$. For TD the model underestimates the *r*-value.

Work has been done by Kasemer, Falkinger and Roters [19] on AA6016 and AA5182 in T4 and O tempers respectively. For both materials the predictions are generally good but the model seems to overestimate the *r*-value for $\alpha = 45$, 67.5 and 90°, largest $\delta \sim 0.15$. The materials have higher elongation compared to AA3003-H14 so the *r*-value could be determined in the range 8 - 12%.

Heat treatable alloys AA6016 and AA7021 have been studied in the work by Engler [14]. Materials were studied in T4, T61, T6, T76 and T73 tempers. Good agreement between experiment and simulation was obtained for AA6016, for example in T4 temper $\delta < 0.1$ for at $\alpha = 0°$, 45° and 90°. For the AA7021 similar trend is observed as shown in figure 2, i.e. reasonable agreement in RD and TD but discrepancy at $\alpha = 45°$. For $\alpha = 45°$, $\delta \sim 0.2 - 0.3$ depending on the temper, i.e. somewhat better compared to the data in figure 2 but the same trend as a function of α. The variation in *r* for different tempers varies in the range 0.05 - 0.1 and the authors explain this in terms of back-stresses generated during ageing.

In summary, typically $\delta < 0.1$ indicates good accuracy in the simulations. However, larger discrepancy is found both for non-heat treatable and heat treatable alloys. A trend is that materials with fcc rolling texture appears to be most challenging to model and discrepancy is not only at $\alpha = 45°$ for these materials.

3.2 Effect of hardening parameters in the polycrystal model
While model calibration is complex and crucial for materials deforming by multiple deformation systems, our experience on pure octahedral slip, as in that case of aluminium, is that the choice of hardening parameters has relatively small influence on the results. The work by Kasemer, Falkinger and Roters [19] is an extensive parametric study comprising 1600 simulations using texture data from AA6016 and AA5182. Varying the hardening assumption, strain rate sensitivity and the hardening rate

show statistically significant effects but very small compared the differences exemplified in figure 2. The largest sensitivity was found for latent hardening, but the largest spread in *r*-value was only 0.03. Hence, the discrepancy found for certain aluminium grades cannot be explained by the choice of hardening parameters.

3.3 Effect of existing substructure

Substructure can potentially have large effects on strength and also anisotropy. The most extreme example found is from the work by Davis and Hutchinson [20] on steel, where the initial *r*-value of a deep drawing steel increases from ~1.7 to ~6 in RD after tensile straining of only a few percent in TD. In the current work, the *r*-value of strain hardened AA3003-H14 was measured between 1-2% so this may well play a role. It is however difficult to justify why this effect would be pronounced at 45° relative to the rolling direction after pre-straining by rolling. Moreover, in view of literature on different materials and tempers it was found that materials with aged and recovered microstructures also display this phenomenon, i.e. not only strain hardened materials. Hence, it is not likely that strain hardening and associated back stresses are responsible for the rather large discrepancy between model and experiment as exemplified in figure 2. It is however likely that it has some effect, at least at small-to-moderate strains. Possibilities to systematically study strain path changes with recent elasto-viscoplastic formulations that includes back-stresses will be of great interest [9-11].

3.4 Effect of texture

Texture is the microstructural feature with largest influence on the *r*-value and arguably very important input parameter to the model. A range of simulations were made by systematically modifying the input texture to study the effect of different rolling texture components. This was done by random sampling 100 orientations within 15° from selected ideal rolling components and adding these new orientations to the original dataset containing 1000 orientations. *r*-value calculations using datasets modified by adding either Cu:$\{11\bar{2}\}\langle 111\rangle$, S:$\{12\bar{3}\}\langle 634\rangle$, Br:$\{1\bar{1}0\}\langle 112\rangle$ or Cube:$\{001\}\langle 100\rangle$ is given in figure 3. Cu, S as well as Br all increase the *r*-value compared the reference EBSD dataset, particularly for $\alpha = 45°$. Cube texture on the other hand reduces the *r*-value for the same angle $\alpha = 45°$. Hence, the *r*-value appears to be particularly sensitive to rolling texture components at $\alpha = 45°$. As for the Cube texture, the trend in simulated *r*-values from $\alpha = 0°$ to $\alpha = 45°$ are in line with experiments, i.e. linearly increasing and not showing the characteristic peak as exemplified in figure 2. However, the simulations are offset by ~0.1 - 0.2.

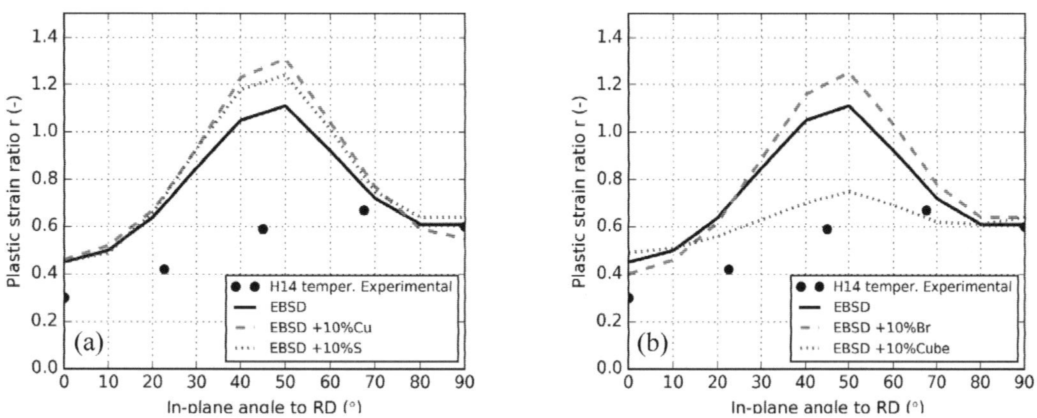

Figure 3. Calculated *r*-values using different input textures. Black line corresponds to the data shown in figure 2 corresponding to 1000 orientations obtained by EBSD. Dashed lines correspond to simulations made with the same dataset, but having additionally 100 orientations of Cu, S, Br or Cube.

IOP Conf. Series: Materials Science and Engineering 1249 (2022) 012025 doi:10.1088/1757-899X/1249/1/012025

EBSD allows collection of high-quality orientation data, but uncertainties still remain, primarily with regard to hit rate of heavily cold-worked materials and spatial variations of the texture. Spatial variations in thickness of aluminum sheet, in particular for the Goss component, have been pointed out by Ghosh, Miroux and Kestens [21] with large potential errors for the *r*-value in TD. They recommend collecting texture data through the complete thickness. In the current work ~30 mm^2 of material was covered from the complete thickness of the material. Despite this, relatively large discrepancies are observed, possibly due to texture variations in the material on larger length scales.

Additional simulations were made to further explore the uncertainty in the texture and its effect on *r*. This was done by using subsets of texture data, sampled from the original dataset, as outlined in figure 4. Results using this method for the texture data collected for AA3004-H14 by sampling 20 datasets, each with 500 orientations from the total 1000 orientations is given in figure 5. The uncertainty is not constant with α, the standard deviation in *r* is about five times higher for $\alpha = 45°$ compared to $\alpha = 0°$. The maximum spread between largest and smallest *r* is about 0.4 for $\alpha = 45°$, which is large compared to the other effects discussed. Small variations in the texture clearly have large effects on the texture at $\alpha = 45°$. However, at $\alpha = 0°$ variations in the texture have a smaller effect, which is also seen in figure 3. Reproducing the experimentally determined *r*-values at $\alpha = 0°$ is not possible simply by varying the texture within reasonable ranges. Hence, uncertainty in the texture is likely a main source of discrepancy between simulation and experiments at $\alpha = 45°$, but achieving say $\delta < 0.05$ will require more than a better description of the texture.

A final comment of the discrepancy at $\alpha = 45°$ shown in figure 2 is that the simulations, presented here and in literature, always tend to always overestimate the *r*-value. This needs further investigations, but it indicates that simply measuring larger areas is not sufficient. Systematic non-indexed orientations in the EBSD analysis or regions of different texture and strength giving rise to non-homogeneous deformation are two possible explanations. Using texture data obtained from x-ray diffraction (XRD) on stacked cross-sections of material could be one way to assess a more statistically representative texture and should be evaluated for this purpose.

4. Conclusions

In view of recent literature on *r*-value determination using crystal plasticity, the difference between simulation and experiment is typically < 0.10, indicating high accuracy of the method. Discrepancies between simulation and experiment do however exist and aluminium sheet with rolling texture appears to be particularly challenging for $\alpha = 45°$ relative the rolling direction. This is at least partially explained by uncertainties in the texture determination as all major texture rolling texture component largely influences the *r*-value in this direction. A simple method to assess uncertainty from texture is to perform batches of calculations on subsets of the overall texture data. Texture alone is however not sufficient to explain the overall discrepancy observed between simulation and experiment. Assumptions in the model itself and effect from existing substructure are two possible sources of discrepancy between model and experiment that requires further investigations.

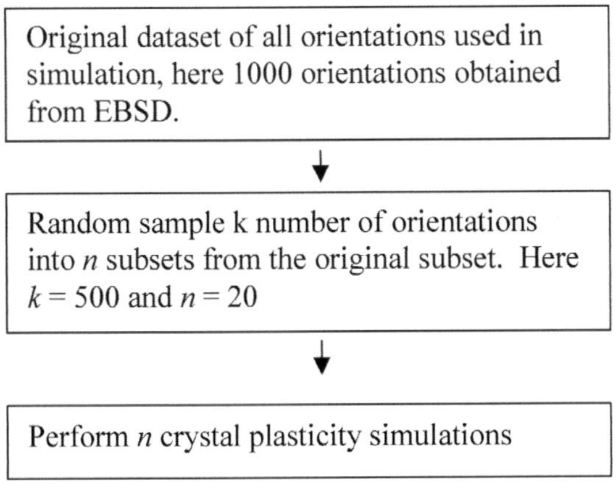

Figure 4. Method to assess angular-dependent uncertainty in texture data.

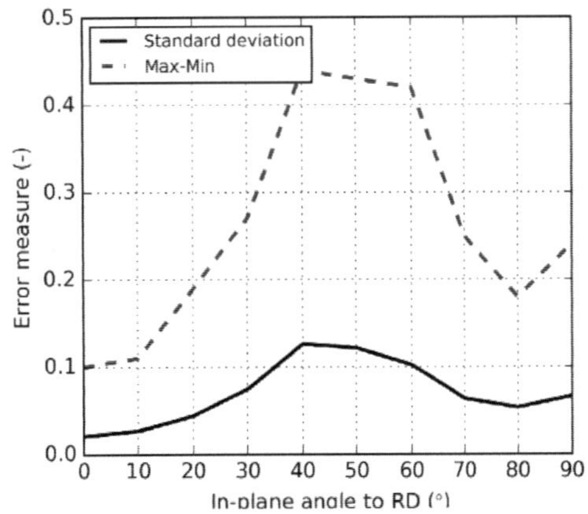

Figure 5. Variation in r-value from 20 simulations with different texture input data as described in figure 4.

Acknowledgements
Ning Wang at Gränges Finspång AB is acknowledged for performing EBSD analysis. Reem Rahmatalla, Anders Oskarsson, Emmanuel Hersent at Gränges Finspång AB and Bevis Hutchinson at Swerim are acknowledged for providing material fruitful discussions.

References

[1] Lebensohn R A 2001 *Acta Mater.* **49** 272337
[2] Roters F et. al. 2019 *Comp. Mat. Sci.* **158** 420-78
[3] Van Houtte P, Delannay L and Samajdar 1999 I *Textures Microstruc.* **31** 109-49
[4] Van Houtte P, Li S, Seefeldt M and Delannay L 2005 *Int. J. Plast.* **21** 589-624
[5] Crumbach M, Pomana G, Wagner P and Gottstein G 2001 *Proc. 1st Int. Conf. Recrytallisation and grain growth* 1053
[6] Mánik T and Holmdal B 2013 *Mater. Sci. Eng. A* **580** 349-54
[7] Zecevic M, Pantelon W, Lebensohn R A, McCabe R J and Knezevic M 2017 *Acta Mater.* **140** 398-410
[8] Zecevic M, Lebensohn R A, McCabe R J and Knezevic M 2018 *Int. J. Plast.* **109** 193-211
[9] Zecevic M and Knezevic M 2015 *Int. J. Plast.* **72** 200-17
[10] Zecevic M and Lebensohn R A 2020 *Int. J. of Solids Struc.* **202** 434-53
[11] Daroju S, Kuwabara T, Sharma R, Fullwood D T, Miles M P and Knezevic . 2022 *Int. J. Plast.* **153** 103292
[12] Bunge H and Roberts W T 1969 *J. Appl. Cryst.* **2** 116-28
[13] Hutchinson B, Jain J and Barnett M R 2012 *Acta Mater.* **60** 5391-98
[14] Engler O 2022 *Mat. Sci. Eng. A* **830** 142324
[15] Bachman F, Hielscher R and Schaeben 2010 *Solid State Phenom.* **160** 63-8
[16] Lebensohn R and Tomé C N 1993 *Acta Metallurgica et Materialia* **41** 2611-24
[17] Engler O and Aegerter J 2014 *Mat. Sci Eng. A* **618** 663-71
[18] Engler O 2015 *Mat. Sci Technol.* **31** 1058-65
[19] Kasemer M, Falkinger G and Roters F 2020 *Model. Simul. Mat. Sci. Eng.* **28** 085005
[20] Hutchinson W B and Davis T 1982 *Proc. 4th Int. Conf. on the Behaviour of Materials* 1227-33
[21] Ghosh M, Miroux A and Kestens L A 2015 *J. Alloys and Compd.* **619** 585-91

IOP Conf. Series: Materials Science and Engineering 1249 (2022) 012026 doi:10.1088/1757-899X/1249/1/012026

Influence of shielding gas flow on the uniformity of additively manufactured martensitic stainless steel

Z Pan, V K Nadimpalli, T L Christiansen, S A Andersen, M B Kjer, O V Mishin and Y Zhang

Department of Civil and Mechanical Engineering, Technical University of Denmark, 2800 Kgs., Lyngby, Denmark

E-mail address: yubz@dtu.dk

Abstract. Shielding gas flow is essential to the additive manufacturing (AM) process, and the effects of argon shielding gas flow variation on the macroscopic homogeneity of additive manufactured stainless steel parts have been studied using an open-architecture AM system. Such a variation manifests itself layer-by-layer within one part and part-by-part across the build plate. Within one build, a combination of balling behavior, conduction melting and keyhole melting is observed across the build plate using the same laser parameters. Quantitative characterization of the melt pool shapes show that the melt pool width and the penetration depth exhibit the largest variations. Possible relations between the gas flow condition and macroscopic structure variations are discussed and guidelines for improved design of a gas flow system as well as future research directions are suggested for achieving macroscopically uniform metal AM.

1. Introduction

Laser powder bed fusion (LPBF) is a layer-by-layer additive manufacturing (AM) technique, which enables the fabrication of complex shapes by taking advantage of its increased design freedom [1]. The process operates in a closed build chamber, and shielding gas such as nitrogen, argon or helium [2] is constantly introduced and circulated to generate a shielding gas flow across the build platform. The gas maintains an inert atmosphere in the chamber and prevents oxidation during manufacturing. The gas flow can at the same time remove undesired by-products, such as plume, condensate/soot and spatters, generated when the laser interacts with powders [3,4], as shown in figure 1. These by-products can further give rise to the attenuation and defocusing of the laser beam [5,6]. Moreover, they might redeposit onto the surface and thus affect the quality of the printed part, e.g. causing a significant increase in the layer thickness and inducing lack of fusion [7]. Therefore, good shielding gas flow conditions are essential for controlling the quality of the parts and the repeatability of the LPBF process.

Despite its importance, the impact of gas flow variation on the printing uniformity is only reported in a few studies. Ferrar et al. [8] found that variations in gas flow uniformity impacted the density and compression strength of a porous structure. Ladewig et al. [7] evaluated the uniformity of the gas flow based on the width of single welds and found that the weld lines were wider close to the gas outlet than at the gas inlet. The morphology of the melt pool and defects influenced by variations in the gas flow have not been well studied yet.

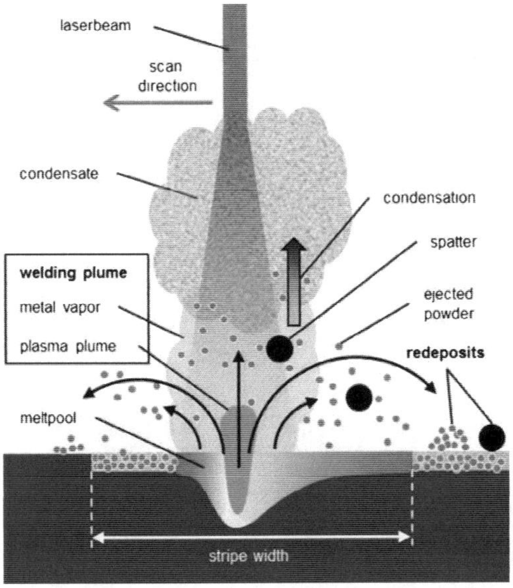

Figure 1. Schematic representation of possible process by-products [5].

In the present study, the effects of gas flow conditions on both the macroscopic and microscopic printing uniformity were examined in an open-architecture AM system. Cubes and single tracks were arranged and manufactured at different positions with respect to the shielding gas flow on the build plate using identical laser parameters. The relative density of the cubes, the morphology of the single tracks and defects were characterized to reveal the structure variations of the printed parts.

2. Experimental set-up and characterization

The AM process was conducted in a new open-architecture AM system equipped with a single mode SPI diode pumped fiber laser (250 W, 100 µm beam diameter) [9]. The process was performed in argon atmosphere with an oxygen level below 200 ppm. Figure 2(a) presents a schematic overview of the gas flow system in the chamber. The system consists of a cross flow fan, an outlet nozzle and a suction fan to achieve laminar flow across the build platform in the chamber. The gas flow system and the build plate were symmetric with respect to the designed flow direction. The build plate was lowered 9.5 mm away from the build platform each printing cycle after recoating to reach the focal plane. The material used was gas-atomized martensitic stainless steel powder (Fe-13Cr-0.1Mo-0.2Mn-0.6Si-0.2C, wt.%, according to the manufacturer) with a 37 µm median diameter. Nine parts in the form of cubes (12 mm x 12 mm x 6 mm) were manufactured and named #1~9; the position of the cubes on the build plate is shown in figure 2(b). The cubes were printed at a laser power of 200 W and a scanning speed of 350 mm/s. The layer thickness and the hatch distance were 0.03 mm and 0.1 mm, respectively. The scanning was unidirectional, with a 67° scan rotation between every layer. Additionally, four single tracks were printed on the top of each cube to investigate the full geometry of the melt pool.

The morphology of porosities and single tracks were characterized using a Zeiss Axio optical microscope. The specimens were polished using SiC grinding papers and oxide polishing suspensions, and etched using Kalling's reagent for 30 s. The relative density of the nine cubes were determined using the Archimedes method. The topography of the second single track close to the outlet nozzle (see figure 2(a)) on each cube was examined using a LEXT OLS4000 3D laser measuring microscope and analyzed using SPIP™ software. The morphology of the melt pools was measured using ImageJ.

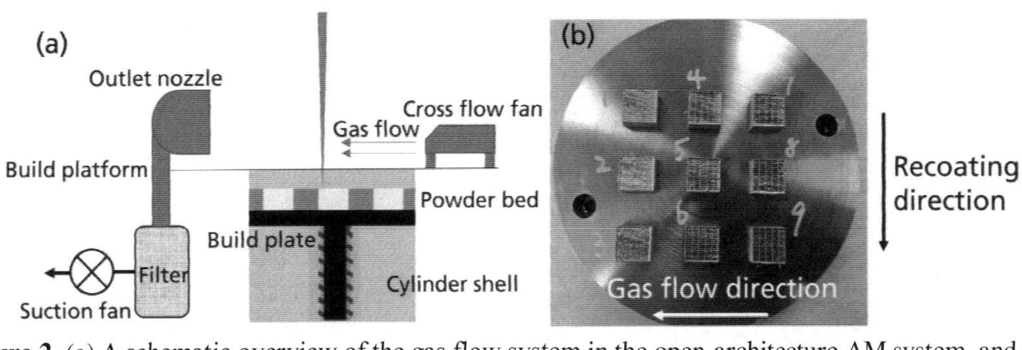

Figure 2. (a) A schematic overview of the gas flow system in the open-architecture AM system, and (b) the arrangement of the cubes and single tracks with respect to the recoating direction and the gas flow direction.

IOP Conf. Series: Materials Science and Engineering 1249 (2022) 012026 doi:10.1088/1757-899X/1249/1/012026

3. Results

The relative density of the nine printed cubes is shown in figure 3a, where a complex gradient pattern is revealed. Cubes #2 and #4 have the highest relative density, whereas cube #9 has the lowest relative density. Nevertheless, the difference in the relative density between the cubes is small (see figure 3a). Optical micrographs in figure 3b show that the material has a martensitic structure. Evidence of keyhole melting is seen in cubes #1 and #5, where keyhole porosity dominates, while evidence of the conduction mode is seen in cube #9, where pores caused by the lack of fusion dominate. In cube #4, the majority of the melt pools are characteristic of the conduction mode, although keyhole melting occurs occasionally. Another important observation is that the number of visible layers per unit length along the build direction is highest for cube #9 and lowest for cube #1. At the same time, the layer-to-layer variation is largest for cube #1 and smallest for cube #9 (see figure 3b).

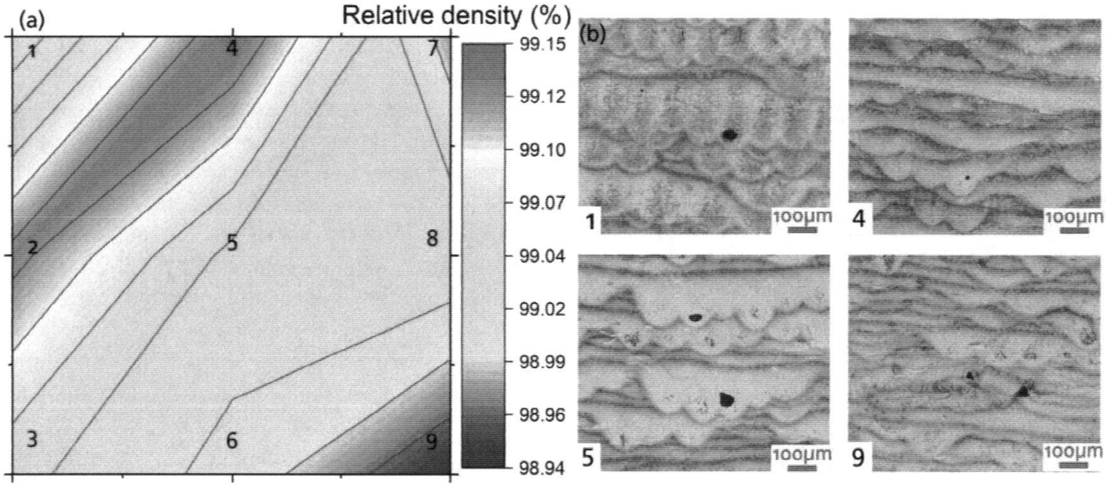

Figure 3. (a) A contour map of relative density of the nine cubes measured using the Archimedes method, and (b) optical micrographs for cubes #1, #4, #5 and #9.

Figure 4 presents the morphology and topography of the single tracks on the top of the 9 cubes. It is evident that the dimension of these tracks and the melting mode vary significantly depending on the printing position. Note that there is only a small difference between the four single tracks on the top of each cube. Except for cube #1, there is no significant difference between melting modes observed in the single tracks and those in the bulk (see figure 3b). The single tracks on cube #1 exhibit balling behavior but not keyhole melting (see figure 4a). The average cap height for the single tracks on cube #1 is 160 μm. The average cap heights for the other single tracks are much smaller, varying from 50 μm to 80 μm. An apparent keyhole is seen on top of cubes #3 and #5, where the maximum penetration depth is ~200 μm and the maximum penetration depth to melt pool width (P/W) ratio is 1.3. The melt pools of the single tracks on cubes #2, #4, #6, #7, #8 and #9 show characteristics of the conduction mode [10]. Moreover, spatters of 20 ~ 50 μm in diameter, e.g. those marked by white arrows in figure 4b, are dispersed on the surface as well as on the single tracks, with many spatters observed on both sides of the single tracks.

Figure 5 quantifies variations in the single-track dimensions. A decrease in the melt pool width is observed from the right side (inlet) to the left side (outlet) of figure 5a, i.e. along the gas flow direction. A reverse trend is seen along the gas flow direction in the cap height map (figure 5b). It is also evident that cube #1 stands out in terms of both the melt pool width and the cap height, which is a result of its balling behavior. A more complex gradient pattern is seen for the penetration depth, with the maximum depth observed in cube #5 (figure 5c). It is also seen in these figures that the penetration depth is much smaller for the cubes close to the powder stock, e.g. cubes #1, #4 and #7.

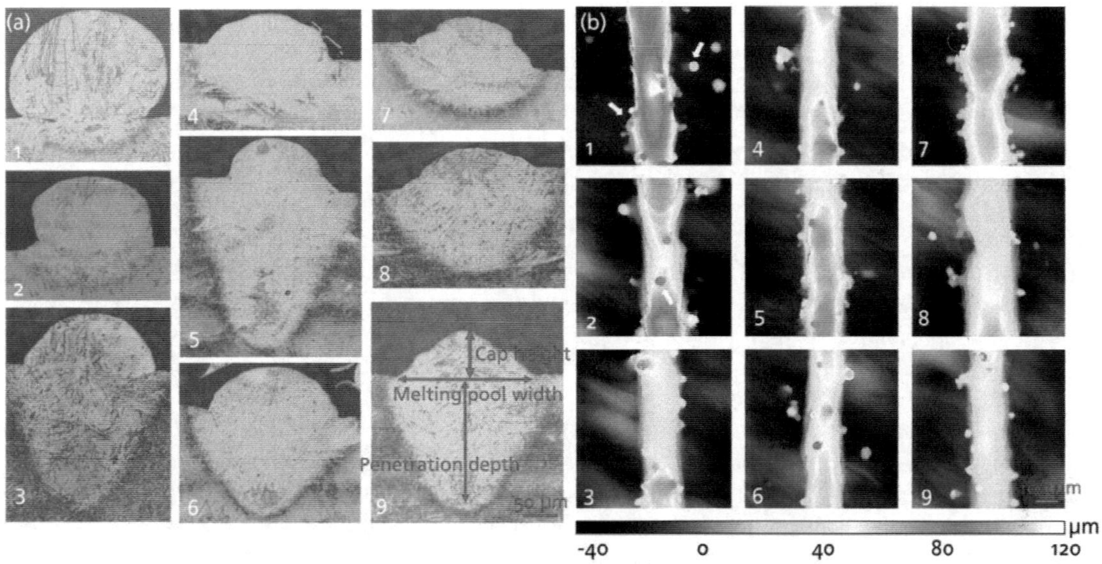

Figure 4. (a) The cross section of the single tracks on top of the nine cubes, and (b) the topographical information of the nine single tracks.

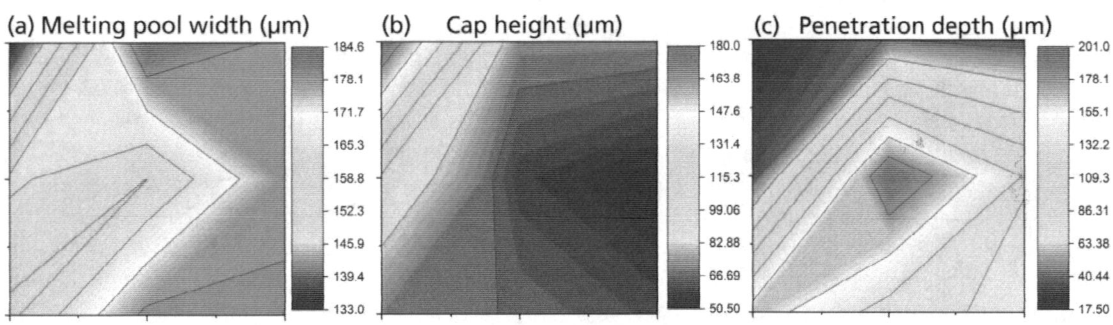

Figure 5. Contour maps: (a) melt pool width; (b) cap height; (c) penetration depth constructed in the same manner as figure 3a. Each map represents data averaged over four single tracks on top of each cube.

4. Discussion

The results show that the shielding gas flow has significant influence not only on the quality but also on melt pool geometry and layer-to-layer variation of the martensitic stainless steel used in this study. Quantitative measurements of the melt pool geometry show an apparent increase in the melt pool width from outlet to inlet. This result is in contrast to the findings by Ladewig et al. [7]. The reason for the difference is likely to be related to the relative position of the beam focal plane with respect to the laminar gas flow. In [7] the gas flow was close to the surface of the build platform/focal plane, while for the present system, the focal plane is about 35 mm below the cross flow fan, which is much larger than the typical height of the vapor plume and soot. This huge difference in height between the focal plane and the cross flow fan is expected to lead to a shadowing effect for the cubes close to the gas inlet (i.e. cubes #7, 8, 9). The vapor plume and soot generated on the top of the melt pool remain there during printing, which will additionally attenuate and defocus the laser, thus leading to a larger melt pool width and smaller cap height. This shadowing effect becomes weaker at the outlet side, and a larger portion of the vapor plume and soot can be transported away. The laser beam is therefore less affected by these by-products when printing the cubes close to the outlet (i.e. cubes #1, 2 and 3), resulting in a small melt pool width and larger cap height.

In general, a more focused laser beam is expected to produce a smaller melt pool width and a larger penetration depth. However, no simple correlation between the melt pool width and penetration depth is observed in the present experiment. A trend along the recoating direction rather than along the gas flow direction is seen for the penetration depth. This may suggest that there are some variations in the powder layer thickness along the recoating direction.

The melt pool geometry affects not only the quality of the printed cubes (in term of porosity), but also the local layer-to-layer heterogeneities. For example, the cap height observed in the single tracks for cube #1 is about 160 μm, which is about 2-3 times greater than that for the other cubes. With such a large layer thickness increase, it is possible that very little or even no powder will be deposited for the next layer. In this case, the laser will directly remelt the previously printed layer and likely lead to keyhole melting. This can explain why differences in the melting behavior are observed between the single tracks on cube #1 and the bulk of this cube, as well as large layer-to-layer thickness variations (see figure 3b).

Whereas the melting modes for the different cubes can be macroscopically related to the gas flow velocity, it is difficult to understand the microscopic layer-to-layer variation observed within the individual cubes, especially for cube #1. This result may suggest that local gas flow turbulence exists and that it is more severe at locations close to the outlet, which is similar to simulation data [11]. More experimental characterization and simulation of the overall gas flow pattern are required for a better understanding of this behavior.

To increase the reproducibility of single- or multi-part printing, the optimum gas flow should be completely uniform across the build platform, as well as being sufficient for removing the process by-products whilst leaving enough recoated powder. These conditions can be difficult to achieve, and sometimes a compromise has to be made when the gas flow system is designed. For example, Wirth et al. [11] found that a larger inlet height generated a less turbulent gas flow and a more homogeneous gas flow velocity field across the powder bed. However, soot was less efficiently removed due to the lower gas flow velocity compared to the velocity enabled by a smaller inlet height [11]. Based on the present findings, several suggestions can be made to optimize the gas flow: (1) setup geometry should be carefully examined to ensure the alignment of the gas flow system with other sub-systems, e.g. the laser or the recoating system; (2) a properly designed channeled gas rail can contribute to the uniformity of the gas flow [8]; (3) a purge gas flow from the top can be introduced to suppress the vortex at the rear wall, so that soot particles can be more efficiently removed [11].

5. Summary and outlook

The variation in the argon shielding gas flow uniformity and its effect on the powder layer thickness in an open-architecture AM system are found to have significant influence on the morphology of the melt pool and the relative density in the printed parts of the investigated martensitic stainless steel. Such a variation causes non-uniform geometry of the melt pools, with balling behavior, conduction melting and keyhole melting observed in different cubes printed on one build plate. In addition to variation across the build plate, it is possible that there are also variations across the build height. The variation in melting modes generates different types of defects. The severe turbulence induced by the non-uniform gas flow contributes to the layer-by-layer non-uniformity of the melt pool. It is expected that these findings can help improve open-architecture LPBF systems for optimized metal AM. Several recommendations are given to optimize the gas flow system in order to more efficiently remove by-products and improve the uniformity of the printed parts across the build plate, such as optimizing the AM setup geometry and inlet design, and introducing purge gas flow. A further study should involve measurements/simulations of the actual gas flow and the impact of different shielding gases both on the morphology of the melt pool and on the quality of the printed part.

Acknowledgement

VKN and SAA acknowledge the support from the 'Open AM initiative' from the Poul Due Jensen Foundation. YZ acknowledges the financial support from the European Research Council (ERC) under the European Union's Horizon 2020 research and innovation programme (M4D-grant agreement No. 788567).

References

[1] DebRoy T, Wei H L, Zuback J S, Mukherjee T, Elmer J W, Milewski J O, Beese A M, Wilson-Heid A, De A and Zhang W 2018 Additive manufacturing of metallic components – Process, structure and properties *Prog. Mater. Sci.* **92** 112–224

[2] Kurzynowski T, Chlebus E, Kuźnicka B and Reiner J 2012 Parameters in selective laser meltingfor processing metallic powders *High Power Laser Mater. Process. Lasers, Beam Deliv. Diagnostics, Appl.* **8239** 823914

[3] Fabbro R, Slimani S, Doudet I, Coste F and Briand F 2006 Experimental study of the dynamical coupling between the induced vapour plume and the melt pool for Nd-Yag CW laser welding *J. Phys. D. Appl. Phys.* **39** 394–400

[4] Liu Y, Yang Y, Mai S, Wang D and Song C 2015 Investigation into spatter behavior during selective laser melting of AISI 316L stainless steel powder *Mater. Des.* **87** 797–806

[5] Greses J, Hilton P A, Barlow C Y and Steen W M 2004 Plume attenuation under high power Nd:yttritium–aluminum–garnet laser welding *J. Laser Appl.* **16** 9–15

[6] Grünberger T and Domröse R 2015 Direct Metal Laser Sintering *Laser Tech. J.* **12** 45–8

[7] Ladewig A, Schlick G, Fisser M, Schulze V and Glatzel U 2016 Influence of the shielding gas flow on the removal of process by-products in the selective laser melting process *Addit. Manuf.* **10** 1–9

[8] Ferrar B, Mullen L, Jones E, Stamp R and Sutcliffe C J 2012 Gas flow effects on selective laser melting (SLM) manufacturing performance *J. Mater. Process. Technol.* **212** 355–64

[9] Andersen S A 2020 *Open Architecture Laser Powder Bed Additive Manufacturing* (Technical University of Denmark)

[10] Rai R, Elmer J W, Palmer T A and DebRoy T 2007 Heat transfer and fluid flow during keyhole mode laser welding of tantalum, Ti-6Al-4V, 304L stainless steel and vanadium *J. Phys. D. Appl. Phys.* **40** 5753–66

[11] Wirth F, Frauchiger A, Gutknecht K and Cloots M 2021 *Industrializing Additive Manufacturing* (Springer International Publishing)

IOP Conf. Series: Materials Science and Engineering 1249 (2022) 012027 doi:10.1088/1757-899X/1249/1/012027

Grain orientation dependence of deformation microstructure evolution and mechanical properties in face-centered cubic high/medium entropy alloys

S Yoshida[1,2*], R Fu[3], W Gong[4], T Ikeuchi[1], Y Bai[2,5], Z Feng[3], G Wu[3], A Shibata[2,6], N Hansen[7], X Huang[3], and N Tsuji[1,2]

[1] Department of Materials Science and Engineering, Kyoto University, Kyoto, Japan
[2] Elements Strategy Initiative for Structural Materials (ESISM), Kyoto University, Kyoto, Japan
[3] International Joint Laboratory for Light Alloys (MOE), College of Materials Science and Engineering, Chongqing University, Chongqing, China.
[4] J-PARC Center, Japan Atomic Energy Agency, Tokai, Ibaraki, Japan
[5] School of Materials Science and Engineering, Dalian University of Technology, Dalian, China
[6] Research Center for Structural Materials, National Institute for Materials Science, Tsukuba, Ibaraki, Japan
[7] Technical University of Denmark, Risø campus, Roskilde, Denmark

E-mail: yoshida.shuhei.5s@kyoto-u.ac.jp

Abstract. This study revealed characteristics of the deformation behavior in high/medium entropy alloys (HEAs/MEAs) with face-centered cubic (FCC) structure. A $Co_{60}Ni_{40}$ alloy and a $Co_{20}Cr_{40}Ni_{40}$ MEA having low and high friction stresses (fundamental resistance to dislocation glide in solid solutions), respectively, but similar in other properties, including their stacking fault energy and grain sizes, were compared. The MEA exhibited a higher yield strength and work-hardening ability than those in the $Co_{60}Ni_{40}$ alloy at room temperature. Deformation microstructures of the $Co_{60}Ni_{40}$ alloy were composed of coarse dislocation cells (DCs) in most grains, and a few deformation twins (DTs) formed in grains with tensile axis (TA) nearly parallel to <111>. In the MEA, three microstructure types were found depending on the grain orientations: (1) fine DCs developed in TA~//<100>-oriented grains; (2) planar dislocation structures (PDSs) formed in grains with other orientations; and (3) dense DTs adding to the PDSs developed in TA~//<111>-oriented grains. The results imply difficulty in cross-slip of screw dislocations and dynamic recovery in the MEA, leading to an increase in the dislocation density and work-hardening rate. Our results suggest that FCC high-alloy systems with high friction stress inherently develop characteristic deformation microstructures advantageous for achieving high strength and large ductility.

1. Introduction

High entropy alloys (HEAs) and medium entropy alloys (MEAs) are novel classes of solid solution alloys having five or more and four or fewer constituent elements, respectively, with near equiatomic fractions [1,2]. Among such systems, HEAs and MEAs with face-centered cubic (FCC) structures have been known to exhibit an exceptional balance of strength and ductility in a wide range of temperatures (from room temperature to cryogenic temperatures) [3,4]. Our previous studies revealed that FCC HEAs

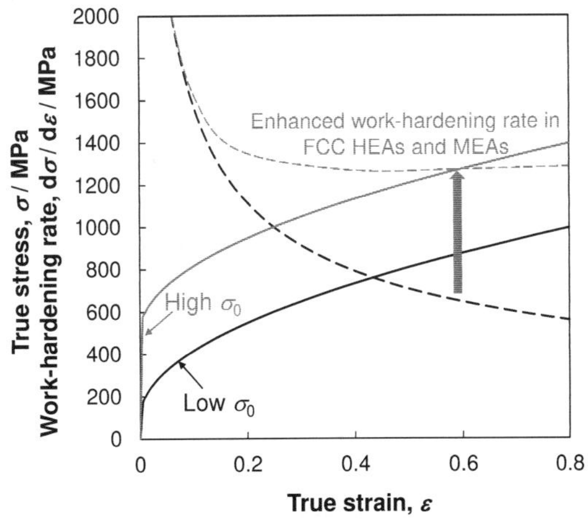

Figure 1. True stress-true strain curves of virtual alloys with high (red solid curve) and low (black solid curve) friction stresses, σ_0. Their work-hardening rate curve as a function of true strain is also shown by a black dashed curve. The red dashed curve indicates the typical work-hardening rate of FCC HEAs and MEAs observed experimentally.

and MEAs showed significantly high friction stress, σ_0, (fundamental resistance to dislocation motion in solid solutions), contributing to their high yield strength and high tensile strength [5–7]. It remains a question as to whether the excellent strength-ductility balance of FCC HEAs and MEAs can be explained simply based on the enhanced strength by their high σ_0. To examine this question, we now conduct a simple thought experiment. Figure 1 shows tensile true stress-true strain curves of virtual FCC alloys with low σ_0 (black solid curve) and high σ_0 (red solid curve). It is assumed that the same deformation mechanisms (e.g., activated slip systems and/or deformation twinning systems) operate in these two alloys, and they have identical elastic properties. Accordingly, the work-hardening rate curves of these two alloys are identical (dashed black curve). In general, uniform elongation of strain-rate insensitive materials can be determined by the plastic instability condition [83], expressed by the Considère equation [8]:

$$\frac{d\sigma}{d\varepsilon} \leq \sigma. \tag{1}$$

When the work-hardening rate becomes equal to the flow stress (corresponding to the crossover of the true stress-true strain curves and the work-hardening curve), macroscopic necking is initiated according to equation (1), leading to fracture. From figure 1, we can understand that the uniform elongation of the alloy with high σ_0 is smaller than that of the alloy with low σ_0 under the identical work-hardening rate. Therefore, it can be concluded in this thought experiment that an increase in the friction stress leads to degradation of the strength-ductility balance, unless the work-hardening rate increases. This means that the strength-ductility balance of FCC HEAs and MEAs (with high σ_0) is worse than that in other conventional dilute FCC alloys (with low σ_0), contradictory to the excellent mechanical properties observed in reality [3,4]. Thus, we hypothesize that, in FCC HEAs and MEAs, there exists a mechanism to enhance their strength and ductility simultaneously (i.e., enhanced work-hardening ability as indicated by the red dashed curve in figure 1). To validate this idea, we have investigated the deformation behavior of a $Co_{20}Cr_{40}Ni_{40}$ MEA with high σ_0 (280 MPa) and a $Co_{60}Ni_{40}$ binary model alloy with low σ_0 (52 MPa) where these alloys are similar in other materials properties, such as elastic constants, lattice constants, melting points, and stacking fault energies (SFEs) [9,10].

2. Experimental methods

Ingots of the $Co_{60}Ni_{40}$ alloy and $Co_{20}Cr_{40}Ni_{40}$ MEA were cast by a vacuum arc-melting of high purity elements (99.9 wt.%). The ingots were then homogenized at 1100 °C for 24 h and rapidly cooled by water quenching. The homogenized $Co_{60}Ni_{40}$ alloy and $Co_{20}Cr_{40}Ni_{40}$ MEA were cold-rolled by a reduction of 92% in thickness and subsequently heat-treated at 750 °C for 120 s and 850 °C for 3.6 ks, respectively. This produced fully-recrystallized microstructures with mean grain sizes of approximately 3 μm, including annealing twin boundaries. To evaluate their mechanical properties, tensile specimens with a gauge dimension of 2 mm (length) × 1 mm (width) × 0.5 mm (thickness) were cut from the recrystallized materials, and tensile tests were carried out at a quasi-static strain rate of 8.3×10^{-4} s^{-1} at room temperature. The surface of the specimens was painted with white and black ink to produce random dot patterns. The displacement of the gauge section was precisely measured by tracking the patterns using a CCD video camera extensometer, and the strain was calculated using the digital image correlation (DIC) technique [11]. We have previously confirmed that tensile tests using such small specimens with DIC measurements give stress-strain data equivalent to those obtained using standard-sized specimens [5,6]. The tensile deformation was interrupted at different strain levels, and the deformation microstructures of the materials were characterized by transmission electron microscopy (TEM) equipped with bright-field (BF) and annular dark-field (ADF) detectors for scanning transmission electron microscopy (STEM). The TEM/STEM observations were conducted at an acceleration voltage of 200 kV under two-beam conditions as described in Ref. [12]. The crystallographic orientations (crystal direction) parallel to the tensile axis in each grain were determined based on Kikuchi patterns obtained by convergent beam electron diffraction. In each material, more than 20 grains from two thin foils were observed, confirming the reproducibility of the results.

3. Results and discussion

Figure 2 shows true stress-true strain curves obtained for the $Co_{60}Ni_{40}$ alloy and $Co_{20}Cr_{40}Ni_{40}$ MEA. The MEA exhibited higher yield strength (375 MPa) than the $Co_{60}Ni_{40}$ alloy (140 MPa) owing to the higher friction stress of the MEA. Based on the slopes of the true stress-true strain curves, the work-hardening rate of the MEA was found to be higher than the $Co_{60}Ni_{40}$ alloy, leading to their fracture at almost the same strain levels. Therefore, it was concluded that the MEA exhibited a better strength-ductility balance.

To reveal the reasons for the difference in deformation behaviors in those two alloys, we investigated the evolution of their deformation microstructures. As demonstrated in previous studies [13,14], the evolution of deformation microstructures in FCC metals is grain-orientation dependent. Therefore, we classified the types of deformation microstructures depending on the grain orientation against the tensile axis (TA). Figure 3 (a) shows an example of the deformation microstructures showing dislocation cells (DCs) in the $Co_{60}Ni_{40}$ alloy. Coarse DCs with a size of several hundred nanometers developed in most grains of the $Co_{60}Ni_{40}$ alloy regardless of grain orientation. In a small number of grains oriented to TA ~// <1 1 1>, a few deformation twins were found, as shown in figure 3 (b), in addition to DCs slightly finer than those in figure 3 (a). In the $Co_{20}Cr_{40}Ni_{40}$ MEA, three distinct types of deformation

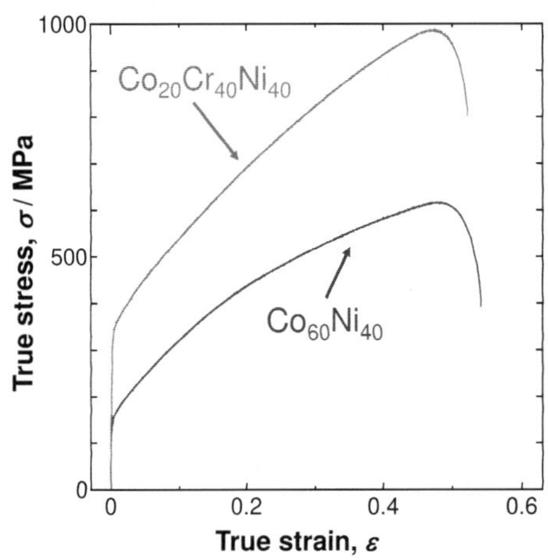

Figure 2. True stress-true strain curves for the $Co_{60}Ni_{40}$ alloy and $Co_{20}Cr_{40}Ni_{40}$ MEA obtained by tensile tests at room temperature.

IOP Conf. Series: Materials Science and Engineering 1249 (2022) 012027 doi:10.1088/1757-899X/1249/1/012027

microstructures were found to be developed. Figure 4 (a) shows an example of fine DCs with a size of several tens of nanometers in grains oriented to TA ~// <1 0 0>. In the case of grains with other orientations, characteristic planar dislocation structures (PDSs) parallel to specific {1 1 1} planes were formed, as shown in figure 4 (b). This suggests that cross-slip of screw dislocations is severely suppressed in the MEA. In addition, among the grains with PDSs, many deformation twins were additionally observed in <1 1 1>-oriented grains, as presented in figure 4 (c).

The results described above indicate that, although the two alloys have different friction stress levels and equivalent materials properties, including SFEs, dynamic recovery (i.e., annihilation of dislocations with opposite signs through cross-slip of screw dislocations) is greatly inhibited in the MEA. One possible reason for the inhibited dynamic recovery is the effect of variation in stacking fault widths due to the local chemical environment in the MEA. Because different elements are in different lattice positions in HEAs and MEAs, SFE can be locally different, resulting in variations of stacking fault width depending on the local chemistry [15,16]. Generally, stacking faults need to shrink in the original slip plane and extend again in the cross-slip plane to allow cross-slip of screw dislocations. Therefore, it is possible that the parts of stacking faults with wider widths may be difficult to shrink, and the shrinkage of such wider parts controls the process of cross-slip of screw dislocations in HEAs and MEAs.

Another possibility is variations in stacking fault widths due to stress levels. As the SFE of the alloys used in the present study were as low as 30 mJm^{-2}, dislocations could dissociate into two Shockley partial dislocations. Furthermore, under uniaxial loading, the resolved shear stresses acting on the

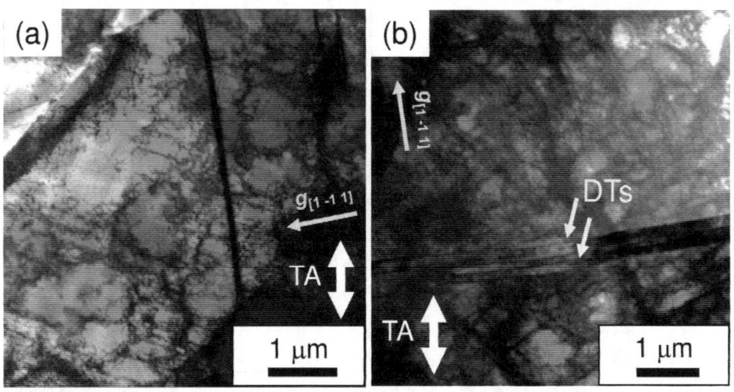

Figure 3. Examples of dislocation microstructures in the $Co_{60}Ni_{40}$ alloy showing (a) DCs (at a nominal strain of $e \sim 30\%$, TA ~// < 1 2 2>) and (b) DTs ($e \sim 50\%$, TA ~// <1 1 1>). TA and g-vector ($g_{[h\,k\,l]}$) are indicated in each image.

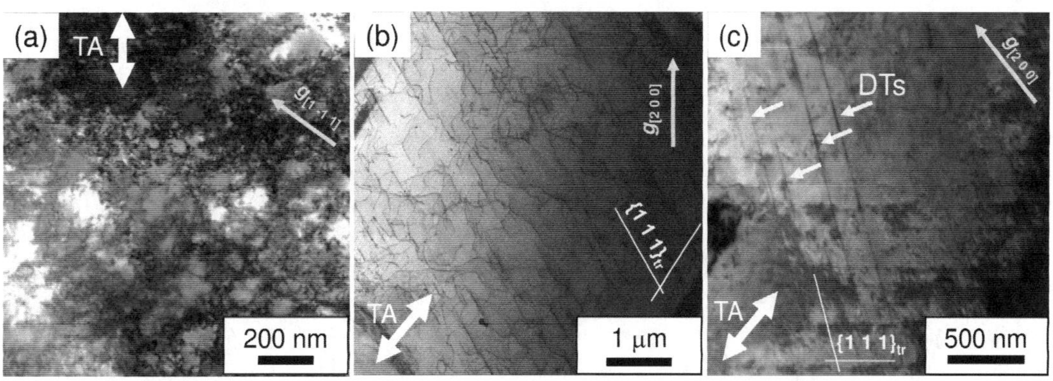

Figure 4. Examples of dislocation microstructures in the $Co_{20}Cr_{40}Ni_{40}$ MEA showing (a) DCs (at a nominal strain of $e \sim 30\%$, TA ~// <1 0 0>), (b) PDSs ($e \sim 10\%$, TA ~// <1 1 2>), and (c) DTs ($e \sim 10\%$, TA ~// <1 1 1>). TA and g-vector ($g_{[h\,k\,l]}$) are indicated in each image. The yellow lines are traces of {1 1 1} planes.

leading and trailing partial dislocations are different because they have different Burgers vectors. As a result, stacking faults can expand or shrink upon loading, and apparent SFEs (effective SFEs) can change depending on the grain orientation. Such orientation-dependent variation of effective SFEs (γ_{eff}) has been formulated by Copley *et al.* [17] as follows:

$$\gamma_{\mathrm{eff}} = \gamma_0 - \frac{m_{\mathrm{L}}-m_{\mathrm{T}}}{m_{\mathrm{L}}+m_{\mathrm{T}}}\sigma b_{\mathrm{p}}, \quad (2)$$

where γ_0 is the stacking fault energy without external stress, σ is the applied tensile stress, and b_{p} is the magnitude of the Burgers vector of the partial dislocations, and where m_{L} and m_{T} are the Schmid factors of the leading and trailing partial dislocations, respectively. According to equation (2), in the MEA with higher σ_0 (i.e., higher flow stress, σ), changes in effective SFEs under uniaxial stress can be more significant than in the $Co_{60}Ni_{40}$ alloy. For example, in the case of TA // <1 1 1> grains, the Schmid factor of the leading partial dislocations (m_{L} = 0.314) belonging to the primary slip system (the a/2<1 1 0>{1 1 1} slip system with the highest Schmid factor) is larger than that of the corresponding trailing partials (m_{T} = 0.157). Thus, stacking faults expand upon tensile loading, and the effective SFEs of the $Co_{60}Ni_{40}$ alloy and $Co_{20}Cr_{40}Ni_{40}$ MEA are calculated at their yield points as 23.1 $\mathrm{mJm^{-2}}$ and 11.6 $\mathrm{mJm^{-2}}$, respectively. Therefore, cross-slip of screw dislocations in the MEA is more inhibited than in the $Co_{60}Ni_{40}$ alloy, resulting in the formation of the PDSs in these grain orientations. In TA // <1 0 0>-oriented grains, the situation is the opposite: the Schmid factor of the leading partial dislocations belonging to the primary slip system (m_{L} = 0.236) is smaller than that of the trailing partials (m_{T} = 0.471), and stacking faults shrink under tensile stress. The effective SFEs of the $Co_{60}Ni_{40}$ alloy and $Co_{20}Cr_{40}Ni_{40}$ MEA are calculated at their yield point as 36.8 $\mathrm{mJm^{-2}}$ and 48.3 $\mathrm{mJm^{-2}}$, respectively. These SFEs might be high enough to allow cross-slip of screw dislocations forming DCs. Consequently, the orientation dependence of effective SFEs can explain the difference in the observed deformation microstructure evolution in the alloys.

As dynamic recovery was more inhibited in the $Co_{20}Cr_{40}Ni_{40}$ MEA than in the $Co_{60}Ni_{40}$ alloy, the dislocation density in the MEA can quickly increase during tensile deformation, resulting in an increase in the work-hardening rate. In addition, in grains with TA ~// <1 1 1>, dense DT boundaries introduced during deformation could have acted as further obstacles for dislocation motion and dynamically reduced the mean free path of dislocations. Although the number fraction of such grains was less than 10%, this may also have been a contributing factor to the increase in the work-hardening rate, the so-called dynamic Hall-Petch effect. Thus, we conclude that, for these reasons, the $Co_{20}Cr_{40}Ni_{40}$ MEA with high friction stress exhibited higher work-hardening ability and superior strength-ductility balance compared with the $Co_{60}Ni_{40}$ alloy. We believe that FCC high-alloy systems with high friction stress inherently develop characteristic deformation microstructures that are advantageous for realizing better mechanical properties than those in conventional FCC metals and alloys.

4. Conclusion

In the present study, we compared the deformation behavior of a $Co_{60}Ni_{40}$ alloy and $Co_{20}Cr_{40}Ni_{40}$ MEA with low and high friction stresses, respectively, but having otherwise similar materials properties, such as SFEs and elastic constants, and investigated the deformation microstructure evolution during tensile deformation at room temperature. The $Co_{20}Cr_{40}Ni_{40}$ MEA exhibited a higher yield strength and a higher work-hardening rate than the $Co_{60}Ni_{40}$ alloy. In the $Co_{60}Ni_{40}$ alloy, we observed coarse DCs regardless of the grain orientation, and a small number of DTs were found in a few grains oriented to TA ~// <1 1 1>. In the $Co_{20}Cr_{40}Ni_{40}$ MEA, we found fine DCs in TA ~// <1 0 0>-oriented grains, while PDSs were observed in others. Dense DTs were formed in grains oriented to TA ~// <1 1 1> in addition to the PDSs. The results indicate that cross-slip of screw dislocations (i.e., dynamic recovery during deformation) is inhibited in the MEA. This can lead to a rapid increase in dislocation density and an increase in the work-hardening rate of the MEA compared with the $Co_{60}Ni_{40}$ alloy. Therefore, characteristic deformation microstructures developed in the $Co_{20}Cr_{40}Ni_{40}$ MEA with high friction stress are beneficial for simultaneously achieving high strength and large tensile ductility.

Acknowledgments
This work was financially supported by the Elements Strategy Initiative for Structural Materials (ESISM, No. JPMXP0112101000), the Grant-in-Aid for Scientific Research on Innovative Area "High Entropy Alloys" (No. JP18H05455), the Grant-in-Aid for Early-Career Scientists (No. JP22K14501), the Grant-in-Aid for Research Activity Start-up (No. JP21K20487), and the Grant-in-Aid for JSPS Research Fellow (No. JP18J20766), all through the Ministry of Education, Culture, Sports, Science and Technology (MEXT), Japan. All the supports are greatly appreciated. The authors also thank Dr. Tianbo Yu and Ms. Gitte Christiansen of Technical University of Denmark (Denmark), Prof. Mitsuhiro Murayama of Virginia Polytechnic Institute and State University (USA), and Mr. Akira Yasuhara of JEOL Ltd. (Japan) for the technical supports on the transmission electron microscopy experiments.

References
[1] Tsai M H and Yeh J W 2014 High-Entropy Alloys: A Critical Review *Mater. Res. Lett.* **2** 107–23
[2] Miracle D B and Senkov O N 2017 A critical review of high entropy alloys and related concepts *Acta Mater.* **122** 448–511
[3] George E P, Raabe D and Ritchie R O 2019 High-entropy alloys *Nat. Rev. Mater.* **4** 515–34
[4] George E P, Curtin W A and Tasan C C 2019 High entropy alloys: A focused review of mechanical properties and deformation mechanisms *Acta Mater.* **188** 435–74
[5] Yoshida S, Bhattacharjee T, Bai Y and Tsuji N 2017 Friction stress and Hall-Petch relationship in CoCrNi equi-atomic medium entropy alloy processed by severe plastic deformation and subsequent annealing *Scr. Mater.* **134** 33–6
[6] Yoshida S, Ikeuchi T, Bhattacharjee T, Bai Y, Shibata A and Tsuji N 2019 Effect of elemental combination on friction stress and Hall-Petch relationship in face-centered cubic high / medium entropy alloys *Acta Mater.* **171** 201–15
[7] Yin B, Yoshida S, Tsuji N and Curtin W A 2020 Yield strength and misfit volumes of NiCoCr and implications for short-range-order *Nat. Commun.* **11** 2507
[8] Considère M 1885 L'emploi du fer de l'acier dans les constructions, Memoire no 34 *Ann. des Ponts Chaussées* **9** 574–775
[9] Köster E H, Thölén A R and Howie A 1964 Stacking fault energies of Ni–Co–Cr alloys *Philos. Mag.* **10** 1093–5
[10] Beeston B E P P, Dillamore I L and Smallman R E 1968 The Stacking-Fault Energy of Some Nickel-Cobalt Alloys *Met. Sci. J.* **2** 12–4
[11] Schreier H, Orteu J J and Sutton M A 2009 *Image correlation for shape, motion and deformation measurements: Basic concepts, theory and applications* (Boston: Springer US)
[12] Zhu Y, Ophus C, Toloczko M B and Edwards D J 2018 Towards bend-contour-free dislocation imaging via diffraction contrast STEM *Ultramicroscopy* **193** 12–23
[13] Huang X 1998 Grain Orientation Effect on Microstructure in Tensile Strained Copper *Scr. Mater.* **38** 1697–703
[14] Huang X and Hansen N 1997 Grain orientation dependence of microstructure in aluminium deformed in tension *Scr. Mater.* **37** 1–7
[15] Smith T M, Hooshmand M S, Esser B D, Otto F, McComb D W, George E P, Ghazisaeidi M and Mills M J 2016 Atomic-Scale Characterization and Modeling of 60° Dislocations in a High Entropy Alloy *Acta Mater.* **110** 352–63
[16] Rao S I, Woodward C, Parthasarathy T A and Senkov O 2017 Atomistic simulations of dislocation behavior in a model FCC multicomponent concentrated solid solution alloy *Acta Mater.* **134** 188–94
[17] Copley S M and Kear B H 1968 The dependence of the width of a dissociated dislocation on dislocation velocity *Acta Metall.* **16** 227–31

IOP Conf. Series: Materials Science and Engineering 1249 (2022) 012028 doi:10.1088/1757-899X/1249/1/012028

Lattice rotations during channel-die compression at high strain rates in copper single crystals of {110}<112>, {112}<111> and {123}<634> orientations

I Mania, H Paul and R Chulist
Polish Academy of Sciences, Institute of Metallurgy and Materials Science, 25 Reymonta St., 30-059 Krakow, Poland

Abstract. The microstructure and texture evolutions of high purity copper single crystals of Br(1-10)[112], C(112)[11-1] and S(346)[63-5] initial orientations were investigated to evaluate the influence of high strain rate on the intensity of twinning and shear banding. The samples were deformed up to 60% in a channel-die at a strain rate of 7×10^5 s^{-1} using explosive energy to propel the punch. The microstructure was characterized by optical microscopy as well as scanning and transmission electron microscopy. High strain rates of deformation of C and S-oriented single crystals lead to massive deformation twinning. Despite the identification of three families of compact clusters of twins of two generations, the microstructure and texture of both single crystals are practically dominated by the twinning on only one co-planar slip plane. The common rotation of twin-matrix layers, within the narrow area, combined with deformation twinning in the re-oriented matrix are responsible for final SB texture formation. This mechanism contributes to the formation of orientations from the neighborhood of the {110}<100> orientation. The {110}<112> orientation, stable during deformation in channel-die in the range of 'conventional' strain rates, lost its stability giving rise to intense twinning and deformation bands across the deformed specimen.

1. Introduction

The microstructure and texture evolutions during quasi-static loading of single crystals of face-centered cubic (fcc) metals have been intensively studied in the past [1-5]. However, corresponding changes that occur as result of deformation at high strain rates are still poorly investigated. The strain rate sensitivity of metals is known to increase dramatically when the strain rate rises above ~ 10^3 s^{-1} and leads to the specific microstructure/texture formation, different from those observed at conventional strain rates [6]. Therefore, the effect of high strain rates on deformation mechanisms is of fundamental interest to those who analyze dynamic loading during technological processes. In early works, an increase of the strain rate sensitivity has been interpreted as a transmission in deformation mechanism from thermal activation control at low strain rate to dislocation drag control at high strain rate [7]. However, later results show that the material response dependence on temperature is much smaller than that of the strain rate in the region of extreme deformation rates. This leads to the conclusion that change in temperature does not lead to a noticeable change in the material response [8].

Yaghoobi and Voyiadjis [9] and Fan et al [10] showed in the case of fcc metals subjected to extreme strain rates that the interaction between dislocations is the governing mechanism of material strength, and the contribution of thermal activation mechanism is negligible. The thermal softening effect is particularly relevant in terms of the so-called adiabatic shear band formation. The core issue, when analyzing this phenomenon, is the description of the mechanism responsible for shear bands nucleation, thickening and propagation across the homogeneously deformed structure of metals. Another issue is the role of shear bands in preserving material integrity [11].

Most previous experimental works have examined microstructural changes in terms of high strain rates using the so-called 'hat-shaped' samples. This specific sample geometry force shearing even in materials that do not tend to form shear bands spontaneously during rolling. However, using hat-shaped, polycrystalline samples the disclosure of influence of high strain rates on deformation texture is difficult, especially when attempting to relate the observed textural changes in the hat-shaped samples to the analysis of transformations that occur in the rolled sheets.

Therefore, the purpose of this paper is to describe the microstructure and texture evolutions of Cu single crystals of three different initial orientations, that constitute the main components of the copper-type rolling texture. Using explosive energy to propel the punch, the single crystalline samples were subjected to the compression in a channel-die at strain rates larger than 10^5 s^{-1}. A detailed characterization of the microstructure and texture was conducted using scanning (SEM) and transmission (TEM) electron microscopy.

2. Experimental

High purity (99.998%) copper single crystals of three orientations, that constitute the main components of the copper-type rolling texture were used to analyse the influence of high strain rate on slip, twinning and micro-/macro- scale shear banding. The measured orientations of the initial crystal used here were 'non-ideal' and deviate a few degrees from the ideal position. These are: (346)[63-5] (denoted hereafter as S) which is close to the one variant of the {123}<634> (S) orientation, (112)[11-1] (C or copper) orientation and (1-10)[112] (Br or brass) orientation. High strain rate deformation of 60% in terms of plane state of strains in a channel-die was conducted at ZTW Explomet (Opole, Poland). Figure 1 presents schematically the set-up used. A 10 mm thick steel driving (flyer) plate and a spacing of 15 mm between the punch of the channel-die and the flyer plate, were applied. Modified ANFO explosive charge with a thickness of 110 mm was used to initiate detonation (impact) velocity of 470 m/s. This means that the single crystalline samples were subjected to strain rates, greater than 10^5 s^{-1}. The dimensions of single crystalline samples were: 7 (width) x 10 (height) x 8 (length) mm^3. An Explomet-Fo-2000 measuring instrument (Kontinitro) was used to measure the vertical velocity of the flyer plate.

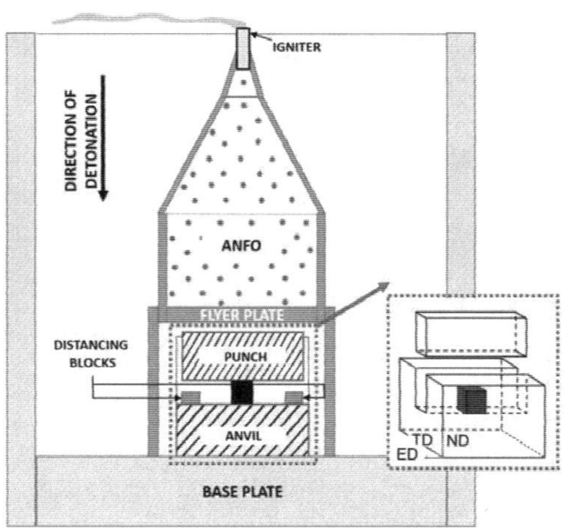

Figure 1. Schematic presentation of the compression test in a channel-die using an explosive energy to propel the punch.

The samples for micro-/nano- scale observations were cut along the normal direction (ND)/extension direction (ED) section from the as-deformed samples using a precision wire saw. Meso- scale analysis was based on an FEI Quanta 3D microscope equipped with the electron backscattered diffraction (EBSD) facility aiming at describing the microstructure and texture evolutions. Diffraction pattern acquisition and indexing were performed using TSL OIM Analysis 5 software. Orientation maps were measured in beam scanning mode using a step size ranging between 50 and 200 nm. As-deformed samples were mechanically ground and then polished using 1 μm diamond suspension and colloidal silica, and electropolished for microstructure/texture investigations. Nano-scale analyses were carried out using TEM techniques with a FEI Technai Super TWIN G2 FEG microscope (200 kV). Analyses were performed on thin foils cut-off from bulk samples along ND/ED plane, i.e. perpendicular to TD (where: ND, ED, TD normal, extension, and transverse directions, respectively) using the wire saw and then standard electro-polishing technique. Initial and post-deformation crystal orientations were additionally checked by X-ray diffraction measurements.

3. Results and discussion

3.1. Global orientation changes

Figure 2 shows the {111} pole figures of samples deformed up to 60%. A strong scattering of initial orientations was observed in all analysed crystallites. This applies not only to unstable crystallites with C and S orientations but also to crystallites with Br orientation supposed to be stable during plane strain deformation at conventional strain rates [2, 3]. Detailed analysis confirms that the formation of new texture components is due to the crystal lattice rotation around axes close to TD and the formation of deformation twins.

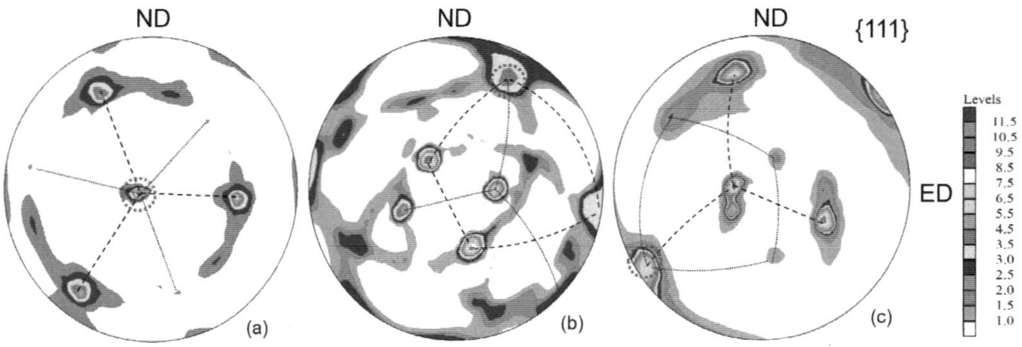

Figure 2.{111} pole figures showing texture of Br, C and S-oriented samples deformed up to 60% in channel-die. Red circle shows twinning plane normals. X-ray diffraction.

The scattering tendencies of the initial orientations are particularly well-visualized in the {111} pole figures measured on the ND-ED plane. In the sample of Br orientated sample, the {111} pole figure shows (figure 2a) dominant scattering *via* ~ 40° crystal lattice rotation around axes located near the TD (in this initial orientation the TD coincides with the <111> crystallographic direction). There is also observed a slightly marked tendency for a twin orientation to appear, with the twinning plane located perpendicular to the TD. In C and S-oriented crystallites, the most pronounced orientation changes result from the appearance of twinning on the (111) plane (figures 2b, c) on which operate co-planar (CP) – (111)[-101] + (111)[0-11] pair of slip systems. Other twinning systems can be also identified however, with much lower levels of intensities. In both unstable orientations, the dominant and minor twinning systems are supplemented by strong scattering of the {111} poles due to the crystal lattice rotation around axes located in the vicinity of the TD.

3.2. Deformation twins formation. Twinning systems

Regardless of crystallite orientation, the formation of deformation twins is practically not observed in poly- and single crystalline Cu deformed over a wide range of conventional strain rates. This is due to the relatively high stacking fault energy (SFE) of Cu [12]. Nevertheless using extremely high strain rates and the channel-die compression, twins appeared in all three analysed orientations. However, there is strong variation in the intensity of twinning depending on the crystal orientation. The crystallites also differ in prevalence of different twin families, i.e. the occurrence of twinning on different planes of the {111}-type, as well as in the possibility of the occurrence of different generations of twins, i.e. twinning of previously twinned areas.

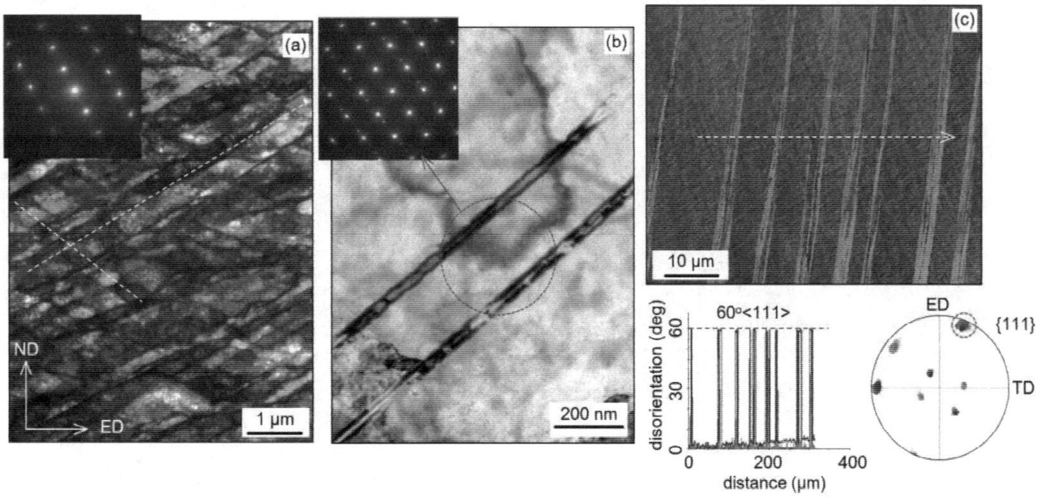

Figure 3. Microstructural features in Br-oriented single crystal. (a) Structure composed of two crossing families of microbands with dislocation walls situated along traces of the {111} planes, (b) thin platelets of twins observed on one of the most privileged for dislocation slip planes and (c) twinning on plane perpendicular to the compression plane (normal of twinning plane is lying close to ED). (a, b) TEM observation and (c) SEM/EBSD measurements in the ND-ED section.

The microstructure of Br-orientated single crystal is dominated by an arrangement of two families of microbands (figure 3a), where one often dominates in a given area. In the ND-ED section, the traces of dislocation walls coincide with the traces of the {111} planes on which the most favored slip systems operate. This behavior is analogous to that observed in crystallites of Br orientation deformed in the range of conventional strain rates [3, 4]. However, in the range of ultra-high strain rates, a strong tendency for strain localization through the formation of deformation bands was also observed. The bands were located along the traces of one of the (1-11) or (1-1-1) planes, and usually exhibited disorientation exceeding 40°, with respect to the neighboring less deformed matrix. These bands formation can be linked with the single slip system operating on one of the {111}-type planes since the disorientation axis (in relation between the band area and the surrounding matrix) coincides with the <112>-type axis connected with one of the systems operating on (1-11) or (1-1-1) plane). As expected, deformation twins are observed relatively rarely in the microstructure of Br-orientated crystallites. They mostly appear as very thin platelets situated along the traces of one of the {111} plane on which the most privileged slip systems operate (figure 3b). However, in this orientation, twins also appeared as a result of a transformation on the (111) plane situated perpendicular to the compression plane, with the normal to the twinning plane located near the ED (figure 3c).

For C and S-oriented crystallites, the microstructural changes are essentially similar for both orientations. The microstructures are dominated by twinning on the CP(111) plane (figure 4a). In areas with orientations close to the initial orientations the traces of the twinning planes are inclined at an angle of about 20° with respect to the compression plane. However, significant volumes near both ends of the samples, show an increased inclination of the CP(111) plane with respect to the rolling plane, which results in (+)TD rotation of initial crystal lattice orientation. This situation is presented in figures 4a, c. The microstructures consist of compact clusters of T-M platelets inclined at an angle of about 27° with respect to the ED. The rotation from the position of C orientation towards (100)[011] orientation increases the preference for slip (and twinning) to occur on the (-1-11) plane, situated symmetrically to the CP(111) plane (relative to the plane perpendicular to the ED). In matrix platelets characterized by orientation close to the (100)[011], the values of Schmid factors for systems operating on (111) and (-1-11) planes are the same. This opens the possibility for additional twinning on the (-1-11) plane (figure 4b); in some areas of the sample, this variant of twinning can occupy a significant volume of the crystallite. Simultaneously, the occurrence of twinning in the interior of the original twinned platelets was also observed (figure 4c).

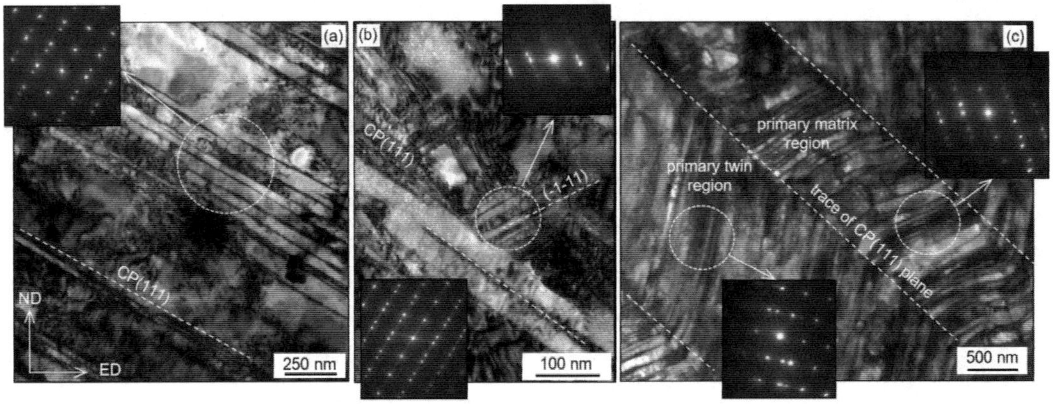

Figure 4. Twinning in C-oriented single crystal from primary matrix on (a) (111) CP plane, (b) on (111) and (-1-11) planes. (c) Twinning in primary matrix region on (-1-11) plane and in primary twin regions (twins from twins).

3.3. Shear banding vs. orientation changes

In Br orientation, strong orientation changes are associated not only with locally generated deformation twins but also with the formation of deformation bands, as described shortly in section 3.2. For the analyzed strain rates, no strain localization in the form of shear bands was observed in crystallites with Br orientation.

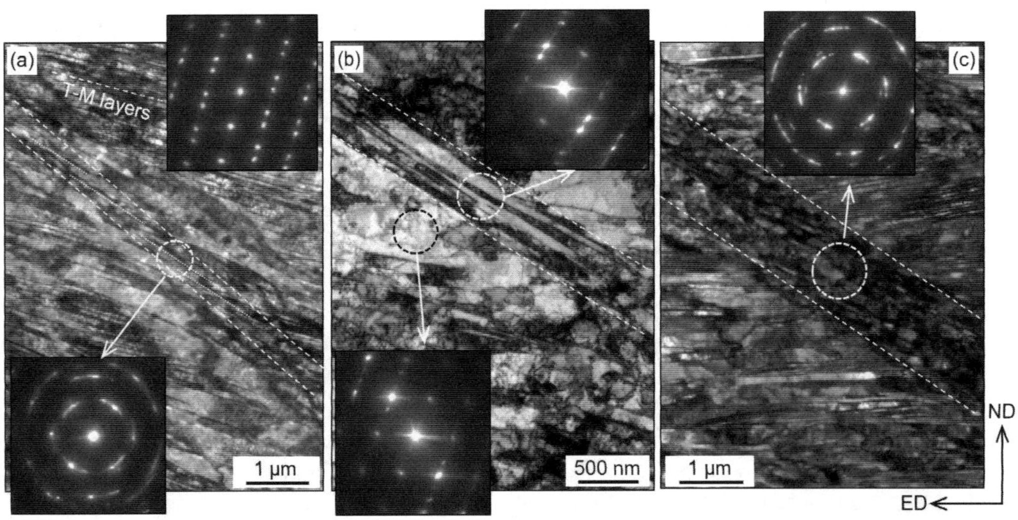

Figure 5. Shear bands formation in S-oriented single crystals. (a) Kink-type bands formation inside T-M layers, (b) twinning in re-oriented matrix inside the band region, (c) well-developed shear bands developed against the background of compact clusters of twins situated parallel to the compression plane.

As expected, the shear bands form very intensively in twinned structures of C and S orientated crystals. The mechanism of shear bands formation is analogous to that previously observed in twinned structures of face-centered cubic (fcc) metals of low stacking fault energy (SFE) [11, 13] or fcc metals of medium SFE but deformed at 77 K [14-16]. The initial stages of the shear band formation are inherently related to the global rotation tendency of the layered structure of twin-matrix (T-M) platelets towards the positions of twinning planes parallel to the compression plane. This global rotation tendency enhances the intensity of the texture components located close to two complementary {111}<112>-type components. However, the rotation observed in regions of strain localization is opposite to that observed globally (as described above). The initial stage of the loss of stability of the layered structure of T-M platelets is associated with the rotation of the T-M platelets (within the narrow region) in the 'direction' which increases the inclination of twinning planes with respect to the ED (figure 5a); this rotation aims at locating the twinning plane parallel to the shear band plane. As a result of this rotation, the orientation of the twin platelets moves towards the position of Goss (110)[001] orientation. On the other hand, the orientation of the retained primary matrix platelets rotates towards the (001)[110] orientation. In both cases, the rotation increases the value of the Schmid factor of co-planar systems operating on the twinning plane. Simultaneously, it increases the tendency of the remaining un-twinned matrix for further twinning (figure 5b). This leads to a complete disappearance of the orientations associated with the original matrix in the shear band region and an enhancement of the density of positions near the Goss (110)[001] orientation. The above mechanism for the formation of shear bands as a result of strain localization within narrow regions enables strain accommodation at the macro- (sample) scale, as shown in a series of earlier works [13, 16].

Once the strain accommodation limit is reached in one family of macro- shear bands, the process of strain localization begins in the second family of macro- shear bands (MSB^2) which is located symmetrically to the first one. This process is analogous to that observed during the formation of the first family of macro- shear bands (MSB^1). This is illustrated in figure 5c. Also in the case of MSB^2 formation, the deflection of compact clusters of the T-M platelets inside the strain localization region is connected with the situation of the twinning planes along the shear band plane. Again, in the strain localization region, the deflection of the compact clusters T-M lamellas is associated with an increase in the inclination of the twinning planes with respect to ED.

In the shear band formation mechanism described above, the twin-matrix deflection within some narrow regions begins with kink-type bands formation, which are precursors of shear bands. This local kinking in the strain localization region leads to the re-orientation of the crystal lattice in the T-M platelets and this strictly 'directed' re-orientation of the crystal lattice leads to the situation of the twinning plane parallel to the shear band plane.

4. Summary

In this study, microstructure and texture changes in the channel-die compressed (up to 60%) copper single crystals with (1-10)[112] (Br), (112)]11-1] (C), and (346)[63-5] (S) orientations, in the range of ultra-high strain rates, were analyzed. It was found that the deformation behaviour of the crystallites is different from that observed during deformation in the range of 'conventional' strain rates. This is mainly due to the intense deformation twinning as well as deformation banding and shear banding. It was found that all analyzed single crystals do not demonstrate microstructural and textural stability during deformation at extremely high strain rates. Twinning in all three orientations, especially intensive in C and S orientations, and the rapid appearance of shear bands in crystallites with C and S – oriented crystals allows to conclude that the evolution of the microstructure and texture of pure Cu crystallites during deformation at extremely high strain rates is analogous to the behavior of Cu deformed at 77K [14-16] or its low SFE fcc alloys.

The Br orientation, stable during channel-die deformation at 'conventional' strain rates, lost its structural and textural stability. In particular, the microstructure of Br-oriented crystallites shows the formation of thin platelets of twins and deformation bands, leading to strong scattering of the initial orientation. The formation of twins was mainly related to the twinning transformation on planes strongly favored for the occurrence of slip during the deformation process. However, to a lesser extent twinning was also observed on the {111} plane perpendicular to the compression plane with the normal situated close to ED.

The evolution of the microstructure and texture of crystallites with C and S initial orientations was associated with intense twinning and shear banding. The microstructure and texture of both single crystals was dominated by the twinning on the co-planar slip plane in spite of identification of three families of compact clusters twins of two generations. This dominant structure of T-M layers favors the formation of the plastic flow instabilities in the form of shear bands. The common twin-matrix layers' rotation, within the narrow area, combined with deformation twinning in the re-oriented matrix is responsible for the formation of orientations from the neighborhood of the {110}<100> orientation.

Acknowledgments
This research was supported by Polish National Centre of Science (NCN) within the project no.: 2018/31/B/ST8/00942 (Opus). Authors wishing to acknowledge assistance to Prof. L. Litynska-Dobrzynska (IMMS PAS) for conducting TEM observations and Mr A. Gałka (ZTW Explomet) for providing the samples deformed using explosive energy.

References

[1] Wagner P, Engler O and Lücke K 1995 Formation of Cu-type shear bands and their influence on deformation and texture of rolled f.c.c. {112}<111> single crystals *Acta Met. Mat.* **43** 3799

[2] Godfrey A, Jull Jensen D and Hansen N 1998 Slip pattern, microstructure and local crystallography in an aluminium single crystal of brass orientation {110}<112> *Acta Mater.* **46** 823

[3] Miszczyk M, Paul H, Driver JH and Maurice C 2015 New orientation formation and growth during primary recrystallization in stable single crystals of three face-centred cubic metals *Acta Mater.* **83** 120

[4] Wrobel M, Dymek S, Blicharski M and Gorczyca S 1994 Dislocation microstructure and texture development in rolled copper single crystals *Z. Metallkd.* **85** 415

[5] Paul H, Maurice C and Driver JH 2010 Microstructure and microtexture evolution during strain path changes of an initially stable Cu single crystal *Acta Mater.* **58** 2799

[6] Huang M, Rivera-Díaz-del-Castillo PE, Bouaziz O and van der Zwaag S 2009 A constitutive model for high strain rate deformation in FCC metals based on irreversible thermodynamics *Mech. Mater.* **41** 982

[7] Follansbee PS 1987 The rate dependence of structure evolution in copper and its influence on the stress-strain behavior at very high strain rates *Materials of Int. Conf. IMPACT'8*

[8] Follansbee PS 1985 High strain rate deformation in FCC metals and alloys *Int. Conf. on metallurgical applications of shock wave and high strain-rate phenomena Materials of Int. Conf. EXPLOMET '85*

[9] Yaghoobi M and Voyiadjis GZ 2018 The effects of temperature and strain rate in fcc and bcc metals during extreme deformation rates *Acta Mater.* **151** 1

[10] Fan H, Wang Q, El-Awady JA, Raabe D and Zaiser M 2021 Strain rate dependency of dislocation plasticity *Nat. Commun.* **12** 1845

[11] Hong CS, Tao N R, Huang X and Lu K 2010 Nucleation and thickening of shear bands in nano-scale twin/matrix lamellae of a Cu–Al alloy processed by dynamic plastic deformation *Acta Mater.* **58** 3103

[12] Murr LE 1975 *Interfacial phenomena in metals and alloys* (Addison-Wesley, Reading MA)

[13] Paul H, Driver JH and Jasieński Z 2002 Shear banding and nucleation of recrystallization in a Cu-2%Al alloy single crystal *Acta Mater.* **50** 815

[14] Paul H, Driver JH, Maurice C and Piatkowski A 2007 The role of shear banding on deformation texture in low stacking fault energy metals as characterized on model Ag crystals *Acta Mater.* **55** 575

[15] Paul H, Driver JH , Maurice C and Jasieński Z 2003 Shear band microtexture formation in twinned face centred cubic single crystals *Mater. Sci. Eng. A* **359** 178

[16] Paul H, Morawiec A, Driver JH and Bouzy E 2009 On twinning and shear banding in a Cu-8 at.% Al alloy plane strain compressed at 77 K *Int. J. Plast.* **25** 1588

3D characterization of a nanostructured Al-Cu-Mg alloy

Wanquan Zhu[1,2*], Ran Yang[1,3], Zongqiang Feng[1,2], Guilin Wu[1,2,4], Andrew Godfrey[5] and Xiaoxu Huang[1,2]

[1]International Joint Laboratory for Light Alloys (MOE), College of Materials Science and Engineering, Chongqing University, Chongqing, 400044, China
[2]Shenyang National Laboratory for Materials Sciences, Chongqing University, Chongqing, 400044, China
[3]ZJU-Hangzhou Global Scientific and Technological Innovation Center, No.733 Jianshe San Road, Xiaoshan District, Hangzhou, Zhejiang, 311200, China
[4]Beijing Advanced Innovation Center for Materials Genome Engineering, University of Science and Technology Beijing, Beijing, 100083, China
[5]Laboratory of Advanced Materials (MOE), School of Materials Science and Engineering, Tsinghua University, Beijing, 100084, China

E-mail: wanquanzhu@cqu.edu.cn

Abstract. Three-dimensional (3D) characterization of variations in crystallography and chemistry of nanostructured metals will provide vital information to understand their mechanical and thermal behaviours. This study applied a surface sliding friction treatment (SSFT) at liquid nitrogen temperature to produce nanostructured surface layers in a peak-aged Al-Cu-Mg alloy. The nanostructured surface was characterized by means of 3D orientation mapping in the transmission electron microscope (3D-OMiTEM) and atom probe tomography (APT). 3D-OMiTEM results revealed a lamellar structure with an average lamellar boundary spacing of 26 nm at the topmost surface layer (depth < 20 μm), which is much finer than normally achievable in commercial purity Al deformed to high strain levels. Based on the 3D-OMiTEM data, a five-parameter grain boundary character analysis was carried out. It was found that low angle grain boundaries dominate the nanoscale structure and that the grain boundary plane distribution of high angle lamellar grain boundaries shows a preference around {101}. APT analysis showed segregation of Cu and Mg atoms at lamellar boundaries, which is believed to play a role in stabilizing the boundaries and enhancing the structural refinement during SSFT.

1. Introduction

Altering the properties of nanomaterials can be realized by tailoring their crystallographic character [1] and chemical distribution [2] which are conventionally revealed by orientation mapping technique and atom probe tomography (APT). Although present nanoscale 2D orientation mapping techniques, e.g., precession electron diffraction (PED) [3] and transmission Kikuchi diffraction [4], have been utilized to characterize the crystallography of nanostructured materials, they are still restricted by the grain overlap problem [5] and cannot offer complete crystallographic information directly, i.e., the grain boundary plane can only be calculated by a statistical method based on a stereological approach. Recently, it has been proved that 3D orientation mapping in the transmission electron microscope (3D-OMiTEM) [6] is a powerful tool for revealing the five-parameter grain boundary character of nanocrystalline materials

[7]. In the present study, we use 3D-OMiTEM and APT to reveal the 3D crystallographic character and element distribution of a nanostructured Al alloy.

2. Experimental

A rolling plate (100 × 100 × 10 mm^3) of commercial Al alloy 2024, with a composition of Al-4.2 Cu-1.5 Mg-0.6 Mn-0.5 Fe-0.5 Si (wt.%), was solution treated for 1 h at 495 °C, water quenched and then aged for 9 h at 195 °C to achieve peak-aged state. One surface of the peak-aged plate was processed by surface sliding friction treatment (SSFT) at liquid nitrogen temperature to achieve structural refinement. The SSFT was realized by iterative sliding of a 10 mm diameter WC-Co ball on the sample surface for 100 cycles with a load of 100 N [8, 9]. The sliding direction and the direction parallel with the plate normal are defined as SD and ND, respectively. The sample after SSFT processing is designated as PA-SSFT. A conventional thin foil TEM specimen was prepared from the SD-ND cross-section of PA-SSFT sample by cold ion milling, with special care taken to ensure accurate determination of the depth of the observed region from the top-surface of sample [9]. 3D-OMiTEM characterization was performed for a volume centered at a depth of around 16μm from the top-surface, using a JEOL JEM-2100 TEM equipped with a TVIPS high-speed camera. The source dataset of 3D orientation mapping contains dark-field conical scanning images for 8 diffraction rings with a tilt angle range of -60° – +53°. The electron beam scans at 2° intervals for each considered diffraction ring. More than 160,000 dark-field images consisting of 1024 × 1024 pixels with a resolution of 0.92 nm/pixel were used to reconstruct **g** vectors of every voxel in 3D real space. Orientation indexing within a volume of 1024 (SD) × 250 (TD) × 1024 (ND) voxels was realized by a dictionary-based forward indexing algorithm [10]. Post analysis and visualization of grain boundary character were carried out by using a metric-based grain boundary plane distribution filter built in Dream3D software [11, 12] and in-house code.

Needle-shaped APT samples for the topmost surface layer of PA-SSFT sample with axis parallel to ND were prepared by Zeiss Auriga FIB-SEM dual beam system. CAMECA LEAP 4000 HR instrument was utilized to perform atom probe tomography at a temperature of 20 K under a voltage pulsing with a voltage fraction of 20% and a target evaporation rate of 1%. Analysis and visualization of the APT data were performed using the IVAS 3.6.12 software.

3. Results and discussion

3.1. 3D-OMiTEM observations

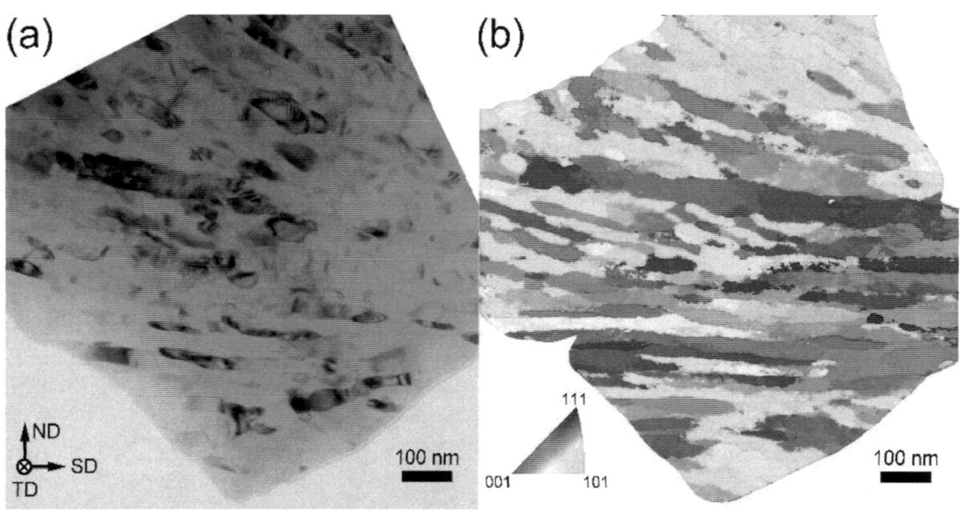

Figure 1. (a) TEM bright field image at a depth of 16 μm in the PA-SSFT sample. (b) Plane view of the reconstructed 3D orientation map. The coloring refers to the crystal direction parallel to the sliding direction during SSFT of the sample.

Figure 1 shows the microstructure and the three-dimensional orientation map of the PA-SSFT sample. The bright-field image and orientation map both reveal a lamellar structure in the PA-SSFT sample. The 3D orientation map reveals that in many areas a single grain occupies the foil thickness (varying from 36 to 97 nm in the mapped volume), while in some other places 2-3 grains overlap along the foil thickness. The average lamellar boundary spacing calculated from different 2D slices of the 3D orientation map along TD is 26 nm which is finer than high-strained commercial purity Al [13, 14] and close to the results for gradient alloys with similar composition and depth obtained by Xu et al. [15] and Feng et al. [16] using PED. Compared with the solution treated sample [9], finer lamellae are observed in regions deeper from the surface, which indicates that the precipitation introduced by aging treatment before deformation can effectively inhibit dynamic recovery of the alloy during deformation. Shuai et al. [17] observed the pinning phenomenon of Al_2Cu particles to triple junctions in Al-0.3wt.%Cu by in-situ electron microscopy, indicating that the second phase particles can improve the stability of the alloy. This stabilization mechanism also plays an important role in the deformation process of the samples in the aging state.

Figure 2a shows three typical cross-sections (ϕ_2 = 0°, 45° and 65°) of the orientation distribution function (ODF) calculated from the 3D orientation mapping data shown in figure 1b. Compared with the corresponding ideal texture component position in the ODF of face centered cubic (FCC) crystals (figure 2b), strong texture components of the nanostructure at a depth of 16 μm include the E {111}<011> and rotated cube {001}<110> texture components. Moreover, minor intensities can be detected at the rotated copper {112}<011> texture component and the F {111}<112> texture component. Similar shear texture components have also been observed in other surface-treated Al alloys [16, 18] and in the surface regions of metals such as Cu and Ni [19, 20]. The dominance of shear texture components in the near-surface region of the sample is in agreement with the repeated shear deformation of the plate surface.

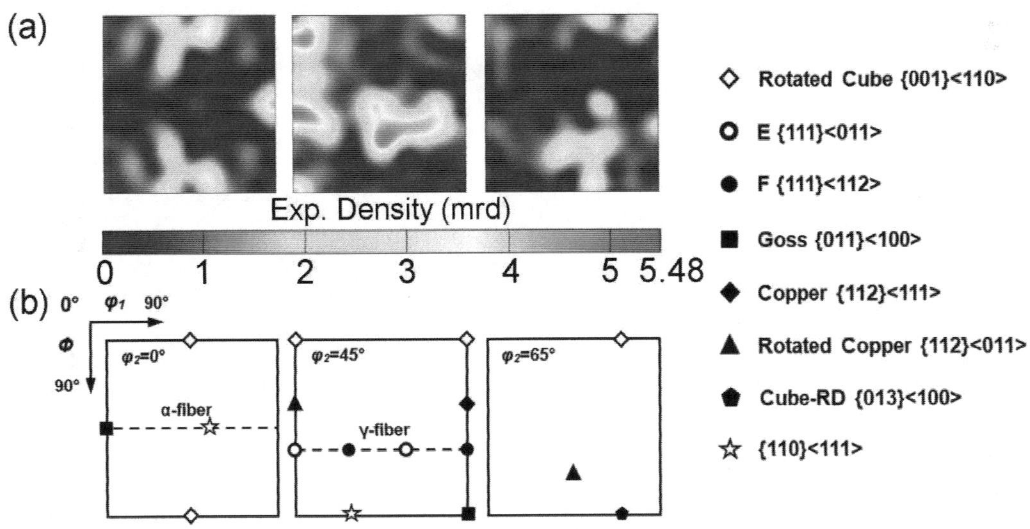

Figure 2. (a) ODF sections (φ_2 = 0°, 45° and 65° from left to right) of the PA-SSFT sample at a depth of 16 μm and (b) corresponding ideal texture components in the same sections of (a) [16].

Figure 3 shows the five-parameter grain boundary character distribution of the PA-SSFT sample. A complex 3D lamellar structure is observed, as shown in figure 3a. Some lamellar boundaries are not edge-on along the ND and hence appear with a certain width in the SD-TD plane view. The lamellar boundaries are almost high-angle grain boundaries (with misorientation angles larger than 15°). A large number of low-angle boundaries are found between the lamellae. Based on the present angular step size of dark-field conical scanning mode, boundaries with misorientation angles lower than 2.5° cannot be detected with confidence and were omitted during data processing. Nevertheless, as shown in figure 3c, the area fraction of low-angle grain boundaries is about 48% higher than observed in the near-surface layer of gradient Al–4 wt.%Cu alloy [15]. It was also found that 21% of high-angle grain boundaries are CSL boundaries ($\Sigma \leq 51$) according to the Brandon criterion [21]. The fractional areas measured in 3D for the $\Sigma 3$, $\Sigma 11$ and $\Sigma 17b$ boundaries are 6.1%, 1.2% and 1.1%, respectively.

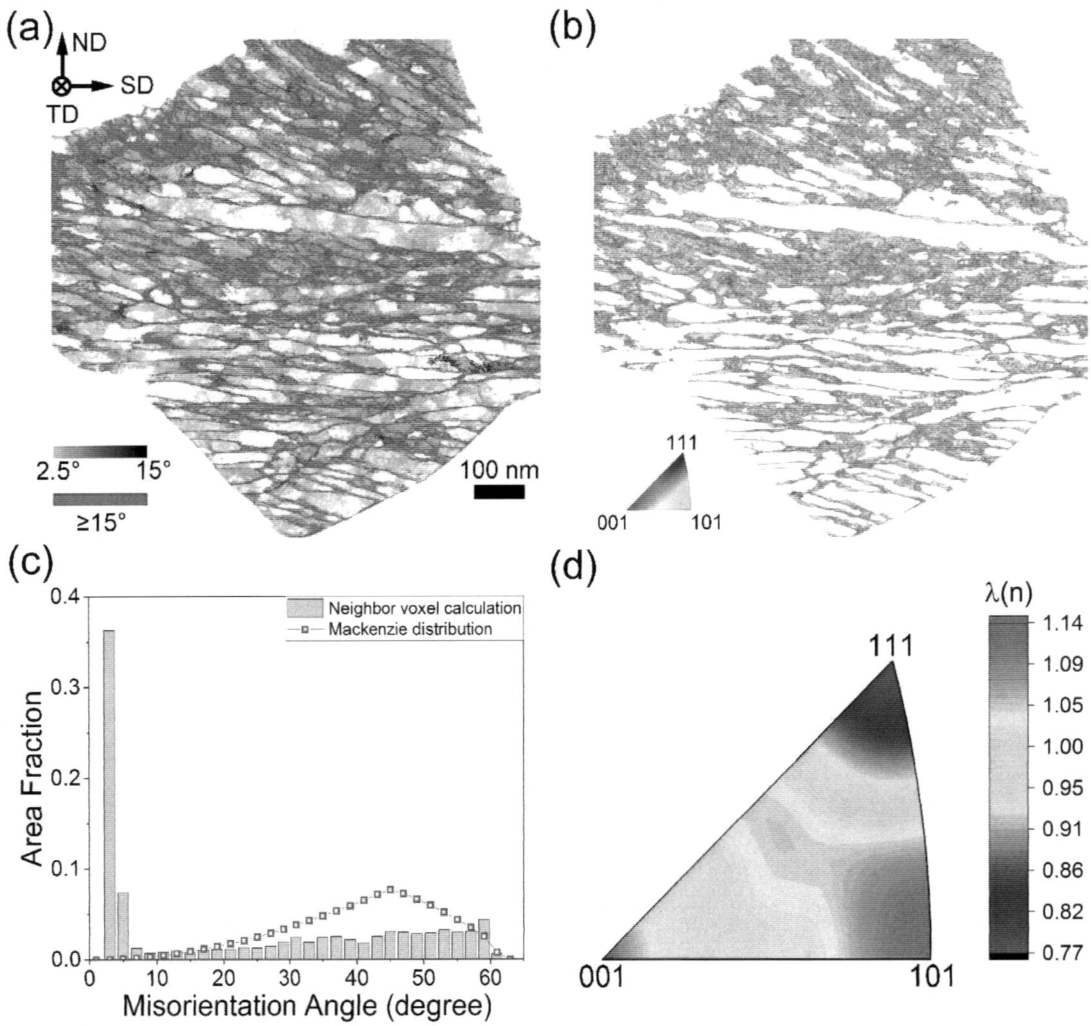

Figure 3. Plane view of the 3D reconstruction of the grain boundaries of the PA-SSFT sample, where grain boundaries are colored according to (a) boundary misorientation type and (b) grain boundary plane normal as indicated in the legend. Shown also are corresponding distributions of (c) grain boundary misorientation angle and (d) grain boundary plane from the 3D mapping in (a) and (b).

With the advantage of 3D orientation mapping, the normal direction of each grain boundary can be determined. Figure 3b shows the three-dimensional reconstruction of the grain boundaries in the PA-SSFT sample using a 5° misorientation angle for grain reconstruction, rendering using colours referred to the grain boundary plane normal in the crystal reference frame. Combined with figure 3a, the five-parameter character of individual grain boundaries can be obtained. The relative population for all grain boundary plane normals, $\lambda(n)$, where n is the normal to the grain boundary plane, is represented in one fundamental zone of the crystal reference frame (figure 3d). The distribution is represented in multiples of random distribution (MRD), where values >1 mean that planes are observed more frequently than anticipated in a random distribution. It is seen that the boundary planes of all high-angle grain boundaries have a preference to lie on {101} planes, while the boundary intensity for {111} planes is lower than that of random distribution, which is different from the previous results for coarse-grained commercial Al [22] and extruded Al alloy 6061 [23].

3.2. *APT analysis*

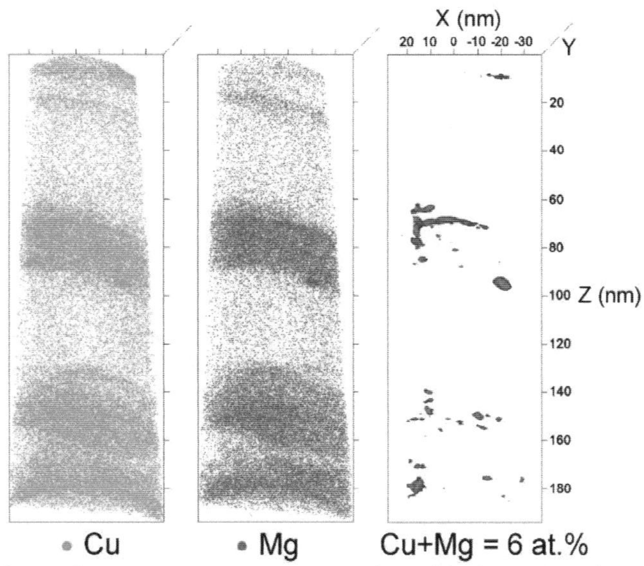

Figure 4. Distributions of Cu atoms, Mg atoms together with the Cu-Mg isosurface at 6 at.% for the PA-SSFT sample.

Figure 4 shows a typical example of reconstructed atom probe images for the topmost surface layer of the PA-SSFT sample. Heterogeneous co-segregation of Cu and Mg atoms is observed. Two kinds of enrichment can be identified. Some Cu and Mg atoms (in the top part of the data) are present in narrow enrichments, believed to correspond to solute segregation at the grain boundaries. Other parts show a diffuse elemental enrichment region of 20–30 nm width, which is suggested to be Cu and Mg atoms from the decomposed and dissolute S phase under the action of the severe plastic deformation [8]. Due to the low diffusion capability at low deformation temperature, these atoms still remain around the decomposed precipitates. Segregation and precipitation in the peak-aged sample can stabilize the grain boundaries from both a thermodynamic and kinetic perspective, and restrain dynamic recovery in the deformation process [17], which leads to thinner lamellar spacing compared with solute-treated SSFT samples [9]. APT can provide abundant 3D chemical information but limited crystallographic knowledge, while 3D-OMiTEM is an effective complementary technique. Further studies on correlating element distribution and crystallography by carefully combining these two characterization methods are still required.

4. Conclusions

In summary, the crystallographic texture, five-parameter grain boundary character distribution and chemical distribution of a nanostructured surface layer in a peak-aged Al-Cu-Mg alloy processed by SSFT have been revealed by 3D-OMiTEM and APT. Severe plastic deformation resulted in the formation of a strong shear texture and a complex three-dimensional lamellar nanostructure in the near-surface region of the specimen. The lamellar structure was composed mostly of high misorientation angle grain boundaries with low-angle dislocation grain boundaries between the lamellae. It was found that the grain boundary plane of lamellae had a preference of {101}. A heterogeneous distribution of solute atoms is observed, improving the stability of the present nanostructure. The ability to collect 3D orientation data and chemical distribution is expected to further contribute to a comprehensive understanding of the relationship between the microstructure and properties of nanomaterials.

Acknowledgements
The present work was supported by the National Key Research and Development Program of China (2021YFB3702101), the National Natural Science Foundation of China (No. 52071038 and No. 51971043).

References
[1] Li X Y, Jin Z H, Zhou X and Lu K 2020 *Science* **370** 831-836.
[2] Hu J, Shi Y N, Sauvage X, Sha G and Lu K 2017 *Science* **355** 1292.
[3] Wu G, Zhu W, He Q, Feng Z, Huang T, Zhang L, Schmidt S, Godfrey A and Huang X 2020 *Nano Mater. Sci.* **2** 50.
[4] Niessen F, Burrows A and Fanta A 2018 *Ultramicroscopy* **186** 158.
[5] Kobler A and Kubel C 2017 *Ultramicroscopy* **173** 84.
[6] Liu H H, Schmidt S, Poulsen H F, Godfrey A, Liu Z Q, Sharon J A and Huang X 2011 *Science* **332** 833.
[7] Zhu W, Wu G, Godfrey A, Schmidt S, He Q, Feng Z, Huang T, Zhang L and Huang X 2022 *Scr. Mater.* **214** 114677.
[8] Feng Z, Luo X, Chen Y, Chen N and Wu G 2019 *J. Alloy. Compd.* **773** 585.
[9] Yang R, Feng Z, Huang T, Wu G, Godfrey A and Huang X 2022 *Scr. Mater.* **206** 114240.
[10] Larsen P M and Schmidt S 2017 *J. Appl. Crystallogr.* **50** 1571.
[11] Glowinski K and Morawiec A 2014 *Metall. Mater. Trans. A* **45** 3189.
[12] Groeber M A and Jackson M A 2014 *Integr. Mater. Manuf. Innov.* **3** 56.
[13] Xu W, Liu X C and Lu K 2018 *Acta Mater.* **152** 138.
[14] Yu T, Hansen N and Huang X 2011 *P. Roy. Soc. A-Math. Phy.* **467** 3039.
[15] Xu W, Liu X C, Li X Y and Lu K 2020 *Acta Mater.* **182** 20.
[16] Feng Z, Chen Y, Zhang W and Wu G 2019 *Philos. Mag. Lett.* **98** 410.
[17] Shuai L, Huang T, Yu T, Wu G, Hansen N and Huang X 2021 *Acta Mater.* **206** 116595.
[18] Sheikh H 2011 *Scr. Mater.* **64** 556.
[19] Cai W, Mabon J and Bellon P 2009 *Wear* **267** 485.
[20] Liu X C, Zhang H W and Lu K 2013 *Science* **342** 337.
[21] Brandon D G 1966 *Acta Metall. Mater.* **14** 1479.
[22] Saylor D M, El Dasher B S, Rollett A D and Rohrer G S 2004 *Acta Mater.* **52** 3649.
[23] Zhao Y, Niverty S, Ma X, Liu X and Chawla N 2020 *Mater. Charact.* **170** 110716.

IOP Conf. Series: Materials Science and Engineering 1249 (2022) 012030 doi:10.1088/1757-899X/1249/1/012030

Challenges related to tomographic reconstruction of 3D intragranular orientation fields in the presence of orientation relationships

Z Liu[1,2], N Viganò[2] and W Ludwig[1,2]

[1] MATEIS, INSA Lyon, UMR5510 CNRS, 69621 Villeurbanne, France
[2] ESRF, Grenoble, France

E-mail: zheheng.liu@esrf.fr

Abstract. X-ray diffraction contrast tomography (DCT) is a near-field diffraction imaging technique to characterize the 3D shape and crystallographic orientation of grains within polycrystalline samples. The presence of orientation relationships as such encountered between $\sum3^n$ annealing twins found in Cu and Ni lead to systematic diffraction spot overlap, since a significant fraction of the lattice planes is shared between the parent and twin crystal lattices. If not correctly addressed, these overlaps will lead to artifacts in the individual grain reconstruction. In this paper we introduce a strategy for joint, tomographic reconstruction of Twin Related Domains (TRD) and cost function weights into DCT to improve the grain reconstruction quality. A large-grained polycrystalline sample made from pure Ni is used for testing this approach and its ability to reveal intragranular orientation gradients related to plastic strain localization at the onset of plastic deformation in this type of materials.

1. Introduction

Strain localization is a plastic instability process before fracture. There are many studies on that subject, and some examples can be found in [1, 2]. During initial stages of plastic deformation in metallic alloys, strain localizes in the form of slip bands. The observation of these slip bands at the sample surface, using a combination of scanning electron microscopy techniques (SEM, EBSD, DIC) is an important method to study plasticity at the microscale. In this paper we study the feasibility to use full-field X-ray diffraction imaging techniques to characterize orientation gradients and slip bands in three-dimensions in the bulk of metallic polycrystals.

X-ray Diffraction Contrast Tomography (DCT) [3] is a non-destructive X-ray imaging technique to reconstruct the microstructure of the grains in polycrystalline sample volumes. The initial approach, hereafter termed 3D-DCT, assumes all the voxels of the grain have the same orientation and applies to well recrystallized microstructures with small intragranular orientation spread. 6D-DCT [4] on the other hand can reconstruct both the shape and intragranular orientations of the grain. It is based on a discretization of 3D orientation space, so that the 3D position space is expanded into a 6D position-orientation space. In single grain reconstructions, only a small sub-region of 3D orientation space (centered on the grain average orientation) is sampled. This concept can be extended to *cluster* reconstructions [5], i.e. sampling of multiple, disjoint regions of orientation space, enabling the joint (simultaneous) reconstruction of grains exhibiting orientation relationships. This in turn allows to correctly handle the presence of shared reflections between different domains of the cluster, which can now be included in the optimization process without violating the consistency of the tomographic reconstruction problem.

In this study, we apply the above outlined 6D-DCT cluster reconstruction to the case of a Twin Related Domain (TRD) in a miniature tensile sample, made from pure Ni. The reconstructed intragranular orientation gradients and signatures of strain localization, visible in the diffraction topographs are discussed.

2. Material and method

The miniature tensile sample made from pure Ni with a cross-section of 400×400 μm is extracted and subsequently annealed to produce a fully recrystallized grain microstructure with an average grain size of ~ 100 μm. The sample is mounted in the NANOX [6] stress rig and scanned at two different applied loads of ~ 2N and 20N, corresponding to ~ 0 and 0.2% applied strain. Each of the DCT scans comprises 7200 images with 0.05° rotation increment, using an X-ray beam energy of 43.5 keV. The detector pixel size is 1.63 μm.

The average orientations and the positions of the grains are computed in a Friedel-pair-based grain indexing step [7]. Then, the shapes and the intragranular orientations of the grains are reconstructed from its corresponding diffraction spots [4]. Usually, the reconstructions are performed grain by grain individually. However, complications may arise for materials exhibiting orientation relationships between neighboring grains. A prominent example are annealing twins in metallic materials with FCC lattice structure like Cu and Ni, for which the parent and twin orientation share subsets of common lattice directions. For example, if the grain has a $\sum 3$ twin, half of the {2 2 0} and {3 1 1} lattice planes of the grain have common reflections with the twin. These common reflections lead to spots with overlapping intensity distributions, which would have to be excluded from the reconstruction of the parent / twin domains in order to keep consistency of the tomographic reconstruction problem. However, these materials often exhibit extended neighborhoods of TRD, containing higher order $\sum 3^n$ twins, which further reduce the number of reflections which can unambiguously be assigned to an individual domain. The quality of the individual grain reconstruction can be strongly compromised, if the presence of resulting diffraction spot overlaps is not taken into account.

To solve this problem, we propose a generalization of the joint reconstruction approach outlined in [4]. Instead of reconstructing grains one by one, the entire material volume of the TRD is reconstructed at once. Firstly, the TRD recognition [8] is applied. Then, the mappings are created between the grains and the diffraction spots. Afterwards, the TRD is jointly reconstructed based on a modified version of the 6D-DCT algorithm [9].

The reconstruction problem is formulated as (1). The column vector $\mathbf{x}$ contains the unknown variables in 6D position-orientation space. $\mathbf{y}$ contains the pixel intensities of the corresponding diffraction spots. $\mathbf{A}$ is the (sparse) projection matrix, which specifies the contribution of each of the elements of x to each of the detector pixels $\mathbf{y}$. $\lambda\|\mathbf{H}(\mathbf{x})\|_1$ is the regularization term. Here, the spatial 3D intensity volume (i.e. the sum over the reconstructed orientation space contributions) is regularized by the total variation norm, promoting solutions exhibiting piecewise constant intensity (i.e. similar diffraction power for each voxel inside a grain). W is a diagonal matrix containing the cost function weights.

$$\mathbf{x}^* = \arg\min_{\mathbf{x}} \frac{1}{2} \left\| \mathbf{W}^{\frac{1}{2}}(\mathbf{y} - \mathbf{A}\mathbf{x}) \right\|_2^2 + \lambda \|\mathbf{H}(\mathbf{x})\|_1, \text{ subject to } \mathbf{x} \geq 0 \tag{1}$$

The weights are inversely proportional to the noise variance. The pixels of the detector are assumed to have a Poisson noise and the variance of this Poisson noise scales with the intensity of the diffracted beams. Moreover, an estimate of σ_{dark}, the variance of the thermal noise of the detector, is calculated from the images without beam. Therefore, the weights are formulated as (2).

$$\mathbf{W} = \sigma_{\mathrm{dark}}[\mathrm{diag}(\sigma_{\mathrm{dark}} + \mathbf{y})]^{-1} = \begin{bmatrix} \frac{\sigma_{\mathrm{dark}}}{\sigma_{\mathrm{dark}}+y_1} & & \\ & \frac{\sigma_{\mathrm{dark}}}{\sigma_{\mathrm{dark}}+y_2} & \\ & & \ddots \end{bmatrix} \tag{2}$$

3. Results

As a first step, 3D-DCT is used to reconstruct an approximate grain map of the sample under the load of 20N. A first refinement is obtained by switching to the 6D-DCT algorithm, yielding improved grain shape reconstructions for this material, which shows non-negligible intra-granular orientation spread, inherited from previous processing and sample handling steps. By segmenting the reconstructed grain intensity and dilating the segmented grain masks by 5 pixels, the grain map of the sample volume is created. Figure 1 shows a slice in the y-z plane of the sample. Next, using the TRD recognition procedure outlined in [8], the TRD map of the grain volume is obtained. Figure 2 shows the corresponding slice of the TRD map. Comparing these two figures, the grains A, B and C indicated in figure 1 belong to the same TRD.

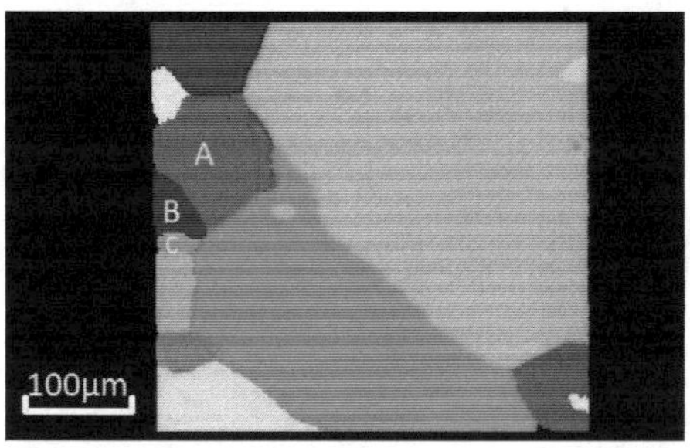

Figure 1. 2D slice (y-z plane) of the grain map in reconstructed sample volume.

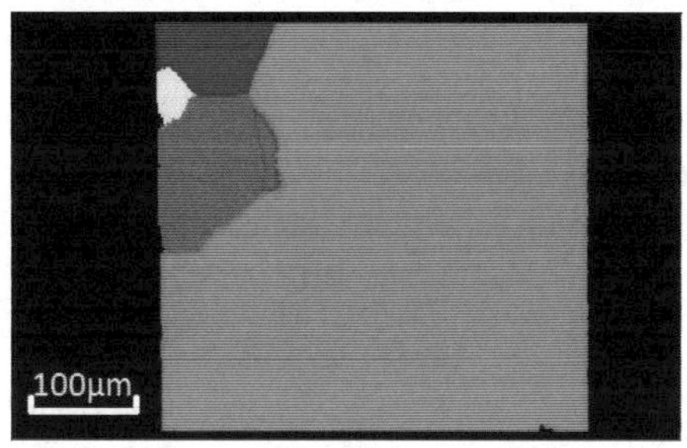

Figure 2. 2D slice (y-z plane) of the TRD map in reconstructed sample volume.

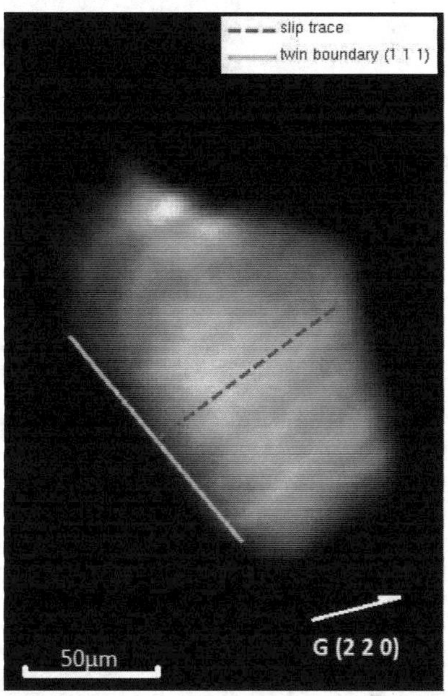

Figure 3. Diffraction spot of grain A.

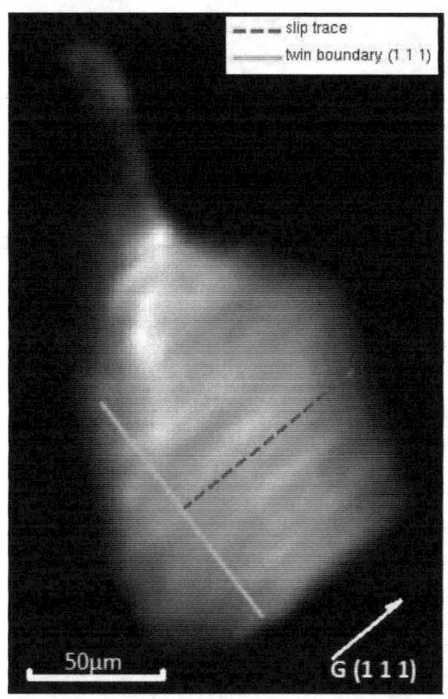

Figure 4. Overlapped diffraction spots of grain A and B.

Figure 3 shows the (2 2 0) diffraction spot of grain A, integrated over the rotation. Its average ω and η are ~ 54° and 255°, where ω is the rotation angle of the sample and η is the azimuthal angle along the diffraction ring. Figure 4 shows an example of an overlapped diffraction spot, corresponding to the (1 1 1) reflection, common to grains A and B (their twin boundary is (1 1 1) plane). Its average ω and η are ~ 23° and 232°. These two spots have similar ω and η, so the spots show the similar structures of grain A, such as the edges and the band-like intensity contrast. In all the figures, the direction of load is upward, which is close to the crystal orientation <2 $\bar{1}$ 2> of grain A. Interestingly, the band-like image contrast does not correspond to any of the close-packed (1 1 1) planes, which are the expected slip planes in this type of material.

Figure 5 shows the slices of the individually reconstructed grains A, B and C. The inclusion of shared diffraction spots into individual grain reconstructions gives rise to ghost images, as can be observed in figure 5(a), where the increased intensity in the bottom left corner corresponds to the region occupied by the twin orientation. Automated identification and removal of shared diffraction spots can be a solution for such small clusters, containing limited number of twin related orientations. For larger clusters which may contain several tens of different twin related orientations (including higher order $\sum 3^n$ twins), this approach becomes impractical, since the number of "valid" (not shared) diffraction spots decreases quickly.

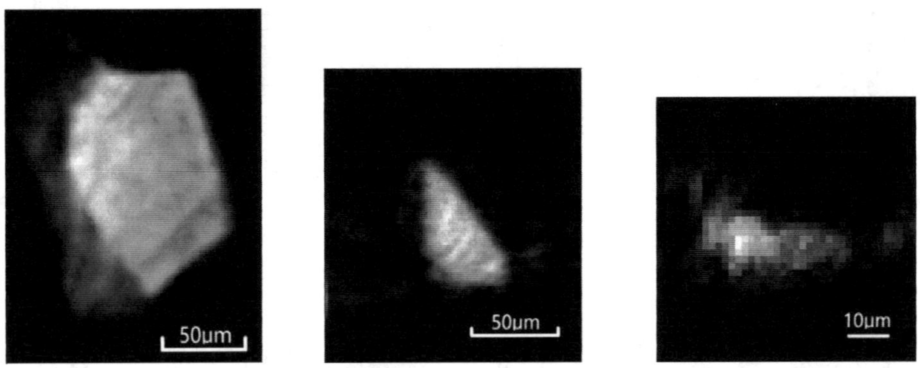

(a) grain A (b) grain B (c) grain C

Figure 5. 2D slices of Individual grain reconstructions of grains A, B and C.

Figure 6(d) shows a slice of the same grain cluster, this time reconstructed using the TRD cluster reconstruction approach. Sub-figures 6(a), 6(b) and 6(c) are the slices of the grains A, B and C in the reconstructed TRD, whereas figure 6(d) is the sum of figures (a), (b) and (c), indicating that the reconstructed twin boundaries of the grains match each other.

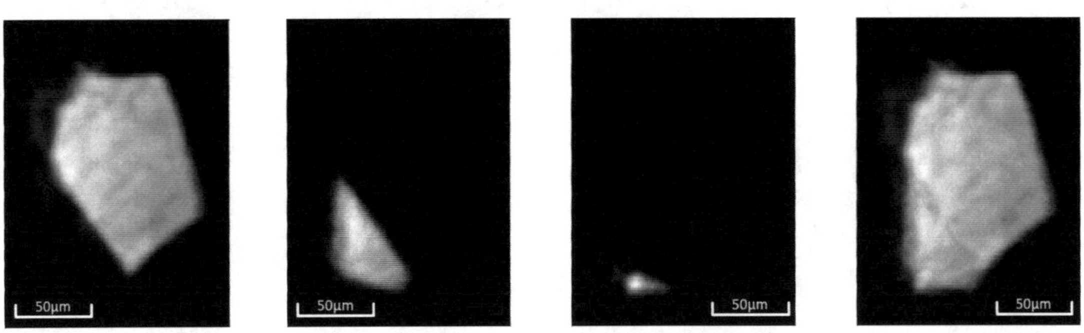

(a) grain A (b) grain B (c) grain C (d) TRD

Figure 6. 2D slices of the volumes of TRD joint reconstructions.

Comparing the results of two reconstruction methods, the TRD joint reconstruction result has less artifacts. After the selection of reflections, there are 81, 58 and 58 reflections used for the individual grain reconstructions (figure 5) of the grains A, B and C respectively, while the TRD reconstruction in figure 6 uses a total of 288 reflections, making better use of the diffraction spot data.

Figures 7 and 8 show the Kernel Average Misorientations (KAM) and the Intragranular Misorientations (IGM) of the TRD at applied load of ~ 2N. There is pre-existing orientation spread, which can be caused by imperfect sample preparation. The IGM of the voxel indicates the misorientation angle between voxel orientation and the average grain orientation. The KAM indicates the average of the misorientation angles between the voxel and its 3×3×3 neighborhood.

Figure 7. 2D slice of KAM.

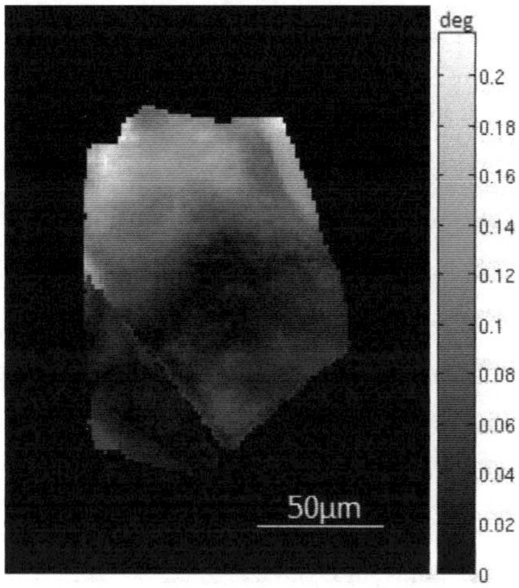

Figure 8. 2D slice of IGM.

Figures 9 and 10 respectively show the slices of the KAM and the of the reconstructed TRD at applied load of ~ 20N. From the images, we can see the band like structures intersecting the twin boundaries, which matches the contrast visible in the diffraction spots in figures 3 and 4. The orientation of these bands do not correspond to any of the {1 1 1} planes and we attribute them as signature of lattice misorientations created by kink bands which are known as an alternative mode of plastic strain localization in fcc materials [10].

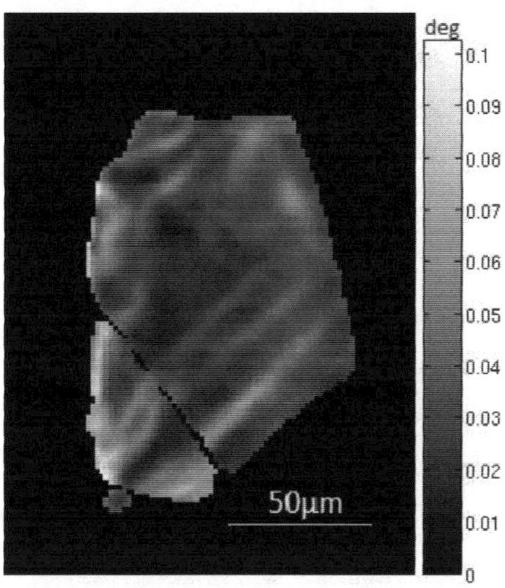

Figure 9. 2D slice of KAM.

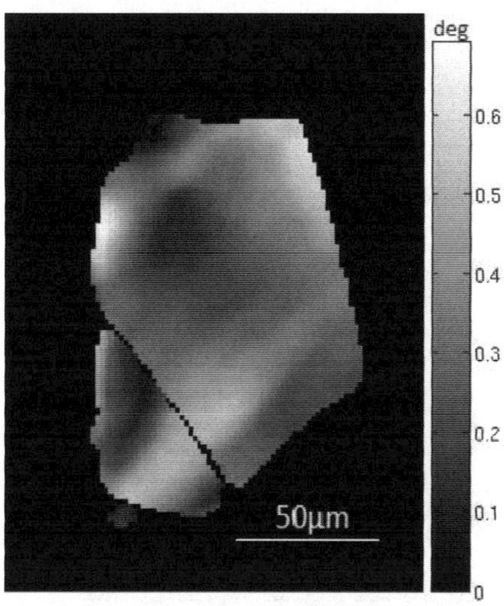

Figure 10. 2D slice of IGM.

4. Conclusions

The presence of shared reflections, as commonly observed in twin related domains in fcc materials, can lead to reconstruction artifacts when not handled appropriately. To address this problem, we propose a joint TRD (grain cluster) reconstruction approach. After grain indexing and an initial grain reconstruction, the TRDs are recognized as a first step. Based on the TRD recognition, the domain occupied by the TRD is reconstructed in a joint (simultaneous) reconstruction, making optimal use of the available diffraction data. According to experiment results, compared to the individual reconstruction, the joint reconstruction and proposed modification of the cost function weight improve the reconstruction quality. The methd has been applied to a Ni polycrystal, deformed to about 0.2% plastic strain. The grain projections, and reconstructed orientation field reveal band-like structures, aligned parallel to the $(1\ \bar{2}\ 1)$ lattice planes.

Acknowledgments
Many thanks to Henry Proudhon and Muhammad Fakhry Hatta for preparing the samples. We thank ESRF for beam time under proposal MA4761. Agence nationale de la recherche is acknowledged for funding the PhD of ZL under the project 3DiPolyPlast.

References

[1] A saro R J and Rice J 1977 *J. Mech. Phys. Solids* **25** 309–38
[2] Antolovich S D and Armstrong R W 2014 *Prog. Mater. Sci.* **59** 1–160
[3] Ludwig W, Reischig P, King A, Herbig M, Lauridsen E, Johnson G, Marrow T and Buffiere J Y 2009 *Rev. Sci. Instrum.* **80** 033905
[4] Vigano N, Ludwig W and Batenburg K J 2014 *J. Appl. Crystallogr.* **47** 1826–40
[5] Vigano N, Nervo L, Valzania L, Singh G, Preuss M, Batenburg K J and Ludwig W 2016 *J. Appl. Crystallogr.* **49** 544–55
[6] Guéninchault N, Proudhon H and Ludwig W 2016 *J. Synchrotron Radiat.* **23** 1474–83
[7] Reischig P, King A, Nervo L, Viganó N, Guilhem Y, Palenstijn W J, Batenburg K, Preuss M and Ludwig W 2013 *J. Appl. Crystallogr.* **46** 297–311
[8] Reed B W, Minich R W, Rudd R E and Kumar M 2004 *Acta Crystallogr.* A **60** 263–77
[9] Vigano N R 2015 *Full-field X-ray orientation imaging using convex optimization and a discrete representation of six-dimensional position-orientation space* Ph.D. thesis INSA de Lyon; Université de Antwerp
[10] Marano A, Gélébart L and Forest S 2019 *Acta Mater.* **175** 262–75

IOP Conf. Series: Materials Science and Engineering 1249 (2022) 012031 doi:10.1088/1757-899X/1249/1/012031

Finite element modelling of indentation testing in an AA1050 microstructure

S Goutianos[1], **H Pan**[2,3] **and X Zhang**[2]

[1] Department of Manufacturing and Civil Engineering, Norwegian University of Science and Technology, 2815 Gjøvik, Norway

[2] Department of Mechanical Engineering, Technical University of Denmark, Kongens Lyngby, 2800, Denmark

[3] Department of Materials Science and Engineering, Kunming University of Science and Technology, Kunming 650093, China

E-mail: stergios.goutianos@ntnu.no

Abstract. Heterogeneous microstructures can be produced in structural metals from heterogeneous deformation, and they can also be purposely designed to meet specific requirements in high-demanding engineering applications. When designing a heterogeneous microstructure in metallic materials, it is important to know the local mechanical properties such as the stress-strain behavior. However, due to the local varying microstructure, direct measurement of the local mechanical properties is not feasible either by bulk testing or by miniature testing. In this work, an iterative approach, based on the finite element method and under the assumption that the local mictrostructure is homogeneous, will be presented as an example to extract the local mechanical properties of cold-rolled AA1050 from indentation tests as an example. The uncertainties of the proposed method are discussed.

1. Introduction

Plastic deformation of metals typically refines the microstructure and results in high strength [1]. Several approaches have been developed to introduce high strains in bulk samples in order to obtain high strength metals. Alternatively, an increased strength can be also obtained by limiting the hard nanostructure only close to the surface region e.g. by shot-peening [1, 2, 3]. In this case, the resultant microstructure is not homogeneous. Similarly, graded microstructures can be generated in the surface and its neighboring region by surface deformation such as friction and wear [4, 5]. In both cases, the microstructures extend only few micrometers to a few tens of micrometers into the bulk material [5, 6], and therefore it is difficult to determine its stress-strain behaviour.

The importance of being able to assess the mechanical properties at such small scales and in parallel the improvements in indentation equipment, have led in a increased interest in instrumented indentation [7]. Several methods have been developed to extract the mechanical properties such as the Young's modulus, E, the yield stress, σ_y and the strain hardening exponent, n, from the indentation load-displacement curves [8, 9]. In general, a reverse analysis, typically using the finite element method, is performed to solve for the three unknowns mentioned above and thus the elasto-plastic material properties can be determined.

The present work follows to a large extend the existing numerical approaches for reverse analysis of indentation experiments. Although, the final aim is to characterize nanostructured graded microstructures, here the numerical method will be illustrated on microstructures in cold-rolled AA1050 (see Fig. 1) and more specifically in 70% cold-rolled AA1050. The present works focuses on the numerical approach, and therefore the experimental details will not be discussed here. However, this method will be incorporated with the experimental methods in near future, to remove the uncertainties in the modelling.

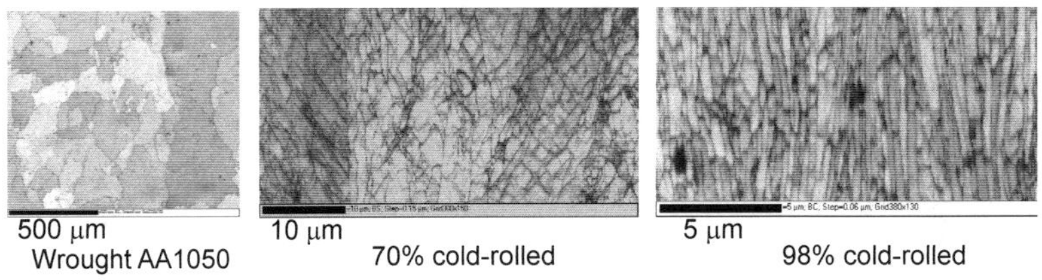

Figure 1. Typical wrought AA1050 microstructure and for 70% and 98% cold-rolling levels.

2. Finite element analysis of micro-indentation

Fig. 2a shows schematically the Berkovich indenter used in the indentation tests. The angle ψ was equal to 65.3^o. A three-dimensional finite element model was built using the finite element program Abaqus. Since a three-dimensional model is computationally expensive even if one sixth of the indenter is modelled due to the three symmetry planes, it is not practical for reverse analysis. Therefore an equivalent axisymmetric finite element model was also built. The equivalent axisymmetric model assumes that the indenter is conical (Fig. 2b) and the angle θ, is given by [10] in order to have the same projected area:

$$\theta = \tan^{-1}\left(\sqrt{\frac{3\sqrt{3}}{\pi}\tan\psi}\right) = 70.3^o \tag{1}$$

A second axisymmetric finite element model with θ equal to ψ [10] was also built. In both the three-dimensional model and the two axisymmetric models, the indenter was assumed to be rigid (modelled as analytical surface). Hard frictional contact was modelled between the indenter and the indented material allowing finite sliding but not interpenetration. The contact accounts for the indenter pile-up or sink-in.

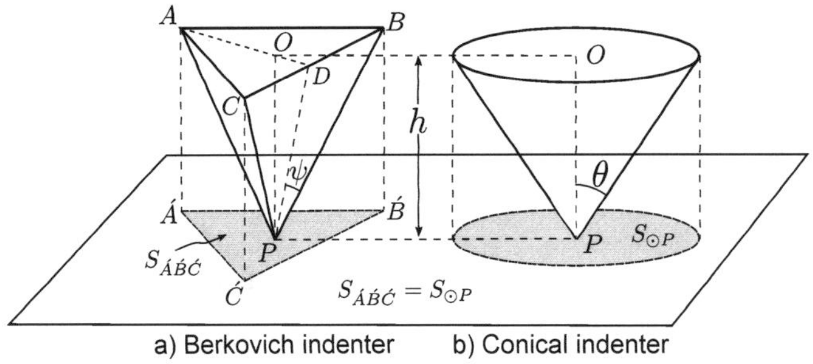

Figure 2. Schematic diagram of the Berkovich indenter with indenter angle ψ, and its equivalent conical indenter with indenter angle θ.

Fig. 3 shows schematically the axisymmetric finite element model with the boundary and loading conditions. The length, L, and the height, H, were large to avoid edge effects. Displacement boundary conditions were applied to reference point (RP), which controls the movement of the indenter in the z-direction during loading and unloading. Since the exact tip geometry of the Berkovich indenter was unknown or difficult to be measured, several tip geometries were modelled: a) tip with sharp edges, and b) tip with a fillet of radius r as shown in Fig. 3. The boundary and loading conditions for the three-dimensional model are analogous and they are not presented here.

A finite element model to simulate and extract the mechanical properties of the bulk cold-rolled aluminum was built (see Section 3.1).

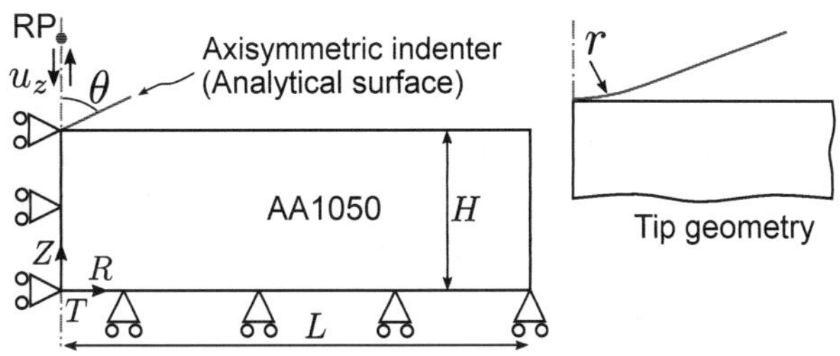

Figure 3. Schematic illustration of the axisymmetric finite element model with boundary and loading conditions.

It was assumed that the indented material can be described by classical isotropic J2 plasticity theory, and that the stress-strain behaviour can be approximated by a power law description:

$$\sigma = \begin{cases} E\varepsilon, & \text{for } \sigma \leq \sigma_y \\ \sigma_o \varepsilon^n, & \text{for } \sigma > \sigma_y \end{cases} \tag{2}$$

3. Results and Discussion

3.1. Tensile stress-strain behaviour

Fig. 4 shows a typical stress-strain curve of 70% cold-rolled AA1050. From the slope of the initial part of the stress-strain curve, the Young's modulus, E, was calculated equal to 63.44 GPa. The tensile experiment was modelled in the Abaqus finite element model assuming that the material follows an elastic-plastic response described by ten points (stress σ, plastic strain ε_{pl}). The first point represents the yield stress, σ_y, of the bulk sample. Then, the Isight optimisation platform was used, in combination with Abaqus, to optimise the modelled elastic-plastic material law. It was found that the yield stress was equal to 79.72 MPa. The elastic modulus and the ten points (σ, ε_{pl}) from the optimisation were used as the starting values in the indentation finite element models e.g. to obtain the initial power law (Eq. 2). This elastic-plastic material law is called here the "reference" material law.

3.2. Indentation response

First, a comparison between the three-dimensional model using a Berkovich indenter and the axisymmetric models (conical indenter) was performed using the "reference" material law (Section 3.1). As can be seen from Fig. 5, when the angle θ is calculated from Eq. 1, the load-displacement curve from axisymmetric model agrees well with the load-displacement curve from the three-dimensional model. As can be seen in Fig. 5 a mesh sensitivity analysis was also performed to increase the confidence in the obtained results.

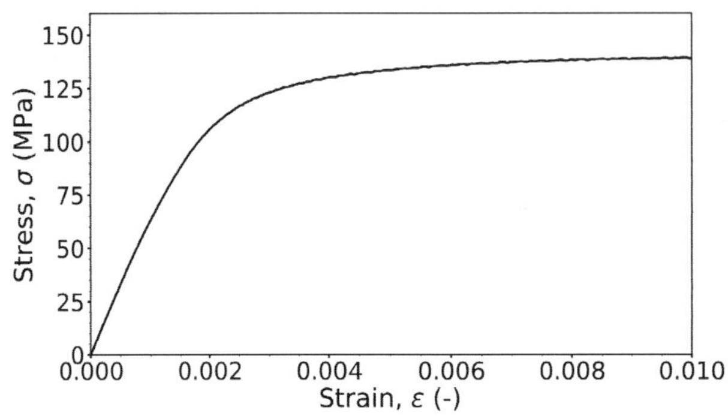

Figure 4. Experimental tensile stress-strain curve of 70% cold-rolled AA1050.

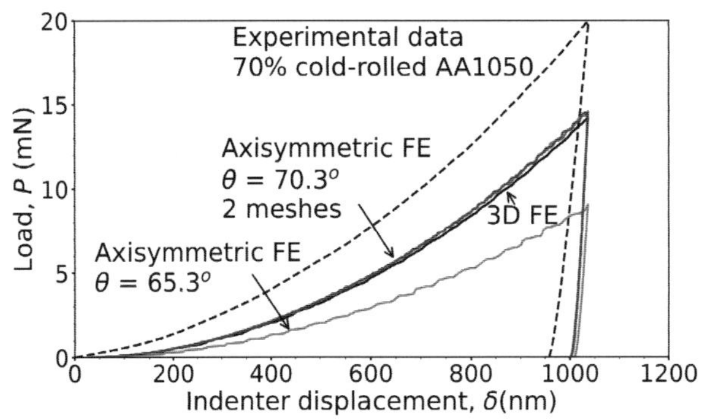

Figure 5. Comparison between 3D finite element (FE) predictions, with a Berkovich indenter, and axisymmetric FE predictions with two different indenter angles θ. The friction coefficient, μ, is equal to 0.1 and the indenter has sharp edges. Fig. 4 was used to extract the elastic-plastic behaviour of AA1050.

It can be seen from Fig. 5 that in all cases, the predicted load-displacement curves, using the "reference" material law, are lower compared to the experimentally obtained load-displacement curve. It should be mentioned that in these simulations the indenters had sharp edges and that the coefficient of friction, μ, was equal to 0.1.

Once it was determined that θ should be equal to 70.3^o, several indentation simulations were performed by varying the elastic-plastic response of the cold-rolled AA1050 in order to have a better agreement with the experimental response. In all cases, it was not possible to match the initial part (for indentation depths approximately up to 200 nm) of the load-displacement curve. A better agreement was found for indenters with rounded tip and as can be seen from Fig. 6. When r is equal to 2 μm the predicted initial part of the load-displacement curve agrees well with the experimental data. It should be mentioned that for larger indentation depths, r should not influence the load-displacement curve.

The effect of the friction coefficient is shown in Fig. 7 for indenters with both sharp and rounded edges. As can be seen the effect of friction is negligible.

Having determined the fillet radius, r, and setting mu equal to 0.1, the elastic-plastic response

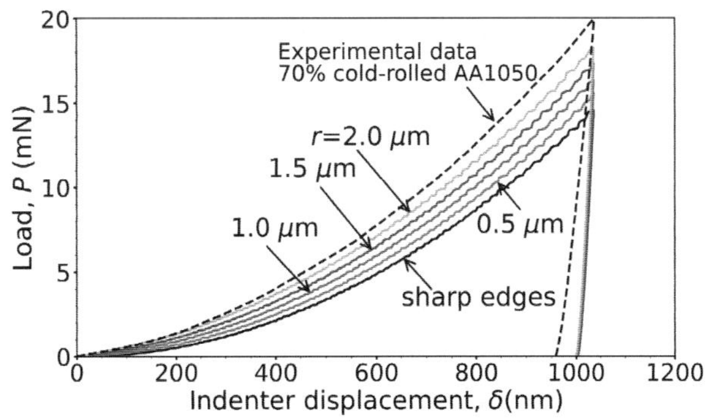

Figure 6. The effect of filler radius, r, on the indentation response using the axisymmetric FE model of Fig. 3 with θ equal to 70.3^o. The friction coefficient, μ, is equal to 0.1. Fig. 4 was used to extract the elastic-plastic behaviour of AA1050.

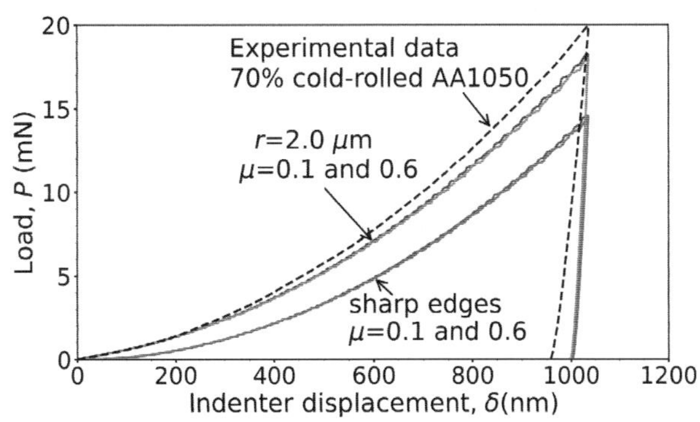

Figure 7. The effect of friction, μ, on the indentation response using the axisymmetric FE model of Fig. 3 with θ equal to 70.3^o. The indenter has sharp edges or a rounded tip with a fillet of radius, r, equal to 2 μm. Fig. 4 was used to extract the elastic-plastic behaviour of AA1050.

of the 70% cold-rolled AA1050 was iteratively modified until a good agreement was found between the numerical predictions and the experimental results. Fig. 8 shows some typical results. It can be seen that when σ_y is equal to 111.2 MPa and n is equal to 0.155, a good agreement is found.

The iterative numerical process showed that other material property combinations showed also a good agreement with the experimentally measured load-displacement curve e.g. for E=63.44 GPa, σ_y=109.24 MPa and n=0.16. Even, when E=60.0 GPa, σ_y=86.83 MPa and n=0.22 the agreement with the experimental response is reasonable. These findings are in agreement with the findings that the elasto-plastic parameters, E, σ_y and n cannot be uniquely obtained from the load displacement curves [7, 11] unless for example two or more indenters with different geometry are used. Here, by assuming that E is the same or close to the Young's modulus of the bulk samples, σ_y and n can be estimated with a certain degree of confidence. However, the procedure presented above has to be repeated in order to verify this claim. At

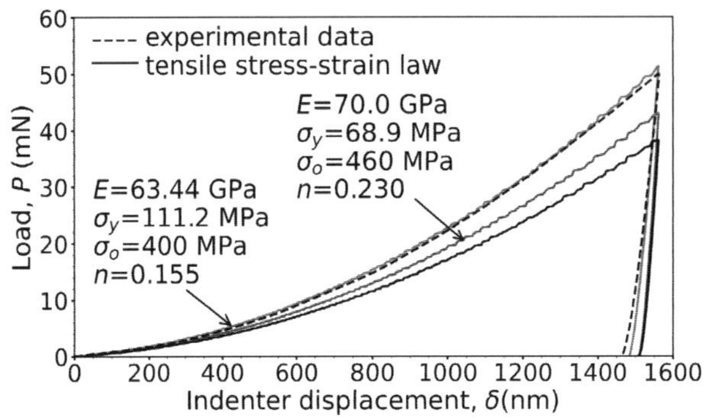

Figure 8. Indentation response using two different stress-strain curves and comparison with the experimentally measured indentation response. The friction coefficient, μ, is equal to 0.1, the filler radius, r, is equal to 2 μm and the indenter angle, θ is equal to 70.3^o.

the same time, the experimental methods to determine the local stresses and strains [12, 13]are also to be further explored, taken into account the internal microstructure and external testing effects.

4. Conclusions

In the present work a numerical iterative approach was presented in order to obtain the elasto-plastic parameters of metals at a small scale from indentation experiments. With a certain degree of confidence, the plastic parameters were estimated. However, using load-displacement curves from the indentation tests using a single indenter geometry does not lead to a unique solution. It is suggested that indenters with different geometries should be used or obtain additional topographic information form the experiments.

Acknowledgements

H. Pan and X. Zhang acknowledges the support from the European Research Council under the European Union Horizon 2020 research and innovation program (grant agreement No 788567-M4D).

References

[1] Zhang X, Hansen N, Gao Y and Huang X 2012 *Acta Materialia* **60** 5933–43
[2] Lu K and Lu J 2004 *Materials Science and Engineering: A* **375-377** 38–45
[3] Dai K and Shaw L 2008 *International Journal of Fatigue* **30** 1398–1408
[4] Hughes D A and Hansen N 2001 *Physical Review Letters* **87**(13) 135503
[5] Rigney D, Naylor M, Divakar R and Ives L 1986 *Materials Science and Engineering* **81** 409–425
[6] Tao N R, Wang Z B, Tong W P, Sui M L, Lu J and Lu K 2002 *Acta Materialia* **50** 4603–16
[7] Tho K K, Swaddiwudhipong S, Liu Z S, Zeng K and Hua J 2004 *Journal of Materials Research* **19** 2498–2502
[8] Giannakopoulos A E and Suresh S 1999 *Scripta Materialia* **40** 1191–98
[9] Zeng K and Chiu C H 2001 *Acta Materialia* **49** 3539–51
[10] Shi Z, Feng X, Huang Y, Xiao J and Hwang K C 2010 *International Journal of Plasticity* **26** 141–8
[11] Cheng Y T and Cheng C M 1999 *Journal of Materials Research* **14** 3493–96
[12] Zhang X, Nielsen C V, Hansen N, Silva C M A and Martins P A F 2019 *International Journal of Plasticity* **115** 93–110
[13] Pan H, He Y and Zhang X 2021 *Materials* **14** 1012

Effect of inclusion type on the rotating bending fatigue properties of a high carbon chromium bearing steel

Zhiyue Shi[1,2], Wenquan Cao[1]*, Chengjia Shang[2] and Xiaodan Zhang[3]

[1] Central Iron and Steel Research Institute (CISRI), Beijing 100081, China
[2] School of Materials Science and Engineering, University of science and technology of Beijing, Beijing 100083, China
[3] Section of Manufacturing Engineering, Department of Mechanical Engineering, Technical University of Denmark, 2800 Kgs, Lyngby

E-mail: caowenquan@nercast.com

Abstract. The use of the double vacuum melting route (conventional second refining + vacuum arc refining, CSR+VAR) and the electroslag remelting route (vacuum induction melting + electroslag remelting, VIM+ESR)) has been investigated for the SAE52100 (100Cr6) bearing steel. The tensile properties, impact toughness, hardness and RCF life of the CSR+VAR steel and the steel prepared using the VIM+ESR route are similar. The number and size of TiN inclusions in the VIM+ESR steel are higher and larger than those in the CSR+VAR steel, providing an explanation for the observation of better fatigue properties for the CSR+VAR steel. Two major types of inclusions, magnesium aluminates/Al2O3-CaO-CaS and TiN, are located in different areas in a map of stress intensity factor versus rotatory bending fatigue cycles (N_{RBF}). The negative impact of TiN inclusions on fatigue properties are greater than those of magnesium aluminates/Al2O3-CaO-CaS inclusions.

1. Introduction

Bearings work as key precision components in a broad scope of machinery and gearboxes, and are vital machine elements that constrain relative motion to a desired error and reduce friction between moving parts [1]. During their service life, bearings are always subjected to complex multi-axial stress states that change periodically [2]. For this reason, the fatigue properties of bearing steel are directly related to the service life of bearings. Bearing steel fatigue can be mainly divided into three stages: crack initiation, crack propagation and instability propagation [3,4], where crack initiation and propagation are the key processes [5,6]. Based on this, many researchers have sought to improve the fatigue properties of bearings by improving the metallurgical treatment and yield strength, and by controlling the surface state and size effect of components [7]. From the view point of materials design, controlling the number and size of inclusions, surface carburizing (or nitriding) and refining the grain size are the most effective methods to improve the fatigue properties of bearing steels [8,9]. Our previous studies have shown that reducing the average inclusion size from 27 μm to 13 μm through changing the refining process from basic oxygen furnace + ladder furnace + RH furnace to vacuum induction melting + vacuum arc refining can improve the rotating bending fatigue strength by about 8.5% and the rolling contact fatigue life by 4 times [10]. Different kinds of inclusions are present in bearing steel, mainly oxides, sulfides and TiN, but previous studies have regarded these inclusions as the same in terms of crack sources and have not classified them [11]. The shape, hardness and elastic modulus of different types of inclusions are different, and their influences on their fatigue properties are also different [12].

In this study, efforts have been made to produce the standard SAE 52100 bearing steel with different amounts of cleanliness, by using two different processing routes, namely conventional second refining plus vacuum arc melting (CSR+VAR), and vacuum induction melting plus electroslag remelting route (VIM+ESR). Rotatory bending fatigue (RBF) tests and inclusion characterization were carried out on samples for the two differently-processed SAE52100 steels to study the effect of different types of inclusions on fatigue properties.

2. Material and experimental procedure

High carbon SAE52100 bearing steel was fabricated either by CSR+VAR or VIM+ESR smelting and refining routes with the compositions listed in Table 1 (oxygen and nitrogen are determined by the pulse heating inert gas fusion-infra-red absorption method, and other elements are determined by spark discharge atomic emission spectroscopy). The oxygen, nitrogen and titanium content are about 3 ppm, 39 ppm and 9 ppm for the CSR+VAR steel, and 12 ppm, 53 ppm and 14 ppm for the VIM+ESR steel, indicating a different cleanliness. The oxygen, nitrogen and titanium content in the CSR+VAR steel are significantly lower than those in the VIM+ESR steel.

Table 1. Chemical compositions in wt.% of SAE52100 fabricated by the CSR+VAR and VIM+ESR processes

Steel	C	Si	Mn	P	S	Cr	Ti	N	O
CSR+VAR	1.04	0.23	0.25	0.007	0.004	1.51	0.0009	0.0039	0.0003
VIM+ESR	1.02	0.25	0.35	0.009	0.005	1.50	0.0014	0.0053	0.0012

The same heat treatment and mechanical tests have been applied to these two steels. All specimens for RBF tests and mechanical tests, such as tensile tests and hardness measurement, were austenized at 840°C for half an hour, followed by oil quenching, and finally by tempering at 170°C for 3 hours. Details regarding the fatigue tests, tensile testing and hardness measurement are reported elsewhere [10]. The fracture surfaces were examined by scanning electron microscopy (SEM, Hitachi S4300) with an energy dispersive X-ray (EDX) analyzer to investigate the corresponding property-related microstructural features and surface morphology. Inclusion parameters, such as the inclusion type, size and distribution, are quantitatively determined using an ASPEX inclusion analyzer [13].

3. Results

3.1. Mechanical properties and fatigue behaviour

The mechanical properties after heat treatment of the SAE52100 steels prepared using the CSR+VAR and VIM+ESR melting routes are shown in Table 2. It can be seen that the yield stress ($R_{p0.2}$), impact toughness (*Aku*) and Rockwell hardness (HRC) of these two steels are roughly the same, around 1770 MPa, 10 J and 59 HRC respectively. The ultimate tensile strength (R_m), the tensile elongation (A) and the area reduction (Z) of the SAE52100 CSR+VAR steel are better than those of the VIM+ESR steel.

Table 2. Hardness, impact toughness and mechanical properties of SAE52100 steels

Steel	Rockwell hardness (HRC)	Ak_u (J)	R_m (MPa)	$R_{p0.2}$ (MPa)	A (%)	Z (%)
CSR+VAR	59.3±1.0	9.7±2.0	2503±29	1767±17	3.3±1.0	5.3±1.0
VIM+ESR	59.5±1.0	10.0±2.0	2273±29	1775±1.0	1.2±1.0	4.7±1.0

The stress amplitude (σ_a) as a function of number of rotatory bending fatigue cycles to failure (N_{RBF}) of two groups of SAE52100 steels are plotted in figure 1, together with their relative inclusion sizes and types analyzed from the RBF fracture surfaces. It shows that the value of N_{RBF} increases with decreasing stress amplitude as a common trend, and the fatigue strength (σ_{-1}) of CSR+VAR and VIM+ESR steels measured and calculated by the up and down method are 1131 MPa and 1085 MPa, respectively. The inclusions and surfaces of the two groups of samples are the crack sources for fatigue failure. Through EDS analysis, it is found that the inclusion-induced crack initiation of these two groups of samples are mainly magnesium aluminates/ Al_2O_3-CaO-CaS inclusions and TiN inclusions. For the VIM+ESR steel, 9 specimens exhibited failure related to TiN inclusions, but for the CSR+VAR steel only one TiN-related failure was observed. For the CSR+VAR steel, under the stress amplitude of 1400MPa, 11 samples failed due to cracks initiated at surface defects, accounting for 69% of the total number of samples. This is understandable as higher stress intensity leads to preferred crack initiation and propagation from surface defects.

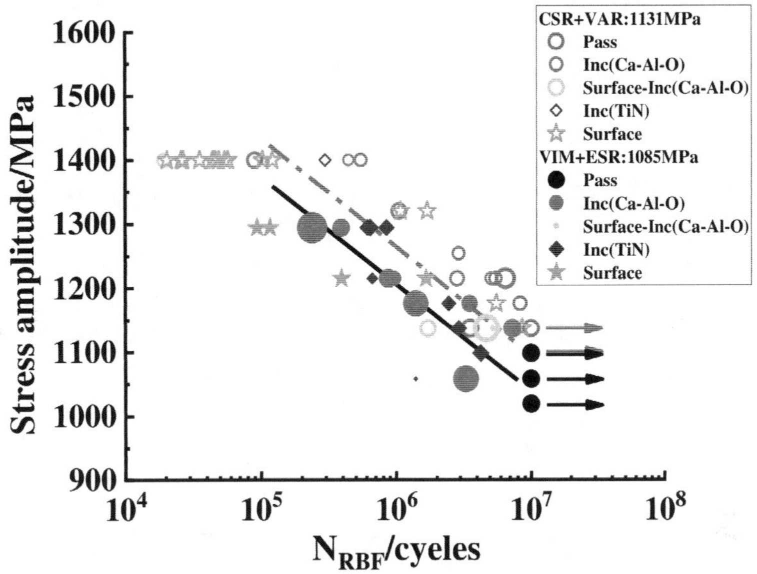

Figure 1. Stress amplitude as a function of rotatory bending cycles to failure for RBF. Red circles represent failures from magnesium aluminates/Al_2O_3-CaO-CaS inclusions, green circles represent failures from magnesium aluminates/Al_2O_3-CaO-CaS inclusions on the surface, boxes represent failures from TiN inclusions. In all cases the label size represents the relative size of inclusions.

3.2. Inclusion characterization and fracture topography

Typical inclusions of RBFed fracture samples are shown in figure 2 for both CSR+VAR and VIM+ESR steels. Figure 2(a) and 2(b) show the fatigue crack initiation caused by magnesium aluminates/Al_2O_3-CaO-CaS inclusions in the CSR+VAR steel, and figure 2(c) and 2(d) show the fatigue crack initiation caused by TiN inclusions in the VIM+ESR steel. Each fatigue fracture was characterized in detail by SEM. A RBF fracture surface can be divided into four characteristic zones from inside to outside, namely the inclusion zone, the fisheye zone, the open fracture zone, and the final fracture zone [14]. A schematic illustration of these four different zones and the definition of related parameters on the fracture surface of specimens are shown in figure 3(a). Values of representative parameters of the fracture surface zones, together with the RBF strength, are given in table 3. It is found that the inclusions for crack initiation are mainly the magnesium aluminates/Al_2O_3-CaO-CaS inclusions for the CSR+VAR steel, with sizes varying from 8.3 μm to 17.8 μm, and are mainly magnesium aluminates/ Al_2O_3-CaO-CaS and TiN inclusions for the VIM+ESR steel, with sizes varying from 3.4 μm to 25.6 μm. Interestingly, the average size of inclusions in the two groups of samples is similar, around 13 μm, but the difference in fatigue strength is 46 MPa. In order to set up the relationship between the parameters, D_{In}, d_{Fish}, and d_{open} are shown as a function of H_{Dep} in figure 3(b). It can be seen that the diameter of the fisheye zone increases linearly with the inclusion depth. For the inclusion on the surface or subsurface of the specimens, the specimens have no, or only a small fisheye, morphology. With increasing H_{Dep}, the value of d_{Fish} increases gradually until the inside of the fisheye coincides with d_{open}. It is interesting that: (i) the slope of the d_{Fish} line of the two groups of samples is basically the same; (ii) there are many inclusions near the surface of the sample, and most of them are TiN inclusions for the CSR+VAR steel; (iii) the size of TiN inclusions near the surface is smaller than magnesium aluminates/ Al_2O_3-CaO-CaS inclusions. Although the average inclusion size of the two groups of samples is the same, there is a big difference in fatigue strength, which may be related to the influence of different types of inclusions and their volume/area fraction on fatigue failure. This will be discussed in detail later.

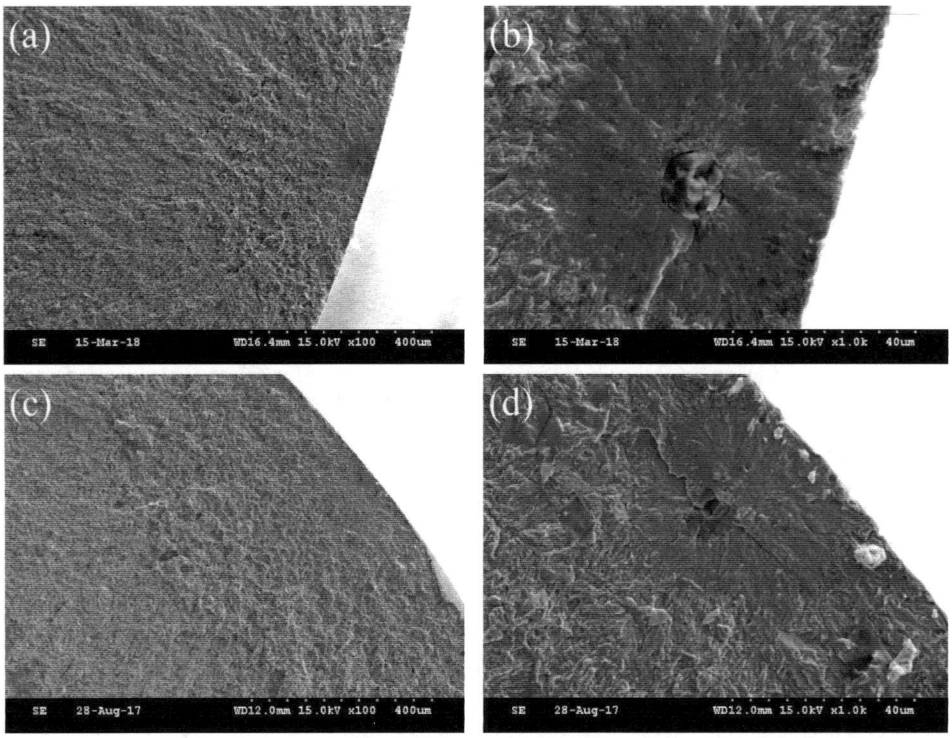

Figure 2. Fracture topography of RBF-failed specimens for both the CSR+VAR and VIM+ESR steel samples. (a) and (b) show the lower and higher magnification fracture topographies of the CSR+VAR failed specimen after 8,312,400 cycles under a stress amplitude of 1176 MPa; (c) and (d) show the lower and higher magnification fracture topographies of the VIM+ESR failed specimen after 665,700 cycles under a stress amplitude of 1215 MPa.

In order to understand the characteristics of different types of inclusions in the matrix, the inclusion density were examined using the Aspex Explorer microanalysis system by reconstruction diagrams of inclusion size, types and location, as shown in figure 4 (a) and (b). It can be seen that four types of inclusions are found in both the CSR+VAR and VIM+ESR steels, namely Al_2O_3-CaO-CaS and magnesium aluminates, titanium nitrides and other unclassified oxides. However, the size, number and distribution of the inclusions are very different. The total inclusion number in a measured area of 120 mm^2 is about 280 for the CSR+VAR steel and 948 for the VIM+ESR steel. Furthermore, compared with the VIM+ESR steel, the CSR+VAR steel has much fewer TiN inclusions. The statistical number density and area density of Ca-Al-O and TiN inclusions are shown in Table 4, where it is seen that the number density of TiN inclusions in CSR+VAR and VIM+ESR steels are 0.14 and 0.34 per square millimeter, respectively, and their corresponding area densities are 0.52×10^{-4} % and 6.08×10^{-4} % respectively. This differences can explain why the fatigue properties of the CSR+VAR steel are better.

Because the O and Ti content in VIM+ESR steel is higher than that in CSR+VAR steel, the number of TiN inclusions increases.

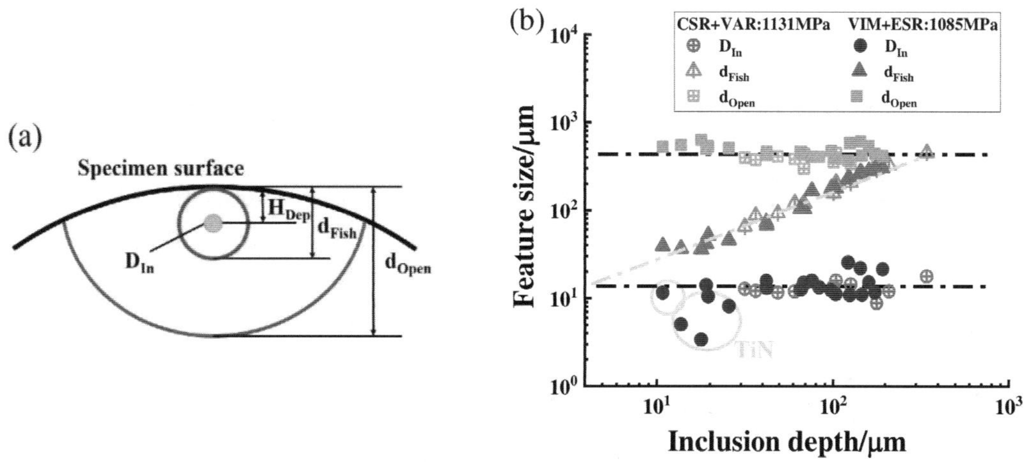

Figure 3. (a) Schematic illustration of RBF-failed specimens in showing the four zone parameters of fracture surface morphology as a function of inclusion depth; (b) variation of these parameters with inclusion depth for the CSR+VAR steel and the VIM+ESR steel.

Table 3. RBF-strength and fracture parameters measured from fracture surfaces

Steel	σ_{-1} (MPa)	D_{In} (μm)	H_{Dep} (μm)	d_{Fish} (μm)	d_{Open} (μm)
CSR+VAR	1131	8.3 – 17.8 (13.7±2.4)	31.5 – 344.1 (113.0±58.0)	64.8 – 453.7 (184.6±101.1)	308.3 – 456.3 (437.7±69.2)
VIM+ESR	1085	3.4 – 25.6 (13.3±5.2)	10.8 – 193.1 (83.2±58.0)	36.0 – 307.0 (150.5±101.1)	373.0 – 634.0 (482.4±74.2)

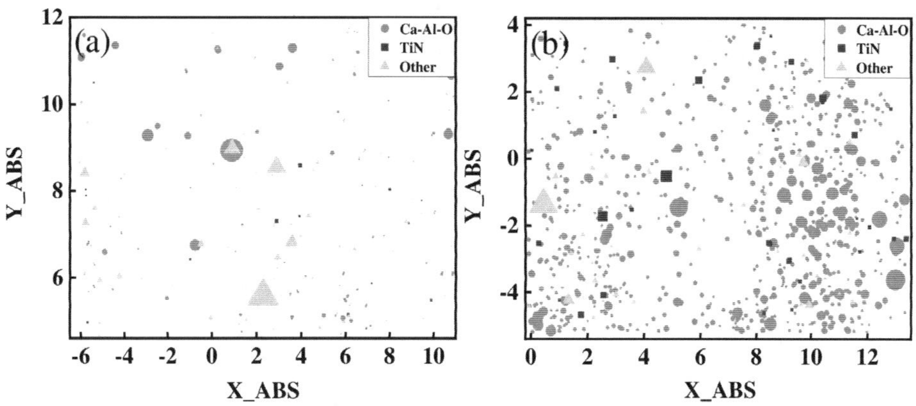

Figure 4. Characterization of inclusions in CSR+VAR and VIM+ESR SAE52100 steels using the Aspex Explorer: (a) the CSR+VAR steel, (b) the VIM+ESR steel.

Table 4. Statistics of number density and area density of CSR+VAR and VIM+VAR steels in an area of 120 mm^2 from the Aspex Explorer

steel	Number density (number/mm2)	Area density (%)

	Ca-Al-O	TiN	Ca-Al-O	TiN
CSR+VAR	1.16	0.14	2.14×10^{-3}	0.52×10^{-4}
VIM+ESR	7.16	0.34	8.76×10^{-3}	6.08×10^{-4}

4. Discussion

Through RBF fracture topography observation by SEM and EDX, it is found that although the average inclusion size of the two groups of samples is about 13 μm, better fatigue properties can be achieved from a lower volume/area fraction of inclusions. It is also interesting that the effects of different types of inclusions causing fatigue fracture may also be different. In order to study the effect of inclusion type on fatigue properties, the stress intensity factor (SIF) at inclusions were evaluated using the following equation [15]:

$$\Delta K = F\sigma\sqrt{\pi a}, \tag{1}$$

where F is the crack shape coefficient ($F = 0.65$ for surface defects and $F = 0.5$ for internal defects), σ is stress amplitude, a is the crack length ($a = \sqrt{area_{Inc}}$).

Figure 5 shows the SIF range for magnesium aluminates/Al2O3-CaO-CaS and TiN inclusions for the CSR+VAR and VIM+ESR steels. It can be seen that no matter what type of inclusion, ΔK_{Inc} decreases with increasing value of N_{RBF}. However, the two inclusion types of magnesium aluminates/Al_2O_3-CaO-CaS and TiN are not found in the same area, and TiN inclusions lie below magnesium aluminates/Al2O3-CaO-CaS inclusion. In other words, TiN inclusion can cause crack initiation and propagation under a smaller SIF. Combined with the morphology of these two types of inclusions observed on fatigue fracture and the Aspex Explorer observations, magnesium aluminates/ Al_2O_3-CaO-CaS inclusions are mainly spherical, while TiN inclusion are mainly square with obvious edges and corners, which is more likely to lead to stress concentration and fatigue crack initiation [16]. A more detailed and systematic investigation of the crack initiation and early propagation will be investigated in future using advanced in-situ SEM and TEM mechanical testing methods [17], with a focus on the interactions between dislocations and boundaries [18]. From the perspective of the smelting process, in addition to reducing the content of O in the matrix, it is also more important to reduce the content of Ti and N in the matrix for the improvement of fatigue properties.

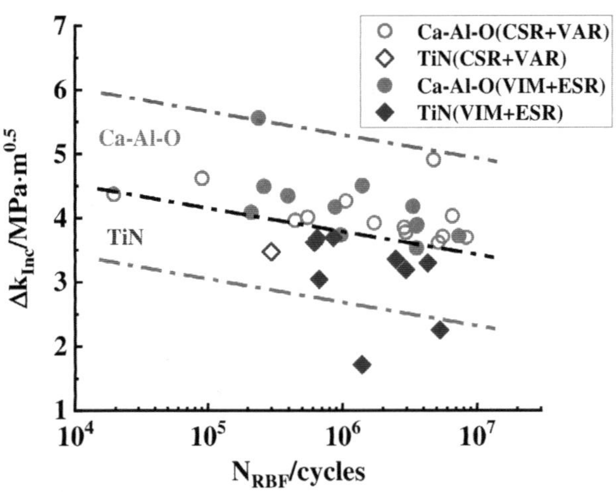

Figure 5. Experimental relationships between the stress intensity factors of D_{In} and the number of cycles to failure, N_{RBF}

5. Conclusions

Two groups of SAE52100 bearing steels have been prepared by the CSR + VAR and the VIM + ESR routes, separately. The fatigue behaviour and mechanical properties of these two steels have been tested, and the inclusions have been analysed by SEM and the Aspex Explorer system. The conclusions can be drawn as follows.

- The O, Ti and N contents of the SAE52100 bearing steel melted by the CSR+VAR process are 0.0003%, 0.00039 and 0.0009% respectively, which are significantly less than the values of 0.0012%, 0.00053 and 0.0014% in the VIM+VAR steel.
- Compared with the VIM+ESR route, the tensile properties, impact toughness, hardness and RCF life of the CSR+VAR steel are similar. At the same time, the RBF fatigue strength (σ_{-1}) of the CSR+VAR steel is 1131 MPa, 46 Mpa higher than the VIM+ESR steel.
- The number and size of inclusions in the VIM+ESR steel are larger than those in the CSR+VAR steel, which can explain why the fatigue properties of the CSR+VAR steel are better.
- Two major types of inclusions, magnesium aluminates/Al2O3-CaO-CaS and TiN are located in the different areas of a map of stress intensity factor versus N_{RBF}. The negative impact of TiN inclusions on fatigue properties is higher than that of magnesium aluminates/Al_2O3-CaO-CaS inclusions.

Acknowledgements
X. Zhang acknowledges the support from the European Research Council under the European Union Horizon 2020 research and innovation program (grant agreement No 788567-M4D).

References
[1] Bhadeshia H K D H 2012 *Prog. Mater. Sci.* **57** 268–435
[2] Campbell J 2017 *Res. Int.* **88** 1–13
[3] Ugi U E S 1988 *Isij Int.* **28** 893–9
[4] Sippel J P and Kerscher E 2020 *Appl. Sci.* **10** 1–27
[5] Muramoto S, Matsunaga H, Moriyama S and Endo M 2008 *Key Eng. Mater.* **387** 449–52
[6] Lu L and Shiozawa K 2011 *Fatigue Fract. Eng. Mater. Struct.* **42** 781–90
[7] Bomas H, Bacher-Hoechst M, Kienzler R, Kunow S, Loewisch G & M 2008 *Fatigue Fract. Eng. Mater. Struct.* **40** 7–11
[8] Cao Z, Liu T, Yu F, Cao W, Zhang X and Weng Y 2020 *Int. J. Fatigue* **131** 105351
[9] Cao Z, Shi Z, Yu F, Sugimoto K ichi, Cao W and Weng Y 2019 *Int. J. Fatigue* **128** 105176
[10] Cao Z, Shi Z, Liang B, Zhang X, Cao W and Weng Y 2020 *Int. J. Fatigue* **140** 105854
[11] Sakai T, Oguma N and Morikawa A 2015 *Fatigue Fract. Eng. Mater. Struct.* **38** 1305–14
[12] Fu H, Rydel J J, Gola A M, Yu F, Geng K, Lau C, Luo H and Rivera-Díaz-del-Castillo P E J 2019 *Int. J. Fatigue* **129** 104899
[13] An X, Shi Z, Xu H, Wang C, Wang Y, Cao W and Yu J 2021 *Metals (Basel).* **11**
[14] Shi Z, Wang H, Gao Y, Wang Y, Yu F, Xu H, Zhang X, Shang C and Cao W 2022 *Fatigue Fract. Eng. Mater. Struct.* 1995–2009
[15] Wu X 2017 *Scr. Mater.* **133** 113
[16] Wang P, Wang B, Liu Y, Zhang P, Luan Y K, Li D Z and Zhang Z F 2022 *Scr. Mater.* **206** 0–5
[17] Zhang X, Godfrey A, Winther G, Hansen N and Huang X 2012 *Mater. Charact.* **70** 21–7
[18] Pan H, He Y and Zhang X 2021 *Materials (Basel).* **14** 1–48

IOP Conf. Series: Materials Science and Engineering 1249 (2022) 012033 doi:10.1088/1757-899X/1249/1/012033

Synthesis, characterization, and dealloying of Al-Cu alloys prepared using spark plasma sintering

S Sun, C F Zhang, W Q Gao and A Godfrey

Key Laboratory of Advanced Materials (MOE), School of Material Science and Engineering, Tsinghua University, Beijing 100084, P.R. China

E-mail: awgodfrey@mail.tsinghua.edu.cn

Abstract. A series of Al-Cu alloys were prepared using spark plasma sintering by adjusting the initial composition ratio and powder sizes. The resulting phase and microstructure in these samples were investigated by X-ray diffraction, as well as energy-dispersive spectroscopy and electron back-scatter diffraction of samples examined in the scanning electron microscope. Both the initial composition ratio and powder size affect the phase composition and phase distribution of the sintered samples. Smaller size powders lead to formation of near-equilibrium phase structures, while larger initial powders lead to metastable microstructures containing phases that deviate significantly from the phase diagram. Dealloying of a metastable microstructure was observed to result in the formation of hierarchical skeleton nanoporous Cu structure.

1. Introduction

In recent years there has been increasing interest in nanoporous metals prepared by dealloying of a precursor alloy, in which one phase or element is selectively removed by a controlled reaction. The three-dimensional bi-continuous structure of such metals provides high surface area, and high specific strength together with good thermal and electrical performance [1]. Applications for nanoporous metals include high-efficiency catalysts [1,2], surface-enhanced Raman scattering (SERS) [3, 4], and actuators [5, 6]. In addition to applications in the functional field, it has also been found that as the ligament size approaches the nanoscale, the mechanical properties deviate from the Gibson-Ashby relationship associated with traditional porous materials, with the strength of single ligaments postulated to approach theoretical strengths as the ligament size is reduced [7,8]. Among the many possible nanoporous metal systems, Cu has been widely studied due in part to the relatively low cost of precursor alloys, and to its high catalytic potential [9, 10]. Precursor systems for nanoporous copper that have been studied include Al-Cu [11,12], Mn-Cu [13,14], and Zn-Cu [15,16]. Typically, these dealloying precursors are prepared by a process of melting and homogenization annealing [13,16,17]. There is, however, still a lack of research on the dealloying of metastable microstructures.

Spark plasma sintering (SPS) can be used to obtain densified bulk metal through application of a pulsed high amperage direct current to a powder compact under a relatively low pressure. This technique is environmentally friendly allows some control of the final grain size of the obtained material after sintering [18]. Some studies have recently been carried out on the metastable Al-Cu alloy microstructure that can be achieved by SPS [19,20]. The influence of the choice of initial powder sizes was not considered, however, in these studies. Additionally, although it has been shown that dealloying of SPS-prepared Al-Cu alloys can give nanoporous structures with high porosity and adequate compressive properties [21], many details regarding the range of possible sintered microstructures and their evolution during the dealloying process remain unknown. In this study, we explore various equilibrium and non-equilibrium Al-Cu microstructures prepared by SPS by changing the initial powder size and the composition ratio of the Al and Cu powders. The microstructures of these samples have been characterized by a variety of methods, and an initial study has been carried out to explore the relationship between the sintered microstructure and the dealloying process.

2. Experimental

Spherical Al powder and Cu powder (average diameters of 1 μm and 30 μm for both metals) were used to prepare samples. The powders were mixed using a high-efficiency resonance mixer (LABSEM-1, CASC, China). Three samples using 1 μm Cu powder and 1 μm Al powder were prepared at mass ratios of Cu powder of 35%, 50% and 65%. In addition, a further three samples were prepared at a fixed Cu mass percentage of 65% using different powder size combinations (specifically 1 μm Al-30 μm Cu, 30 μm Al-1 μm Cu, and 30 μm Al-30 μm Cu). The mixed powders were sintered by SPS (SPS-1050, Sumitomo Heavy Industries Ltd., Japan). After an initial heating ramp a pressure of 30 MPa and a temperature of 500 °C were maintained for 30 min to allow full sintering of the sample and to allow diffusion mixing of the Al and Cu. After sintering, the sample was slowly furnace-cooled. During the sintering process, a vacuum below 40Pa was maintained.

X-ray diffraction (XRD) patterns for the sintered Al-Cu samples were obtained using an X-ray diffractometer (Empyrean, Malvern PANalytical, Netherlands). Mechanically polished samples were also characterized in a scanning electron microscope (SEM; Mirca 3LMH, Tescan, Czechia) and the composition of the samples was analyzed by energy dispersive spectroscopy (EDS). Flat surfaces of the samples were also prepared by cross-section polishing for electron back-scatter diffraction (EBSD) characterization. Dealloying of the 30 μm Al-30 μm Cu (65wt% Cu) alloy was carried out on mechanically polished samples using 0.5 mol/L sodium hydroxide solution. Dealloying was continued until no bubbles could be seen escaping from the sample surface. The surface and cross-section of the dealloyed sample was observed in the SEM. A cube of 3 mm side length was cut from the sample, then dealloyed before being compressed at a rate of 0.1 mm/min using the deformation stage attached to an Xradia 620 instrument (Carl Zeiss AG).

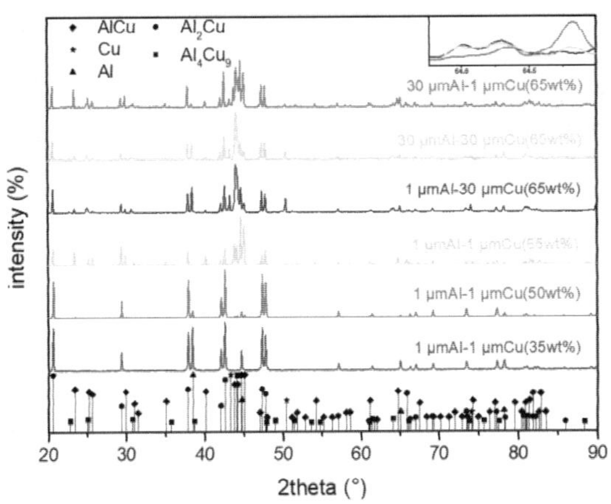

Figure 1. XRD profiles of the six SPS Al-Cu samples with black markers showing the positions of peaks corresponding to different phases (positions indicated on the bottom row). All peak intensities in the profiles are normalized.

3. Results and discussion

The XRD results of the SPS samples prepared from the Al-Cu powder combinations are shown in figure 1. Both the 1 μm Al-1 μm Cu (35wt%) and 1 μm Al-1 μm Cu (50wt%) samples contain Al and Al_2Cu, with higher Al_2Cu peak intensity in the latter. The sample prepared from the 1 μm Al-1 μm Cu (65wt%) mixture shows peaks only for the Al_2Cu and AlCu phases. For samples prepared at a fixed powder composition with at least one 30 μm powder peaks corresponding to Al, Cu, Al_2Cu, AlCu and

IOP Conf. Series: Materials Science and Engineering 1249 (2022) 012033 doi:10.1088/1757-899X/1249/1/012033

Al_4Cu_9 phases were detected. The Al and Cu peaks for the 30 μm Al-1 μm Cu (65wt%) sample are significantly weakened compared to those sintered using the 30μm copper powder, indicating a decrease in the proportion of these phases. The upper right inset of figure 1 shows that for the fixed mass fraction samples the Al_4Cu_9 peak intensity at 64.2° decreases as the AlCu peak intensity at 64.8° increases, indicating that part of the Al_4Cu_9 phase is replaced by AlCu.

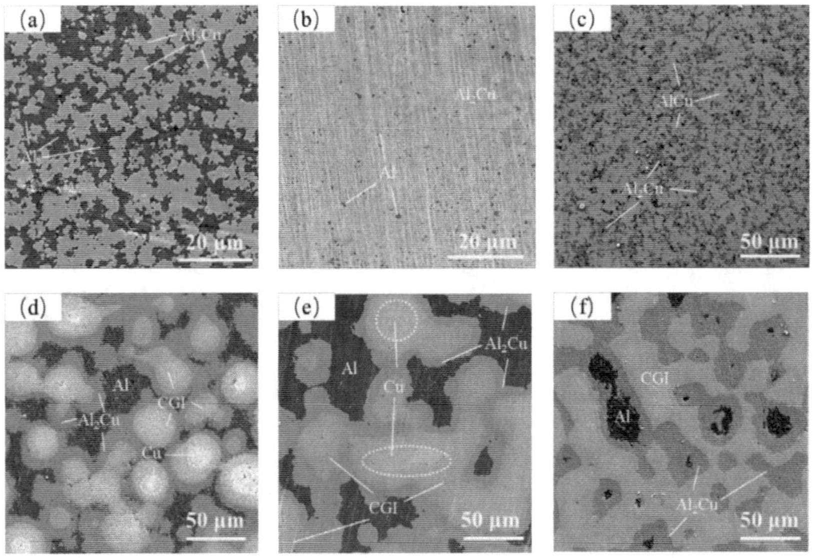

Figure 2. SEM micrographs showing the morphology of the sintered samples: (a) 1 μm Al powder and 1 μm Cu powder (35wt% Cu); (b) 1 μm Al powder and 1 μm Cu powder (50wt% Cu); (c) 1 μm Al powder and 1 μm Cu powder (65wt% Cu); (d) 1 μm Al powder and 30 μm Cu powder (65wt% Cu); (e) 30 μm Al powder and 30 μm Cu powder (65wt% Cu); (f) 30 μm Al powder and 1 μm Cu powder (65wt% Cu).

Table 1. Composition values (or ranges) as determined from EDS point scan measurements at positions based on SEM image contrast. (Values are given as the weight percentage of Cu)

Sample	Cu-rich	CGI layer	Al_2Cu	Al-rich
35wt% Cu/1μmAl-1μmCu	-	-	55%	7-27%
50wt% Cu/1μmAl-1μmCu	-	-	56%	8-32%
65wt% Cu/1μmAl-1μmCu	-	72%	55%	-
65wt% Cu/1μmAl-30μmCu	98%	70-84%	54%	3-11%
65wt% Cu/30μmAl-30μmCu	98%	76-86%	56%	4-15%
65wt% Cu/30μmAl-1μmCu	-	72-78%	56%	5-17%

Figure 2 shows the morphology of the sintered Al-Cu powder mixtures. The sintered microstructures of samples sintered from 1 μm powders in different mass ratios are shown in figure 2a-c. Due to the large difference between the atomic numbers of Al and Cu, the phases were identified based on a combination of composition contrast, combined with EDS point scans (values given in table 1) and the XRD results (figure 1). The Al-rich FCC phase and Al_2Cu phase are alternately distributed in the 35wt% Cu sample in approximately equal proportion. Due to the initial composition gradient, the Cu content in individual regions of the Al-rich phase can reach 27wt%, which is higher than the solid solubility of Cu in Al, such that the resulting Al phase is a supersaturated solid solution. In the 50wt% Cu sample, the area of the Al_2Cu phase is further increased, with a small amount of spherical Al-rich phase still present. The EDS measurements show that the degree of supersaturation in Al-rich areas is also further increased. In contrast, in the 65wt% Cu sample (prepared from 1 μm powders) an alternating arrangement of Al_2Cu and AlCu phases are present. According to the results of the Al-Cu phase diagram [22], except for some areas of inhomogeneous powder mixing, the phase composition obtained by SPS of 1 μm powders is close to the expected steady-state structure, albeit with some variations in solute content.

By comparing the results in figure 2c-f, the influence of the initial powder diameter on the sintered structure can be analyzed for a fixed nominal composition (65wt% Cu). Regions with four distinctive greyscale intensity are seen in the 1 μm Al-30 μm Cu sample (figure 2d). Combining the XRD results (figure 1) with the EDS measurements (table 1), these have been identified the Al-rich FCC phase (darkest), the Al_2Cu phase, intermetallics with a gradient in Cu content (Cu-gradient intermetallics, CGI), and the Cu-rich FCC phase (lightest). The same phases are found in the 30 μm Al-30 μm Cu sample (figure 2e), though the contrast between the Cu and CGI phases is weaker. Kim et al. [20] sintered Al-Cu to obtain kinetically preferentially nucleated and grown Al_2Cu and Al_4Cu_9 phases. In our experiments a longer sintering time was used, promoting diffusion, and resulting in the formation of the CGI layer, and the AlCu phase, as detected in the XRD measurements. According to the Kirkendall effect, the faster diffusion rate of Al leads to the formation of intermetallic compounds in the Cu layer during reaction diffusion. The formation of intermetallic compounds reduces the compositional gradient, further slowing the diffusion of Al. In addition, for the 1 μm Al-30 μm Cu (65wt%) sample the native oxide film on the fine Al powder surface hinders diffusion of Al atoms into Cu so that the Al atoms cannot fully penetrate into the Cu-rich area and Cu-rich region surrounded by a CGI layer is obtained.

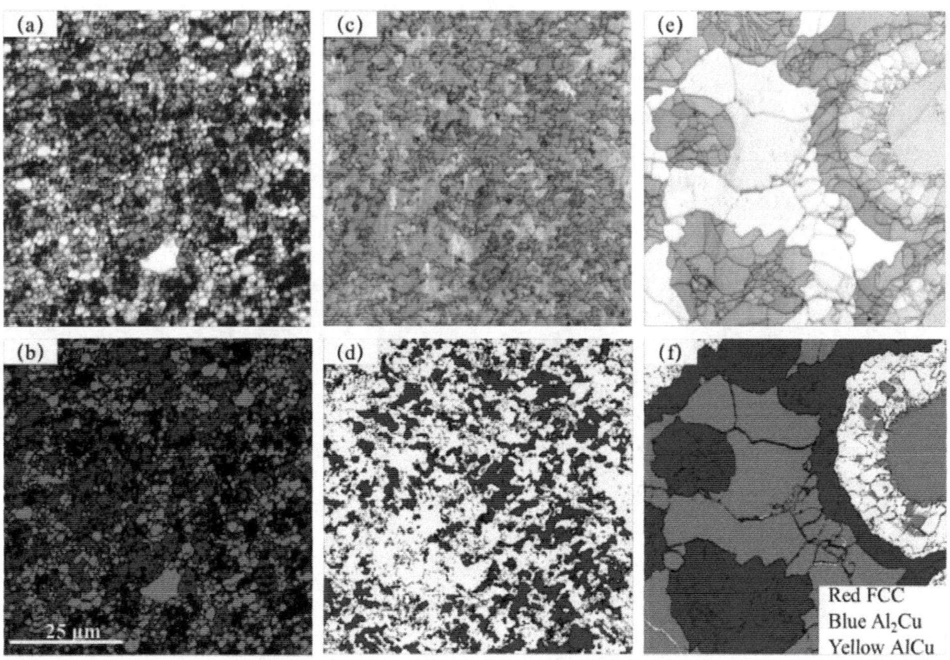

Figure 3. EBSD maps for selected Al-Cu mixed powder samples after sintering. The first row shows EBSD band contrast; the second row shows phase distribution maps: (a,b) 1 μm Al powder and 1 μm Cu powder (35wt% Cu); (c,d) 1 μm Al powder and 1 μm Cu powder (65wt% Cu); (e,f) 30 μm Al powder and 30 μm Cu powder (65wt% Cu).

For 30μm Al-1μm Cu (65wt%) sample, only regions with contrast corresponding to the Al-rich FCC region, the Al_2Cu phase, and the CGI region were observed in the SEM image (figure 2f). Although XRD showed that the Cu phase was still present in this sample, this phase could not be identified under

SEM examination. As Al atoms have a higher diffusion rate, the finer Cu powder results in shorter diffusion distances in this sample. Al atoms are also more prone to short-circuit diffusion through grain boundaries, resulting in a significantly reduced fraction of copper-rich FCC regions and a decrease in copper content in the CGI regions (table 1). This variation corresponds to the decreases in peak intensity in the Cu phase and Al_4Cu_9 phase seen in the XRD results (figure 1). The more dispersed copper powder results also in the formation of near-spherical Al-rich regions surrounded by an Al_2Cu layer. The short-circuit diffusion of Cu atoms additionally results in a reduction in the proportion of Al-rich FCC phases, while the relatively low diffusion rate and long diffusion distance of Cu leads to the retention of Al-rich regions. Accordingly, using 30 μm powder to sinter Al-Cu samples results in a non-equilibrium structures that deviate from those expected from the phase diagram.

Examples of the sintered samples as characterized by EBSD are shown in figure 3 based on both EBSD band contrast and indexing-based phase contrast. These maps also allow the grain structure within each phase to be probed. For the 1 μm Al powder and 1 μm Cu powder sample (35wt% Cu), the Al phase and Al_2Cu phase are alternately distributed with size and shape of the grains remain basically similar to the powder size (figure 3a,b). For the 1 μm Al powder and 1 μm Cu powder (65wt% Cu) an alternating distribution of Al_2Cu and AlCu phases is obtained, where both the AlCu and Al_2Cu grains are significantly larger than the powder size and are no longer equiaxed (figures 3c,d). In previous studies [23,24], grains both in pure Al and pure Cu have been shown to exhibit high thermal stability (resistance to grain growth) during the SPS process. It is speculated therefore that the formation of intermetallic compounds may lead to decreased grain boundary stability, though this should be studied further. Maps for the 30 μm Al powder and 30 μm Cu powder (65wt% Cu) sample are shown in figure 3e,f. The possible phases selected prior to EBSD scanning are FCC (Al, Cu), Al_2Cu and AlCu. Although the CGI layer is most likely not completely AlCu phase (in a preliminary study it was not possible to distinguish using EBSD between AlCu and Al_4Cu_9) the overall phase distribution and grain size can nevertheless be obtained by EBSD. Although both initial powders were 30 μm, the grains in the Al_2Cu phase and the CGI region are relatively small, with smallest grain size in the layer closest to the Cu phase. The results suggest that Al_2Cu nucleates and grows at the two-phase interface. With the progress of diffusion, the grains gradually grow toward the Cu-rich FCC region, and a composition gradient results in the formation of new Cu-rich intermetallic compound grains comprising the CGI layer.

Figure 4. Hierarchical skeleton nanoporous Cu structure obtained from dealloying of the 30 μmAl-30 μm Cu (65wt% Cu) sample: (a) dealloyed sample surface, (b,c) sample cross-section.

Sintered samples prepared from 30 μm Al powder and 30 μm Cu powder (65wt% Cu) were also dealloyed to obtain a hierarchical skeleton nanoporous structure, the morphology of which is shown in figure 5. The flat surface after dealloying is shown in figure 4a. Due to the removal of the Al-rich phase, a channel with a width of about 20 μm was formed. Under the dealloying conditions used here both the Cu-rich region and the CGI layer are not dealloyed. Therefore, these regions were retained, forming a spherical skeleton structure. The Al_2Cu phase was, however, dealloyed resulting in the formation of nanoporous copper. In figure 4b,c a large number of ≈ 1 μm diameter nanoporous copper spheres can be observed attached to the Cu-CGI skeleton, wherein the ligament size of the nanoporous Cu is in the range of 40 nm. A cube with a side length of 3 mm was prepared from this sample for compression testing. The maximum compressive strength of the sample was found to reach 45MPa with a compression strain before failure of 8%. Future experiments will explore further dealloying conditions as well as thermal coarsening with the goal of improving the compressive strength and ductility.

4. Summary and conclusions

A series of samples with different phase distribution microstructures were prepared under the same SPS process conditions by varying the initial size and the composition ratio of mixed Al and Cu powders. The initial powder size affects both the connectivity of the mixed powders and grain size, as well as the diffusion of atoms. In this study it was found necessary to consider the influence of the diffusion distance of each atom species, short-circuit diffusion along grain boundaries, reaction diffusion to form intermetallic compounds, and the relative diffusion rates of the two elements, as well as the effect in some cases of oxide films. It is worth noting that for metal sintering of multi-component or completely miscible systems, initial powder with too large a diameter will also lead to insufficient diffusion, resulting in a metastable microstructure. The present study also has some relevance for such systems. The use of metastable microstructures to obtain a hierarchical skeleton nanoporous structure provides a new idea for the design of nanoporous materials and will be explored further in future studies.

References

[1] Zhang Y, Sun X, Nomura N and Fujita T 2019 *Small* **15** e1805432
[2] Wittstock A, Zielasek V, Biener J, Friend C M and Baumer M 2010 *Science* **327** 319–22
[3] Chen L Y, Yu J S, Fujita T and Chen M W 2009 *Adv. Funct. Mater.* **19** 1221–6
[4] Li R, Liu X J, Wang H, Wu Y, Chu X M and Lu Z P 2014 *Corros. Sci.* **84** 159–64
[5] Jin H J and Weissmuller J 2011 *Science* **332** 1179–82
[6] Roschning B and Weissmüller J 2020 *Adv. Mater. Interfaces* **7** 24
[7] Biener J, Hodge A M, Hayes J R, Volkert C A, Zepeda-Ruiz L A, Hamza A V and Abraha F F 2006 *Nano Lett.* **6** 2379–82
[8] Xiang Y H, Liu L Z, Shao J C and Jin H J 2020 *Acta Mater.* **186** 105–15
[9] Huang J E et al 2021 *Science* **372** 1074–8
[10] Behrens M et al 2012 *Science* **336** 893–7
[11] Zhang Q and Zhang Z 2010 *Phys. Chem. Chem. Phys.* **12** 1453–72
[12] Liu W, Zhang S, Li N, Zheng J and Xing Y 2010 *J. Electrochem. Soc.* **157** D666
[13] Hayes J R, Hodge A M, Biener J, Hamza A V and Sieradzki K 2011 *J. Mater. Res.* **21** 2611–6
[14] Wang L, Xie H, Shao J-C, Yang H, Feng P-Z and Jin H-J 2021 *Scripta Mater.* **203** 114114
[15] Egle T, Barroo C, Janvelyan N, Baumgaertel A C, Akey A J, Biener M M, Friend C M, Bell D C and Biener J 2017 *ACS Appl. Mater. Inter.* **9** 25615–22
[16] Cheng I C and Hodge A M 2012 *Adv. Eng. Mater.* **14** 219–26
[17] Zandersons B, Lührs L, Li Y and Weissmüller J 2021 *Acta Mater.* **215** 116979
[18] Ali R et al 2021 *J. Alloy. Compd.* **889** 161531
[19] Kim D, Kim K and Kwon H 2021 *Materials* **14** 226
[20] Kim K, Kim D, Park K, Cho M, Cho S and Kwon H 2019 *Materials* **12** 1546
[21] Kong Q, Lian L, Liu Y and Zhang J 2014 *Mater. Lett.* **127** 59–62
[22] Funamizu Y and Watanabe K 1971 *Trans. Jpn. I. Met* **12** 147–52
[23] Le G M, Godfrey A, Hansen N, Liu W, Winther G and Huang X 2013 *Acta Mater.* **61** 7072–86
[24] Zhu K N, Ruan Q and Godfrey A 2015 *IOP Conf. Series: Mater. Sci. Eng.* **89** 012060

IOP Conf. Series: Materials Science and Engineering 1249 (2022) 012034 doi:10.1088/1757-899X/1249/1/012034

Laboratory-scale gas atomizer for the manufacturing of metallic powders

D C del Rio, D Juul Jensen, T Yu and N S Tiedje

Department of Civil and Mechanical Engineering, Technical University of Denmark, DK-2800 Kgs. Lyngby, Denmark

E-mail: dcdr@mek.dtu.dk

Abstract. Metallic powders for additive manufacturing (AM) processes are primarily produced by gas atomization, which consists of three steps: melting, atomization and cooling. In the present work, we report on the refurbishing of a laboratory-scale gas atomizer. The equipment facilitates small-scale atomization, useful for developing powders tailored specifically to metal AM processes (e.g. binder jetting, laser powder-bed fusion and direct energy deposition). The refurbished atomizer is operated by an in-house measurement and control system, fully equipped with pressure, oxygen, gas-flow and temperature sensors that allow the user to experiment with the input parameters, and thus, understand how they affect the physical and chemical properties of powders. In this paper, the working principle of the laboratory-scale gas atomizer is presented and the main characteristics of the newly refurbished equipment are described.

1. Introduction

The development of powder-based metal manufacturing processes such as metal injection molding and AM has a strong impact on the metallic powder production market [1,2]. Different methods for powder production exist. Most of the production techniques are based on atomization due to the ease with which the molten metal could be broken up into fine droplets, which will solidify into powder particles. Primary types of atomization processes are: jet atomization, melt spinning centrifugal atomization, rotating electrode and plasma transferred arc [1]. For powders used as feedstock for AM processes, jet atomization is the most used production process [3]. The atomization step during powder production in jet atomization, could be achieved using either gas, water or plasma torches as jets. Each process produces powder that might be different in size distribution and shape. Gas and plasma atomized powder particles have spherical shapes while water atomization generally leads to non-spherical shapes. Most metallic alloys available to AM are developed by gas atomization [4,5]. Spherical powders have higher packing density compared to non-spherical ones [1,6]. A higher packing density is often beneficial, as it will improve the theoretical density of the final component [7]. Along with higher packing density, spherical powders provide better flowability. Since gas and plasma atomization are usually performed under inert atmospheres, the presence of oxides in the powder is remarkably reduced compared to water atomization.

Particle size distribution (PSD) obtained by gas atomization is in the size range of 20–500 μm [8,9], which fulfills the PSD requirements for different AM methods such as binder jetting, laser powder-bed and laser powder-fed systems. Although gas atomization as a production process for metal powders, fulfills the requirements needed for being used as feedstock for AM powder-based processes, more efforts are needed to tailor powder properties for AM and not many grades of steel powders are commercially available. The varying quality of the powders currently produced, may be one of the reasons for the observed microstructural variations in metal AM components. In the present work, we refurbish a lab-scale gas atomizer with the purpose of creating a well-controlled atomization process, where the main factors affecting the process are measured and registered.

2. Experimental Set-up

Gas atomization consists of three main steps: melting, atomization and solidification. The melting takes place in an induction furnace located at the topmost of the equipment (figure 1). Once liquid phase is reached, the molten metal is allowed to flow down the melting chamber through a ceramic tube. Subsequently, the metal stream is broken into droplets by interaction with a high-velocity gas flow. The droplets are afterwards cooled and solidified. Atomization and solidification steps occur in the atomization chamber. Each sub-process affects the quality of the produced powder. The PSD of the powder are determined by two-phase turbulent effects between melt and gas flows and the heat transfer during solidification [10].

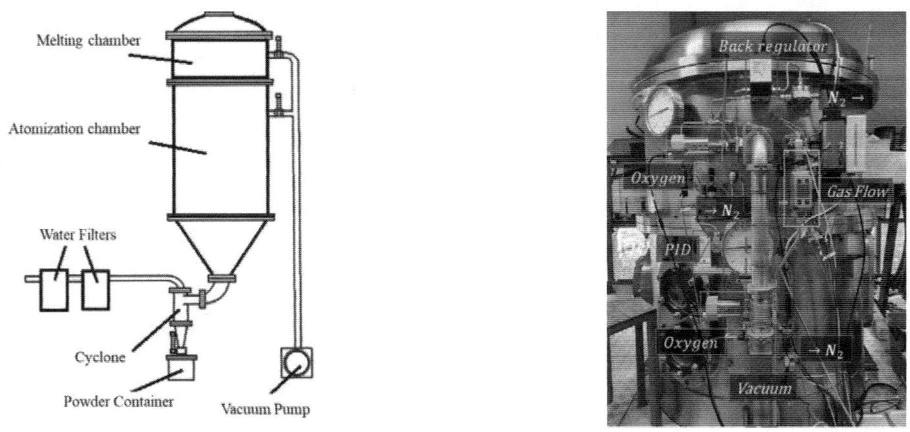

a) b)

Figure 1. Lab-scale gas atomization equipment. a) Schematic *and* b) Photograph with an indication of where the measurement sensors are installed to ensure efficient process control.

The median volume-weighted particle size diameter has been correlated with the opening of the ceramic pouring tube $'d_{melt}'$, as [11,12]:

$$d_{50} = d_{melt}K\sqrt{\frac{\eta_{melt}}{\eta_{gas}We}\left(1+\frac{J_{melt}}{J_{gas}}\right)} \quad (1)$$

where K is a numerical constant specific to the atomizer; η_{melt} is the kinematic viscosity of the molten metal; η_{gas} is the kinematic viscosity of the atomizing gas; J_{melt} and J_{gas} are the molten metal and atomizing gas flow rates and We is the Weber number defined by:

$$We = \frac{\rho_{melt}{v_{gas}}^2 d_{melt}}{\sigma_0} \quad (2)$$

where ρ_{melt} is the density of the molten metal; v_{gas} is the velocity of the atomizing gas at the moment of impingement with the liquid; and σ_0 is the surface tension of the melt. An schematic of the droplet formation proposed by See et al. [13] is shown in figure 2. After ligaments breakdown into droplets, the cooling and solidification step starts.

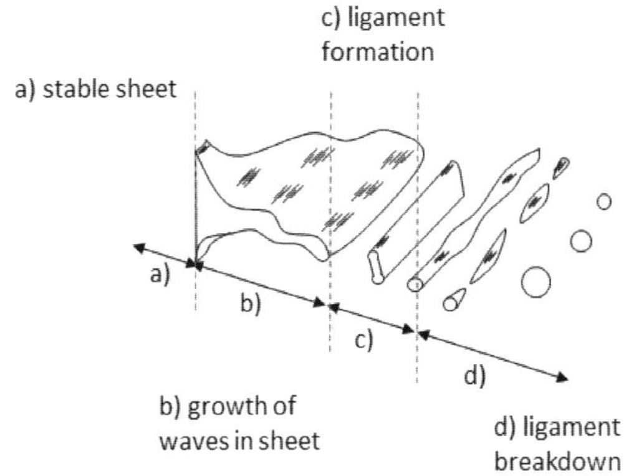

Figure 2. Disintegration of a liquid sheet by a high velocity jet of gas. (Adopted from See et al. [14]).

Based on the parameters that govern droplet formation, the atomizer is constructed such that d_{melt}, J_{melt}/J_{gas} and v_{gas} are monitored and adjustable.

Ceramic pouring tubes are cast in-house. This gives flexibility in varying d_{melt}. The length of the pouring tube is also adjustable. This affect the melt velocity and the mass flow, giving a longer distance of laminar stream before turbulence [12].

The kinematic viscosity, η_{melt}, and surface tension of the melt, σ_0, are parameters that can also be modified in order to affect the droplet formation. However, measuring those properties during the atomization process is rather complex. It is known that surface active elements, such as S and O_2, have a marked effect on the surface tension and viscosity of the melt, even at ppm levels [15,16]. During printing, recent studies documented a temperature dependent surface tension coefficient and its influence on the melt pool flow during laser AM processes [17,18]. These are reasons why measuring S and O_2 in the powder is important. For our experiments, the S content is analyzed in the material before melting by the infrared absorption method after combustion in an induction furnace (LECO Analyzer). Same measurement is performed for the atomized powder. The O_2 content in the melt and powder is affected by the atmosphere in both the melting and atomization chambers. Zirconium oxide flow-based O_2 sensors are installed (A19-P Angst+Pfister). One sensor is located in each chamber with a measurement range between 1 and 1000 ppm and a resolution of 1 ppm. The sensor locations are shown in figure 1b. Low O_2 concentrations are achieved by two means: Initially, a 0.01 mbar absolute pressure in the equipment is attained by pumping. Subsequently, a non-reactive gas is used to fill the chambers and to create an inert atmosphere, usually N_2 or Ar, which helps in dissipating O_2. In the described system, N_2 is used. The O_2 levels are measured and recorded during the whole atomization process. We use the oxygen concentration as a conditional limit for starting the atomization process. The program is set to start only if the O_2 levels are below 0.1%.

The mass flow rate ratio between the molten metal and the gas, J_{melt}/J_{gas}, can be adjusted in the system either by using a different atomizer and nozzle design or by increasing the gas pressure. The latter will also affect the velocity of the atomizing gas, v_{gas}.

The two commonly used atomizer designs are free-fall and confined/close-coupled design (figure 3). It is possible to obtain higher yields in small-to-medium sized particles (20–60 µm). By using a confined design, d_{50} will be smaller as J_{melt}/J_{gas} is reduced since the gas velocity at the interaction with the metal stream is maximized. The equipment is currently using the confined design. However, a free-fall configuration could be attained by changing crucible. This is possible thanks to the modular approach design of the atomizer. The same applies for the gas nozzle, where v_{gas} is affected by the type and shape of the nozzle hence different gas velocities and flows could be reached.

Gas flow J_{gas}, is measured at the pipe that supplies gas to the nozzle. The sensor installed is an inline flow meter that uses the thermal mass flow measurement principle for calculating the gas entering the system in m^3/min (EE741 E+E). A 93 m^3 battery pack at 200 bar is used to supply the gas needed for atomization. The pressure at the gas nozzle is measured by a pressure transducer positioned at the nozzle. The gas used to fill the melting and atomization chambers comes from the gas battery. The N_2 pressure at the melting chamber is controlled using a PID electronic pressure valve that regulates the gas supplied to the chamber. It contains a proportional valve with a flow restrictor that allows the user to set different N_2 partial pressures during melting. The overpressure is controlled by a back pressure regulator system that enables release of N_2 from the melting chamber and maintain the set point.

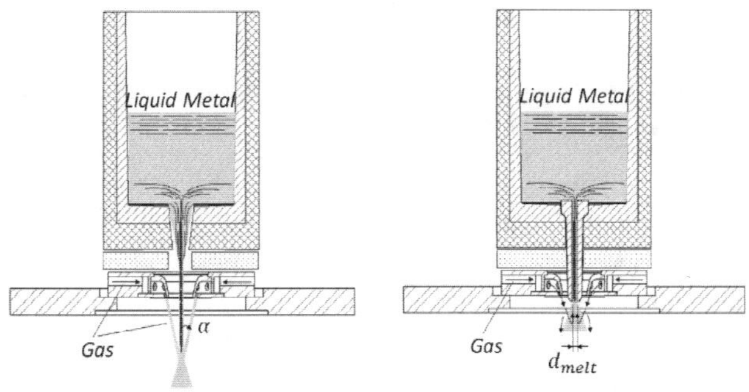

Figure 3. Atomizer designs: a) free-fall and b) confined/close-coupled design. Design characteristics: α, angle formed by free-falling molten metal and gas; d_{melt}, diameter of molten metal nozzle.

Pressure transducers are installed at each chamber; readings are collected and displayed by the control program during the entire process. Several thermocouples are mounted including one inside the stopper rod (alumina tube that prevents the melt to flow down to the atomization chamber before a pre-defined temperature is reached) so the melt temperature is known at any time. Additional melt temperature measurements are taken with an optical pyrometer located at the top of the melting chamber.

Another relevant parameter in the atomization process, is the melt superheat. Melt superheat influences the determination of the powder particle size as the melt surface tension, σ_0, is affected by the melt temperature [16]. During solidification, cooling rate, solidification time and nucleation are affected by the superheat in the metal droplet. The melt temperature is controlled in the process at the induction furnace (50 kW power unit), where the power and heating time are defined. The thermocouple located inside the stopper rod is continuously measuring the melt temperature so atomization is done when the predefined superheat is achieved. By being able to adjust the solidification of the particles, the powder microstructure can be tailored.

Liquid droplets form after ligament breakdown (figure 2). The cooling and solidification step starts during the droplet trajectory before being collected at the powder container. The heat from the spherical droplet is transferred towards the gas flow field and is determined by convection, radiation and release of latent heat (ΔH_f) arising from the solidification [19]. The first term in the right-hand-side of equation (3) indicates convection heat extraction whereas the second term address radiation cooling. The heat balance equation for an individual metallic particle is described as follows:

$$\frac{\partial Q}{\partial t} = \rho_{melt} V \left(c_{melt} \frac{\partial T}{\partial t} - \Delta H_f \frac{\partial f_s}{\partial t} \right) = A \left(h\left(T_{drop} - T_{gas}\right) + \varepsilon \sigma_{SB} \left(T_{drop}^4 - T_{surr}^4\right) \right) \tag{3}$$

where $\frac{\partial Q}{\partial t}$ is the heat flux; ρ_{melt} is the density of the droplet; V is the volume of the droplet; c_{melt} is the heat capacity of the droplet; $\frac{\partial T}{\partial t}$ is the cooling rate; $\frac{\partial f_s}{\partial t}$ is the solidification rate; A is the droplet surface area; h is the heat transfer coefficient between the droplet and the gas; T indicates temperature; ε indicates the emissivity of the droplet surface and σ_{SB} is the Stefan-Boltzmann constant.

The heat loss due to radiation cooling is as much as two orders of magnitude lower compared to convection [19]. Thus, heat loss at the droplet due to radiation can be neglected. Convective cooling is mostly affected by the heat transfer coefficient h given by:

$$h = \frac{k_{gas}}{d}\left(2.0 + 0.6\sqrt{\frac{d\rho_{gas}(v_{gas} - v_{drop})}{\eta_{gas}}}\sqrt[3]{\frac{\eta_{gas}c_{gas}}{k_{gas}}}\right) \quad (4)$$

where k_{gas} is the thermal conductivity of the atomization gas (W/(m·K)); d is the droplet diameter (m); ρ_{gas} is the density of the atomization gas (kg/m^3); v_{gas} and v_{drop} are the gas and droplet velocity (m/s); η_{gas} is the dynamic viscosity of the gas (mPa·s) and c_{gas} is the heat capacity of the gas (J/(kg·K)). From equation (4), the only parameters that are not related to the gas are the droplet diameter d and the relative gas-droplet velocity $v_{gas} - v_{drop}$. Therefore, cooling of droplets conditions can also be monitored by the sensors installed in the equipment as discussed above.

The gas atomizer uses a 1.5 L clay-bonded graphite crucible with alumina liners that prevent C diffusion towards the melt. It is possible to melt up to 10 kg of ferrous alloys per run. Absolute N_2 pressure of 2.5 bar could be reached inside the chambers. The process time for the complete process depends on the material to be produced, for 316L stainless steel, the process time is around 90 min. Once the atomization process is completed, the powder is collected in the powder container at the bottom of the atomizer (figure 1a). A cyclone is located between the atomization chamber and the powder container so the gas and big particles are removed. The gas is released to the atmosphere and the big particles are collected in the water filter. The powder container is sealed and handled inside a glovebox to avoid any contamination. Further categorization may be performed by sieving to obtain specific particle size ranges.

3. Conclusions

A lab-scale vacuum induction gas atomization facility was refurbished with the purpose of having full control over the atomizing parameters. The sensors and actuators installed allow the users to control and acquire data from the main parameters that influence powder quality. Establishing stable atomization conditions is a key factor to seek high quality metal powders tailored for AM processes. The equipment will provide experimental results that will help understanding the influence of the powder's physical and chemical properties on the printed components. The small batch/lab-scale equipment furthermore allows to carry alloy design experiments with close-to production conditions without the need of using large amounts of raw material. A modular lab-scale gas atomizer for metal powders is now available and open to collaborations.

Acknowledgement

This research work was supported by DFF-FTP Danmarks Frie Forskninsfond as the project, REPOWDER-Design of metal powders for 3D printing – from scrap to resource. Contract No. 9041-00335B. We thank Benny Edelsten, Lars Lorentzen and Bo Holdt-Simonsen for their support during construction of the equipment.

References

[1] Kassym K and Perveen A 2020 Atomization processes of metal powders for 3D printing *Mater Today Proc.* **26** 1727–33

[2] Baufeld B, Biest O Van der and Gault R 2010 Additive manufacturing of Ti-6Al-4V components by shaped metal deposition: Microstructure and mechanical properties *Mater. Des.* **31** S106–11

[3] Herzog D, Seyda V, Wycisk E and Emmelmann C 2016 Additive manufacturing of metals *Acta Mater.* **117** 371–92

[4] Sun Y, Hebert R J and Aindow M 2018 Non-metallic inclusions in 17-4PH stainless steel parts produced by selective laser melting *Mater. Des.* **140** 153–62

[5] Fayazfar H, Salarian M, Rogalsky A, Sarker D, Russo P, Paserin V and Toyserkani E 2018 A critical review of powder-based additive manufacturing of ferrous alloys : Process parameters , microstructure and mechanical properties *Mater. Des.* **144** 98–128

[6] Hausnerova B, Mukund B N and Sanetrnik D 2017 Rheological properties of gas and water atomized 17-4PH stainless steel MIM feedstocks: Effect of powder shape and size *Powder Technol.* **312** 152–8

[7] Koseski R P, Suri P, Earhardt N B, German R M and Kwon Y S 2005 Microstructural evolution of injection molded gas- and water-atomized 316L stainless steel powder during sintering *Mater. Sci. Eng. A* **390** 171–7

[8] Kassym K and Perveen A 2020 Atomization processes of metal powders for 3D printing *Mater. Today Proc.***26** 1727-1733

[9] Lawley A 1994 *Encyclopedia of Materials: Science and Technology.* ed Lawley A (Oxford: Elsevier) 387–93

[10] Antipas G 2012 Gas atomization of aluminium melts: Comparison of analytical models *Metals (Basel).* **2** 202–10

[11] Lubanska H 1970 Correlation of spray ring data for gas atomization of liquid metals *J Met.* **22** 45–9

[12] Rai G, Lavernia E and Grant N J 1985 Powder size and distribution in ultrasonic gas atomization *Jom* **37** 22–6

[13] Bruce See J and Johnston G H 1978 Interactions between nitrogen jets and liquid lead and tin streams *Powder Technol.* **21** 119–33

[14] Neikov O D 2019 *Atomization and Granulation* (Oxford: Elsevier Ltd.) 125-185

[15] Li Z, Mukai K, Zeze M and Mills K C 2005 Determination of the surface tension of liquid stainless steel *Journal of Materials Science.* **40** 2191–5

[16] Mills K C and Brooks R F 1994 Measurements of thermophysical properties in high temperature melts *Mater. Sci. Eng. A* **178** 77–81

[17] Aucott L, Dong H, Mirihanage W, Atwood R, Kidess A, Gao S, Wen S, Marsden J, Feng S, Tong M, Connolley T, Drakopoulos M, Kleijn C R, Richardson I M, Browne D J, Mathiesen R H and Atkinson H V. 2018 Revealing internal flow behaviour in arc welding and additive manufacturing of metals *Nat. Commun.* **9** 1–7

[18] Leung C L A, Marussi S, Atwood R C, Towrie M, Withers P J and Lee P D 2018 In situ X-ray imaging of defect and molten pool dynamics in laser additive manufacturing *Nat. Commun.* **9** 1–9

[19] Henein H, Uhlenwinkel V and Fritsching U 2017 *Metal Sprays and Spray Deposition* (Cham: Springer)

In-situ study of microstructural evolution and strain distribution during single axis tension of 316L stainless steel prepared by selective laser melting

W Y Wang, W Q Gao, S Sun and A Godfrey

Key Laboratory of Advanced Materials (MOE), School of Material Science and Engineering, Tsinghua University, Beijing 100084, P.R. China

E-mail: awgodfrey@mail.tsinghua.edu.cn

Abstract. A combined digital image correlation (DIC) and electron backscatter diffraction (EBSD) study of the multi-scale strain distribution and microstructure evolution during tensile loading has been carried on 316L stainless steel samples prepared by selective laser melting (SLM). The initial material shows a range of microstructural heterogeneity. Macro- and micro-DIC results reveal a non-uniform distribution of plastic strain over several length scales, including the formation of localized slip-bands at the intra-grain level. No obvious correlation is found between the heterogeneity in plastic strain and either grain size, grain morphology or the orientation of the cellular structures formed during SLM. Intragranular orientation subdivision and the evolution of local misorientation have also been investigated by EBSD during tension, where a complex pattern of heterogeneity in both microstructure and plastic strain is found.

1. Introduction

Metal additive manufacturing is a rapid prototyping technology using metal powders together with a 3D model, and has a wide potential for use in industry [1-3]. Near net-shape metal parts with near full density can be directly prepared by fully melting metal powders using a laser beam, followed by solidification of the melt pools in a layer-by-layer manner. One specific metal additive manufacturing technique is selective laser melting (SLM), characterized by a high temperature gradient (~10^5 K/m), high cooling rate (~10^6 K/s), and small size of melt pools (~100 μm) [4]. The high laser scanning speed (usually over 400 mm/s), together with the high solidification and cooling rates during the printing process typically results in violent melt pool convection, and as such the rapid solidification process during SLM is very different from that of traditional casting processes. Metals prepared by SLM therefore possess complex heterogeneous microstructures covering a large range of length scales, including the melt pools, a micro-scale grain structure, dislocations, solidification dendrites, and nano-scale precipitates [5,6]. Understanding and controlling the evolution of the complex microstructures formed during metal SLM is essential to allow manufacture of components with good properties.

SLM-prepared 316L stainless steel has been demonstrated to exhibit high strength without losing plasticity compared to 316L stainless steel manufactured by conventional methods [5-7]. Many studies have attributed the improved mechanical properties of SLM-prepared 316L stainless steel to the presence of the characteristic cellular structure found in many SLM-prepared metals [8,9], on the basis that segregated elements and tangled dislocations at the cellular structure boundaries are both important factors affecting the mechanical properties. Previous studies have proposed various explanation for the simultaneous improvement of strength and ductility, including interaction of dislocations with other dislocations, accumulation of dislocations at the boundaries, interaction of dislocations with precipitates, and the stability of the dislocation cellular structure under large deformation strains [5,6,8-12]. The micro-scale deformation mechanisms are, however, not clear yet, as there are still only a limited number of studies relating the deformation behavior with the evolution of microstructures over different length scales. In this paper we investigate the evolution of multi-scale microstructures and strain distribution simultaneously in SLM 316L stainless steel during mechanical deformation and better understand the relationship between microstructural heterogeneity and local deformation heterogeneity.

2. Methods

A SLM Solutions instrument (SLM 280) was used for preparation of the samples. The metal powder was spherical 316L stainless steel, produced by gas atomization with 90% particle size in the range of 10-45μm. After comparing a series of printing parameters, using the density of the printed samples as criterion, the optimal printing parameters were determined as laser power of 180 W, laser scanning speed of 600 mm/s, powder layer thickness of 30μm, and rotation angle between layers of 67°. The density of the printed samples was 99.9%, as measured using the Archimedes method. A bulk sample of size of 18x7x20 mm^3 was printed, and then dog-bone shape tensile samples were cut from the bulk sample by electron-discharge machining as shown in figure 1a. The width, gauge length and thickness of the tensile samples were 1.8, 5 and 1 mm, respectively (see figure 1b for full specimen geometry).

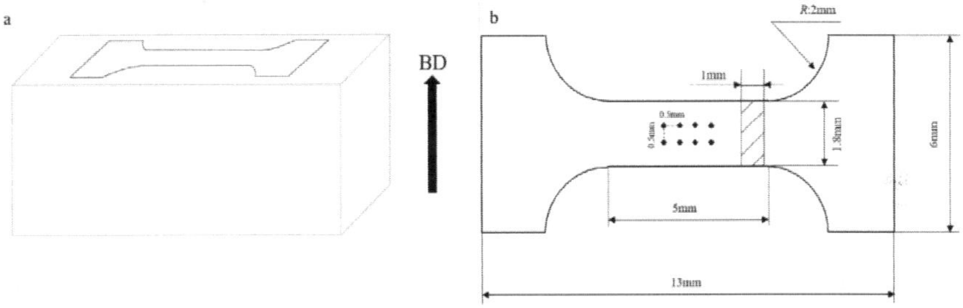

Figure 1. (a) Orientation of the tensile samples cut from the as-printed bulk sample, (b) dimensions for the tensile sample. BD indicates the build direction.

The macroscopic mechanical properties and strain distribution of the tensile sample were investigated by tensile testing combined with a macro-digital image correlation (DIC) system. Before tensile testing, samples were first ground to 2000-grit sandpaper, with the sample thickness reduced to about 0.9 mm. White and black oil paint were then sprayed on the surface of the sample as the markers for macro-DIC. Tensile tests were carried out on an AG-X tensile machine under displacement control at a rate of 0.5mm/min (initial strain rate of about 1.6 x 10^{-3} s^{-1}) until failure.

In-situ tensile deformation was carried out using a self-made push-to-pull stage, equipped with a load cell to record the instantaneous force during tensile loading. Samples were first ground up to 2000 grit paper to reduce the thickness to about 0.4 mm (due to the limited capacity of the load cell). Samples were then electrochemically polished in a 10% perchloric acid and 90% alcohol solution to achieve a flat polished surface free of mechanical damage and residual deformation. For micro-DIC experiments, surface patterning was achieved using silica nanoparticles [13]. Colloidal silica solution containing particles with an average size of 50 nm was first given an ultrasonic treatment for 90 minutes, after which one drop was placed onto the sample surface. Hardness indents were added as shown in figure 1(b) to allow easy identification of the same area for observation under SEM and EBSD.

The in-situ observations were conducted using a TESCAN MIRA 3 LMH scanning electron microscope (SEM), equipped with a CMOS-based electron back-scatter diffraction (EBSD) detector from Oxford Instruments and operated at 20 kV. Secondary electron (SE) images were used for the micro-DIC measurements, with a tracked region for micro-DIC of initial size 69.2 x 69.2 $μm^2$, recorded in a single image at a microscope magnification of x3000 at a resolution of 4096 x 4096 pixels. To ensure full overlap of the tracked region at each deformation stage, EBSD maps with a size of 75.2 x 75.2 μm² were collected using a step size of 0.1 μm. SE images for micro-DIC measurements and EBSD maps for analysis of microstructure evolution were collected before tensile testing (0% strain) and at tensile strains of 3.5% and 10.3%. For micro-DIC calculations the Ncorr package was used [14].

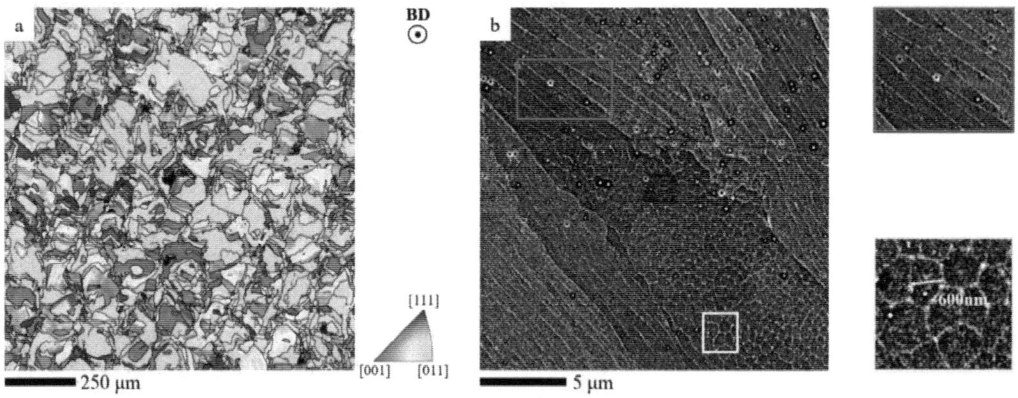

Figure 2. (a) EBSD map (inverse pole-figure coloring corresponding to the build direction, BD) of the as-printed sample on the cross-section perpendicular to the build direction (black lines show grain boundaries of ≥15°), (b) SE image showing cellular structures on the cross-section perpendicular to the build direction.

3. Results and discussion

A distinctive and typical microstructure is seen in the cross-section perpendicular to the build direction (figure 2a) where strip-like features can be seen corresponding to the laser scanning path and the resulting melt tracks (this pattern is only seen weakly here as a result of the 67° rotation between powder layers). This results a strongly non-uniform grain size distribution as seen in this section, consisting both of large columnar grains and regions of smaller equiaxed grains between the columnar grains. The average grain size was determined as approx. 20 μm from the EBSD data using an area reconstruction method and a 15° grain boundary definition, however a large fluctuation in values is present with maximum grain diameter of up to 140 μm.

The growth of grains in as-printed metal samples is mainly influenced by factors such as epitaxial growth, heat conduction in the melt pool, and local temperature gradients, thus forming a columnar grain morphology with the preferred growth direction along the direction of the local temperature gradient [15]. The temperature gradient at the center of each melt track is parallel with the build direction. The morphology of the grains at the center of melt tracks as seen in figure 2a represents therefore the cross-section of the columnar grains in the short axis direction, resulting in the appearance of a fine equiaxed grain morphology. In contrast, at the edge of the melt tracks, the direction of local temperature gradient is perpendicular to the tangential direction of the melt pool boundary. Local grains will grow along the temperature gradient from the edge of the melt pool to the center of the melt track. Hence in the section perpendicular to the build direction the columnar grains are seen in the long axis direction, and appear as columnar grains with larger size.

Figure 2b is a SEM image showing cellular structures in the as-printed samples. Cell morphologies in this section of both equiaxed and columnar shape are seen, attributed to different sectioning of a three-dimensional prism-like solidification structures. The reason for such diverse morphology in solidified structures in SLM 316L stainless steel is mainly related to the instability of the solidification interface caused by rapid solidification during the SLM process [16,17]. The characteristic diameter of the cellular structure is about 600 – 800 nm, which is approximately the primary dendrite arm spacing calculated from the expected temperature gradient and solidification rate during the SLM process [18], suggesting that the cellular structure is likely to be related to a solidification structure.

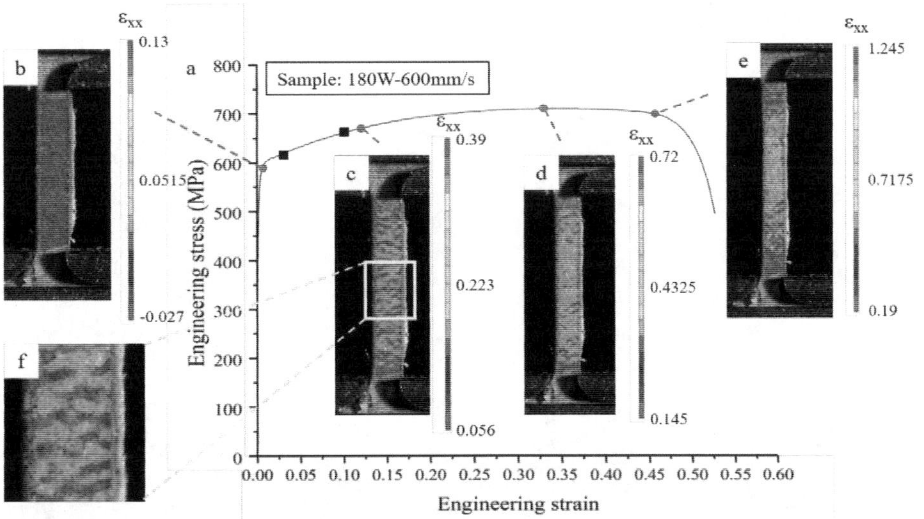

Figure 3. (a) Engineering strain-stress curve of as-printed 316L stainless steel, (b-e) strain distribution maps under strains of 1%, 12%, 32% and 45%, respectively, (f) local magnification of strain distribution at 12% strain (macro-DIC measurements).

The engineering stress-strain curve of the as-printed sample obtained from tensile testing is shown in figure 3a. The curve does not exhibit a yield plateau, indicating that the as-printed samples have good strain hardening ability. The yield strength, ultimate tensile strength, uniform elongation and elongation at fracture of this sample are 592 MPa, 710 MPa, 34% and 53%, respectively, all of which are much higher than 316L stainless steel prepared by conventional casting, and similar to values reported elsewhere for SLM of 316L stainless steel [6,7,9]. Figure 3(b-e) shows the tensile strain distribution (ε_{xx}) over the sample gauge as determined from macro-DIC measurements after strains of 1%, 12%, 32% and 45% (indicated by red dots in figure 3a), where ε_{xx} is given a Green-Lagrangian strain [14]. The strain distribution over the entire gauge length remains relatively uniform during loading, and no obvious macroscale inhomogeneous strain distribution is observed until necking occurs. However, as seen in the enlarged observation of the strain distribution map at a strain of 12% (figure 3f), a characteristic pattern of non-uniform strain distribution is present at smaller length scale. As this ripple-like pattern representing a mesoscale inhomogeneity of strain distribution was observed at several strain stages, further experiments based on micro-DIC were carried out to study the relationship between the deformation behavior and different length scales of microstructures.

The two black squares in figure 3a show the conditions at which micro-DIC data collection was carried out. A good agreement is observed between these data (taken from the load-cell and displacement of hardness markers) and the load curve obtained using the laboratory testing frame. The results, shown in figure 4a-c, reveal a pattern of increasingly complex heterogeneity in the micro-scale strain distribution during the tensile deformation process. In addition to the significant variation in strain within each grain a variation over a larger length scale covering several grains is also observed, with, in this case a higher average strain on the left side of the tracked region compared to the right side. Such pattern is consistent with the uneven ripple-like strain distribution seen in the macro-DIC measurements (figure 3). This pattern appears at low strain and is maintained throughout the deformation process. At present, however, the origin of this variation is unknown, as the length scale is larger than any parameter related directly to the SLM conditions. At the microscale a pattern of evident strain localization in fine bands is observed in many grains, albeit with varying extent of strain concentration. The direction of these strain localization bands is consistent with that of the slip bands observed in SEM images as shown in figure 4d-f. From the inset in figure 4f, it can be observed that the slip bands are not related to the local orientation of the cellular structure in the as-printed sample, but are instead consistent with the direction of the trace of the slip plane corresponding to the slip system with the highest Schmid factor under tension. As such the dislocations in the cell boundaries in the as-printed samples have only an indirect influence on the strain distribution and local deformation behavior.

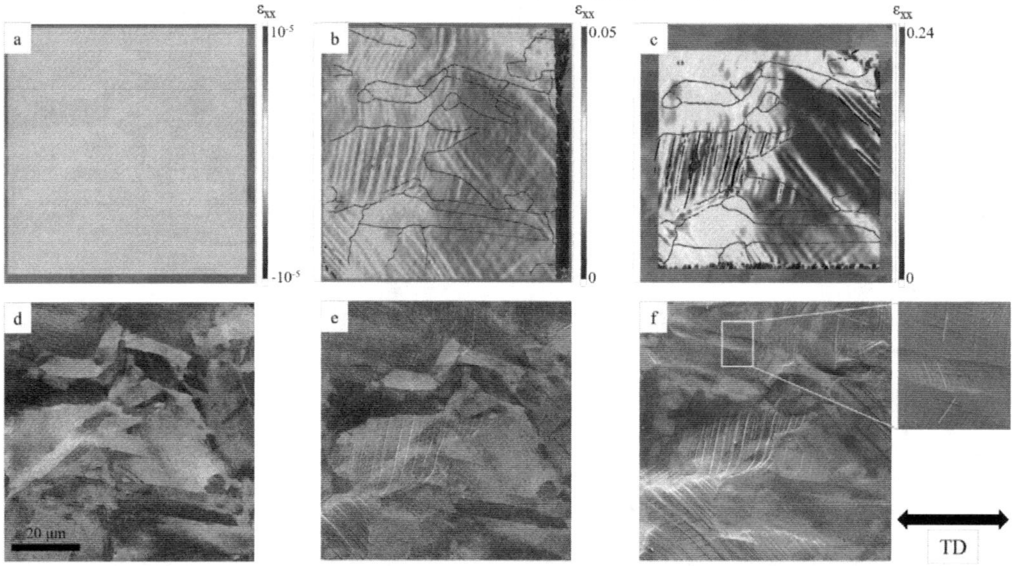

Figure 4. (a-c) Micro-DIC strain distribution maps at tensile strains of 0%, 3.5%, 10.3%, (d-f) SE images of the tracked region; red lines in (f) represent the direction of cellular structures, while yellow lines represent the direction of the slip trace associated with the highest Schmid factor based on tensile deformation of the orientation as determined from the EBSD data (TD indicates the tensile loading direction)

A key advantage of using silica particles as markers for micro-DIC is that these particles do not degrade the EBSD signal, thus enabling simultaneous characterization and analysis of local crystal orientation evolution during deformation. Figure 5a-c show EBSD images of the tracked region, where it is seen that grain orientations do not change significantly during the deformation process, local grain subdivision takes place with an increase in intragranular orientation spread, and in slip band width. This pattern of grain subdivision is consistent with that already present in the initial as-printed sample, indicating some continuity in the evolution of microstructural heterogeneity during deformation. Maps of kernel average misorientation also show a complex pattern in the evolution of local crystal misorientation, tracking in some grains the fine scale local strain variation seen in the micro-DIC data related to slip bands, though also showing many variations not seen in the DIC local strain data.

4. Summary

Samples of 316L stainless steel with near full density were fabricated by SLM and the microstructure and deformation behavior under tension characterized over different length scales. As-printed samples exhibit a complex heterogeneous microstructure, mainly related to the extreme solidification and cooling processes in the SLM process. Investigation of local strain distribution based on macro- and micro-DIC investigations show a heterogeneous pattern of strain distribution at the mesoscopic and microscopic length scales, where this pattern does not correlate significantly with either the grain or cellular structures. The use of colloidal silica markers for micro-DIC measurements also allows EBSD data to be collected over the tracked region, allowing the evolution of local crystal orientations and rotations to be compared with the deformation behavior. It is found that the original in-grain pattern of local misorientation and the intragranular subdivision that develops during tensile loading have only a limited effect on the strain distribution. Further detailed investigations into the evolution of microstructure and strain distribution during deformation of SLM prepared metals should be carried out to better understand the relationship between microstructural heterogeneity and local deformation mechanisms.

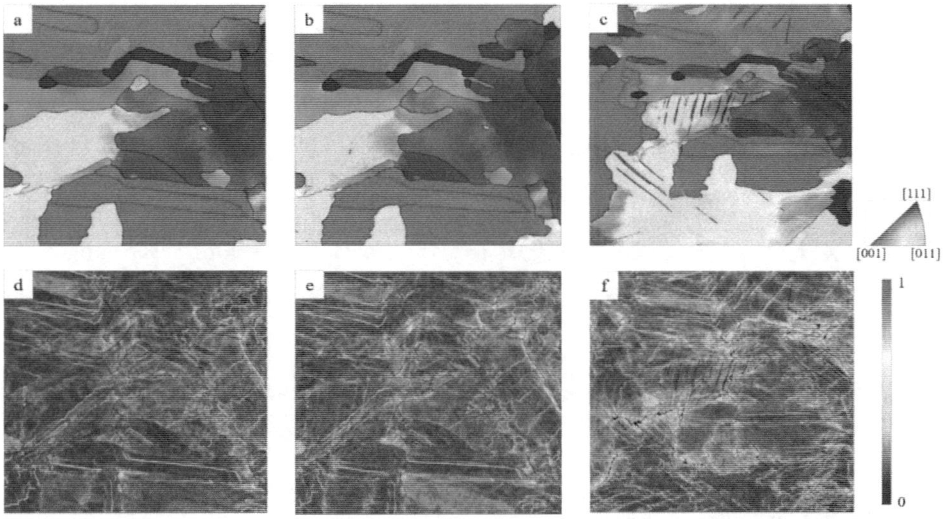

Figure 5. EBSD data covering the same surface region tracked in the micro-DIC investigation at strains of (a,d) 0%, (b,e) 3.5%, and (c,f) 10.3% : (a-c) inverse pole figure coloring (based on tensile loading direction – see figure 4), (d-f) 3 x 3 kernel average misorientation maps.

Acknowledgements
Financial support from the National Natural Science Foundation of China (project number 51971117) and from the Tsinghua-Imperial Research and Innovation Seed Fund is gratefully acknowledged.

References

[1] Frazier W E 2014 *J. Mater. Engin. Perf.* **23** 1917–28
[2] Gu D D, Meiners W, Wissenbach K and Poprawe R 2013 *Inter. Mater. Rev.* **57** 133–64
[3] Gorsse S, Hutchinson C, Goune M and Banerjee R 2017 *Sci. Technol. Adv. Mater.* **18** 584–610
[4] King W E, Anderson A T, Ferencz R M, Hodge N E, Kamath C, Khairallah S A and Rubenchik A M 2015 *Appl. Phys. Rev.* **2** 041304
[5] Liu L, Ding Q, Zhong Y, Zou J, Wu J, Chiu Y-L, Li J, Zhang Z, Yu Q and Shen Z 2018 *Mater. Today* **21** 354–61
[6] Wang Y M *et al* 2018 *Nat. Mater.* **17** 63–71.
[7] Qiu C, Kindi M A, Aladawi A S and Hatmi I A 2018 *Sci. Rep.* **8** 7785
[8] Barkia B, Aubry P, Haghi-Ashtiani P, Auger T, Gosmain L, Schuster F and Maskrot H 2020 *J. Mater. Sci. Technol.* **41** 209–18
[9] Voisin T, Forien J-B, Perron A, Aubry A, Bertin N, Samanta A, Baker A and Wang Y M 2021 *Acta Mater.* **203** 116476
[10] Bertsch K M, Meric de Bellefon G, Kuehl B and Thoma D J 2020 *Acta Mater.* **199** 19–33
[11] Li Z, He B and Guo Q 2020 *Scr. Mater.* **177** 17–21
[12] Pham M S, Dovgyy B and Hooper P A 2017 *Mater. Sci. Engin.* A **704** 102–11
[13] Gao W Q and Godfrey A 2022 *Mater. Lett.* **308** 131272
[14] Blaber J, Adair B and Antoniou A 2015 *Exp. Mech.* **55** 1105–22
[15] DebRoy T, Wei H L, Zuback J S, Mukherjee T, Elmer J W, Milewski J O, Beese A M, Wilson-Heid A, De A and Zhang W 2018 *Prog. Mater. Sci.* **92** 112–224
[16] Kurz W, Giovanola B, Trivedi R 1986 *Acta Metall.* **34** 823–30
[17] Pham M S, Dovgyy B, Hooper P A, Gourlay C M, Piglione A 2020 *Nat. Commun.* **11** 749
[18] Fisher D J and Kurz W 1980 *Acta Metall.* **29** 11–20

Heterogeneous lamellar-structured high manganese steel with excellent tensile properties

J C Xiong[1], J F Zhao[1], L Kong[1], H K Yang[2], X D Zhang[3] and Y H Wang[1]*

[1] National Engineering Research Center for Equipment and Technology of Cold Strip Rolling, Yanshan University, Qinhuangdao 066004, P. R. China
[2] Hong Kong Productivity Council (HKPC), Hong Kong SAR, 999077, P. R. China
[3] Department of Civil and Mechanical Engineering, Technical University of Denmark, DK-2800 Kgs. Lyngby, Denmark

Email: yhwang@ysu.edu.cn

Abstract. A single-phase Fe-30Mn-0.11C steel with a heterogeneous lamellar structure composed of micron-sized recrystallized grains (MRGs) and deformation bands (DBs) was successfully produced by cold rolling to 90% in thickness reduction and annealing at 550 °C for 120 min. The tensile properties of samples cut out along the rolling direction (RD) and transverse direction (TD) were investigated. The results indicate that the samples along the RD and TD both exhibit excellent tensile properties, with a yield strength of approximately 600 MPa and a total elongation of more than 30%.

1. Introduction

Many studies have reported that the strategy of designing heterogeneous structures can achieve superior mechanical properties for metallic materials. Heterogeneous structure materials such as heterogeneous lamellar structure, gradient structure and dual phase structure materials generally exhibit better mechanical properties than conventional homogeneous structure materials [1-5], and have attracted more and more attentions in recent years.

For single-phase metallic materials, it has been confirmed that heterogeneous lamellar structures can be obtained by cold rolling and subsequent annealing [6, 7]. Generally, the soft domain is the recrystallized structure, and the hard domain is the deformed structure has been observed in as heterogeneous lamellar-structured Fe-34.5Mn-0.04C steel [6] and pure Ti [8]. A heterogeneous lamellar structure was also obtained in 316L stainless steel by 85% cold rolling and annealing at 750 °C [7]. In addition, much work has focused on the strain hardening and plastic deformation mechanisms of heterogeneous lamellar-structured materials [9, 10].

Generally, multi-phase laminated composites are very sensitive to the loading direction, as a result of the inhomogeneous distribution of voids, cracks, oxides, intermetallic compounds and other impurities in the interfaces or the different degree of plastic deformation in the lamella [11-16]. However, there are few studies on mechanical properties of single-phase heterogeneous lamellar-structured alloys along the different directions, especially in high manganese austenitic steel.

In this work, a heterogeneous lamellar structure composed of MRGs and DBs was successfully obtained in the Fe-30Mn-0.11C steel by a combination of cold-rolling to 90% thickness reduction and annealing at 550 °C for 120 min. The tensile properties of samples cut out along RD and TD were investigated.

2. Experimental

The steel used has a nominal composition of Fe-30Mn-0.11C (wt.%). The steel was melted using the vacuum induction technique. The ingot was heated to 1000 °C and held for 1 h, and then forged at 800~1000 °C to produce a 21 mm thick plate. The hot forged plate was then cold rolled to a thickness reduction of 90%, followed by annealing at 550 °C for 120 min. Subsequently, the samples were water-quenched. The Fe-30Mn-0.11C steel has a stable austenitic structure after 90% cold rolling, which was characterized in our previous work [17-18].

Microstructure analyses were carried out using electron microscopy. The microstructure observations were all carried out on the longitudinal section along the rolling direction (RD) and the normal direction (ND). A field emission scanning electron microscope (FE-SEM) equipped with a backscatter electron (BSE) detector and an electron backscatter diffraction (EBSD) system operated at an accelerating voltage of 20 kV was used for microstructural characterization. The microstructure was also characterized using a FEI Talos F200s transmission electron microscope (TEM) operated at 200 kV. Tensile specimens with a gauge length of 10 mm, a width of 2 mm and a thickness of 2 mm were prepared from the annealed plates. Tensile tests were carried out using a Zwick Z050TEW testing machine with a strain rate of 10^{-3} s^{-1} at room temperature. The tensile testing was repeated 3 times.

3. Results and discussion

Figure 1 shows the microstructure of the cold rolled Fe-30Mn-0.11C steel. The EBSD IPF+IQ map of the cold-rolled sample in figure 1a, indicates that the microstructure is elongated along the RD. There are clear DBs along RD. The microstructure was significantly refined via cold rolling. The ECCI image (figure 1b) shows that the cold rolled microstructure contains DBs (white dashed area) and nano-twin bundles (NTBs) (blue dashed area) which were confirmed by TEM images. The microstructure after cold rolling is mainly a nano-lamellar structure (NLS) delineated by geometrically necessary dislocation boundaries (GNBs), as shown in figure 1c. Selected area electron diffraction (SAED) is ring-shaped, which matches the nanocrystalline structure. Figure 1d shows that there are NTBs in the microstructure, which are surrounded by the NLS. The above results show that the microstructure after 90% cold rolling mainly consist of NLS, DBs and NTBs.

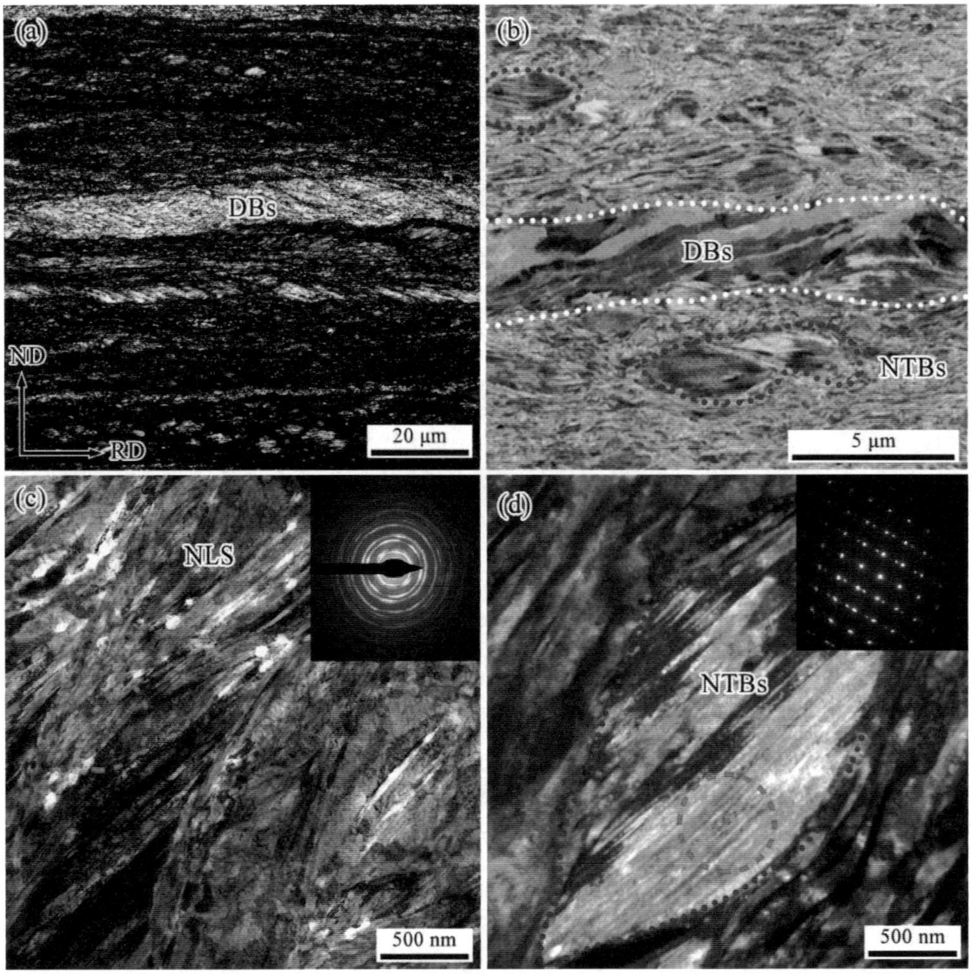

Figure 1. Microstructure of the cold rolled Fe-30Mn-0.11C steel (a) EBSD IPF+IQ image (b) electron channel contrast image (ECCI), (c) TEM image of NLS, the inset is the selected area diffraction pattern (SAED) of NLS, (d) TEM image of NTBs, the inset is the SAED of NTBs. The regions of SAEDs are circled by red dotted lines.

Figure 2 shows the heterogeneous lamellar structure consists of the MRGs and DBs after cold rolling and annealing at 550 °C for 120 min. The initial NLS has recrystallized and the recrystallization area is composed of MRGs (see figure 2b). The average grain size calculated by EBSD is 2.0 μm. Figure 2c shows DBs surrounded by MRGs. The DBs consist of nanocrystals delineated by dislocation boundaries, which is confirmed by SAED. Figure 2d shows that some NTBs still exist in the annealed structure.

Tensile tests were carried out along the RD and TD, as sketched in figure 3a. The engineering stress-strain curves (see figure 3b) of the samples show a discontinuous yielding behavior, which was also observed in the Fe-34.5Mn-0.04C steel [19]. This is attributed to a lack of dislocation sources in these fine-grained samples, similar to the behavior observed in many other materials with grain sizes in the near-micrometer regime, resulting also in an additional strengthening due to the lack of dislocation sources [20-21]. The yield drop phenomenon can be eliminated by additional slight cold rolling which introduce extra dislocations [19]. The yield strength and total elongation of the samples along RD and TD are 583 MPa, 37% and 625 MPa, 30%, respectively.

Figure 2. Microstructure of Fe-30Mn-0.11C steel after cold rolling and annealing at 550 °C for 120 min (a) EBSD IPF+IQ image (b) TEM image of recrystallized microstructure, (c) TEM image of a region with DBs, the inset is SAED of DBs (d) TEM image of a region with NTBs. The regions of SAEDs are circled by red dotted lines.

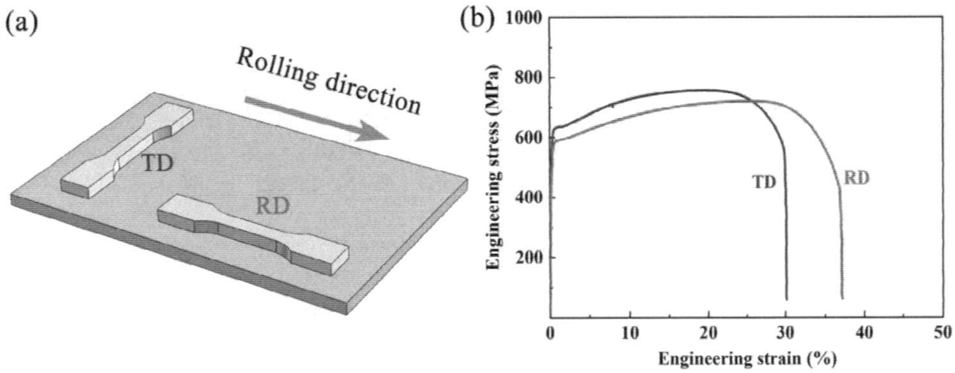

Figure 3. (a) Schematic diagram (b) engineering stress-strain curves of Fe-30Mn-0.11C steel annealed at 550 °C for 120 min.

Figure 4 shows the fractography of the RD sample. The fracture is cup-cone-shaped (see figure 4a) and the fracture appearance is composed of large and deep dimples (see figure 4b), indicating a ductile fracture mode.

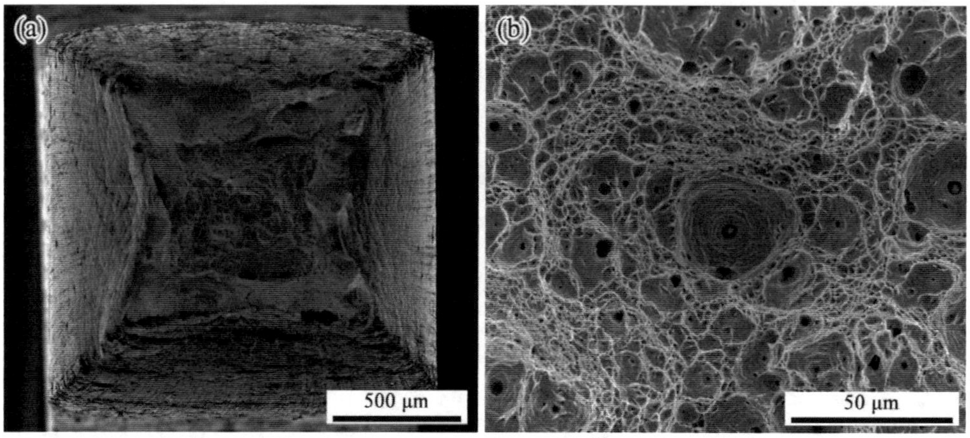

Figure 4. The fractography of the sample along the RD.

Generally, multi-phase laminated composites exhibit an obvious anisotropic mechanical behavior due to the inhomogeneous distribution of voids, cracks, oxides, intermetallic compounds and other impurities in the interfaces or different degree of plastic deformation in the lamella [11-16]. The Ti/Al laminated composites exhibited a significant anisotropy due to the difference in the microstructure and microhardness of the Ti and Al layer [12]. Moverare and Odén [11] reported load partitioning between austenitic and ferritic phase in cold-rolled duplex stainless which was dependent on the loading directions, leading to an obvious anisotropy in the mechanical properties.

However, it is worth noticing that the heterogeneous lamella structure of the Fe-30Mn-0.11C steel exhibit unexpected tensile properties along RD and TD, which is different from multi-phase laminated composites (such as duplex stainless [11], Ti-Al [12], Mg-Al [13]). This may be related to the following two aspects: i) the large difference in the grain size between the MRGs (2.0 μm) and DBs/NTBs (73 nm); ii) the remaining DBs/NTBs zones are well distributed. The detailed mechanisms for the observed excellent mechanical properties need to be investigated systematically, taking into account the strengthening contributions from different zones [21] and the dislocation-boundary interactions [22].

4. Summary

In this work, a heterogeneous lamellar structure composed of MRGs and DBs/NTBs was successfully produced in a single-phase Fe-30Mn-0.11C steel by cold-rolling to 90% thickness reduction followed by annealing at 550 °C for 120 min. Excellent tensile properties both along RD and TD were achieved: the yield strength and total elongation of the samples along RD and TD are 583 MPa, 37% and 625 MPa, 30%, respectively.

Acknowledgments

The authors gratefully acknowledge the support from the National Natural Science Foundation of China (Grant No. 51871194) and the National Natural Foundation of Hebei Province, China (Grant No. E2021203011, 226Z1003G). X. Zhang acknowledges the support from the European Research Council under the European Union Horizon 2020 research and innovation program (grant agreement No 788567-M4D).

References

[1] Zhu Y, Ameyama K, Anderson P M, Beyerlein I J, Gao H, Kim H S, Lavernia E, Mathaudhu S, Mughrabi H, Ritchie R O, Tsuji N, Zhang X and Wu X 2021 *Mater Res Lett* **9** (1) 1-31
[2] Bhattacharjee T, Wani I S, Sheikh S, Clark I T, Okawa T, Guo S, Bhattacharjee P P and Tsuji N 2018 *Sci. Rep.* **8** (1) 3276
[3] Wu X, Zhu Y and Lu K 2020 *Scripta Mater.* **186** 321-5

[4] Liu Y, Cao Y, Mao Q, Zhou H, Zhao Y, Jiang W, Liu Y, Wang J T, You Z and Zhu Y 2020 *Acta Mater.* **189** 129-44
[5] Zhang X, Hansen N, Gao Y and Huang X 2012 *Acta Mater* **60** 5933–43
[6] Wang Y H, Kang J M, Peng Y, Wang T S, Hansen N and Huang X 2018 *Scripta Mater.* **155** 41-5
[7] Li J, Cao Y, Gao B, Li Y and Zhu Y 2018 J. *Mater. Sci.* **53** (14) 10442-56
[8] Wu X, Yang M, Yuan F, Wu G, Wei Y, Huang X and Zhu Y 2015 *PNAS* **112** (47) 14501-5
[9] Li J, Zhang Q, Huang R, Li X and Gao H 2020 *Scripta Mater.* **186** 304-11
[10] Moering J, Ma X, Malkin J, Yang M, Zhu Y and Mathaudhu S 2016 *Scripta Mater.* **122** 106-9
[11] Moverare J J and Odén M 2002 *Metall Mater Trans A* **33** (1) 57-71
[12] Huang T, Song Z, Chen F, Guo J, Pei Y, Xing B, Xiang N and Song K 2020 *Materials* **13** (16) 3556
[13] Wang T, Li S, Ren Z, Jia Y, Fu W, Guo M, Ma X and Han J 2019 *Metals* **9** (6) 690
[14] Ng H P, Przybilla T, Schmidt C, Lapovok R, Orlov D, Höppel H and Göken M 2013 *Mater. Sci. Eng. A* **576** 306-15
[15] Lee K S, Lee S E, Sung H K, Lee D H, Kim J S, Chang Y W, Lee S and Kwon Y N 2013 *Mater. Sci. Eng. A* **583** 177-81
[16] Wang T, Wang Y L, Bian L P and Huang Q X 2019 *Mater. Sci. Eng. A* **765** 138318
[17] Kang J M, Wang Y H, Chen X M, Zhang C L, Peng Y and Wang T S 2019 *Results Phys.* **13** 102247
[18] Wang Y, Zhang Y, Godfrey A, Kang J, Peng Y, Wang T, Hansen N and Huang X 2021 *Commun Mater* **2** (1)
[19] Wang Y, Kang J, Peng Y, Wang T, Hansen N and Huang X 2018 *J. Mater. Sci. Technol.* **34** (10) 1939-43
[20] Huang X 2009 *Scripta Mater.* **60** (12) 1078-82
[21] Huang X, Hansen N and Tsuji N 2006 *Science* **312** 249-51
[22] Zhang X, Nielsen CV, Hansen N, Silva CMA and Martins PAF 2019 *Int. J. Plast.* **115** 93–110
[23] Pan H, He Y and Zhang X 2021 *Materials* **14** 1012

IOP Conf. Series: Materials Science and Engineering 1249 (2022) 012037 doi:10.1088/1757-899X/1249/1/012037

Effects of alloying, microstructure and texture on the strength–ductility trade-off in layered aluminum

Y P Xia[1,3], H Wu[2], K S Miao[2], L Geng[1], G H Fan[2], T Yu[3] and D Juul Jensen[3]

[1]School of Materials Science and Engineering, Harbin Institute of Technology, Harbin, 150001, China

[2]Key Laboratory for Light-weight Materials, Nanjing Tech University, Nanjing, 211816, China

[3]Department of Civil and Mechanical Engineering, Technical University of Denmark, DK-2800 Kgs. Lyngby, Denmark

E-mail: genglin@hit.edu.cn (L. Geng); ghfan@njtech.edu.cn (G.H. Fan);

Abstract. Tailoring constituent layers can lead to improved mechanical properties of layered metals. In this work, we designed and fabricated two samples of layered Al. One sample with AA3003 and AA1060 layers and the other with only AA1060 layers. The effects of these designs on the deformation behaviors are revealed by in-situ monitoring of the local strain evolution. It was found that larger microstructural and textural variations can lead to a larger difference in transverse compressive strain between constituent layers under uniaxial tension, which indicates more significant defomation incompatibility and thus a larger strain gradient caused by the constraint effect. The effects of the texture on the transverse compressive strain levels was analysed by the Lankford values calculated by the visco-plastic self-consistent method. Furthermore, the fracture behaviors of the two samples are compared and discussed based on the cross-sectional fracture morphologies. This work illustrates the potential of tailoring layered metals for a superior strength–ductility combination.

1. Introduction

For metallic materials, it is generally found that an improvement in strength or ductility invariably sacrifices the other, which is known as the strength–ductility trade-off dilemma [1]. In recent years, it has however been realized that layered structural design may be a promising strategy to break the strength–ductility trade-off [2]. The layered structural design contains two main aspects, the structural parameters and the variations between constituent layers. Concerning the structural parameters, many investigations suggest that both the layer thickness and layer thickness *ratio* have significant effects on the tensile properties. It has been reported that the optimal strength–ductility combination in layered Cu can be achieved when the interface affected zones (IAZs) just overlap [3]. A number of studies have shown that tailoring layer thickness ratio can tune the mechanical properties by altering the deformation mechanisms [4] and fracture modes [5].

However, it is still an open question how an optimal or at least an improved strength–ductility combination can be achieved by controlling *variations* between constituent layers. Recent work on this topic centered on the discrepancies in the work hardening rate between two constituent layers [6]. It was suggested that a larger discrepancy in work hardening rate leads to a steeper strain gradient with more geometrically necessary dislocations (GNDs), resulting in higher work hardening capacity [7]. However, the Ti/Ti_3Al metallic/intermetallic multilayered composite [8], possesses a large work hardening discrepancy, but has unsatisfactory tensile properties at ambient temperature. This implies that the effect of the work hardening discrepancy on the strength–ductility combination may have to be reconsidered. Furthermore, the difference in plastic anisotropy, namely the difference in the ratio of width to thickness strain during uniaxial tension (Lankford values), can also lead to significant deformation incompatibility between constituent layers. The difference in plastic anisotropy may therefore also be crucial for tailoring the mechanical properties of layered metals [2,9].

The different mechanical properties of constituent layers, such as yield strength, work hardening rate, and plastic anisotropy, are dependent on material parameters such as alloy composition, microstructure and texture. In this study, we designed and fabricated two kinds of layered Al with different microstructural and textural variations between constituent layers. The effects on both the deformation behavior and the strength–ductility combination are analyzed and discussed based on in-situ mapping of the local strain distribution and on fracture observation.

2. Experiments and methods

One of the present samples is composed of layers of unannealed and annealed Al foils (AA1060), each with a thicknesses of 200 μm. The unannealed AA1060 foils have a typical rolling texture with an average grain size of 2.7 μm. The annealing was performed at 500 °C for 1 h to achieve a complete recrystallization. After stacking the unannealed and annealed foils alternately, the laminate was hot-pressed at 400 °C and 45 MPa for 1 h. Then the hot-pressed sheet was rolled at 300 °C to a thickness reduction of ~65%, followed by 1 h annealing at 250 °C. The other layered sample was fabricated by aumulative rolling bonding (ARB) with 1.5-mm-thick AA3003 and AA1060 alloy sheets as raw material. The AA3003 and AA1060 sheets were annealed at 610°C and 550°C for 1h, respectively. Then, the sheet surfaces were cleaned in a ethanol bath and roughened by steel brushing. The rolling was carried out at 200 °C with a thickness reduction of 50%, followed by a cut at the middle of roll-bonded sheet. This procedure was repeated 4 times without changing rolling direction (RD). Finally the layered Al produced by ARB was annealed at 320 °C for 1 h. In the following, the layered Al sample with only AA1060 layers is termed as A11 while the other with AA3003 and AA1060 alloys is termed as A31. In addition, a reference sample was fabricated with only unannealed AA1060 foils by the same process as that for A11, termed as FG1. Reference samples with only annealed AA3003 sheets (termed as FG3) and with only annealed AA1060 sheets (termed as CG) were also prepared using the same ARB–annealing process as that for A31, resulting in similar microstructures to those of constituent layers in A31.

Tensile tests were carried out using an Instron-5569 universal testing machine. Dog-bone-shaped specimens were equipped with a 10 mm extensometer and tensioned along the transverse direction (TD) at a strain rate of $\sim 1\times10^{-3}$ s^{-1}. The local mechanical properties of the constituent layers were characterized by nano-indentation tests (Nano-Indenter XP machine). The depth of the nano-indentations were 400–500 nm. A scanning electron microscope (SEM, ZEISS MERLIN COMPACT) equipped with an HKL-EBSD system was utilized to characterize the microstructures.

In-situ observations of the strain evolution were performed with a commercial micro-tensile rig mounted in a TESCAN S8000 SEM. The evolution in representative areas on the surface containing the loading direction (LD) and the normal direction (ND) were recorded during tensile deformation. The results were analyzed based on these SEM micrographs using the commercial software (VIC-2D). The cross-sectional fracture morphologies were also captured in SEM for analyzing the fracture behavior of the two samples.

3. Results and discussion

The microstructures of the two layered Al samples are shown in figure 1. They both comprise of fine-grained and coarse-grained layers (FLs and CLs) with similar layer thicknesses of ~75 μm. The samples have different microstructures and textures in the constituent layers (figure 1(a-b)). Figure 1(c-d) shows the distributions of grain sizes (equivalent spherical diameter, determined using a cutoff angle of 15°). The FLs and CLs in A31 have average grain sizes of 1.1 μm and 9.3 μm, respectively, i.e. an eight-fold difference. The average grain sizes of FLs and CLs in A11 are 4.2 μm and 13.0 μm, respectively, i.e.

three-fold difference. The FLs of both A31 and A11 have typical rolling textures (strong S texture {123}<634> and brass texture {110}<112>). The CLs of A31 have a shear texture ({100}<011>) with a weak cube texture ({100}<001>) [10]. The microstructures of the layers are recrystallized and contain a high fraction of high-angle grain boundaries (f_{HAGB}). The CLs of A11 have a weak S and cube texture (figure 1(e-f)) with a low f_{HAGB}, and a lamellae microstructures elongated along LD. The larger microstructural and textural variations in A31 are mainly caused by different alloys. The addition of Mn in AA3003 can lead to Al-Mn eutectic phases embedded in the Al matrix as well as a higher recrystallization temperature compared to AA1060.

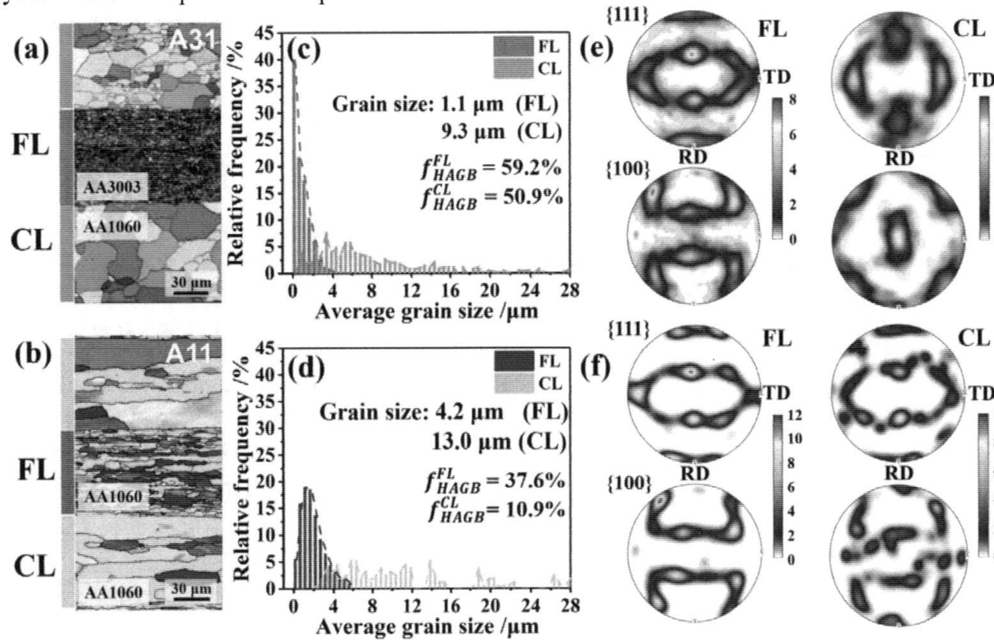

Figure 1. The microstructures of the two layered Al samples: IPF maps for (a) A31 and (b) A11; grain size (equivalent spherical diameter) distributions and proportions of high-angle grain boundaries (f_{HAGB}) of FLs and CLs in (c) A31 and (d) A11; {111} and {100} pole figures of FLs and CLs in (c) A31 and (d) A11.

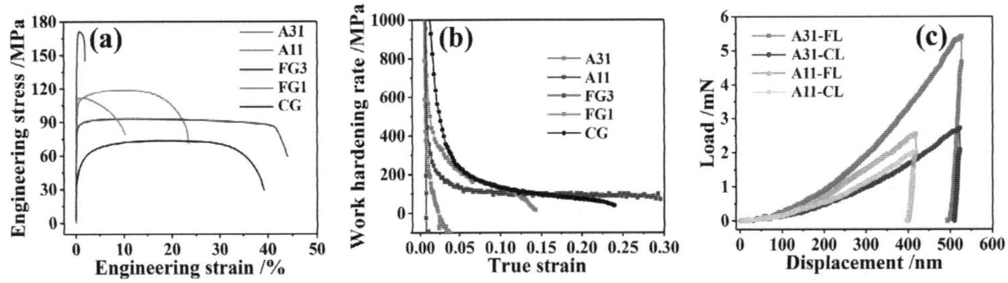

Figure 2. The mechanical properties of the layered Al sample compared to different homogeneous Al samples (FG3, FG1 and CG): (a) Engineering stress–strain curves, (b) work hardening curves. (c) Load-displacement curves of the layers in two Al samples from nano-indentation test.

Compared to monolithic fine-grained AA3003 and AA1060 Al with similar microstructures to FLs in A31 and A11 (termed as FG3 and FG1, respectively), the layered Al sample have a better strength–ductility combination. The strengths of the two layered Al samples basically follow the law of mixture. The sample, A31, with larger variations of microstructures and textures between constituent layers, exhibits a higher work hardening rate compared to A11, especially before the true strain reaches ~0.125.

After that, the strain hardening rate of A31 drops rapidly to ~100 MPa, and even below those of A11. The more steady strain hardening rate of A11 implies that A11 has a better resistance to plastic instability, hence a better elongation to fracture than A31. The elongation is even superior to the monolithic coarse-grained AA1060 (CG). These results support the general observation that layered structures can pose a positive effect on the strength–ductility trade-off by increasing the work hardening rate and resistance to plasticity instability. Furthermore, the present results reveal that the microstructural variation between constituent layers can affect these improvements. Figure 2(c) shows the nano-indentation loading curves of the different constituent layers in the two layered samples. The mechanical response difference between FLs and CLs in A31 is larger than that in A11, which agrees well with the larger microstructural variation.

Figure 3 compares the local strain along LD (ε_{yy}) in A31 and A11 at the similar average strains. The average strain, $\bar{\varepsilon}_{yy}$ is calculated from the analyzed regions. The local high strain fluctuations with red and blue colored band-shaped domains reveal strain localization. The strain localization occurs in both constituent layers of A31 and A11 (figure 3(a-b)), which suggests some local deformation incompatibilities within the constituent layers. The ε_{yy} strain distributions along ND (figure 3(c-d)) is largely consistent across the layer at all the investigated average strains, implying that the FLs and CLs keep an iso-strain condition along LD.

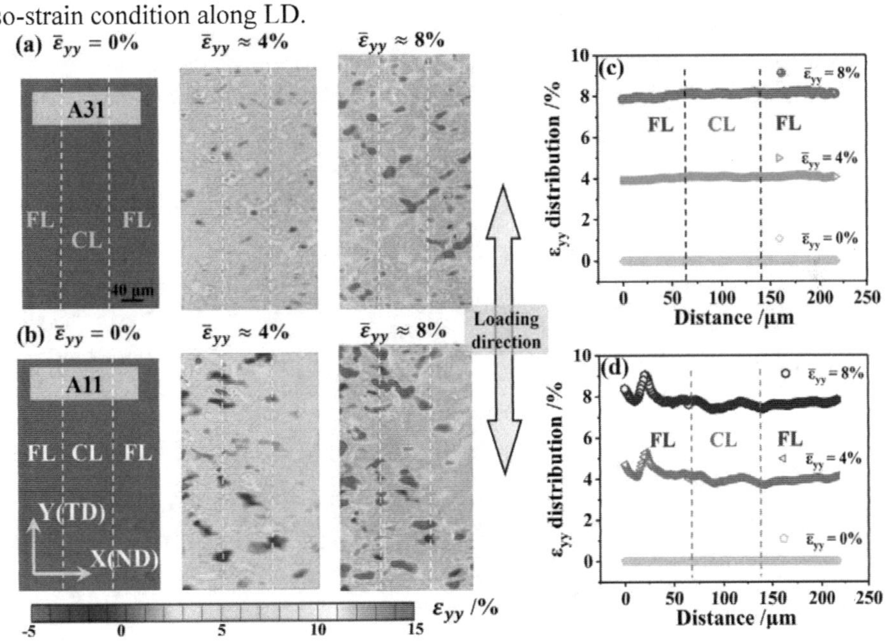

Figure 3. The distributions of local strain along LD, ε_{yy}, at average strains $\bar{\varepsilon}_{yy}$ of 0%, ~4% and ~8%: (a) A31 and (b) A11. The ε_{yy} averaged along LD as a function of distance along ND: (c) A31 and (d) A11.

Figure 4 shows the distribution of local strain along ND (ε_{xx}). As the tensile deformation increases, different strain partition emerged in the two Al samples (figure 4(a-b)). More specifically, the strain levels of ε_{xx} are significantly different between the FLs and CLs in both A31 and A11. The average strains of FLs and CLs are about -4.3% and -7.9% in A31, yet about -5.8% and -3.8% in A11. The larger strain difference in A31 (~3.6%) compared to that in A11 (~2.0%) indicates that more deformation accommodation between constituent layers is required in A31 than in A11 [6,9]. This result also suggests that more significant deformation incompatibility between the constituent layers can exist in the layered sample with the larger microstructural and textural variations as well as a difference in chemical components. Due to the deformation continuity requirement, more deformation accommodation in A31 can be expected, resulting in a change of slip activities and a higher work hardening rate in the soft AA1060 layers near the interface [11].

Previous work also reported steeper strain gradients in samples with larger microstructural variations [6], and quantified the relation between strain gradients and GNDs accumulation based on the strain gradient theory [12]. However, the deformation incompatibilities were always attributed to the difference of grain sizes and work hardening capacities in the previous work [6,7]. In the present work, it was found that the average compressive strain along ND (ε_{xx}) in FLs of A31 is less than that in CLs. On the contrary, the FLs bear larger compressive strain (ε_{xx}) than CLs in A11 (figure 4(c-d)). In other words, the strain gradient directions are opposite in these two samples. The work hardening rate can only be related to the onset of plastic instability, and cannot explain the opposite strain gradient between constituent layers in our present samples. The significance of plastic anisotropy should be considered. For rationalizing effects of plastic anisotropy on different compressive strain levels, the visco-plastic self-consistent (VPSC) simulations [13] with Affine scheme was performed to calculate the Lankford values [14] of the individual constituent layers. The textures characterized by EBSD (figure 1(e-f)) were imported into VPSC. It was found that the FLs with S and brass texture in A31 and A11 have Lankford values of 0.3 and 0.33, respectively. However, the Lankford value of CLs in A31 is only 0.11 due to the strong shear texture. The slight cube texture in CLs of A11 is also important, leading to a Lankford value of 0.85. A higher Lankford value means more compressive strain along the thickness direction (i.e. ND) for the same tensile strain. Accordingly, the VPSC-predicted Lankford values of the constituent layers agree well with the difference of ε_{xx} strain levels observed on the free surface of A31 and A11. In the bulk interior, more uniform strains are expected due to 3D constraint, and thus the slip activities are altered and the work hardening rate is increased compared to free deformation, especially for the soft CLs.

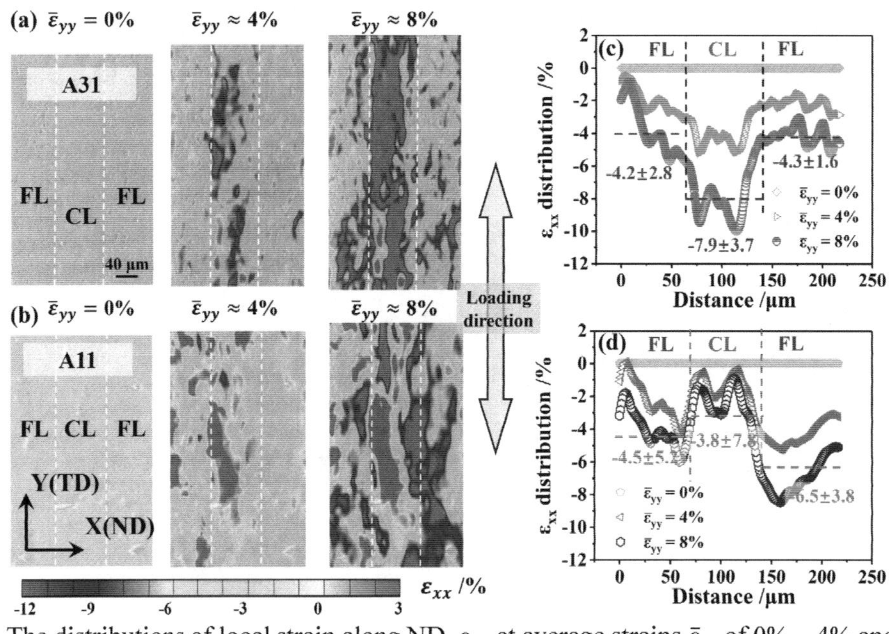

Figure 4. The distributions of local strain along ND, ε_{xx}, at average strains $\bar{\varepsilon}_{yy}$ of 0%, ~4% and ~8%: (a) A31 and (b) A11. The ε_{xx} averaged along LD as a function of distance along ND: (c) A31 and (d) A11.

Figure 5 compares the fracture morphologies of A31 and A11. A large amount of interfacial microcracks can be found in both layered Al samples, indicated by magenta arrows in figure 5(a-d). However, some microcracks are also located within FLs of A31, indicated by red arrows in figure 5(a-b). The backscattered electron image (figure 5(b)) further verified that the intra-layer microcracks dominantly initiated at the interface between Al matrix and Al-Mn eutectic phase. The intra-layer microcracks can facilitate the transverse propagation of interfacial microcracks across FLs in A31, which is detrimental to the resistance to plastic instability, and leads to the formation of primary cracks.

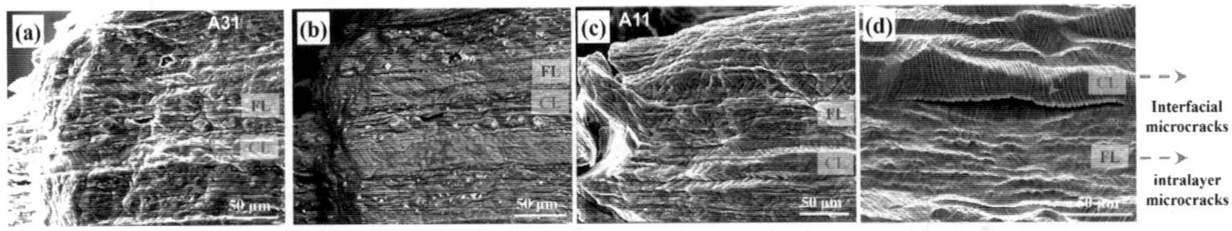

Figure 5. Fracture morphologies of A31: (a) secondary electron image, (b) backscattered electron image, and (c-d) A11. Red and magenta arrows indicate interfacial and intra-layer microcracks.

4. Summary

We fabricated two samples of layered Al with different compositions, which are shown to have different tensile properties. Different transverse strain levels observed by DIC, and Lankford values predicted by VPSC imply the effects of microstructural and textural variations on the deformation incompatibility between constituent layers. Compared to only interfacial microcracks in A11, the additional microcracks initiating within the hard FLs of A31 lead to plastic instability. An important insight from this study is that the microstructural and textural variations are closely related to the work hardening capacity, plastic anisotropy and fracture modes of constituent layers, hence providing an feasible approach for regulating the deformation incompatibility between constituent layers and achieving superior mechanical properties.

Acknowledge

This work was financially supported by the National Key Research & Development Program of China (Grant No. 2020YFA0405900), the National Natural Science Foundation of China (Grant Nos. 51927801, 51971075), Natural Science Foundation of Jiangsu Province (Grant No. BK20202010). Y P Xia would also like to acknowledge the support from China Scholarship Council (No. 202106120142). T Y and D J J thank the European Research Council (ERC) under the European Union's Horizon 2020 research and innovation programme (M4D – grant aggreemet No. 788567).

References

[1] Ovid'ko I A, Valiev R Z and Zhu Y T 2018 *Prog. Mater. Sci.* **94** 462-540
[2] Wu H and Fan G 2020 *Prog. Mater. Sci.* **113** 100675
[3] Huang C X, Wang Y F, Ma X L, Yin S, Höppel H W, Göken M, Wu X L, Gao H J and Zhu Y T 2018 *Mater. Today* **21** 713-9
[4] Misra A, Hirth J P and Hoagland R G 2005 *Acta Mater.* **53** 4817-24
[5] Xia Y, Wu H, Miao K, Li X, Xu C, Geng L, Xie H and Fan G 2022 *J. Mater. Sci. Technol.* **111** 256-67
[6] Wan T, Cheng Z, Bu L and Lu L 2021 *Scripta Mater.* **201** 113975
[7] Wang Y F, Huang C X, Fang X T, Höppel H W, Göken M and Zhu Y T 2020 *Scripta Mater.* **174** 19-23
[8] Wu H, Fan G, Huang M, Geng L, Cui X and Xie H 2017 *Int. J. Plasticity* **89** 96-109
[9] Ashby M F 2006 *Philos. Mag.* **21** 399-424
[10] Suwas S and Ray R K 2014 *Crystallographic Texture of Materials,* pp 143-77
[11] Xia Y, Miao K, Wu H, Geng L, Xu C, Ku C-S and Fan G 2022 *Int. J. Plasticity* **152** 103240
[12] Gao H J, Huang Y, Nix W D and Hutchinson J W 1999 *J. Mech, Phys. Solids* **47** 1239-63
[13] Lebensohn R A and Tomé C N 1993 *Acta Metall. Mater.* **41** 2611-24
[14] Inoue H and Takasugi T 2007 *Mater. Trans.* **48** 2014-22

Microplasticity in inhomogeneous alloys

Yash Pachaury[1], Tomohisa Kumagai[2] and Anter El-Azab[1*]

[1]Purdue University, West Lafayette, IN 47906, USA
[2]Energy transformation laboratory, Central Research Institute of Electric Power Industry, Kanagawa, Japan

* E-mail: aelazab@purdue.edu

Abstract. We report on a preliminary modelling effort to understand the influence of compositional inhomogeneity on alloy microplasticity from a dislocation dynamics perspective. We tackle this problem by multiscale simulations in three steps: (1) analysis of the 3D composition morphology in alloys with tendency to undergo spinodal instability both thermally and under irradiation, with bcc FeCrAl alloys as a model system, (2) atomistic simulation of the dislocation mobility as a function of the local alloy composition, and (3) using dislocation dynamics simulations to understand the impact of composition inhomogeneity on microplasticity. The dislocation dynamics model takes into consideration the coherency stress associated with composition inhomogeneity when computing the forces driving the dislocation motion and on cross slip. Our preliminary investigation shows that the stress-strain response of the alloy and the dislocation density evolution depend on the wavelength of the composition fluctuations. Our investigation also shows that the alloy inhomogeneity may alter the cross-slip activity, which, in turn, influences the dislocation density evolution. The dependence of the dislocation mobility and coherency stress on local composition and its variation, as well as the altered cross slip rates, cause the dislocation microstructure to differ relative to that in the homogeneous alloy of the same average composition.

1. Introduction

Plasticity in random solid solutions is a consequence of collective dynamics of dislocations and their interactions with solute atoms. The microscopic stress due to solute atoms and the localized changes in the elastic properties affect the local velocities of the dislocations under external load, hence giving rise to different microplasticity of the alloy that affects the macroscale plasticity response. Additionally, the composition of an alloy may become non-uniform due to processing of alloys prior to or during their intended application. Aged [1], irradiated [1,2], and additively manufactured [3] alloys typically exhibit this non-uniform distribution of the solutes which are often characterized as spinodal/spinodal-like composition fluctuations [1,2]. Formation of gradual composition-rich and composition-deficient regions within an alloy is a characteristic feature of this phenomena. The associated composition fluctuation wavelengths may range from 10s to 100s of nanometers all the way to the micrometer scale, hence affecting the mechanical behavior of such alloys at the mesoscale.

Significant experimental evidence exists which suggests that the composition fluctuations affect the mechanical behavior of the alloys [4–8]. Along the modeling front, the first attempt to explain the hardening in spinodally decomposed alloys was made by Cahn [9] and later followed by Kato et al. [10–12]. Their theories are based on predicting critical resolved shear stress (CRSS) due to the motion of single (screw, edge (Cahn [9]) or mixed ([10])) dislocation within sinusoidally modulated spinodal composition fields. However, in order to truly understand plasticity in alloys exhibiting propensity towards spinodal/spinodal-like composition fluctuations, both yielding, and work-hardening (post-yielding) response must be investigated. The deformation regime where the work hardening is significant is governed by the collective motion of dislocations and their short- and long-range interactions. Utilizing mesoscale modeling tools, such as discrete dislocation dynamics (DDD), provide a very important avenue for exploring the mesoscopic deformation response of the spinodally-decomposed alloys where the fluctuations in composition are of the order of the dislocation interaction distances. However, studies involving collective dynamics of dislocations have not been pursued before due to lack of a proper computational framework except recently. Sudmanns and El-Awady [13] carried out DDD for multi-principal element FCC alloys (MPEA) to predict the mechanical behavior. Solute clustering was modeled based on Dirichlet distribution and hardening and increased cross-slip activity was reported for alloy containing solute clusters. In this paper, we present a more comprehensive and an alternative computational methodology for studying the collective dynamics of dislocations using DDD which is capable of taking into consideration the correlation in the composition data representative of the fluctuations. We present a DDD framework involving (1) stochastic reconstruction of spinodal-like composition fields, (2) molecular dynamics for determining stress, composition, line-orientation and temperature dependent velocities of the dislocations, and (3) internal coherency stresses due to elastic inhomogeneity. We present some preliminary results on the effects of the composition fluctuation wavelengths on the mechanical behavior of alloys by choosing FeCrAl alloy as a model material system.

2. Computational method

We are interested in understanding the bulk deformation response of FeCrAl alloys in the presence of composition fluctuations. To achieve this, we have utilized a representative volume element (RVE) framework with a domain of size 2.43 μm × 2.45 μm × 2.45 μm for investigating bulk plasticity in the alloy. Figure 1 shows the modelling framework adopted for studying the mesoscopic influence of the composition fluctuations on the deformation behaviour of the FeCrAl alloys. Virtual composition fields with probability distributions representing composition data from a neutron irradiated FeCrAl (C35M) alloy, with 13wt% and 5wt% nominal Cr and Al concentrations, respectively, are generated using a stochastic reconstruction scheme presented in Pachaury et al. [14]. Three different wavelengths/covariance relaxation length scales, λ, of 100 nm, 200 nm and 300 nm are utilized and the dislocation mediated plasticity in these alloys are compared to an alloy with homogeneous composition for C35M representing a uniform distribution. The composition fields are also utilized to determine composition, stress, temperature, and line direction dependent velocities of dislocations using molecular dynamics (MD). Readers are referred to Kumagai et al. [15] for more details regarding the determination of the dislocation velocities in FeCrAl alloys. Reconstructed composition maps are utilized to determine the coherency stresses within the material. These coherency stresses represent the internal stress field in the alloy due to fluctuating lattice parameter for the fluctuating composition fields. The coherency stresses are determined by solving an elastic boundary value problem (BVP) considering the domain to be an RVE of the bulk crystal. Consequently, a strain periodicity framework is implemented similar to one of the co-author's earlier work on continuum dislocation dynamics (CDD) [16]. The difference in this work and the earlier work is the formulation of the eigenstrain for the fluctuating composition fields which is modelled as $\boldsymbol{\varepsilon}^* = \frac{1}{3}\frac{\Delta V}{V_{nom}}\delta_{ij}\boldsymbol{e}_i \otimes \boldsymbol{e}_j$, where ΔV is the change in the volume of the lattice unit cell with respect to the nominal volume V_{nom}, δ_{ij} is the Kronecker delta, and $\boldsymbol{e}_i$ and $\boldsymbol{e}_j$ are the basis vectors. The implementation of the elastic BVP for the coherency stresses is carried out in MOOSE by sampling grid/voxel dependent eigenstrains [17]. 3D composition fields, coherency stresses, and a parametric form of the dislocation velocities are coupled to DDD. For simulating the collective dynamics of dislocations, Mechanics of Defect Evolution Library (MoDELib) is utilized [18]. We utilize RVE boundary conditions for DDD to simulate bulk plasticity in the alloy [19–22]. Constant elastic properties with elastic modulus and Poisson's ratio determined from Field et al. [7] have been utilized for the coherency stresses and DDD calculations at room temperature. Initial dislocation density consists of randomly populated circular dislocation loops of radius 2.5 μm with density of approximately 4 μm^{-2}. Only glide on $\{1\bar{1}0\}\langle 111\rangle$ slip system is considered for the BCC alloy. A phenomenological cross-slip law of Chaussidon et al. [23] is implemented for screw segments. It is worthwhile to note that we do not distinguish between the twin and anti-twin sense of deformation for determining the probabilities of cross-slip from the primary slip system to the conjugate slip system due to unavailability of cross-slip behavior in BCC alloys. All calculations are carried out at a constant strain rate of 100 s^{-1} along $[001]$ direction where the crystal is oriented as $[120] \times [\bar{2}10] \times [001]$ to minimize self-annihilations. Initial relaxation of the dislocation configuration for all the alloys is carried out under the presence of the coherency stresses (for alloys with composition fluctuations) and self-stresses of the dislocations.

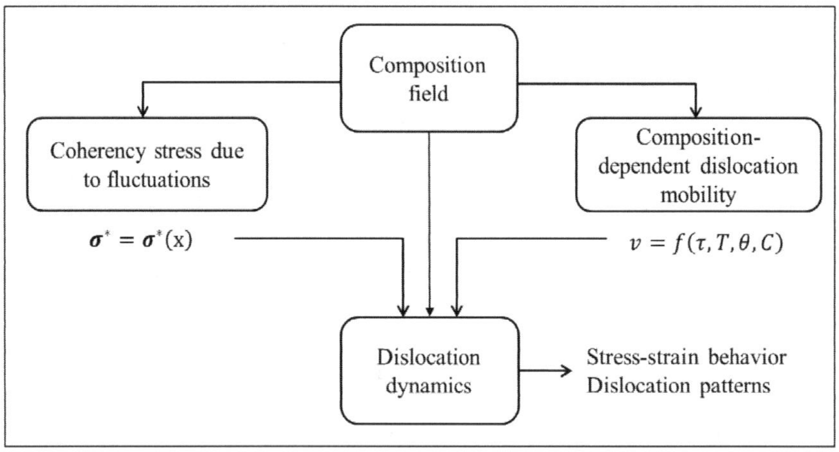

Figure 1. The modeling framework illustrating the coupling of stochastic composition reconstruction, molecular dynamics calculations for determining stress, temperature, line direction and composition dependent mobility law for FeCrAl alloys, and coherency stresses to DDD. $\boldsymbol{\sigma}^*$ represents the elastic coherency stress field which depends on spatial location.

3. Results and discussion

Figure 2 (a) shows the Cr-composition field with wavelength of 100 nm reconstructed using the stochastic process. Similar composition fields are also obtained for the composition of Fe, Cr and Al for the three wavelengths used in this study. To confirm the increasing wavelengths of the composition fluctuations, covariance of the Cr-Cr composition is also plotted in figure 2 (b) for all the reconstructed alloys. More details regarding the determination of covariance can be found in Pachaury et al. [14]. With the increase in the wavelength of the composition fields, the wavelength of the fluctuating coherency stress increases. This along with the composition dependent velocities are expected to facilitate/impede the motion of the dislocations as they glide, hence affecting the microplasticity in the alloys. The combined effects manifest by altering the macroscopic response of the alloys. To study this, we now turn our attention towards our understanding of the alloy plasticity from a DDD perspective.

IOP Conf. Series: Materials Science and Engineering 1249 (2022) 012038 doi:10.1088/1757-899X/1249/1/012038

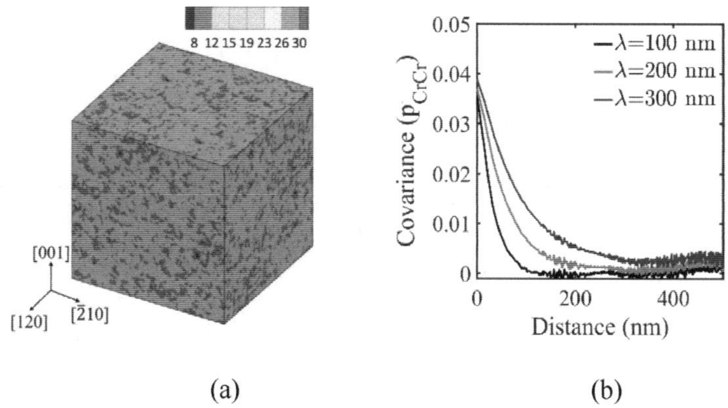

Figure 2. Reconstructed composition map using stochastic process, (a) Cr-composition field (wt%) with 100 nm fluctuating wavelength, and (b) CrCr covariance for the three reconstructed composition maps.

Figure 3 shows the smoothed stress-strain curve, (a), and the evolution of the dislocation density, (b), for the homogeneous and decomposed alloys. The smoothed stress-strain curves are plotted after taking moving average in the plastic regime of the stress-strain response. All decomposed alloys exhibit increased hardening response as compared to the homogeneous alloy consistent with the experimental observations. With the increase in the composition fluctuation wavelengths, the yield strength of the alloys increases. Differences are also observed in the evolution of the dislocation densities in the decomposed alloys as compared to the homogeneous alloy. During the initial (elastic) loading stage of the crystal, the dislocations do not get activated below a threshold strain which is representative of the friction stress barrier for the BCC FeCrAl alloys. For more details regarding the threshold stress, the readers are referred to Kumagai et al. [15]. During the plastic deformation of the alloy, the dislocations tend to evolve more rapidly with strain for decomposed alloys as compared to the homogeneous alloy. Plateaus (such as marked by point A in figure 3 (b)) are indicative of stick-slip motion of the dislocations for all the alloy configurations tested. For the homogeneous alloy, the density increases monotonically with plateaus whereas some fluctuations are observed in the evolution of the densities for decomposed alloys where the increase of the density is not perfectly monotonic. Snapshots of dislocation densities in homogeneous and decomposed alloys at a strain of ~0.38% are shown in figure 3(c) and (d), respectively. It is evident that the arrangement of dislocations differs significantly in the decomposed alloys than in homogeneous alloy due to localized composition variations which alter the mobility of the dislocations and coherency stresses. The coherency stresses is also expected to alter the local cross-slip frequency of the screw segments, hence changing the macroscopic behavior of the crystal manifested in the form of different evolution of the dislocation densities. It is expected that the evolution of the slip-system densities is also different for the decomposed alloys as compared to the homogeneous alloys. However, a detailed investigation of the cross-slip and slip-system densities is beyond the scope of the current work and will be a subject of a future study. Moreover, relatively straight dislocation segments are observed for the homogeneous alloy whereas wavy dislocation lines are observed for the decomposed alloys. This observation is consistent with the experimental observations of Kato et al. [12] for CuNiSn alloys.

IOP Conf. Series: Materials Science and Engineering 1249 (2022) 012038 doi:10.1088/1757-899X/1249/1/012038

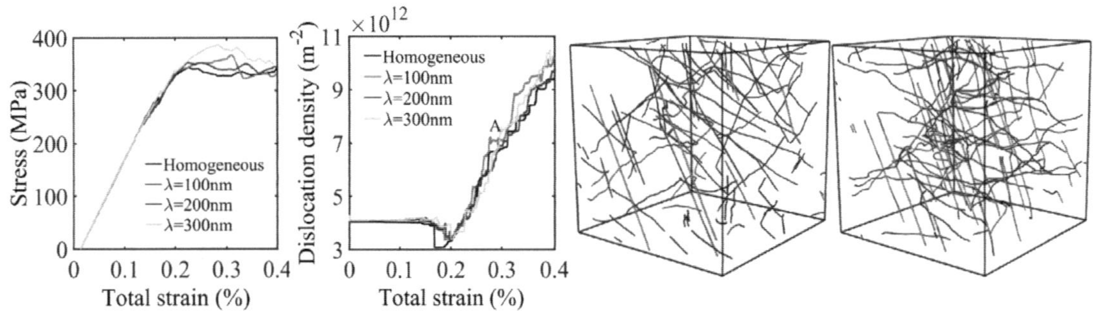

(a) (b) (c) (d)

Figure 3. DDD results. (a) Stress vs total strain, (b) evolution of the total dislocation density as a function of total strain for the 4 runs with different composition fluctuation wavelengths, and (c) and (d) dislocation density snapshots at 0.38% strain for the homogeneous and decomposed alloys, respectively.

4. Concluding remarks

The question of how the plastic strength of alloys depends on composition is critical to alloy design. Numerous classical works have tackled this question in the past. Yet, the models available to date primarily focus on the strength of alloys at the onset of yielding and seldom address the role of alloy composition in the hardening and dislocation microstructure evolution regime. The above question becomes even more important in situations in which the alloys are compositionally nonuniform at the mesoscale, as in spinodally decomposed alloys, high entropy alloys, and additive-manufactured alloys. In this work, we have presented a data-driven multiscale framework based on DDD for investigating plasticity in alloys exhibiting spinodal/spinodal-like composition fluctuations. Specifically, a 3-way coupling involving stochastic reconstruction for composition fields, MD for dislocation velocities, and an elastic BVP for coherency stresses, has been implemented in DDD simulations. The developed framework is capable of predicting the hardening in decomposed alloys which is dependent on the wavelength of the composition fluctuations. Additionally, it is expected that the composition fluctuations change the cross-slip activity in the decomposed alloys which results in different dislocation microstructures in decomposed alloys as compared to the homogeneous alloy. Hence, microplasticity in decomposed alloys results in different macroscopic response of the decomposed alloys as compared to the homogeneous alloy.

Acknowledgements

This project is supported by the US Department of Energy, Office of Nuclear Energy, contract DE-NE0008758 and by the Office of Science, Division of Materials Sciences and Engineering, through award number DE-SC0017718 at Purdue University.

References

[1] Garner F A, McCarthy J M, Russell K C and Hoyt J J 1993 Spinodal-like decomposition of Fe-35Ni and Fe-Cr-35Ni alloys during irradiation or thermal aging *J. Nucl. Mater.* **205** 411–25

[2] Garner F A, Brager H R, Dodd R A and Lauritzen T 1986 Ion-induced spinodal-like decomposition of Fe-Ni-Cr invar alloys *Nucl. Instrum. Methods Phys. Res. B: Beam Interact. Mater. At.* **16** 244–50

[3] Ostovari Moghaddam A, Shaburova N A, Samodurova M N, Abdollahzadeh A and Trofimov E A 2021 Additive manufacturing of high entropy alloys: A practical review *J Mater Sci Technol* **77** 131–62

[4] Brager H R, Garner F A and Hamilton M L 1985 The influence of composition on microstructural evolution and mechanical properties of irradiated Fe-Ni-Cr ternaries *J. Nucl. Mater.* **133–134** 594–8

[5] Chen W-Y, Miao Y, Gan J, Okuniewski M A, Maloy S A and Stubbins J F 2016 Neutron irradiation effects in Fe and Fe-Cr at 300 °C *Acta Mater.* **111** 407–16

[6] Field K G, Hu X, Littrell K C, Yamamoto Y and Snead L L 2015 Radiation tolerance of neutron-irradiated model Fe–Cr–Al alloys *J. Nucl. Mater.* **465** 746–55

[7] Field K G, Snead M A, Yamamoto Y and Terrani K A 2018 Handbook on the Material Properties of FeCrAl Alloys for Nuclear Power Production Applications *(FY18 Version: Revision 1) (Oak Ridge National Lab. (ORNL), Oak Ridge, TN (United States))*

[8] Massey C P, Zhang D, Briggs S A, Edmondson P D, Yamamoto Y, Gussev M N and Field K G 2021 Deconvoluting the Effect of Chromium and Aluminum on the Radiation Response of Wrought FeCrAl Alloys After Low-Dose Neutron Irradiation *J. Nucl. Mater.* **549** 152804

[9] Cahn J W 1963 Hardening by spinodal decomposition *Acta Metall* **11** 1275–82

[10] Kato M, Mori T and Schwartz L H 1980 Hardening by spinodal modulated structure *Acta Metall* **28** 285–90

[11] Kato M, Horie S and Lee T C 1986 Evaluation of Critical Resolved Shear Stress for Systems with General Periodic Stress Fields *Phys. Status Solidi A* **98** 209–14

[12] Kato M, Katsuka S, Okamine S and Sato A 1986 Deformation behaviour and microstructure of Cu-10Ni-6Sn spinodal alloy single crystals *Mater. Sci. Eng.* **77** 95–102

[13] Sudmanns M and El-Awady J A 2021 The effect of local chemical ordering on dislocation activity in multi-principle element alloys: A three-dimensional discrete dislocation dynamics study *Acta Mater.* **220** 117307

[14] Pachaury Y, Kumagai T, Wharry J P and El-Azab A 2022 A data science approach for analysis and reconstruction of spinodal-like composition fields in irradiated FeCrAl alloys. *Acta Mater.* **234** 118019. https://doi.org/10.1016/j.actamat.2022.118019

[15] Kumagai T, Pachaury Y, Maccione R, Wharry J and El-Azab A 2021 An atomistic investigation of dislocation velocity in body-centered cubic FeCrAl alloys *Materialia* **18** 101165

[16] Lin P and El-Azab A 2020 Implementation of annihilation and junction reactions in vector density-based continuum dislocation dynamics Modelling Simul. Mater. Sci. Eng. **28** 045003

[17] Permann C J, Gaston D R, Andrš D, Carlsen R W, Kong F, Lindsay A D, Miller J M, Peterson J W, Slaughter A E, Stogner R H and Martineau R C 2020 MOOSE: Enabling massively parallel multiphysics simulation *SoftwareX* **11** 100430

[18] Po G, Mohamed M S, Crosby T, Erel C, El-Azab A and Ghoniem N 2014 Recent Progress in Discrete Dislocation Dynamics and Its Applications to Micro Plasticity JOM **66** 2108–20

[19] El-Azab A 2000 The boundary value problem of dislocation dynamics Modelling Simul. Mater. Sci. Eng. **8** 37–54

[20] Pachaury Y, Po G and El-Azab A 2022 Discrete dislocation dynamics for crystal RVEs. Part 1: Periodic network kinematics *J Mech Phys Solids* **163** 104861

[21] Bulatov V, Rhee M and Cai W 2002 Periodic Boundary Conditions for Dislocation Dynamics Simulations in Three Dimensions *MRS Symp Proc* **653**

[22] Marian J, Fitzgerald S and Po G 2020 Discrete Dislocation Dynamics Simulations of Irradiation Hardening in Nuclear Materials *Handb. Mater. Model.* pp 2243–71

[23] Chaussidon J, Robertson C, Rodney D and Fivel M 2008 Dislocation dynamics simulations of plasticity in Fe laths at low temperature *Acta Mater.* **56** 5466–76

3D grain mapping by laboratory X-ray diffraction contrast tomography implemented on a conventional tomography setup

H Fang[1, 2], R Granger[1, 3], W Ludwig[2, 4] and P Lhuissier[1]
[1]Université Grenoble Alpes, Grenoble INP, CNRS SIMaP, 38000 Grenoble, France.
[2]European Synchrotron Radiation Facility, BP 220, 38043 Grenoble, France.
[3]Université Grenoble Alpes, Grenoble INP, CNRS 3SR, 38000 Grenoble, France.
[4]Université de Lyon, INSA-Lyon, CNRS MATEIS, 69621 Villeurbanne, France.

E-mail: haixing.fang@grenoble-inp.fr

Abstract. Non-destructive 3D characterization of grain orientations, shapes and sizes, i.e. grain mapping, offers immense opportunities for studying microstructural evolution in polycrystalline materials. In addition to a number of well-established grain mapping techniques available at synchrotron facilities, a polychromatic variant - laboratory diffraction contrast tomography (LabDCT) - using lab-based x-rays, has been developed and commercialized. Yet, the product is bounded to a specific instrument and requires a commercial license, which limits the use on widely available laboratory instruments. To promote the availability of LabDCT, we have developed a grain reconstruction method and implemented it on a conventional X-ray tomography setup at the SIMaP laboratory for LabDCT grain mapping. First, we tested the grain reconstruction algorithm by comparing an input virtual grain structure and a reconstructed volume using the forward simulated diffraction projections from the input structure. Then, we experimentally characterized an AlCu alloy sample using LabDCT and validated the grain mapping result by a grain reconstruction from synchrotron DCT measurement. Last, perspectives on further development of generalizing LabDCT technique on conventional tomography setups are discussed.

1. Introduction

A detailed characterization of grain orientations, morphologies and grain boundary characters is often required for improving properties of metals and alloys. X-ray based grain mapping techniques offer such possibility to characterize 3D grain structures in a non-destructive manner over reasonably large fields-of-view (up to mm^3) and are thus highly complementary to established (surface / thin foil) electron diffraction techniques [1]. A number of grain mapping techniques have been well established at large synchrotron facilities, such as differential aperture X-ray microscopy (DAXM) [2], three-dimensional X-ray diffraction microscopy (3DXRD) [3], diffraction contrast tomography (DCT) [4] and dark-field X-ray microscopy [5]. To broaden the use of grain mapping and overcome the limited access to synchrotron facilities, lab-based diffraction contrast tomography (LabDCT), using a conical polychromatic beam from laboratory source, has been developed [6] and commercialized [7]. The former approach was based on Friedel pair based indexing, followed by tomographic reconstruction and can only deal with moderate number of grains in the illuminated sample volume due to overlap of diffraction spots. The latter product is based on a forward modelling strategy [7], bounded to a specific instrument and requires a commercial license, limiting the use on widely available laboratory instruments.

To promote the availability of LabDCT, we have developed a forward modelling based grain reconstruction method and implemented it on a conventional X-ray tomography setup at the SIMaP laboratory for grain mapping. Multiple algorithms for grain reconstruction have been developed based on both forward and back calculations. Testing of these algorithms was done by comparing an input virtual grain structure and a reconstructed volume using the forward simulated diffraction projections from the input structure. After the reconstruction algorithms were carefully assessed, we experimentally characterized an AlCu alloy sample using LabDCT and validated the grain reconstruction result by the one from synchrotron DCT measurement. With regard to the flexibility in geometrical configuration and choices of X-ray sources and detectors, we discuss perspectives on further development of generalizing the LabDCT technique on conventional tomography setups.

2. Experimental setup

LabDCT was implemented on a conventional tomography setup (EASYTOM XL Nanofocus tomography manufactured by RX solutions) at the SIMaP laboratory. This tomography instrument has two sources: i) a micro source generating high-flux beam with a source size from 1 to 20 μm, suitable for scanning large objects; ii) a nano source providing a less powerful beam but with a smaller source size, ranging from 0.3 to 3 μm, suitable for scanning small samples to obtain high resolution (Figure 1). For each source there are three modes for tuning the source size (being small, middle and large) according to specific experimental demands. Different combinations of targets (W or Mo) and filaments (W or LaB6) are also available. This setup is adaptable to three detectors: a flat panel detector with 1536×1920 pixels and a pixel pitch of 127 μm, a CCD detector with 2040×2040 pixels and a pixel pitch of 24 μm, and a direct photon counting CdTe detector equipped with two modules. Each module contains 512×402 pixels with a pixel pitch of 62 μm. The large range of choices in sources and detectors offers a flexible solution for different tasks with different samples.

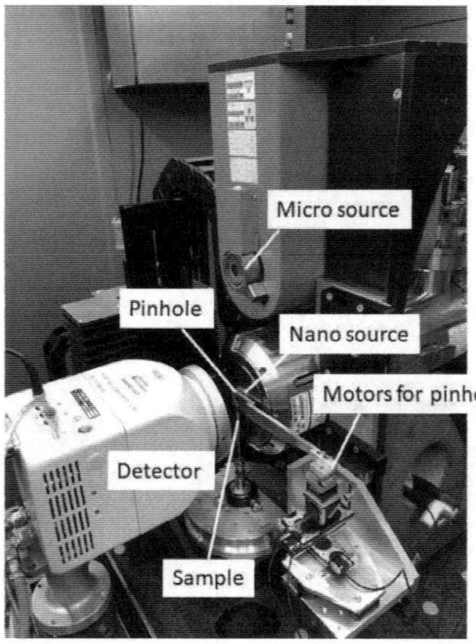

Figure 1. Tomography setup for LabDCT grain mapping.

In diffraction mode for grain mapping, the incoming beam was confined by a pinhole placed in between the sample and the source. As shown in Figure 1, a customized pinhole support was machined and three micro motors (enable movements in x, y and z directions) were used to position the pinhole with a precision of several nanometers. Pinholes with different sizes ranging from Ø50 μm to 2 mm are available for LabDCT measurements. Circular beamstops with different diameters were made of 0.5 mm Pb plates. For diffraction tomography measurement, a beamstop is placed right in front of the detector to block the direct transmitted beam, and thus to enhance diffraction signals recorded by the outer area of the detector. For absorption tomography, the beamstop is removed.

3. Materials and Methods

An AlCu alloy (8 wt.% Cu) sample with a cylindrical diameter of ~650 μm and a height of ~160 μm was cut from a heated treated rod, in which the copper enriched eutectic phase has precipitated at grain boundaries. A conical polychromatic beam emitted from a tungsten target of the nano source (80 kV, 1.3 W and in middle-size mode) was used to illuminate this sample. 2D projections behind the sample were collected by a CCD detector with 2040×2040 pixels and a pixel pitch of 24 μm. The direct transmitted beam was shielded by a beamstop to enhance the diffraction signal. Sample-to-source (L_{ss}) and sample-to-detector (L_{sd}) distances were 6.1 and 52.9 mm, respectively. In total, 121 projections were acquired at a rotation step of 3° and each projection was averaged over 6 frames with an exposure time of 60 s (maximum exposure time allowed for this detector). The experimental parameters selected here were based on exercises to balance the signal-to-noise ratio and the scanning time. A subsequent tomographic scan was also performed to reconstruct a sample shape that serves as a volume mask for grain mapping. The whole diffraction and tomography scan took about 13 h. To reconstruct grain maps, we developed multiple computational efficient methods and the result reported here was obtained using one of these methods (which will be presented in a forthcoming publication). The reconstruction time for this sample was about 6 h using a computing node equipped with a NVIDIA Tesla V100-PCIE-32GB GPU on the ESRF cluster.

This sample was also characterized by synchrotron DCT at beamline ID11 of European Synchrotron Radiation Facility (ESRF). A parallel monochromatic beam with an energy of 43.6 keV was used to illuminate the sample. Transmitted diffraction signals were recorded by a sCMOS (Andor Marana) camera with 2048×2048 pixels, coupled to a 10 μm thickness transparent luminescent screen via a 7.5× objective lens, resulting in an effective pixel size of 1.22 μm. The distance between the rotation axis and the detector was 11 mm. A DCT scan contains 3600 equally spaced projections over 360° sample rotation. Exposure time for each projection was 0.1 s and a total scan took about 10 min. Subsequent grain reconstruction was performed using the method described in [4, 8]. The obtained grain map was verified by the good agreement with the tomography volume, where the Cu-enriched eutectic phase was resolved at grain boundaries. Therefore, we directly use this grain map to check the grain structure reconstructed by LabDCT.

To test the grain reconstruction algorithm for LabDCT, simulated projections using a forward projection model [9] were created from a virtual Fe grain structure, containing 6 grains with an average size of 100 μm in a cylindrical volume (a diameter of 150 μm and a height of 300 μm). The simulations were performed in the same conditions as the experiment described above. A grain structure was reconstructed based on the simulated projections. The grain reconstruction time was about 20 min using the same computer as described above.

4. Results and discussion

4.1. Reconstruction of the virtual sample

Figure 2 compares grain structures between the ground truth input and the reconstructed result. In total 6 grains have been reconstructed, which is the same as the input. 3D visualizations (Figure 2a and b) show that all the reconstructed grains are very similar in shape with respect to the input. Examination on 2D slices show that grain boundary positions are largely coincident (Figure 2c). 95% of the voxels are fully matched and more than 98% deviate no more than 2 pixels. The maximum deviation is found to be 4 pixels only. As shown in Figure 2d, comparison of grain size confirms that the grains were accurately reconstructed and the relative grain size difference is 1.7% ± 0.8% (mean and standard deviation). The disorientation of grain pairs between the input and the reconstructed grain maps is 0.05 ± 0.01°. The grain center-of-mass is 0.89 ± 0.46 pixels. All these numbers indicate that the reconstruction algorithm performs well and the reconstructed grain map is in very good agreement with the input one.

To further assess the performance of the reconstruction algorithm, we used the forward projection model to check the positions and the shapes of the spots from the reconstructed grain map on the initially simulated LabDCT projections, i.e. regarded as "experimental" projection. Figure 3a shows that the outlines of the simulated spots match very well with the "experimental" ones. To quantify the difference, we calculated the difference in Euclidian distance of center-of-mass for each spot pair (Δ_{dis}). Statistics of the spot pairs present in all the 121 projections are summarized in Figure 3b. The figure shows that $\Delta_{dis} = 1.7 \pm 1.3$ pixels, approximately 2 times the deviation of grain centroids.

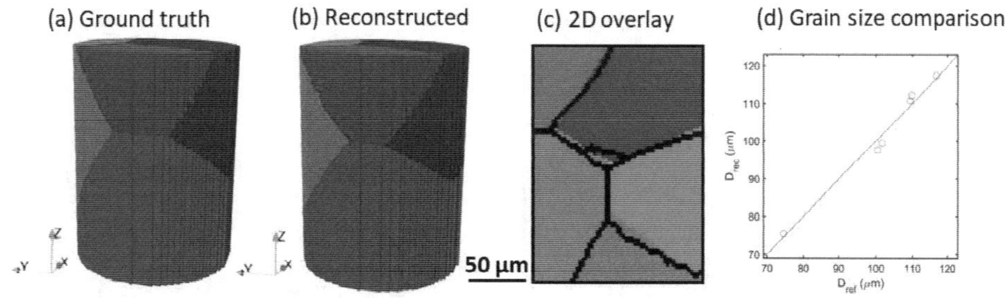

Figure 2. Comparison of reconstructed grain structure by LabDCT with the ground truth input. (a) Ground truth input and (b) reconstructed structures visualized in 3D; (c) overlay of grain boundaries in the reconstructed grain map onto the corresponding XZ slice of the ground truth; (d) grain size comparison between the ground truth and the LabDCT reconstruction. Grains are colored in IPF-Z direction. Voxel size is 2.5 μm.

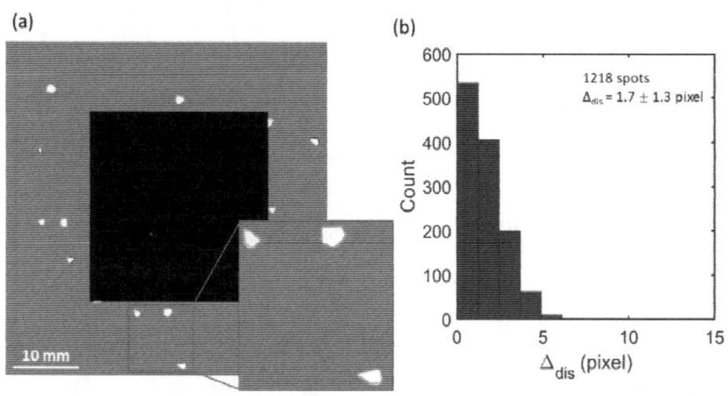

Figure 3. Verification by forward simulated spots. (a) Overlay of forward spots from the reconstructed grain structure onto the LabDCT projection simulated for the initial ground truth input; (b) statistics of Δ_{dis} between the reconstructed and the input structures. Outlines of the forward projected spots are colored according to {*hkl*} families: red-{011}; green-{002}; blue-{112}; yellow-{022}; cyan-{013}.

4.2. Reconstruction of the experimental AlCu alloy sample

Figure 4 shows grain maps reconstructed from the synchrotron radiation DCT (SR-DCT) and the LabDCT measurements. A full volumetric grain map was reconstructed from SR-DCT, which contains 329 grains with an average grain size of 52.5 ± 20.5 μm. At the time of the LabDCT measurement performed, the current on the target of the X-ray source was rather limited, thereby a relatively low flux was used for scanning this sample. As a result, diffraction signals from relatively small grains (mostly < 40 μm) could not be distinguished and properly segmented from the background noise in the projection images. To exclude the effect of the undetected signals on the comparison of grain maps between SR-DCT and LabDCT, we used the forward model to compute values of two parameters for each grain centroid voxel: 1) completeness (*Comp*), i.e. the number of intersected spots between forward simulation and experiment divided by the theoretically expected number of spots on the detector; 2) median distance (D_{med}), median value of distances between each forward simulated spot and its nearest experimental one. During the LabDCT reconstruction, indexing of voxel orientation was accepted only when the voxel has $Comp \geq 0.3$ as well as $D_{med} < 20$ pixels. Therefore, the same condition was used to filter the grain map of SR-DCT and a partial map was thus obtained as shown in Figure 4b.

Figure 4c shows the grain map reconstructed from LabDCT. Compared to Figure 4b, the same number of grains was obtained and each grain has similar shapes and locations as its counterpart in SR-DCT. By pairing all the grains, we compared the disorientation and the center-of-mass distance between corresponding grains. Results show that the disorientation of grain pairs is 0.06 ± 0.03° and the grain center-of-mass distance is 3.96 ± 1.35 pixels. This is less well, compared to the result of the reconstruction for the synthetic dataset described above. Since the diffraction signals were relatively low because of the low flux, a complete grain indexing and reconstruction cannot be obtained for this sample volume. Additionally, many spots can be under-segmented, restricting the proper voxel growth of the corresponding grains. Examples of these grains are indicated by the red arrows in Figure 4c, showing that they are significantly smaller compared to their counter parts in Figure 4b. The difference in average grain size further confirms a high degree of spot under-segmentation, where the average grain size is 75.9 ± 8.8 μm for the filtered grain map of SR-DCT and it is 64.7 ± 19.7 μm for the grain map of LabDCT.

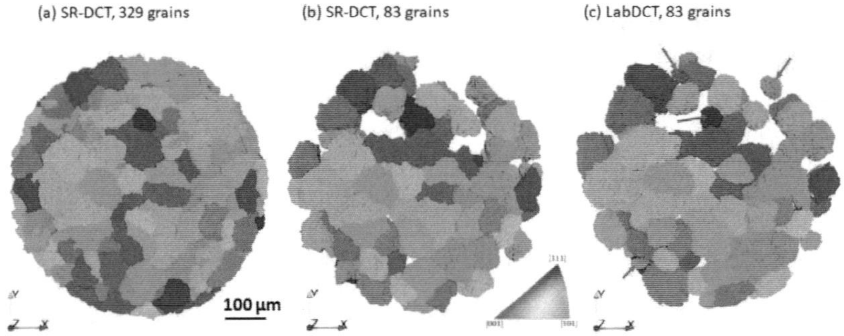

Figure 4. Grain maps reconstructed from synchrotron radiation DCT (SR-DCT) and LabDCT for an AlCu alloy sample. (a) full grain map from SR-DCT contains 329 grains; (b) a filtered grain map contains 83 grains, whose completeness (*Comp*) ≥ 0.3 and median distance (D_{med}) < 20 pixels based on forward calculation result on the LabDCT projections; (c) grain map reconstructed from LabDCT contains 83 grains with *Comp* ≥ 0.3 and D_{med} < 20 pixels. Grains are colored in IPF-Z direction. Voxel size is 2.7 μm.

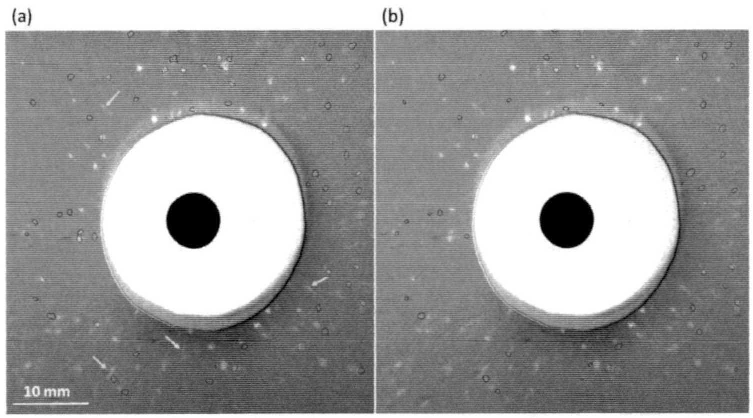

Figure 5. Overlay of forward projected spots from the reconstructed (a) SR-DCT and (b) LabDCT grain maps onto the LabDCT projections. Yellow arrows indicate examples of streaking spots. Outlines of the forward spots are colored according to {*hkl*} families: red-{111}; green-{002}; blue-{022}.

To verify both reconstructions, we computed the diffraction spots using the forward projection model and overlaid them onto the experimental projection. As shown in Figure 5a and b, most of the forward spots superimposed well onto experimental spots. Some may slightly shift while some others may be too big or too small compared to their overlaid experimental spots. Statistics of Δ_{dis} for more than 1400 spots for all the 83 grains show that Δ_{dis} = 5.2 ± 3.1 pixels for SR-DCT and Δ_{dis} = 5.7 ± 3.3 pixels for LabDCT. This suggests that the LabDCT reconstruction is only slightly less performed compared to SR-DCT. The reason for not having a smaller value of Δ_{dis} for SR-DCT is that some grains may contain non-negligible strain as indicated by some streaky spots (see Figure 5). Although the lattice strain was not considered during SR-DCT reconstruction, the quality of the SR-DCT reconstruction was less affected compared to LabDCT as a monochromatic parallel beam was used for the SR-DCT measurement.

4.3. Discussion
The comparisons of different grain maps demonstrate that grain mapping by LabDCT has been implemented successfully on a conventional laboratory instrument and a computational efficient grain reconstruction method has been established. Notably, this method is general and it can be applied to reconstruct grain maps for diffraction data acquired on other tomography instruments in both Laue-focusing and non-Laue-focusing geometries.

The experimental result also shows that a weak X-ray beam ultimately limits the capability of LabDCT to reconstruct small grains. As a result of unsuccessful reconstruction for small grains, shape reconstruction for already-indexed grains may be affected and may not be accurate due to over-growth. Therefore, increasing the photon flux as well as higher detective quantum efficiency of the detector system are very important to further improve the performance of LabDCT in addition to improvements in spot segmentation and reconstruction algorithm [10, 11]. To these aims, we suggest the following measures: 1) allow higher current on the target of the source using thicker material for the anode or improving the heat conduction around the target; 2) use direct photon counting detectors such as the CdTe 'color' detectors, which are expected to produce low background noise, have better detective quantum efficiency at high X-ray energies and offer the capability to define an X-ray energy range for detecting signals. On the other hand, there is a need to reduce scanning time for LabDCT experiments, thereby enabling 4D (space + time) characterization. This requires fast counting detectors, such as flat panel with large pixel pitch or direct photon counting detector equipped with CMOS chip, as well as a generous geometrical constraint for optimizing different settings of L_{sd} considering the area and the pixel pitch of each different detectors.

5. Conclusions
In this study, we demonstrate that grain mapping by LabDCT was made available on a conventional tomography setup at the SIMaP laboratory. A computational efficient grain reconstruction algorithm has been developed and a pipeline to reconstruct grain maps has been established. Verification by a virtual Fe sample indicates a good performance of the current method. The differences in disorientation and positions of grain center-of-mass are $0.05 \pm 0.01°$ and 0.89 ± 0.46 pixels, respectively, and the relative grain size difference is $1.7\% \pm 0.8\%$. The comparison of grain maps for an AlCu alloy sample reconstructed from SR-DCT and LabDCT further validates the reconstruction method. Suggestions for further developing the LabDCT technique towards better detection limit and faster acquisition were presented.

Acknowledgements
This work is part of the project Advanced Laboratory X-ray Microtomography funded by the Agence Nationale de la Recherche (ANR-18-CE42-0005).

References
[1] Shahani A J, Xiao X, Lauridsen E M and Voorhees P W 2020 *Mater. Res. Lett.* **8** 462
[2] Larson B C, Yang W, Ice G E, Budai J D and Tischler J Z 2002 *Nature* **415** 887
[3] Poulsen H F 2020 *Curr. Opin. Solid. St. Mater. Sci.* **24** 100820
[4] Ludwig W, Schmidt S, Lauridsen E M and Poulsen H F 2008 *J. Appl. Cryst.* **41** 302
[5] Simons H, King A, Ludwig W, Detlefs C, Pantleon W, Schmidt S, Stöhr F, Snigireva I, Snigirev A and Poulsen H F 2015 *Nature Comm.* **6** 6098
[6] King A, Reischig P, Adrien J and Ludwig W 2013 *J. Appl. Cryst.* **46** 1734
[7] Bachmann F, Bale H, Gueninchault N, Holzner C and Lauridsen E M 2019 *J. Appl. Cryst.* **52** 643
[8] Reischig P, King A, Nervo L, Viganó N, Guilhem Y, Palenstijn W J, Batenburg K J, Preuss M and Ludwig W 2013 *J. Appl. Cryst.* **46** 297
[9] Fang H, Juul Jensen D and Zhang Y 2020 *Acta Cryst.* A **76** 652
[10] Fang H, Juul Jensen D and Zhang Y 2021 *IUCrJ* **8** 559
[11] Fang H, Hovad H, Zhang Y, Clemmensen L K H, Kjaer Ersbøll and Juul Jensen D 2021 *IUCrJ* **8** 719

Separation of XRD peak profiles in single-phase metals with bimodal grain structure to analyze stress partitioning

E Sjögren-Levin[1], W Pantleon[2], A Ahadi[1], Z Hegedüs[3], U Lienert[3], N Tsuji[4], K Ameyama[5] and D Orlov[1]

[1] Department of Mechanical Engineering Sciences, Lund University, Lund, Sweden.
[2] Department of Civil and Mechanical Engineering, Technical University of Denmark, Kongens Lyngby, Denmark.
[3] Deutsches Elektronen-Synchrotron (DESY), Hamburg, Germany.
[4] Department of Materials Science and Engineering, Kyoto University, Kyoto, Japan.
[5] Department of Mechanical Engineering, Ritsumeikan University, Kusatsu, Japan.

E-mail: elis.sjogren@material.lth.se, dmytro.orlov@material.lth.se

Abstract. Materials with bimodal grain size distributions have an attractive combination of strength and ductility. Harmonic structure materials are a category of bimodal-structure materials with a specific microstructure design. The deformation mechanisms of such novel materials during the early stages of deformation are not well understood. Thus, we deformed nickel with harmonic structure in tension until a true strain of 0.04 while recording powder diffraction patterns with high-energy synchrotron X-rays. Line profile analysis based on such data enables quantification of stress states and lattice defect densities in different phases in multi-phase materials. Bimodal size distributions in single-phase materials add extra complexity due to the absence of differences in composition and crystal structure causing the diffraction peaks from fine and coarse grains to appear at the same diffraction angles. Therefore, prior to any meaningful line profile analysis, the respective diffraction profiles need to be separated. A general method for automatically separating profiles originating from different grain fractions in bimodal materials is presented in this work.

1. Introduction

A rapidly increasing number of studies over the past decade has demonstrated possibilities for overcoming the strength – ductility balance in single-phase materials [1]. This can be achieved by careful tailoring of heterogeneous microstructures with regions of coarse and fine grains [2, 3]. Harmonic structure (HS) materials have a special kind of heterogeneous microstructure where regions of fine grains create a continuous network (skeleton) encompassing islands of coarse grains [4, 5].

Several recent experimental [6-8] and computational [3, 9] studies have been focused on investigating the early stages of plastic deformation in HS materials to understand the mechanics of fine and coarse grains. These studies revealed that stresses and strains partition between the grain fractions from the early stages of elastic-plastic transition. However, a direct quantitative measurement of stresses was not performed. *In situ* X-ray diffraction experiments allow following the evolution of stress-state in the constituent phases of composites and alloys during mechanical tests, cf. [10]. A dedicated *in situ* X-ray diffraction experiment can capture the average elastic behavior within each fraction of coarse and fine grains, allowing direct determination of stress-partitioning in addition to microstructural features.

In the case of multi-phase materials, diffraction peaks from different phases are often easy to distinguish as they appear at different scattering angles in a diffraction pattern. For single phase materials with heterogeneous grain sizes, the separation of diffraction peaks from coarse and fine grain fractions is more complicated since they appear at the same diffraction angle. The mean grain diameters of both grain fractions are substantially higher than 1 μm in HS wherefore no significant difference in size broadening is expected between the fractions. Therefore, peak broadening-based methods such as presented in [11] cannot be applied. Furthermore, it is not possible to distinguish the different fractions from specific texture components as in [12].

In this study, we outline a procedure for tailoring the collection of X-ray diffraction data for materials with heterogeneous grain sizes along with an algorithm for separating average diffraction signals from grains of different sizes. The elastic strains of the respective grain fractions are determined and the accuracy of the algorithm is evaluated.

2. Experimental details

2.1. *Material fabrication and characteristics*

Nickel samples were prepared through a route of powder metallurgy by mechanical milling of nickel powder particles with average particle size of 150 μm followed by spark plasma sintering to achieve a harmonic-structure [8]. Samples for characterization by electron backscatter diffraction (EBSD) were prepared by grinding with SiC paper down to #4000 grit followed by polishing with colloidal silica. The typical characteristics of HS materials with a continuous network of fine grains surrounding islands of coarse grains is shown in figure 1a where high angle boundaries (with misorientation angles above 15°) are indicated. A reasonable threshold for separating the grain size fractions into coarse and fine is an equivalent circular diameter of 5 μm (which we will elaborate on in a future paper). Assuming such a threshold, the area fraction of fine grains below 5 μm is determined to be 24%.

Dog-bone shaped, flat tensile specimens with 7 mm gauge length and 1.5 mm width were prepared using electric discharge machining and subsequently mechanically ground to a thickness of 0.8 mm.

2.2. *Experimental set-up at Petra III*

In situ synchrotron X-ray data were acquired at the synchrotron radiation facility Petra III, beamline P21.2, using a beam energy of 52 keV and a beam cross-section of 100x100 μm^2 until a true strain 0.04. The sample was mounted in a custom-made screw-driven load frame equipped with a 1 kN load cell. Markers were placed at the ends of the gauge length for the measurement of macroscopic strain with an optical camera. The load frame was placed with the tensile axis horizontally on top of a diffractometer. The diffractometer allowed for translations as well as ω-rotation around the vertical axis, as indicated in figure 1b. A position-sensitive area detector (VAREX XRD4343CT) with pixel size 150x150 μm^2 and array size 2880x2880 was placed at approximately 1.5 m downstream of the specimen to cover the first six Debye-Scherrer rings of nickel. The detector position and tilt were calibrated with a NIST SRM660c standard LaB_6 powder using Fit2D software.

2.3. *Experimental procedure*

Before commencing tensile loading, the specimen was translated such that the position of a reference grain in the gauge section of the specimen coincided with the rotation center of the diffractometer. Furthermore, this grain was fully illuminated by the beam and had its 400 reflection parallel with the tensile axis. The tensile loading was conducted stepwise with 10-20 MPa stress increments. During loading, the detectors were capturing images continuously for keeping the selected reference grain in the beam. This ensured the illumination of approximately the same specimen volume throughout the experiment. After each load step, the reference grain was re-centered. For data acquisition, the specimen was rotated 1° in 50 intervals of 0.02° around the ω-axis perpendicular to the load axis. Optical images were acquired after each loading step for subsequent digital image correlation to determine the strain.

The beam cross-section of 100x100 μm^2 was selected based on two criteria. The beam needs to be sufficiently large to fully illuminate grains over the entire range of sizes, but small enough to not capture a too high number of coarse grains. Figure 1c depicts one ω-interval of a diffraction pattern corresponding to the Debye-Scherrer ring 111. It can be seen that the pattern constitutes of a continuous ring of low intensity and many overlaying high-intensity diffraction spots. The ring has an appearance similar to that of partially recrystallized cold rolled samples [12-14]. In those studies, the diffraction spots were associated with growing grains, and the continuous ring with a deformed matrix. Thus, in the present study we assume that diffraction spots and continuous rings represent coarse and fine grain fractions, respectively. The use of a larger beam size would result in a diffraction pattern of mostly high-intensity spots. To evaluate the mechanical behavior of the two grain fractions separately, their respective diffraction patterns need to be separated. To capture full diffraction spots from individual grains and allow them to be separated, the data are acquired in small ω-intervals while rocking over ω.

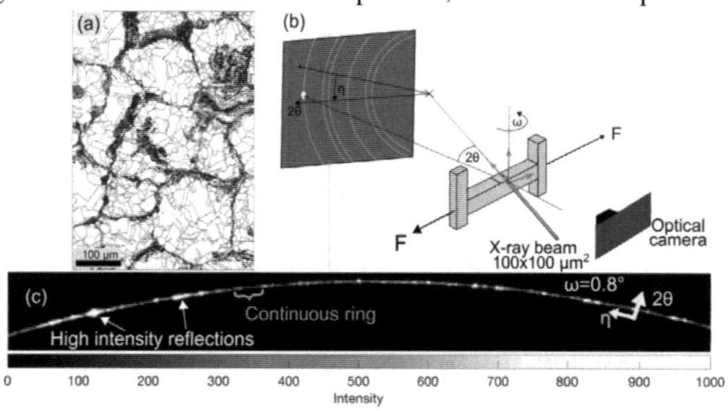

Figure 1: (a) High angle boundary map obtained by ESBD showing a representative overview of harmonic structure nickel. (b) Schematic of the experimental set-up. (c) Typical section of the first Debye-Scherrer ring for one specific ω-interval showing an arc in η of about 30°. Note that on the Debye-Scherrer ring, high-intensity diffraction spots originate from coarse grains, while the continuous ring of low intensity originates from fine grains.

3. Data analysis

3.1. Algorithm for peak separation

The algorithm for separating the diffraction patterns according to the fractions of different grains was developed and realized as a MATLAB® script. The algorithm merely defines a threshold in intensity between the diffraction spots and the continuous Debye-Scherrer rings in a range of $[2\theta, \eta, \omega]$. The recorded intensity may vary among diffraction patterns acquired at different load increments due to variation in beam intensity and acquisition time. Moreover, the recorded intensity varies with peak position (on the detector) depending on Lorentz factor. Therefore, the separation procedure was carried out on the diffraction patterns acquired at each load increment and diffraction ring separately. The entire separation procedure for the 111 ring is presented as an illustrative example. Note, only grains with <hkl> along the tensile axis within a range of ± 15° in η-direction (figure 2a) are of interest in this study.

Step 1: From the recorded diffraction pattern, a 3D intensity distribution is extracted for each ring in an interval $[2\theta_{hkl}\text{-}2\theta_{full}\ldots 2\theta_{hkl}+2\theta_{full}, -15°\ldots15°, 0°\ldots1°]$ with large $2\theta_{full}$ to separate hkl rings (figure 2a).

Step 2: A 1D azimuthal distribution along η is obtained by integrating 2θ and ω (figure 2b). From this, high-intensity peaks are identified and ranked according to their maximum intensity. Simultaneously their positions of maximum intensity η_{max} and their η ranges are determined from the intensity minima between them. The most intense peak at η^{1}_{max} having a range $\eta^{1}_{max}\text{-}\eta^{1}_{low}\ldots\eta^{1}_{max}+\eta^{1}_{high}$ is shown in figure 2c.

Step3: To optimize further analysis, the 2θ range is now limited based on the most intense peak. Azimuthal integration is performed in η and ω in the range $[\eta^{1}_{max}\text{-}\eta^{1}_{low}\ldots\eta^{1}_{max}+\eta^{1}_{high}, 0°\ldots1°]$ to obtain a radial profile (still not background corrected). The full width at half maximum intensity (FWHM) of the profile in 2θ is determined and used to limit the 2θ range to $[2\theta_{hkl} - 10*2\theta_{FWHM}\ldots 2\theta_{hkl} + 10*2\theta_{FWHM}]$. This smaller 2θ range (indicated by dashed box in the inset in figure 2c) still comprises the most intense peak while keeping the region to investigate smaller.

Step 4: For further quantification the intensity background is determined as average in the combined 2θ range $[2\theta_{hkl} - 15*2\theta_{FWHM}\ldots 2\theta_{hkl} - 10*2\theta_{FWHM}]$ and $[2\theta_{hkl} + 10*2\theta_{FWHM}\ldots 2\theta_{hkl} + 15*2\theta_{FWHM}]$ and subtracted from the 3D intensity distribution $[2\theta_{hkl} - 10*2\theta_{FWHM}\ldots 2\theta_{hkl} + 10*2\theta_{FWHM}, -15°\ldots15°, 0°\ldots1°]$.

Step 5: For each peak identified from the azimuthal distribution along η, a 1D azimuthal distribution along the angle ω is obtained by integrating 2θ over $[2\theta_{hkl}-10*2\theta_{FWHM}\ldots2\theta_{hkl}+10*2\theta_{FWHM}]$ and η over $[\eta_{max}-\eta_{low}\ldots\eta_{max}+\eta_{high}]$. The intensity distribution varies significantly along both azimuthal directions due to the individual coarse grains giving rise to high intensity peaks at certain (η, ω) positions. The contribution of the finer grains is rather smooth and assumed to be an almost constant function in η and ω. When normalizing the intensity by the integration range ($\Delta\eta = \eta_{high} - \eta_{low}$, $\Delta\omega = 0.02°$), fine grains are assumed to contribute a constant normalized intensity level.

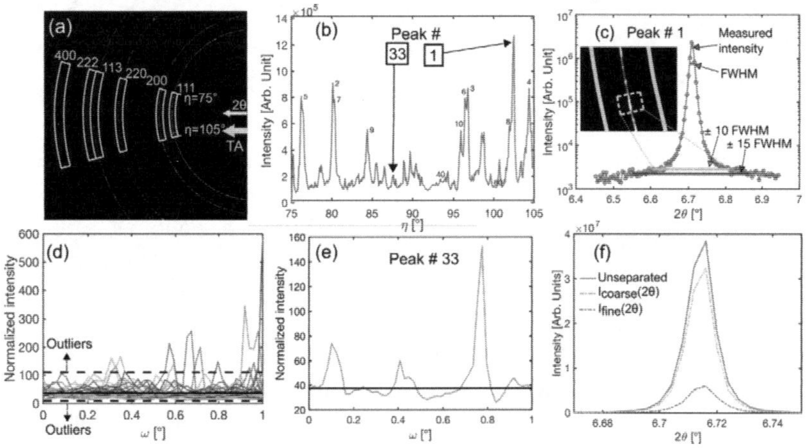

Figure 2: Illustration of the critical steps explained in the main text of the separation of high-intensity spots from the continuous ring in diffraction data.

Step 6: The constant normalized intensity level to be used as threshold is defined by selecting the 25 peaks recognized along η with lowest maximum integrated intensity to represent regions outside pronounced high intensity peaks. These 25 peaks correspond to rather small grains. Consequently, their average normalized intensity $I_{norm,thr}$ defines a suitable threshold for distinguishing between fine grains contributing to continuous rings with low variation in η and ω and coarse grains giving rise to individual high intensity peaks. The 25 profiles in ω (figure 2d) may still contain high-intensity peaks. For removing them, the profiles are inspected individually, and recognized outliers removed from analysis.

Step 7: Based on the threshold value for the normalized intensity $I_{norm,thr}$, an azimuthal threshold $I_{thr} = I_{norm,thr} * \Delta\eta * \Delta\omega$ is established. With that azimuthal threshold, the azimuthal range for each ring is separated in two regions depending on whether the (background corrected) intensity exceeds the threshold value (region A) or not (region B). Figure 2e displays the ω-profile of a peak with low intensity but still high enough to be part of the region of pronounced high intensity peaks. Intensities higher than the azimuthal threshold indicated by the horizontal line correspond to region A and lower to region B.

Step 8: The intensity distribution of the latter region B (for which the radially integrated intensity does not exceed the threshold) is attributed to solely fine grains. The corresponding radial profile $I_B(2\theta)=I_{fine}(2\theta)$ from these azimuthal regions is derived and analyzed. The intensity distribution of region A (i.e. from all the regions exceeding the threshold) contain diffraction signal from coarse grains and fine grains of e.g. same orientation diffracting at the same azimuthal angles. The determined radial profile $I_A(2\theta) = f_{A,coarse}\ I_{coarse}(2\theta) + (1-f_{A,coarse})\ I_{fine}(2\theta)$ is a superposition of two profiles from fine and coarse grains which can be separated presuming that the fine grains contribute with the same profile $I_{fine}(2\theta)=I_B(2\theta)$ as in region B $I_{coarse}(2\theta) = (f_{A,coarse})^{(-1)}\ (I_A(2\theta) - (1-f_{A,coarse})\ I_B(2\theta))$ (figure 2f). The fraction of coarse grains $f_{A,coarse}$ contributing in region A is obtained from the ratio between the integrated intensities above the threshold and the entire integrated intensity. The radial profile of the fine grains in the full azimuthal region (η, ω) [-15°... +15°, 0°...1°] is given by $I_{fine}(2\theta) = I_B(2\theta) + (f_{A,coarse})\ I_B(2\theta)$ (figure 2f).

In this manner, the average radial profiles of the grain fractions are obtained which are of interest for the present study. If analysis of individual coarse grains is attempted, their corresponding individual high-intensity peaks within the intensity distribution (of region A) are to be identified by projecting the intensity along 2θ on the azimuthal plane (η, ω). The relevant ranges in the azimuthal angles of

IOP Conf. Series: Materials Science and Engineering 1249 (2022) 012040 doi:10.1088/1757-899X/1249/1/012040

individual high-intensity peaks can be determined and individual radial profiles constructed. However, details of the analysis of individual coarse grains are outside of the scope of this paper.

3.2. Radial profiles

The separated and unseparated radial profiles are characterized in terms of average peak positions and total intensities. Intensities above 1/50 of the maximum intensity constitute the corresponding peaks, and the average peak position ($2\theta_{mean}$) is calculated as the weighted average based on these intensities. The peak position relates to the average spacing d_{hkl} of diffracting {hkl} lattice planes and the wavelength λ through Bragg's law $\lambda = 2d\sin(\theta)$. The total intensities in the separated profiles correspond to the total diffracting volume of coarse and fine grains.

4. Results & Discussion

Diffraction patterns were acquired during interruptions of the tensile test at certain macroscopic stress values as indicated in figure 3a. Strain values at each loading step were obtained by digital image correlation from optical images using the GOM Correlate commercial software. The outlined algorithm was applied to separate diffraction signals from coarse and fine grains. The ratio between the total intensity from the fine grains and the total intensity from the entire diffracting volume, i.e. the volume fraction of fine grains, for the first four Debye-Scherrer rings at the unloaded state were 17.4% <111>, 18.7% <200>, 24.9% <220> and 26.7% <311> with an average fraction of fine grains of 22%. As the used threshold I_{thr} value is based on the average intensity level of the continuous ring (or better the 25 identifiable peak of lowest integrated intensity) it is *a priori* unknown to which threshold value d_{thr} in terms of grain size this corresponds. Comparing the obtained volume fraction of fine grains with the finding from orientation maps obtained by EBSD on the same material that grains smaller than 5 µm contribute an area fraction of 24 %, it is reasonable to presume that the algorithm discriminates grains larger and smaller than 5 µm into coarse and fine grain fraction, respectively. The threshold value d_{thr} of 5 µm and its stability is further evaluated by checking the fraction of fine grains with increasing strain.

The lattice strains $\varepsilon_{hkl}=(d_{hkl}-d^0_{hkl})/d^0_{hkl}$ of grain families with four different <hkl> along the tensile direction are calculated based on the first four Debye-Scherrer rings. The stress-free $2\theta_{mean}$ position ($2\theta^0_{mean}$) was evaluated at 0% strain for the unseparated and separated profiles, and the $2\theta_{mean}$ was calculated for the subsequent stress increments. Figure 3b shows the variation of lattice strain with true stress for grains with tensile axis along <111> obtained from the first Debye Scherrer ring. The three different curves represent the average lattice strain from the unseparated profile and for the two different grain fractions separately. With increasing macroscopic stress and strain, all elastic lattice strains increase, but the fine grain fraction develops larger elastic strains then the coarse-grained fraction indicating the development of internal stresses and stress partitioning. Where forward stresses occur in the fine grains and back stresses develop in the coarse grains. The details of the stress partitioning and the yielding of the harmonic structure will be presented in a forthcoming paper.

Figure 3c presents the evolution of the fraction of fine grains with macroscopic strain based on the ratio of integrated intensity of fine grains to the total intensity. For all grain families, the fraction of fine grains increases with strain. This might be a consequence of coarse grains getting subdivided by deformation-induced dislocation boundaries due to plastic yielding. The developing mosaicity leads the diffraction spots to spread in the azimuthal directions (η, ω). This causes the smooth continuous ring to comprise of a combination of intensity from fine grains and parts of high-intensity spots from subgrains in deformed coarse grains. At approximately 0.01 true strain, the continuous ring intensity starts to increase relative to the strongest high-intensity spots, which results in a higher f_{fine} and an increase of d_{thr}. Furthermore, the superposition of intensities of fine grains and subgrains from coarse grains causes the lattice strains of the fine-grained fraction to be underestimated. It is also noted, that the fraction of fine grains is always largest for grains with one of their <311> directions along the tensile axis followed by <220>, <200> and <111>. This might not be a real effect, but an artefact of the separation algorithm and the used threshold for separating fine and coarse grains. In particular rings with a high multiplicity will have a plentitude of high-intensity peaks, which makes their discrimination difficult. Overlapping peaks will be attributed to the smooth intensity distribution causing a higher fraction of fine grains.

In summary, the algorithm separates the diffraction signals from coarse and fine grains nicely at the targeted threshold d_{thr} of 5 μm until 0.01 true strain. Afterwards, the effective threshold d_{thr} increases and as a consequence, the lattice strains of the initial fine-grained fraction are not monitored properly anymore as part of the coarse grains become merged into the fine-grained fraction.

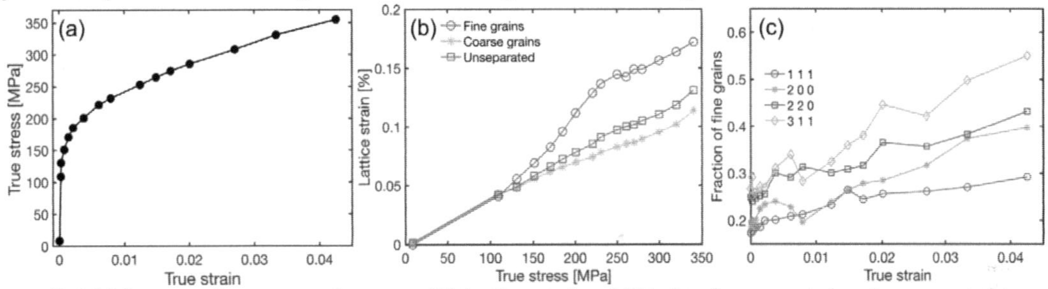

Figure 3: (a) Macroscopic stress strain curve. (b) Lattice strain of 111 ring for separated and unseparated profiles with true stress. (c) Fraction of fine grains with true strain according to separation algorithm.

5. Conclusion

In situ X-ray diffraction was carried out during tensile deformation of harmonic structured nickel with bimodal grain size distribution until 0.04 true strain. The settings for acquisition were adjusted such that azimuthal segments from coarse and fine grains were distinguishable. An algorithm for automatic separation of radial profiles of respective grain fractions was developed. The fraction of fine grains (with diameters below 5 μm) is accurately monitored until approximately 0.01 true strain. Then the diffraction signal from the coarse grains partially merges with the signal from fine grains. The average elastic strains of the two grain fractions differ during tensile deformation revealing stress partitioning.

Acknowledgements

The authors thank the financial support by the Swedish Research Council through research grant # 2016-03811, the Swedish Foundation for International Cooperation in Research and Higher Education (STINT) grants # SJ2017-7432 and MG2019-8450, and Interreg program “ESS & MAX IV: Cross Border Science and Society”, grants DTU007 and LU070. We acknowledge DESY (Hamburg, Germany), a member of the Helmholtz Association HGF, for the provision of experimental facilities. Parts of this research were carried out at PETRA III. Beam time at beamline P21.2 was allocated for proposal I-20190991 EC and has been supported by the project CALIPSOplus under the Grant Agreement 730872 from the EU Framework Programme for Research and Innovation HORIZON 2020.

References

[1] Y. Zhu *et al* 2021 *Mater. Res. Lett.* **9** 1-31
[2] X. Wu and Y. Zhu 2017 *Mater. Res. Lett.* 1-6
[3] A. Shokry, A. Ahadi, P. Ståhle and D. Orlov 2021 *Sci. Rep.* **11** 17445
[4] K. Ameyama *et al* 2022 *Mater. Res. Lett.* **10** 440-71
[5] D. Orlov and K. Ameyama 2020 *Mater. Sci. and Technol.* **36** 517-26
[6] D. Orlov, H. Fujiwara and K. Ameyama 2013 *Mater. Trans.* **54** 1549-53
[7] H.K. Park *et al* 2018 *Mater. Res. Lett.* **6** 261-7
[8] D. Orlov *et al* 2019 *IOP Conf. Ser. Mater. Sci. Eng.* **580** 012019
[9] D. Orlov, R. Kulagin and Y. Beygelzimer 2020 *Mater. Lett.* **275** 128126
[10] H. Van Swygenhoven and S. Van Petegem 2013 *Mater. Charact.* **78** 47-59
[11] Z. Matěj *et al* 2014 *Powder Diffr.* **29** S35-S41
[12] F.X. Lin *et al* 2015 *Philos.Mag.* **95** 2427-49
[13] E.M. Lauridsen, D.J. Jensen, H.F. Poulsen and U. Lienert 2000 *Scr. Mater.* **43** 561-6
[14] Y.D. Wang *et al* 2002 *J. Appl. Crystallogr.* **35** 684

IOP Conf. Series: Materials Science and Engineering 1249 (2022) 012041 doi:10.1088/1757-899X/1249/1/012041

Characterization of local deformation and fracture behavior in ferrite + martensite dual-phase steels having different grain sizes

M H Park[1], Y Tagusari[1] and N Tsuji[1, 2]

[1] Department of Materials Science and Engineering, Kyoto University, Yoshida Honmachi, Sakyo-ku, Kyoto 606-8501, Japan
[2] Elements Strategy Initiative for Structural Materials (ESISM), Kyoto University, Yoshida Honmachi, Sakyo-ku, Kyoto 606-8501, Japan

E-mail: park.myeongheom.8r@kyoto-u.ac.jp

Abstract. Low carbon dual-phase (DP) steels composed of soft ferrite and hard martensite have been widely used in the automotive industry due to their good strength-ductility balance and large strain hardening ability. DP steels have a wide variation in mechanical properties depending on several microstructural features such as grain size, phase fraction and distribution. Among them, the grain refinement of DP steels is known to be an effective option for enhancing mechanical performance in strength and ductility (especially post-uniform elongation). However, the exact reason for the significant improvement of post-uniform elongation by grain refinement has not been fully understood. It is considered that the characterization of local deformation behavior and micro-void formation/growth behavior in connection with microstructures is an essential approach for understanding the enhanced post-uniform elongation realized in the fine-grained DP specimen. In the present study, we prepared two kinds of DP specimens with mean ferrite grain sizes of 14.9 μm (coarse-grained DP) and 7.1 μm (fine-grained DP), and carefully investigated local strain distribution of tensile specimen and micro-void formation/growth behavior using digital image correlation (DIC) analysis and SEM observations. The fine-grained DP specimen exhibited a gradual strain localization after necking and had sufficient strain capacity that could endure against fracture. The fine-grained DP structure had a great number of micro-voids in the necked region, but almost all the micro-voids maintained a very small size, which was contrasted with the case of coarse-grained DP specimen containing very large-sized micro-voids. Such a significant difference in micro-void size/number characters between two kinds of DP specimens would be one possible reason for exhibiting greatly different post-uniform elongation behavior.

1. Introduction

Steels used for structural materials have been increasingly required to manage both high strength and large ductility for the safety of structural applications and the achievement of high fuel-efficiency through the reduction of body weight of automobile components. Dual-phase (DP) steels consisting of soft ferrite phase and hard martensite phase are widely used steels due to their good mechanical balance between strength and ductility. Over several decades, many studies of DP steels have been achieving much better mechanical performance by changing microstructural factors such as phase distribution, grain size, phase strength (hardness) and phase fraction [1-3]. Recently, it has been found that the grain refinement for DP steels is a promising strategy to simultaneously improve strength and ductility (especially post-uniform elongation) [2, 4]. However, the underlying reason for the enhancement of post-uniform elongation realized by grain refinement is still not fully understood. The characterization of local deformation behavior including micro-void evolution can be one effective approach for understanding the overall deformation-fracture nature of DP structures [5, 6]. In the present study, we characterize the local deformation behavior of DP tensile specimens by means of the digital image correlation (DIC) technique providing strain distribution maps, and then investigate the micro-void size and its number by SEM observation for tensile-fractured DP specimens having different grain sizes.

2. Experimental procedures

A low carbon steel having a chemical composition of Fe-2Mn-0.1C (mass%) was used in the current study. Ae3 and Ae1 temperatures of the present alloy were calculated to be 804 °C and 655 °C, respectively, using Thermo-Calc software with TCFE7.0 thermodynamic database. Specimens were austenitized at two different temperatures of 950 °C and 810 °C for 3 hr, followed by furnace cooled (F.C.) in a vacuum condition for obtaining ferrite + pearlite microstructures having different ferrite grain sizes. Afterward, ferrite + martensite DP structure was obtained from ferrite + austenite intercritical annealing at 750 °C for 2 hr, followed by water quenching (W.Q.), during which austenite was transformed to martensite. Microstructure observation was carried out using a field-emission scanning electron microscope (FE-SEM, JEOL JSM 7100F) equipped with an electron back-scattered diffraction (EBSD) system at an accelerated voltage of 15kV. The specimen for SEM observation was prepared through mechanical polishing with 400-4000 grid-sized emery papers and subsequent electrolytic polishing in a chemical solution of 90 vol% of CH_3COOH and 10 vol. % of $HClO_4$ with a voltage of 22 V for 30 s at room temperature. Tensile properties were evaluated using a uniaxial tensile testing machine (Shimadzu, AG-100kN Xplus) with an initial strain rate of 8.3×10^{-4} s^{-1} at room temperature. Tensile specimen with a gauge length of 10 mm, gauge width of 5 mm and thickness of 1 mm was cut from the heat-treated sheets. Tensile displacement (or strain) of gauge and strain distribution of tensile specimen were accurately measured by the digital image correlation (DIC) technique. For DIC strain analysis, a speckle pattern was introduced on the surface of the tensile specimen with white and black inks prior to the tensile test. The images including tensile specimen were captured by a CCD camera (SVS-VISTEK) with a constant speed of 5 fps during the tensile test, and they were provided for DIC analysis software (Correlated solutions Inc., Vic-2D) to obtain a strain distribution map. The size and number of micro-voids were investigated by careful SEM observation with the tensile-fractured DP specimens, for which a dedicated image-processing software (ImageJ) was utilized.

3. Results and discussion

3.1. Microstructural characteristic and tensile properties

Figure 1 exhibits SEM-EBSD image quality (IQ) maps and corresponding kernel average misorientation (KAM) maps of the DP specimens obtained from the heat treatments of (a) 950 °C-3hr-F.C. + 750 °C-2hr-W.Q. and (b) 810 °C-3hr-F.C. + 750 °C-2hr-W.Q. Contrast on IQ map reflects diffraction pattern quality of Kikuchi line, and colors on KAM map indicate KAM value according to the key KAM color bar. Since the martensite containing a high density of lattice defects lead to lower IQ value (darker contrast in IQ map) and higher KAM value, ferrite and martensite can be distinguished with such differences in IQ map and KAM map. Capital letters **F** and **M** on the IQ map are ferrite and martensite, respectively. The specimen (a) and specimen (b) in figure 1 exhibited different mean ferrite grain size of 14.9 μm and 7.1 μm, respectively, but they had almost the same martensite fraction (f_M = 45% for the specimen (a) and f_M = 46% for the specimen (b)) and similar martensite distribution with chain-shaped martensite surrounding ferrite grains, each of which will be referred to as "coarse-grained DP" and "fine-grained DP" hereafter.

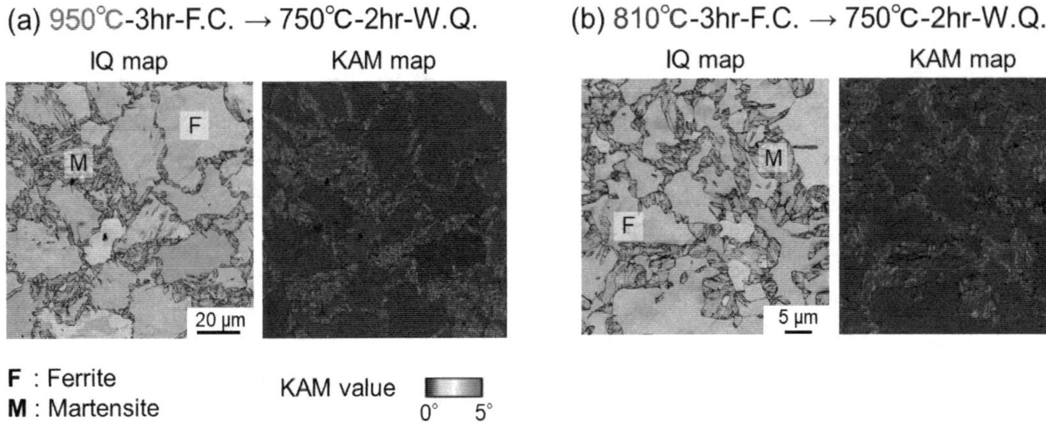

Figure 1. EBSD image quality (IQ) maps and corresponding kernel average misorientation (KAM) maps of DP specimens obtained from the heat treatments of (a) 950 °C-3hr-F.C. + 750 °C-2hr-W.Q. and (b) 810 °C-3hr-F.C. + 750 °C-2hr-W.Q. **F** and **M** on IQ map denote ferrite and martensite, respectively, and colors on KAM map indicate KAM value according to the key color bar.

Figure 2 displays nominal stress-strain (S-S) curves of the coarse-grained DP specimen (red) and the fine-grained DP specimen (blue). The circle symbol on each S-S curve indicates S-S point at uniform elongation. The fine-grained DP specimen exhibited a higher yield strength (0.2% proof stress) of 410 MPa and ultimate tensile strength (UTS) of 801 MPa than the coarse-grained DP specimen having a yield strength of 374 MPa and a UTS of 734 MPa. The fine-grained DP specimen exhibited much better ductility with a uniform elongation of 0.107 and a post-uniform elongation of 0.104 than the coarse-grained DP specimen (uniform elongation of 0.106 and post-uniform elongation of 0.072), indicating that grain refinement of DP structure greatly enhanced post-uniform elongation as well as high strength. We further investigated the local deformation behavior of tensile specimens in the period of post-uniform elongation using DIC strain analysis.

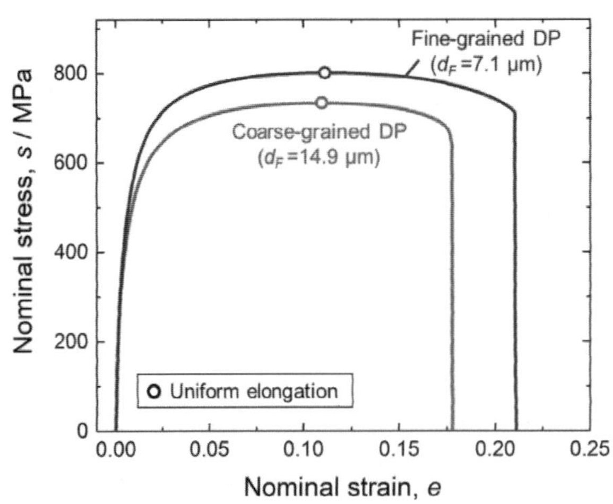

Figure 2. Nominal stress-strain curves of coarse-grained DP specimen (red) and fine-grained DP specimen (blue). Circle symbol on the curve corresponds to uniform elongation of each DP specimen.

Figure 3 shows the DIC strain distribution maps of (a) coarse-grained DP and (b) fine-grained DP tensile specimens at specific global tensile strains (nominal strain). The tensile direction is parallel to the vertical direction of images, and the colors superimposed on the tensile specimens correspond to the value of normal strain in the tensile direction (ε_{yy}), according to the key strain color bar. It was found that two kinds of tensile specimens showed rather homogeneous strain distributions in the gauge part until the global tensile strain of 10%. At a global tensile strain of 15%, a clear strain-localized band was observed in both DP tensile specimens, and a sharp difference in the magnitude of strain localization was recognized at the further tensile-strained condition of 17.8%. The maximum local strain value in the necked region of the coarse-grained DP specimen was 0.65 at the global tensile strain of 17.8%, whereas that of the fine-grained one was 0.44 at the same global tensile strain. The fine-grained DP specimen could be further tensile-deformed to the global tensile strain of 21.1%, in which the maximum local strain in the necked region was reached to 0.90.

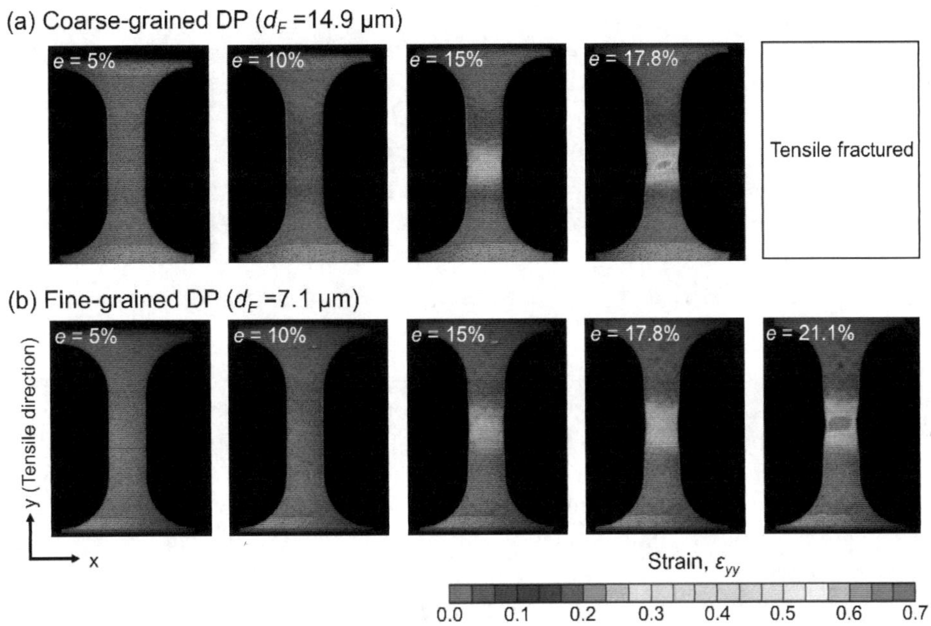

Figure 3. DIC strain distribution maps of (a) coarse-grained DP and (b) fine-grained DP tensile specimens at the nominal strain of 5%, 10%, 15%, 17.8% and 21.1%. Colors superimposed on the tensile specimen correspond to the normal strain value in tensile direction (ε_{yy}), according to key strain color bar.

Figure 4 displays local strain (ε_{yy}) changes at the strain-inspection point as a function of global tensile strain (i.e., nominal strain). The local strain data were extracted from the DIC strain distribution results in figure 3, and the strain-inspection point was set on the most strain-localized region. An X-mark on the graph indicates tensile-failure, and a broken line is a one-to-one correspondence line of global tensile strain and local strain (i.e., homogeneous deformation line). In both DP specimens, the local strain linearly increased along the homogeneous deformation line until around 10% global tensile strain. After that, in both specimens, the local strain showed an exponential increase, greatly deviating from the homogeneous deformation line. The fine-grained DP specimen exhibited a more gradual increase of local strain, and eventually had a much larger strain capacity that could endure against tensile failure. It is believed that such a significantly different strain localization manner between coarse-grained DP and fine-grained DP specimens can be understood by investigating microstructural deformation behavior including micro-void evolution in the strain-localized region.

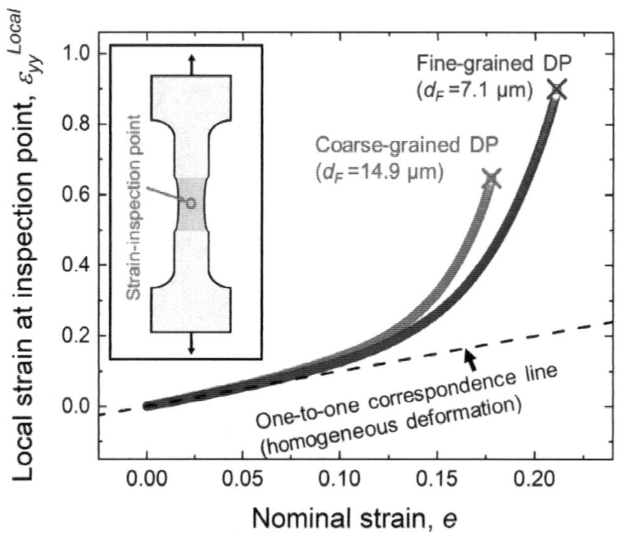

Figure 4. Local strain change at the strain-inspection point in coarse-grained DP (red) and fine-grained DP (blue) tensile specimens as a function of nominal strain. The Strain-inspection point was set on the most strain-localized region.

3.2. Micro-void characterization in the strain-localized region

Figure 5 shows micro-void characterization results investigating (a) the size of each micro-void and (b) the number of micro-voids in the observed area at a tensile-directional distance from the fracture surface of tensile-fractured DP specimens. The schematic illustration of the observed position is inserted in both graphs. Note that, in figure 5 (a), each data point indicates the size for an individual micro-void, and the size is evaluated as an area unit (μm^2) of each micro-void. It was found that in figure 5 (a) that almost all the micro-voids observed in both DP specimens had a quite small void size less than 10 μm^2 even in the vicinity of the fracture surface. It was noteworthy that large-sized micro-voids ranging from 20 μm^2 to 70 μm^2 were observed only in the coarse-grained DP specimen, which intensively occurred near fracture surface. This suggests that several large-sized micro-voids formed in the coarse-grained DP specimen give more critical chances to lead to relatively early tensile failure. In the result of Figure 5 (b) investigating the number of micro-voids, it was found in both DP specimens that a great number of micro-voids were observed in the vicinity of the fracture surface, and the number of them were significantly decreased as a distance from the fracture surface. On the other hand, the absolute number of micro-voids in the vicinity of the fracture surface was apparently different between coarse-grained DP and fine-grained DP specimens. It is intriguing that the fine-grained DP specimen exhibiting larger post-uniform elongation has five times more micro-voids (n_{void} = 62) than the coarse-grained DP specimen (n_{void} = 12) at the position of 150 μm from the fracture surface. A great number of micro-voids in the fine-grained DP specimens just before fracture might be related to its very large local strain value in the necked region of the tensile specimen. Considering that the fine-grained DP specimen exhibits a gradual strain increase even after necking as shown in figure 3, profusely distributed micro-void formation sites such as grain boundaries and interface may lead to delocalizing microscopic strains. We believe that the reason for such a large number of micro-void formations in the fine-grained DP specimen can be understood by characterizing microscopic deformation behavior through the microscopic DIC (μ-DIC) technique. The finding(s) combined with μ-DIC microstructure strain result will show a great synergy for the global view of the deformation nature of DP structure in connection with global mechanical response, leading to effective material design strategy that can realize a simultaneous improvement of strength and ductility in DP structures.

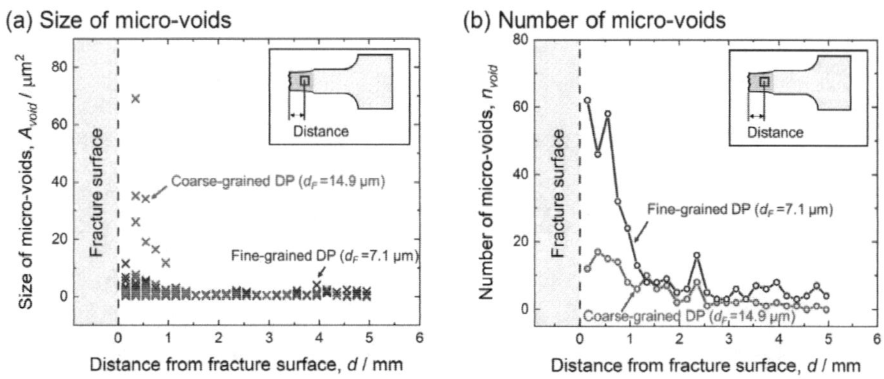

Figure 5. Characterization of micro-voids in tensile-fractured DP specimens having a mean ferrite grain size of 14.9 μm (red) and 7.1 μm (blue): (a) Size and (b) number of micro-voids as a function of distance from the fracture surface.

4. Conclusions

In the present study, local deformation behavior and micro-void characterization of DP specimens having different ferrite grain sizes were investigated using DIC analysis and SEM observation. DP specimen having a mean ferrite grain size of 7.1 μm (fine-grained DP) exhibited an excellent mechanical balance between strength and ductility (especially post-uniform elongation), compared to the coarse-grained DP specimen having a mean ferrite grain size of 14.9 μm. From the result of the DIC strain distribution of tensile specimen, the fine-grained DP specimen showed a slow strain localization and large strain localization capacity against fracture, which might be one of the possible reasons for the enhancement of post-uniform elongation of fine-grained DP specimen. In microstructure observation, a great number of small-sized micro-voids were observed in the fine-grained DP specimen, which indicated that micro-void growth can be effectively inhibited by grain refinement in DP structure.

References

[1] Delincé M, Bréchet Y, Embury J D, Geers M G D, Jacques P J and Pardoen T 2007 Structure-property optimization of ultrafine-grained dual-phase steels using a microstructure-based strain hardening model, *Acta Mater.* **55** 2337–50
[2] Park M H, Shibata A and Tsuji N 2016 Effect of Grain Size on Mechanical Properties of Dual Phase Steels Composed of Ferrite and Martensite, *MRS Adv.* **1** 811–6
[3] Mazinani M and Poole W J 2007 Effect of martensite plasticity on the deformation behavior of a low-carbon dual-phase steel, *Metall. Mater. Trans. A* **38** 328–39
[4] Calcagnotto M, Adachi Y, Ponge D and Raabe D 2011 Deformation and fracture mechanisms in fine- and ultrafine-grained ferrite/martensite dual-phase steels and the effect of aging, *Acta Mater.* **59** 658–70
[5] Azuma M, Goutianos S, Hansen N, Winther G and Huang X 2012 Effect of hardness of martensite and ferrite on void formation in dual phase steel, *Mater. Sci. Technol.* **28** 1092–1100
[6] Saeidi N, Ashrafizadeh F, Niroumand B, Forouzan M R, Mohseni M S and Barlat F 2015 Void coalescence and fracture behavior of notched and un-notched tensile tested specimens in fine grain dual phase steel, *Mater. Sci. Eng.* A **644** 210-7

IOP Conf. Series: Materials Science and Engineering 1249 (2022) 012042 doi:10.1088/1757-899X/1249/1/012042

In-situ X-ray microtomography of interface between additively manufactured aluminium bronze and H13 tool steel

Håkon Linga[1], Yubin Zhang[2], Vegard Brøtan[3], Xiaobo Ren[4], Ida Westermann[1] and Bjørn Holmedal[1]

[1] Department of Materials Science and Engineering, Norwegian University of Science and Technology, Sem Sælands vei 14, 7034, Trondheim, Norway.
[2] Department of Civil and Mechanical Engineering, Technical University of Denmark, Produktionstorvet, 425, 2800 Kgs. Lyngby, Denmark.
[3] SINTEF Manufacturing AS, S.P. Andersens vei 3, 7031, Trondheim, Norway.
[4] SINTEF Industry, Richard Birkelandsvei 2B, N-7465 Trondheim, Norway

E-mail: hakon.linga@ntnu.no

Abstract. Directed energy deposition (DED) is an additive manufacturing process category where material is melted as it is deposited, often powder melted with a focused laser. In this work aluminium bronze was deposited onto H13 tool steel using this technique, forming a mixing zone between the two metals. This mixing zone was examined with X-ray microtomography to obtain a three dimensional perspective that is unrealizable with conventional microscopy. Both the shape of the melt pools, the microstructure within, and processing defects could be discerned due to varying absorption of the radiation. In addition to characterizing the microstructure, the sample was also strained in three steps; the first step was to approximately 2.5 % strain, the second to 10.5 % and finally until fracture of the sample. The sample was scanned between each step, including a scan of the fracture surface. The ultimate tensile strength was found to be approximately 850 MPa and the fracture was observed to originate from cracks between the H13 substrate and the mixing zone. These cracks appeared to form in the second step of the straining. Additionally, local strains were estimated by utilizing pores in the sample as tracking points.

1. Introduction

Additive manufacturing (AM) allows combinations of different metals and alloys in the same part, which is difficult to achieve in more traditional production methods such as welding or casting. Uses for such multi-material solutions are abundant, but in general the idea is to utilize distinct properties for each material. One example is to use steel for its strength or wear resistance, while using copper or its various alloys for thermal- or corrosion properties. An application could be in injection moulding of plastic where thermal conduction is highly important in addition to a need for wear resistance in the tools[1]. Several studies have been done on deposition of copper on steels, but many of them focus on hardness or wear rate, and not on the adhesion between the metals[2, 3]. In this work, aluminium bronze deposited onto H13 tool steel is examined with X-ray microtomography, with gradually increasing loads. The motivation of using X-ray microtomography for additively manufactured parts is that it can give a three-dimensional view on defects such as lack of fusion and pores[4]. Both types of defects are distinguishable with this technique, as voids results in lower absorption of radiation. Additionally, the mixing areas between metals are often anisotropic and complex so it is difficult to get a full understanding of the microstructure and its impact on mechanical properties from micrographs alone.

2. Experiment

The sample was made by first producing a block of H13 tool steel from powder with a composition of 5.40 wt% Cr, 0.38 wt% C, 1.38 wt% Mo, 1.18 wt% Si, 1.09 wt% V, 0.41 wt% Mn and Fe as balance. This was deposited by a laser with spot size of 1000 μm at a power of 500 W. 5 grams of powder was fed per minute, with a deposition speed of 1000 mm/min. After this, aluminium bronze powder with 8.5-10.75 wt% Al, 0.5-2.0 wt% Fe and Cu as balance was deposited onto the H13 steel with a laser spot size of 800 μm at 500 W, 3.8 grams of powder/minute and a laser speed of 1000 mm/min. A bi-directional laser scanning pattern was used throughout the build. The mixing zone between the two alloys was formed where the deposition of aluminium bronze remelted the previously deposited steel. From the as-built block a small tensile test sample, with a cross section of 0.5 x 1.0 mm and a gauge length of 1.25 mm, was made with electric discharge machining. This sample consisted of approximately 0.1 mm of aluminium bronze and a thin layer of H13 with a mixing zone in between the two alloys. The mixing zone with a thickness of 0.3 - 0.4 mm was roughly at the center of the gauge volume and parallel to the loading direction, as shown in Figure 1 a) and d). It would be sensible to have the interface perpendicular to the loading direction to test the strength here, but this could be problematic as it would be possible for the bronze to fail before the mixing zone[5]. This is also reinforced by preliminary microhardness results that show a hardness of approximately 500 HV3 in the bulk H13, 170 HV3 in the bulk aluminium bronze, and around 230 HV3 in the mixing zone, meaning that the bulk aluminium bronze should be the weakest part.

The absorption contrast was recorded with ZEISS Xradia 520 Versa, at 20 seconds exposure time at 3000 incident angles. The source to sample distance was 18 mm, while the sample to detector distance was 80 mm. An optical magnification of 4x was used, and the detector was set to a binning of 2. The reconstructed volume had a voxel size of 1.24 μm. Between scanning the sample, the sample was strained with a crosshead speed of $2 \cdot 10^{-3}$mms^{-1} using a specially designed rig [6]. While the sample was strained, applied force was recorded and closely monitored.

3. Results

Figure 1 a) shows the tensile test specimen prior to straining. The sample contains a region with aluminium bronze, a small region of H13 tool steel, and around half of the samples volume consists of the mixing zone between the two alloys. An example slice from the unstrained volume is displayed in Figure 1 d), while Figure 1 c) shows an optical micrograph within the mixing zone etched with Murakami's etchant. The mixing zone mainly consists of spherical iron particles suspended in the aluminium bronze. However, only larger particles can be clearly distinguished in the CT data, as seen in Figure 1 d). The line deposition direction refers to the direction of the deposition tracks of the mixing layer. The contour of these tracks of previous melt pools can be observed by the collection of iron particles along the two diagonal lines in Figure 1 d).

The samples were strained to fracture through three loading steps and the last load step corresponds to fracture. In the first two straining steps the engineering strain was estimated by measuring the change in distance between selected pores in opposite ends of the scanned volume. The loading data from the experiment is shown in Figure 1 b), where each strain step is separated by a dashed vertical line. Since the sample was analyzed with x-ray microtomography in between strain steps, approximately a day passed between each straining step. This time is not shown in Figure 1 b). Additionally, some small pauses were done in the first straining step, and are the cause of the discontinuity in the plot around 50 to 60 seconds. The engineering strain along the loading axis was estimated to be around 2.5 % and 10.5 % in the first and second steps, respectively. The ultimate tensile strength was approximately 850 MPa.

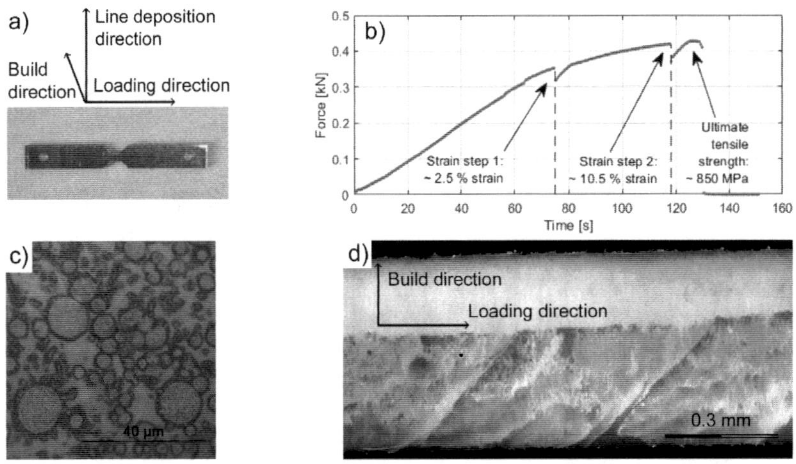

Figure 1. a) Tensile test specimen. b) Applied load plotted against time. c) Optical micrograph from the mixing zone. d) A slice from the unstrained volume. Red lines show melt pool contours.

By visual inspection the microstructure does not change much between straining steps; the iron particles can be observed in the same relative locations. However, pores are observed to emerge and grow in size as shown in Figure 2. This shows minimum intensity projections, where the minimum recorded absorption of the X-rays into the plane are displayed. The lowest intensity comes from pores in the material, so all pores throughout the sample are presented. To get a more complete view, the projection is shown at three different angles, along the loading direction, 45° rotated about the build direction, and along the line deposition direction in the mixing zone. The contour of the melt pool can clearly be seen from the projection along the line deposition direction, similarly to what can be seen in Figure 1 d), as particles of iron are accumulated here throughout the whole volume.

It should also be noted that the projections shown of the unstrained, 2.5 % strained, and 10.5 % strained sample were cropped to avoid any of the volume outside to interfere with the projection. This is not done in the fractured sample, as this approach would not allow the fracture surface to be revealed. Therefore the outside volume was instead detected and set to the maximum possible intensity. The apparent increase in pores around the fracture surface is therefore due to errors in detecting the fracture surface. The vertical lines show the corners of the volume that is projected.

Five different pores are marked in Figure 2. Pore 1 can very faintly be observed in the unstrained volume, but is clearly visible in the volume strained to 2.5 %. This pore continues to grow with increasing strains. Pores 2 and 3 are also seen in the previous volumes, but are only distinctly revealed in the volume strained to 10.5 %. All these contrast the behaviour of the majority of the pores that are already present from the unstrained volume, that do not seem to grow as the sample is strained. Pore 4 shows a crack that is formed when the sample is strained to 10.5 %. This does not appear at all in previous volumes, and closer inspection shows that the crack is positioned at an interface between the H13 steel and the aluminium bronze. This is seen in the large square in the bottom right corner of Figure 2, and is a slice from the volume where the crack observed from the projection is present. Pore 5 shows a similar formation of pores in the material close to the contour of the melt pool, but is instead a series of smaller unconnected pores. It should be noted that the five pores shown here do not make an exhaustive list of pore formation with increasing strain in the volume and are only highlighted as examples.

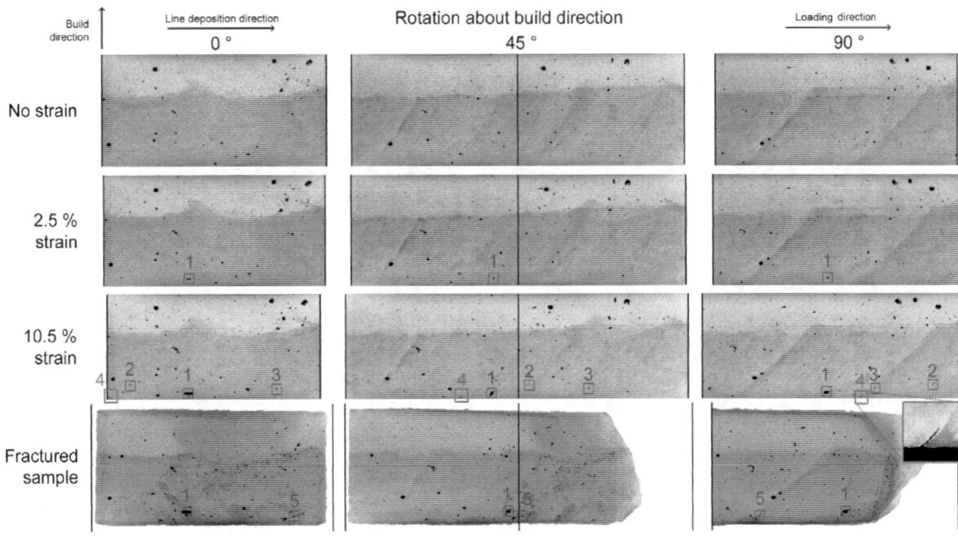

Figure 2. Minimum intensity projections at various strains and projection angles.

Closer examination of the location of the crack shown in Figure 2 reveals that the crack is not only positioned close to the contour of the melt pool, but is also aligned with the fracture surface. This can be observed better in Figure 3 a). The fracture surface is shown as a white line overlaid with slices from the volume strained to 10.5 %. The fracture surface is shown to closely follow the contour of the melt pool in some areas, but deviates from this contour and follows nearby pores in other areas as shown in Figure 3 b). However, a general observation is that the fracture initially follows the interface between the steel and bronze before deviating, as seen in Figure 3 c).

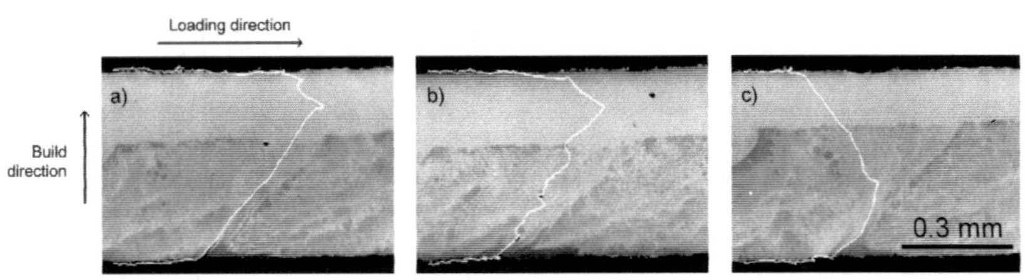

Figure 3. Fracture surface shown as white line overlapped with slices from the 10.5 % strained volume.

To further understand the plastic deformation behavior, the local strains were analyzed using pores as markers. The spatial coordinates were found for pores in both the unstrained and 10.5 % strained volumes. A selection of 80 out of around 250 pores were manually verified to be identical between the two volumes and were used to create a mesh using Delaunay triangulation. This resulted in just over 400 tetrahedrons with an average vertex distance of approximately 140 μm. The four vertexes in each tetrahedron were used to estimate the deformation gradient tensor, and from this the Green-Lagrange strain was found and placed in the incenter of each

IOP Conf. Series: Materials Science and Engineering 1249 (2022) 012042 doi:10.1088/1757-899X/1249/1/012042

tetrahedron. Figure 3 a) displays the normal strains compared to the location along the build direction and shows uniform strain with some local deviations throughout the build. In Figure 3 b) the axial strain is plotted against the radius ratio of the tetrahedrons. R is the radius of the circumscribed sphere and r is the radius of the inscribed sphere of the tetrahedrons, meaning an equilateral tetrahedron has a radius ratio of one[7]. An increased variance in axial strain is observed with decreasing radius ratio. The shear strains were also found, with an uniform distribution around zero with some small local variations.

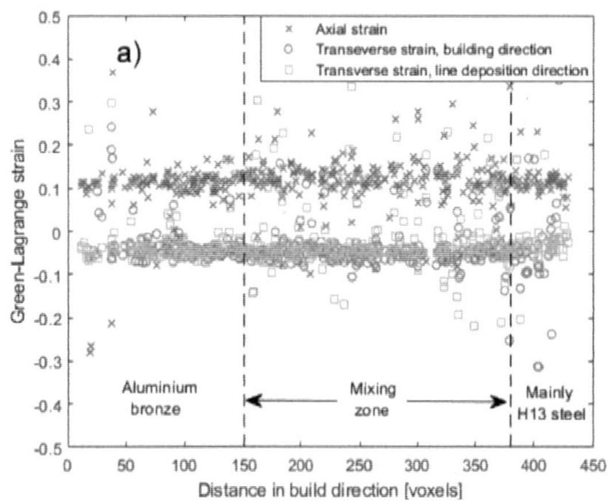

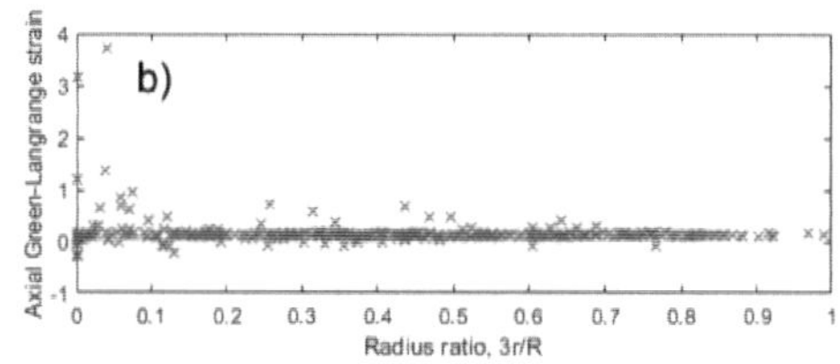

Figure 4. a) Normal strains plotted against distance in build direction. b) Axial strain plotted against radius ratio of tetrahedrons.

4. Discussion

The calculations done shows that the local strains were almost uniform throughout the sample with some local variations. There could be several reasons for these local strain variations. The first is related to the local microstructural variation, where one could envision variation depending on location within the melt pools in addition to variation along the building direction, where the local composition changes considerably. Such correlation could however not be observed, but is likely due to the large average tetrahedron size. An alternative explanation of the observed variation in strain is that it originates from inaccuracies in the calculation of the strain. This could come from two different errors. The first is faults in the meshing. The pores can have very irregular spacing, making the Delaunay triangulation occasionally create tetrahedrons with sharp angles. This makes even exceedingly small errors in localizing the pore coordinates cause large errors in the calculated strains. This is confirmed in Figure 3 c), where the largest deviations are observed at low radius ratios, and thus unfavorably shaped tetrahedrons. The second type of possible error is shifting of indexed coordinate due to pore growth. As seen in Figure 2, some pores grow in size with increasing strain, making this a possible source of deviation. However, the growth of certain pores but not others suggests that the strain is not uniform. A finer mesh of tracked points would make the correlation between strain and microstructure clearer, and using a larger fraction of the pores would certainly be a step in the right direction.

Another possibility for improving both the accuracy and the resolution of the local strain calculations is to use the iron particles as tracking points. These are decidedly more abundant in the volume than pores, and more tracking nodes would allow a finer mesh. Digital volume correlation is an alternative method of calculating local strains[8]. This would eliminate the need of identifying and matching individual pores and particles in the microstructure, since this is based only on intensity variation. More work will be done along these lines to refine the strain calculation and to relate them with local microstructure.

When it comes to the subject of the crack formation shown in Figure 2, it is evident that pore 4 is in fact the initiation point, considering how perfectly it aligns with the fracture surface. However, one could argue that in a real scenario more of the H13 steel substrate would be used, and that the crack could simply be a result of surface imperfections. This could very well be the case, but either way large interfaces between H13 steel and aluminium bronze are weak compared to the rest of the material. This is evident from the fracture consistently aligning with this interface, even far away from the fracture initiation position. One possibility is that the crack could be a result of lack of fusion between the two metals. However, this is improbable as the boundary would in that case have a lower absorption due to a small gap between the metals. This would be present already in the unstrained volume. Although this cannot be observed here, it is plausible that the resolution is not high enough to discern it. Since the mixing zone is formed by rapid cooling after the powder is melted, high thermal stresses are also expected. Therefore it is not unlikely that residual stress built at the bronze-steel interface contributes to the failure.

5. Conclusions

The ultimate tensile strength of the aluminium bronze - H13 tool steel sample tested was approximately 850 MPa, and the axial engineering strain just prior to fracture was estimated to around 10.5 %. The local strain calculated from displacement of pores in the material showed similar results. Both the normal- and shear strains were close to uniform throughout the volume, independent on microstructural features. Some pores are shown to grow in size with increasing levels of strain, while others are seemingly unaffected and remain approximately the same size. A correlation between pore growth and local strain could however not be established. The fracture is likely to have been initialized in a crack that formed in an interface between H13 steel and aluminium bronze in the sample surface. Furthermore the crack is shown to propagate in interfaces between the two metals in addition to passing nearby pores. For further improvement of the fracture toughness in the mixing zone between aluminium bronze and H13 tool steel, the deposition parameters need to be optimized such that the segregation of iron along melt pool edges is reduced.

Acknowledgments

This research is based on the research activites within the center for research-based innovation, SFI Manufacturing in Norway and is partially funded by the Research Council of Norway (NFR) under contract number 237900. A special thanks to Nordic Additive Manufacturing AS and the national research project IPN NextMoldMaker hosted by Skriverform AS (Project No.: 281967, NFR) that have contributed with test parts, case description and AM parameter development. YZ acknowledges the support from the European Research Council (M4D ERC grant agreement, project number 788567).

References

[1] Wang Y, Gao Y, Li Y, Zhai W, Sun L and Zhang C 2019 *Emerg. Mater. Res.* **8** 538–51
[2] Schmidt M, Kolleck R, Grimm A, Veit R and Bartkowiak K 2010 *CIRP Ann. Manuf. Technol.* **59** 211–4
[3] Samodurova M, Shaburova N, Radionova L, Zakirov R, Pashkeev K, Myasoedov V, Erdakov I and Trofimov E 2020 *Mater.* **13** 461–71
[4] Sames W, Medina F, Peter W, Babu S and Dehoff R 2014 *Effect of Process Control and Powder Quality on Inconel 718 Produced Using Electron Beam Melting* (John Wiley & Sons, Ltd) pp 409–23
[5] Liu L, Zhuang Z, Liu F and Zhu M 2013 *Int. J. Adv. Manuf. Technol.* **69** 2131–2137
[6] Kobayashi M, Zhang Y, Ishikawa H, Sun J, Oddershede J, Jensen D and Miura H 2021 *Exp. Mech.* **61** 817–28
[7] Liu A and Joe B 1994 *BIT Numer. Math.* **34** 268–87
[8] Xu C, Andriollo T, Zhang Y, Hernando J, Hattel J and Tiedje N 2020 *Scr. Mater.* **178** 463–7

IOP Conf. Series: Materials Science and Engineering 1249 (2022) 012043 doi:10.1088/1757-899X/1249/1/012043

Stress relief during annealing of railway wheel steel characterized by synchrotron X-ray micro-diffraction

Yubin Zhang[1,*], Casey Jessop[2], Dimitrios Nikas[2,3], Tianbo Yu[1], Wenjun Liu [4] and Johan Ahlström[2,*]

[1] Department of Civil and Mechanical Engineering, Technical University of Denmark, 2800 Kgs. Lyngby, Denmark
[2] Department of Industrial and Materials Science, Chalmers University of Technology, 412 96 Gothenburg, Sweden
[3] Department of Engineering and Physics, Karlstad University, 651 88 Karlstad, Sweden
[4] Advanced Photon Source, Argonne National Laboratory, Argonne, 60439-4800 Illinois, USA

*E-mail: yubz@dtu.dk; johan.ahlstrom@chalmers.se

Abstract. Railway wheels in service experience rolling contact fatigue loading, but also need to resist frictional heating on braking, yielding temperatures up to 500 °C. The combination of mechanical and thermal loads leads to changes in the mechanical properties of the material. The focus of this study is to investigate the effect of annealing on local microstructure and residual stresses in railway wheel pearlitic steel (medium carbon steels, ~0.55 wt.% C) using synchroton X-ray Laue micro-diffraction. It is found that the local residual stress releases to a large extent after annealing at 500 °C. The stress formation and relief mechanisms and their relationship to the local microstructure are discussed.

1. Introduction

Medium carbon steels (with approximately 0.55 wt.% C) in railway wheel materials typically has a pearlitic microstructure with a small amount (5-10%) of pro-eutectoid ferrite decorating the previous austenite grain boundaries [1,2]. The manufacturing of railway wheels involves heat treatments in order to achieve desirable characteristics in mechanical properties and residual stress distributions. Investigations presented in literature show that the temperature in the wheel tread can rise to over 500 °C for wheels on block-braked freight trains [3]. The high temperatures reached during operation can, cause degradation of the microstructure, e.g. pearlite spheroidisation, which has been found to affect both the mechanical properties and fatigue behavior [2,4]. At the same time, the local residual stresses in the microstructure may relax. How the local residual stresses evolve upon annealing and how they may affect the degradation of the pearlite structure are essential information for optimization of the processing route and mechanical properties of railway pearlitic steel as well as for accurate constitutive modelling of the mechanical behaviour on different scales, and thus need to be quantified.

In this study, an advanced synchrotron X-ray technique, differential aperture X-ray microscopy (DAXM) [5,6], is used to quantify the changes in microstructure and residual plastic/elastic strain after annealing of the R8T steel grade used for railway wheels. This technique allows 3D non-destructive characterization of both crystallographic orientations and elastic strain with high precision (0.01° and 1×10^{-4}, respectively) [7]. By correlating the microstructural and strain changes upon annealing and comparing with other materials systems, the local stress relief mechanisms can be better understood.

2. Material and experimental procedures

The material studied in this investigation is the railway wheel steel R8T, which complies to the ER8 grade specified in the standard EN 13262-1 [8], and the specific batch examined in this work has a chemical composition as shown in Table 1. Samples were extracted from unused ('virgin') wheels approximately 15 to 20 mm below the wheel tread surface; this region is of interest since it becomes exposed to elevated temperatures and mechanical loading during operation. The typical microstructure of R8T steel in the selected position is shown in figure 1. The fraction of these so-called free ferrite grains is about 10%. The samples were annealed for 4 hours at temperatures at 300° C and 500 °C. More information about the samples can be found in [9].

Table 1. Chemical composition of R8T wheel material, in wt.%

C	Si	Mn	Mo	S	Cr	Ni	V	P	Fe
0.59	0.35	0.78	0.05	0.006	0.13	0.17	<0.005	0.005	Bal.

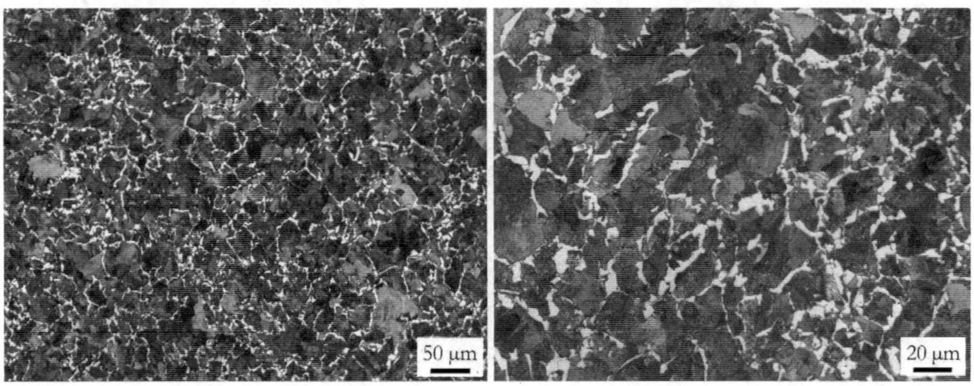

Figure 1. Optical micrograph of typical microstructure of R8T steel: free ferrite grains (white regions) and pearlite colonies (non-white regions).

2.1 Synchrotron X-ray diffraction

The DAXM experiments were conducted at beam line 34-ID-E at the Advanced Photon Source (APS), Argonne National Laboratory. First, a focused polychromatic microbeam (with size of ~0.3 μm) was used to determine the orientations of the grains. The X-ray microbeam scans the sample, which is mounted on a holder at an inclination angle of 45° to the incoming beam. A flat panel detector mounted in a 90° reflection geometry 510.3 mm above the specimen was used to record the Laue diffraction patterns. A Pt knife-edge, scanning along the sample surface at a distance of 250 μm, was used as a differential aperture for resolving diffraction patterns from different depths penetrated by the microbeam. The Laue patterns at each depth were reconstructed and indexed using the LaueGo software available at beamline 34-ID-E. More details about the experimental set-up can be found in [7].

A monochromatic beam was subsequently used for measurements of elastic strain within selected grains. By scanning the X-ray energy around the calculated value (calculated from the indexed Laue pattern using an ideal lattice parameter) for a selected Laue diffraction spot, the intensity distribution as a function of the diffraction vector $Q = 2\pi/d$ (Q-distribution) can be determined for the selected diffraction spot at different depths. At each depth, the Q-distribution is fitted using a Gaussian function and the center of the distribution, Q_C, is used to determine the absolute lattice spacing, d. The elastic strain, ε, can then be determined using equation (1):

$$\varepsilon = \frac{(d-d_0)}{d_0} \quad (1)$$

where d_0 is the lattice spacing of the sample in the stress-free state. For the present study, d_0 was calculated, according to the chemical composition in table 1 with the lattice parameter $a_0 = 2.86734$ Å. More detailed information about the calculation procedure can be found in [10,11]. For the present study, the strain component along the normal direction of the polished surface was measured for three bulk grains for each sample.

3. Results

The microstructure of the annealed samples characterized using polychromatic beam DAXM are shown in figures 2 and 3. As can be seen from the inverse pole figure (IPF) maps (figures 2a and 3a), most of the volume within approximately 40 μm from the surface is indexed while the indexing rate is reduced at further depth. It is anticipated that the indexed grains are free ferrite grains and favorably oriented pearlite colonies, while the black areas most likely correspond to pearlite colonies, which cannot be indexed either due to unfavorable orientation of ferrite and shielding cementite lamellae, or due to large orientation gradients [4,9].

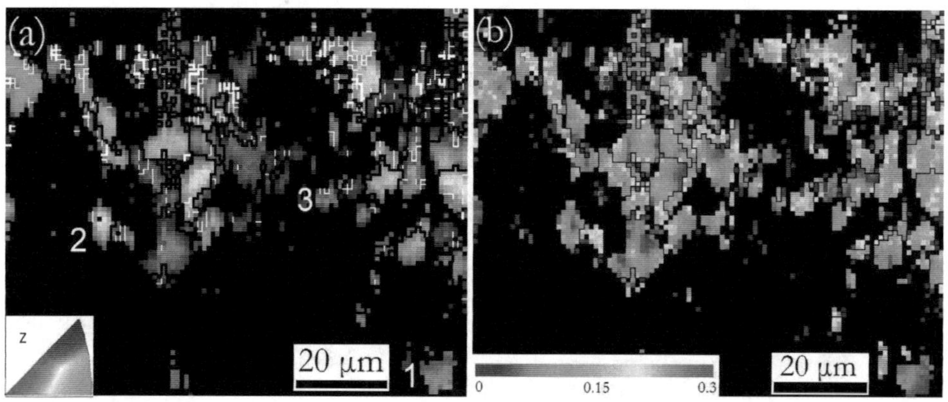

Figure 2. Distribution of crystallographic orientations in the sample annealed at 300 °C for 4h. (a) IPF map overlaid on top of a grey contrast from the number of indexed spots (in the range of 6-20), and (b) KAM map showing misorientation below 0.3° within the grains. The numbers 1-3 in (a) mark three grains selected for monochromatic energy scans. White and black lines in (a) mark boundaries with misorientations of >0.3° and 3°, respectively. (The sample surface is at the top of the figures)

The small misorientation range (0-0.3°) in the kernel average misorientation (KAM) map (figure 2b) emphasizes the local orientation variation within the free ferrite grains as well as in the ferritic phase of favorably oriented pearlite colonies. Such small misorientations would not be detected using 'standard' EBSD. The KAM value for a given pixel is defined as the average misorientation angle of that pixel to its 8 nearest neighbors, calculated with the proviso that misorientation angles above a cut-off angle (0.3° for this map) are not counted in the averaging process. The mean KAM value within the indexed ferrite grains selected from approx. 0.1 mm depth (to increase the chance that only free ferrite grains are included in the calculation) from sample surface is about 0.08° for both samples, which corresponds to a geometrically necessary dislocation (GND) density of 1.2×10^{13} m^{-2}, estimated using the method described in [12]. This result implies that there are no dramatic dislocation activities in the free ferrite grains in the temperature range 300-500°C.

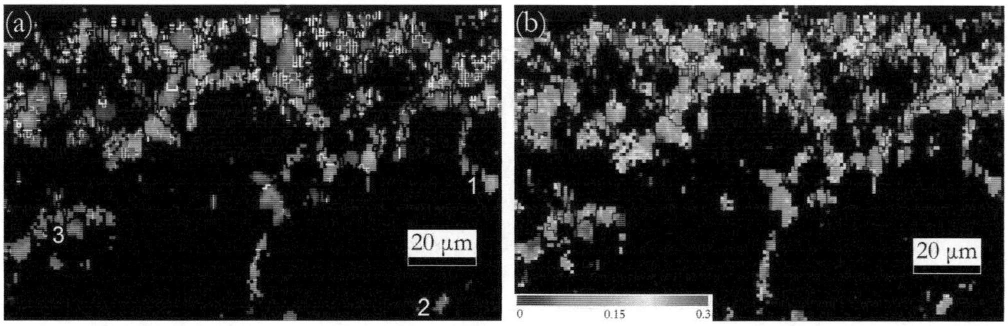

Figure 3. Distribution of crystallographic orientations in the sample annealed at 500 °C for 4h. (a) IPF map overlaid on top of a grey contrast from the number of indexed spots (in the range of 6-20), and (b) KAM map showing misorientation below 0.3° within the grains. The numbers 1-3 in (a) mark three grains selected for monochromatic energy scans. White and black lines in (a) mark boundaries with misorientations of >0.3° and 3°, respectively. (The sample surface is at the top of the figures)

In addition to the crystallographic orientation/misorientation information presented in the IPF and KAM maps, monochromatic energy scans were used to determine the elastic strain in certain grains in the samples annealed at 300 °C and 500 °C (numbered in figure 2a and figure 3a, respectively). The strains measured in three selected grains in each of the samples are shown in figure 4. The elastic strain values vary spatially within each grain and between grains, and therefore no systematic trend is seen to conclude if the strains close to the grain boundaries are higher or lower than those in the grain interiors. The overall strain variation is approximately 9×10^{-4} (from -5×10^{-4} to 4×10^{-4}) in the 300 °C annealed sample, and 3×10^{-4} (from -2×10^{-4} to 0.7×10^{-4}) in the 500 °C annealed sample. Despite the limited number of grains measured, it appears that a smaller reduced strain variation is observed as annealing temperature increases. The distribution of strains is narrower and the values are close to zero in the 500 °C annealed sample, indicating that the strain within the grains is largely released during the annealing. It is anticipated, based on the morphology and the ability to release residual stress at such low temperature (i.e. before severe spheroidisation occurs [9]), that these selected grains are free ferrite grains.

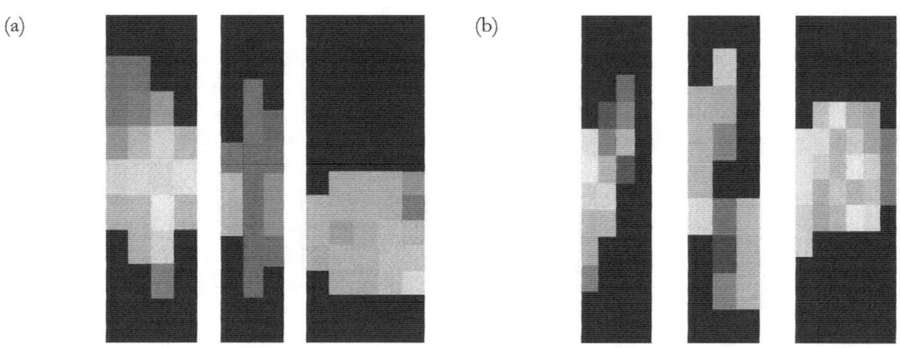

Strain (x10^{4})	Grain 1	Grain 2	Grain 3
Min.	-5.24	-2.43	-2.54
Max.	-1.67	3.94	1.43
Range	3.57	6.36	3.98

Strain (x10^{4})	Grain 1	Grain 2	Grain 3
Min.	-0.565	-0.650	-2.38
Max.	0.683	0.157	0.00572
Range	1.25	0.807	2.44

Figure 4. Strain range measured for 3 grains from samples annealed at (a) 300 °C, and (b) 500 °C. Note that the color scales are according to the minimum and maximum strain values, which are different for the different grains. All strains are $\times 10^{-4}$.

4. Discussion

The results show the relief of the local residual stresses in the present medium carbon R8T pearlitic steel up on annealing. The local residual elastic strain variation developed during manufacturing is $1.5\text{-}2\times10^{-3}$ in the virgin material [13]. These residual strains/stresses are developed mainly during phase transformation because of the orientation and lattice differences between the ferrite and cementite [13,14]. Since the local orientation and size of the pearlite grains around individual free ferrite grains are different, the local residual stresses within and between individual free ferrite grains thus vary (see e.g. figure 4). The stress due to the thermal expansion seen in other steels [15] should be small in this material, as the thermal expansion coefficients for cementite and ferrite are similar in the temperature range 300-500 °C [16,17]. The present results show that the measured local residual stresses are largely released upon annealing at 500 °C, which further supports this argument, as thermal stresses are typically difficult to be removed.

Upon annealing at 300 °C, there is nearly no structural change in the pearlitic colonies [2,9]. A reduction of the GND density from 1×10^{14} m^{-2} [13] in the free ferrite grains in the virgin material to 1.2 $\times10^{13}$ m^{-2} after annealing at 300 °C for 4h corresponds well to the reduction of elastic strain variation, from $1.5\text{-}2\times10^{-3}$ to 9×10^{-4}. Upon annealing at the higher temperature of 500 °C, there is a further decrease in elastic strain in the free ferrite grains, although the GND density there is largely unchanged. At this stage, the stress relief should mainly be related to the recovery of the pearlitic colonies and the initial stages of spheroidization of the cementite [2,9].

The local residual strains/stresses in the present pearlitic steel can be compared with partially recrystallized pure iron [18], for a better understanding of the residual stress formation mechanisms. After annealing at 500-550 °C to ~10% recrystallization, a variation in residual elastic strain of 4×10^{-4} has been observed in the recrystallized grains of size in the range 5-25 μm. In those grains, the GND density is very low, in the order of 10^{-12} m^{-2}. The large difference in the GND density but similar variation in residual elastic strain between the iron and the present steel results suggests that the residual elastic strain variation within individual free/recrystallized ferrite grains is largely independent of GND density, but depends more on the surrounding pearlite colonies/deformed matrix of the steels. The development of medium- to long-range residual stresses from dislocation boundaries in deformed materials has been discussed and quantified extensively in the past (e.g. [19,20]). The presence of a significant residual stress variation in the free/recrystallized ferrite grains, with size significantly larger than the typical range affected by the long-range residual stresses from dislocation boundaries (0.5-1 μm) in the deformed materials, suggests that the formation mechanisms of long-range residual stresses have to be reconsidered taking the new results into account.

5. Summary

In the present study, the microstructure and elastic strain in annealed wheel steel samples were measured using differential aperture synchrotron X-ray microscopy. Only free ferrite grains and favorably oriented pearlite colonies with low orientation gradients were possible to index and it is difficult to differentiate between them in the region close to the surface, where indexing rates are high. A large variation in the elastic strain is seen within individual free ferrite grains as well as between free ferrite different grains. Both the local dislocation density and the residual stresses in the free ferrite grains decrease during annealing. The stress relief in the the free ferrite grains is mainly affected by the spheroidization of cementite in and recovery of the surrounding pearlitic colonies.

Acknowledgments

This work is part of the research activities within CHARMEC (Chalmers Railway Mechanics, www.charmec.chalmers.se). It was partly financed within the European Horizon 2020 research and innovation programme in the projects In2Track, In2Track2 and last in the In2Track3 project under grant agreement 101012456. YZ acknowledges the support by the Danish Council for Independent Research Fund (8022-00340B) and TY acknowledges the support by the European Research Council (ERC) under the European Union's Horizon 2020 research and innovation programme (grant No 788567, M4D). Use of the Advanced Photon Source, an Office of Science User Facility operated for the U.S. Department of Energy (DOE) Office of Science by Argonne National Laboratory, was supported by the U.S. DOE under Contract No. DE-AC02-06CH11357.

References

[1] Cvetkovski K, Ahlström J and Karlsson B 2011 Monotonic and cyclic deformation of a high silicon pearlitic wheel steel *Wear* **271** 382–7
[2] Nikas D, Ahlström J and Malakizadi A 2016 Mechanical properties and fatigue behaviour of railway wheel steels as influenced by mechanical and thermal loadings *Wear* **366–367** 407–15
[3] Peters C J and Eifler D 2009 Influence of service temperatures on the fatigue behaviour of railway wheel and tyre steels *Mater. Test.* **51** 748–54
[4] Nikas D and Ahlström J 2014 Thermal deterioration of railway wheel steels *36th Riso international symposium of material science* pp 411–20
[5] Larson B C, Yang W, Ice G E, Budai J D and Tischler J Z 2002 Three-dimensional X-ray structural microscopy with submicrometre resolution *Nature* **415** 887–90
[6] Levine L E, Larson B C, Yang W, Kassner M E, Tischler J Z, Delos-Reyes M A, Fields R J and Liu W 2006 X-ray microbeam measurements of individual dislocation cell elastic strains in deformed single-crystal copper. *Nat. Mater.* **5** 619–22
[7] Larson B C and Levine L E 2013 Submicrometre-resolution polychromatic three-dimensional X-ray microscopy *J. Appl. Crystallogr.* **46** 153–64
[8] Anon 2009 *European standard for Wheels EN 13262 - Product requirements*
[9] Nikas D, Zhang Y, Jessop C and Ahlström J 2022 Effect of annealing on microstructure in railway wheel steel *IOP Conf. Ser. Mater. Sci. Eng. This conference.*
[10] Zhang Y B, Andriollo T, Fæster S, Liu W, Hattel J and Barabash R I 2016 Three-dimensional local residual stress and orientation gradients near graphite nodules in ductile cast iron *Acta Mater.* **121** 173–80
[11] Dhar S, Zhang Y, Xu R, Danielsen H K and Juul Jensen D 2017 Synchrotron X-ray measurement of residual strain within the nose of a worn manganese steel railway crossing *IOP Conference Series: Materials Science and Engineering* vol **219** p 012016
[12] Zhang Y and Barabash R 2019 High resolution mapping of orientation and strain gradients in metals by synchrotron 3D X-ray Laue microdiffraction *Quantum Beam Sci.* **3** 1–14
[13] Yildirim C, Jessop C, Ahlström J, Detlefs C and Zhang Y 2021 3D mapping of orientation variation and local residual stress within individual grains of pearlitic steel using synchrotron dark field X-ray microscopy *Scr. Mater.* **197** 113783
[14] Bramfitt B L and Marder A R 1973 A transmission-electron-microscopy study of the substructure of high-purity pearlite *Metallography* **6** 483–95
[15] Zhang Y B, Andriollo T, Fæster S, Barabash R, Xu R, Tiedje N, Thorborg J, Hattel J, Juul Jensen D and Hansen N 2019 Microstructure and residual elastic strain at graphite nodules in ductile cast iron analyzed by synchrotron X-ray microdiffraction *Acta Mater.* **167** 221–30
[16] Yershov V M 1981 High-Temperature X-Ray Investigation of Thermal Expansion of the Cementite Lattice. *Phys. Met. Metallogr.* **52** 90–4
[17] Bhadeshia H K D H 2020 Cementite *Int. Mater. Rev.* **65** 1–27
[18] Zhang Y, Yu T, Xu R, Thorborg J, Liu W, Tischler J Z, Godfrey A and Juul Jensen D 2022 Local residual stresses and microstructure within recrystallizing grains in iron *Submitted for Publication.*
[19] Levine L E, Geantil P, Larson B C, Tischler J Z, Kassner M E, Liu W, Stoudt M R and Tavazza F 2011 Disordered long-range internal stresses in deformed copper and the mechanisms underlying plastic deformation *Acta Mater.* **59** 5803–11
[20] Mughrabi H 1983 Dislocation wall and cell structures and long-range internal stresses in deformed metal crystals *Acta Metall.* **31** 1367–79

IOP Conf. Series: Materials Science and Engineering 1249 (2022) 012044 doi:10.1088/1757-899X/1249/1/012044

Multiscale Exploration of Texture and Microstructure Development in Recrystallization Annealing of Heavily Deformed Ferritic Alloys

C Yildirim[1], N Mavrikakis[2], P K Cook[1], R Rodriguez Lamas[1], H F Poulsen[3], C Detlefs[1] and M Kutsal[3]

[1] European Synchrotron Radiation Facility, 38043 Grenoble, France
[2] OCAS, Pres. J.F. Kennedylaan 3, BE-9060, Zelzate, Belgium
[3] Department of Physics, Technical University of Denmark, 2800 Kgs. Lyngby, Denmark

E-mail: can.yildirim@esrf.fr, mkutsal@fysik.dtu.dk

Abstract. We present a multi-scale study of recrystallization annealing of an 85% cold rolled Fe-3%Si-0.1%Sn alloy using a combination of dark field X-ray microscopy (DFXM), synchrotron X-ray diffraction (SXRD), and electron backscatter diffraction (EBSD). Grains of interest from high stored energy (HSE) regions in a 200μm-thick sample are studied using DFXM during isothermal annealing. The intra-granular structure of the as deformed grain reveals deformation bands separated by $\approx$ 3–5°misorientation. Geometrically Necessary Dislocation evolution during recrystallization and growth is investigated. These findings are supported by a quantitative non-destructive texture analysis using SXRD in terms of pole figures and orientation distribution functions. Although no significant macroscopic texture change is observed up to 50% recrystallization, the calculated texture index indicates different nucleation and growth processes at various stages of annealing. Our results show that zones of local misorientation in the HSE regions are decisive for the formation and growth of recrystallized grains.

1. Introduction

Metal processing often includes steps during which the material is plastically deformed and new defects are introduced. The density and distribution of these defects not only determine the hardening in the deformed state, but also constitute the main driving force for annealing phenomena. Understanding the nucleation and growth of recrystallization in deformed materials therefore is of great industrial significance.

The experimental understanding of strengthening mechanisms, in the deformed state, and recrystallization has improved drastically over the past 30 years. Notably, with the introduction of electron backscatter diffraction (EBSD) in scanning electron microscopes, fast and automated data collection providing morphological and crystallographic information can be employed. Studies in the last decades have shown that the distribution of the recrystallized (RX) grains is highly influenced by the local deformation microstructure [1, 2]. However, although these studies improved our understanding of the dislocation structures, electron-based techniques have limited possibilities of dynamic studies within bulk material, due to their sample preparation requirements.

Within the past 20 years, synchrotron-based techniques such as 3D X-ray diffraction, or High Energy Diffraction Microscopy (3DXRD or HEDM) [3, 4], Diffraction Contrast Tomography (DCT) [5] and Differential Aperture X-ray Microscopy (DAXM) [6] have been used to investigate recrystallization and growth phenomena in 3D and 4D (x, y, z, t). These studies have provided fascinating insights on the recrystallization and growth of near perfect grains within bulk materials. Yet, these methods fail to capture the multi-scale relation between the new nuclei and the highly deformed matrix due to limited spatial resolution and the peak overlap problem.

A relatively new synchrotron-based method called Dark Field X-ray Microscopy (DFXM) provides an alternative approach to the above-mentioned challenges. DFXM is a recently-developed diffraction imaging method for probing 3D nanostructures with their associated strain and orientation in bulk materials, with better angular resolution than its electron counterparts [7, 8]. An important feature of the microscope is that it can be coupled with coarser grain mapping methods such as 3DXRD and DCT without having to dismount the sample. Although at high deformation levels grain indexing is not possible, a quantitative texture analysis can still be carried out using the observed intensities from Debye-Scherrer rings. It has been shown that XRD is a well-suited tool for non-destructive texture characterization [9, 10]. Combining synchrotron XRD (SXRD) texture information with DFXM data provides not only a volume-averaged understanding of the sample, but also the fine details of the intricate physics of a grain of interest (GOI) in micro scale within bulk.

Here, we explore the multi-scale links between the deformed state, recrystallization and grain growth using DFXM, SXRD and EBSD upon annealing of a highly deformed ferritic alloy. We employ a texture analysis using pole figures (PFs) and orientation distribution functions (ODFs) from the SXRD data to get an overall picture of the texture evolution. We investigate the evolution of microstructure in terms of GND density using the DFXM data.

2. Experimental

2.1. DFXM

The DFXM experiments were carried out at beamline ID06 at the European Synchrotron Facility (ESRF) [11]. A Si monochromator was used to select 33 keV photons. A set of 1D compound refractive lenses (CRLs) were used to obtain a line-focused beam. A far-field FReLoN CCD was positioned to image the (110) reflection of the grain. An X-ray objective comprising 2D CRLs was used to magnify the diffracted signal from the 2D slices of a GOI, which was projected onto the far-field detector with $\approx$90nm effective pixel size. The DFXM images showing the 3D structure of the GOI were then obtained by moving the sample in the z direction across the 1D condensed beam (figure 1(a)). The local crystallographic orientation, i.e. local 110 pole figure, was measured by scanning one (ω) or two (ω, χ) sample axes at a fixed scattering angle (2θ) (figure 1(a)). More details about the experimental setup can be found in [12]. The collected DFXM data were used to generate center-of-mass (COM) maps using the intensity at a pixel coordinate as function of the goniometer angles across the scan [13].

2.2. SXRD

For the SXRD experiments (figure 1(a), a FReLoN CCD 2D detector was used having 47.3 μm pixel size positioned at 170 mm from the sample, at beamline ID06. $500 \times 500\,\mu m^2$-sized box beam with 33keV photon energy illuminated the sample while ω was rotated 360° in 1° steps. This configuration enabled complete coverage of the first two Debye-Scherrer rings, 110 and 200, and partial coverage (65° out of 360°) of the third ring, 211. The collected SXRD patterns were pre-processed for background subtraction and distortion correction. The respective experimental geometries were calibrated using pyFAI software [14]. In order to derive a polar intensity distribution of the observed diffraction signal, the processed SXRD patterns were averaged in 5° bins in ω-direction. The ω-averaged images were further divided to 5° slices in azimuthal direction, η, and subjected to $1D$ azimuthal integration with pyFAI software. The resulting polar intensity distributions were corrected for Lorentz and polarization factors [9, 15]. These distributions were used to derive pole figures (PFs) and orientation distribution functions

(ODFs) using MTEX software [16]. For the ODF derivation, the resolution of the associated Fourier transform was set as 5°. The derived ODFs were further utilized for re-evaluation of the PFs and for calculation of inverse pole figures (IPFs).

2.3. EBSD

Global recrystallization tendencies were followed with EBSD in the as deformed condition and after interrupted annealing in a dilatometer at 600°C. More details about the EBSD measurements can be found in [12].

2.4. Samples and Sample Environments

We studied Fe-3%Si-0.1%Sn samples that were hot rolled and annealed having a final average grain size of 100 μm [17]. These samples were then cold rolled at a Von Mises true strain of about $\epsilon = 2$, corresponding to 85% reduction in size. The samples remain ferritic due to the high Si content of the alloy. Details of sample preparation is given elsewhere [12, 17]. Different samples from the same batch were used for the EBSD and synchrotron experiments. A hot air blower positioned 5 mm from the sample was used to perform the isothermal annealing steps for the synchrotron annealing studies. A total of 1432 s of isothermal heating at ≈ 610°C was applied in 6 intermittent steps, achieving a global recrystallization amount of ≈50% [18]. All X-ray measurements were then done at room temperature after air cooling.

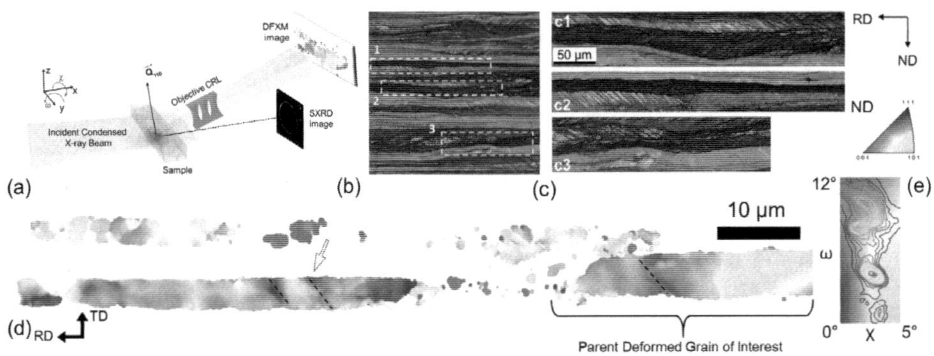

Figure 1. (a) Schematics of DFXM and SXRD setup. The GOI is shown darker.(b) EBSD Normal Direction-Inverse Pole Figure (ND-IPF) map of the initial microstructure of the as deformed state through thickness.(c) Zoom in on regions of high dislocation density. (d) DFXM mosaicity map of a selected layer of the deformed state. (e) The local 110 pole figure color key of the two sample tilts.

3. Microstructure of the Deformed State

Metals that are highly deformed typically exhibit a strongly textured microstructure and heterogeneous stored energy. Figure 1(b-c) shows the EBSD normal direction (ND) IPF superimposed on the image quality (IQ) mappings of the as deformed state. Deformation features such as transition bands (i.e. shear bands, deformation bands) are present. The deformed microstructure shows a rolling texture manifested by the marked appearance of α-fibre ($\{hkl\}\langle 110\rangle$), blue-red-magenta colored regions in (ND-IPF) and γ-fibre ($\{111\}\langle uvw\rangle$), blue regions in ND-IPF) [19]. In conventional cold rolled bcc metals, $\{111\}\langle uvw\rangle$ regions have the highest Taylor factor, with high stored energy (HSE) [20]. As a result, these HSE regions typically produce the first RX grains during the early stages of annealing [21–23]. We, therefore, focused on these HSE regions and mapped the 3D orientation of one using DFXM. An example of a typical DFXM mosaicity map of a given layer of the GOI in the deformed state is shown in figure 1(d). High spatial and angular resolution of DFXM uncovers the fine details of the deformed microstructure of GOI, located in the HSE regions. These details include deformation bands (e.g. yellow arrow) and cells having <1° misorientation across the stretched grain. The orientation gradients along the rolling direction separates zones with different substructure.

Notice that DFXM maps show intensity of the diffracting lattice, and the overall angular spread of the measured deformed grains exceeds 10°. Parts of the sample where the orientation falls outside of the scan range appear empty in the mosaicity maps.

4. Texture Analysis at the Deformed State

Figure 2(a) shows the SXRD-measured PFs of 110, 200 and 211 rings. A stronger signal in 110 and 200 PFs was observed with respect to 211 PF due to limited azimuthal coverage of the 211 ring. In order to compare the obtained SXRD PFs to its EBSD-counterpart, the derived SXRD PFs re-aligned to the EBSD geometry by rotating them with the appropriate angular relations between EBSD and SXRD/DFXM experimental geometries. Therefore, the directions with respect to rolling measured by SXRD correspond to that of the EBSD results. TD and RD IPFs are shown in figure 2(b), displaying 111 and 110 tendencies, respectively. Figure 2(c) shows sections of derived ODFs. The expected cold rolled bcc texture is observed (with α- and γ-fibre) when comparing the 45° ODF section to previous findings [24, 25]. In order to validate our

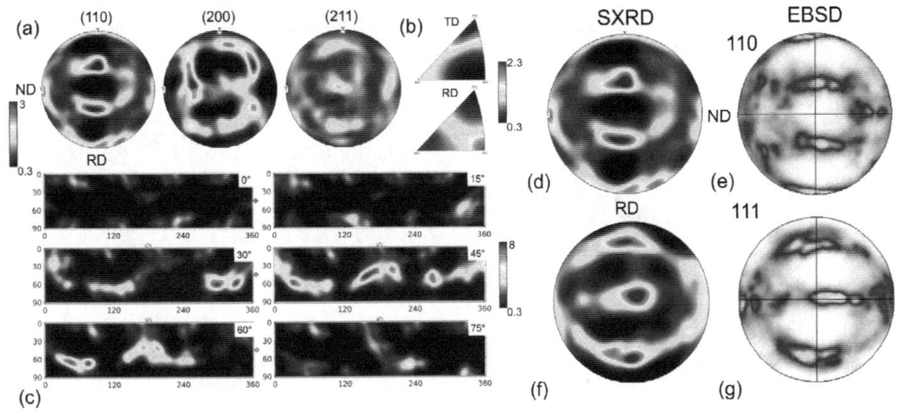

Figure 2. Texture in the as deformed state shown by (a) SXRD Pole Figures. (b) Inverse Pole Figures (c) Orientation distribution functions (ODFs) serial sections of 15° at ϕ_2 of the 3D Euler space (Bunge's notation). 110 and 111 pole figures obtained by SXRD (d, f) and EBSD (max. at 12 mrd) (e, g) measurements. All SXRD pole figures have the same color scale.

SXRD results, we compare the texture of the as deformed state with the EBSD PFs. Fig 2(d-g) shows this comparison of the SXRD and EBSD measured 110 and 111 PFs. For both methods, the observed poles, peak position and relative intensities are in good agreement. Discrepancies between these maps can be attributed to the probed volume differences (i.e. limited penetration depth of EBSD compared to SXRD). A detailed comparison of EBSD and XRD texture analysis in general is discussed elsewhere [24].

Looking at the 111 PFs (figure 2 (f-g)), one can notice that a 111 pole lies almost in the center. The other 111 poles have angles of 60 to 90° between them. According to the angles between 111 planes, the dominant texture can be found as $\{111\}\langle 112\rangle$-type texture [25], which is a common rolling texture of ferritic alloys. Both SXRD and EBSD results show similar orientation relations with different intensity levels.

5. Texture Evolution Upon Annealing

Having established the texture in the deformed state, we can now focus on the effect of isothermal annealing on texture. Figure 3(a-f) show the SXRD 111 PFs at different stages of annealing at 610°C. At a first glance, the overall orientation relations remain the same even at 50% recrystallization, in line with our previous work showing variance maps of the diffraction rings [12]. Nevertheless, 111 PFs show subtle changes upon annealing. In particular, a noticeable decrease of intensity is observed when PFs of the as deformed state and the later stages of annealing are compared. Moreover, a gradual rotation of the texture peak is observed by $\approx 22°$ counter clockwise for one of the 111 poles, figure 3(f) (noticeable at the upper hemisphere of PFs).

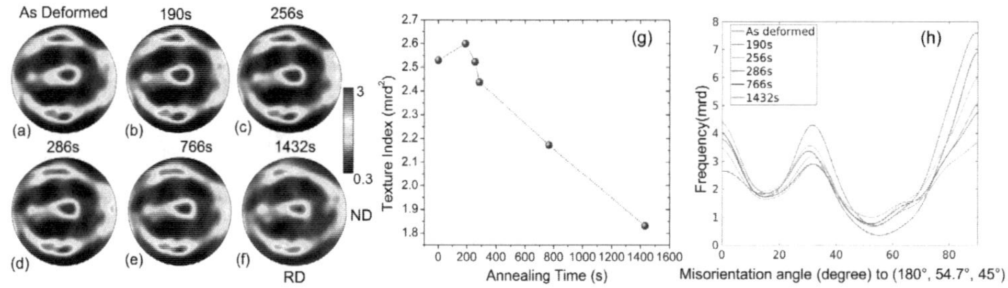

Figure 3. Evolution of 111 Pole Figures measured using SXRD (a-h). Intensity scales are identical for each pole figure. Calculated texture index as a function of annealing time at 610°C (g). Texture density along the γ-fibre (180-270°, 54.7°, 45°) with annealing time (h).

Interestingly, the 111 pole along the ND, towards $\langle 112 \rangle$ increases intensity up to 2.5 mrd from the as deformed state to 286s of annealing. Further annealing results in a continuous decrease of this texture component. Figure 3(g) shows the calculated texture index as a function of annealing time. The texture index is calculated using the following relation:

$$t = \int f(g)^2 dg \tag{1}$$

where $f(g)$ is the ODF [26]. There is an increase in the texture index at the early stages of annealing. Upon further annealing texture index decreases in two fashions: from 190s to 286s, a slope of -0.0016 mrd^2/s is read. Further annealing results in a slope change to -0.00053 mrd^2/s, indicating the different effects of simultaneous recrystallization and grain growth.

From the calculated ODFs, we can focus on a specific texture component of the system and monitor its evolution upon annealing. Figure 3(h) shows the texture density along the γ-fibre with annealing. The peak position at around 30° remains almost the same with annealing. However, there is an increase of frequency of this texture component from the as deformed state to the early stages of annealing, at 190s, similar to what is observed in evolution of the texture index. This may be explained as the HSE regions give birth to many new nuclei during the early stages of annealing before the lower stored energy (LSE) regions can recrystallize. Since these new grains have similar orientation to their parents [12], they contribute to the overall HSE-zone texture, thus increasing the texture index. As recrystallization proceeds, the frequency of the γ decreases continuously, in line with the observed 111 PF intensity drop. This is because the LSE regions start to recrystallize. Hence, the advancement of recrystallization results in an overall texture softening.

6. GND Evolution Upon Annealing

After having established the meso-scale texture evolution, we now zoom into the parent GOI located in a HSE region shown in Fig 1(d), and monitor the GND evolution upon recrystallization and grain growth. Upon plastic deformation, GND density increases accommodating the lattice curvature and manifesting itself with a measurable local intragranular misorientation [27–29]. From mosaicity maps (Fig 1(d)), one can calculate the intragranular misorientation. The misorientation is defined by $\Delta\gamma = \sqrt{(\Delta\omega)^2 + (\Delta\chi)^2}$ [12]. Here, $\Delta\omega$ and $\Delta\chi$ are the differences between the local sample tilt COM and their grain averages. Considering the case of a lattice deformation proposed by [30], one can calculate the GND density using the following equation:

$$\rho_{GND} = \frac{\alpha \Delta \gamma}{bx} \tag{2}$$

where γ is the misorientation angle, α is a constant ($\alpha = 2$ for tilt misorientation [31]), b is the Burger's vector, and x is unit length.

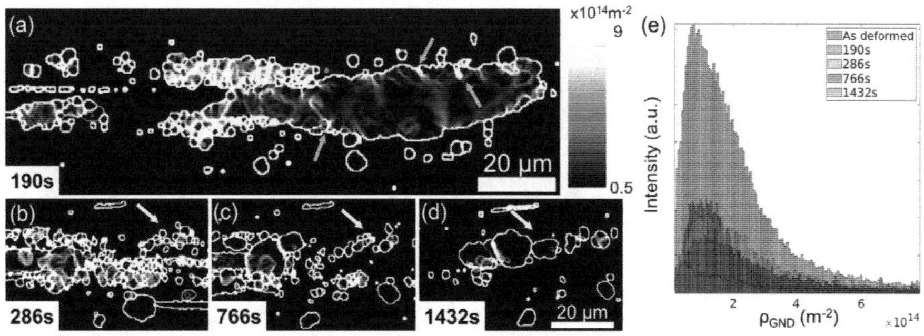

Figure 4. Generated GND density maps during different steps of annealing. (a) shows the early stages of the recrystallization, annealed at 190s at 610°C. (b-d) are zoomed-in views of the same area on the sample showing the recrystallization and growth competition along the further stages of annealing – all figures have the same color scale. (e) GND density distribution histograms at different heating steps.

Figure 4(a-d) shows the generated GND maps of given layers of the GOI at different annealing times. At 190s, we can observe a partially RX structure with new nuclei having formed within and around the deformed parent grain. The deformed parent grain still displays an angular spread larger than 8°, with considerably recovered zones. The regions with higher GND density corresponds to the junction points of the deformation bands (green arrows), and the boundaries of the newly-formed grains, reaching up to values $9 \times 10^{14}\text{m}^{-2}$. Similar values were reported for the same material [17] and on 92% colled-rolled Tantalum using EBSD [31]. The new nuclei, however, show ultra-fine structure with orientation spread $< 0.05°$. The lower orientation spread is manifested by the low dislocation density, below 10^{13}m^{-2}. The new grains form around the bands and the prior grain boundaries of the as deformed parent grain. This is in line with the literature that the misorientation across the boundaries provides the driving force for recrystallization [32]. These regions are marked as high-strain regions in the as deformed structure, reaching up to measured axial strain values above 0.002 with elastic stored energy values larger than 10^4J/m^3 (not shown here). Further annealing causes grain growth (GG) of the RX grains. They grow at the expense of the remaining deformed matrix. The deformed matrix is progressively consumed by the new nuclei, while some of the already-nucleated grains grow. In particular, the growth competition between some RX grains can be monitored: at 286s, more than 15 RX grains having a size of $< 2\,\mu$m are observed (yellow arrow), whereas this number decreases down to 8 grains at 1432 s. GND density continues to decrease during this process. The growth rates of selected grains from the mosaicity maps were analyzed between 286 and 1432s of annealing. The growth rate appears to be size dependent, with values between 0.001 and 0.006 μms^{-1}. This difference in growth rate can be attributed to grain impingement; certain individual grains consume the non-RX matrix as they grow freely, while some impinge with other nuclei. The DFXM-calculated growth rates were compared with previous global observations of the same material [18]. These observations were done using optical microscopy (OM) and EBSD analyses after interrupted annealing. There, the mean growth rates were estimated as 0.015 μms^{-1} for 650°C, 0.11 μms^{-1} for 700°C, 0.71 μms^{-1} for 750°C, respectively. The DFXM growth values during recrystallization correlate well with that of OM/EBSD considering the grain boundary velocity is proportional to the annealing temperature.

Finally, a general picture of the evolution of average GND density distribution upon annealing of the GOIs is shown in Fig 4(e). The large intensity drop between the as deformed and 190s of annealing is due to the combined recovery and recrystallization effect, as the dislocations annihilate during recovery process. Further annealing results in a continuous decrease of the GND density with a slower rate, combining the effects of recrystallization and grain growth.

7. Conclusions and Outlook

We presented a multi-scale investigation of the microstructure and texture evolution of a heavily deformed (85%) ferritic alloy upon isothermal annealing, using a combination of DFXM, SXRD and EBSD methods. The SXRD texture analyses agree well with the EBSD results, showing a typical rolling texture observed in ferritic alloys. Although there are no significant texture changes up to 50% recrystallization, SXRD results reveal quantifiable subtle, but meaningful, variations at different steps of annealing. The evolution of the GND structure with annealing illustrate the coupled effects of recovery, recrystallization and grain growth. Combined with SXRD, DFXM can be a powerful technique to resolve subgrain dynamics and along with texture information can offer a new perspective in the study of recrystallization.

Acknowledgments

The authors acknowledge the ESRF for provision of beamtime at ID06. N.M. acknowledges the financial support of the National Association of Research and Technology (ANRT – Project no.1073–2015). HFP acknowledges support from ERC Advanced Grant nr 885022 and from the Danish ESS lighthouse on hard materials in 3D, SOLID. M.K. thanks Dr. M. Hazar Şeren for his valuable discussions on the development of the SXRD data analysis routine.

References

[1] Xu W, Quadir M and Ferry M 2009 *Metall. Mater. Trans. A* **40** 1547–56
[2] Zhang Y, Juul Jensen D, Zhang Y, Lin F, Zhang Z and Liu Q 2012 *Scr. Mater.* **67** 320–323
[3] Schmidt S, Nielsen S F, Gundlach C, Margulies L, Huang X and Jensen D J 2004 *Science* **305** 229–32
[4] Hefferan C M e a 2012 *Acta Mater.* **60** 4311–18
[5] Sun J, Yu T, Xu C, Ludwig W and Zhang Y 2018 *Scr. Mater.* **157** 72–75
[6] Xu C, Zhang Y, Godfrey A, Wu G, Liu W, Tischler J Z, Liu Q and Juul Jensen D 2017 *Sci. Rep.* **7** 1–7
[7] Simons H *et al.* 2015 *Nat. Commun.* **6** 6098
[8] Poulsen H F, Jakobsen A, Simons H, Ahl S R, Cook P and Detlefs C 2017 *J. Appl. Cryst* **50** 1441–56
[9] Yuan H, Chen Z, Buslaps T, Honkimäki V and Borbély A 2018 *J. Appl. Cryst.* **51** 883–94
[10] Sinclair C W *et al.* 2003 *Adv. Eng. Mater.* **5** 570–74
[11] Kutsal M *et al.* 2019 *IOP Conf. Mat. Sci. and Eng.* **580** 012007
[12] Yildirim C *et al.* 2022 *Scr. Mater.* **214** 114689
[13] Ferrer Garriga J *et al.* 2022 *arXiv preprint: 2205.05494*
[14] Ashiotis G *et al.* 2015 *J. Appl. Crystallogr.* **48** 510–519 ISSN 1600-5767
[15] Poulsen H 2004 *Three-Dimensional X-Ray Diffraction Microscopy* (*Springer Tracts in Modern Physics* vol 205) (Germany: Springer Berlin Heidelberg)
[16] Bachmann F, Hielscher R and Schaeben H 2010 *Solid State Phenom.* **160** 63–68
[17] Mavrikakis N e a 2019 *Acta Mater.* **174** 92–104
[18] Mavrikakis N *et al.* 2018 *IOP Conf. Ser.: Mater. Sci. Eng.* vol 375 (IOP Publishing) p 012016
[19] Hoelscher M, Raabe D and Luecke K 1991 *Steel Res* **62** 567–75
[20] Dillamore I L and Katoh H 1974 *Metal Science* **8** 73–83
[21] Doherty R D *et al.* 1998 *Materials Today* **1** 14–15
[22] Inagaki H 1994 *ISIJ International* **34** 313–321
[23] Hutchinson W B 1984 *Int. Metals Rev.* **29** 25–42
[24] Wright S I, Nowell M M and Bingert J F 2007 *Metall. Mater. Trans. A* **38** 1845–55
[25] Fu J 2021 *J. Mater. Res. Technol.* **14** 1284–91
[26] Hielscher R and Schaeben H 2008 *J. Appl. Cryst.* **41** 1024–37
[27] Ashby M 1970 *Phil. Mag.* **21** 399–24
[28] Nye J F 1953 *Acta Metal.* **1** 153–162
[29] Gao H and Huang Y 2003 *Scr. Mater.* **48** 113–18
[30] Kubin L and Mortensen A 2003 *Scr. Mater.* **48** 119–125
[31] Moussa C, Bernacki M, Besnard R and Bozzolo N 2015 *IOP Conf. Ser.: Mater. Sci. Eng.* vol 89 p 012038
[32] Humphreys F J and Hatherly M 2004 *Recrystallization And Related Annealing Phenomena* 2nd ed (United Kingdom: Elsevier Science)

Recent advances of lab-based diffraction contrast tomography – reconstruction speed benchmark testing and validations

J Sun[1], F Bachmann[1], J Oddershede[1] and E Lauridsen[1]

[1]Xnovo Technology ApS, Galoche Allé 15, 1st floor, 4600 Køge, Denmark
E-mail: jsun@xnovotech.com

Abstract. Lab-based diffraction contrast tomography (DCT) enables the user to reconstruct 3D grain maps of polycrystalline materials non-destructively. For each grain, the morphology and crystallographic orientation, as well as derived properties such as grain boundaries, can be determined. Through two application examples this paper demonstrates the data acquisition and reconstruction speed of the current implementation and validates the resulting grain maps. Firstly, for a conventional Laue focusing scan of an AlCu sample comprising 340 grains, major performance enhancements of the reconstruction algorithm have reduced the reconstruction time from half a day to half an hour. The second example highlights an advanced scan with projection geometry of an oriented electrical steel sheet containing 7,800 grains. While the data collection time is around a day for each of the data sets, the boosted reconstruction of the advanced acquisition data takes half a day and gives the full grain map without the need for stitching. After the major algorithm speed enhancements, grain map qualities are comparable with misorientations below 0.02° and grain boundary distances less than a voxel for both samples.

1. Introduction

Imaging the 3D grain microstructure of polycrystalline materials non-destructively is key to a better understanding of the material performance. Characterizing such microstructures through DCT was first introduced at the large-scale synchrotron facilities [1,2] before it migrated to the lab-based X-rays sources [3] and was commercialized as the LabDCT™ solution for ZEISS Xradia Versa X-ray microscopes [4,5] powered by the GrainMapper3D™ reconstruction engine [6].

The development history of lab-based DCT is briefly outlined in Figure 1. In 2015, when LabDCT was first introduced, GrainMapper3D v.1 only gave cube plots of grain centroids. With v2.0 came full grain morphologies, followed by a series of optimizations of algorithm speed, map fidelity and user friendliness (v2.1-v2.3). The recent v3.0 takes lab-based DCT to the next level of throughput, versatility and representative volume mapping. This is done by introducing advanced scanning schemes [7], enabling seamless data acquisition and reconstruction of larger, irregularly-shaped sample volumes like dogbones and plate geometries, and by the release of CrystalCT™, a dedicated microCT system for 3D grain mapping. The current GrainMapper3D v3.1 implementation features a major enhancement in the reconstruction algorithm and the introduction of reconstruction boosting.

Reconstruction benchmark tests comparing the different versions of GrainMapper3D, performed both for a conventional and an advanced scan, will be presented in the following. The different acquisition strategies and reconstruction details will be outlined in Section 2, while the results of the reconstruction benchmark tests will be presented, validated, and discussed in Section 3, and Section 4 will round off with summary and outlook.

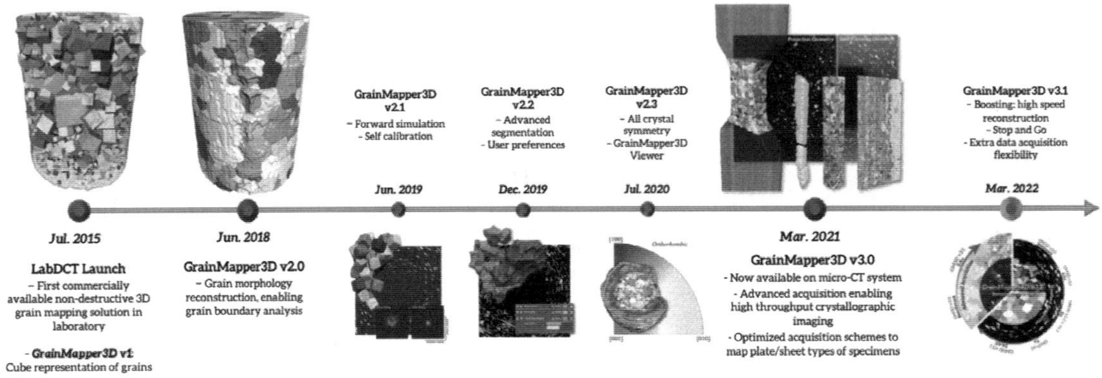

Figure 1. Schematic overview of the development history of lab-based DCT highlighting GrainMapper3D versions.

2. Reconstruction benchmark testing

Two scans, performed on a Zeiss Xradia 520 Versa with a flat panel extension and LabDCT Pro module, have been selected for the reconstruction benchmark tests: 1) A conventional DCT scan of an AlCu cylinder previously used to validate the GrainMapper3D v2.0 implementation [6], and 2) An advanced DCT scan of an oriented electrical steel sheet also mapped along with a number of other electrical steels to demonstrate the capabilities of the advanced scanning schemes for this type of samples [8]. Table 1 shows the main sample, data acquisition and reconstruction details of relevance for the benchmark tests.

Table 1. Sample, data acquisition and reconstruction parameters for DCT reconstruction benchmark testing.

	Conventional DCT	**Advanced HP-HART DCT**
Sample		
Material	Al-5%wt Cu	Fe-3.1%wt
Geometry	AlCu cylinder	Oriented electrical steel sheet
Specimen size	1.4 mm diameter	4 cm (RD) × 2 cm (TD) × 0.4 mm (ND)
Acquisition		
Geometry	Laue focusing	Projection
Source		
Voltage / Power / Distance	160 kV / 10 W / 14 mm	120 kV / 10 W / 60 mm
Detector		
Type / Binning / Distance	DCT4X / bin1 / 14 mm	Flat panel / bin2 / 245 mm
Number of projections / Exposure time / Total time	181 / 420 s / 22 h	6877 / 5 s / 20 h
Data file size	1.5 GB	20.5 GB
Reconstruction		
Number of voxels	200 × 180 × 80	840 × 40 × 1200
Voxel size	5 μm	25 μm

2.1. *Data acquisition*

While conventional DCT data is collected in Laue focusing geometry on a high-resolution detector with scintillator optics, the current advanced HP-HART DCT data was collected in projection geometry on a flat panel detector. Parallel efforts have explored reducing the number of projections [9] and using projection geometry on high-resolution detectors [10]. However, the lower detection efficiency of the high-resolution detectors combined with the increased source to detector distance leads to inconviently long acquisition times. The current approach relies on efficient medium resolution flat panel detectors and high geometrical magnification, which enables high spatial resolution while still maintaining the same number of projections per volume element within reasonable acquisition times.

The term *conventional* DCT scan refers to the traditional way of collecting DCT data (rotation, no translations) limiting the sample width to fit inside the aperture field of view (FOV) at all rotation angles, and the sample region of interest (ROI) height to be defined by the beam height. The *advanced* DCT scanning schemes introduced in GrainMapper3D v3.0 combine sample rotations and translations to map a sample ROI larger than the aperture FOV. In the present study a HP-HART (Helical Phyllotaxis High Aspect Ratio Tomography) advanced scan was used. Between every projection of a helical phyllotaxis scan the sample is rotated by 137.5° and translated vertically by a few microns in order to uniformly cover a sample ROI taller than the aperture FOV. To cover a sample ROI that is also wider than the aperture FOV, the helical phyllotaxis motion can be coupled with a horizontal translation, either in a HP-Raster scan where the number of horizontal steps is equal for each projection angle, or in an adaptive manner for high aspect ratio tomography of plate-like samples using the HP-HART scan. For more details please refer to [7].

Example projection data from the conventional and advanced HP-HART scans can be seen in Figure 2a and b, respectively. Note that the total data collection times of the two scans are almost the same because the advanced HP-HART scan has a substantial overhead due to the larger sample rotations (137.5° versus 2° for the conventional scan) and additional translations between projections.

2.2. *Data reconstruction*

Within GrainMapper3D, indexing is first performed at certain super-voxels or seeds, after which the seed orientation is grown to adjacent voxels. This approach is significantly faster than indexing voxels one-by-one. During reconstruction several (often only slightly different) seed orientations may attempt to grow into the same sample region. GrainMapper3D v3.1 introduces boosting to reduce the effort spent on growing similar solutions. The growth is governed by the completeness (ratio of observed to expected reflections matching a given orientation) tolerances and boosting constrains the growth of new regions such that only seeds with a completeness higher than a given percentage (the boosting factor) of the existing completeness will grow. For the current study, boosting factors of 95% and 90% were used for the GrainMapper3D v3.1 reconstructions of conventional and advanced HP-HART data, respectively.

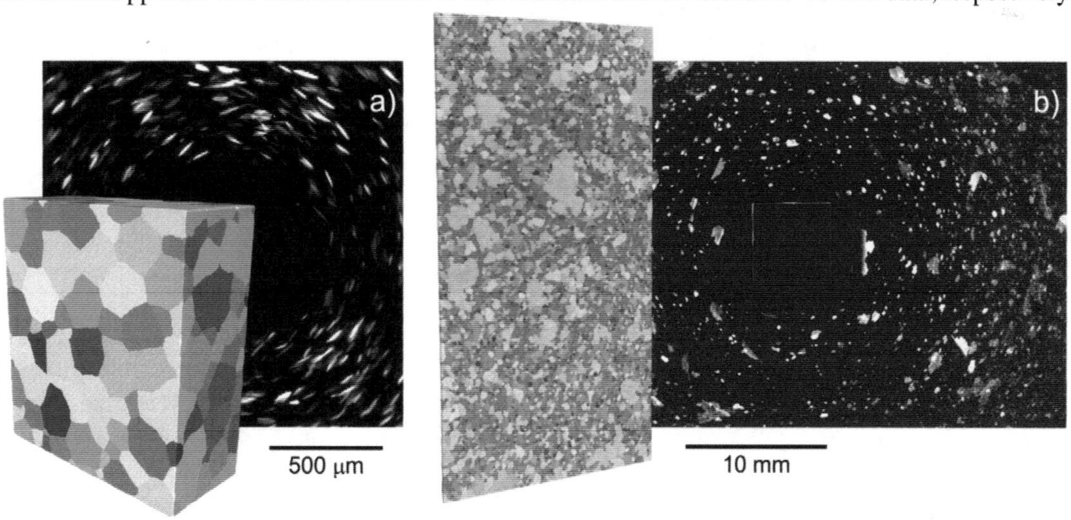

Figure 2. Example data projections and reconstructed 3D grain maps from the speed benchmark testing. a) Conventional Laue focusing scan of AlCu. b) Advanced projection geometry HP-HART scan of oriented electrical steel.

Box-shaped sub-volumes as shown in Figure 2 (number of voxels and voxels sizes given in Table 1) of both samples were selected and used throughout the reconstruction benchmark test. For all tests of a given data set with the different versions of GrainMapper3D, parameters affecting indexing speed were kept the same, *i.e.*, the number of *hkl* families, levels (seed grid size), and completeness tolerances. For more details on how these parameters are defined and affect the indexing please refer to [6]. Likewise the same instrumental parameters were used for all benchmark tests as these affect both indexing speed and grain map fidelity [11].

For the conventional DCT data the reconstruction benchmark tests can be done all the way back to GrainMapper3D v2.0, while reconstruction of advanced DCT data was not enabled until v3.0, so here comparisons can only be made between v3.0 and v3.1. The speed benchmark tests were run on a workstation equipped with an AMD Threadripper 3970X CPU (32 cores, 3.7 GHz) and 256 GB RAM.

3. Results and discussion

3.1. Reconstruction benchmark tests

Table 2 summarizes the results of the reconstruction benchmark tests. Both application examples, representing vastly different data types, show a 3.4× speed up in reconstruction time between GrainMapper3D v3.0 and v3.1 because of the major enhancement in the reconstruction algorithm and the introduction of reconstruction boosting. Furthermore, the conventional DCT reconstruction showed a 24× speed up between GrainMapper3D v2.0 and v3.1, basically reducing the reconstruction time to be measured in the unit of hours rather than days.

3.1.1. Data file size. As can be seen from Table 1, the data file size of advanced DCT scans with many projections soon becomes large to the point where it gets challenging to handle, even on a powerful workstation. Keep in mind here that not only the raw data, but also the result map (more than 40 million voxels for the advanced HP-HART DCT reconstruction), takes up a significant amount of memory. The current recommendation is to keep the number of projections below 2,500 for bin1 data and 10,000 for bin2 data (the advanced HP-HART DCT data is bin2).

Table 2. Reconstruction benchmark test results. *ESD > 20 μm (4voxels). **ESD > 100 μm (4 voxels)

GrainMapper3D	Conventional DCT			Advanced HP-HART DCT		
	Time [min]	Speed up	Number of grains*	Time [min]	Speed up	Number of grains**
v2.0	828		314			
v2.1	186	4.5×	322			
v2.2	160	5×	321			
v2.3	166	5×	321			
v3.0	119	7×	343	2478		7870
v3.1	34	24×	338	734	3.4×	7767

3.2. Grain map validations

Table 2 also gives the number of indexed grains with estimated sphere diameter (ESD) larger than 4 voxels for each reconstruction. When validating the different reconstructions against each other the maps will be compared pairwise within a center-of-mass distance criteria of 4 voxels and a misorientation criteria of 0.1°. The minimum and maximum number of matching grain pairs within these criteria will be given along with the average center-of-mass distances and misorientations.

3.2.1. Conventional DCT. Not surprisingly, the best match of AlCu conventional DCT reconstructions is found between GrainMapper3D v3.0 and v3.1 (330 grains with average deviations of 1.3 μm and 0.005°), while the poorest match is between v2.0 and v3.0 (284 grains with average deviations of 7.7 μm and 0.014°). The unmatched grains are mainly small, such that they fall just below the 20 μm size limit in one map but not in the other, and located at the sample surfaces.

As an example of a more detailed validation, Figure 3 shows a slice through the AlCu reconstruction performed using GrainMapper3D a) v2.1 and b) v3.1, respectively. For these two maps 317 grains match with average deviations of 2.6 μm and 0.006°. In order to quantify the very faint differences between the maps, the voxel-by-voxel comparison of c) misorientation (scale max 0.1°) and d) grain boundary distance (scale max 6 μm) are also shown in Figure 3. The average grain boundary distance is found to be 1.2 μm, while 95% of the boundaries are within 4.5 μm, which is less than a voxel. For the v3.0 versus v3.1 comparisons the corresponding average grain boundary distance is 0.7 μm, while 95% of the boundaries are within 3.5 μm.

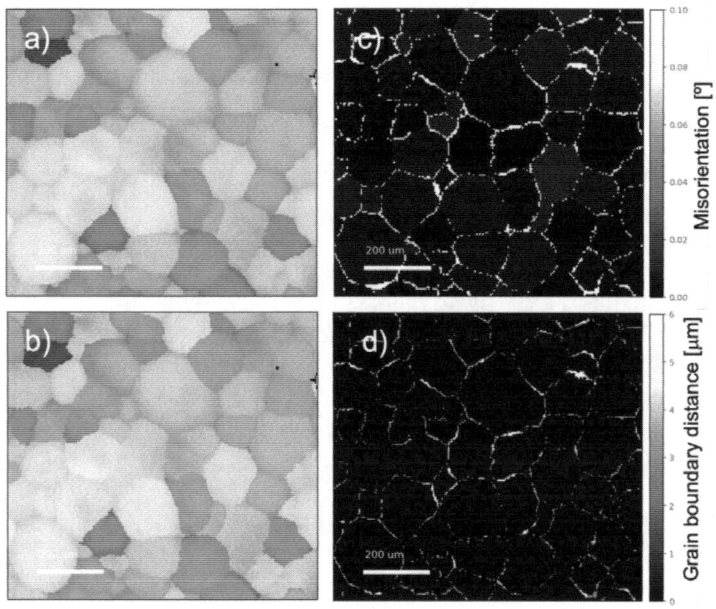

Figure 3. Slice through reconstructions of conventional DCT scans of AlCu done with GrainMapper3D a) v2.1 and b) v3.1 and voxel-by-voxel comparisons of c) misorientation and d) grain boundary distance.

When looking at the number of indexed grains as a function of GrainMapper3D version there seems to be a tendency that v2.0 has fewest grains, v2.1-v2.3 have slightly more and very similar numbers of grains, and then v3.0 and v3.1 have the most grains and are again very similar in numbers. The pairwise voxel-by-voxel comparisons also confirm that the v2.1, v2.2 an v2.3 maps are equally similar as the v3.0 and v3.1 maps. A closer look of the additional 21 grain indexed with v3.1 compared to v2.1 reveal that nine of these are at the very surface of the reconstructed volume, another nine have misorientations to larger neighboring grains just above the 0.1° misorientation criteria used to define grains and the remaining three are genuine new grains. These three new grains have average completeness values in the range 72-78% as compared to the average map completeness of 86%. They are 30-60 μm in size, located in low completeness grain boundary junctions of the v2.1 map, and also indexed in the v3.0 reconstruction. The increased number of grains in the v3.x maps compared to v2.x probably originate from the new background correction and noise reduction options, introduced with GrainMapper3D v3.0 to enhance diffraction signal-to-noise ratios and used for the v3.x reconstructions. This makes more of the weaker diffraction spots (typically originating from smaller grains) available for the indexing routine, and hence some of these grains no longer fall below the minimum completeness limit. Similar observations were made in a recent study applying deep learning algorithms for background subtraction and segmentation of lab-based DCT data to improve the grain map reconstructions [12]. To wrap up it should also be mentioned that the five extra grains larger than 20 μm found with v3.0 compared to v3.1 are all associated with the 0.1° misorientation criteria used to define grains, which is slightly too small to consistently merge all sub-grains of the AlCu sample [6].

3.2.2. Advanced HP-HART DCT. The number of grain pairs matching between the two oriented electrical steel reconstructions performed with GrainMapper3D v3.0 and v3.1 are 7656. The average center-of-mass distance of these is 13.5 μm, and the misorientation 0.007°. The average grain boundary distance is found to be 25 μm (1 voxel), while 95% of the boundaries are within 88 μm (3.5 voxels). While the misorientations are very similar to those observed for the conventional DCT reconstruction comparisons, the center-of-mass and grain boundary distances are somewhat larger. For the center-of-masses this can mainly be attributed to the fivefold larger voxel size. For the grain boundary distances the difference is more than a factor of five, although still comparable to previous reproducibility studies [13] when taking the voxel size into account. It would be natural to also attribute the difference in grain boundary accuracy to the differences in data collection modalities, *i.e.*, conventional versus advanced scan and Laue focusing versus projection geometry, but this is generally not observed when using the same sample for all scans.

4. Summary and outlook

The present paper has demonstrated the current capabilities of lab-based DCT in terms of data acquisition and reconstruction speed. It was validated that the qualities of the resulting grain maps are unaffected by the recently implemented speed enhancements. The access to significant improvements in grain statistics within the same acquisition time when moving from a conventional Laue focusing to an advanced projection geometry DCT scan has been highlighted. New encouraging results show that reconstruction times for conventional and advanced DCT scans should now be counted in hours and days, respectively, rather than in days and weeks. With the shift towards larger samples and improved grain statistics, the data volumes increase accordingly. This puts new demands on the reconstruction algorithms, and for this reason a better memory management to handle larger data files is on the roadmap for GrainMapper3D.

Acknowledgement
The authors thank Jules Dake, Ulm University, Germany for providing the AlCu sample and Yang Li, University of Science and Technology Beijing, China for providing the oriented electrical steel sheet.

References

[1] King A, Johnson G, Engelberg D, Ludwig W and Marrow J 2008 *Science* **321** 382–5
[2] Ludwig W, Reischig P, King A, Herbig M *et al.* 2009 *Rev. Sci. Instrum.* **80** 033905
[3] King A, Reischig P, Adrien J, Peetermans S and Ludwig W 2014 *Mater. Charact.* **97** 1–10
[4] McDonald S A, Reischig P, Holzner C, Lauridsen E M *et al.* 2015 *Sci. Rep.* **5** 14665
[5] Holzner C, Lavery L, Bale H, Merkle A, McDonald S *et al.* 2016 *Micros. Today* **24** 34–43
[6] Bachmann F, Bale H, Gueninchault N, Holzner C and Lauridsen E M 2019 *J. Appl. Crystallogr.* **52** 643–51
[7] Oddershede J, Bachmann F, Sun J and Lauridsen E 2022 *Integr. Mater. Manuf. Innov.* **11** 1–12
[8] Sun J, Petryshynets I, Meng L, Zhang N, Li Y, Bachmann F and Lauridsen E 2021 *Integr. Mater. Manuf. Innov.* **10** 551–8
[9] Lindkvist A, Fang H, Juul Jensen D and Zhang Y 2021 *J. Appl. Crystallogr.* **54** 99–110
[10] Fang H, Juul Jensen D and Zhang Y 2021 *IUCrJ* **8** 559–73
[11] Niverty S, Sun J, Williams J, Bachmann F, Gueninchault N *et al.* 2019 *JOM* **71** 2695–704
[12] Fang H, Hovad E, Zhang Y, Clemmensen L K H, Ersbøll B K and Juul Jensen D 2021 *IUCrJ* **8** 719–31
[13] Sun J, Oddershede J, Bachmann F, Bale H and Lauridsen E 2019 *IOP Conf. Ser. Mater. Sci. Eng.* **580** 012046

IOP Conf. Series: Materials Science and Engineering 1249 (2022) 012046 doi:10.1088/1757-899X/1249/1/012046

Advanced time-resolved characterization of Stress Assisted Grain Boundary Oxidation of 718 Ni superalloy

P D Resende[1,2], N Viganò[1], D Bardel[3], J Réthoré[4], and W Ludwig[1,2]

[1] European Synchrotron Radiation Facility (ESRF), Grenoble, France
[2] MATEIS, INSA Lyon, Université de Lyon, UMR5510 CNRS, France
[3] Framatome, 2 rue Professeur Jean Bernard, 69007 Lyon, France
[4] GeM - Génie Civil et Mécanique - UMR 6183 CNRS - Centrale de Nantes - Université de Nantes
E-mail: pedro.damas-resende@esrf.fr

Abstract. Nickel superalloys are used for harsh condition application cases as they have high chemo-thermomechanical stability. However, they can suffer from embrittlement due to stress corrosion cracking. This effect is difficult to observe as it can take place at long time scales. Here we propose a feasible experiment to study stress assisted grain boundary oxidation, a phenomenon that has similar mechanisms involved that can take place in laboratory compatible time scales. We observed the event using phase and diffraction contrast tomography while applying monotonic loading of the sample at 650 °C. This initial analysis shows that the experimental setup is a good candidate for the study of such degradation mechanism.

1. Introduction

Nickel superalloys (SA) are widely known for withstanding harsh environments while maintaining good chemo-mechanical stability, key properties for applications in nuclear reactors, turbines and others [1–3]. The IN718 alloy is mainly composed by Ni-Cr-Fe with additions of elements like Nb and Mo. This alloy is constituted by a γ matrix that is hardenable by γ'/γ" precipitates. The crystal structure of the matrix is an FCC solid solution, while the γ' has a ordered cubic structure of type $L1_2$ with approximate composition of $Ni_3(Al,Ti)$ and γ" has an ordered BCT structure of type $D0_{22}$ [2,4] and approximate composition of Ni_3Nb. Both phases are coherent with the matrix, while γ" is the main hardening precipitate [4,5]. The occurrence of δ phase is also possible with an orthorhombic structure [6]. MC carbides can also be found in the microstructures [4].

Regarding the application in nuclear power plants, premature failure of components made with 718 alloy was found in literature due to the Stress Corrosion Cracking (SCC) phenomenon [1]. The main mechanism of failure in reactor conditions (~350 °C, mechanical loading, and irradiation) is Stress Corrosion Cracking (SCC). Leonard *et al.* [1] studied leaf springs that were subjected to a commercial reactor exposure for 2 years. In their study it was found that the samples showed extensive grain boundary (GB) cracking in spite of its low content of δ phase. They also found that cracks tend to propagate preferentially along random high angle GBs.

SCC is caused by a coupling between oxygen diffusion and mechanical loading [7,8] that leads to embrittlement. This susceptibility of the material to embrittlement in such application is related to crystallographic orientation relationships within the grain structure of the material [9,10]. In the literature the most common type of study relies on observation of *post-mortem* samples, and/or surface-limited observations. There is a clear need to complement this type of surface analysis with in-situ bulk observations in order to gain a better understanding of the kinetics and to account for the three-dimensional character of the damage process.

Observing SCC *in-situ* is a difficult task as the phenomena can take a long time and never reach failure [11]. Stress Assisted Grain Boundary Oxidation (SAGBO) observation arises as a sound method to mimic SCC within laboratory consistent time-scales. SAGBO occurs at conditions with combined high temperature, oxygen partial pressure and mechanical loading over a limited range of strain rates [7,8,12]. Andrieu, Max and Viguier showed in their work [8] that at high strain rates ($\sim 10^{-2}$ s^{-1}) intragranular cracking and embrittlement is not observed even at higher temperatures in air atmosphere. For lower strain rates ($< 10^{-4}$ s^{-1}), it is possible to see a clear transition from the transgranular fracture regime, accompanied by Portevin-Le Chatelier effect, to an intergranular failure, depending on the temperature. Therefore, a condition to carry out SAGBO observation can be defined around 650 °C and strain rates in the order of $\sim 10^{-5}$ s^{-1}.

In the present work we present an experimental setup for *in-situ* observation of SAGBO that allows to correlate 3D grain orientation maps with the evolution of a crack network by means of combining phase contrast and diffraction contrast tomography. Whereas the former allows to visualize the 3D crack network during in-situ testing at high temperature, the former can be used to map the 3D microstructure of the material prior to the in-situ experiment. For heat treatments promoting precipitation of the delta phase, it may also be possible to use the phase contrast data to calculate local strain fields by means of digital volume correlation (DVC) [13,14]. However, the weak difference in electron density implies poor image contrast between matrix and the precipitate phase and careful optimization of the scan acquisition and processing steps is required to provide suitable contrast for this type of analysis. Hereafter we discuss incipient processing details to achieve such image quality.

2. Experimental procedures

The *in-situ* experiment was conducted at the ID11 Materials Science beamline at ESRF. Miniaturized samples of IN718 alloy were cut in dog-bone shape specifically designed for the Nanox [15] tensile rig. The rig was modified to accommodate a longer shaft allowing SAGBO *in-situ* observations inside a tubular furnace by means of phase contrast tomography (PCT). The samples, already mounted in the stress-rig, were initially grain-mapped by means of diffraction contrast tomography (DCT) before straining and/or heating. All measurements were carried out on the 3DXRD endstation of ID11.

2.1. *Diffraction contrast tomography (DCT)*

The data acquisition for the grain map of the undeformed sample was performed using a 2048 x 2048 pixels sCMOS detector system with effective pixel size of 2.44 µm and sample-detector distance, $d = 7$ mm. 50.25 keV x-rays ($\lambda = 0.24673$ Å) were used with an incident beam of 0.7 x 0.3 mm in cross-section. The diffraction spots images were measured with angular steps ($\Delta\omega = 0.1°$) and exposure time of 0.05 s over a full turn of the sample. The DCT data processing pipeline is implemented in Matlab is based on finding and indexing Friedel pairs among the segmented spots and then selecting the spots for shape reconstruction [16,17].

2.2. *Phase Contrast Tomography (PCT)*

The PCT imaging was performed with effective pixel size of 0.61 µm and similar beam conditions. The reconstructions were performed using an implementation from ESRF's data analysis group. Denoising and ring removal routines were also applied to assure the best conditions for digital volume correlation (DVC).

In-situ data acquisition. The cylindrical furnace was placed over the Nanox sample using a motorized translation stage. The temperature was controlled by measurements from a thermocouple near the sample region. The dwell temperature was set to 923 K and remained stable for at least 1 h before the test. The force applied to the sample at a strain rate of order of $\sim 10^{-6}s^{-1}$ was recorded through the experiment allowing the determination of the tensile behavior as function of time.

PCT data processing pipeline. The PCT data was mainly processed using ESRF's Nabu software for tomographic reconstruction [18] using a filtered back-projection (FBP) algorithm. The experimental setup in tomographic acquisitions can introduce image artifacts, such as rings, streaks and error-motion originated defects. Different methods have been proposed to suppress rings [19–22]. In this work we chose to apply the ring corrections suggested by Vo, Atwood and Drakapoulos in their paper [21,23]. The error motion was calculated using a method based on the tomographic consistency premise [24] and the resulting translation vectors were applied before final reconstruction.

3. Results and Discussion

3.1. Initial reconstructions

Figure 1 depicts a vertical slice reconstruction of different time steps of the SAGBO test without corrections. One can notice the evolution of shape/features in the middle slice as the sample is deformed, this is due to the relative movement imposed by the tensile rig together with cross-section diminishing. It is also possible to notice that the dark spots on the slices, which correspond to MC carbides, present self-similar star-like artifacts. These artifacts are due to small movements of the sample relative to the rotation axis, caused by periodic temperature fluctuations during a full rotation of the sample in our experimental setup. Ring artifacts are prominent in all the images and high frequency noise can be observed.

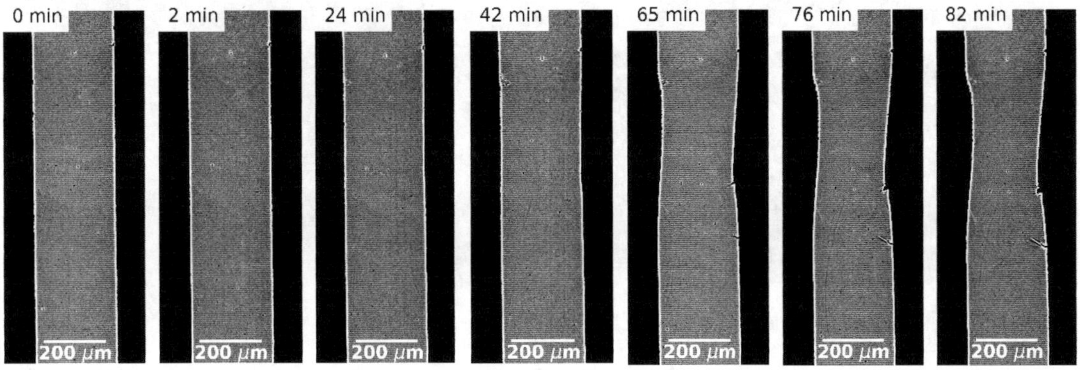

Figure 1. Vertical slice reconstruction of 7 different time steps along the in-situ SAGBO test.

3.2. Deringed and error-motion corrected reconstructions

Figure 2 shows a separation between the effects of error-motion and error-motion + ring correction steps in the 2 minutes acquisition. Comparing figure 2(a) and (b) it's possible to see that the amplitude of the star-like artifacts is reduced. In figure 2(c) one can see that finer rings are removed, although some weaker, new ones have been introduced at locations which seem correlated with the presence of larger carbides.

In figure 3 we present the ring and error-motion corrected slices. It's possible to note that smaller carbide precipitates have enhanced contrast and boundary definition in the reconstructed slice, while bigger particles maintain residual streak artifacts, most likely due to strong phase contrast at the operating distance. The ring removal procedure added/enhanced streak artifacts on the reconstruction as previously reported in [20] potentialized by interaction with cracks in latter time steps.

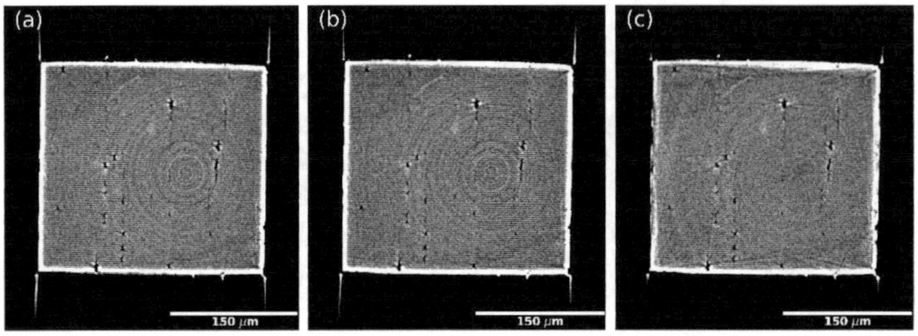

Figure 2. Comparison between slices with (a) no corrections, (b) error-motion correction, and (c) ring and error-motion correction.

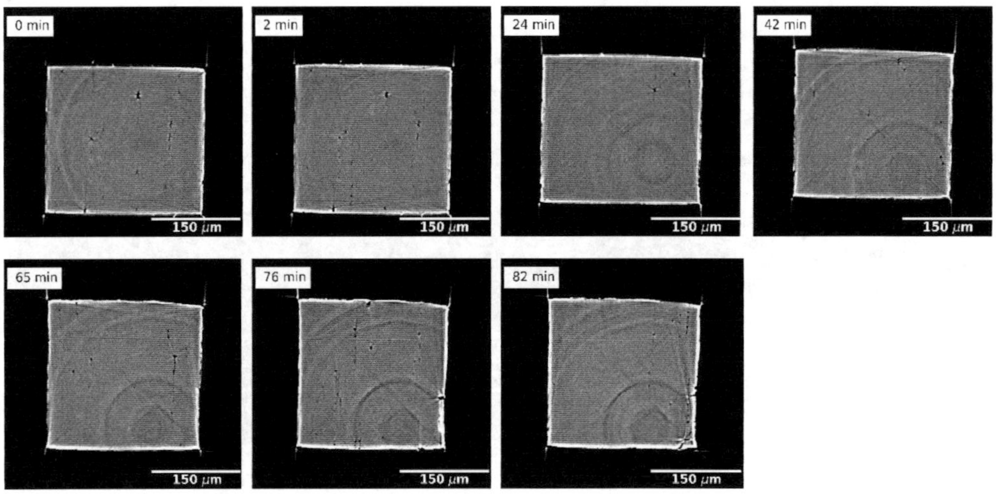

Figure 3. Middle slice reconstruction of 7 different time steps along the in-situ SAGBO test with ring and error-motion correction.

3.3. Effects of filtering on the quality of FBP reconstructed slice

In order to use tomography data to calculate 3D strain fields it is necessary that the contrast found on the images are related to local characteristics of the sample. DVC is based on the assumption of optical flow, in which each point of the sample produces a level of grey within its neighbourhood that will be maintained as the sample is strained. High frequency noise, as well as ring artifacts, is detrimental to DVC, as it introduces random contrast that does not represent, nor follows, the deformation of the sample. In figure 4 we present different frequency cut-off values used with a Hann filter [25] for the reconstruction of the slices. A divergent colormap was chosen to accentuate the differences in pixel value. It is possible to note that even though the filtering produced an overall smoothing of the image the internal gray level fluctuation remain unchanged.

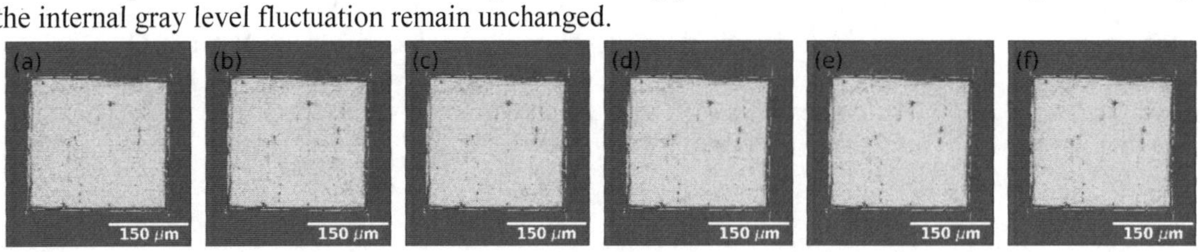

Figure 4. Slices obtained by using different frequency cut-off values of (a)0.5 to (f)1.0 with increments of 0.1 between pictures. Blue color represents lower intensity values in the reconstructed image.

3.4. *Diffraction contrast tomography*

In figure 5 we present the DCT reconstruction of three mid-height centered superimposed on the respective radiograph of the same view of the undeformed sample. It is possible to notice that the sample has a comparatively large grain size with respect to its lateral dimensions, with one grain extending over the full cross-section of the sample on the upper half of the image.

On the referred grain it is possible to see a twin forming an angle of 49.6 degrees with the loading axis of the sample. On the right-hand side of the image a radiograph of the deformed sample, prior to big crack opening, is presented. One can infer from this image that the deformation of the sample is dictated by the rotation of the big grain (purple) in the clockwise direction. On the radiograph it is also possible to see faint extinction spots, with orientation angle of 54.2 degrees, that corroborate this observation.

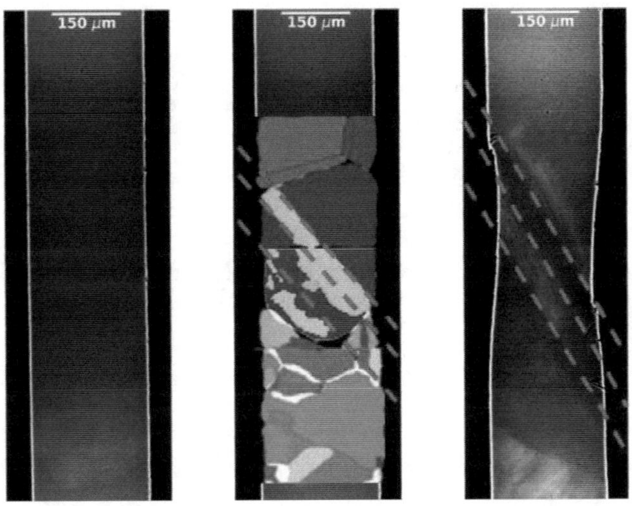

Figure 5. DCT map of the middle slice through the thickness superposed to the radiograph of the sample.

4. Conclusions

In this work we presented an overview of a first attempt of *in-situ*, time-lapse observation of SAGBO in 718 Ni superalloy. We showed that a combination of phase and diffraction contrast tomography can be used to follow the evolution of damage during a continuous tensile test and we succeeded in the direct observation of SAGBO while monotonically loading the sample at appropriate strain rates. However, the challenging experimental conditions give rise to various artifacts in the reconstructed images. Further improvement of the experimental conditions is required to reach the level of data quality required for quantitative analysis of the evolving strain fields by means of DVC. Nevertheless, our preliminary results confirm the feasibility of joint PCT & DCT characterization, enabling crystallographic analysis of the damage initiation sites.

Acknowledgments

This project has received funding from the European Union's Horizon 2020 research and innovation programme under the Marie Skłodowska-Curie grant agreement No 847439.

References

[1] Leonard K J, Gussev M N, Stevens J N and Busby J T 2015 Analysis of stress corrosion cracking in alloy 718 following commercial reactor exposure *J. Nucl. Mater.* **466** 443–59

[2] Villalón N O, Covarrubias O and Colás R 2013 Heat treating of a 718Plus superalloy *Mater. Perform. Charact.* **2** 153–62

[3] Firoz R, Basantia S K, Khutia N, Bar H N, Sivaprasad S and Murthy G V S 2020 Effect of

microstructural constituents on mechanical properties and fracture toughness of Inconel 718 with anomalous deformation behavior at 650 °C *J. Alloys Compd.* **845** 156276
[4] Reed R C 2009 *The physical metallurgy of nickel and its alloys*
[5] Ahmadi M R, Rath M, Povoden-Karadeniz E, Primig S, Wojcik T, Danninger A, Stockinger M and Kozeschnik E 2017 Modeling of precipitation strengthening in Inconel 718 including non-spherical γ" precipitates *Model. Simul. Mater. Sci. Eng.* **25** 055005
[6] Rafiei M, Mirzadeh H and Malekan M 2019 Micro-mechanisms and precipitation kinetics of delta (δ) phase in Inconel 718 superalloy during aging *J. Alloys Compd.* **795** 207–12
[7] Milanese J, Andrieu E, Osio J B, Alexis J and Bardel D 2018 Study of the oxidation assisted intergranular cracking mechanism on a Ni-base superalloy *Miner. Met. Mater. Ser.* **2018-June** 805–15
[8] Andrieu E, Max B and Viguier B 2015 Oxidation Assisted Intergranular Cracking in Alloy 718: Effects of Strain Rate and Temperature *AerospaceLab J.* **9** 1–7
[9] da Cruz Gallo F, Barreto de Azevedo L M, Figueiredo A R, de Vasconcelos Varela A, Araújo L S and de Almeida L H 2021 Effect of increased length fraction of Σ3n special boundaries on OAIC response of cold rolled Ni-based alloy 718 thin sheets *Mater. Sci. Eng. A* **825**
[10] Lu X, Ma Y and Wang D 2020 On the hydrogen embrittlement behavior of nickel-based alloys: Alloys 718 and 725 *Mater. Sci. Eng. A* **792**
[11] Parkins R N and Singh P M 1990 Stress corrosion crack coalesence *Corrosion* **46** 485–99
[12] Ter-Ovanessian B, Villani A, Andrieu É and Forest S 2019 Oxidation-assisted Cracking *Mechanics - Microstructure - Corrosion Coupling* (Elsevier) pp 339–58
[13] Lachambre J 2014 *Développement d'une Méthode de Caractérisation 3D des Fissures de Fatigue à l'aide de la Corrélation d'Images Numériques obtenues par Tomographie X*
[14] Réthoré J 2018 UFreckles
[15] Gueninchault N, Proudhon H and Ludwig W 2016 Nanox: A miniature mechanical stress rig designed for near-field X-ray diffraction imaging techniques *J. Synchrotron Radiat.* **23** 1474–83
[16] Ludwig W, Reischig P, King A, Herbig M, Lauridsen E M, Johnson G, Marrow T J and Buffière J Y 2009 Three-dimensional grain mapping by x-ray diffraction contrast tomography and the use of Friedel pairs in diffraction data analysis *Rev. Sci. Instrum.* **80** 033905
[17] Reischig P, King A, Nervo L, Viganó N, Guilhem Y, Palenstijn W J, Batenburg K J, Preuss M and Ludwig W 2013 Advances in X-ray diffraction contrast tomography: flexibility in the setup geometry and application to multiphase materials *J. Appl. Crystallogr.* **46** 297–311
[18] Paleo P, Payno H, Vigano N and Nemoz C 2021 Nabu 2021.2.0
[19] Mäkinen Y, Marchesini S and Foi A 2022 Ring artifact and Poisson noise attenuation via volumetric multiscale nonlocal collaborative filtering of spatially correlated noise *J. Synchrotron Radiat.* **29** 829–42
[20] Makinen Y, Marchesini S and Foi A 2021 Ring artifact reduction via multiscale nonlocal collaborative filtering of spatially correlated noise *J. Synchrotron Radiat.* **28** 876–88
[21] Vo N T, Atwood R C and Drakopoulos M 2018 Superior techniques for eliminating ring artifacts in X-ray micro-tomography *Opt. Express* **26** 28396
[22] Šalplachta J, Zikmund T, Zemek M, Břínek A, Takeda Y, Omote K and Kaiser J 2021 Complete Ring Artifacts Reduction Procedure for Lab-Based X-ray Nano CT Systems *Sensors* **21** 238
[23] Vo N T, Atwood R C, Drakopoulos M and Connolley T 2021 Data processing methods and data acquisition for samples larger than the field of view in parallel-beam tomography *Opt. Express* **29** 17849
[24] Guizar-Sicairos M, Boon J J, Mader K, Diaz A, Menzel A and Bunk O 2015 Quantitative interior x-ray nanotomography by a hybrid imaging technique *Optica* **2** 259
[25] Toda H 2021 *X-Ray CT* vol 58 (Singapore: Springer Singapore)

IOP Conf. Series: Materials Science and Engineering 1249 (2022) 012047 doi:10.1088/1757-899X/1249/1/012047

Magnetoresistive behaviour of ternary Cu-based materials processed by high-pressure torsion

M Kasalo[1], S Wurster[1], M Stückler[1], M Zawodzki[1], L Weissitsch[1], R Pippan[1] and A Bachmaier[1]

[1] Erich Schmid Institute of Materials Science of the Austrian Academy of Sciences, Jahnstrasse 12, 8700 Leoben, Austria

Abstract. Severe plastic deformation using high-pressure torsion of ternary Cu-based materials (CuFeCo and CuFeNi) was used to fabricate bulk samples with a nanocrystalline microstructure. The goal was to produce materials featuring the granular giant magnetoresistance effect, requiring interfaces between ferro- and nonmagnetic materials. This magnetic effect was found for both ternary systems; adequate subsequent annealing had a positive influence. The as-deformed states, as well as microstructural changes upon thermal treatments, were studied using scanning electron microscopy and X-ray diffraction measurements. Deducing from electron microscopy, a single-phase structure was observed for all as-deformed samples, indicating the formation of a supersaturated solid solution. However, judging from the presence of the granular giant-magnetoresistive effect, small ferromagnetic particles have to be present. The highest drop in room temperature resistivity (2.45% at 1790 kA/m) was found in $Cu_{62}Fe_{19}Ni_{19}$ after annealing for 1 h at 400 °C. Combining the results of classical microstructural studies and magnetic measurements, insights into the evolution of ferromagnetic particles are accessible.

1. Introduction

The change of the electrical resistance in an external magnetic field is termed magnetoresistance (MR) and depending on the material, different MR-effects can occur. Fert et al. and Grünberg et al. independently showed that the change in MR can be huge, for instance ~50% were found in alternating thin layers of Cr and Fe at low temperatures. Due to the large changes in resistance, it was labelled giant magnetoresistance (GMR) [1,2]. The GMR effect is distinguished from the ordinary, anisotropic MR effect by its isotropic behaviour. Currently, research is conducted on e.g. its application for MR-RAM or the detection of influenza A virus by GMR biosensors [3]. Monoclonal antibodies are exchanged for a combination of viral nucleoproteins and magnetic nanoparticles. In the presence of an influenza virus, the magnetic nanoparticles can be bound to the GMR biosensor and the binding rate is proportional to the concentration of the virus. A change in concentration results in a change in resistance, which can be measured in real-time [4].

Alternating ferromagnetic (FM) and nonmagnetic (NM) multi-layered structures are no necessity for GMR. Bulk materials exhibiting small FM particles embedded in a NM matrix can also show a GMR behaviour, called *granular* GMR [5,6]. In general, the evolution of the GMR with increasing magnetic field is connected to the magnetization of the material. For materials featuring the granular GMR, a disadvantage is that the saturation of the hysteresis and thus the full effect of GMR is reached at high magnetic fields. Regarding application, however, 3D-bulk materials show more freedom in application-relevant shaping compared to 2D-multilayered materials.

IOP Conf. Series: Materials Science and Engineering 1249 (2022) 012047 doi:10.1088/1757-899X/1249/1/012047

Typical methods to fabricate granular alloys are mechanical alloying, ion implantation, melt spinning and co-evaporation. It was already demonstrated that materials processed by high-pressure torsion (HPT), a severe plastic deformation technique, also show a GMR effect [7–9]. Fathoming the prospects of HPT together with a proper concatenation of annealing treatments, this work focuses on the granular GMR effect of a variety of ternary Cu-based materials. For the existence of an interface between FM and NM phases, the chosen material systems have to show a pronounced miscibility gap. Fe-Co, due to its increased saturation magnetization, and Fe-Ni, due to its easy processability, were chosen as FM phases and Cu for the NM phase.

2. Experimental

The used powders are Cu (99.9% purity, 170+400 mesh, Alfa Aesar), Co (99.998%, -22 mesh, Alfa Aesar Puratronic), Fe (99.9%, -100+200 mesh, MaTeck), Ni (99.9%, -3+6 mesh, Alfa Aesar). The FM volume fraction was chosen to be 20 vol.%, as it is known from the literature that this ratio leads to the highest granular GMR [10]. In addition, samples containing 40 vol.% of FM material were produced. In the first step, the pure elemental metal powders were mixed to the desired composition. In the second step, the powder blends were pre-compacted in Ar atmosphere to obtain a bulk sample. In the last step, the powder compacts were HPT-deformed at room temperature (RT) at a nominal pressure of 5 GPa, using a rotational speed of 1.28 min^{-1} for 100 rotations. Annealing treatments, followed by water quenching, were performed using a conventional furnace. For long-time annealing treatments (100 h) or treatments at the highest temperature (600 °C) the samples were encapsulated in Ar-filled glass tubes.

For RT resistivity measurements within a magnetic field, the experimental setup described in [8,9] was used. Herein, GMR is represented by the difference in resistivity at zero field and highest field in relation to the resistivity at zero field. Two different orientations of current flow in relation to the magnetic field lines are probed. The exact chemical composition of each material system was determined by electron dispersive X-ray spectroscopy (e-Flash, Bruker) attached to a scanning electron microscope (SEM, Magna, Tescan). All compositions of HPT-samples are given in wt.%. Microstructural investigations in different viewing directions (LEO1525, Zeiss) and X-ray diffraction (XRD) measurements (D2-Phaser, Bruker) of all states were performed.

3. Results and Discussion

3.1. Scanning electron microscopy

As shown in figure 1a, a homogeneous microstructure in the nanocrystalline (NC) regime was reached for $Cu_{78}Fe_{16}Co_6$ after deformation. The low phase-contrast in backscatter mode indicates a good intermixing. The additional annealing treatments for 1 h at 400 °C and 10 min at 500 °C do not increase the grain size significantly. However, an increased number of small, darker particles is observed in figure 1b compared to 1a and 1c. These can be identified as FM particles (Fe and/or Co) due to Z-contrast.

For $Cu_{62}Fe_{19}Ni_{19}$, the as-deformed state and two chosen annealed samples are presented in figure 1d-f. In comparison to the as-deformed state, figure 1d, annealing for 1 h at 300 °C and 400 °C (figure 1e) caused hardly any grain growth. The achieved microstructure is somewhat coarser for higher annealing durations, 100 h at 400 °C, and higher annealing temperatures, 500 °C (figure 1f) and 600 °C. However, a homogeneous NC microstructure was retained in each case, even at the highest annealing temperature. Deduced from the very small changes in grain size, the microstructure of $Cu_{62}Fe_{19}Ni_{19}$ shows remarkable temperature stability up to 600 °C, which corresponds to a homologous temperature $T_{hom,Cu}$ of 0.64. The coarsest microstructure was obtained for the longest annealing duration of 100 h at 400 °C; however, the grain sizes still being in the NC regime.

$Cu_{84}Fe_7Ni_9$ was exposed to the same heat treatments as $Cu_{78}Fe_{16}Co_6$. The microstructure of the as-deformed material is shown in figure 1g. A homogeneous NC microstructure was reached in each state with the subsequent thermal treatments for 1 h at 400 °C (figure 1h) and 10 min at 500 °C (figure 1i) resulting in insignificant grain growth.

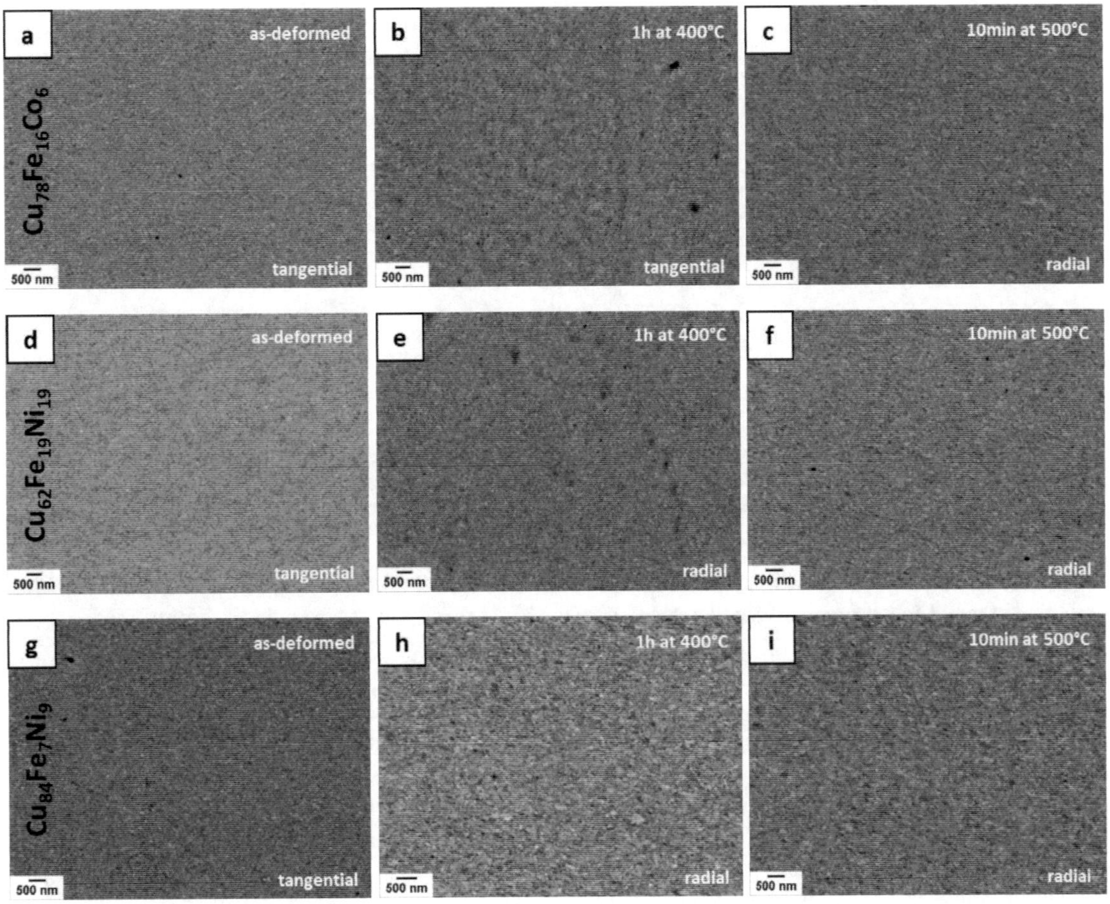

Figure 1. SEM micrographs in backscatter mode of all investigated materials in as-deformed and annealed states. All micrographs were taken at a radius of ~3 mm; the viewing direction with respect to the HPT-disc is indicated.

3.2. XRD

XRD measurements were performed to study the crystallographic phases of all samples. They were carried out on the samples which were used for electrical measurements; thus, the XRD-investigated spots are located at a radius larger than 1.5 mm from the specimen's center. For $Cu_{78}Fe_{16}Co_6$, no Fe-bcc peaks are present in the as-deformed state (figure 2a, lower part), indicating again the formation of a supersaturated solid solution, although Co-Cu and Fe-Cu are immiscible systems at equilibrium conditions [11]. In case of partial supersaturation, another possible explanation for vanishing Fe-peaks could be their low intensity, as bcc-phase content would be strongly reduced in volume after HPT. After annealing the samples, Fe-bcc peaks occurred for $Cu_{78}Fe_{16}Co_6$, indicating Fe precipitation and grain growth. Furthermore, annealing narrows the full width at half maximum (FWHM), which is in good accordance with the BSE micrographs in figure 1a-c.

Figure 2b, shows XRD results of as-deformed and annealed $Cu_{62}Fe_{19}Ni_{19}$. A single-phase fcc structure close to Cu-fcc was reached for the as-deformed state. No peaks appear at the Fe-bcc and Ni-fcc positions. Therefore, one can assume that Fe and Ni are dissolved in the fcc Cu-matrix. The annealing leads to peak narrowing, especially for 100 h at 400 °C and 1 h at 600 °C. This indicates grain growth and correlates with the BSE micrographs in figure 1d-f. In addition, the high-temperature stability of the $Cu_{62}Fe_{19}Ni_{19}$ alloy is confirmed, since no phase separation occurs, even after annealing at 1 h at 600 °C.

For $Cu_{84}Fe_7Ni_9$, a single-phase fcc structure in the as-deformed and annealed states is present (figure 2a, upper part). While $Cu_{78}Fe_{16}Co_6$ samples show a phase separation after annealing, CuFeNi-samples do not show this behaviour, though being exposed to the same annealing parameters.

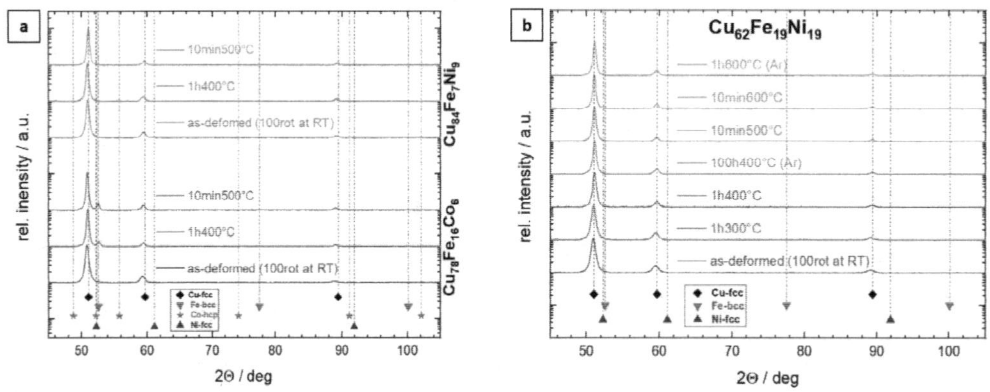

Figure 2. XRD patterns of a) $Cu_{78}Fe_{16}Co_6$ and $Cu_{84}Fe_7Ni_9$, and b) $Cu_{62}Fe_{19}Ni_{19}$ for as-deformed and annealed states. Measurements were performed in the axial direction at a radius > 1.5 mm.

3.3. Magnetoresistive measurements

In figure 3, the decrease of resistivity with increasing applied magnetic field is shown for all as-deformed and annealed states. $Cu_{78}Fe_{16}Co_6$ exhibits an isotropic GMR behaviour in each state; the linear decrease of perpendicular and parallel measurements is almost identical (figure 3a). At the maximum applied field H_{max} of 1790 kA/m, a GMR of about 1.1% was reached for the as-deformed state. In general, for the occurrence of a granular GMR effect, interfaces between FM and NM phases are required. In SEM- and XRD-data, these interfaces are not clearly evident. However, a granular GMR effect was yet measured. Stückler et. al showed with atom probe tomography measurements that supersaturation of Fe (between 7 wt.% and 25 wt.%) in Cu on the atomic scale can be obtained after HPT deformation. Still, some small Fe-rich particles remain in the Cu-matrix [12]. It is evident from MR results in figure 3 that these interfaces; thus the FM particles, must be present, though the particles are so small that they cannot be detected by low energy XRD and SEM.

For $Cu_{78}Fe_{16}Co_6$, subsequent annealing led to an amplification of the GMR amplitude. The highest GMR of ~1.45% at H_{max} was reached after 1 h at 400 °C. Compared to the literature, a GMR of ~15% was observed for ball-milled $Cu_{80}Fe_{14}Co_6$ powder annealed for 1 h at 415 °C [13]. Furthermore, the annealed material reaches even higher values up to 16% at fields of ~4000 kA/m, where saturation is finally reached [13]. It is important to note that these measurements were carried out at 5 K, where higher GMR values are accessible. As can be seen in figure 3a, no saturation in GMR occurs for $Cu_{78}Fe_{16}Co_6$ at H_{max}, even for the annealed samples. This behaviour could be caused by the expected slowly rising hysteresis curve of these materials due to partial supersaturation, i.e. a fraction of FM atoms is diluted in a NM matrix giving rise to a change of the FM long-range interaction [14]. It can be assumed due to literature values [13] that a saturation behaviour of $Cu_{78}Fe_{16}Co_6$ is expected to be reached at much higher applied magnetic fields and/or lower temperatures.

MR measurements of as-deformed and annealed $Cu_{84}Fe_7Ni_9$ are shown in figure 3b. All MR measurements show an isotropic GMR behaviour. For the as-deformed sample, a GMR of 0.6% was reached at H_{max}. Both subsequent annealing treatments, 1 h at 400°C and 10 min at 500 °C, lead to an improvement of the GMR amplitude. The obtained GMR of 1.3% at H_{max} is more than two times higher compared to the as-deformed state. Annealing for 10 min at 500 °C results in a maximum GMR of ~2.1%, which is more than three times the value of the as-deformed sample. Furthermore, this state shows a tendency of a saturation behaviour and the GMR values decrease quadratically with increasing field. Thus, the change in resistivity is expected to correlate well with the squared global relative magnetization according to the relationship already described in [5]. In comparison, the same heat treatment (10 min at 500 °C) resulted only in a weak raise of the GMR amplitude in $Cu_{78}Fe_{16}Co_6$ and a reduction of the GMR amplitude in $Cu_{62}Fe_{19}Ni_{19}$ compared to their as-deformed state. Annealing for 1 h at 400 °C had in each case a positive effect regarding the change in resistivity. Referring to the literature [16], GMR values of 8% were found in $Cu_{80}Fe_{10}Ni_{10}$ melt spun ribbons. An even higher GMR amplitude of 19% was achieved by subsequent annealing for 2 h at 400 °C. However, these values were measured at a low temperature of 50 K and a very high field of ~5570 kA/m.

The as-deformed $Cu_{62}Fe_{19}Ni_{19}$ sample (figure 3c) shows a linear drop in resistivity with increasing magnetic field. Demonstrating reproducibility, three samples of the $Cu_{62}Fe_{19}Ni_{19}$ were HPT-deformed and measured. For the measurement in parallel field alignment, they show perfect agreement of GMR values at the highest field of 1.84%, 1.85%, and 1.80%, respectively. Two annealed samples (1 h at 300 °C and 1 h at 400 °C) show a linear decrease of GMR values without any saturation behaviour. An increase of the GMR effect was achieved by annealing for 1 h at 400 °C. At higher annealing temperatures or for longer times (100 h at 400 °C, 10 min at 500 °C, 10 min at 600 °C, and 1 h at 600 °C) the changes in GMR become smaller. In all cases, a reduction of GMR amplitudes was found. A GMR of ~2.4% was observed at similar conditions (RT measurement up to H ~ 1590 kA/m) in magnetron sputtered and spinodal decomposed $Cu_{60}Fe_{20}Ni_{20}$ films [15]. Higher GMR values of 6.5% were reached for higher fields of ~4770 kA/m, whereby the GMR curve does not saturate even for the highest fields. For the presented HPT-deformed material, it can be assumed that increased GMR amplitudes will be achieved in higher applied magnetic fields.

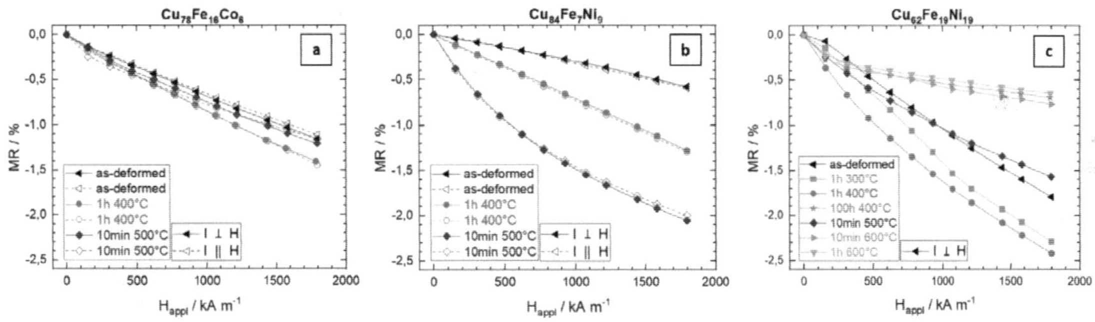

Figure 3. Decrease in resistivity with increasing applied magnetic field of a) $Cu_{78}Fe_{16}Co_6$, b) $Cu_{84}Fe_7Ni_9$, and c) of $Cu_{62}Fe_{19}Ni_{19}$ for as-deformed and annealed states. For a) and b) both measurement directions with current flow parallel (open symbols) or perpendicular (filled symbols) to the applied magnetic field are shown. For the sake of clarity, only the perpendicular measurement direction is presented for $Cu_{62}Fe_{19}Ni_{19}$.

4. Conclusion

The granular GMR effect was measured in three different ternary Cu-based materials processed by HPT deformation. At RT and the highest achievable magnetic field of 1790 kA/m, $Cu_{62}Fe_{19}Ni_{19}$ shows the highest GMR amplitude of all as-deformed states. Subsequent heat treatments (1 h at 400 °C and 10 min at 500 °C) led to higher GMR amplitudes in $Cu_{78}Fe_{16}Co_6$ and $Cu_{84}Fe_7Ni_9$ samples. Generally, the highest GMR of 2.45% at 1790 kA/m was found in $Cu_{62}Fe_{19}Ni_{19}$ after annealing for 1 h at 400 °C. For this material, a weakening of the effect was observed for high annealing temperatures and long durations. The maximum measured GMR of $Cu_{62}Fe_{19}Ni_{19}$ is somewhat higher than the reported GMR value of spinodal decomposed $Cu_{60}Fe_{20}Ni_{20}$ films at similar measuring conditions [15]. In comparison to the literature, $Cu_{78}Fe_{16}Co_6$ and $Cu_{84}Fe_7Ni_9$ show smaller GMR values. However, no saturation in GMR was observed in any Cu-based composition in as-deformed and annealed states. Higher magnetic fields or lower temperatures are expected to lead to a stronger GMR effect. Furthermore, the microstructure of as-deformed $Cu_{62}Fe_{19}Ni_{19}$ exhibits astonishingly high-temperature stability up to 600 °C, which corresponds to a homologous temperature of 0.64 of Cu. Comparing both material systems, CuFeNi exhibits a higher single-phase stability than CuFeCo.

References

[1] Baibich M N, Broto J M, Fert A, Van Dau F N, Petroff F, Etienne P, Creuzet G, Friederich A and Chazelas J 1988 Giant Magnetoresistance of (001)Fe/(001)Cr Magnetic Superlattices *Phys. Rev. Lett.* **61** 2472–5
[2] Binasch G, Grünberg P, Saurenbach F and Zinn W 1989 Enhanced magnetoresistance in layered magnetic structures with antiferromagnetic interlayer exchange *Phys. Rev. B* **39** 4828–30
[3] Ennen I, Kappe D, Rempel T, Glenske C and Hütten A 2016 Giant Magnetoresistance: Basic Concepts, Microstructure, Magnetic Interactions and Applications *Sensors* **16** 904
[4] Krishna V D, Wu K, Perez A M and Wang J-P 2016 Giant Magnetoresistance-based Biosensor for Detection of Influenza A Virus *Frontiers in Microbiology* **7** 400
[5] Xiao J Q, Jiang J S and Chien C L 1992 Giant magnetoresistance in nonmultilayer magnetic systems *Phys. Rev. Lett.* **68** 3749–52
[6] Berkowitz A E, Mitchell J R, Carey M J, Young A P, Zhang S, Spada F E, Parker F T, Hutten A and Thomas G 1992 Giant magnetoresistance in heterogeneous Cu-Co alloys *Phys. Rev. Lett.* **68** 3745–48
[7] Suehiro K, Nishimura S, Horita Z, Mitani S, Takanashi K and Fujimori H 2008 High-pressure torsion for production of magnetoresistance in Cu–Co alloy *J. Mat. Sci.* **43** 7349–53
[8] Wurster S, Weissitsch L, Stückler M, Knoll P, Krenn H, Pippan R and Bachmaier A 2019 Tuneable magneto-resistance by severe plastic deformation *Metals* **9** 1188
[9] Wurster S, Stückler M, Weissitsch L, Müller T and Bachmaier A 2020 Microstructural Changes Influencing the Magnetoresistive Behavior of Bulk Nanocrystalline Materials *Appl. Sci.* **10** 5094
[10] Wang J-Q and Xiao G 1994 Transition-metal granular solids: Microstructure, magnetic properties, and giant magnetoresistance *Phys. Rev. B* **49** 3982–96
[11] Kormout K S, Pippan R and Bachmaier A 2017 Deformation-Induced Supersaturation in Immiscible Material Systems during High-Pressure Torsion *Adv Eng. Mat.* **19** 1600675
[12] Stückler M, Weissitsch L, Wurster S, Felfer P, Krenn H, Pippan R and Bachmaier A 2020 Magnetic dilution by severe plastic deformation *AIP Adv.* **10** 015210
[13] Lardé R and Le Breton J M 2005 Influence of the milling conditions on the magnetoresistive properties of a Cu80(Fe0.7Co0.3)20 granular alloy elaborated by mechanical alloying *J. Mag. Mag. Mat.* **290–291** 1120–2
[14] Stückler M, Krenn H, Kürnsteiner P, Gault B, De Geuser F, Weissitsch L, Wurster S, Pippan R and Bachmaier A 2020 Intermixing of Fe and Cu on the atomic scale by high-pressure torsion as revealed by DC- and AC-SQUID susceptometry and atom probe tomography *Acta Mat.* **196** 210–9
[15] Chen L H, Jin S, Tiefel T H and Wu T C 1994 Giant magnetoresistance in spinodally decomposed Cu–Ni–Fe films *J. App. Phys.* **76** 6814–6
[16] Martins C S and Missell F P 2000 Annealing dependence of giant magnetoresistance in CuFeNi alloys *J. App. Phys* **87** 4840–2

Funding: This project has received funding from the European Research Council (ERC) under the European Union's Horizon 2020 research and innovation programme (Grant No. 757333).

IOP Conf. Series: Materials Science and Engineering 1249 (2022) 012048 doi:10.1088/1757-899X/1249/1/012048

A simple procedure to produce bio-mimic brick-and-mortar microstructure in commercially pure Ti

Chengze Liu*, Jianping Xu, Huan Li and Yusheng Zhang

Xi'an Rare Metal Materials Institute Co. Ltd, 96 Weiyang Road, Xi'an 710016, China

E-mail: czliu14s@alum.imr.ac.cn

Abstract. The extremely high fracture toughness of nacre is known to arise from its brick-and-mortar microstructure and has inspired the development of metals, metallic glasses and composites with similar bio-mimic microstructures. Although various approaches have been proposed to produce materials with a brick-and-mortar microstructure, including powder metallurgy, electro deposition, spray deposition, etc., the production of large-scale samples remains to be a challenge. In this research, we present a simple way to produce large samples of commercially pure Ti with a brick-and-mortar microstructure, which includes a pre-heat-treatment followed by deformation in the single-phase temperature region and subsequent annealing. EBSD characterization shows a layered microstructure in the produced samples resembling a brick-and-mortar structure.

1. Introduction

Inspired from the unique brick-and-mortar structural characteristics of nacre, researchers have proved that lots of properties, including mechanical properties [1-3], corrosion resistance [4] and electric performance [5] of titanium alloys, metallic glasses, and Cr/Cu layers, could be potentially improved beyond traditional theoretical constraints. The excellent properties can be largely attributed to the characteristic brick-and-mortar structure, which regulates the initiation and/or propagation of cracks in these materials through modulating strain distribution and avoiding premature cut-through cracks [1-5].

The brick-and-mortar microstructures show a layered feature with near-rectangular grain boundaries, and are challenging to produce via traditional materials preparation procedures. One way of fabricating such materials is a two-step process involving first a layer-by-layer electrophoretic deposition step, to introduce a layered microstructure, and then a rolling step, to bound the interfaces between the layers [4]. However, the deposition step largely limits the efficiency of this approach, making it unfavorable in the large-scale production. Another attempt for fabricating such materials applied a combined additive-subtractive, ultrafast laser fabrication technique, which is also limited to small-sized specimens [5]. Recently, a layered microstructure containing randomly-distributed α-phase particles was produced in a bulk Ti-Ta composite via spark plasma sintering (SPS) followed by plastic deformation and annealing [2]. Strictly speaking, the microstructure was not a brick-and-mortar microstructure, but the results showed the potential of producing metallic materials with such a microstructure using traditional plastic deformation and heat treatment methods.

In this work, using pure titanium as an example, we present a simple method for producing a brick-and-mortar microstructure in Group ⅣB transition metals via controlled plastic deformation and annealing.

2. Experimental procedures

The commercial pure titanium plate (Grade-1, 99.5%) studied in the present work has the size of 50 × 50 × 20 mm^3. In its as-received state the plate had been cast after arc-melting, forged in the β-phase region (1100℃) with thickness reduction of 50%, annealed in the α-phase region (600℃) and finally air cooled. In this work the as-received plate was further cold rolled with a total thickness reduction of 50%, followed by annealing at 600℃ with different durations varying from 15 to 120 minutes. The rolling geometry are referred to as RD (rolling direction), TD (transverse direction) and ND (normal direction), respectively.

Specimens for characterization was prepared by mechanical grinding, mechanical polishing followed by electro polishing. Electro polishing was operated at -40℃ for 2 min in Ethanol, n-butanol and perchloric acid with volume ratio of 6:3:1, using a current of 0.6 A. Electron backscatter diffraction (EBSD) analysis was carried out on SU3500, a scanning electron microscope (SEM), using an acceleration voltage of 15 kV, a working distance of 10 mm and a step size of 0.5 μm. Data analysis and visualization was done using the HKL® Channel 5 software.

3. Results and discussion

3.1. Microstructural characteristics of the as-received and cold-rolled Ti plate

The microstructure of the as-received Ti plate was characterized using SEM. As shown in figure 1, the Ti plate contains a typical lamellar microstructure, which results from its heat-treatment in the β-phase region. The size of the original β grain is in the range of 500-2000 μm. Each β grain is subdivided into regions called colonies. Throughout each colony, nearly parallel α-Ti lamellae distribute homogeneously and are expected to have the Burgers orientation relationship with the β phase, which is well-documented for Ti [7]. The α-Ti lamellae in different colonies orient differently, as shown in figure 1 where three colonies meet. The width of the α-Ti lamellar ranges between 5 and 20 μm.

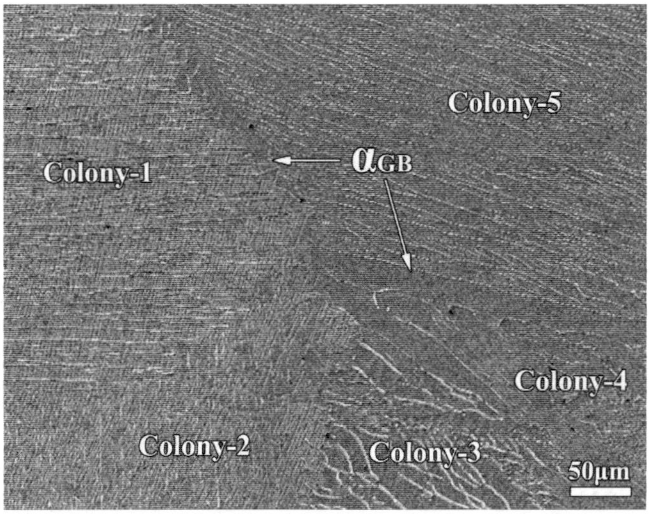

Figure 1. SEM secondary electron image showing the microstructure of the as-received commercial pure Ti plate

As shown in figure 2a, the cold-rolled Ti plate contains fragmented grains, while the phase boundaries of the initial α laths can still be identified. Viewed along ND, these laths are approximately parallel to RD. Within each α lath, high-density-defects as well as substructures are observed, corresponding to a large area fraction of deformed (red) and substructured (yellow) regions shown in figure 2b. The recrystallized (blue) fraction in the cold-rolled Ti plate is only 2.61%, whereas 83.6% of the studied region belongs to deformed microstructure, which is characterized by the high density of low-angle grain boundaries that are typical of cold-rolled samples. These regions were automatically identified through the HKL® Channel 5 software, depending on the density of high-angle grain boundaries (HAGBs) as well as low angle grain boundaries (LAGBs) [10].

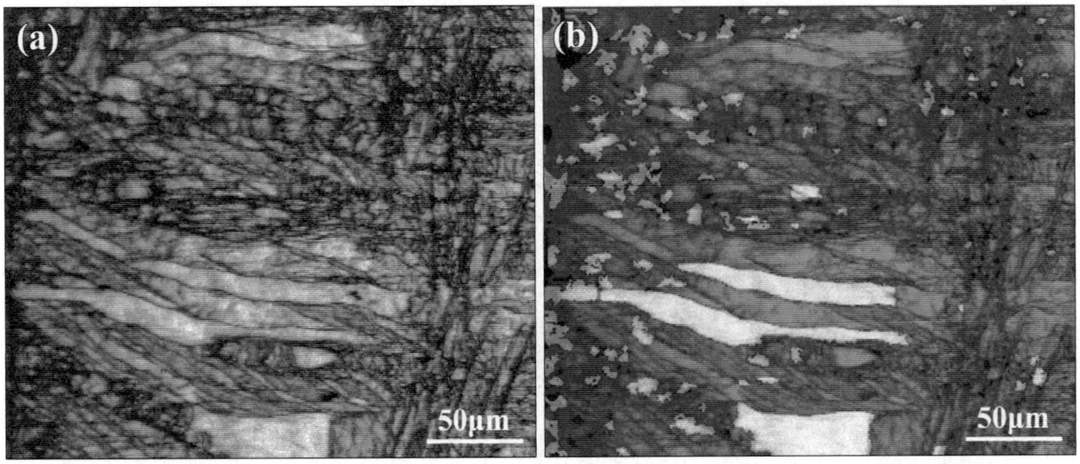

Figure 2. Band contrast map (a) and different regions identified (b) for the cold-rolled Ti plate viewed along ND. The red, yellow and blue colors refer to deformed, substructured and recrystallized regions, respectively. The horizontal direction is along RD.

3.2. Microstructural characteristics of the Ti plates with a brick-and-mortar microstructure

After cold rolling with total reduction of 50%, all the Ti plates were annealed at 600℃ with different durations varying from 15 to 120 minutes. As shown in figure 3, with increasing annealing durations, the grains recrystallized into different morphologies. In the plate annealed for 15 min, the grains are arranged along RD, and some regions show characteristics of a brick-and-mortar microstructure. After annealing for 30 min, the Ti plate shows an ideal brick-and-mortar microstructure, where the planar distributed grains are packed like "bricks" parallel with RD. In addition, many grain boundaries lie at approximately 90 degrees to the planar boundaries, very similar to the microstructure of nacre. The widths of the 'bricks' are between 5 and 20 μm, consistent with the initial α-Ti lamellar width. During cold rolling, the α laths of prior β grains were curved and broken along RD, and high density of dislocations were generated within the α laths [8]. During the subsequent annealing, these laths were recovered and partially recrystallized. However, due to the hindering effect of grain boundaries [9], the boundaries of initial α laths were inherited. As the annealing duration increases to 60 and 120 min, these rectangular grain boundaries gradually evolve towards a more equiaxed morphology to lower the interfacial energy. Interestingly, lots of annealing twins are observed in the 60 min annealed sample (figure 2c), which could result from the residual stress generated during the grain morphology transformation [8].

Table 1. Area fraction of different regions of heat-treated Ti plates

Annealing duration (min)	Deformed fraction (%)	Substructured fraction (%)	Recrystallized fraction	Brick-and-mortar microstructure
15	13.0	11.7	75.3	No
30	3.26	32.0	64.8	Yes
60	10.2	49.2	40.6	No
120	4.08	31.5	64.4	No

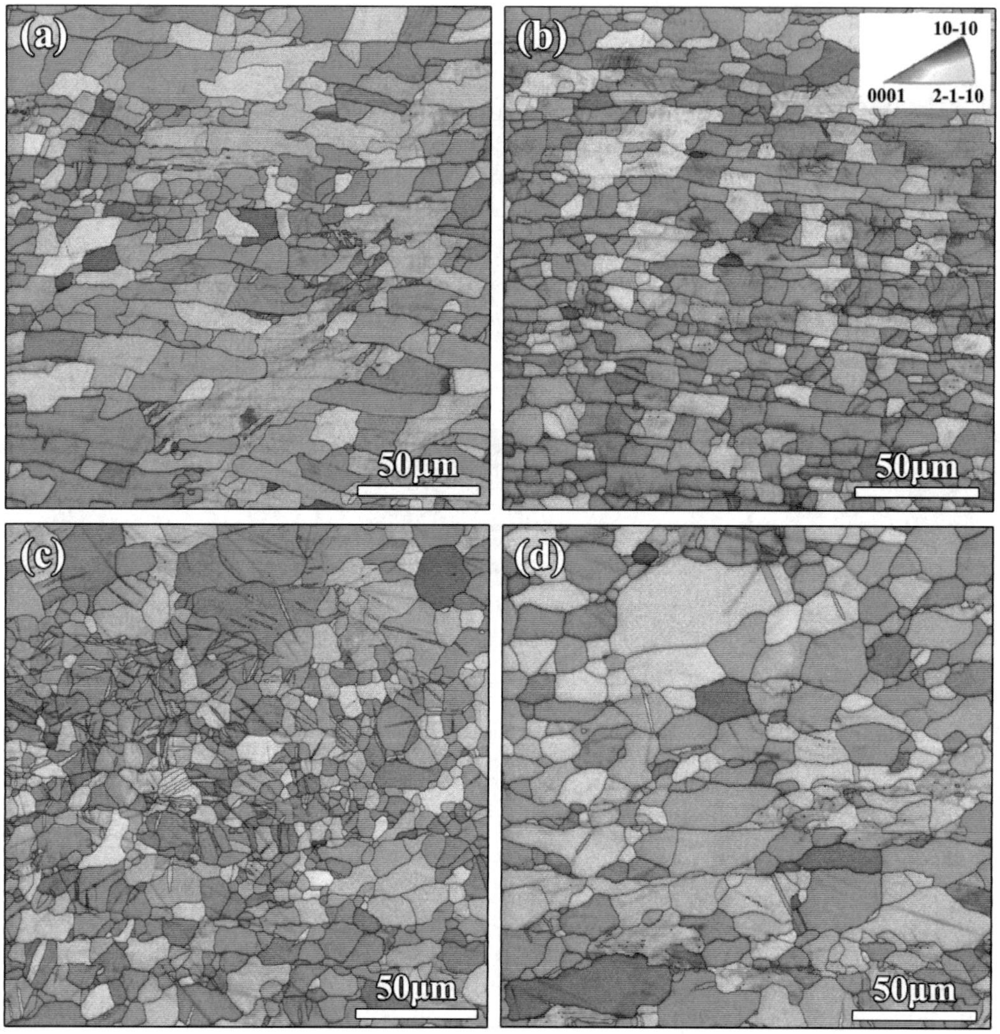

Figure 3. IPF maps of the Ti plate heat treated for 15 (a), 30 (b), 60 (c) and 120 (d) minutes, respectively. Viewed along ND. The horizontal direction is RD.

As shown in figure 4 (a), after 15 min annealing of the cold-rolled Ti plate, 75.3% area fraction has been recrystallized. However, due to the microstructural inhomogeneities in the cold-rolled samples, some regions still contain a deformed structure (in red color). The deformed and substructured regions were determined depending on the density of geometrically necessary dislocations (GNDs) or LAGBs [10]. As the annealing duration reaches 30 min, the volume fraction of deformed structure reduces to 3.26%, being the smallest of all samples (see table 1). After annealing for 60 min, 49.2% area fraction is detected as substructured regions. It is observed in figure 4c that most of the substructured grains are associated with annealing twins. Therefore, these twins were considered to be thermodynamically unstable because large residual strains were expected in the substructured regions adjacent to them, which generated enough energy for detwinning during heating and cooling [8]. This hypothesis was supported by the significantly reduced number of annealing twins in the sample annealed for 120 min (figure 4d). The recrystallization fraction fluctuated with the annealing time, which was due to the heterogeneity [8] and the limited size of characterization area. Thus, measurement error was not taken into account in the above discussions. Since this research mainly introduce a simple way to produce brick-and-mortar microstructures, the effect of annealing parameters on the microstructure was not listed in the conclusion part.

It is suggested that the initial microstructure be important in the production of a brick-and-mortar microstructure. In the present case the fully lamellar microstructure transformed from the β phase can provide the framework for forming the brick-and-mortar microstructure. The approach for fabricating a brick-and-mortar microstructure has been proven to work in Zr, Hf and their alloys [11]. The mechanical properties of these materials with novel microstructures would be further studied in our upcoming researches.

Figure 4. Different regions identified in the Ti samples heat treated for 15 (a), 30 (b), 60 (c) and 120 (d) minutes, respectively. Viewed along ND. The red, yellow and blue colors refer to deformed, substructured and recrystallized regions, respectively. The horizontal direction is RD.

4. Conclusions

The microstructures of a commercially pure Ti processed by cold rolling and subsequent annealing were examined using SEM-EBSD. Annealing at 600 °C for different duration results in different extent of recrystallization and microstructural morphologies. The layered microstructure in the sample annealing at 600 °C for 30 min highly resembles a brick-and-mortar microstructure in morphology, showing that a brick-and-mortar microstructure can be introduced into commercially pure Ti using a relatively simple approach of cold rolling following by annealing.

Acknowledgement

This work was financially supported by Natural Science Foundation of Shaanxi Province (2020JC-50), Qinchuangyuan Project of Shaanxi Province (2021QCY1-01), Xi'an Science and Technology Plan Project (2020YZ0028), and RMMI research project (Y1905S). The authors thank Dr Chuanshi Hong as well as reviewers for comments on the manuscript.

References

[1] Suyue Yuan, Xinxiang Song and Paulo S Branicio 2020 *Scr. Mater.* **186** 69–73
[2] Shenghang Xu, Meng Du, JiaLi, Kun Yan, Biao Cai, Quanfeng He, Qihong Fang, Oxana Magdysyuk, Bin Liu, Yong Yang and Yong Liu 2019 *Mater.* **8** 100463
[3] K Z Motomichi, M M Wang, D Ponge, D Raabe, K Tsuzaki, H Noguchi and C C Tasan 2017 *Science* **355** 1055–57.
[4] Shu-sen Peng, Xing-yun Chen, Yong-cun Ma and Ya-lin Feng 2019 *Sur. Tech.* **48** 170–8 (in Chinese)
[5] Bin Feng, Xin Jiang, Guisheng Zou, Wengan Wang, Tianming Sun, Heng Yang, Guanlei Zhao, Mingye Dong, Yu Xiao, Hongwei Zhu and Lei Liu 2021 *Adv. Funct.* **31** 2102359
[6] Qingsong Pan, Liangxue Zhang, Rui Feng, Qiuhong Lu, Ke An, Andrew Chihpin Chuang, Jonathan D Poplawsky, Peter K Liaw and Lei Lu 2021 *Science* **374** 984–9
[7] Chengze Liu, Geping Li, Fusen Yuan, Muhammad Ali, Fuzhou Han, Yingdong Zhang and Hengfei Gu 2019 *Mater. Chem. Phys.* **223** 75–7
[8] Jianping Xu, Chengze Liu, JinpingWu, Boyi Qi, Yusheng Zhang, Yongqing Zhao and Lian Zhou 2021 *Mater. Sci. Eng.* A **819** 141508
[9] Chengze Liu, Geping Li, Fusen Yuan, Fuzhou Han, Muhammad Ali, Yingdong Zhang, Wenbin Guo and Hengfei Gu 2020 *Scr. Mater.* **185** 170–4
[10] Charbel Moussa, Marc Bernacki, Rémy Besnard and Nathalie Bozzolo 2015 *IOP Conf. Ser.: Mater. Sci. Eng.* **89** 1
[11] Chengze Liu, Jianping Xu, Jinping Wu, Xing Zhao, Pei Cui and Fan Yang 2022 *Chinese Patent*, 2021115857578

IOP Conf. Series: Materials Science and Engineering 1249 (2022) 012049 doi:10.1088/1757-899X/1249/1/012049

Registration of data from a multimodal investigation of particle stimulated nucleation in aluminum

E F F Knipschildt[1], X Lei[2,3], Y B Zhang[1], T Yu[1], S Fæster[4], W Liu[5], R E Sanders[3] and D Juul Jensen[1]

[1] Department of Civil and Mechanical Engineering, Technical University of Denmark, DK-2800, Kgs. Lyngby, Denmark.
[2] Key Laboratory for Light-weight Materials, Nanjing Tech University, Nanjing 210009, China
[3] International Joint Laboratory for Light Alloys (Ministry of Education), Department of Materials Science and Engineering, Chongqing University, Chongqing, People's Republic of China.
[4] Department of Wind Energy, Technical University of Denmark, Risø Campus, Roskilde, 4000, Denmark.
[5] Advanced Photon Source, Argonne National Laboratory, Lemont, IL, 60439-4800, USA.

E-mail: effkn@mek.dtu.dk

Abstract. Nucleation of recrystallization at large second phase particles during annealing is essential in many industrially relevant aluminum alloys. To adequately quantify the spatial distribution of such particles and their impact on nucleation in bulk materials, 3D investigations are needed. This study presents a multimodal characterization method to investigate particle stimulated nucleation in a heavily deformed AA5182 alloy. To map the spatial distribution of large second phase particles, conventional laboratory absorption contrast X-ray tomography is used, while the recrystallization nuclei are revealed by Laue micro-diffraction using synchrotron X-rays. The registration of data obtained by the two different non-destructive high-resolution experimental techniques is described. Finally, some preliminary results are presented.

1. Introduction

Particle stimulated nucleation (PSN) is of utmost importance in many industrial relevant materials undergoing thermomechanical processing [1–3]. When a metal is deformed, orientation gradients are created around un-deformable large second phase particles [4–7], these volumes are called 'deformation zones' [8]. Nucleation of recrystallization at such sites has been found to be a dominating nucleation mechanism in many studies, e.g. [9, 10]. Even though PSN is a widely studied nucleation mechanism, it is not known how the particle size and shape, strain and surrounding matrix orientations affect the nucleation behavior and it is therefore not known if nucleation takes place at every large second phase particle [11].

The majority of investigations preformed to date on PSN has been carried out using 2D or destructive 3D techniques (e.g. [12]). When using light optical or electron microscopy, only the sample surface or a thin surface slice of the sample is investigated [2]. Some 2D techniques have been expanded to cover the third dimension by sectioning of the sample, yet this destructive approach leaves several complications as damage by the sectioning method of choice, lack of knowledge of in-between layers as well as limiting dynamic studies [2]. Laboratory diffraction contrast tomography (LabDCT) can be used to map 3D microstructures of recrystallized samples [13]. However, the present spatial resolution limit of $\sim$ 10 μm [14] makes LabDCT less suitable for investigations of the early stages of recrystallization in highly strained samples.

In this study, the efficiency of PSN was investigated quantitatively by two different types of 3D mapping. The material was an AA5182 sample cold-rolled to 75 % reduction. This alloy was chosen because the aluminum AA5xxx series alloys have good formability, good weldability, excellent corrosion resistance and high strength, making these alloys of interest in e.g. building, automotive and chemical industries [15]. More importantly for the present work, PSN is assumed to be the dominating nucleation mechanism in highly strained AA5182. Synchrotron X-ray Laue micro-diffraction [16] was used to map the microstructure, while conventional laboratory absorption contrast X-ray tomography (CT) [14] was used to map the second phase particles. In this paper, data registration of the two characterized volumes is presented. It is essential to get the registration very precise to allow a proper analysis of PSN.

2. Experimental

2.1. Materials

The material investigated is commercial AA5182 alloy sheet supplied by Novelis Inc., where it was hot rolled to 4 mm and then cold-rolled to 1 mm (75% reduction) on industrial rolling mills. The sheet was annealed at 250°C for 160 minutes to achieve an initial recrystallization state.

2.2. Synchrotron X-ray Laue micro-diffraction

The Laue micro-diffraction data were collected at beamline 34-ID-E at the Advanced Photon Source (APS) at Argonne National Laboratory in USA by white-beam differential-aperture X-ray microscopy (DAXM), also called the X-ray microbeam Laue diffraction technique [16]. The focused polychromatic beam had a Lorentzian profile with a full width half maximum of $\sim$ 0.3 μm. The sample was mounted on a sample holder 45° relative to the incident beam. A Perkin-Elmer flat panel detector, positioned in 90° reflection geometry 510.3 mm above the sample, collected Laue diffraction patterns from the volume illuminated by the incident microbeam. A knife-edge was utilised as diffraction aperture and scanned at a distance $\sim$ 250 μm from the sample surface; this was done to resolve different gauge volumes below the surface, i.e. to give depth resolved information. Depth resolved patterns were reconstructed from peak detection and indexing by ray-tracing using *LaueGo*, which was available at the APS beamline. Only the aluminum peaks were indexed. The diffraction patterns were obtained from individual voxels to a depth of 360 μm into the sample with a step size of 0.75 μm. The crystallographic orientations were indexed with an orientation resolution of 0.01°. The microbeam was scanned across the volume repeating the wire scan at each position and thereby mapping a volume of $360 \times 60.75 \times 49.5\ \mu\text{m}^3$.

A fluorescence scan was made of the sample surface including the surface area of the mapped gauge volume.

2.3. Conventional absorption contrast tomography

A Zeiss Xradia 520 Versa X-ray microscope was used for the CT data acquisition. The CT scan was performed with a source voltage of 50 kV, tube current of 79 μA along with source-sample and sample-detector distances between 8 and 60 mm. In total, 4500 projections were collected with an exposure time of 45 seconds per projection. With a 4x objective and camera binning of 2, the reconstructed CT volume had a pixel size of 0.40 μm.

CT data were used to find positions of large second phase particles in the gauge sample volume. Noise was filtered by various masks including median filtering and morphological top-hat filtering. The step size of the CT data was scaled to fit the step size of the Laue micro-

IOP Conf. Series: Materials Science and Engineering 1249 (2022) 012049 doi:10.1088/1757-899X/1249/1/012049

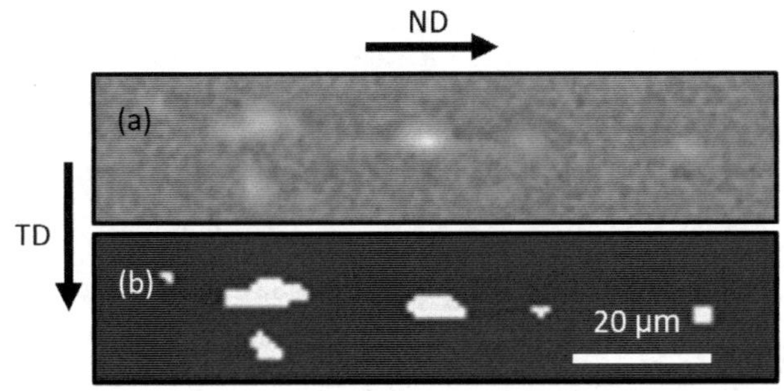

Figure 1. A 2D slice of the CT data perpendicular to the rolling direction (RD) before **(a)** and after threshold and filtering **(b)**.

diffraction data. A threshold was chosen to a level where all particles were still represented while single pixels were filtered out, see figure 1.

3. Results and discussions

3.1. Data registration based on particle distribution

The 3D microstructure of the scanned gauge volume acquired from Laue micro-diffraction is shown in figure 2**a**. In this figure, the larger uni-colored volumes are recrystallized grains, while the many voxels of different colors are the deformed matrix. Figure 2**b** shows the CT data which is covering a volume larger than that mapped by Laue micro-diffraction. Here, the brighter and darker areas are the second phase particles.

An optical micrograph including the surface region of the Laue micro-diffraction acquisition is shown in figure 3**a**. Figure 3**b** shows a fluorescence map, including the area shown by the red dashed rectangle in figure 3**a**, which reveals iron-rich particles of the surface region of interest. Based on the pattern of the particle distribution, the sample surface region mapped by synchrotron X-rays was identified in the CT data, see figure 3**c**. The large particles and the string of particles, seen in figure 3**c**, were utilized to make a first estimation of the data alignment between the two data sets. To further improve alignment accuracy, particle information was obtained based on the diffraction images. Particle phases were not indexed by *LaueGo*. By looking at the raw patterns it can however be seen that they typically produce more spots than aluminum. Therefore, the number of detected spots were used as a measure for second phase particles. The counts of spots were filtered by a median filter and noise was removed. Then 2D slices were plotted in real space. Here, areas of high number of spots resemble second phase particles to some extend. All the raw Laue patterns from large particle-like positions of high diffraction spot count were gathered and all the positions constituting the particle-like areas were found in the real space coordinate system to fit the full data set. After the pre-alignment by the large particles on the surface, the positions of the high diffraction spot count from the Laue micro-diffraction data and the positions of the particles from the CT data were compared manually in 3D. 21 positions of high diffraction spot counts in the Laue micro-diffraction data were related to particles found in the CT data. A translation and a rotation matrix were created as well as a scaling between the data sets was found; representing the whole transformation. The translation is found to be close to the presumed translation obtained from the sample surface, while the rotation matrix is close to unity. The influence of the rotation was found by comparing the use of the rotation matrix as oppose to not allowing any rotation; it was found that the position of high diffraction spot count in the Laue microdiffraction furthest away from the sample surface was shifted by 4 pixels using the rotation matrix. Additionally to the transformation matrices, the scaling of ~ 2 % of the Laue micro-diffraction data, which was found to optimize the registration, was allowed. The Laue micro-diffraction data were transformed by the matrices and the scale found.

Data registration was assumed to be successful if the trends of large particle positions, seen

IOP Conf. Series: Materials Science and Engineering 1249 (2022) 012049 doi:10.1088/1757-899X/1249/1/012049

in figure 3, could be related to the sample surface region of the cropped volume of the CT data corresponding to the volume of the Laue micro-diffraction data. This cropped surface region from the CT data is shown in figure 4. The raw CT data in figure 4**a** also shows the surrounding large particles, while figure 4**b** only shows the cropped area equivalent to the Laue micro-diffraction data after alignment. The threshold of 4**b** is the same as in figure 1.

The uncertainty of the data registration was defined as the mean of the difference between the positions found in the two volumes and was found to be ~ 2 voxels after rotation, translation and scaling.

The true uncertainty of the data registration might however be significantly larger if all the chosen positions are being misaligned be equal amounts in the same directions. This is a possibility, as the relations between the two data sets were heavily based on the pre-alignment by the sample surfaces, as described earlier, and on an assessment of each pair of positions in regard to distance off-set and particle visibility in the CT data.

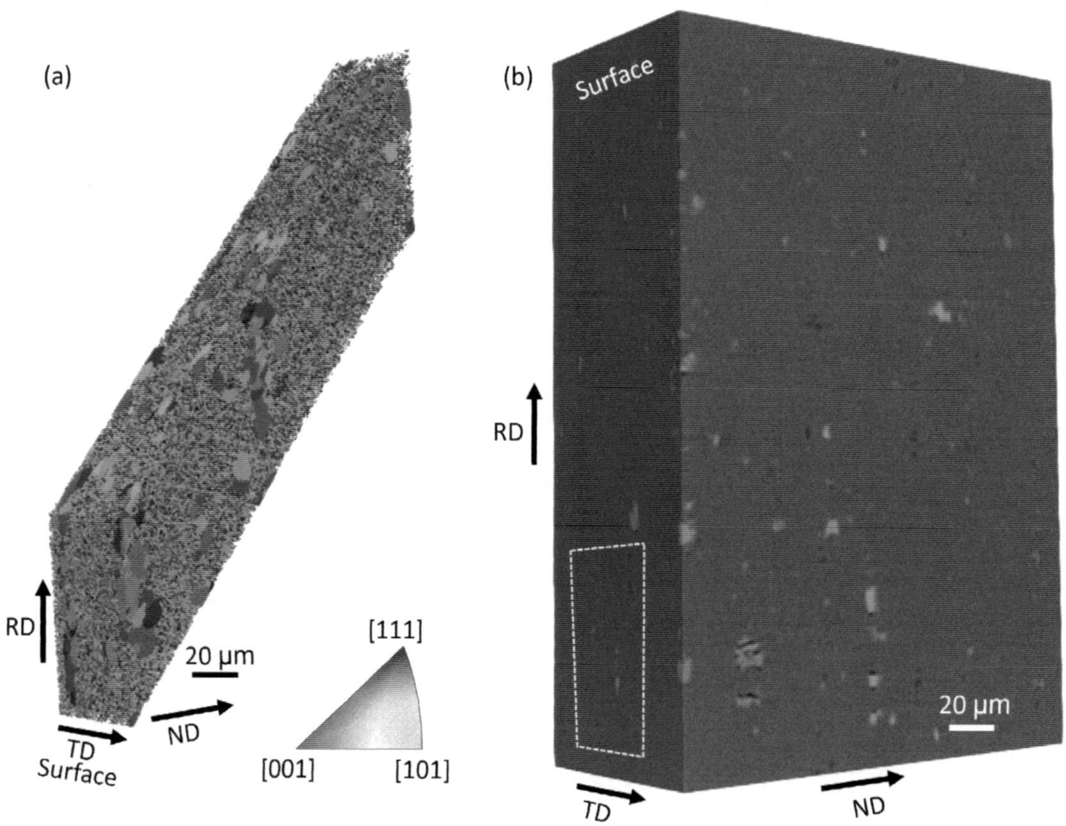

Figure 2. **(a)** Inverse pole figure (IPF) map of the partly recrystallized microstructure of aluminum AA5182 sample acquired from Laue micro-diffraction. **(b)** Visualization of the CT data of the same sample containing the volume seen in **a**. The yellow dashed rectangle indicates the surface area approximately corresponding to the sample surface in **a**. The normal direction (ND), transverse direction (TD) and RD in the two figures are parallel. The planes showed on the surfaces of both data sets are $\sim 10\ \mu m$ below the sample surface.

IOP Conf. Series: Materials Science and Engineering 1249 (2022) 012049 doi:10.1088/1757-899X/1249/1/012049

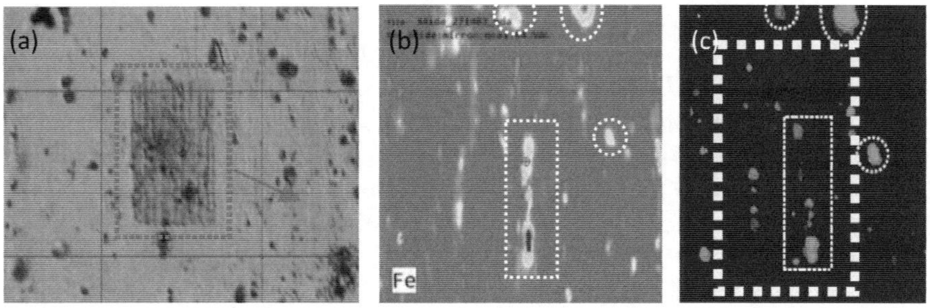

Figure 3. **(a)** An optical micrograph of the sample surface after mapping at the synchrotron source. The green squares are 100 × 100 μm^2. **(b)** Fluorescence map including the inspected area shown in **a**. Large particles are mapped by dashed lines. **(c)** A reconstruction of the CT data representing the sample surface. The yellow dashed rectangle indicates the expected surface area mapped by Laue micro-diffraction.

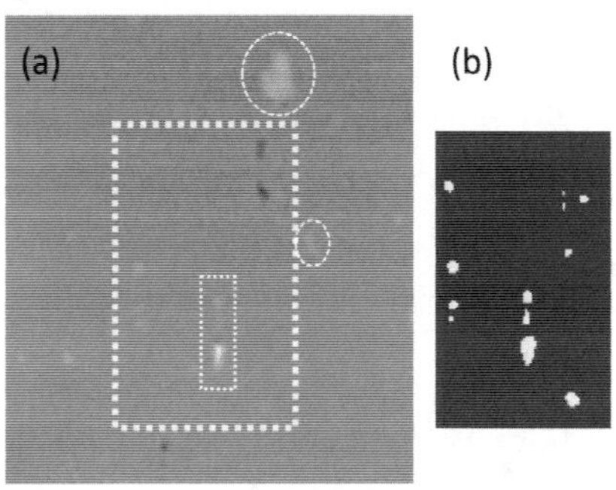

Figure 4. Planes perpendicular to the ND and ~ 6 μm from the first layer of the CT data set. **(a)** Shows the area of interest, shown by the yellow dashed rectangle, and the large particles around this. **(b)** shows the area of interest with the threshold mentioned earlier.

3.2. Preliminary quantification of particle stimulated nucleation
Recrystallized grains were mapped in 3D. Recrystallized grains were assumed to have a volume larger than 8 voxels, which in this case corresponds to grains larger than ~ 3.4 μm^3. The recrystallized grains were dilated by 2 voxels, assuming a disk shape, representing the uncertainty in the data registration. It was assumed that if the voxels in the CT data, constituting particles, touch each other, these belong to the same particle or cluster of particles. Particles and particle clusters are expected to have the same effect on PSN [9]. Hereafter, grains, observed in the Laue micro-diffraction data, touching particles, observed in the CT data, were counted as being PSN grains. The PSN result might be overestimated, using this definition, since some nuclei might have formed away from the particle and only grow to become near [9].

Using the mentioned assumptions, it was found that ~ 76% of particles larger than 5 μm stimulated nucleation. It should be noticed that only second phase particles of high peak count in the CT data, assumed to be Fe-containing particles, were included in this preliminary analysis. Areas of low peak count were observed and can possibly represent second phase particles of different chemical composition which might also stimulate nucleation.

4. Summary and outlook
Data registration of two 3D data sets acquired by two experimental techniques was made possible by comparing high diffraction spot counts in Laue micro-diffraction data and particle sites in CT data using in-house *MATLAB* scripts. This new method, using excess diffraction spots to find low symmetry second phase particle positions, makes it possible to quantitatively investigate PSN behavior in 3D with high spatial resolution.

From the preliminary analysis, $\sim 76\%$ of the large particles ($> 5\ \mu$m) were found to stimulate nucleation. A further analysis of the PSN results is ongoing.

In our future work, we will vary the threshold of the CT data and the dilation of the recrystallized grains to qualify how accurate the data registration is. It would also be of interest to include Mg particles to see if soft particles influence PSN.

If the large second phase particles can be indexed based on synchrotron X-ray Laue micro-diffraction data, the alignment of such a data set and a CT data set would be simpler. This would furthermore add the possibility of investigating potential differences in PSN behavior based on the chemical composition of the second phase particle. Furthermore, it would contribute to the accuracy of the alignment and to the determination of the data registration uncertainty.

Funding

E F F Knipschildt, Y B Zhang, T Yu and D Juul Jensen acknowledge the financial support by the European Research Council (ERC) under the European Union's Horizon 2020 research and innovation program (M4D - grant agreement No. 788567).

References

[1] Humphreys F J and Hatherly M 2004 *Recrystallization and Related Annealing Phenomena* (Oxford: Elsevier)
[2] Juul Jensen D and Zhang Y B 2020 *Curr. Opin. Solid. State. Mater. Sci.* **24** 2
[3] Mandal D and Baker I 1997 *Acta. Mater.* **45** 453–61
[4] Humphreys F J and Juul Jensen D 1986 *Proc. of the 7th Risø Int. Symp. on Mater. Sci.* (Roskilde: Risø Nat. Lab.) pp 93–106
[5] Engler O, Kong X W and Yang P 1997 *Scr. Mater.* **37** 1665–74
[6] Chan H M and Humphreys F J 1984 *Met. Sci.* **18** 527–30
[7] Hirsch J and Engler O 1995 *Proc. of the 16th Risø Int. Symp. on Mater. Sci.* (Roskilde: Risø Nat. Lab.) pp 49–62
[8] Humphreys F J 2004 *Mater. Sci. Forum* vol 467 (Switzerland: Trans Tech Publ) pp 107–16
[9] Lei X, Zhang Y B, Sun J, Bachmann F, Yang X, Sanders R and Juul Jensen D 2021 *Mater. Res. Lett.* **9** 65–70
[10] Humphreys F J 1977 *Acta metall.* **25** 1323–44
[11] Humphreys F J, Ferry M, Johnson C and Paillard P 1995 *Proc. of the 16th Risø Int. Symp. on Mater. Sci.* (Roskilde: Risø Nat. Lab.) pp 87–104
[12] Weiland H, Rouns T N and Liu J 1994 *Int. J. Mater. Res.* **85** 592–7
[13] Lindkvist A, Fang H, Juul Jensen D and Zhang Y 2021 *J. Appl. Crystallogr.* **54** 99–110
[14] Oddershede J, Sun J, Gueninchault N, Bachmann F, Bale H and Holzner Cand Lauridsen E 2019 *Integr. Mater. Manuf. Innov.* **8** 217–25
[15] Engler O, Kuhnke K and Hasenclever J 2017 *J. Alloys Compd.* **728** 669–81
[16] Liu W, Zschack P, Tischler J, Ice G and Larson B 2011 *The 10th Int. Conf. on X-ray Microscopy* vol 1365 (Chicago: AIP Conference Proceedings) pp 108–11

IOP Conf. Series: Materials Science and Engineering 1249 (2022) 012050 doi:10.1088/1757-899X/1249/1/012050

Recovery of small orientation gradients in a recrystallized grain observed in 3D during ex-situ annealing

Chuanshi Hong[1,2,*], Yubin Zhang[1], Tianbo Yu[1], Wenjun Liu[3] and Dorte Juul Jensen[1,*]

[1] Department of Civil and Mechanical Engineering, Technical University of Denmark, DK-2800 Kgs Lyngby, Denmark
[2] Shenyang National Laboratory for Materials Science, Institute of Metal Research, Chinese Academy of Sciences, Shenyang 110016, China
[3] Advanced Photon Source, Argonne National Laboratory, Argonne, Illinois, 60439, USA

*E-mail: cshong@imr.ac.cn; doje@mek.dtu.dk

Abstract: We report the observation of a small orientation gradient (~0.5° over a few micrometers) in a recrystallized grain formed at the sample surface in pure Al and the elimination of this orientation gradient during subsequential ex-situ annealing. The observation is in 3D and done using synchrotron white-beam differential-aperture X-ray microscopy with high angular resolution of 0.01°. The elimination of the orientation gradient during subsequent annealing are analyzed and discussed.

1. Introduction

During recrystallization, nuclei form in a deformed microstructure and grow by grain boundary migration [1,2]. The nuclei, and thus the recrystallizing and recrystallized grains, are generally considered to be perfect, or at least almost perfect, single crystals. However, if for example the nucleation mechanism is strain induced boundary migration (SIBM) [3], crystallographic misorientations are present in the part of the nucleus that are within the deformed grain from which it originated, i.e., in the part of the nuclei which has not been swept by a migrating recrystallization boundary. An example of this is reported in [4], where misorientations as large as 6° within nuclei formed by SIBM at triple junctions in 40% cold rolled pure aluminum were observed for the early stages of recrystallization. Those measurements were done by electron back scattering diffraction (EBSD) which has limited angular resolution. Recent developments of advanced synchrotron X-ray techniques have enabled 3D non-destructive mapping of both crystallographic orientations and residual strains with very high angular resolution as well as good spatial and strain resolutions [5–9]. By such synchrotron measurements, it was revealed that a network of very low angle (<0.1°) dislocation boundaries and local residual strains as large as 1×10^{-3} may exist within recrystallizing grains [10–12]. It is thus of interest to study interior misorientations and strains in recrystallizing grains in more detail using these advanced experimental methods, to understand effects hereof on the nucleation and growth of recrystallizing grains, and to follow the evolution during the recrystallization anneal.

The present work is based on data we got as a 'byproduct' of a 3D experiment in which we focused on i) the boundary migration and its relations to the deformation microstructure during recrystallization [13], and ii) recovery within the remaining deformed matrix material during annealing [14]. The experiment however contained high resolution orientation data *also* for the recrystallizing grain, which is what we focus on in the present work. Sample preparation had introduced small misorientations in a part of the recrystallized grain at the sample surface, and the aim of the present paper is quantifying the orientational evolution of this misoriented part during the course of recrystallization.

2. Material and methods

High purity Al (99.996 wt.%) was used in this study. A fully annealed Al plate of 94 × 46 × 3 mm^3 was rolled to the thickness reduction of 12% at room temperature. A slice of ~8 × 1 × 3 mm^3 from the transverse section (containing the rolling direction and the normal direction) of the plate was mechanically ground and polished, electro-polished, indented with 1 kg Vickers hardness load and then annealed to stimulate recrystallization nucleation. A recrystallized grain was obtained near the bottom of the indent after annealing at 335 °C for 10 min.

The evolution of this grain was followed in an ex-situ experiment using white-beam differential-aperture X-ray microscopy (DAXM) at 34-ID-E at the Advanced Photon Source (APS), Argonne National Laboratory, USA [15]. For this experiment, the indented surface was furthermore mechanically ground with care using #4000 SiC paper to remove about half the depth of the indent and then electro polished. This sample preparation was performed *after* the nucleation but *before* the first DAXM scan.

After the first DAXM scan (referred to as Scan 0), the sample underwent three additional sets of annealing treatments outside of the synchrotron facility, with a DAXM scan carried out after each annealing treatment. The additional annealing treatments were at 335 °C for 15 min, at 335 °C for 90 min, and at 375 °C for 30 min, corresponding in this work to DAXM Scans 1, 2 and 3, respectively. For DAXM, a focused polychromatic X-ray beam with a full width at half maximum of ~0.3 μm was used, and the scanning step size was 1.5 μm. The open-source software packages, Dream3D and Paraview [16], were used to process and visualize the 3D volumetric results.

Figure 1 illustrates the geometry of the DAXM experiment. A cuboid volume inclined at 45° to the sample surface was scanned and reconstructed (figure 1a). The volume contains part of the recrystallized grain and the recovered deformation microstructure (figure 1b). The boundary migration of the recrystallized grain and the recovery of the deformation microstructure have been analyzed in previous work [13,14]. In this work, we focus on the evolution of the small orientation gradients found inside the recrystallized grain.

3. Results and discussion

When plotted using misorientation relative to the mean orientation of the recrystallized grain, it was found that the grain contains small orientation gradients which evolve during subsequent annealing. This microstructural evolution is shown in figure 2, which present the internal microstructure of the recrystallized grain in Scans 0, 1 and 2.

As shown in figure 2a_1–a_3, for the state annealed at 335 °C for 10 min, the grain contains a shallow region of about 35 × 15 × 5 μm^3 which is disoriented from the matrix grain orientation. This region is extending from the intersection between the indented surface and the polished sample surface all the way to the migrating recrystallization boundary. The maximum misorientation relative to the mean orientation of the grain is about 0.5°. This value is significant compared with the angular resolution of crystallographic orientation measured by DAXM (~0.01° [15]) and the fluctuations in orientation within the bulk of the grain (< 0.1°). The orientation gradient may be introduced during mechanical polishing and not fully removed by the electropolishing.

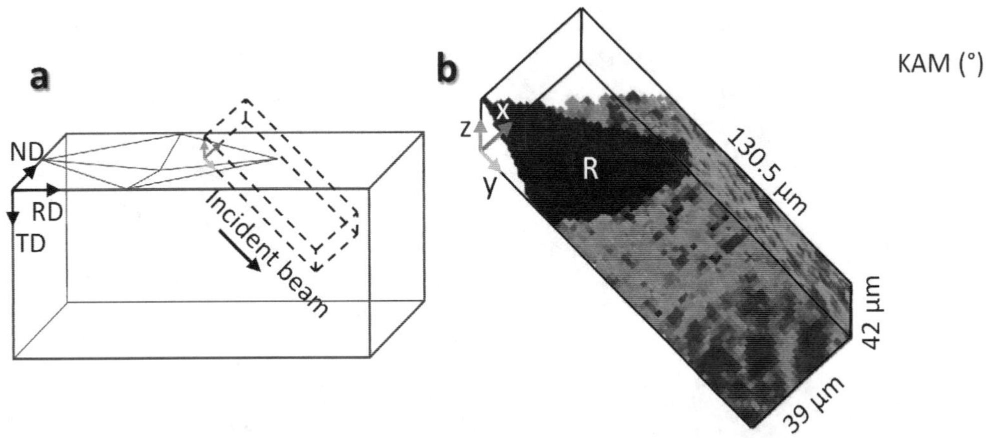

Figure 1. (a) Geometry of the indented sample surface and the DAXM experiment. RD, TD and ND stand for the rolling, transverse and normal directions, respectively. The volume scanned by DAXM is outlined by the dashed cuboid frame. Note that the sketched sample frame and indentation are not in proportion. (b) The reconstructed 3D volume of Scan 0 colored according to the kernel average misorientation (KAM). The recrystallized grain is marked by "R". (The figure is adapted from [13]).

After further annealing at 335 °C for 15 min, as shown in figure 2b_1–b_3, the maximum misorientation in the recrystallized grain decreases to ~0.3°. The size of the disoriented region also decreases by a few microns in the lateral direction along the sample surface and by ~1.5 µm (i.e. ~1 voxel) in the depth direction. It is furthermore interesting to note that the average of the misorientations relative to the mean orientation of the grain increases from 0.02° in Scan 0 to 0.03° in Scan 1 for the part of the grain away from the surface region, which has an orientation difference to that of the major part of the grain.

After further additional annealing at 335 °C for 90 min, as shown in figure 2c_1–c_3 for Scan 2, the disoriented region in the recrystallized grain disappears completely. The entire grain shows homogeneous orientation with occasional fluctuation at the level of 0.06°. Because the microstructure in the recrystallized grain has already been cleaned up in Scan 2, the microstructure in Scan 3 contains little additional information and is thus not shown in this work.

As previously observed for nucleation by SIBM, recrystallizing grains may contain significant orientation gradients, in particular during the early stages of recrystallization. In the work by Sabin et al. [4], nucleation in 40% cold rolled pure aluminum, from the same batch as the present sample, was studied in 2D by EBSD. Like the present investigation, interior misorientations were recorded and followed during ex-situ annealing. For approximately half of the nuclei, large orientation variations were found across them of a magnitude considerably larger than the error in measurements, which was reported to be around 1°. During subsequent annealing the nuclei were observed to grow – both those with and those without interior orientation variations, and during the growth the interior misorientations were observed to decrease or even disappear completely [4]. The present grain seems to follow the same evolution pattern, although the observed misorientations in the present work are introduced at the surface by the sample preparation (and thus is not a 'natural' misorientation). The present misorientations are also far smaller and the scale over which they extend is larger, than those observed in the work by Sabin et al. [4]. During extended annealing, the microstructure within the recrystallizing grain 'cleans up'. The rate of this,

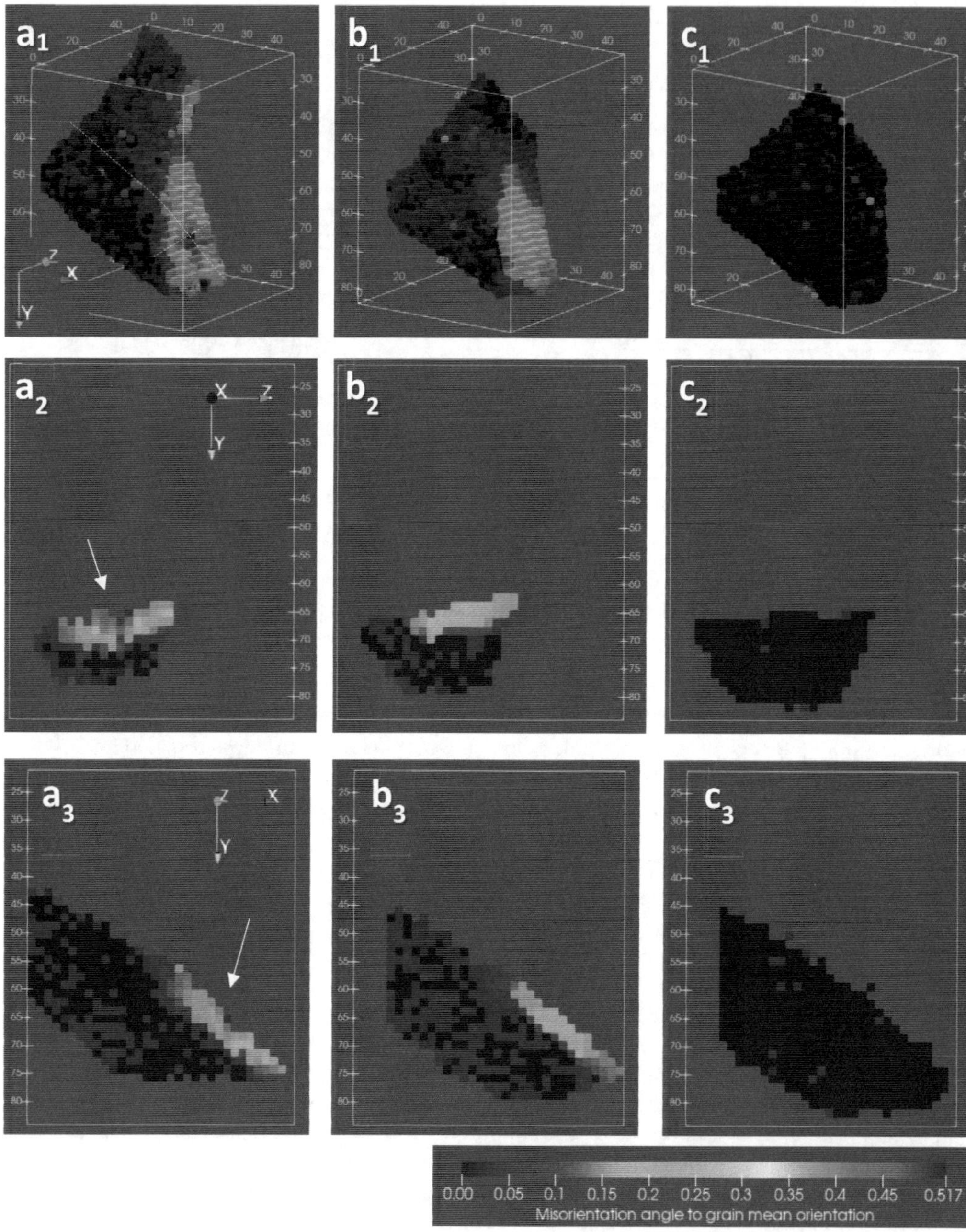

Figure 2. Evolution of internal microstructure of the recrystallized grain. (a_1–a_3) Scan 0 after annealing at 335 °C for 10 min. (b_1–b_3) Scan 1 after additional annealing at 335 °C for 15 min. (c_1–c_3) Scan 2 after further additional annealing at 335 °C for 90 min. The first row of images, a_1, b_1, c_1, show 3D views. The second and third rows of images show y-z and x-y sections from positions marked by dashed lines in a_1 and the same positions in b_1 and c_1. White arrows in a_2 and a_3 mark the sample surface. The scales on the axes are in micrometers.

is significantly faster than that observed in the remaining non-recrystallized parts of the present sample [14], even though the misorientations (i.e. stored energy) are smaller. Reasons are likely to be related to the nearness to the free sample surface, and possibly also to less well-organized dislocation microstructure of the investigated misoriented region within the recrystallized grain than in the deformed/recovered regions of the sample.

Neglecting the misoriented region discussed above, it is furthermore of interest to notice that the present recrystallizing grain does not contain an interior ultra-low-angle 'cellular' microstructure such as that observed by Ahl et al. [10]. In that previous work several possibilities were suggested to explain the presence of the intragranular dislocations organized in a cell structure: dislocation formation caused by small stresses due to the volume changes occurring when the sample is recrystallized, residual dislocations from the deformed structure, which possesses cell structures on a similar length scale as that observed for the intragranular substructure, as well as dislocations formed or pinned due to the presence of impurities [17]. The present aluminum is purer than that used in the work by Ahl et al. [10], reducing the likelihood of dislocation pinning. The deformed microstructure at the present nucleation site is expected to be more complex with a less regular dislocation boundary arrangement, due to the combined effect of a hardness indent and the 12% cold rolling, than that in the previous work, where the aluminum was cold rolled to a reduction of 40%.

Whereas continuous 'clean-up' is observed everywhere within recrystallizing grains upon annealing in Ahl et al. [10], an increase in internal misorientation is seen locally for the present grain in Scan 1. It is unknown whether this may be related to the dislocations moving away from the orientation gradient region close to the surface or to dislocations from the surrounding deformation matrix. However, a better understanding of this process is likely to be important for the understanding of migration mechanisms of recrystallization boundaries. *In-situ* work to follow the growth of individual recrystallized grains with high temporal resolution is required. Concerning residual stresses, which have not been mapped in the present work, they may as well be of high importance. Other works have reported very large residual stress even within recrystallizing grains [11,12]. It is thus suggested to include strain mapping in future studies of recrystallization, as residual strains are likely a missing link in identifying nucleation sites and understanding local migration phenomena.

4. Conclusions

The present work has demonstrated the ability of recrystallizing gains to recover interior misorientations. It is shown that misorientations as large as 0.5° in regions larger than $35\times15\times5\mu m^3$, introduced artificially during sample preparation into a recrystallizing grain, are 'cleaned up' within the angular uncertainty of <0.01° upon further annealing. This recovery within the recrystallizing grain is completed long before that in the bulk non-recrystallized parts of the sample. The present data together with previous results on i) reduction of interior misorientations in recrystallizing grains formed by SIBM, and ii) elimination of very low angle dislocation boundary networks and reduction of residual strains within recrystallizing grains suggest that while the general assumption of recrystallized grains being single crystals may not be valid during the early stages of recrystallization, it is a good assumption for the grains in almost fully recrystallized samples.

Acknowledgement

This work is supported by funding from the European Research Council (ERC) under the European Union's Horizon 2020 research and innovation program (grant agreement No 788567, M4D). The authors thank Dr. Chaoling Xu for providing the as-rolled sample for this study.

References

[1] Humphreys F J, Rohrer G S and Rollett A 2017 *Recrystallization and related annealing phenomena* (Elsevier)

[2] Doherty R D, Hughes D A, Humphreys F J, Jonas J J, Juul Jensen D, Kassner M E, King W E, McNelley T R, McQueen H J and Rollett A D 1997 Current issues in recrystallization: a review *Mater. Sci. Eng. A* **238** 219–74

[3] Beck P A and Sperry P R 1950 Strain induced grain boundary migration in high purity aluminum *J. Appl. Phys.* **21** 150–2

[4] Sabin T, Winther G and Juul Jensen D 2003 Orientation relationships between recrystallization nuclei at triple junctions and deformed structures *Acta Mater.* **51** 3999–4011

[5] Larson B C, Yang W, Ice G E, Budai J D and Tischler J Z 2002 Three-dimensional X-ray structural microscopy with submicrometre resolution *Nature* **415** 887–90

[6] Jakobsen B, Poulsen H F, Lienert U, Almer J, Shastri S D, Sørensen H O, Gundlach C and Pantleon W 2006 Formation and subdivision of deformation structures during plastic deformation. *Science* **312** 889–92

[7] Zhang Y B, Budai J D, Tischler J Z, Liu W, Xu R, Homer E R, Godfrey A and Juul Jensen D 2017 Boundary migration in a 3D deformed microstructure inside an opaque sample *Sci. Rep.* **7** 4423

[8] Simons H, Haugen A B, Jakobsen A C, Schmidt S, Stöhr F, Majkut M, Detlefs C, Daniels J E, Damjanovic D and Poulsen H F 2018 Long-range symmetry breaking in embedded ferroelectrics *Nat. Mater.* **17** 814–9

[9] Hayashi Y, Setoyama D, Hirose Y, Yoshida T and Kimura H 2019 Intragranular three-dimensional stress tensor fields in plastically deformed polycrystals. *Science* **366** 1492–6

[10] Ahl S R, Simons H, Zhang Y B, Detlefs C, Stöhr F, Jakobsen A C, Juul Jensen D and Poulsen H F 2017 Ultra-low-angle boundary networks within recrystallizing grains *Scr. Mater.* **139** 87–91

[11] Zhang Y and Barabash R 2019 High resolution mapping of orientation and strain gradients in metals by synchrotron 3D X-ray Laue microdiffraction *Quantum Beam Sci.* **3** 1–14

[12] Zhang Y, Yu T, Xu R, Thorborg J, Liu W, Tischler J Z, Godfrey A and Juul Jensen D 2022 Local residual stresses and microstructure within recrystallizing grains in iron *Submitted for publication.*

[13] Hong C, Zhang Y, Lindkvist A, Liu W, Tischler J, Xu R and Juul Jensen D 2021 Recrystallization boundary migration in the 3D heterogeneous microstructure near a hardness indent *Scr. Mater.* **205** 114187

[14] Yu T, Hong C, Zhang Y, Lindkvist A, Liu W, Tischler J and Juul Jensen D 2022 Recovery of deformation microstructure in the bulk interior revealed by synchrotron X-ray micro-diffraction *Submitted for publication.*

[15] Larson B C and Levine L E 2013 Submicrometre-resolution polychromatic three-dimensional X-ray microscopy *J. Appl. Crystallogr.* **46** 153–64

[16] Groeber M A and Jackson M A 2014 DREAM.3D: A Digital Representation Environment for the Analysis of Microstructure in 3D *Integr. Mater. Manuf. Innov.* **3** 5

[17] Humphreys F J and Martin J W 1968 The effect of dispersed phases upon the annealing behaviour of plastically deformed copper crystals *Philos. Mag.* **17** 365–403

IOP Conf. Series: Materials Science and Engineering 1249 (2022) 012051 doi:10.1088/1757-899X/1249/1/012051

Corrosion Performance of Additively Manufactured Stainless Steel by Binder Jetting

P H Kaae[1] and E Z Eikeland[2]
[1] Material Technology and Support, Grundfos, 8850 Bjerringbro, Poul Due Jensens Vej 7, Denmark
[2] Material laboratory, Grundfos, 8850 Bjerringbro, Poul Due Jensens Vej 7, Denmark

E-mail: Pkaae@grundfos.com

Abstract. The influence of microstructural defects on the corrosion resistance of AISI 316L additively manufactured with Metal Binder Jetting in high Cl^- environments were studied herein. A comprehensive metallographic examination was conducted to identify microstructural defects followed by a corrosion study with electrochemical techniques with the aim of finding correlations between impurities in the microstructure and preferential nucleation sites. Despite, the high number of inclusions in the microstructure of the studied samples, none of the conducted corrosion tests showed a clear indication that they resulted in a decreased corrosion resistance as compared to areas with few inclusions. However, a significant higher pitting potential was observed on the as-printed samples compared to samples grinded with 600 grit SiC paper which will be discussed further.

1. Introduction

Austenitic stainless steel is a widely used industrial material due to its excellent corrosion resistance, formability and weldability. The microstructure is controlled by the chemical composition and the manufacturing routes from melt to final components. With the implementation of additive manufacturing, it is possible to obtain different microstructures and thereby properties compared to the wrought and casted alloys. This has resulted in increased interest among the research communities, evaluating the link between process parameters, powder properties, microstructures, mechanical properties and corrosion resistance. Previously the attention has mainly been on laser power bed fusion (L-PBF) printing technology which is also the most widely used. However, L-PBF comes with several disadvantages such as low production volumes and the need of support material and rough down skin, which has resulted in an increased interest in the additive manufacturing Metal Binder Jet (MBJ) technology.

MBJ is an additive manufacturing technique originally developed at MIT in the early 1990s [1]. It offers the possibilities of larger volume production compared to the L-PBF techniques. The process works by dispensing a liquid binder on powder to form a two-dimensional layer pattern. The process is replicated a number of times to form a three-dimensional component. Additionally, the component needs to undergo several post processing steps before reaching the final stage incl.; curing at low temperature for polymerization of the binder, de-powdering with the aim of extracting the green body, de-binding for complete removal of the binder agent and finally a sintering at high temperature to activate a strong bonding of the metal powder driven by diffusional forces [2]. Even though the MBJ process was developed already in the 1990's, the research activity has been low and resulted in MBJ being the least matured powder bed AM technology, having a Technology Readiness Level (TRL) between 1 and 2, compared to other AM technologies having a TRL of 6-7 [1].

Prior study highlights that the corrosion resistance of sintered stainless steel is expected to be lower compared to cast and wrought equivalents due to crevice corrosion mechanism taking place in the pores [3]. The authors emphasize the importance of increasing the density by adjusting the sintering temperature, but also state that the increase in sintering temperature can lead to a higher degree of delta ferrite and precipitation of other unwanted phases during cooling resulting in lower corrosion resistance [3].

In this work, the microstructure of AISI 316L stainless steel additively manufacturing by MBJ has been analysed and correlated to the corrosion performance. The aim is to test the corrosion resistance in high saline environments and to understand the nucleation mechanism in MBJ samples and compare it with wrought stainless steel equivalents.

2. Materials and Methods

Test and analysis were performed on wrought EN 1.4404 and AISI 316L additively manufactured by Binder Jetting. The wrought test samples were cut out of a cold-rolled coil with a thickness of 1mm. delivered with a 2B surface finish (cold rolled, annealed, pickled and lightly rolled). Its chemical composition was measured by optical emission spectroscopy, see table 1 for elemental composition in weight percentage. The AISI 316L MBJ samples were printed on a production scale MBJ machine with parameters defined by the supplier. Hereafter, the samples were cured at low temperatures to obtain polymerization of the binder followed by a de-powdering, de-binding and finally sintering at a temperature between 1300-1400°C for more than 2 hours applying a hydrogen atmosphere. The AISI 316L powder was studied by the use of a Scanning Electron Microscope (SEM) FEI Quanta 600 at 20kV voltage and a spot size of 5.2. The aim of the investigation was to evaluate the particle size distribution, geometry and inhomogeneity. The chemical composition of the printed MBJ samples can be seen in table 1. To obtain a well-defined surface finish before corrosion testing, all samples were grinded with a SiC250 grit paper followed by SiC600 grit paper, unless stated otherwise. Samples for metallographic examination were cold moulded in epoxy, grinded to 220 SiC paper and polished with diamond suspension (9µm, 3µm, 1µm). The samples were cleaned in an ultrasonic batch for 5 min in acetone and 5 min in ethanol.

2.1. Metallographic examinations

To reveal the microstructure, the polished AISI 316L MBJ samples were electrolytically etched in two different solutions: 10% oxalic acid for 50s and 40% NaOH for 40s. The microstructural analysis was carried out on two optical microscopes (Keyence VHX-7000 and Zeiss Axio Observer) and an SEM equipped with an energy dispersive X-ray detector (EDX) (SEM - FEI Quanta 600 and EDX Noran System Six model no. 7752C-3UUS-SN from Thermo Scientific). The data was analyzed with NSS software from Thermofisher Scientific. The SEM-EDX analysis was conducted in High vacuum ($9e^{-5}$ mbar) with a voltage of 25kV and a spot size of 5.2.

2.2. Corrosion experiments

The Cyclic Potentiodynamic Polarization (CYP), Potentiodynamic Step-by-Step (PStep) and Single Pulse (SP) measurements were all performed using a Gamry Reference 600 potentiostat, with a standard three electrode and Avesta cell setup [4]. The solution media for the Avesta cell consist of 250 mL of artificial seawater (ASW) produced according to ASTM D1141-98 although not including the stock solution No. 3 and pH adjustment steps. The chlorinity of the ASW is 19.4 g. All steel samples were cleaned in acetone and ethanol for five minutes, using an ultrasonic bath, before being mounted to the bottom of the Avesta cell and tightened with a torque of 20 Nm exposing a sample area of 1 cm^2. In order to ensure pitting corrosion conditions, demineralized water was used to flush a filter paper ring, with a 1 cm^2 hole, located between the sample and Avesta cell. The demineralized water flow was set to 0.4-0.5 mL/min using a peristaltic pump. The exposed sample acted as the working electrode, a Pt mesh was used as the counter electrode, while a saturated calomel electrode (SCE) was used as the reference electrode.

The CYP measurements (T = 25°C) started by measuring the open circuit potential (OCP) for 1 h before conducting a cyclic sweep from −0.25V vs. OCP to 1V vs the SCE with a scan rate of 0.167mV/s. The scan was set to reverse when the current density reached 5 mA/cm^2 or the potential reached 1V vs. the SCE. The reverse scan rate was 0.167mV/s back to −0.25V vs. OCP. The breakdown potential (E_{break}) was approximated as the potential when reaching a current density of 0,01 mA/cm^2.

The PStep measurements (T = 25°C) started by measuring the OCP for 1 h before increasing the potential to 0.025 V vs. OCP and holding the potential for 3 hours. Subsequently the potential was increased 0.025 V every 3 hours until high currents of 1-10 mA were reached, or the potential reached 1 V. The sharp current increase indicates the onset of pitting corrosion, and the given potential is noted E_{pit}.

The SP measurements (T = 40°C) were started by holding the potential at the repassivation potential for 2 s before increasing the potential for 1 s or 4 s to Epit + 300mV before lowering the potential to the repassivation potential for 2 s while measuring the current response. After the SP measurements a selective number of samples were examined (before and after electrolytic etching in 10% oxalic acid for 90 s) with the purpose of investigating the pit density, morphology, and nucleation. The analyses were carried out by optical microscopy and SEM. To obtain a better understanding of the magnitude of corrosion, the area below the current/time curves were integrated and by utilizing Faradays law the mass loss determined (see supplementary data S3).

3. Results and discussion

The measured chemical composition showed that both the wrought EN 1.4404 and MBJ AISI 316L were in accordance with the standard (ISO/EN10088-2 and ASTM A240/240M), showing similar compositions of the main elements Cr, Ni and Mo, see Table 1.

Table 1. Chemical composition in wt% of wrought EN 1.4404 and MBJ AISI 316L

Elements	C	Si	Mn	P	S	Cr	Mo	Ni	N	Fe
Wrought EN1.4404	0.03	0.46	1.31	0.035	0.003	16.62	2.02	10.02	0.035	Bal.
MBJ AISI316L*	0.01	0.73	0.84	0.023	0.003	16.54	2.14	10.57	0.007	Bal.

* Uncertainties in results due to porosities in analysed sample.

Physical powder properties influence the quality of MBJ due to the effect on flowability and density [3]. The flowability of the powder is of significant importance for MBJ due to the low external forces for packing of the powder. The flowability and shape of the powder is therefore directly impacting the density of the green body. The powder used for MBJ is commonly gas atomized due to the spherical shape which improve the flowability and limiting the particle friction [3]. Another important parameter is the particle size distribution, where small powders are limiting the flowability, but lead to good packing density together with larger particles [3]. The SEM analysis performed of the powder showed a particle size distribution of d10 = 4 µm, d50 = 9.9 µm and d90 = 20.8 µm and revealed spherical powders with few satellites. The powder and process parameters enabled a relative density of 98.8% measured by Archimedes' principle.

IOP Conf. Series: Materials Science and Engineering 1249 (2022) 012051 doi:10.1088/1757-899X/1249/1/012051

3.1 Microstructure characterization

The as-printed MBJ AISI 316L has an austenitic matrix with approximately 2-5% delta-ferrite, see figure S1 c) (supplementary data). The level of delta-ferrite is due to the high temperature under sintering. It is generally found that up to 8% delta-ferrite has negligible effect on the corrosion resistance [3]. However, the BCC phase of delta-ferrite is rich in chromium and molybdenum, which can result in the formation of σ-phase if the cooling rate from the high sintering temperature is too slow in the temperature region 600-100°C [5]. The presence of σ-phase has a significant impact on the corrosion resistance due to the segregation of chromium and molybdenum leading to depletion in the surrounding regions. The metallographic examination of the samples showed σ-phase in the delta-ferrite, see figure S1 d) (supplementary data). However, σ-phase was not detected close to the surface, so it is not expected to impact the corrosion resistance. Additionally, the microstructural observations revealed a microstructure with twins and equiaxed grains with a mean size of ~30μm, see figure S1 e) (supplementary data). Figure 1 shows the presence of circular "black" spots mostly less than 5 μm in diameter with a high concentration in the center of the sample. The literature is defining these as gas porosities which are commonly observed in sintered parts [6]. However, other studies find these circular "black" spots not exclusively to be porosities, but also Cr/Mn, Si, Mn/Si/Cr rich inclusions [1]. To evaluate the circular black spots in the MBJ samples, an EDX analysis was conducted confirming the presence of inclusions rich in Mn, Si, Cr as presented in figure S2 and table S2 (supplementary data).

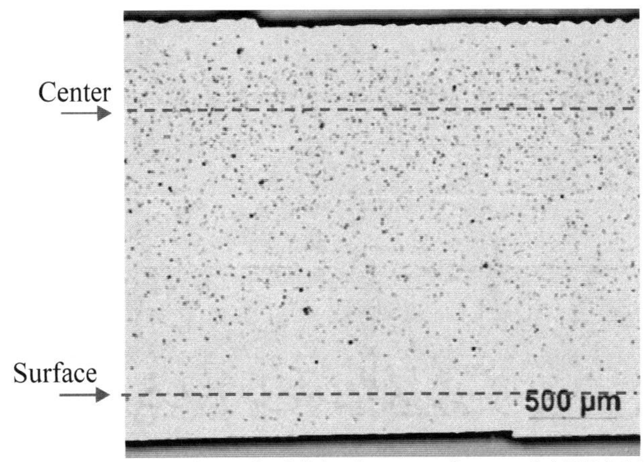

Figure 1. Metallographic observation of additively manufactured AISI 316L by MBJ – unetched: showing black circular spots.

Deposited MnO has been found in large quantities on the surfaces inside the ovens, ultimate resulting in flakes of MnO tearing off inside. The critical aspect of Mn containing steels in sintering has been addressed extensively in the literature, explaining that Mn is used as a so-called "self-cleaning" element against oxygen in the atmosphere, protecting the highly oxygen sensitive powder [7]. The self-cleaning effect of Mn is related to the high volatility of Mn that leads to significant sublimation at elevated temperatures. Thereby, the Mn is preventing the specimen from subsequent oxidation. The literature also states that Mn sublimates at relatively low temperatures and its partial pressure reaches about 10^{-3} Pa at 700°C, where thermodynamic conditions favor oxidation. Based on the conducted analysis of the microstructure the hypothesis is the following: for the areas close to the surface Mn will diffuse and leave the surface, while the Mn in the center will be trapped due to the rapid shrinkage rate in the temperature range 1000-1050°C.

The microstructural defects in the form of porosities, σ-phase, Mn/Cr or O/Si/Mn inclusion are potential nucleation points for pitting corrosion and will be examined in the following section.

3.2 Cyclic potentiodynamic polarization

The cyclic potentiodynamic polarization curves of wrought EN1.4404 2B and MBJ 316L as-printed are shown in figure 2. Both samples are exhibiting self-passivation characteristics, meaning that they spontaneously form a passive Cr_2O_3 layer at the surface [8]. Additionally, figure 2 shows a successful repassivation when reversing the potential. More so, figure 2. presents large variations in pitting potential for MBJ AISI 316L (~400mV – transpassive region). Therefore, it was decided to replicate the test for a total of 11 times, resulting in 7 out of 11 samples reaching 1000mV before breakdown of the passive layer. However, on 3 occasions breakdown was observed on the return from the transpassive region. Wrought EN1.4404 showed a pitting potential of 530mV ±20 (n=3). Besides the lower pitting potential of EN1.4404 2B the results also show a higher instability in the passive region - indicating occurrence of metastable pitting - compared to MBJ AISI 316L where the majority show a passive region without metastable pits, see figure 2.

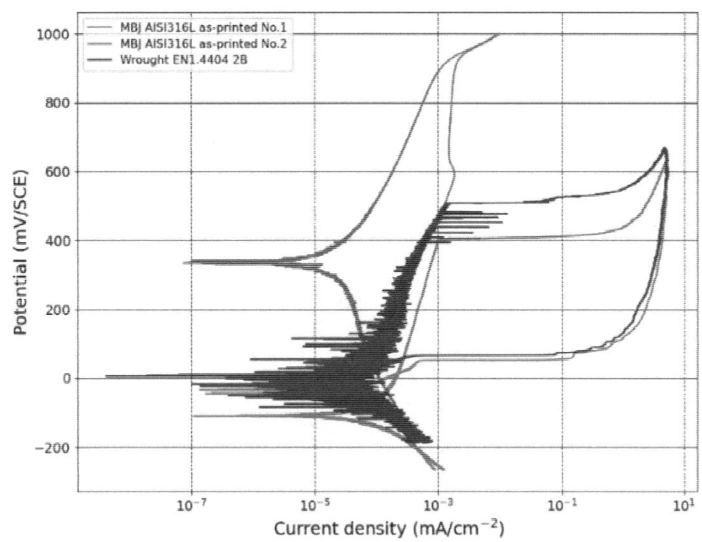

Figure 2. Cyclic potentiodynamic polarization curves of MBJ AISI 316L (two samples, one showing transpassive behavior and one showing a pitting potential of 400mV) and wrought EN1.4404 2B.

3.3 Potentiostatic Step-by-Step

Figure 3 summarizes the results from the potentiostatic step-by-step tests conducted on the following samples: EN 1.4404 2B, EN 1.4404 SiC600 grit paper, MBJ AISI 316L as-printed, MBJ AISI 316L SiC600 grit paper (surface & center; see figure 1 for definition).

The MBJ AISI 316L as-printed samples did not experience any breakdown of the passive layer despite reaching a potential of 1000-1400mV and all tests were therefore interrupted (n=3). The grinded MBJ samples (both center and surface) on the other hand exhibit E_{pit} values in the range of 370-420mV. The potentiostatic step-by-step test reveals a trend of lowering E_{pit} when mechanically removing a few microns, while an unchanged behavior when removing additionally millimeters despite the high level of O/Si/Mn rich inclusions in the center of the samples. The outstanding corrosion resistance of the as-printed surface is assumed to be a result of sintering in a reducing atmosphere, leading to an outer surface free from impurities. However, tests have not been conducted to support this hypothesis, but XPS could be used to obtain a better understanding of the as-printed surface.

The results presented in figure 3 confirm that MBJ AISI316L as-printed has a significant higher E_{pit} compared to EN1.4404 despite the better surface finish (2B and SiC600, see supplementary data S1) as also observed in the cyclic potentiodynamic polarization tests. Analysis of variance (ANOVA) and Fisher's Least Significant Difference (LSD) methods were applied to evaluate if a significant difference could be found between wrought EN1.4404 (2B and grinded) and MBJ AISI316L grinded. The ANOVA resulted in a p-value of 0.0006, leading to the conclusion that at least one of the tested samples differs from the rest. Fishers LSD methods were calculated with 95% confidence interval, identifying that E_{pit} for wrought EN1.4404 (2B and SiC600 grit paper) are significantly different from the MBJ AISI316L grinded surfaces (surface and center). Additionally, it can be stated that there is no significant difference between the MBJ AISI 316L SiC600 grit paper surface and center based on the performed experiments.

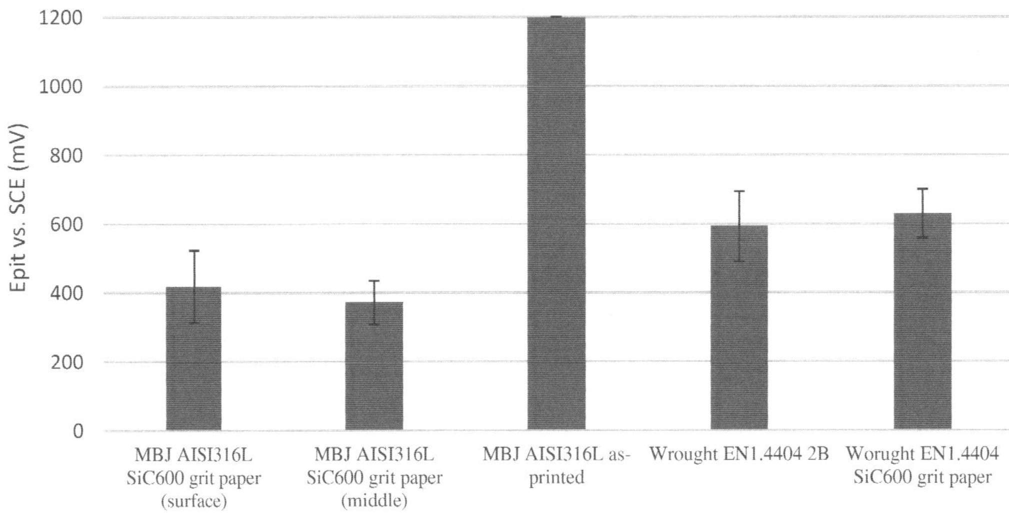

Figure 3. Summary of E_{pit} from potentiostatic step-by-step test. Showing a significant higher corrosion resistance of MBJ AISI316L as-printed which did not experience corrosion despite applying a potential of 1000-1400mV in the tested conditions. The tests have been replicated (n=3-6).

3.4 Single potentiostatic pulse tests

The single potentiostatic pulse tests were conducted on wrought EN 1.4404 SiC600 and MBJ AISI 316L SiC600. The E_{pit} and E_{rep} was found from the cyclic potentiodynamic polarization (performed at 40°) for the respective samples, and the potentials for the single potentiostatic pulse tests were chosen according to E_{pit}+300mV and E_{rep}. A potential of 695mV vs SCE was selected for wrought EN 1.4404 and a potential of 700mV vs SCE was selected for MBJ AISI 316L. The results after the 1s and 4s test shows that the current density response for wrought EN 1.4404 and MBJ AISI 316L is very similar showing a fast increase in current density when the potential is applied, followed by a decrease in current density and a steady increase until a momentary increase followed by a sudden decrease, see figure S3 (supplementary data). Prior studies performed on SLM samples presented significantly different behavior when applying the potential, showing only minor increase in current density over time, although in this study a different solution medium was used (4 mol.l^{-1} at 50°C) [9].

The total mass loss after the single potentiostatic pulse tests for both 1s and 4s was calculated by Faraday's law showing a higher weight loss of wrought EN1.4404 compared to MBJ AISI316L independent of testing time, see Table. S3 (supplementary data).

SEM observations after 1s and 4s showed a clear increase in pit size going from 1s to 4s applied potential for both alloys, see figure 4. Additionally, by comparing figure 4 e), f), g), h) a tendency can be seen, showing the average pit size of MBJ being slightly larger compared to wrought EN 1.4404. However, figure 4 a), b), c), d) shows a considerable higher mean pit density (pits/area) for wrought EN1.4404.

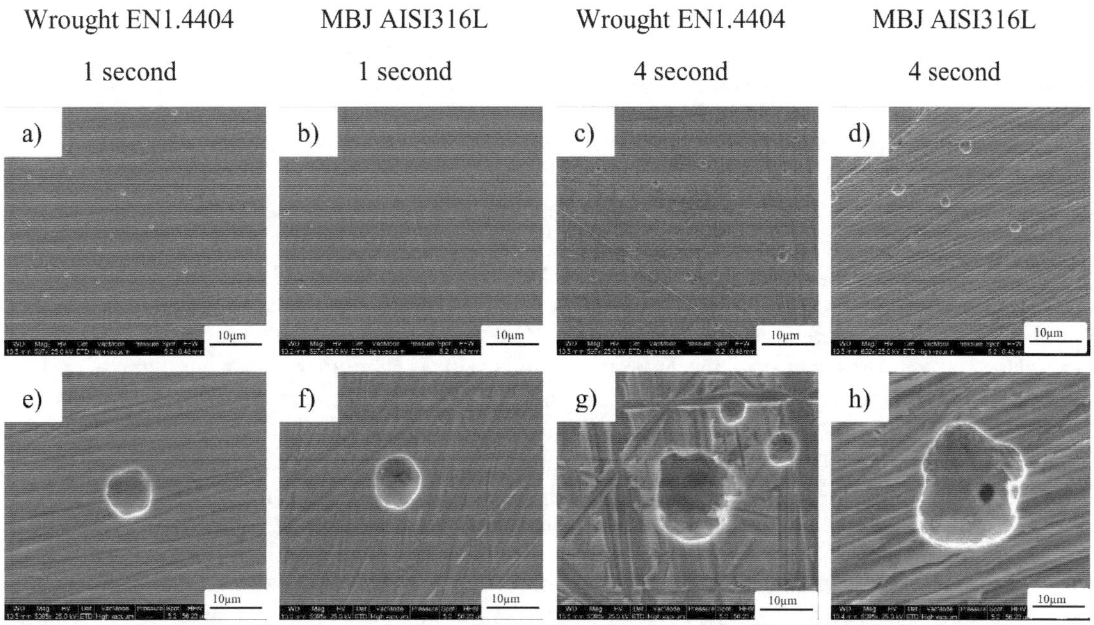

Figure 4. SEM images of pitting corrosion after applying a potential of E_{pit}+300mV for 1s and 4s. a) wrought EN 1.4404 after 1s applied potential, b) MBJ AIS I316L after 1s applied potential, (c) wrought EN 1.4404 after 4s applied potential, d) MBJ AISI 316L after 4s applied potential, e) wrought EN 1.4404 after 1s applied potential, f) MBJ AISI 316L after 1s applied potential, g) wrought EN 1.4404 after 4s applied potential, h) MBJ AISI 316L after 4s applied potential.

SEM microscopy of pits electrolytically etched in 10% oxalic acid shows a tendency for pits to nucleate in grain boundaries or triple grain boundaries of wrought EN1.4404 figure 5. Other studies are presenting similar results estimating that 68% of the pits initiate in grain boundaries including 26% being triple grain boundaries. The results are explained by the presence of chromium and molybdenum rich phases close to the grain boundaries, which thereby changes the properties of the passive layer at the interface between the Cr/Mo rich phase and the austenitic matrix [9]. The same tendency is not observed in MBJ AISI316L, where only few of the examined pits seem to nucleate in grain boundaries. However, porosities are commonly found to be present in the pits. It is unclear if the porosities are the initiation point or a Mn/Si/O rich inclusion has been present and dissolved, leaving a pore behind. Further analyses need to be conducted to clarify the actual nucleation of pits in MBJ samples, but the examination illustrates different mechanisms for pit nucleation of the two alloys.

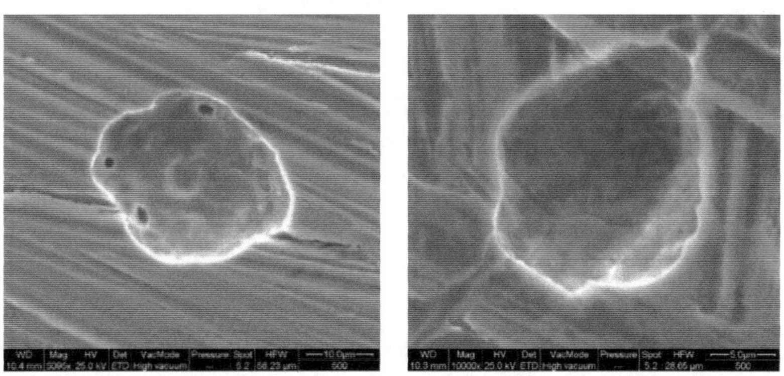

Figure 5. Example of SEM images of pits after electrolytic etching in 10% oxalic acid. Left - MBJ AISI 316L showing porosities inside the pit, right - wrought EN 1.4404 showing triple grain boundaries.

4. Conclusion

A detailed analysis of the microstructure of AISI 316L additively manufactured by Metal Binder Jet has been conducted. Furthermore, a comprehensive corrosion study utilizing electrochemical techniques in high saline environments was performed. The presented work was able to reveal the following information:

1. The microstructure of MBJ AISI 316L contains a large number of Mn/Si/O rich inclusions concentrated in the centre of the samples due to manganese's high volatility which leads to sublimation under sintering condition, where manganese and oxygen react and precipitates inside the porosities. Despite the high concentration of inclusions none of the conducted experiments show a clear tendency that these inclusions lowered the corrosion resistance by acting as nucleation sites for pitting corrosion.
2. The corrosion resistance of the MBJ as-printed AISI 316L outperformed wrought EN1.4404. However, the analysis showed that even minor removal of the surface layer resulted in a drastic decrease of the pitting potential under the given test conditions.
3. Whilst for the wrought EN 1.4404 grain boundaries were found to be a possible nucleation site, it was shown that porosities could be more important as nucleation sites for samples produced with MBJ.

Acknowledgements
The authors would like to acknowledge Dr S. Okoro, Mrs. A. Riis and Mrs. D. Primdahl for their assistance and support throughout the project.

References

[1] Lecis N, Beltrami R and Mariani M 2021 Binder jetting 3D printing of 316 stainless *J. La Metallugia Italiana* **113** 31-41
[2] AMPOWER Insights, *Design guideline for sinter-based Additive Manufacturing*, Version 1.0 (2021)
[3] Mirzababaei S and Pasebani S 2019 A Review on Binder Jet Additive Manufacturing of 316L Stainless Steel *J. Manuf. Mater. Process.*, **3** A 82
[4] Qvarfort R 1988 New Electrochemical Cell for Pitting Corrosion Testing *J. Corros. Sci.* **28** 135-40
[5] Hsieh C and Wu W 2012 Overview of Intermetallic Sigma (σ) Phase Precipitation in Stainless Steels *J. ISRN Metallurgy*
[6] Zwiren A 2018 Comparison of Binder Jetting Additive Manufacturing to Press and Sinter 316L Stainless Steel *Sementic Scholar*
[7] Hryha E, Dudrova E and Nyborg L 2010 Critical Aspects of Alloying of Sintered Steels with Manganese *J. Metallur. Mater. Trans* **41** 2880-98
[8] Sander G, Cruz V, Bhat N and Birbilis N 2020 On the in-situ characterisation of metastable pitting using 316L stainless steel as a case study *J. Corros. sci.* **177**
[9] Racot A, Aubert I, Touzet M, Thiebaut S and M. Demesy 2022 Statistical analysis of the pitting corrosion induced by potentiostatic pulse tests of wrought and SLM 316L stainless steels *J. Corros. sci.* **197**

IOP Conf. Series: Materials Science and Engineering 1249 (2022) 012051 doi:10.1088/1757-899X/1249/1/012051

SUPPLEMENTARY DATA

Corrosion Performance of Additively Manufactured Stainless Steel by Binder Jetting

P H Kaae and E Z Eikeland

S1. Surface quality

The surface topography is largely impacting the corrosion resistance, due to the formation of microenvironments where passive film breakdown can be easily achieved. One of the disadvantages of additive manufacturing is the high roughness in the as-printed conditions. However, MBJ shows significantly lower roughness compared to L-PBF technologies, especially on the downskin side of the parts. Table S1 shows the measured roughness of wrought EN 1.4404 2B, MBJ AISI 316L, Wrought EN 1.4404 SiC600 grit paper and MBJ AISI 316L SiC600 grit paper.

Table S1. Surface roughness measured by optical profilometry of Wrought EN 1.4404 2B, MBJ AISI 316L as-printed, Wrought EN 1.4404 SiC600, MBJ AISI 316L SiC600.

Elements	Ra (perp. to BD)	Rz (perp. to BD)	RSm (perp. to BD)	Sa	Sz	Ra/RSm
Wrought EN1.4404 2B	0.14μm	2.67μm	33.00μm	0.15μm	3.26μm	0.004μm
MBJ AISI316L as-printed	1.352μm	15.29μm	50.69μm	8.57μm	71.32μm	0.026μm
Wrought EN1.4404 SiC600	0.037μm	0.338μm	58.69μm	0.048μm	1.87μm	0.0006μm
MBJ AISI316L SiC600	0.120μm	0.784μm	102.1μm	0.207μm	2.96μm	0.0012μm

From the listed values in Table S1, it can be seen that MBJ in as-printed condition has higher roughness values compared to wrought material. However, previous internal studies on L-PBF showed Sa values in the magnitude of 15.82-31.52μm, which emphasize the better surface quality of MBJ compared to the L-PBF. Based on the Ra/RSm values, the wrought EN1.4404 should have the best surface finish for optimizing the corrosion resistance.

S2. Microstructure characterization

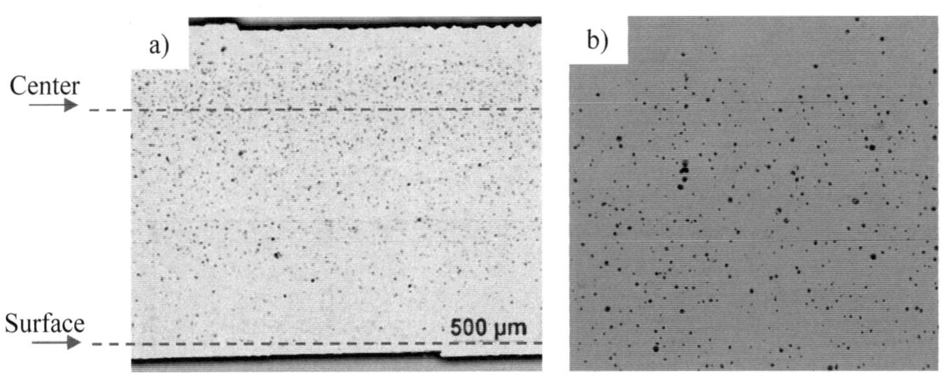

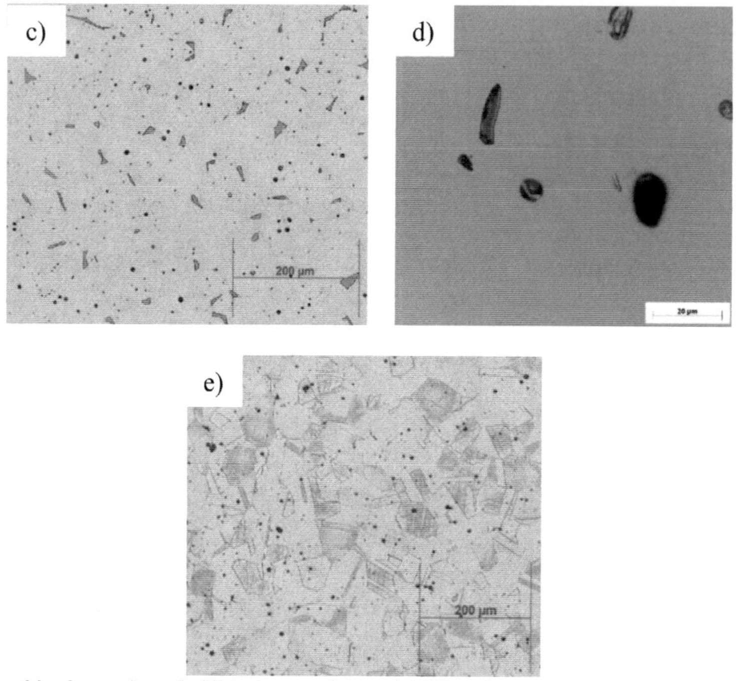

Figure S1. Metallographic observation of additively manufactured AISI 316L by MBJ. a) Unetched, showing black spots, b) Unetched, showing black spots at higher magnification, c) Etched in 40% NaOH showing austenitic matrix with delta-ferrite, d) Etched in NaOH showing σ-phase within the delta-ferrite, e) Etched in 10% oxalic acid showing grains and twinning.

The SEM images in figure S2 show some of the different inclusions that are present within the MBJ microstructure, in addition to austenite, ferrite and porosities. They were analyzed by EDX and found to be rich in Mn, Si, Cr, and O as presented in Table 3. Additionally, EDX mappings showed no segregation inside the inclusions, which means that the elements are uniformly distributed inside the inclusion.

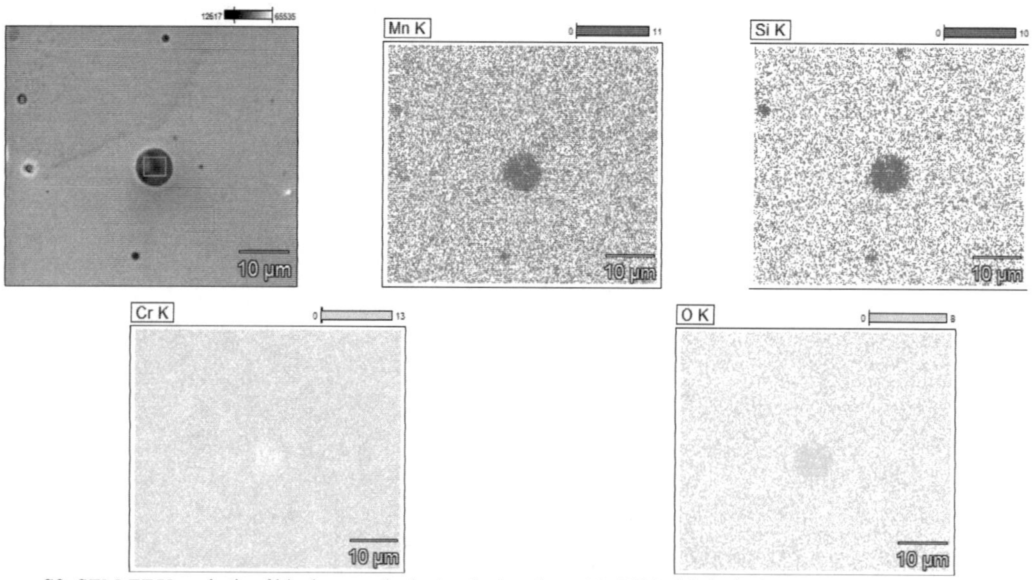

Figure S2. SEM-EDX analysis of black spots (inclusions), showing a Mn/Si/O rich inclusion.

Table S2. Elemental analysis by EDX of phases identified in figure S2 of additively manufactured AISI 316L by MBJ.

Elements	C	Si	Mn	O	S	Cr	Mo	Ni	N	Fe
Base material (P4)	1.33	0.6	1.5	-	-	16.9	1.9	9.7	-	Bal.
P1 – O/Si/Mn	1.0	14.2	23.9	21.7	~1.1*	9.0	-	2.4	-	Bal.
P2 – Cr/Mn	0.5	1.1	2.9	1.1	-	18.1	0.8	7.7	-	Bal.
P3 – O/Si/Mn	0.7	7.5	9.9	15.5	-	12.8	1.7	5.9	-	Bal.

*The EDX was performed with 25kV to be able to distinguish Mo and S. However, further analyzes were not performed to clarify if the high concentration of S were present in the microstructure.

S3. Corrosion experiments – Mass loss

Current density response from Single potentiostatic pulse tests after 1 and 4s of applied potential.

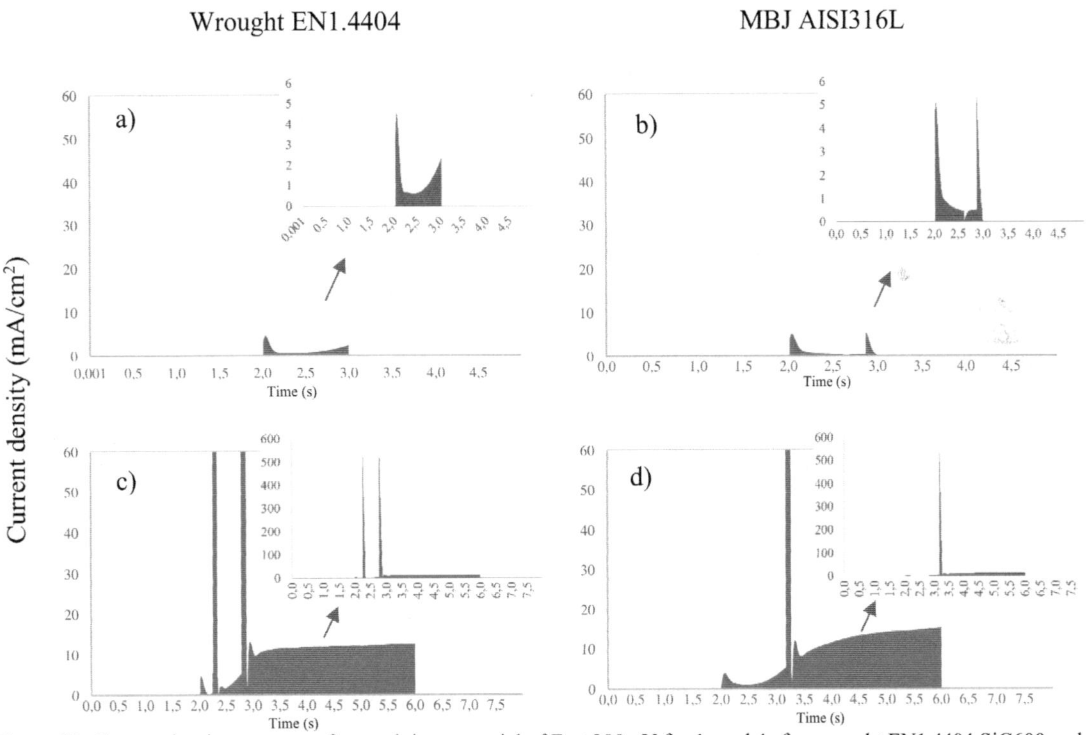

Figure S3. Current density response after applying potential of E_{pit}+300mV for 1s and 4s for wrought EN1.4404 SiC600 and MBJ AISI316L SiC600. a) and b) show results after 1s while c) and d) show the results after 4s.

The total mass loss after the single potentiostatic pulse tests for both 1s and 4s was calculated by Faraday's law as shown in equation 1.

$$mass\ loss = \frac{current * time * molar\ mass}{number\ of\ electrons * Faraday's\ constant} \quad (1)$$

The current and time were calculated by integrating the area below the current vs. time curves using the Gamry Echem Analysist software. The molar mass was calculated according to the following chemical composition containing approximately 16 wt% Cr, 10.5 wt% Ni, 2 wt% Mo and the remaining being balanced by Fe (71.5wt%), resulting in a molar mass of 56.33 g/mol. Using the same composition, the number of electrons was calculated to z=2,2. Table S3 summarizes the calculated mass loss, which shows an increase in mass loss as a result of increased holding time from 1 to 4 s. Additionally, it can be seen that the weight loss after 1s for both alloys are very similar, while wrought EN1.4404 after 4s shows a considerably higher mass loss compared to MBJ AISI316L.

Table S3. Average mass loss after single potentiostatic pulse test for 1s and 4s calculated by Faraday's law. The calculation is based on 3 measurements.

	Wrought EN1.4404, 1 second	Wrought EN1.4404, 4 seconds	MBJ AISI316L, 1 second	MBJ AISI316L, 4 seconds
Mass loss (g)	$1.459e^{-6} \pm 9.871e^{-7}$	$1.679e^{-5} \pm 8.068e^{-6}$	$2.752e^{-6} \pm 4.027e^{-6}$	$6.770e^{-6} \pm 5.865e^{-6}$

IOP Conf. Series: Materials Science and Engineering 1249 (2022) 012052 doi:10.1088/1757-899X/1249/1/012052

Texture in cold rolled and high-temperature annealed Cu foils

J Guo[1*], X Li[2,3], G L Wu[1,4], R Q Zhang[2,3], T L Huang[1], H F Shi[2,3], C L Zhang[5], D Jensen[5] and X Huang[1]

[1] International Joint Laboratory for Light Alloys (MOE), College of Materials Science and Engineering, Chongqing University, Chongqing 400044, China
[2] Chongqing Key Laboratory of Multi-Scale Manufacturing Technology, Chongqing Institute of Green and Intelligent Technology, Chinese Academy of Sciences, Chongqing, 400714, P.R. China
[3] University of Chinese Academy of Sciences, No.19(A) Yuquan Road, Shijingshan District, Beijing, 100049, P.R. China
[4] Beijing Advanced Innovation Center for Materials Genome Engineering, University of Science and Technology Beijing, Beijing, 100083, P.R. China
[5] Department of Civil and Mechanical Engineering, Technical University of Denmark, Kgs. Lyngby DK-2800, Denmark

E-mail: guo_jing@cqu.edu.cn

Abstract. Thick single crystal Cu foil substrates with specific crystallographic orientations such as {111} and {110} planes parallel to the rolling plane are desired for epitaxial growth of two-dimensional materials such as graphene. In this study, annealing experiments of a 46 μm thick cold-rolled foil were carried out at temperatures in the range of 300-1040 °C. The texture evolution during annealing was followed by electron backscatter diffraction (EBSD) orientation mapping. It was found that the texture evolution may be divided into three stages: at relatively low annealing temperatures, a strong cube texture develops with a small population of grains having relatively larger deviations from the cube orientation; at medium temperatures, these grains with large deviations grow abnormally; and at the highest temperature, annealing twins form in the abnormally growing grains giving rise to new orientations including some with the {111} planes parallel to the rolling plane.

1. Introduction

High-quality single-crystalline two-dimensional (2D) materials such as graphene have attracted much attention recently due to their superior properties such as excellent mechanical properties, thermal and electrical conductivities, as well as transparency [1-4]. Graphene can grow epitaxially on a single-crystal metal substrate and domains stitch seamlessly to synthesize single-crystals by chemical vapor deposition (CVD) [5]. The larger the size of the single-crystal metal substrate is, the larger the single-crystal graphene can be [6-9]. Recrystallization of cold-rolled thin foils is an obvious approach to produce such substrates. It is well-established that annealing of cold-rolled Cu normally generates a polycrystalline grain structure that is characterized by a strong cube texture with many low angle grain boundaries [10-15]. Because of the orientation pinning effect [16], it is challenging to produce single crystal Cu foils by normal grain growth. Recently, it was found that large single-crystal Cu foils can be obtained by annealing at high temperatures leading to abnormal grain growth [17]. At the same time, the orientations of these foils deviate from the cube orientation, and more favorable surface orientations for the growth of graphene, such as {111} and {011} planes parallel to the rolling plane can be obtained [17-20]. For simplicity, grains with e.g. {111} planes parallel to the rolling plane are denoted as {111} grains in the following. However, the mechanisms leading to a change from {001} grains to {111} or {011} grains are not clear. The aim of this paper is to study the evolution of texture and grain structure of a Cu foil during annealing with focus on the evolution during high-temperature annealing.

2. Material and experimental details

The material used in this study is a commercial Cu foil with a purity of 99.94%, which was cold rolled to a thickness of 46 μm corresponding to a reduction of approximately 90%. X-ray diffraction measurements revealed that the Cu foil has a typical rolling texture (see figure 1), composed of S {123}<634> (34.6%), Cu {112}<111> (16.6%), Brass {011}<211> (19.7%) and Goss {011}<001> (5.8%). Orientation distribution functions (ODFs) were calculated from three pole figures using the so-called Arbitrarily Defined Cells (ADC) method [21]. The volume fraction of each component was calculated by LaboTex allowing a tolerance angle of 10°.

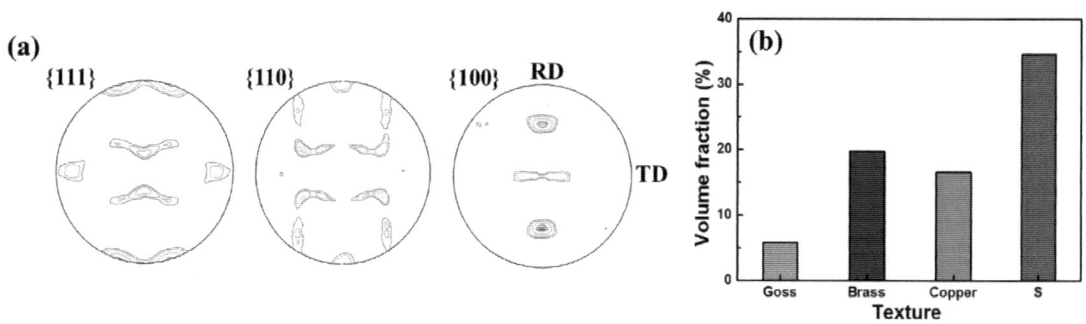

Figure 1. The texture of the initial cold rolled Cu foil. (a) Recalculated pole figures using the ADC method [21] based on XRD data and, (b) Volume fractions of the different texture components.

Rectangular shaped samples with a size of 80 mm × 100 mm were cut from the foil. The samples were placed on a flat quartz plate and then inserted into a quartz tube within a vacuum furnace, with the rolling direction of the foil perpendicular to the quartz tube axis. The samples were annealed for 1 h at various temperatures ranging from 300 °C to 1040 °C. The annealing was conducted within a protective atmosphere of 40 standard cubic centimeter per minute (sccm) H_2 and 500 sccm Ar. The annealed samples were subsequently oxidized in air at 190 °C on a hot plate for 2-3 min, to enable optical visualization of the grains. The orientations and microstructures of annealed samples were characterized by electron backscatter diffraction (EBSD) in a JOEL JEM-7800F scanning electron microscope coupled with an Oxford Aztec detector. Then the EBSD data for the annealed samples were analysed by Channel 5. During the whole experiment, the samples were carefully protected to avoid surface damage, and in all cases, it is the rolling plane at a position near the foil surface, which is inspected.

3. Results and Discussion

Partial recrystallization is observed after annealing for 1h at 300 °C and 500°C. The microstructure consists of recrystallizing grains with cube orientation, annealing twins, and some deformed grains (see figure 2 a, b). In the inset of figure 2a, the {122} twin component of the {100} grains can be clearly seen. The boundaries are mainly high-angle boundaries with misorientations > 15° and twin boundaries (see figure. 2d, e). The frequency of low-angle boundaries increases with increasing temperature from 300 °C to 500°C, whereas the twin boundary frequency remains almost constant. After annealing at 900 °C, the sample is fully recrystallized and substantial grain growth has occurred (figure 2c). The microstructure is dominated by large grains with cube orientation, and most grain boundaries are low angle boundaries. The frequency of twin boundaries is only 0. 025 after annealing at 900°C (approximately one tenth of that at 500°C). The inverse pole figure (IPF) coloring orientation maps reveal that there are dense twins within the cube grains.

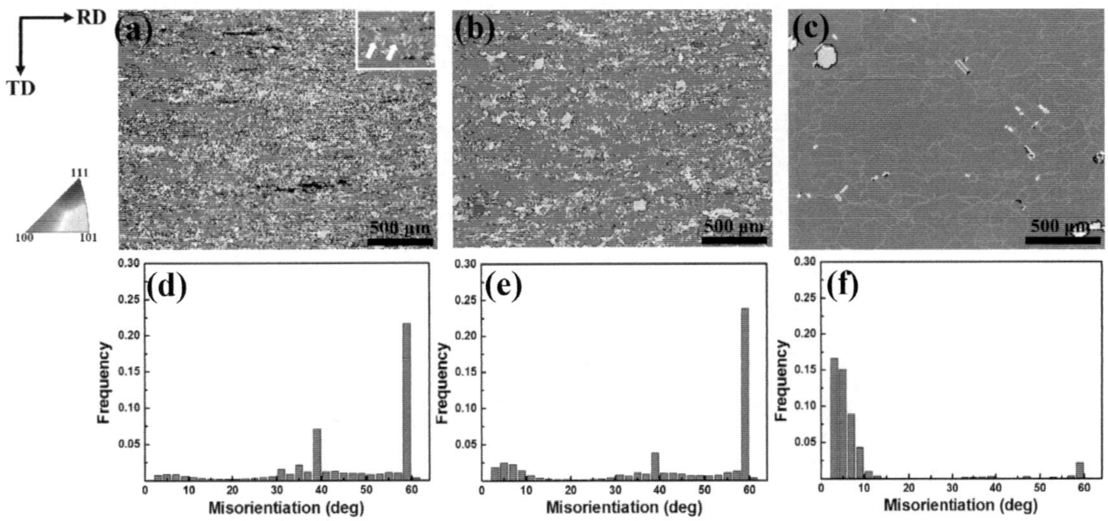

Figure 2. (a-c) Inverse pole figure (IPF) coloring orientation maps (ND colored) and (d-f) misorientation distributions of the Cu foils annealed at different temperatures for 1 h. (a, d) 300 °C, (a magnified map is inserted in the top right-hand corner of figure 2a to show the twins); (b, e) 500 °C; (c, f)) 900 °C. The grey, black and white lines in (a-c) represent boundaries with misorientations of >2°, >15° and twin boundaries, respectively. Black points are unindexed pixels. The white arrows point to the {122} grains, i.e., the twins of {100} grains.

Figure 3(a, c) show a photograph and an orientation map of the sample after annealing at 1000 °C. It is obvious that significant abnormal grain growth has occurred during annealing at this temperature, and the size of the abnormally grown grains reaches centimeters. The orientation of the abnormally grown grain seen to the right in figure 3c deviates from the cube orientation. Only a few twins are left in the grains. Figure 3(b, d) show a photograph and an orientation map of the sample after annealing at 1040 °C. It can be seen that centimeter-sized grains with an average size of 1.6 cm have developed in the sample. Furthermore, a large number of {111} twins appear, and the size of these twins is larger than that of the {122} cube-twins observed in the samples annealed at 300-1000 °C.

Figure 4 shows pole figures after annealing at different temperatures. After annealing at 300 °C, the texture consists of a weak rolling texture, superimposed with a cube and {122} cube-twin components (see figure 4a). At higher annealing temperatures the rolling texture disappears and the intensity of cube texture increases. The cube-twin texture becomes weaker after annealing at 500 °C - 900 °C (figure 4b-e). When the annealing temperature reaches 1040 °C, the orientation of the grains spreads widely. It is important to notice that only ten grains are included here due to the large grain size, see figure 4f.

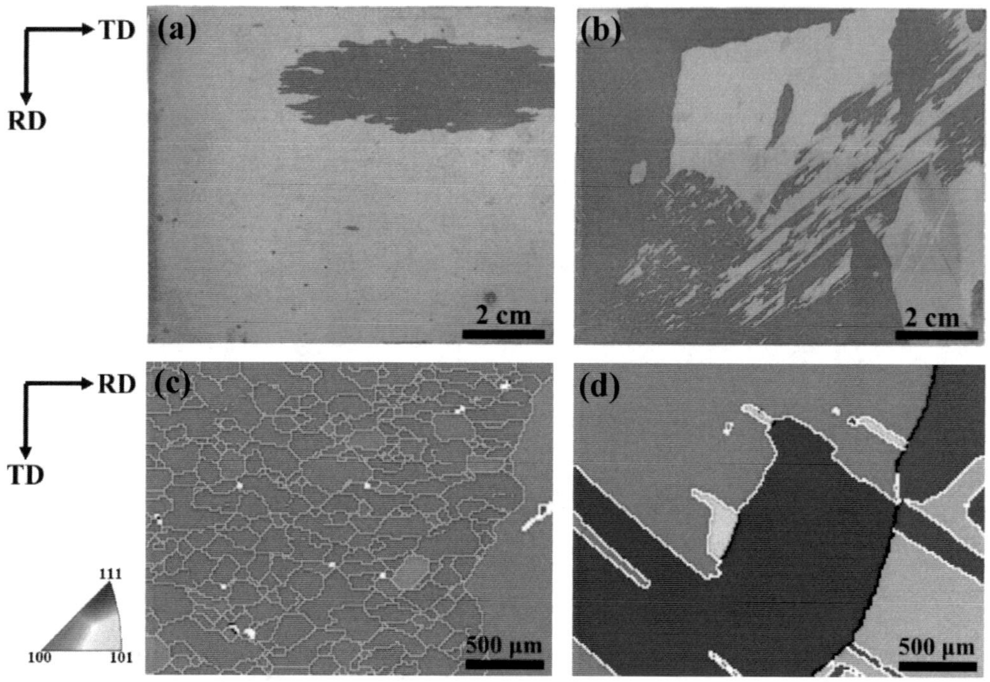

Figure 3. (a-b) Photograph and (c-d) IPF coloring orientation maps (TD colored) of the Cu foil annealed at (a, c) 1000 ℃, (b, d) 1040 ℃ for 1 h. The grey, black and white lines in (c,d) represent boundaries with misorientation of >2° and >15° and twin boundaries, respectively.

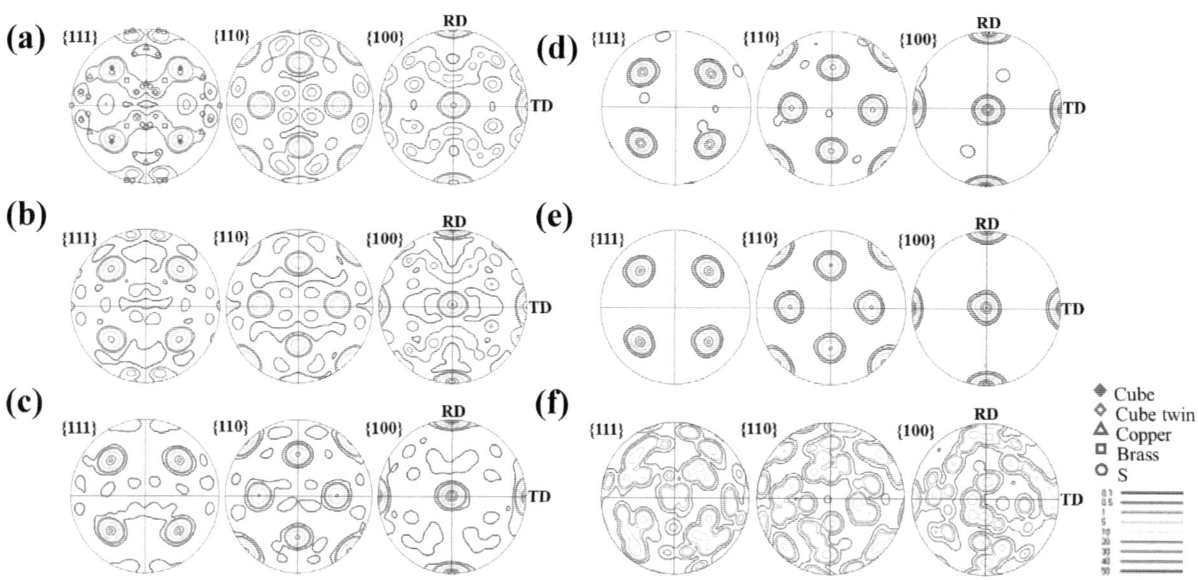

Figure 4. Pole figures (obtained from EBSD data - Channel 5) of the Cu foil annealed at different temperatures for 1 hour. (a) 300 °C and a set of symbols corresponding to the different texture types in the {111} pole figure, (b) 500 °C, (c) 700 °C, (d) 900 °C, (e) 1000 °C and (f) 1040 °C (please note that only 10 grains are included in the measurements at this temperature).

Over all, the texture evolution can be divided into three stages: i) the rolling texture transforms to cube texture during recrystallization; ii) abnormal grain growth leads to orientations that deviate from the cube orientation at higher annealing temperatures; iii) annealing twins appear leading to new orientations, including {111} twins.

After high-temperature annealing, the average grain size is much larger than the thickness of the Cu foil (46 μm) used in the present study. For example, the average grain size reaches 167 μm after annealing at 900 °C. Therefore, the surface energy has a significant effect on the annealing and thus the annealing texture [22]. Since the surface energy of cube grains, is lower than that of {122} cube-twin components [23], the number of cube twins continuously decrease as the annealing progresses.

The most interesting part of the microstructure and texture evolution is related to the high annealing temperatures. When the temperature is around 1000°C, the cube texture is very strong and the grains are separated mostly by low angle boundaries. As such boundaries have quite low mobilities, the grains with orientations deviating from the cube may be energetically favorable and therefore grow abnormally, in agreement with the general observation that a strong recrystallization texture facilitates abnormal grain growth [24, 25].

In stage iii), at 1040 °C, the grains are so large that surface energy totally dominates [26,27]. In order to further reduce the energy, {111} annealing twins appear because this close-packed plane in the FCC structure has the lowest surface energy [28]. It is suggested that further annealing of the Cu foil for a longer time or at a slightly higher temperature will lead to further growth of the {111} twins into a large-size single-crystal.

4. Summary

The texture evolution of a cold-rolled 46-μm thick Cu foil during annealing has been investigated using the electron backscatter diffraction technique. It is found that the texture evolution can be divided into three stages:

- At 300-900°C, a strong cube texture develops with a small population of grains having relatively large deviations from this orientation;
- At 1000 °C, grains with orientations deviating from the cube orientation grow abnormally;
- At 1040 °C, {111} annealing twins are observed in the abnormally grown grains. It is suggested that further annealing for a longer time or at a slightly higher temperature may lead to further growth of these {111} twins.

Acknowledgments

The authors acknowledge the financial support from the National Key Research and Development Program of China (2021YFB3702101). This paper was finalized while the corresponding author Jing Guo was visiting the Technical University of Denmark as a guest PhD student. For this part of the work we acknowledge the support from the European Research Council (ERC) under the European Union's Horizon 2020 research and innovation programme (Grant agreement No.788567).

References

[1] Novoselov K S, Fal'ko V I and Colombo L 2012 *Nature.* **490** 192-200
[2] Chae H K, Diana Y and Siberio-Pérez 2004 *Nature.* **427** 523-27
[3] Andrea C F, Francesco B and Fal'ko V 2015 *Nanoscale.* 7 4598
[4] Lee C, Wei X and Kysar J W 2008 *Science.* **321** 385-88
[5] Muñoz R and Gómez-Aleixandre C 2013 *Chem. Vap. Depos* **19** 297-22
[6] Dong J, Zhang L and Dai X. *Nat. Commun.* **11** 5862
[7] Nguyen V L, Shin B G and Duong D L 2015 *Adv. Mater.* **27** 1376-82
[8] Xu X, Zhang Z, and Dong J 2017 *Sci Bull.* **62** 1074-80
[9] Shi X, Yin Q and Wei Y 2012 *Carbon.* **50** 3055-63
[10] Baldwin W H 1942 *Metals Tech. Tech. Pub.* 1455
[11] Beck P A 1951 *JOM: The journal of the Minerals, Metals & Materials Society* **3** 474-75

[12] Barrett C S 1952 *Structure of Metals, 2rd Ed*
[13] Beck P A 1954 *Adv Phys-X* **3** 245-24
[14] Sindel M, Kölhoff G D and Lücke K 1990 *Textures & Microstructures.* **12** 37-46
[15] Hammelrath H Bulter J F and Juul Jensen D 1991 *Textures & Microstructures* **14** 687-99
[16] Juul Jensen D 1995 *Acta metal. mater.* **43** 4117-29
[17] Li Y, Sun L and Chang Z 2020 *Adv. Mater.* **32** 2002034
[18] Jin S, Huang M and Kwon Y 2018 *Science.* **362** 1021-25
[19] Wang L, Xu X and Zhang L 2019 *Nature.* **570** 91-96
[20] Li X, Wu G L and Zhang L 2022 *Nat. Commun.* **13** 1773
[21] Pawlik K 1986 *Phys. Stat. Sol. (b)* **134** 477-83
[22] Cotterill P, Mould P R 1976 *R Surrey University Press*
[23] Wang S G, Tian E K and Lung C W 2000 *J Phys Chem Solids.* **61** 1295-00
[24] Rollett A D, Srolovitz D J and Anderson M P 1989 *Acta Metall.* **37** 1227-1240
[25] Grest G S, Anderson M P and Srolovitz D J 1990 *Scripta Metal. Mater.* **24** 661-65
[26] Mao W M 1994 *Metallurgical Industry Press*
[27] Carl V T and Roland C 1996 *J Mech Phys Solids.* **44** 657-73
[28] Zhang J M, Fei M and Xu K W 2003 *Surf Interface Anal.* **35** 1082-90.

IOP Conf. Series: Materials Science and Engineering 1249 (2022) 012053 doi:10.1088/1757-899X/1249/1/012053

Effect of Aging Treatment on the Microstructure and Mechanical Properties of Fe-Mn-Al-C Low Density Steel

G Y Li[1], L Kong[1], E Z Liu[1], X D Zhang[2] W Q Cao[3] and Y H Wang[1*]

[1] National Engineering Research Center for Equipment and Technology of Cold Strip Rolling, Yanshan University, Qinhuangdao 066004, P. R. China
[2] Department of Civil and Mechanical Engineering, Technical University of Denmark, DK-2800 Kgs. Lyngby, Denmark
[3] Central Iron and Steel Research Institute (CISRI), Beijing 100081, China

E-mail: yhwang@ysu.edu.cn

Abstract. Fe-Mn-Al-C steel has excellent weight reduction potential. In this study, the microstructure and mechanical properties of Fe-25Mn-11Al-1.0C steel after solution treatment at 1050 °C and aging at 450~650 °C are studied. The results show that with the increase of aging temperature, the strength increases first and then decreases. The plasticity and strength show a trade-off relationship. With the increase of the aging temperature, the plasticity first decreases and then increases. The above-mentioned changes in mechanical properties are closely related to the changes in the morphology and distribution of κ-carbides during the aging process. In general, the precipitation of κ-carbides is beneficial to obtain higher strength, but has a detrimental effect on plasticity.

1. Introduction

Fe-Mn-Al-C steel has a great potential for weight reduction due to the addition of Al and C element, and at the same time, it has an excellent combination of strength and plasticity, which gives it a great application potential in the automobile lightweight design [1-5]. In addition, when the added Al content is low, the deformation mechanism of Fe-Mn-C-Al steel is generally twin induced plasticity (TWIP) or transformation induced plasticity (TRIP) [6-8], and a higher Al content may change the deformation mechanism.

The addition of Al and C also promotes the precipitation of κ-carbides in Fe-Mn-Al-C steel, which affects the mechanical properties. According to the result of Chen et al. [9], κ-carbides can precipitate only when the Al content is higher than 7 wt.% and the C content is higher than 0.8 wt.%. According to the location of precipitation, κ-carbides can be divided into two categories (intragranular κ-carbides and intergranular κ-carbides), which have different effects on mechanical properties [9,10]. Chih et al [10]. studied the microstructure and mechanical properties of Fe-30Mn-8.5Al-2.0C after aging at 550 °C for 0~7 h, and found that with the increase of the aging time, intragranular κ-carbides precipitated first and then transformed to intergranular κ-carbides. Feng et al [11]. studied the impact toughness and tensile properties of Fe-24.8Mn-7.3Al-1.2C aged at 550 °C for 0~13 h, and found that the best combination of impact toughness and strength could be obtained after aging for 1 h. Most of the previous studies have studied the precipitation process of κ-carbides and its effect on mechanical properties by aging at a certain temperature for different times. However, few studies have paid attention to the effect of aging temperature on the precipitation and mechanical properties, especially at higher aging temperatures. In this study, the effects of aging at different temperatures on the microstructure and mechanical behavior of Fe-25Mn-11Al-1C steel will be studied.

2. Experimental

The nominal chemical composition of the experimental steel is Fe-25Mn-11Al-1.0C. The ingot is produced via a 50 Kg vacuum melting furnace. The smelted ingot was forged at 1100 °C to form a φ15 bar, then the solution treatment (ST) was carried out at 1050 °C for 2 h to homogenize the structure, and then the solid solution sample was aged at 450~650 °C for 4 h.

The tensile test was carried out on the Zwick/Roell tensile universal testing machine, and the strain rate was 10^{-3} s^{-1}. The tensile specimen gauge has a dimension with a length in 4 mm, a width in 1.5 mm, and a thickness in 0.8 mm.

The microstructure was investigated by X-ray Diffraction (XRD), Scanning Electron Microscopy (SEM), Electron Back-Scattered Diffraction (EBSD) and Transmission Electron Microscopy (TEM). The specimens for SEM and EBSD were mechanically polished and then electropolished in a 1:9 HCLO4:C2H6O solution. The polishing voltage was 15 V, the current was 1 A, and the polishing time was 10 s. The XRD analysis was carried out using a D/max 2500 diffractometer. A Cu-α target was used for the XRD test and the step scan mode was used, with a spacing interval of 2 s at each 0.02° step in a 2θ from 40° to 80°. The working voltage is 40 kV and the working current is 200 mA. Thin foil specimens for the TEM observations were prepared by the twin-jet electro-polishing in a 1:9 HCLO4:C2H6O solution at room temperature.

3. Results and discussion

3.1. Microstructure

The XRD result and EBSD map of the sample after solution treatment are shown in Figure 1. The XRD result shows that only the diffraction peaks of γ and α. The EBSD map shows that the α is distributed on the γ grain boundaries and that its content after the solution treatment is 7.3%.

Actually, the experimental steels with high Al and C content can form κ-carbides through spinodal decomposition during quenching in γ [12,13]. However, the κ-carbides formed by spinodal decomposition are extremely fine and diffuse, so they are difficult to detect by XRD and EBSD.

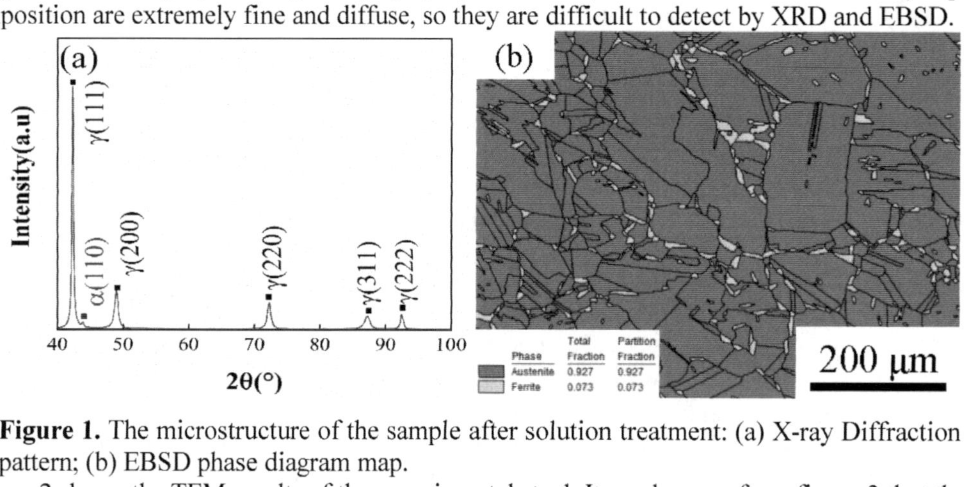

Figure 1. The microstructure of the sample after solution treatment: (a) X-ray Diffraction pattern; (b) EBSD phase diagram map.

Figure 2 shows the TEM results of the experimental steel. It can be seen from figure 3 that there are dispersed and periodically distributed granular precipitates in the austenite matrix. The selected area electron diffraction confirms that the precipitates are κ-carbides. Therefore, the microstructure after solution treatment should be composed of γ, α and nano-sized intragranular κ-carbides.

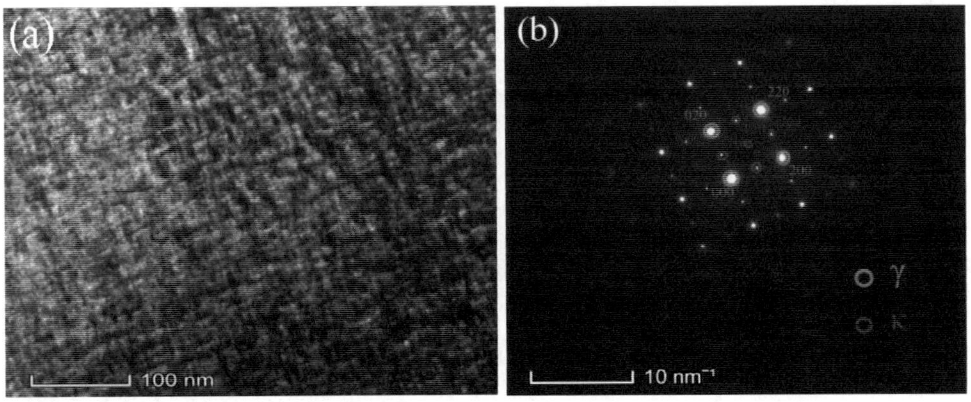

Figure 2. TEM images of the sample after solution treatment: (a) bright field image; (b) selected area electron diffraction.

The SEM results of specimens after solution treatment and aging at 450~650 °C for 4 h are shown in figure 3. No obvious precipitation of κ-carbides was observed in the microstructure after solution treatment and aging at 450 °C for 4 h, as shown in figure 3a and 3b. When the temperature increases to 550 °C, the granular precipitates can be seen on the grain boundary, as shown in figure 3c and 3d. When the temperature increases to 650 °C, the precipitates on the grain boundaries changed from granular to alternating light and dark lamellar structures, and grew into austenite grains, as show in figure 3e and 3f.

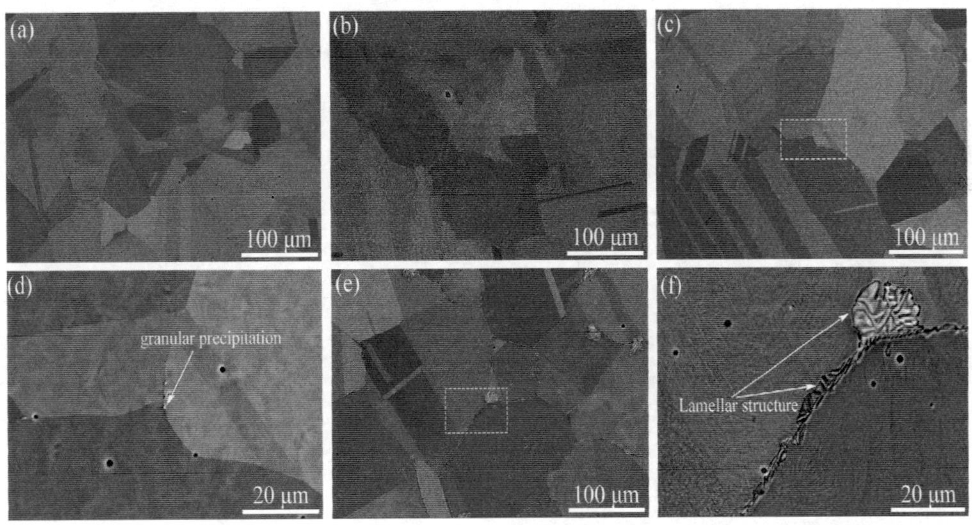

Figure 3. Microstructure after solution treatment and aging at different temperatures for 4 h: (a) solution treatment; (b) 450 °C; (c) 550 °C; (d) enlarged view of the framed are in (c); (e) 650 °C; (f) enlarged view of the framed are in (e)

From the SEM results in figure 3, the precipitates in the aging process have two forms: granular and lamellar. Therefore, to clarify the types of precipitates, the samples after aging were analyzed by the X-ray diffraction, and the results are shown in figure 4. It can be seen that only diffraction peaks of γ and α are observed after solution treatment and aging at 450 °C; when the aging temperature increases to 550 °C, a weak κ-carbides diffraction peak appears, indicating that the granular precipitates on the austenite grain boundaries are κ-carbides. With the further increase of the aging temperature to 650 °C, the diffraction peaks of κ-carbides and α are significantly enhanced at the same time, which is due to the eutectoid reaction of $\gamma \rightarrow \alpha + \kappa$ [14]. The κ-carbides and α transformed from the eutectoid distribute on the austenite grain boundaries in a lamellar structure, which is consistent with the lamellar structure observed in figure 3e and 3f.

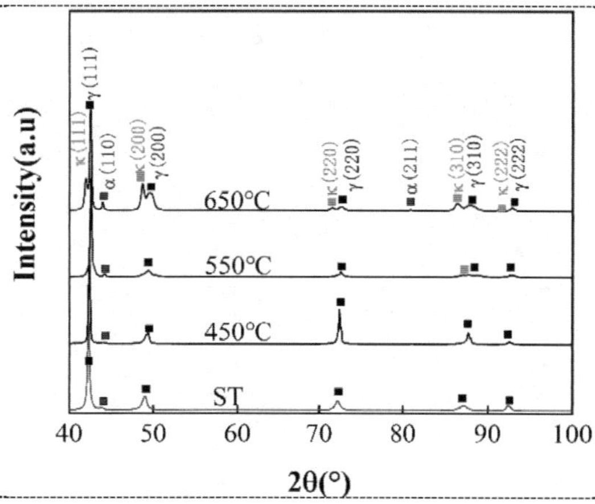

Figure 4. X-ray diffraction patterns of samples aged at different temperatures (for the convenience of comparison, the X-ray diffraction patterns of the sample after solution treatment are also added to the figure).

3.2. Mechanical properties

The mechanical properties are related to the deformation mechanisms. The addition of 11 wt.% Al to the experimental steel reduces the density by about 17.8% while increasing the stacking fault energy to 85.6 mJ/m^2. Research shows that when the stacking fault energy is higher than 45 mJ/m^2, the deformation mechanism is the dislocation slip [15]. Figure 5 shows the EBSD results near the tensile fracture of the sample after solution treatment. It can be seen from figure 5a that the grains are elongated along the tensile direction. From the phase distribution map of the deformed structure, no martensite transformation occurs after deformation, as shown in figure 5b. At the same time, it can be seen from figure 5c, the grain boundary map, no deformation twins are observed. Therefore, it is confirmed that the deformation mechanism of the experimental steel is the dislocation slip. While, the detailed evolution processes of deformation microstructure and mechanisms need to be investigated by advanced in-situ mechanical testing in electron microscopes systematically [16-18].

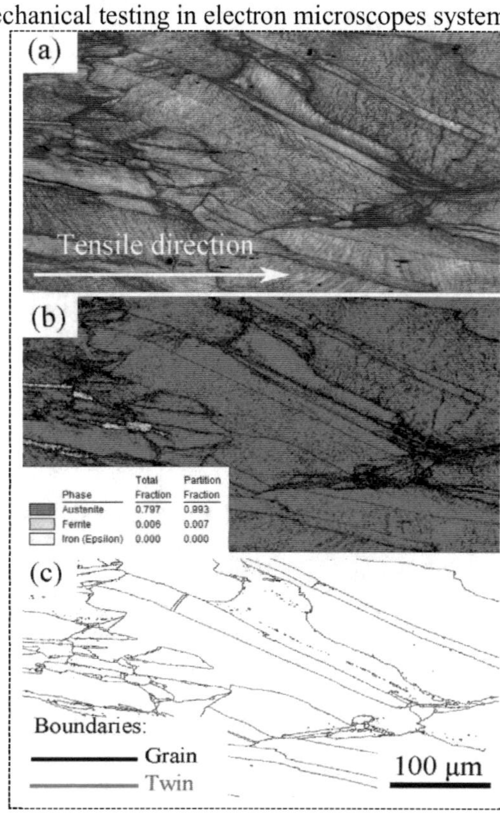

Figure 5. EBSD images of the deformation microstructure near the tensile fracture surface: (a) image quality map; (b) phase distribution map; (c) grain boundary map.

Figure 6 and Table 1 show the tensile properties after solution treatment and annealing at different temperatures. It can be seen from figure 6 that the sample after solution treatment has a large elongation, but its strength is worse than those of any aged samples in this study. Compared with the samples after solution treatment, the yield strength and tensile strength after aging at 450 °C are increased by 152 MPa and 94 MPa, respectively, and the total elongation is almost the same. This is mainly due to the further precipitation of intragranular κ-carbides. However, the size of κ-carbides is still small and cannot be observed by SEM.

When the aging temperature increases to 550 °C, the strength is further increased, but the elongation decreased sharply, which is related to the further growth of intragranular κ-carbides and the precipitation of granular κ-carbides on the grain boundaries. The growth of intragranular κ-carbides reduces the work hardening rate and thus the ductility. When the aging temperature further increases from 550 °C to 650 °C, the transformation of $\gamma \rightarrow \alpha + \kappa$ occurs at the grain boundary, and the content of κ-carbides is further increased. Interestingly, the increase in the κ-carbides content after aging at 650 °C results in a decrease in strength and an increase in elongation. According to the research of Han et al. [19], both intragranular and intergranular κ-carbides are enriched in C. This suggests that the partition of C to intergranular κ-carbides lead to the dissolution of intragranular κ-carbide in the C-depleted austenite. The dissolution of intragranular κ-carbides results in a reduction in strength, especially a rapid decrease in yield strength. In addition, the dissolution of the intragranular κ-carbides leads to the recovery of the work hardening ability, so that the tensile elongation is also recovered, and the elongation is improved.

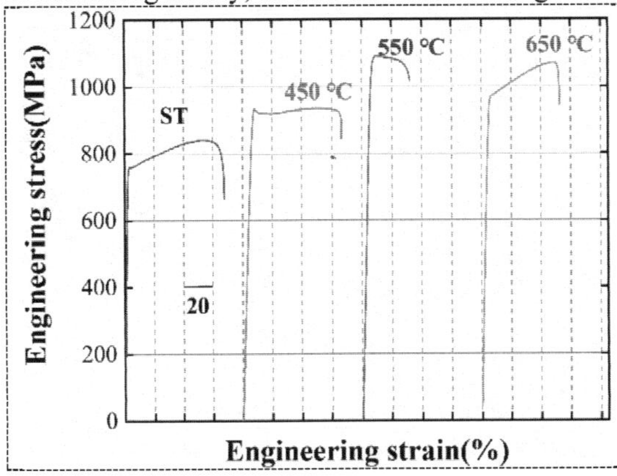

Figure 6. Engineering stress-strain curves after solution treatment and aging at different temperatures.

Table 1. Tensile mechanical properties with different aging temperatures

Sample	Yield Strength (MPa)	Tensile strength (MPa)	Total elongation (MPa)	TS × TE (MPa·%)
ST	726	841	65.2	54833
450 °C	878	935	64.1	59934
550 °C	997	1091	29.0	31639
650 °C	909	1067	49.5	52817

4. Summary

In this study, the Fe-25Mn-11Al-1.0C low-density steel with a weight loss of 17.8% was successfully prepared, and its mechanical properties were optimized by aging at 450~650 °C for 4 h. The main findings are as follows:

- With the increase of the aging temperature, the position and morphology of κ-carbides change, from the intragranular precipitation (450 °C) to the granular precipitation on grain boundaries (550 °C) and further to the κ/α lamellar precipitation on grain boundaries (650 °C).

- Intragranular κ-carbides can significantly increase the strength maintaining excellent plasticity (450 °C); when κ-carbides precipitate in the granular form on the grain boundaries, the strength is further increased, but the plasticity is seriously decreased (550 °C); when the κ-carbide and α are precipitated on the grain boundary in the lamellar structure of κ/α, the strength decreases compared with that after aging at 550 °C, but the plasticity increases significantly. The samples aged at 450 °C for 4 h have the best comprehensive mechanical properties.

Acknowledgments
The authors gratefully acknowledge the support from the National Natural Science Foundation of China (Grant No. 51871194) and the National Natural Foundation of Hebei Province, China (Grant No. E2021203011, 226Z1003G). X. Zhang acknowledges the support from the European Research Council under the European Union Horizon 2020 research and innovation program (grant agreement No 788567-M4D).

References
[1] Hansoo K, Dong-Woo S and Nack J K 2013 *Sci. Technol. Adv. Mater.* **14** 1-9
[2] Sang-Heon K, Hansoo K and Nack J K 2015 *Nature* **518** 77-9
[3] Frommeyer G and Brüx U 2006 Steel *Res. Int.* **77** 627-33
[4] Howell R A and Van Aken D C 2009 *Iron Steel Technol.* **6** 193-212
[5] Joonoh M, Seong-Jun P, Jae Hoon J, Tae-Ho L, Chang-Hoon L, Hyun-Uk H, Heung N H, Jaeeun Lee, Bong Ho Lee and Changhee L 2018 *Acta Mater.* **147** 226-35
[6] Wang Y, Zhang Y, Godfrey A, Kang J, Peng Y, Wang T, Hansen N and Huang X 2021 *Commun. Mater.* 2 (1)
[7] Wang Y, Kang J, Peng Y, Wang T, Hansen N and Huang X 2018 *J. Mater. Sci. Technol.* **34** (10) 1939-43
[8] Wang Y H, Kang J M, Peng Y, Wang T S, Hansen N and Huang X 2018 *Scripta Mater.* **155** 41-5
[9] Chen S P, Rana R, Haldar A and Ray R K 2017 *Prog. Mater. Sci.* **89** 345-91
[10] Lin C L, Chao C G, Bor H Y and Liu T F 2010 *Mater. Tran.* 51 1084-8
[11] Feng Y F, Song R, Pei Z Z, Song R F and Dou G U 2018 *Met. Mater. Int.* **24** 1012-23
[12] Zhang J L, Raabe D and Tasan C C 2017 *Acta Mater.* **141** 374-87
[13] Sung-Won P, Jun Young K, Kyong M C, Jae H J, Seong-Jun P, Joonoh M, Tae-Ho L and Jong-Ho S 2018 *Met. Mater. Int.* **25** 683-96
[14] Han D, Ding H and Liu D 2020 *Mater. Sci. Technol.* **36** 681-9
[15] Li G Y, Wang Y H, Chen X M, Xiong J C, Tian Y Z, Cao W Q and Wang T S 2020 *Heat Treat. Surf. Eng.* **2** 16-21
[16] Zhang X, Godfrey A, Winther G, Hansen N and Huang X 2012 *Mater. Charact.* **70** 21-7
[17] Zhang X, Godfrey A, Winther G, Hansen N and Huang X 2010 *Proc., 31st Risø int., symp., mater., sci.* 489-96
[18] Pan H, He Y and Zhang X 2021 *Materials* **14** 47
[19] Han D, Ding H and Liu D 2020 *Mater. Sci. Technol.* **36** 681-9

3D In-situ Stop Action Study of Recrystallization in Additively Manufactured 316L Stainless Steel: Reconstruction Optimization and Observations

H Liu[1], Y Zhang[2], M Wilkin[2], J-S Park[3], P Kenesei[3], A D Rollett[2] and R M Suter[1]

[1] Department of Physics, Carnegie Mellon University, Pittsburgh, PA 15213, USA
[2] Department of Materials Science and Engineering, Carnegie Mellon University, Pittsburgh, PA 15213, USA
[3] X-ray Science Division, Argonne National Laboratory, Lemont, IL 60439 USA

E-mail: suter@andrew.cmu.edu

Abstract. A volume of an additively manufactured 316L stainless steel sample has been tracked during its recrystallization using near- and far-field High Energy Diffraction Microscopy (HEDM) and absorption tomography at Advanced Photon Source beamline 1-ID. A near-field compatible *in situ* furnace allows monitoring of Bragg diffraction signals as they evolve out of a weak and diffuse background while the sample temperature is $\approx$ 1250° C. The sample is rapidly cooled to room temperature after observation of significant signal evolution and $\sim 0.035\,\mathrm{mm}^3$ is mapped by the near-field method. Four cycles of heat treatment follow the structure from a state of small, isolated grains through impingement of domains to near completion of recyrstallization. Here, the experiment and reconstructions are described, and recrystallized fractions, twin domains, and distributions of grain boundary types are discussed.

1. Introduction

Additive manufacturing (AM) facilitates the rapid manufacture of precision parts in small numbers and made with unusual materials with a drastically abbreviated design-to-production process. Despite the obvious advantages of AM, there are many open questions that deserve attention from the research community. For example, the laser melting and subsequent re-solidification occur orders of magnitude faster than in conventional materials processing and this changes many aspects of the processing-microstructure-properties-performance relationships. [1, 2] Standard aids for processing such as time-temperature-transformation diagrams are no longer valid. Different grain structures, segregation patterns and precipitate structures are obtained, all of which change the material properties and, especially with metals, how the material responds to subsequent heat treatment. [3] Stored defects and consequent stresses can lead to degradation of mechanical properties. Post-build thermal processes have been developed to relieve these stresses. [4]

Here, the heat treatment of a laser powder bed AM sample of 316L stainless steel is investigated. Recrystallization is observed in real time as the sample is heated to $\approx$ 1250°C while mounted for High Energy Diffraction Microscopy (HEDM) measurements at the 1-ID beamline of the Advanced Photon Source (APS) at Argonne National Laboratory. Near-field

HEDM is used to map recrystallized grain unit cell orientations in a measured volume of $V_m = 0.5 \times 0.5 \times 0.14$ mm^3 = 0.035 mm^3. Over four annealing cycles, the sample evolves from a sparse distribution of small, almost randomly oriented, isolated grains to being almost fully recrystallized and filling the measured volume with well ordered grains. The material maintains its face centered cubic crystal structure and this gives rise to extensive twinning with twin-related domains containing up to 20 individual crystals. The final structure is dominated by domains that arose in the first annealing cycle resulting in at most weak recrystallization texture.

2. Procedures

A nf-HEDM compatible *in situ* furnace, developed for these measurements, is shown in figure 1. The system is similar to one currently in use at CHESS. [5] Infrared radiation from two lamps is focused on the sample position and allows rapid heating and cooling. There is no mechanical contact between the furnace and sample. Forming gas (97% N_2, 3%H_2) circulates through the sample chamber to prevent oxidation and cooling water keeps the furnace body close to ambient temperature. To monitor the progress of recrystallization, repeated measurements of near-field diffraction signals integrated over the same 30° interval are carried out with the detector placed as shown. This brings the scintillator to ~ 12 mm from the sample rotation axis. Any illuminated crystalline grain should contribute diffraction in this rotation range. Upon observing subtle but significant changes in the signal, the furnace is turned off and removed and room temperature HEDM and tomography measurements are performed. The sample volume of interest was measured in four states referred to as annealing cycles 1 through 4. Annealing between each state used $T = 1250$°C which was found to generate dynamics that were visible on detector images within a few minutes.

Figure 1: The fully instrumented *in situ* furnace installed at the 1-ID beamline at the APS. The aluminum bar at right holds the furnace and is connected to XYZ translations stages that position the furnace over the sample or remove it for room temperature measurements.

Figure 2 shows the evolution of near-field detector signal over the first three annealing cycles. Initially, from the as-built material, the signal is weak and dispersed with few if any Bragg-like peaks. This indicates that the material is in a deformed state (EBSD measurements are being performed on as built material). In successive annealed states, sharp peaks emerge and grow in size and number.

Microstructure cross-sections are obtained from 1440 images collected at two rotation-axis-to-detector distances (deduced to be $L = 5.614$ and 7.614 mm) integrated over $\delta\omega = 0.25°$ intervals and spanning 180° sample rotation. Standard image analysis is performed on the detector images; this subtracts a median background image which largely eliminates broad, signals from unrecrystallized material, and then extracts sharp signals using the Laplacian of Gaussian "blob" finding algorithm. [6]

Reconstructions of recrystallized grains used the HEXOMAP package (available here on GitHub). This is a python/pyCUDA based GPU accelerated code based on the forward modeling C++ routines from well proven previous generation code. [7, 8] In contrast to the old code, reconstructions are performed on a square array of sample *points* rather than triangular area elements; this makes reconstructions more efficient (fewer redundant projections) and makes importing deduced structures into graphics and modeling codes considerably more straightforward. The volume associated with each reconstruction point is still referred to as a

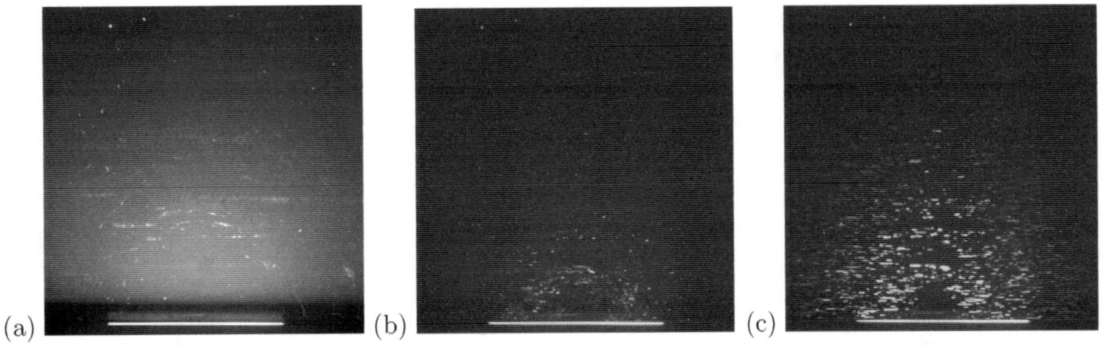

Figure 2: nf-HEDM signal evolution with annealing treatment. (a) Raw image from as-built material with no background subtraction. (b) Background subtracted, integrated image collected while the sample rotated through 30° in the cycle 1 state. (c) Same as (b) but after annealing cycle 3. The clear emergence of sharp Bragg diffraction spots is associated with the growth of recrystallized domains. The bright horizontal line at bottom is the attenuated incident beam. The detector has a 3×3 mm field-of-view.

"voxel;" in the current case, the point mesh has spacings $2 \times 2 \times 4$ μm^3, with the last value being the spacing between layer measurements. For a given unit cell orientation and sample position or voxel, the forward model code projects diffracted beams to a simulated detector space and looks for the orientation that generates maximal overlap with the experimental binarized detector pattern. The overlap is measured by a confidence or completeness parameter, C, which is the fraction of simulated Bragg peaks that strike experimentally observed intensity. The ease of code modification and extension in the python environment allows development of special purpose routines; two such routines were developed in this work and are available for general use:

Boundary position optimization. To accelerate reconstructions, a "flood fill" algorithm is used in which a found orientation is tried as an initial guess for neighboring voxels to avoid *ab initio* searches over the entire zone of possible orientations for each voxel. The found orientation (with local Monte Carlo tweaking) can fill in an entire grain cross-section. However, this has the danger of dilating some grains at the expense of neighbors. To address this, a post-processing algorithm tests all orientations in the neighborhood of a boundary and switches orientations if C increases. Iterations continue until no further changes are made. The small number of test orientations makes this optimization fast.

Twin checking. Small twinned regions generate diffraction signals that are small and potentially weak; more extended parent generated signals may extend into the domain of the twin. By testing all twin variant orientations at each voxel location, weak signals may be found in the detector data. Any increase in C or reduction to within a given threshold can be kept as an existing twin. Again, the small number of test orientations makes this fast. This algorithm was applied here for the FCC twinning case but additional twin variant sets can be generated easily for other cases.

3. Results

Figure 3 shows the evolution of microstructure in a single layer; this layer is in the center of the measured volume which implies that most of the observed grains occur in neighboring layer measurements as well. Close inspection reveals that many of the grains that appear in (a) persist and grow in later cycles. In later states, there are additional small grains that appear and

IOP Conf. Series: Materials Science and Engineering 1249 (2022) 012054 doi:10.1088/1757-899X/1249/1/012054

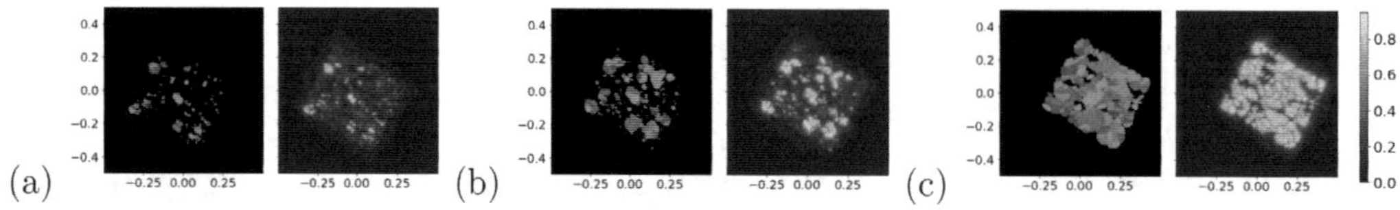

Figure 3: A single cross-section of microstructure after (a) Cycle 1, (b) Cycle 2, (c) Cycle 3 *in situ* annealing treatments. Scale markings are in millimeters. This is layer 17 of the 33 measured. In each case, the color coded orientation map (colors scaled to RGB weights from Rodrigues vector components) is on the left and the confidence parameter, $\mathcal{C}$, is on the right. Note that these images have not been mutually registered, so one can see that the sample did not always return to precisely the same position after heating. The vertical positions of layer measurements were referenced to the top of the sample after each annealing, so should be repeatable to ~ 1 μm.

there are grains that appeared first in other cycles and grew into the layer shown. In addition, some grains that appear in one state are consumed by others through the coarsening process. The maps of $\mathcal{C}$ show the clear contrast between recrystallized ($\mathcal{C} > 0.4$) and unrecrystallized material ($\mathcal{C} < 0.4$). Orientation maps are thresholded to show only voxels with $\mathcal{C} > 0.4$. The smallest grains have reduced values of $\mathcal{C}$ compared to larger grains due to the reduced amount of crystal volume that projects into individual detector pixels and consequent reduced peak intensities, reduced precision in peak shape determination, and more simulated peaks that miss the segmented peaks in the data set. Note, on the other hand, the large dynamic range of the measurement: the smallest grain cross-sections are ~ 16 μm^2 and the largest are $\sim 2 \times 10^4$ μm^2.

The recrystallized grains show resolution limited intra-granular orientation variations, i.e., < 0.1 degree disorientations indicating a high degree of crystalline order and low defect density as observed in recrystallization of more conventionally deformed materials.

Before addressing the details of observed microstructure, the elucidation of which is the motivation for this work, what do the data say about global aspects of recrystallization dynamics? Table 1 shows the progress of recrystallization over the four annealing cycles including anneal times and the fraction of recrystallized volume, $X(t)$, calculated as the measured volume with $\mathcal{C} > 0.35$ in each state divided by the total measured volume, V_m. The time is the duration in seconds of annealing at $T_a = 1250°\text{C}$. The Johnson-Mehl-Avrami-Kolmogorov equation [9], which can be written as

Table 1: Recrystallized volume fractions.

	Time (s)	$X(t)$
Cycle 1	57	0.061
Cycle 2	53	0.201
Cycle 3	150	0.474
Cycle 4*	306	0.915

*Cycle 4 annealing and measurement were performed separately from others and involved removing and remounting the sample.

$$X(t) = 1 - e^{-(t/\tau)^n}, \tag{1}$$

with n a dimensionless exponent expected to be close to three in three dimensions and with nucleation from random sites, [9] and τ is a characteristic time scale where $X(\tau) = 1 - e^{-1} = 0.632$. This equation was proposed as a means to capture two phases of the progression of recrystallization: first, grains nucleate and grow quickly into unrecrystallized deformed material driven by the energy release in crystallization; when domains impinge on each other, growth slows since a large fraction of the advancing fronts are pinned against previously recrystallized material. Of course, the newly formed grain boundaries can continue to migrate to reduce boundary energy, but this has no effect on X. If a simple equation of the form of (1) can be validated and shown to be general, it could be useful for process design in a wide range of contexts. Measurements such as this one present an opportunity to test such expressions.

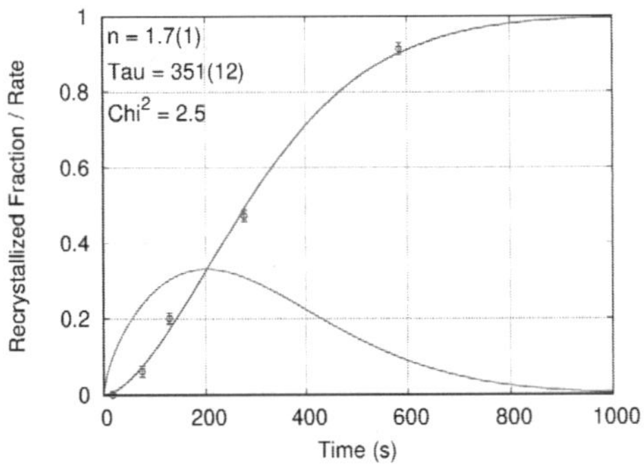

Figure 4: Non-linear least squares fit of the data in Table 1 to (1). The red curve is 150 times the time derivative of (1), giving the recrystallization rate.

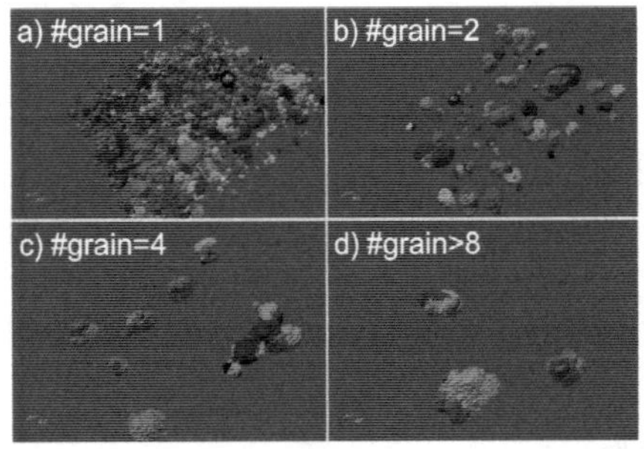

Figure 5: Twin related domains (TRDs) present after annealing cycle 2. Each panel shows only those TRDs composed of (a) 1, (b) 2, (c) 4, and (d) 8 or more twin related grains.

Figure 4 shows a weighted non-linear least squares fit [10] to (1) which results in $n = 1.7(1), \tau = 351(12)$ s, with "(...)" being the uncertainty in the previous digit, and $\chi^2 = 2.5$. Error bars on the X were approximated from sensitivities to variations in the C threshold (strictly, such variations give rise to systematic uncertainty rather than the desired random measurement noise). While the exponent is almost consistent with $n = 2$, the result is clearly inconsistent with $n = 3$. However, as an empirical representation of the data the result is quite good. The rate curve in figure 4 shows the expected behavior with zero rate at $t = 0$, as initial nuclei begin to yield growing crystals, followed by rapid growth and a long tail of slower growth as completion is approached. This fitting exercise is demonstrative but is not robust enough to rigorously support or refute the validity of (1): The values of n and τ have a substantial correlation coefficient of -0.3 and, with only four data points, the data do not clearly define the shape of $X(t)$. Further, while this fit included a data point at $X(18\text{ s}) = 0$ (the result of a brief test heating), the definition of $t = 0$ is unclear and may have a strong effect on the deduced exponent. The reduced value of n may reflect the complexity of growth paths seen in [11] .

Now turning to the grain scale, figure 5 shows three dimensional views of twin related domains (TRD) after two cycles of annealing. No smoothing has been applied to the grain shapes, so the voxels are shown as $2 \times 2 \times 4$ μm^3 boxes which correspond to the reconstruction point spacings. Most grains are not twinned ($\#grain = 1$) but many TRDs contain multiple twins and these numbers grow with further annealing. By cycle 3, the V_m is largely filled with TRDs; domains have as many as 20 twin related grain members, many of which necessarily have the same orientation (given only four twin variant choices) but are not in contact with each other. Multiple examples can be seen in figure 5 in which TRDs contain three and four distinct orientations, some with disconnected grains with the same orientation.

Figure 6 shows the fractional population of different classes of grain boundaries in each measured annealing cycle. Each class is specified by the three parameter disorientation between adjoining grains, so all boundary orientations (relative to the adjoining crystal unit cells) are included in each class. The large fraction of $\Sigma3$ boundaries (red) (disorientation of 60° about a $\langle 111 \rangle$ axis) reflects the extent of twinning in the recrystallization microstructure of this FCC material; this single disorientation represents about 40% of all boundaries. The $\Sigma9$ boundaries (orange) occur at intersections at triple lines involving two twin boundaries. A small number of general boundaries are found within the TRDs (green), reflecting intersections involving at least one non-$\{111\}$ surface. Just over 60% of disorientations are general boundary types that form through impingement between independent TRDs (red) which have essentially random orientations.

4. Conclusion and future work

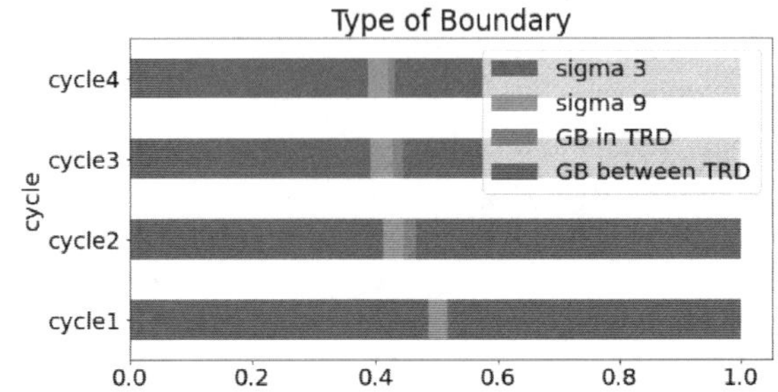

Figure 6: Fractions of grain boundary classes during recrystallization of AM 316L stainless steel.

This brief article has focused on technical developments that were necessary to perform one of the first *in situ* grain mapped measurements in three dimensions of recrystallization and on two examples of unique information extracted. Improvements have been made in the algorithms used to optimize reconstructions of orientation maps making them both fast and accurate and more sensitive to subtle features such as crystallographic twins. [12] It is only with nf-HEDM and analogous high energy x-ray methods that it is possible to gather the statistics to image evolution of material as in figure 5. The procedures applied here can be used in future studies of microstructure evolution processes.

Further studies are needed to determine the generality of the above observations. It would be interesting to compare recrystallization of AM processed 316L to conventional roll-deformed metals, including 316L, other steels and beyond. For example, in this work it is seen that grains that grew out of deformed material in annealing cycle 1 come to dominate the final microstructure. Grains that appear later tend to be consumed due to their small size by the larger cycle 1 grains. Presumably, by controlling the density of cycle 1 grains, one may be able to tailor the final microstructure to be more like that of conventional wrought material which has a smaller grain size than that seen here.

Acknowledgments

The authors thank F. Wagner for informative discussions. This work was supported by the U.S. Department of Energy under NNSA Award DE-NA0003915. This research used resources of the Advanced Photon Source, a U.S. Department of Energy Office of Science user facility at Argonne National Laboratory and is based on research supported by the U.S. DOE Office of Science-Basic Energy Sciences, under contract DE-AC02-06CH11357.

References

[1] Oh S A, Lim R E, Aroh J W, Chuang A C, Gould B J, Bernier J V, Parab N, Sun T, Suter R M and Rollett A D 2021 *JOM* **73** 212–22
[2] Oh S A, Lim R E, Aroh J W, Chuang A C, Gould B J, Amin-Ahmadi B, Bernier J V, Sun T, Pistorius P C, Suter R M and Rollett A D 2021 *Mat. Res. Lett.* **9** 429–36
[3] Zhang F, Levine L E, Allen A J, Stoudt M R, Lindwall G, Lass E A, Williams M E, Idell Y and Campbell C E 2018 *Acta Mat.* **152** 200–14
[4] Pokharel R, Patra A, Brown D W, Clausen B, Vogel S C and Gray G T 2019 *I. J. Plast.* **121** 201–17
[5] Pagan D C, Bernier J V, Dale D, Ko J Y P, Turner T J, Blank B and Shade P A 2018 *Scripta Mat.* **142** 96–100
[6] Lind J, Rollett A, Pokharel R, Hefferan C, Li S and U Lienert R S 2014 IEEE Int. Conf. Image Proc. pp 4877–81
[7] Li S and Suter R 2013 *J. Appl. Cryst.* **46** 512–24
[8] Menasche D, Shade P and Suter R 2020 *J. Appl. Cryst.* **53** 107–16
[9] Humphreys J, Rohrer G and Rollett A 2017 *Recrystallization and Related Annealing Phenomena, Third Edition* (Amsterdam, Netherlands: Elsevier)
[10] Bevington D K and Robinson P R 1992 *Data Reduction and Error Analysis for the Physical Sciences* 2nd ed (McGraw-Hill) ISBN 978-0072472271
[11] Lauridsen E, Poulsen H, Nielsen S and Jensen D J 2003 *Acta Mat.* **51** 4423–35
[12] For HEXOMAP and associated analysis codes, much of which was developed by co-author He Liu for this work, see https://github.com/HeLiuCMU/HEXOMAP.

Heterogeneous microstructure and failure analysis of yaw gear rings

Xiaodan Zhang, Niels Leergaard Pedersen, Peder Klit and Dorte Juul Jensen

Department of Civil and Mechanical Engineering, Technical University of Denmark, Lyngby, Denmark

E-mail: xzha@mek.dtu.dk

Abstract. The microstructure and hardness of two cracked and failed steel (42CrMo4) teeth from two different yaw rings have been investigated. It is shown that the surface quenching and tempering treatments introduce heterogeneous microstructures with a low-temperature tempered hard martensite surface layer and a high-temperature tempered matrix. This is common to both teeth, however, the gradient microstructure and hardness are different for these two yaw rings. It is suggested that this difference in gradient strongly affects cracking and failure. The failure mechanisms for the two teeth are discussed based on the microstructure and hardness analysis.

1. Introduction

Decades of development have made Denmark a frontrunner in wind power – a position that is used to share experiences and help other countries embark on the same green journey [1]. In 2020, onshore and offshore wind turbines produced over 46 percent of Denmark's consumption of electricity [2]. However, an issue often overlooked is that gears and bearings in wind turbines (gear boxes, generators, and yaw systems) still typically fail much earlier than the desired 20 years lifetime. These gears and bearings are made of strong steels. The common methods to manufacture these strong steels include thermomechanical processes, such as casting plus heat treatments, and thermochemical processes, such as carburization [3] and nitridation [4, 5]. These processes produce a heterogeneous microstructure with a hard surface and a soft/ductile interior matrix, to acquire a combination of high wear resistance and good fatigue properties for heavy load-bearing components such as rollers [6-8], gears [9, 10] and bearings [3, 11].

In the present study, the microstructure and hardness of two 42CrMo4 cracked and failed teeth from two different yaw gear rings, which were used to rotate the nacelle at the top of the tower, are investigated. Based on this, the failure mechanisms for the two teeth are discussed.

2. Experimental

The 42CrMo4 steel (EN 10083-3-2006) has the nominal chemical composition as shown in table 1. It is a heat treatable steel which contains at least 0.9%Cr and 0.15%Mo as strengthening elements. After quenching and tempering, the material is characterized by high strength with a typical tensile strength of 900 -1200 MPa and good low-temperature impact toughness.

Table 1. Nominal chemical composition of 42CrMo4 steel (wt. %) [12]

C	Si	Mn	P	S	Cr	Mo
0.38 - 0.45	max 0.4	0.6 - 0.9	max 0.025	max 0.035	0.9 - 1.2	0.15 - 0.3

Two used 42CrMo4 yaw rings were received from a company for the present study, without details about the load conditions they had been exposed to. For both yaw rings, cracks were observed to have initiated from the contact areas between gear pairs; at the same time, the dark heating traces from both sides of the gear teeth can be observed, e.g. as seen in the right figure of figure 1a. Two teeth were selected for the present investigation, each of which being typical for that yaw ring : One, to be named the cracked tooth, has many large cracks but is still intact, while the other, to be named the failed tooth, has a serious fall-off as pitting [9], see figure 1.

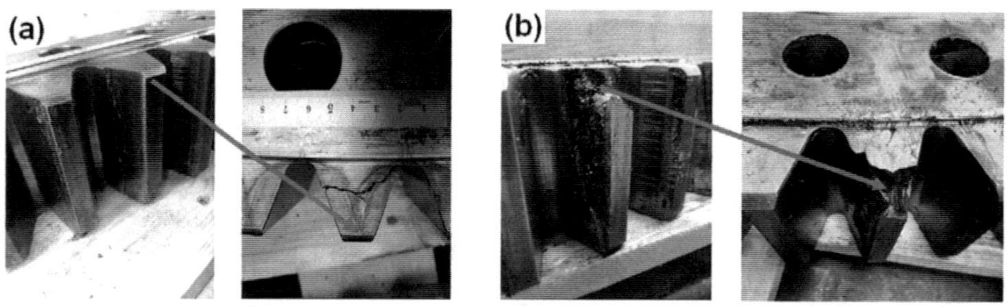

Figure 1. The two low yaw rings; (a) the selected cracked tooth and (b) the failed tooth.

To investigate effects of the heating, quenching and tempering processes on the microstructure and hardness of these two yaw gear rings, the two selected teeth was cut off from the yaw gear rings and sectioned into samples by strong water beams (figure 2). The sectioning revealed extended 3D crack networks in particular in the cracked tooth, which actually fell into pieces due to this crack network upon sectioning (see figure 2a).

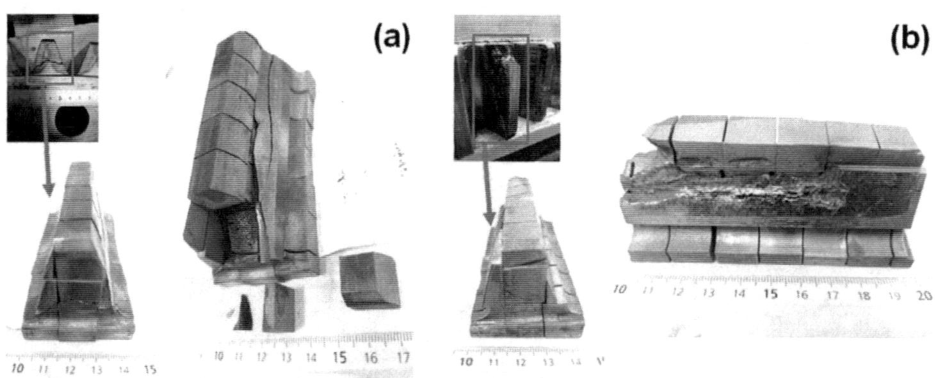

Figure 2. The selected teeth after sectioning: (a) the cracked tooth, and (b) the failed tooth.

The red-framed middle front blocks (figures 2 and 3) were taken from the same positions of both teeth, mounted into epoxy, ground and mechanical polished finally with oxide-polishing solutions (OPS) containing 25% Ammonia and 3% H2O2, to get specimens ready for microstructural characterization and hardness measurements. The microstructural characterization was performed using an optical microscope and a ZEISS Supra 35 scanning electron microscope. The Vickers hardness was performed with a load of 1 kg.

3. Results and discussion

3.1. Hardness

Figure 3 shows the hardness distribution in red-framed middle front blocks from both teeth. For the cracked tooth, the average Vickers hardness of the surface layer is 621 ± 14, while that of the matrix is 297 ± 19. The failed gear tooth has an average Vickers hardness of 600 ± 7 at the surface, while it is 264 ± 7 in the matrix. For both teeth, good correspondence between the surface heat treatment traces and the hardness profiles are observed.

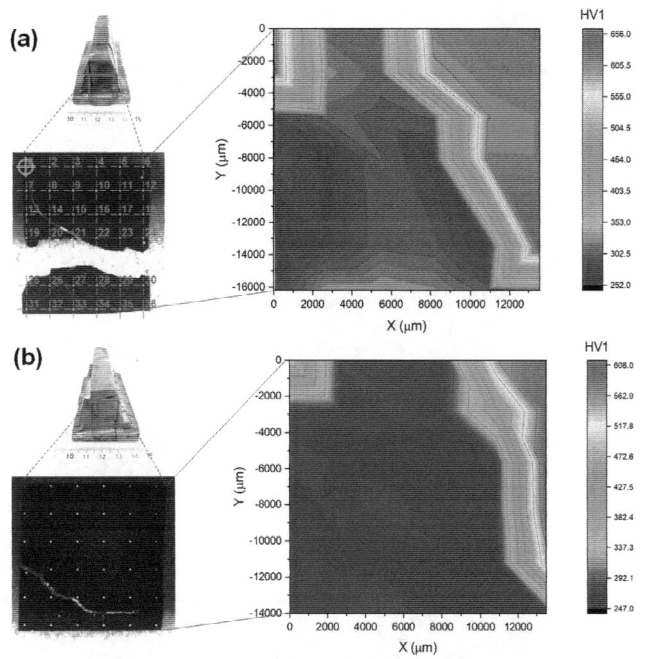

Figure 3. Vickers hardness HV1 distributions in red-framed middle front blocks from (a) the cracked, and (b) the failed tooth.

3.2. Microstructure

Figure 4 shows the averaged hardness profile and microstructure of the soft matrix, the transition zone and the hard surface, normal to the side surface of the cracked tooth. The hard surface layer thickness is around 6 mm. The microstructure in this hard surface layer is composed of thin plate martensite (also called low temperature-tempered martensite, LT-t M). The microstructure of the soft matrix is coarse, high temperature-tempered martensite (HT-t M) with carbide nanoparticles. The microstructure in the transition zone is composed of mixed LT-t M and HT-t M. The width of the transition zone is approximately 100 μm (which is not reflected in the hardness measurements as the step size for these measurements were chosen to be 270 μm).

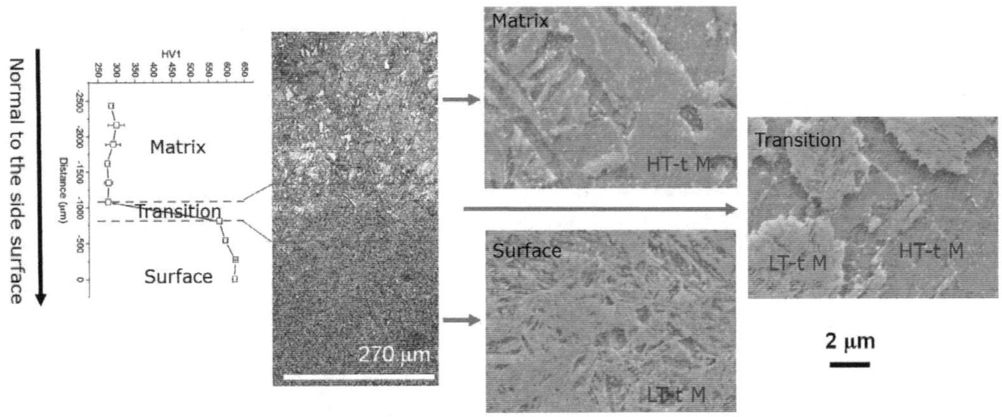

Figure 4. Hardness profile and microstructure of the cracked tooth inspected normal to the tooth's side surface.

Figure 5 shows the same data for the failed tooth. The microstructure in the hard surface layer is thin plate martensite with over-tempered areas. These areas are characterized by a large amount of carbide nanoparticles within a ferrite matrix, with a size comparable to those in the soft matrix. The hard surface layer thickness is around 4 mm. The soft matrix is composed of carbide nanoparticles within a ferrite matrix. The width of the transition zone is approximately 600 μm.

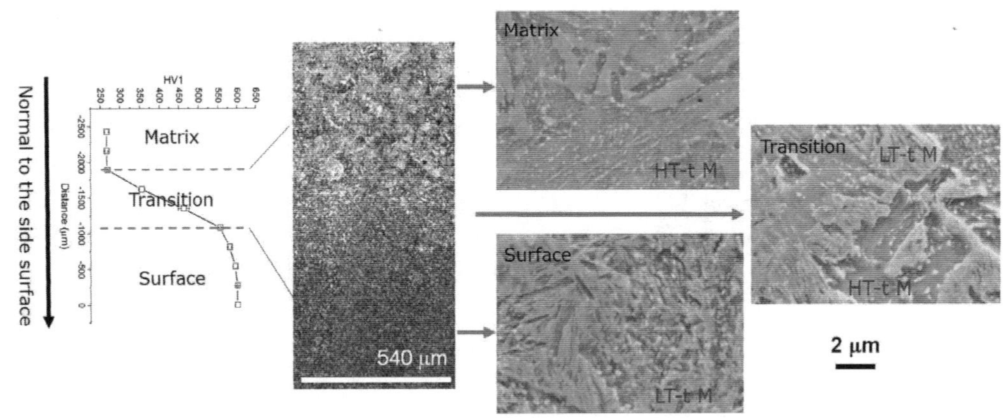

Figure 5. Hardness profile and microstructure of the failed tooth inspected normal to the tooth's side surface.

3.3. Comparison: similarities and differences

The surface quenching and tempering treatment introduce heterogeneous microstructures with a thick, millimetre scale, low-temperature tempered hard martensite surface layer and a high-temperature tempered soft inner matrix. This is common to both teeth. Cracks are seen at the contact areas between gear pairs, and scratches can also be observed in these areas for both the cracked and failed tooth.

The more detailed comparison of the two teeth however give information of critical importance for future gear material microstructure design for longer service life and sustainable development.

i) For the failed tooth, the lower surface hardness and the relatively thin surface layer with the over-tempered microstructure seem to reduce the wear properties. Crack initiation and propagation originate from the soft over-tempered layer with large carbides and ferrite;

ii) For the cracked tooth, the harder and thicker surface layer provides better wear resistance, while the thin transition zone, resulting from fast heating and quenching, is likely to introduce large internal stress, which stimulate development of an internal 3D crack network.

3.4. Failure mechanisms

As found in our previous research [9], cracks and micro pitting is mainly associated with plastic deformation of asperities on contact surfaces of the gear, which induce localized deformation bands. The cracking and micro pitting phenomena may be divided into three stages: running-in, steady progression and degradation. Typically, micro pitting initiates at around several thousands of cycles. In general terms, this is very similar to the observation in the present study, where we see that: i) cracks initiate from the gear pair contact areas; ii) cracks can develop into micro as well as macro pitting (seen in the failed tooth); iii) cracks may however not result in pitting but into development of an extended internal 3D crack network (seen in the cracked tooth).

The answer to why some cracks develop into pitting, while other lead to extended internal 3D crack networks, is suggested to relate to the characteristics of the hard surface layer and the transition zone:

i) when the hard layer is thin, has a relative low hardness and contains "weak areas" such as soft over-tempered martensite, cracks (originating from the surface) will propagate through these areas and develop into micro and macro pitting;

ii) when the surface hard layer is thick, has a relative high hardness and is microstructurally homogeneous, cracks may progress initially only into the transition zone and start to form an internal 3D crack network there. When the transition zone is thin, this evolution is likely to be assisted by tensile residual stresses. The residual stress and its effects are however for future research.

The above comparison indicates that a thick and microstructurally homogeneous hard surface layer, a reasonable wide transition zone between the hard surface layer and the soft matrix, and appropriate annealing to remove internal tensile residual stress may improve gear properties and extend its service life.

4. Conclusion

The microstructure and hardness of two 42CrMo4 teeth from two different yaw rings have been investigated by optical microscopy, scanning electron microscopy and hardness indentation. Surface quenching and tempering treatment are found to introduce heterogeneous microstructures with a thick, millimetre scale, low-temperature tempered hard martensite surface layer and a high-temperature tempered soft inner matrix. When the hard surface layer is thin, has a relative low hardness and contains "weak areas" such as soft over-tempered martensite, cracks may readily propagate through these areas and develop into micro and macro pitting. When the surface hard layer is thick, has a relative high hardness and has a homogeneous microstructure without 'soft spots', cracks tend to stay below the hard surface area and form a 3D crack network. The characteristics of the transition zone between the soft matrix and the hard surface layer is thus of importance, and it is suggested to investigate effects of residual stress in this transition zone, in particular when the zone is thin.

Acknowledgments

The authors wish to acknowledge the financial support from the Energy Technology Development and Demonstration Programme (EUDP 18-II DMDS) (grant No. 60418-0615).

References

[1] Denmark.dk, Denmark is a laboratory for green solutions. Available at: https://denmark.dk/innovation-and-design/green-solutions.
[2] State of green, A record year: Wind and solar supplied more than half of Denmark's electricity in 2020. Available at: https://stateofgreen.com/en/partners/state-of-green/news/a-record-year-wind-and-solar-supplied-more-than-half-of-denmarks-electricity-in-2020/.

[3] Cao Z, Liu T, Yu F, Cao W, Zhang X and Weng Y 2020 *Int. J. Fatigue* **131** 105351.
[4] Yang R, Wu GL, Zhang X, Fu WT and Huang X 2017 *IOP Conf. Ser.: Mater. Sci. Eng.* **219** 012047.
[5] Wang B, Liu B, Zhang X and J Gu 2018 *Surf. Coat. Technol.* **356** 89-95.
[6] Zhang X, Liu W, Sun D and Li Y 2007 *Metall. Mater. Trans. A* **38** 499–505.
[7] Zhang X, Godfrey A, Liu W and Liu Q 2009 *Metall. Mater. Trans. A* **40** 2171–77.
[8] Zhang X D, Liu W, Li Q L, Godfrey A and Liu Q 2010 *Mater. Sci. Technol.* **26** 1177–83.
[9] Mallipeddi D, Norell M, Naidu VMS, Zhang X, Näslund M and Nyborg L 2021 *Tribol. Int.* **156** 106820.
[10] Yang R, Wu G, Zhang X, Fu W, Hansen N and Huang X 2021 *Materials* **12** 3797.
[11] Danielsen H K, Gutiérrez Guzmán F, Fæster S, Shirani M, Rasmussen B H, Linzmayer M and Jacobs G 2022 *Wear* **494-495** 204230.
[12] http://www.steelnumber.com/en/steel_composition_eu.php?name_id=335.

IOP Conf. Series: Materials Science and Engineering 1249 (2022) 012056 doi:10.1088/1757-899X/1249/1/012056

A CPFEM based theoretical analysis of strains resolved by the microstructural feature tracking method

R Li[1*], R Quey[2], Y Zhang[1], M Kobayashi[3], J Oddershede[4] and D Juul Jensen[1]

[1] Department of Civil and Mechanical Engineering, Technical University of Denmark, 2800 Kgs. Lyngby, Denmark
[2] Mines Saint-Etienne, Univ Lyon, CNRS, UMR 5307 LGF, F - 42023 Saint-Etienne, France
[3] Department of Mechanical Engineering, Toyohashi University of Technology, Toyohashi, Aichi 441-8580, Japan
[4] Xnovo Technology ApS, DK-4600 Køge, Denmark

*E-mail: rulia@dtu.dk

Abstract. In this work, it is investigated how grain boundaries influence the local strain determination by the microstructural feature tracking (MFT) algorithm. In this method, tetrahedra are used as the strain calculation unit. We apply the MFT processing procedure on data obtained by a crystal plasticity finite element modeling (CPFEM) simulation to explore the uncertainties in the calculated strains caused by grain boundaries. Effects of tetrahedron types and radius ratios are discussed.

1. Introduction

The strain distribution that develops in polycrystalline materials during plastic deformation is important for understanding their mechanical properties. It has been demonstrated that the microstructural feature tracking (MFT) method is capable of mapping the internal local strain fields within particle-contained bulk metals [1,2], from which the relationships between microstructures and local deformation heterogeneities can be elucidated.

The accuracy of local strain determination based on the MFT method has already been evaluated in terms of particle characteristics in [3], but other factors are still worth investigating. A particular one is related to the spatial resolution of the method and how strong deformation heterogeneities caused by microstructure discontinuities, like grain boundaries, influence the estimation of grain-averaged strains. Previous studies of deformed microstructures have revealed that these discontinuities can cause stress/strain concentrations due to incompatibilities in strain components of neighboring grains [4,5].

The MFT method typically uses a relatively sparse distribution of particles in the microstructure. The particles serve as vertices of the strain analysis tetrahedral units, some of which are distributed within grains, but others cross grain boundaries and effectively cover several grains. The possible large local strain variations induced by boundaries may be smeared out in the MFT analysis. How such strain smearing may influence the MFT results needs to be clarified.

This paper aims to explore the effects of grain boundaries on MFT-measured strains. Instead of using actual experimental data, and to isolate the effect of the spatial resolution associated to the density of the particle distribution, we use a CPFEM simulation to generate model data. Based on the ground truth strain dataset computed via simulation, the effects of the different tetrahedron types and radius ratios on the calculated strains are investigated.

2. Simulation details and analysis method

2.1 CPFEM simulation

The simulation is based on the actual morphology and orientation of an Al-4mass%Cu alloy sample characterized by laboratory-based X-ray diffraction contrast tomography (LabDCT) [3,6]. Figure 1a shows the grain structure of the gauge section of the undeformed sample, which is typical of a well-annealed polycrystalline sample with an average grain size of ~80 μm and a weak texture. The simulation mesh was generated from the experimental data following the procedure detailed in Refs. [7] and [8] and implemented in Neper [8], according to which the grain structure was first approximated by its nearest convex-cell geometry. The resulting tessellation was then meshed into hundreds of elements per grain, leading to a total of 2,673,622 elements. Figure 1d shows the mesh. The FEPX code was employed to simulate the elastoplastic deformation of the virtual polycrystalline aggregates [9]. Tensile loads were applied along the z-axis.

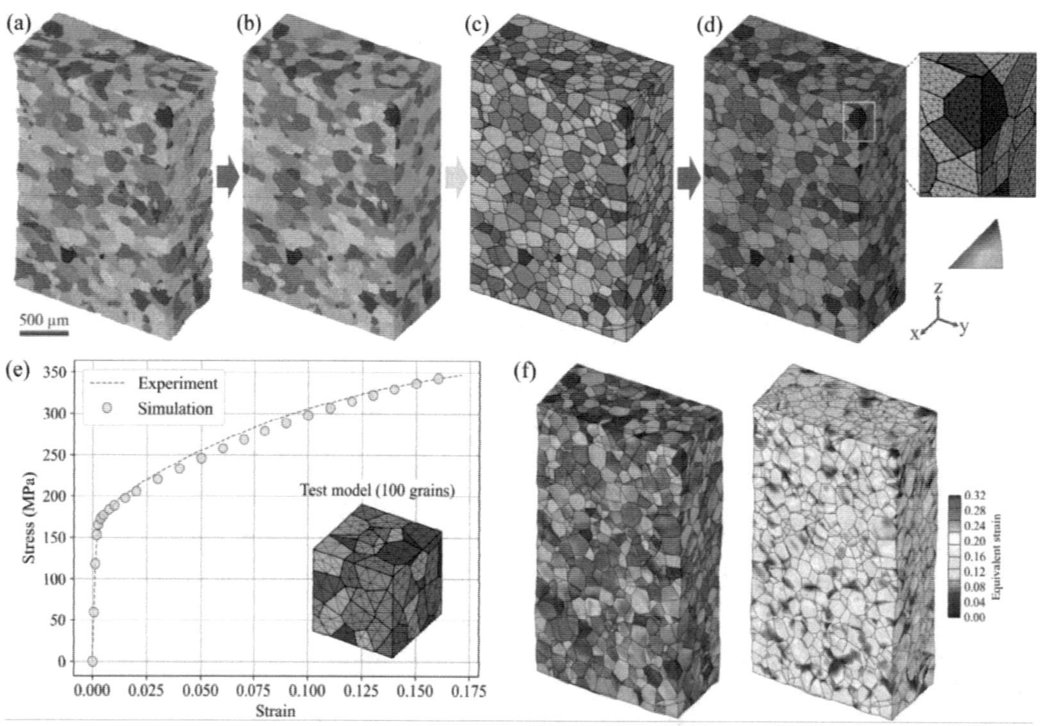

Figure 1. Model configurations. (a) LabDCT measurement, (b) cropped volume, (c) nearest convex-cell approximation, and (d) mesh. (e) Experimental and simulated stress-strain curves. (f) Orientation and equivalent strain maps of the 16% tensile deformed sample.

The material is described via an elastic and a rate-dependent plastic response. The stress (σ) is related to the elastic strain (ε) via Hooke's law: $\sigma = \mathbf{C}\varepsilon$, where $\mathbf{C}$ is the stiffness tensor, defined by the input parameters C_{11}, C_{12}, and C_{44} for cubic materials. The slip kinetics, assumed to be identical for all slip systems, has the form of $\dot{\gamma}^{\alpha} = \dot{\gamma}_0 \left(\frac{|\tau^{\alpha}|}{g^{\alpha}}\right)^{\frac{1}{m}} sgn(\tau^{\alpha})$, where $\dot{\gamma}^{\alpha}$ is the shear strain rate of the slip system α, $\dot{\gamma}_0$ is the fixed-rate strain rate scaling coefficient, g^{α} and τ_{α} are the critical resolved shear stress (CRSS) and current resolved shear stress of slip system α, and m is the rate sensitivity exponent. The slip hardening ($\dot{g}$) is modeled by $\dot{g}^{\alpha} = h_0 \left(\frac{g_s(\dot{\gamma}) - g^{\alpha}}{g_s(\dot{\gamma}) - g_0}\right)^n \dot{\gamma}$, where g_0 and $g_s(\dot{\gamma})$ are the starting and saturation CRSS of slip system α, h_0 is the fixed-state hardening rate scaling coefficient, and n is the non-linear Voce hardening exponent. The slip system saturation strength is calculated at a given state as $g_s(\dot{\gamma}) = g_{s0}\left(\frac{\dot{\gamma}}{\dot{\gamma}_{s0}}\right)^{m'}$, where g_{s0} is the initial slip system saturation strength, m' is the saturation strength rate scaling exponent, $\dot{\gamma}_{s0}$ is the initial saturation slip system shear rate. $\dot{\gamma}$ is calculated as $\dot{\gamma} = \sum_{\alpha} |\dot{\gamma}^{\alpha}|\, g_{s0}\left(\frac{\dot{\gamma}}{\dot{\gamma}_{s0}}\right)^{m'}$. In order to calibrate the crystal-scale material constants, we use a smaller model containing 100 grains (the inset of figure 1e) and fit the measured macroscopic stress. The resulting constants are provided in table 1, and the experimental and simulated stress-strain curves are plotted in figure 1e. Figure 1f shows the simulated orientation and equivalent strain maps of the 16% tensile deformed sample, featuring typical orientation spreading and an inhomogeneous strain distribution.

Table 1. Single-crystal elastic constants (aluminum) and elasto-viscoplastic model constants used in the simulation.

C_{11}, GPa	C_{12}, GPa	C_{44}, GPa	h_0, MPa	g_0, MPa	g_{s0}, MPa	m	m'	$\dot{\gamma}_0$, 1/s	$\dot{\gamma}_{s0}$, 1/s
107.3	60.9	28.32	300.0	85.0	240.0	0.05	0.005	1.0	5×10^{10}

2.2 MFT analysis

We randomly selected 15,000 points in the virtual polycrystal aggregates to serve as virtual marker particles (corresponding to an average of 4 markers per grain). Following the MFT method, tetrahedra were generated from these marker particles by Delaunay triangulation [10], leading to a tetrahedral mesh where particles represent nodes (the tetrahedron "vertices") covering a total of 99,431 tetrahedra (i.e., about 1/27 of those for CPFEM). The local strains were then evaluated within the tetrahedra from the relative displacements of the vertices during straining by the MFT method as detailed in [1]. The von Mises equivalent strain, ε_{eq}, referred to as "strain" in the following, is defined as $\varepsilon_{eq} = \sqrt{\frac{2}{3}\left(\varepsilon_{xx}^2 + \varepsilon_{yy}^2 + \varepsilon_{zz}^2\right) + \frac{1}{3}\left(\varepsilon_{yz}^2 + \varepsilon_{xz}^2 + \varepsilon_{xy}^2\right)}$. The MFT strains and the effective CPFEM strains were then compared. The MFT grain strains, defined as the volume-weighted average strain over tetrahedra or parts of tetrahedra within a grain, were also considered. Similarly, a CPFEM grain strain was defined as the volume-weighted average strain over all elements within a grain.

We classified the strain tetrahedra into five types, depending on their positions relative to grains and grain boundaries. As illustrated in figure 2a, "Tetrahedron-4" represents tetrahedra for which all vertices are located inside the same grain. "Tetrahedron-31 and -22" represent tetrahedra whose vertices are located in two connected grains, which are therefore crossing a grain boundary. Finally, "Tetrahedron-211 and -1111" represent tetrahedra whose vertices are located in 3 or 4 grains, respectively, and which are therefore containing a grain junction. Following this notation, each digit corresponds to the number of vertices inside distinct grains.

The generation of tetrahedral elements from marker particles by Delaunay triangulation often leads to a certain amount of ill-shaped tetrahedra, which may also influence the strain determination. Thus, in the following analysis, we introduce the tetrahedron radius ratio (r) to further classify the tetrahedra. As described in figure 2b, r is defined as $3Ri/Ro$, where Ri and Ro are the radii of the inscribed and circumscribed spheres of a tetrahedron, respectively. The ratio ranges between 0 and 1, and a higher value implies a better quality of the tetrahedron. Figure 2c shows the number of tetrahedra as a function of r, highlighting the existence of some low-quality tetrahedra with extremely low r for the present virtual dataset.

IOP Conf. Series: Materials Science and Engineering 1249 (2022) 012056 doi:10.1088/1757-899X/1249/1/012056

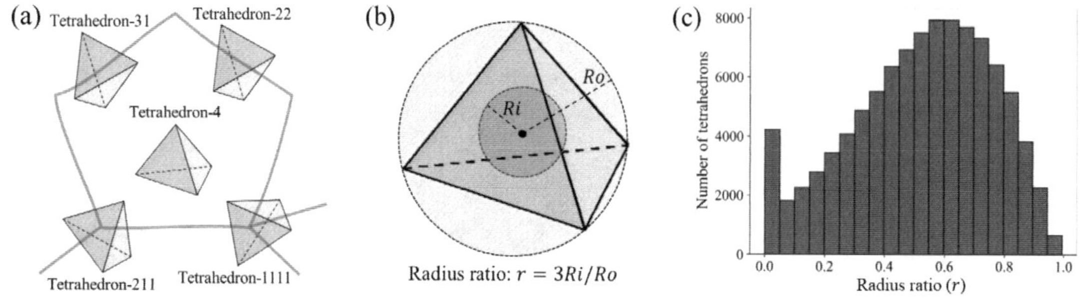

Figure 2. (a) Tetrahedron classification. (b) Tetrahedron radius ratio (r). (c) Frequency distribution of the radius ratios.

3. Results and discussion

We analyzed first MFT tetrahedron strains as a function of tetrahedron type. Since the 'ground truth' strain distribution within each MFT tetrahedron can be acquired from the CPFEM strains within the much refined finite element volumes, a quantitative analysis of the representativeness of the MFT tetrahedron strain can be performed. We therefore first performed the analysis based on the CPFEM strains and then examined the MFT grain strains, based on which suggestions to further improve the accuracy of MFT grain are given.

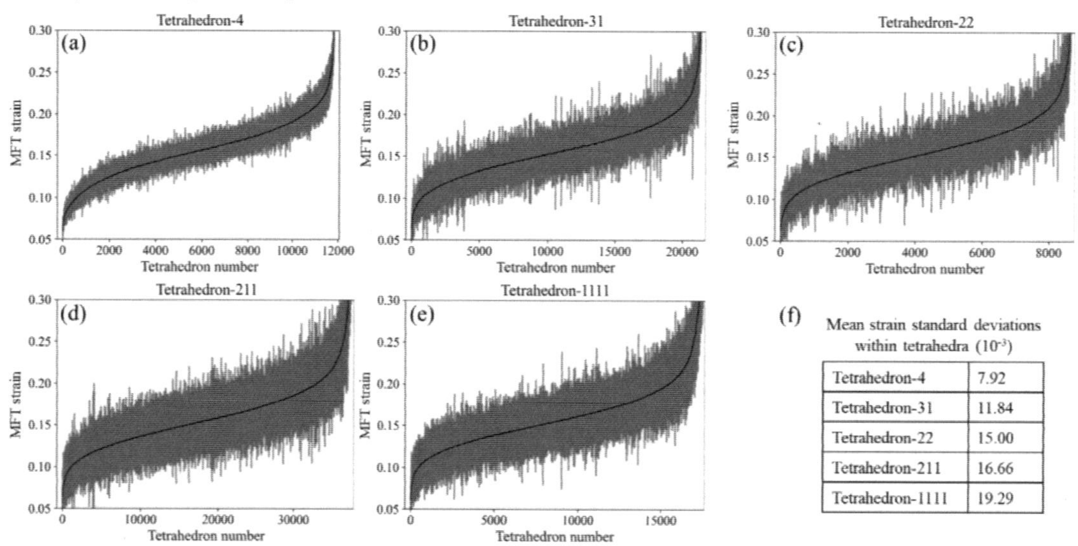

Tetrahedron-4	7.92
Tetrahedron-31	11.84
Tetrahedron-22	15.00
Tetrahedron-211	16.66
Tetrahedron-1111	19.29

Figure 3. MFT strains for different types of tetrahedra. (a) Tetrahedron-4. (b) Tetrahedron-31. (c) Tetrahedron-22. (d) Tetrahedron-211. (e) Tetrahedron-1111. The strain standard deviation for each tetrahedron is shown in green lines. The mean strain standard deviation data for each type of tetrahedra are listed in (f).

Figure 3 shows the MFT strains for different types of tetrahedra. For clarity, strains of the tetrahedra are plotted by increasing strain magnitude as shown by the black line, and the strain standard deviation for each tetrahedron is shown as green lines. Significantly smaller strain standard deviations are noticed in Tetrahedron-4 (figure 3a) than for the other ones (figure 3b-e), see also figure 3f, suggesting that the local strain distributions in Tetrahedron-4 are generally more homogeneous than in other tetrahedra types. The result is understandable because all the other types of tetrahedra contain boundaries or boundary junctions (figure 2a), where the parts belonging to different grains may have significantly different strains, therefore giving a greater probability of recording highly heterogeneous strains.

As the calculation of the MFT grain strain is based on the strains of tetrahedra within or connected to the specific grain, the above-mentioned effects of tetrahedron type may lead to uncertainties in MFT grain strain. To evaluate such uncertainties, the MFT grain strain and CPFEM grain strain for each grain are compared. Furthermore, by selecting different tetrahedra used for MFT grain strain calculation, suggestions on how to optimize such data processing are given.

To maximize accuracy, only grains containing more than 100 vertices are considered. The grain strains calculated by MFT are plotted against those calculated by CPFEM for different r ranges: $r > 0.0$, $r > 0.1$, and $r > 0.5$, in figure 4a, where all tetrahedra are considered. The degree of deviation of the data spots from the identity line indicates how much the MFT grain strains differ from the CPFEM grain strains. As seen from the figure, there are many significantly deviating data points in the $r > 0.0$ plot, but not so many in the $r > 0.1$ and $r > 0.5$ plots. The apparent decrease in the number of deviating grains indicates that the ill-shaped tetrahedra, having lower r values (mostly lower than 0.1), are responsible for the erroneously MFT calculated grain strains. That the strain data are being distributed around lines with slopes significantly lower than 1, suggests that the strain heterogeneities are underestimated by MFT, which may be explained from the averaging effect of boundary-crossing tetrahedra. This is analyzed in the following.

The plots of figure 4b are drawn in the same manner as figure 4a, but for Tetrahedron-4 only. Only large grains, which contain more than 25 Tetrahedron-4 types, are considered. The results clearly show that, after excluding the averaging effect of boundary-crossing tetrahedra, the MFT grain strains are in better agreement with the CPFEM grain strains. This indicates that, for those large grains, the strain averaged over all Tetrahedron-4 is more representative of the effective strain than the MFT strain averaged over all tetrahedra.

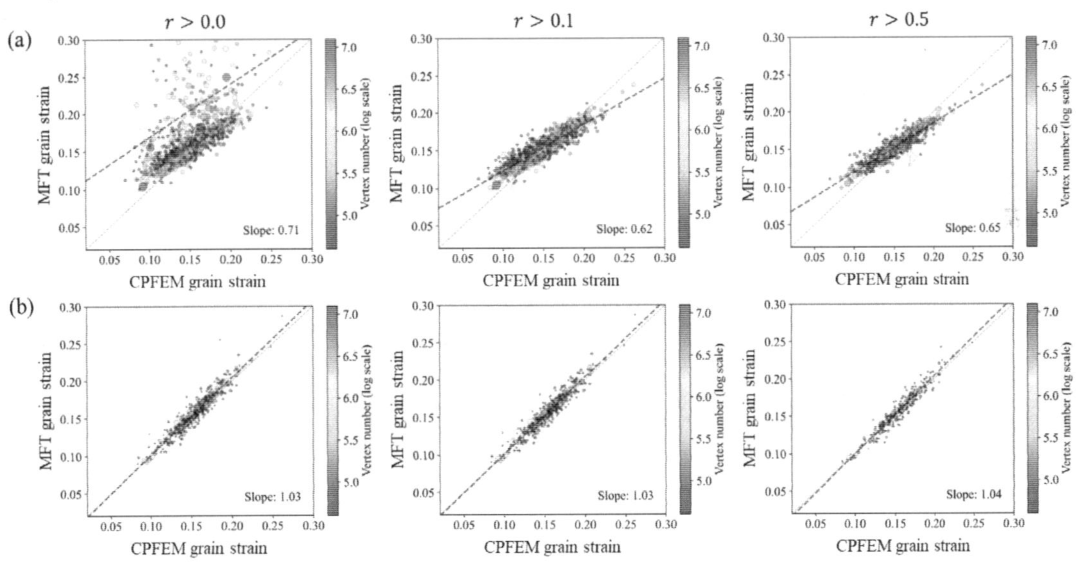

Figure 4. MFT grain strain versus CPFEM grain strain with different r ranges, *i.e.*, $r > 0.0$, $r > 0.1$, and $r > 0.5$. (a) Calculated using all tetrahedra. (b) Calculated using Tetrahedron-4 only. The black dashed line and light dashed line are the linear fit of the data spots and the identity line, respectively.

4. Conclusions

The data processing procedure of the MFT method was evaluated using reference results obtained by CPFEM simulations. The influence of the type and radius ratio of the MFT tetrahedral elements on the computed strains was analyzed. It was found that the strain calculated from the tetrahedra for which all vertices are located inside the same grain is more representative of the local strain than the strain computed using all tetrahedra, including those overlapping grain boundaries. In the MFT method, the strain heterogeneity of boundary-crossing tetrahedra tends to be underestimated. It was also shown that significant uncertainty is induced in the MFT strain calculations when low radius ratio tetrahedra are used. In short, the analysis highlighted the pros and cons of the MFT method for strain measurements, and the results can be used for optimizing the MFT data processing. Additionally, the results may be of value to improve also other methods such as digital image correlation and digital volume correlation.

Acknowledgments
The financial support from the European Research Council (ERC) under the European Union's Horizon 2020 research and innovation programme (M4D-grant agreement No. 788567) is gratefully acknowledged.

References

[1] Toda H, Minami K, Koyama K, Ichitani K, Kobayashi M, Uesugi K and Suzuki Y 2009 Healing behavior of preexisting hydrogen micropores in aluminum alloys during plastic deformation *Acta Mater.* **57** 4391–403

[2] Kobayashi M, Miura H and Toda H 2015 Non-destructive and three-dimensional measurement of local strain development during tensile deformation in an aluminium alloy *IOP Conf. Ser.: Mater. Sci. Eng.* **89** 012030

[3] Kobayashi M, Zhang Y, Ishikawa H, Sun J, Oddershede J, Juul Jensen D and Miura H 2021 Investigation of relationships between grain structure and inhomogeneous deformation by means of laboratory-based multimodal X-ray tomography: strain accuracy analysis *Exp. Mech.*

[4] Bieler T R, Eisenlohr P, Zhang C, Phukan H J and Crimp M A 2014 Grain boundaries and interfaces in slip transfer *Curr. Opin. Solid State Mater. Sci.* **18** 212–26

[5] Kocks U F and Canova G R 1981 How many slip systems, and which? *Deformation of Polycrystals: Mechanisms and Microstructures. 2 nd Riso Int. Symposium on Metallurgy and Materials Science* 35–44

[6] Kobayashi M, Zhang Y, Ishikawa H, Sun J, Oddershede J, Jensen D J and Miura H 2021 Relationships between 3D grain structure and local inhomogeneous deformation: A laboratory-based multimodal X-ray tomography investigation *submitted to Acta Materialia for consideration*

[7] Quey R and Renversade L 2018 Optimal polyhedral description of 3D polycrystals: Method and application to statistical and synchrotron X-ray diffraction data *Comput. Methods Appl. Mech. Eng.* **330** 308–33

[8] Quey R, Dawson P R and Barbe F 2011 Large-scale 3D random polycrystals for the finite element method: Generation, meshing and remeshing *Comput. Methods Appl. Mech. Eng.* **200** 1729–45

[9] Dawson P R and Boyce D E 2015 FEpX--Finite element polycrystals: Theory, finite element formulation, numerical implementation and illustrative examples *arXiv:1504.03296*

[10] Barber C B, Dobkin D P and Huhdanpaa H 1996 The quickhull algorithm for convex hulls *ACM Trans. Math. Softw.* **22** 469–83

Microstructure and properties of cold-drawn Cu and Cu-Fe alloy wires

Fei Yang[1, 2], Xiaodan Zhang[2] and Feng Fang[1*]

[1] Jiangsu Key Laboratory of Advanced Metallic Materials, Southeast University, Nanjing, 211189, China

[2] Department of Mechanical Engineering, Technical University of Denmark, Lyngby, Denmark

E-mail: fangfeng@seu.edu.cn

Abstract. Copper and Cu-Fe alloy wires were cold drawn at room temperature, and the microstructure and properties were characterized and analyzed. The results show that, at the same drawing strain, the strength and ductility of Cu-Fe wires are higher than those of copper wires. At a drawing strain of 4, the tensile elongation of Cu-Fe alloy wires is about 5%, much higher than 2% for pure copper wires. For Cu-Fe wires, with increasing drawing strain, the Fe phase is elongated along the drawing direction and forms a fiber structure with a <110> texture. In addition to the <111> and <100> fiber textures in the copper wire, a high-intensity of texture with a <112> orientation forms in Cu-Fe alloy wires. The ductility of the Cu-Fe alloy wires is greatly preserved due to dynamic recrystallization. Moreover, with the additional refining effect of nano-copper fibers in the Fe phase, the strength of the material is greatly improved.

1. Introduction

Copper is widely used in various fields of electrical transmission due to its excellent electrical conductivity (EC), thermal conductivity, and corrosion resistance. However, pure copper has low strength and hardness, and even through work hardening, the strength and hardness are still insufficient for industrial applications [1]. The development of new high-strength and high-electrical conductivity copper alloys has become one of the current research hotspots. The solid solution of alloying elements will increase the strength of copper, but significantly reduce the electrical conductivity of copper. It is necessary to control therefore the total amount of solid-solution alloying elements [2]. In practical applications, the method of forming second phase strengthening in the copper matrix is generally adopted, where the most common approaches are through precipitation strengthening or dispersion strengthening [3]. In-situ deformation is a new method developed in recent years, which can maintain a high comprehensive matching of electrical conductivity and strength.

At present, research on in-situ deformation of copper matrix composites is mainly focused on the Cu-Ag and Cu-Nb systems, as both Ag and Nb have very low room temperature solubility in copper [4, 5]. The uniform distribution of fibrous Ag or Nb fibers in a matrix close to pure copper can play a significant strengthening role without significantly deteriorating its electrical conductivity. Compared with Cu-Ag and Cu-Nb, the Cu-Fe-based system has not been systematically studied, but it has attracted extensive interest of some researchers due to its low cost. At the same time, Cu-Fe has more prominent advantages [6, 7] including: 1) Fe has a relatively low melting point, and the miscibility between liquid

Fe and Cu is high; and 2) the density of Fe is relatively close to that of Cu. When the material is prepared by a melting and casting method, the specific gravity segregation is small, such that a billet with a larger size can be prepared. Therefore, Cu-Fe in situ composites have more potential in terms of industrial-scale preparation and application. For example, after severe drawing deformation, powder metallurgy formed Cu-15% Fe can obtain a tensile strength of 1200 MPa and an electrical conductivity of 47.6 %IACS[8].

In this work, Cu-20% Fe wires were prepared by cold drawing, and the microstructure and properties at different drawing strains have been investigated. As the copper phase plays a dominant role in electron transport in the Cu-Fe alloy, a drawn pure copper wire was selected as a comparison to analyze the difference in the drawn microstructure of single-phase and dual-phase materials.

2. Material and methods

Commercial available continuously-cast and rolled pure copper rods were selected as the comparison material, with a diameter of 8 mm. For the Cu-Fe alloy a Cu80Fe20 (wt%) ingot was prepared by vacuum induction melting. After hot forging, machining and annealing at 500°C for 1h, a 12 mm diameter bar was obtained. Drawn wires with different strains were obtained through multi-pass drawing at room temperature. The microstructure of the alloy wires was characterized by field-emission scanning electron microscopy (FEI-Sirion). Transmission electron microscopes (Talos F200X and FEI Tecnai G2) were used to observe the nano-structure and precipitates. The tensile properties were tested at room temperature by using a uniaxial tensile testing machine (CMT5105), where the strain rate was kept below 5×10^{-3} s^{-1}. Each group of samples was tested three times, and the average was taken as the final result. The electrical resistivity of the wires was acquired using an ohmmeter (HIOKI-RM3548), and the electrical conductivity was evaluated according to the International Annealed Copper Standard (%IACS).

3. Results

3.1 Microstructural evolution

The cross-section microstructure of the Cu-Fe alloy bars after pre-treatments (casting, hot forging and annealing) is shown in figure 1. The Fe phase in the ingot mainly exists in the form of dendrites, although some Fe particles can also be observed (as shown in figure 1a). The original dendrites are destroyed during subsequent hot forging deformation (figure 1b), and the local dendrites are severely distorted. After annealing at 500 °C for 1h, the Fe particles are more uniformly dispersed in the Cu matrix (figure 1c). A small number of Fe particles are precipitated from the Cu matrix after annealing treatment, with a particle size of about 10 nm, as shown in figure 1d.

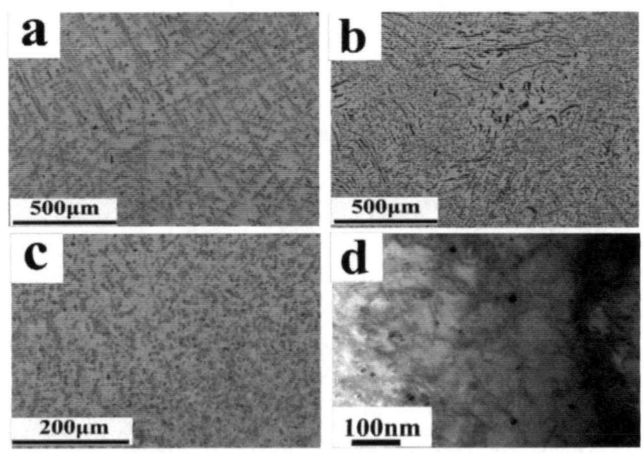

Figure 1. Microstructure of Cu-Fe bar (a) ingot, (b) hot forged, (c) annealed, and (d) precipitation of Fe particles.

The orientation microstructure of copper and Cu-Fe wires at different drawing strains is shown in figure 2, where white regions in figure 2b correspond to the Fe phase. At the low strain of ε = 0.58, the grains in the copper wire are extruded along the drawing direction (figure $2a_2$), and a <111> texture starts to form. At the medium strain of ε = 1.91 (figure $2a_3$), the grains are severely deformed, and grains are completely transformed into <111> and <100> textures. When the drawn strain reaches to ε = 3.47, a clear fiber structure is observed, and grains with <111> and <100> orientation are intersect each other (figure 2a$_4$). In contrast to the fiber structure in the drawn pure copper wires, the Cu phase in Cu-Fe wire is less refined, and the grains are not elongated substantially in the drawing direction or compressed in the transverse direction. Compared with the drawn copper wire, the <111> and <100> textures are not intense in Cu-Fe wires, and there are still a considerable number of grains showing the <112> orientation (the purple area in figure 2 b$_2$-b$_4$). Finally, in both wires, lamellar structures are formed.

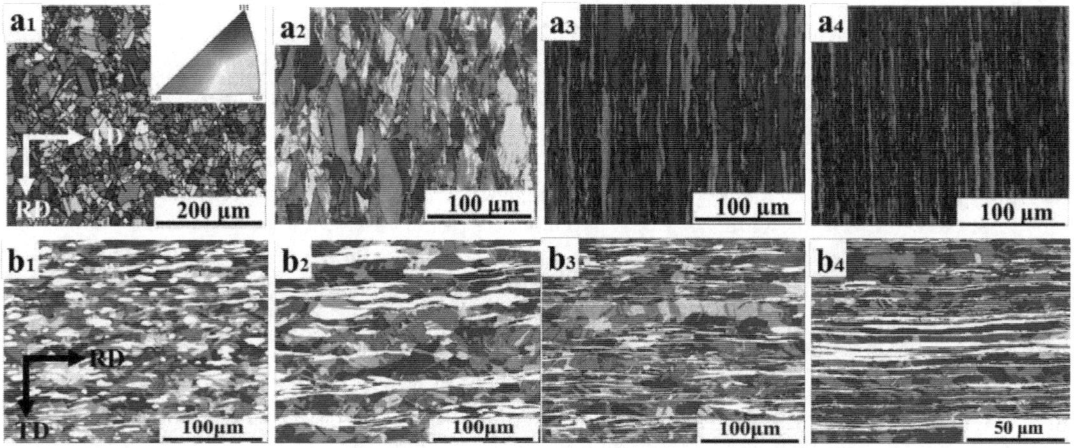

Figure 2. Orientation maps using inverse pole figure coloring of drawn copper wires at (a$_1$) ε=0, (a$_2$) ε=0.58, (a$_3$) ε=1.91, (a$_4$) ε=3.47; and Cu-Fe wires at (b$_1$) ε=0, (b$_2$) ε=0.62, (b$_3$) ε=2.06, (b$_4$) ε=3.32.

Figure 3 shows TEM images of the copper wires and the copper phase in drawn Cu-Fe wires. At a low strain of ε = 0.58, the dislocation density significantly increases in copper wires, the grains in favorable orientations are first elongated along the drawing direction, and a large number of dislocations are present and entangled with each other. As the local dislocation density increases, dislocation substructures are formed (figure 3a$_1$). At a medium strain ε = 1.91, the average grain width is about 500 nm. Due to continuous plastic deformation, dislocations accumulate and annihilate at the dislocation cell walls. In the process of continuous dislocation interaction, the walls gradually transform into sub-grain boundaries with large misorientation, as shown by the white arrows in figure 3a$_2$. At a high strain ε = 3.47 (figure 3a$_3$), the grains form a lamellar structure along the drawing axis, and the grain width continues to be refined. At the same time, a certain number of recrystallized grains were also observed, as shown in figure 3a$_4$. For the microstructure of the Cu phase in Cu-Fe wires, at low strain ε = 0.62, the dislocations slip within the grains, and are pinned at precipitates (shown by yellow arrows). As the drawing strain increases to 2.06, large numbers of dislocation substructures are formed nearby Fe particles. After severe drawing deformation at ε = 3.32, sub-grains are formed with average width about 200 nm (see figure 3b$_3$). It is observed that several Fe particles in the copper matrix have

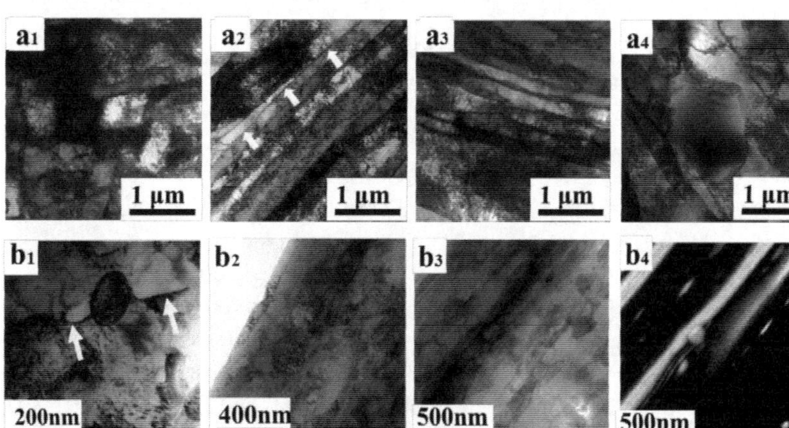

Figure 3. TEM of copper wires at (a$_1$) ε=0.58, (a$_2$) ε=1.91, (a$_3$, a$_4$) ε=3.47; and Cu-Fe wires at (b$_1$) ε=0.62, (b$_2$) ε=2.06, (b$_3$, b$_4$) ε=3.32.

almost no plastic deformation and exist in the form of spheres or ellipsoids (shown by the yellow area in figure 3 b_4). This illustrates that the Fe precipitates with a size of several hundred nano-meters are not easily deformed during the drawing process, and hinder the formation of a fiber structure of the adjacent Cu matrix and Fe phase.

3.2 Mechanical properties

The mechanical properties of drawn copper and Cu-Fe wires are shown in figure 4. For drawing strains exceeding 2, the strength of the copper wire is almost no longer improved, and the tensile strength is stable between 420-430MPa. The stabilization of strength may be related to a balance between work hardening and dynamic recrystallization. The tensile elongation of the copper wire is very small due to the destruction of plasticity by work hardening (figure 4a). The tensile strength of the Cu-Fe alloy wire increases linearly with increasing drawing strain. When the drawing strain reaches 3.32, the ultimate tensile strength (UTS) increases to 802 MPa and the total elongation (TE) is 5.8%. Through the comparison of tensile properties, it can be found that both strength and ductility of Cu-Fe wires are better than those of copper wires at the same drawing strains (UTS < 430 MPa, TE nearly 2%). Different from the drawing texture of pure copper, the Cu-Fe wire shows more <112> texture. The <112> texture was found to be generated in the structure of annealed copper wires [9]. Due to the low recrystallization temperature of copper, more high isolated thermal stress is induced during the drawing deformation of the Cu-20%Fe alloy wires, and results in continuous dynamic recovery or recrystallization of the copper phase, enabling the Cu-Fe wire to exhibit a higher tensile elongation.

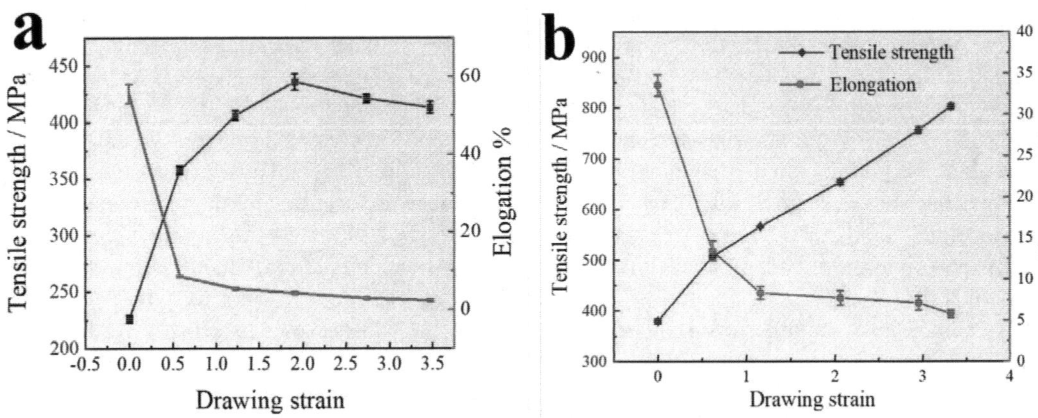

Figure 4. Mechanical properties of cold drawn (a) copper wires, and (b) Cu-Fe wires.

For the Cu-Fe alloy wires at a strain of 3.32, the yield strength is 761 MPa. The strength of the drawn Cu-Fe wire is mainly affected by work hardening, fine grain strengthening and second phase strengthening. Cu-Fe alloys can also be regarded as Fe fiber-reinforced Cu matrix composites, and as such the mixing law can be used to integrate the contribution of each phase, as follows:

$$\sigma_y = 0.22\sigma_{yFe} + 0.78\sigma_{yCu} \tag{1}$$

in which, σ_{yFe} and σ_{yCu} are the yield strengths of the Fe and Cu phases after drawing deformation, where σ_{yCu} is taken from the data for drawn copper wires (σ_{yCu}=425MPa at a strain of 3.47). The volume fraction of Fe and Cu phase are taken to be 0.22 and 0.78, respectively.

According to the report of Feng et al. [10], a similar Hall-Petch relationship can be acquired between yield strength and interlamellar spacing of Fe. As shown in figure 5, there are a lot of copper particles in the Fe phase. With increasing drawing strain, these particles are elongated along the drawing direction, and gradually change from spherical to elliptical, and finally to fibers. The nano-copper fibers greatly refine the structure of the Fe phase (shown in figure 5c). As such the strength of the Fe phase can be expressed by formula (8), as follows:

$$\sigma_{yFe} = \sigma_0 + 5.9Gb \times {d_{Fe}}^{-1} \tag{2}$$

in which, σ_0 is the friction stress, measured from fully recrystallized Fe as about 62MPa, G stands for the shear modulus, about 81.7 GPa, b is the Burgers vector, 0.248 nm, and d_{Fe} is the average lamellar spacing between Fe lamella and copper fibers, which is measured from TEM as about 63 nm. The value of σ_{yFe} is calculated to be 1898 MPa, then the value is substituted into formula (1), and σ_y is estimated to be 749 MPa. The predicted value is in good agreement with the actual yield strength of 761 MPa. Most importantly, it can be concluded that the contribution of Fe fibers to strength is critical for the drawn Cu-20%Fe wires.

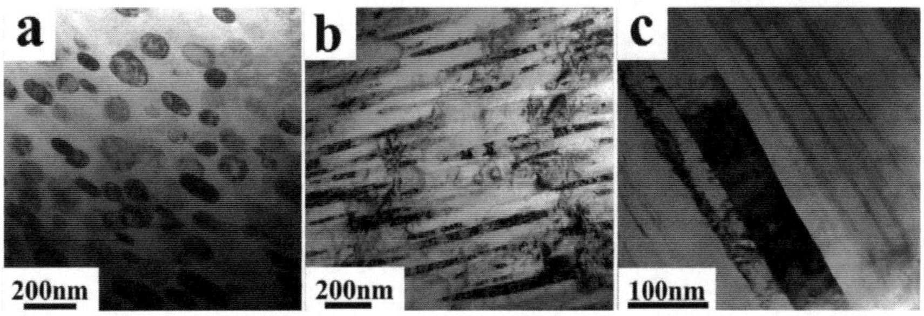

Figure 5. TEM of copper precipitates in the Fe phase(a) ε=0.62, (b) ε=2.06, (c) ε=3.32

3.3 *Electrical conductivity*

In addition to the influence of environmental factors and intrinsic properties of the material, the variation of the electrical conductivity of drawn copper and Cu-Fe wires is mainly related to the microstructure evolution, including dislocations, grain boundaries (GBs), vacancies, precipitated phases and solute atoms. The electrical conductivity of copper and Cu-Fe wires is shown in figure 6.

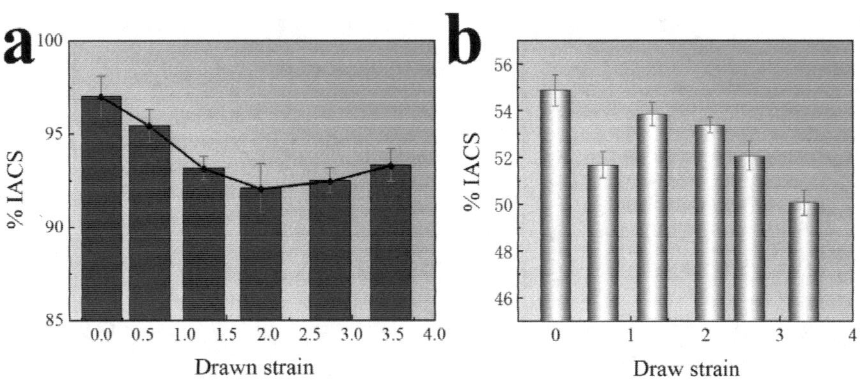

Figure 6. Electrical conductivity of (a) copper wires, and (b) Cu-Fe wires.

The electrical conductivity of copper wires continues to decrease until the drawing strain reaches 1.91, followed by a slight increase. Since there is no second phase and solid solution atoms in the copper grains, the decrease in electrical conductivity can be attributed to the generation of dislocations and the sub-grain boundaries separated by grain refinement during the drawing process. When the cumulative drawing strain is larger than 1.91, the dynamic recovery and recrystallization lead to the elimination of defects, and the electrical conductivity is recovered. Figure 6b shows that the electrical conductivity of Cu-Fe wire decreases almost monotonically with increasing the drawing strain, and drops sharply for strains greater than 2. On the one hand, the smaller the diameter of the drawn wire, the faster the heat dissipation, resulting in a reduction in the time required for dynamic recrystallization. On the other hand, large plastic deformation may result in the diffusion at the phase boundary (as shown in figure 7). The solid solution of heteroatoms in copper will result in serious scattering of electrons and great damage to the conductivity. It has been reported that after being subjected to high-pressure torsion deformation, Fe phase dissolves into the copper matrix and a supersaturated solid solution is formed [12]. The content of solid solution iron in the copper matrix with severe plastic deformation is up to 12at%. Therefore, the sharp decrease of electrical conductivity under large drawing strain may be related to the fragmentation and re-dissolution of Fe fibers.

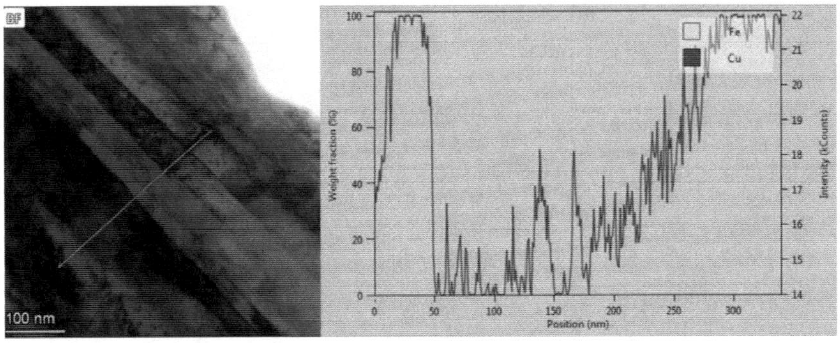

Figure 7. Element distribution at phase boundary

4. Conclusions

The microstructure and properties of copper and Cu-Fe wires have been characterized. Compared with the <111> and <100> textures in the pure copper wire, there is still a considerable amount of <112> texture in the Cu-Fe wire. Due to the dynamic recovery and recrystallization in copper phase, the Cu-Fe wire obtains a higher tensile elongation than copper wires. Due to the fibrillation of the Fe phase in Cu-Fe and the refinement of nano-copper particle.
es in the Fe phase, the ultimate tensile strength of Cu-Fe wires is greatly improved. The sharp decrease in electrical conductivity for drawing strains larger than 2 is mainly attributed to fragmentation of Fe fibers and redissolution of Fe elements into the Cu phase.

Acknowledgments

This work was supported by National Natural Science Foundation of China (No. 52171110). X. Zhang acknowledges the support by a research grant (00028216) from VILLUM FONDEN.

References

[1] Yuan D W, Zeng H, Xiao X P, Wang H, Han B J, Liu B X and Yang B 2021 *Mater. Sci. Eng. A* **812** 141064
[2] Wang M, Yang Q R, Jiang Y B, Li Z, Xiao Z, Gong S, Wang Y R, Guo C L and Wei H G 2021 *Trans. Nonferrous Met. Soc. China* **31** 3039–49
[3] Stepanov N D, Kuznetsov A V, Salishchev G A, Khlebova N E and Pantsyrny V I 2013 *Mater. Sci. Eng. A* **564** 264–72
[4] Zhang B B, Tao N R and Lu K 2017 *Scr. Mater.* **129** 39–43
[5] Spitzig W A, Pelton A R and Laabs F C 1987 *Acta Mater.* **35** 2427–42
[6] Wu Z W, Zhang J D, Chen Y and Meng L 2009 *J. Rare Earths* **27** 87–91
[7] Sauvage X, Wetscher F and Pareige P 2005 *Acta Mater.* **53** 2127–35
[8] Wang F, Wakoh K, Li Y, Ito S, Yamanaka K, Koizumi Y and Chiba A 2017 *Mater. Des.* **126** 64–72
[9] Baudin T, Etter A L and Penelle R 2007 *Mater. Charact.* **58** 947–52
[10] Feng H C, Fang F, Zhou X F, Zhang X D, Xie Z H and Jiang J Q 2020 *Mater. Sci. Eng. A* **796** 140017
[11] Zeng W, Xie J W, Zhou D S, Fu Z Q, Zhang D L and Lavernia E J 2018 *J. Alloys Compd.* **745** 55–62
[12] Quelennec X, Menand A, Le Breton J M, Pippan R and Sauvage X 2010 *Philos Mag.* **90** 1179–95

IOP Conf. Series: Materials Science and Engineering 1249 (2022) 012058 doi:10.1088/1757-899X/1249/1/012058

Microstructure and mechanical properties of a 0.1C5Mn3Al steel fabricated by warm drawing

L Zhang[1], P Cai[1], X Cheng[1], L Zhang[1,2,3], Z Hou[1] and X Huang[1,2,3]

[1] International Joint Laboratory for Light Alloys (MOE), College of Materials Science and Engineering, Chongqing University, Chongqing, 400044, P.R. China

[2] Electron Microscopy Center of Chongqing University, Chongqing University, Chongqing 400044, P.R. China

[3] Shenyang National Laboratory for Materials Science, Chongqing University, Chongqing 400044, P.R. China

E-mail: zhangling2014@cqu.edu.cn

Abstract. In the present study, a recently developed low density 0.1C5Mn3Al dual phase steel was subjected to warm drawing at 600 °C. Microstructure observations and mechanical properties were investigated by scanning electron microscopy (SEM), transmission electron microscopy (TEM) and tensile tests. It was found that carbide precipitates together with laminated ferrite were formed before and after warm drawing. After warm drawing by a Von Mises strain of 1.96, the yield stress (YS) improved from 650 MPa to 1100 MPa and the uniform elongation (UE) dropped to 2%. Further heat treatments were performed to tune the microstructures and mechanical properties. Discontinuous yielding phenomenon was widely observed in the heat treated samples and 15% hot rolling was used to eliminate this discontinuous yielding.

1. Introduction

In view of the demand for energy conservation, emission reduction and high safety, dual phase (DP) steel containing ferrite and martensite has been widely used in the manufacturing industry, especially in the automotive industry because of its good comprehensive mechanical properties. Among them, aluminum containing medium Mn steel has become one of the research hotspots [1-6]. Recently, to further improve the properties of dual phase steel, Kimura and Tsuzaki et al. [7,8] have proposed an effective method for grain refinement involving warm rolling of tempered martensite (tempforming) which results in the elongated ultrafine/nanostructured grains with finely dispersed second phase particles, and allows obtaining a promising combination of mechanical properties in low-alloyed steels.

In this paper, a recently developed 0.1C5Mn3Al laminated dual phase steel was used. After hot rolling, formation of ferrite and martensite dual phase led to a yield stress (YS) of 789 MPa and a total elongation (TE) of about 25%. To further improve the mechanical properties, tempforming [7-13] was tried by warm drawing at 600 °C and followed by annealing to tune mechanical properties. Microstructure observations and mechanical properties were investigated by scanning electron microscopy (SEM), transmission electron microscopy (TEM) and tensile tests.

2. Experimental

The low carbon medium manganese steel has a composition of 0.11 wt% C, 4.91 wt% Mn, 3.11 wt% Al and 0.21 wt% Si. The steel was fabricated in a vacuum induction furnace and cast into 50 kg ingot, then the ingot was hot forged into square slab with a length of 30 mm. After homogenizing at 1200 °C for 2 hours, the slab was hot rolled with the final temperature higher than 850 °C into rods with a diameter of 8 mm and followed by air cooling to room temperature, then re-heated to 600 °C for 1 hour (Referred as WD-0 sample) and drawing to thin rod with a final diameter of 3 mm (a Von Mises strain of 1.96, shorted as WD-1.96 sample). Heat treatments were performed from 550 °C to 800 °C for 5 and 10 minutes.

Dog-bone shaped tensile samples with a gage length of 8 mm, a width of 2 mm and a thickness of 1 mm were used, the tensile direction was parallel to the drawing direction. Tensile tests were conducted at ambient temperature with a rate of $10^{-3}s^{-1}$.

The samples for microstructure analysis were polished and etched in 4% natal for 5 s. Microstructure observations were carried out by scanning electron microscopy (SEM) with a Zeiss Gemini SEM 300 scanning electron microscope and transmission electron microscopy (TEM) with a JEM-2100 electron microscope operated at 200 kV.

3. Results and discussions

3.1. Microstructure and mechanical properties of the warm drawn samples

Figure 1 shows the mechanical properties of the warm drawn samples with different strains. With the increase of strain, the hardness and the strength increase gradually, whereas the elongation decreases greatly. After warm drawing, the hardness increased from 252 HV of the WD-0 sample to 340 HV of the WD-1.96 sample. The YS and UTS of WD-1.96 sample increased to 1235 MPa and 1388 MPa, respectively, with a total elongation (TE) of ~ 6%. It is the highest known strength of the 0.1C5Mn3Al dual phase steel However, with the von Mises strain of 1.96, the ductility is too low for engineering application. In order to restore its elongation, the followed annealing is necessary.

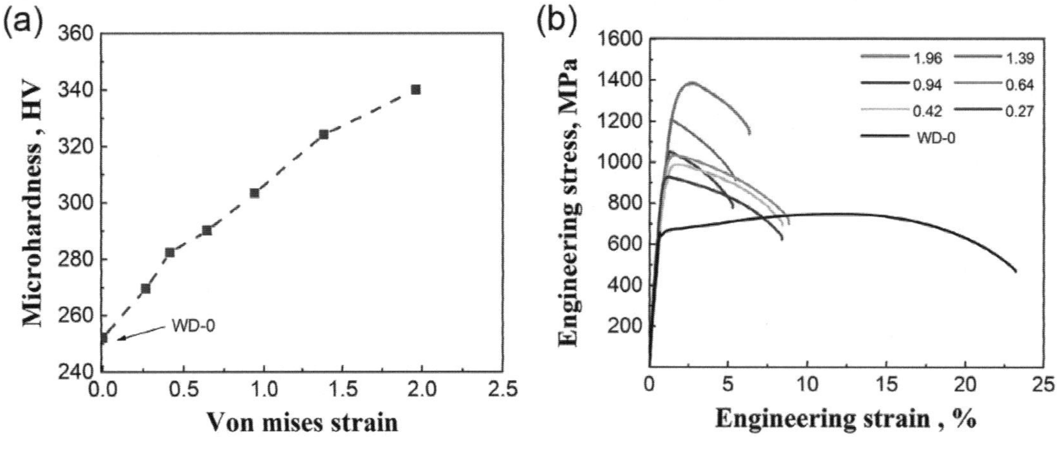

Figure 1. (a) Hardness, (b) engineering stress-strain curves of the warm drawn samples.

As shown in figure 2(a), before warm drawing, the WD-0 sample has a lamellar character containing alternating ferrite and ferrite + carbides layers (transformed from martensite phase and formed during 600 °C annealing, as marked as M'). Both two phases deformed during warm drawing and at a strain of 1.96, only very large carbides can be seen by SEM and the layer thickness decreased to ~ 1.8 μm. The ferrite phase has a characteristic of severely plastic deformed structure with elongated grains/subgrains aligned along the warm drawn direction containing a large amount of dislocations, as shown by figure 2(c). The average thickness is about 340 ± 220 nm. Big carbides can be found occasionally, as indicated by blue arrows, but in the ferrite phase, it is not frequently observed.

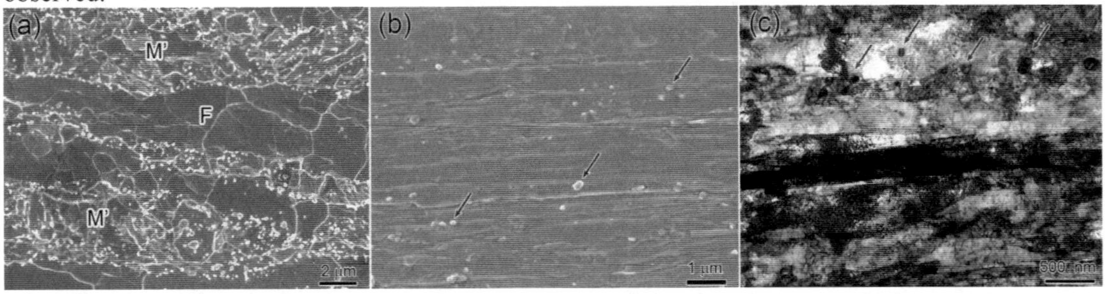

Figure 2. Microstructures of (a) WD-0 sample and (b) WD-1.96 sample observed by SEM; (c) WD-1.96 sample observed by TEM.

3.2. Microstructure and mechanical properties of the annealed samples

Figure 3 shows the engineering stress-strain curves of the warm drawn WD-1.96 sample heat treated for 5 and 10 minutes. It is clear that the strength decreases whereas the elongation gradually increases with the increases of annealing temperature or time. Samples annealed at 750 °C have the best combination of strength and ductility. The UTSs of the 5 min and 10 min heat treated samples are 860 MPa and 649 MPa, respectively, with the UEs are higher than 10%. Furthermore, it is found that most of the stress-strain curves show a discontinuous yielding phenomenon. This yield point can be caused by a number of factors, associated with insufficient mobile dislocations in the sample [14-17]. To understand this phenomenon and to eliminate this yield point, the microstructure of the annealed samples was investigated.

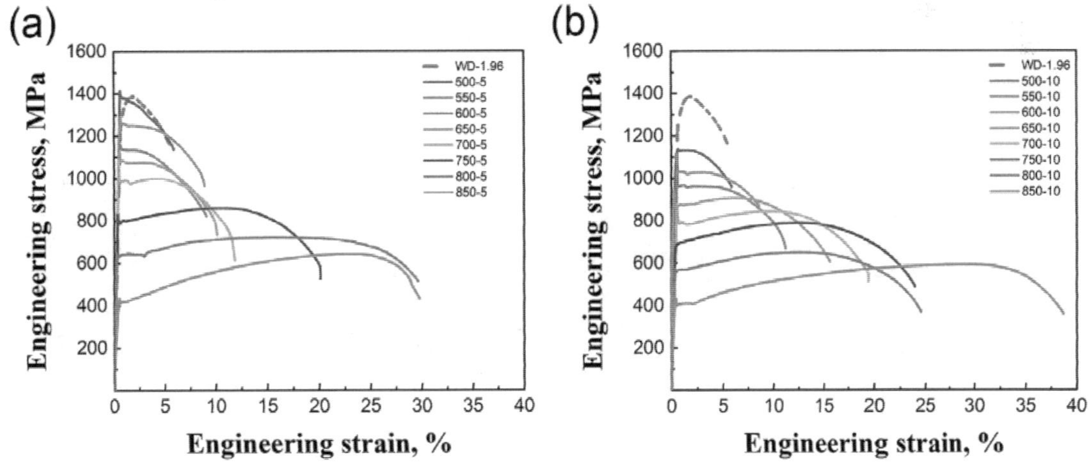

Figure 3. Engineering stress - strain curves of the samples annealed for (a) 5 min and (b)10 min.

The SEM images of the samples annealed at 550 °C to 800 °C for 10 minutes are shown in figure 4. Comparing with unheat-treated WD-1.96 sample, it is seen that the layered structure is reserved for samples annealed at temperatures lower than 700 °C (figures 4(a) - (d)). The ferrite phase underwent recovery at 550 °C, recovery and partial recrystallization at 600 °C, grain growth and coarsening at 700 °C and 750 °C, respectively. Both small and big carbides can be seen in the samples annealed at temperatures lower than 750 °C, as shown in figures 4(a) - (e). The big white carbides are the remained ones and the very small carbides are thought to be newly formed ones. The number density of the carbides gradually decreases with the increasing annealing temperatures. When annealed at temperatures higher than 800 °C, the carbides disappeared and most of the carbides resolved in the high temperature austenite and ferrite phases. This is in agreement with the phase diagram [18].

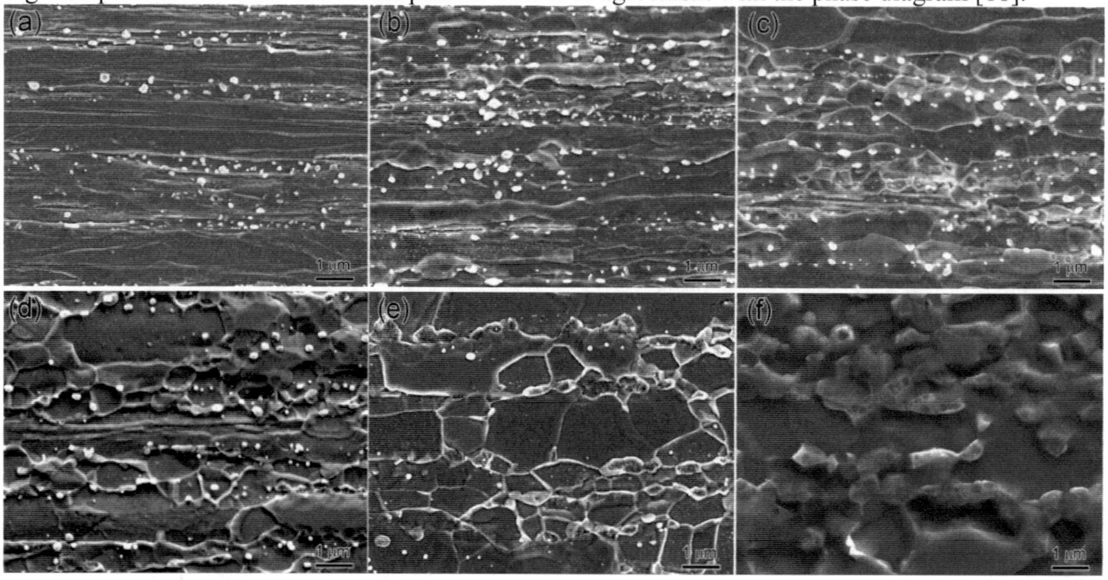

Figure 4. SEM images of the samples annealed at (a) - (f) 550 °C, 600 °C, 650 °C, 700 °C, 750 °C, 800 °C for 10 minutes and followed by air cooling.

Figure 5 is the TEM bright field images of the annealed samples. The microstructure of the sample annealed at 600 °C for 5 minutes is similar to that of the unheat-treated WD-1.96 sample, only recovery occurred during annealing. When annealed at 750 °C for 5 minutes, it is clear that a large amount of grains are recrystallized equiaxed ones. Some of the grains underwent grain growth when annealed at 750°C for 10 minutes. Some carbides were indicated by arrows. The dislocation density decreased significantly after annealing due to interaction between dislocations and occurrence of recrystallization. The reduction of dislocation densities and coarse of grains led to the observed decreasing of the strength. It is thought that the scarce of mobile dislocations may contribute to the discontinuous yielding phenomenon, similar to the reported interstitial free steel and Al [6, 15, 17, 19].

Figure 5. TEM images of the samples annealed at (a) 600 °C for 5 min, (b) 750 °C for 5 min and (c) 750 °C for 10 min.

As given by Ref. [15, 16], the yield drop phenomenon can be removed by introducing deformation. In the present study, the 750 °C - 10 min annealed sample was applied by 15% hot rolling. The corresponding microstructure and stress - strain curves are shown in figure 6. It is clear that the dislocations were successfully introduced in the rolled sample. And the rolled sample shows a continuous yielding with a YS of 800 MPa, UTS of 833 MPa and a TE of 10%.

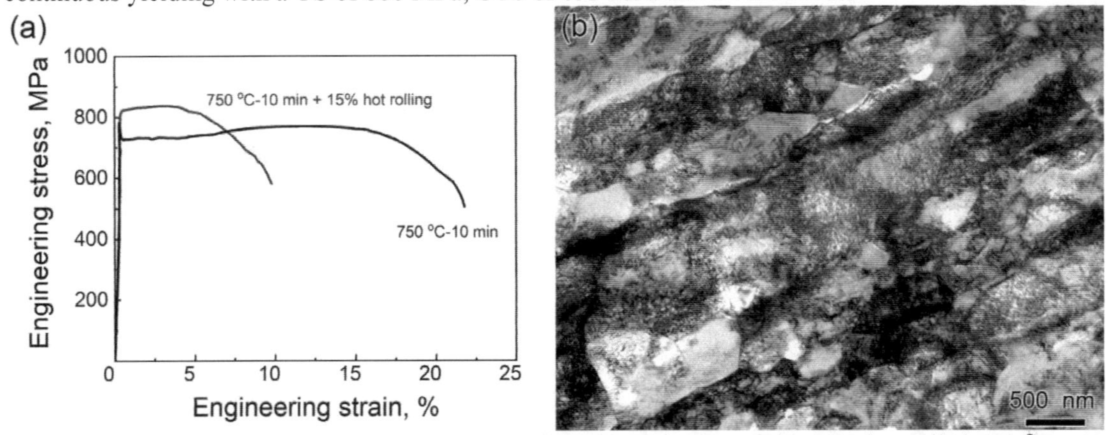

Figure 6. (a) Engineering stress-strain curves and (b) TEM image of the 15% hot rolled 750 °C - 10 min annealed sample.

4. Summary

Warm drawing was applied to 0.1C5Mn3Al dual phase steel. With the increase of the drawing strain, the layer thickness was effectively refined and the tensile strength of the samples gradually increased, but the elongation decreased rapidly. The warm drawn sample with Von Mises strain of 1.96 has a layer thickness of 1.8 μm and a UTS of 1388 MPa. It is the highest known strength of the 0.1C5Mn3Al dual phase steel. Further heat treatments of WD-1.96 sample lead to a decrease in strength and gradual recovering of the elongation, but most of the stress-strain curves exhibit obvious discontinuous yielding phenomenon. Further deformation is effective to remove this discontinuous yielding as illustrated by 750 °C - 10 min annealed sample with 15% hot rolling.

Acknowledgements

This work was supported by the National State Key Research and Development Program of China (2016YFB0700403), the National Natural Science Foundation of China (51671039) and the Fundamental Research Funds for the Central Universities (Project No. 2020CDJDPT001).

References

[1] Zhang M D, Hu J, Cao W Q, Dong H. *Mater. Sci. Eng. A* 2014 **618** 168-175
[2] Zhang Y J, Hui W J., Zhao X L, Wang C Y, Cao W Q, Dong H. *Eng. Fail. Anal.* 2019 **97** 605-616
[3] Zhang M D, Hu J, Cao W Q, Dong H. *Mater. Sci. Eng. A* 2014 **31** 1349-1354
[4] Liu T Q, Cao Z X, Wang H, Wu G L, Jin J J, Cao W Q. *Scr. Mater.* 2020 **178** 285-289
[5] Zhang L J, Liang M Y, Feng Z Q, Zhang L, Cao W Q, Wu G L. *Mater. Sci. Eng. A* 2019 **745** 144-148
[6] Zhang L, Chen Z, Wang Y H, Ma G Q, Huang T L, Wu G L, Jensen D J. *Scr. Mater.* 2017 **141** 111-114
[7] Kimura Y, Inoue T, Tsuzaki K. *J. alloys compd* 2013 **577** S538-S542
[8] Karmakar A, Sivaprasad S, Nath S K, Misra R D K, Chakrabarti D. *Metall. Mater. Trans. A* 2014 **45** 2466-2479
[9] Song R, Ponge D, Raabe D. *Acta Mater.* 2005 **53** 4881-4892
[10] Dolzhenko A, Yanushkevich Z, Nikulin S A, Belyakov A, Kaibyshev R. *Mater. Sci. Eng. A* 2018 **723** 259-268
[11] Dolzhenko A, Kaibyshev R, Belyakov A, *Mater. Lett.* 2021 **303** 130547
[12] Kimura Y, Inoue T. *ISIJ Int.* 2020 **60** 1108-1126

[13] Dolzhenko A, Kaibyshev R, Belyakov A. *Metals* 2020 **10** 1566
[14] Huang X X, Kamikawa N, Hansen N. *Mater. Sci. Eng. A* 2008 **493** 184-189
[15] Huang X X, Hansen N, Tsuji N. *Science* 2006 **312** 249-251
[16] Wang Y H, Kang J M, Peng Y, Wang T S, Hansen N, Huang X X. *Scr. Mater.* 2018 **155** 41-45
[17] Kamikawa N, Huang X X, Tsuji N, Hansen N. *Acta Mater.* 2009 **57** 4198-4208
[18] Zhang M D, Cao W Q, Dong H , Zhu J. *Mater. Sci. Eng. A* 2016 **654** 193-202
[19] Jiang X J, Zhang L J, Zhang L, Huang T L, Wu G L, Huang X X, Mishin O V. *Mater. Sci. Eng. A* 2019 **759** 262-271

IOP Conf. Series: Materials Science and Engineering 1249 (2022) 012059 doi:10.1088/1757-899X/1249/1/012059

Effect of annealing on microstructure in railway wheel steel

Dimitrios Nikas[1,2], Yubin Zhang[3] and Johan Ahlström[1,*]

[1] Department of Industrial and Materials Science, Chalmers University of Technology, 412 96 Gothenburg, Sweden

[2] Department of Engineering and Physics, Karlstad University, 651 88 Karlstad, Sweden

[3] Department of Civil and Mechanical Engineering, Technical University of Denmark, 2800 Kgs. Lyngby, Denmark

*E-mail: johan.ahlstrom@chalmers.se

Abstract. Railway wheels are commonly made from medium carbon steels (~0.55 wt.% C), heat treated to a near pearlitic microstructure with 5–10% pro-eutectoid ferrite. During the operation of freight trains, where block brakes are used, high thermal loads occur together with the high contact stresses, which combined can affect the mechanical properties of the material. In this study, the effects of annealing on local microstructure and mechanical properties in pearlitic railway wheel steel were investigated using electron microscopy and micro-hardness. It is found that after annealing at 650 °C, the room temperature hardness reduces about 25%, accompanied by significant spheroidization of cementite in the pearlitic colonies, though the size and the orientation gradients of the pearlitic colonies have not changed much. The relationship between the microstructural changes and the mechanical properties are discussed.

1. Introduction

Demands on railway wheels have increased in recent years due to the strive for higher train speeds for passenger transport and higher axle loads in freight operation. Especially freight wheels are exposed to very high temperatures in the wheel tread due to block braking systems exposing the wheel tread surface to frictional heating. The material in the tread of the wheel is also exposed to very high stresses from the rolling contact, resulting in cyclic plastic strains. Investigations presented in literature show that the temperature in the wheel tread can rise above 500 °C for wheels on freight trains [1]. Qualification tests of such wheels impose frictional heating of the wheel rim to around 600 °C imitating severe drag braking (50 kW, 45 minutes). This is done to ensure resistance against thermally induced radial cracks as well as to confirm that the shape of the wheel remains within limits necessary for the structural integrity (i.e. keeping the wheelset in shape for the rail gauge). Medium carbon steels (with approximately 0.55 wt.% C) are commonly used due to their cost-effective fatigue and wear properties, and because they can be reasonably well maintained under normal service conditions.

The manufacturing of railway wheels involves heat treatments in order to achieve desirable characteristics in mechanical properties and residual stress distributions. After heat treatments, the forged and rolled wheels have a running surface consisting of fine-lamellar pearlite, a structure of alternating hard cementite and softer ferrite lamellae with a lamellar spacing of around 130 nm, and about 5-10% pro-eutectoid ferrite decorating the previous austenite grain boundaries [2,3]. The high temperatures occurring during operation can however degrade the microstructure, more specifically spheroidize the pearlitic microstructure. The effect of the microstructural degradation has been evaluated in previous studies and found to affect both the mechanical properties and the fatigue behavior [3,4]. Systematic studies of the microstructure after annealing at different temperatures are required to quantify the degradation process, information essential for optimization of the processing route and the mechanical properties of the pearlitic steel. It is also a necessity for accurate material modelling with a multi-scale approach, in order to describe the combined effects of thermal and mechanical loadings on orientation and microstructure degradation, yielding the strain and stress development for wheels in operation.

In this study, we use complementary characterization techniques, including optical microscopy (OM), Scanning Electron Microscopy (SEM) and Electron Back Scattered Diffraction (EBSD), to quantify the changes in microstructure after annealing of the pearlitic R8T steel grade used for railway wheels. With OM we can quantify the fraction of the pro-eutectoid ferrite covering a large area and characterize its morphology, while with SEM/EBSD we can characterize the changes in morphology of the pearlite and the crystallographic orientations within pearlite colonies. A previous study concerned pre-deformed wheel material representative of the near surface region, the first few millimeters below the surface, [5] while the current study concerns undeformed material at larger depths, which was exposed to only thermal loadings. The current study is connected to a parallel manuscript on Differential Aperture X-ray Microscopy (DAXM) done on the same material [6], and is supposed to provide the overview of the microstructure and the correlation to the mechanical properties. By characterization of material exposed to annealing treatments across different length scales, the overall picture of the degradation of pearlitic lamellar structure is obtained and becomes useful for multi-scale modelling of mechanical performance.

2. Material and experimental procedures

The material studied in this investigation is the railway wheel steel R8T which complies to the ER8 grade specified in the standard EN 13262-1. The manufacturing process of this wheel material consists of forging, rolling, heat treatment, rim chilling and annealing. Samples were extracted from unused ('virgin') wheels approximately 15 to 20 mm below the wheel tread surface; this material region is of interest since it will be exposed to elevated temperatures and mechanical loading during operation. Different heat treatments were done for the investigation, to provide a complete understanding of the annealing occurring during operation and enable setting suitable limits for use of the material. The experiments presented here concern virgin wheel material annealed for 4 hours at temperatures in the range 250 – 650 °C with an interval of 50 °C, but with the EBSD investigation focused on the cases also represented in the parallel DAXM study [6].

Vickers hardness values were measured before and after annealing for each sample using a Wolpert 2RC hardness tester and a load of 10 kg (HV10) with a dwell time of 12 s. Indents were positioned randomly on the sample surface and averaged based on three indents.

After heat treatment, the specimens were mechanically ground and polished down to 0.04 μm using a colloidal silica suspension. Gentle etching was done using Nital (3% HNO_3 in ethanol) to gain some topographical contrast. The EBSD measurements were carried out in a LEO 1550 high resolution field emission scanning electron microscope (FEG-SEM). The system was equipped with an EBSD detector (Nordlys, Oxford instruments), a high speed camera for EBSD pattern recording, and software for crystal orientation mapping (AZtecHKL). The samples were tilted to 70° and the SEM was operating at an accelerating voltage of 20 kV. Crystallographic orientation maps were recorded within the pearlite colonies as well as within the pro-eutectoid ferrite, with a step size of 130 nm which is approximately the interlamellar spacing of the pearlite in this material [3]. This step size was chosen for mapping of large areas to resolve the orientation gradient within pearlite colonies, rather than resolve the local pearlitic structure. The material microstructure was evaluated in randomly chosen areas which were selected without focusing on any specific pearlite colony or other microstructural features.

Image analysis of SEM images was performed to quantify the degree of spheroidization at the various annealing temperatures. A Matlab script was developed to process SEM images for each sample at every temperature examined. The first step of the analysis was segmentation using a threshold value in brightness, to determine the areas corresponding to cementite and ferrite. A 'thinning' function was thereafter used to account for the sampling error resulting from cementite lamellae protruding after etching. The cementite areas were then separated into lamellar and globular cementite according to the length over width aspect ratio. Chattopadhyay and Sellars [7] state that regions with aspect ratios lower than 8 should be classified as globular. Using this definition and averaging the results from all (25) images for each sample, the percentage of lamellar and globular cementite was obtained for each temperature. A more detailed explanation of the method is available in [8]. An illustration of the division between lamellar and spheroidal cementite for a strongly spheroidised sample is shown in figure 1.

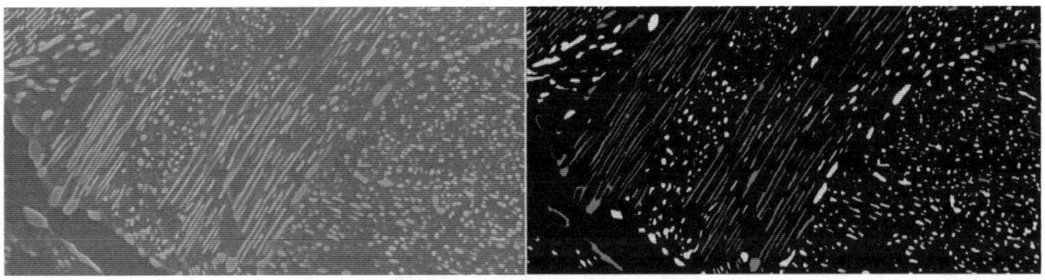

Figure 1. Degree of spheroidization determined from SEM images (from [8]). Original image on the left, processed image on the right (red regions are lamellar cementite defined as L/W>8, green regions are defined as globular).

3. Results

The hardness of the virgin and the annealed samples are shown in figure 2a. The initial hardness of the R8T material in the virgin state is around 260 HV10. The material slightly hardens when annealing around 300 °C, whereas significant softening is observed for higher temperatures. The hardness decrease is approximately 25 % after 4 h annealing at 650 °C.

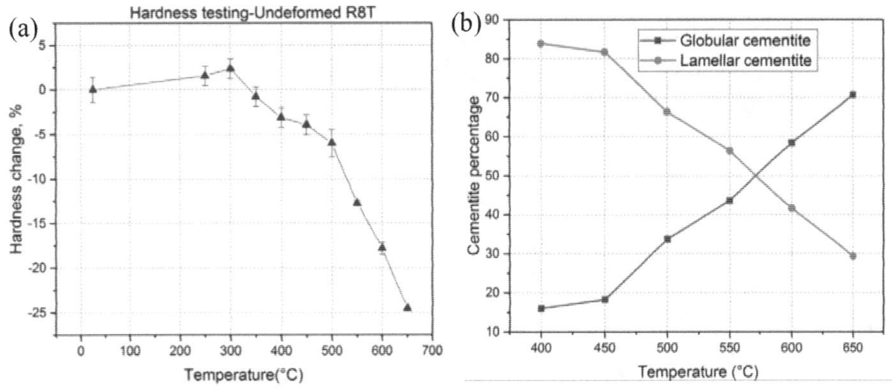

Figure 2. (a) Hardness change after annealing for 4 h relative to the virgin material. (b) Quantification of cementite spheroidization after annealing for 4 h.

The quantified cementite spheroidization fraction determined based on SEM images of the virgin and annealed samples are shown in figure 2b, and figure 3 shows SEM images for some of the examined temperatures. Below 400 °C, no noticeable change is seen in the pearlitic structure. The cementite lamellae start to break up at a temperature around 450 °C, while most of the pearlitic structure remains unchanged. By increasing the annealing temperature, cementite spheroidization becomes evident at 500 °C and gets severe at 650 °C, where ~70% of the cementite is globular (see figure 2b).

An SEM image and an inverse pole figure (IPF) EBSD map of the same field of view are shown in figure 4 for the sample annealed at 300 °C. Due to the small thickness of cementite lamellae, only a very small fraction of the cementite is indexed (~2%). Some grains containing many sub-grain boundaries in the EBSD map and with more contrast in the SEM image are likely pearlitic colonies (see for example the grain marked by P in figure 4a and b). The grains with uniform contrast in the SEM image and no sub-grain boundaries are generally pro-eutectoid ferrite (see for example the grain marked by F in figure 5a and b). Within the pearlite colonies long-range orientation gradients remain (see figure 4c).

Upon annealing at higher temperatures, 500 °C and 650 °C, the grain structures as observed in the EBSD maps do not change much (see figures 5 and 6), even though severe spheroidization of cementite lamellae occurs (figures 2b and 3d). The kernel average misorientation (KAM) map after annealing at 500 °C in figure 5c shows that the majority of the pearlitic colonies (see for example the grain marked by P in figure 5) still have large KAM values, while the pro-eutectoid ferrite grains have very low KAM values, <0.5° (see for example the grain marked by F in figure 5). At 650 °C, the orientation gradients in the ferrite within individual pearlite colonies only slightly redistribute locally (figure 6c), leading to well-defined cell boundaries (figure 6b top, marked P) and an increase in the measured fraction of low angle boundaries (see figure 7). The average low angle boundary spacing (2-15°) changes slightly from 2.8 μm to 2.1 μm after annealing at 300 °C and 650 °C, respectively.

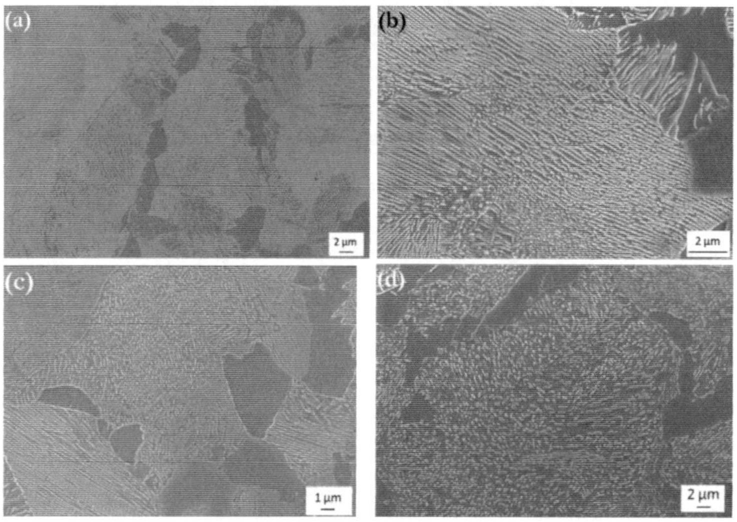

Figure 3. Microstructure of (a) virgin material, representative also for material annealed at 300 °C (b) annealed at 450 °C (c) annealed at 500 °C and (d) annealed at 650 °C. The annealing time was 4 h in all cases.

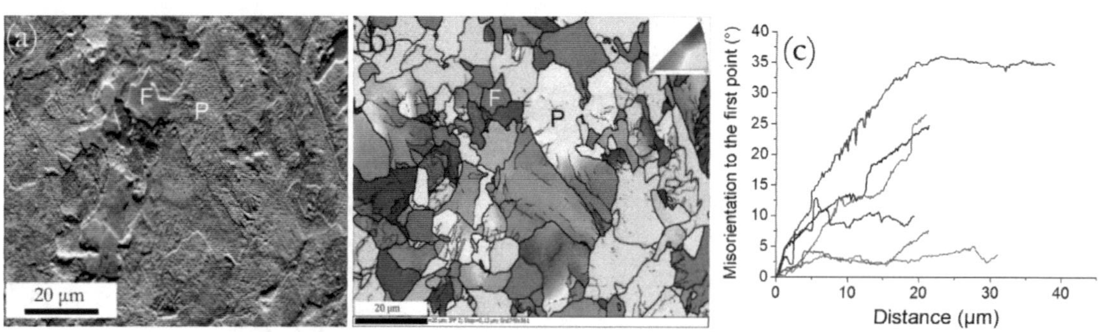

Figure 4. Microstructure and orientations in the sample annealed at 300 °C for 4 h. (a) SEM image, (b) IPF map, and (c) misorientation gradients within 6 randomly chosen pearlite grains. The colors in (b) represent the crystallographic orientation along the measured surface normal direction, see inset. The thin and thick black lines in (b) represent boundaries with misorientation angles >2° and >10°, respectively. Letters F and P mark selected ferrite and pearlite grains, respectively.

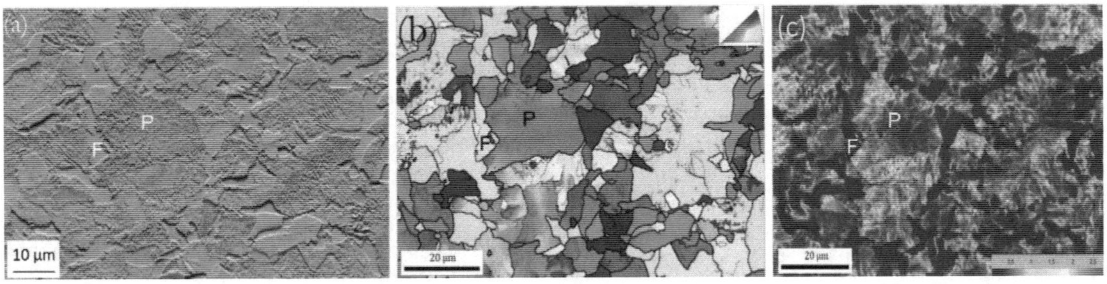

Figure 5. Microstructure and orientations in the sample annealed at 500 °C for 4 h. (a) SEM image, (b) IPF map. The colors in (b) represent the crystallographic orientation along the measured surface normal direction, see inset. (c) KAM map showing orientation variations within grains. KAM value for a given pixel is defined as the average misorientation angle of that pixel to its 8 nearest neighbors, calculated with the proviso that misorientation angles above a cut-off angle (3° for this map) are not counted in the averaging process.

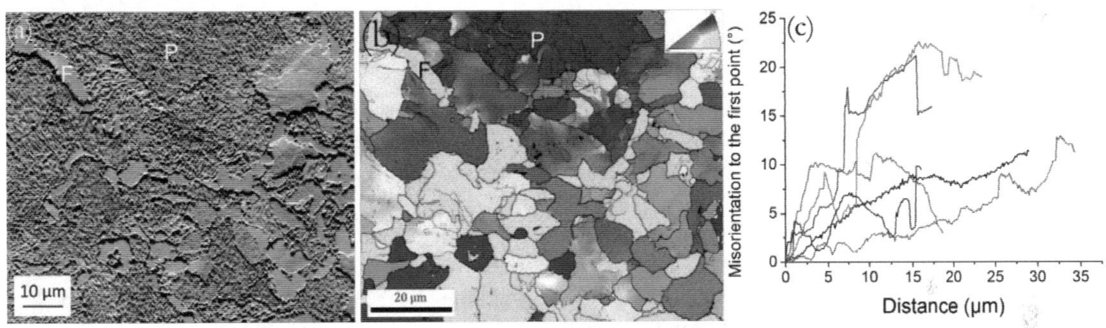

Figure 6. Microstructure and orientations in the sample annealed at 650 °C for 4 h. (a) SEM image, (b) IPF map, and (c) misorientation gradients within 6 randomly chosen pearlite grains. The colors in (b) represent the crystallographic orientation along the measured surface normal direction, see inset.

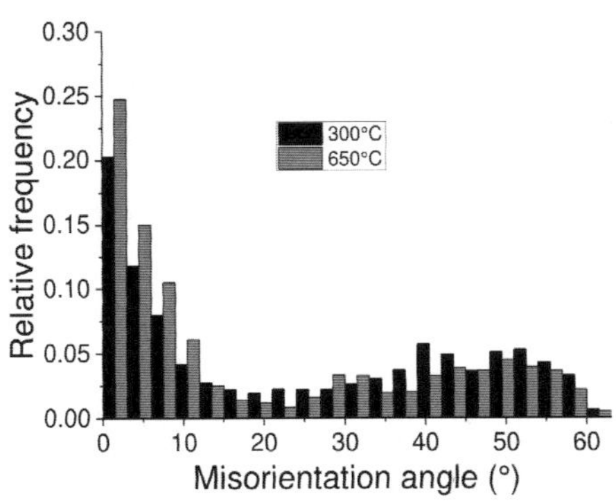

Figure 7. Distribution of misorientations between neighboring pixels of the annealed samples, determined based on the EBSD maps in figures 4 and 6 (step size of 130 nm).

4. Discussion

The results show that upon annealing the cementite lamellae start to break up at 450 °C in the pearlite colonies. Higher annealing temperatures lead to increasing degree of spheroidization. After annealing at 650 °C for 4 h, most of the pearlitic colonies have spheroidized (with cementite aspect ratio < 8). A sub-grain structure has formed in certain spheroidized pearlite colonies, but the long range orientation gradient within less spheroidized pearlite colonies does not change much even after annealing at 650 °C for 4 h. This is in agreement with previous studies reporting long-range orientation gradients remaining since the pearlite was formed in the eutectoid reaction [9]. Only localized changes in subgrain boundary spacing and morphology of the cells occur, which have a rather small effect on average values. This implies that even though the cementite lamellae break up and spheroidize, the neighboring ferrite lamellae re-orient only locally around the cell boundaries, and the total amount of geometrically necessary dislocations (GNDs) that produce the long range orientation gradient within the colonies does not reduce much. This is likely because the dislocations within the (connected) ferrite lamellae are pinned strongly due to the short free mean path between the spheroidized cementite particles and they cannot move long distances to reduce the long range orientation gradients.

Both the reduction in elastic strain and dislocation density may affect the mechanical properties. After annealing at 300 °C for 4h, part of the elastic strain [10] and some dislocations within the free ferrite grains are removed while the cementite and ferrite lamellae remain more or less unchanged. Both the reduction in elastic strain and dislocation density in the free ferrite grains should lead to a reduction of strength of the material, although the fraction of the free ferrite grains is only <10 %. However, this is not the case; the hardness of the annealed sample at 300 °C for 4 h instead increase about 2.5 % compared to the virgin sample. This is due to strain ageing which decreases the mobility of the dislocations [11]. Upon annealing at higher temperature, the elastic strain and dislocation density within the ferrite grains continue to decrease [6], and the cementite lamellae break up and spheroidise. In addition, the total dislocation density (both GNDs and statistically stored dislocations) within the pearlite colonies also reduces as the mean free path between cementite particles increases. All these factors contribute to reduce hardness and strength.

5. Summary

In the present study, the microstructure in annealed wheel steel samples were measured using electron backscatter diffraction in SEM. The microstructure of the unused railway wheel pearlitic steel, R8T, is quite stable upon annealing at 300 °C, while microstructural degradation in term of spheroidization of cementite lamellae in pearlite colonies occurs at temperatures higher than 450 °C. The size of the pearlitic colonies does not change much even after annealing at 650 °C for 4h, though ~70% of the cementite is spheroidized leading to large loss in hardness (-25%).

Acknowledgments

This work is part of the research activities within CHARMEC (Chalmers Railway Mechanics, www.charmec.chalmers.se). It was partly financed within the European Horizon 2020 research and innovation programme in the projects In2Track, In2Track2 and last in the In2Track3 project under grant agreement 101012456. YZ acknowledges the support by the European Research Council (ERC) under the European Union's Horizon 2020 research and innovation programme (grant No 788567, M4D).

References

[1] Peters C J and Eifler D 2009 *Mater. Test.* **51** 748–54
[2] Cvetkovski K, Ahlström J and Karlsson B 2011 *Wear* **271** 382–7
[3] Nikas D, Ahlström J and Malakizadi A 2016 *Wear* **366–367** 407–15
[4] Nikas D and Ahlström J 2014 *Proceedings of the 35th Riso international symposium of material science, Eds S. Faester, et al.* pp 411-420
[5] Nikas D and Ahlström J 2015 *IOP Conf. Series: Mater. Sci. Eng.* **89**
[6] Zhang Y, Jessop C, Nikas D, Yu T, Liu W and Ahlström J 2022 submitted to *42nd Risø international symposium of material science*
[7] Chattopadhyay S and Sellars C M 1977 *Metallography* **10** 89–105
[8] Nikas D 2018 PhD thesis Chalmers Univ Tech
[9] Takahashi T, Ponge D and Raabe D 2007 *Steel Res. Int.* **78** 38–44
[10] Yildirim C, Jessop C, Ahlström J, Detlefs C, and Zhang Y 2021 *Scr. Mater.* **197** 113783
[11] Baird J D 1971 *Metall. Rev.* **16** 1–18

IOP Conf. Series: Materials Science and Engineering 1249 (2022) 012060 doi:10.1088/1757-899X/1249/1/012060

Recrystallization kinetics of the AA5xxx series aluminum alloys - revisited

Roy Vandermeer[1], Fengxiang Lin[2], Xiuchuan Lei[3], Robert Edward Sanders[4] and Dorte Juul Jensen[5]

[1]Consultant, Maryville TN, USA
[2]Department of Engineering and Physics, Karlstad University, 65188 Karlstad, Sweden
[3]Key Laboratory for Light-weight Materials, Nanjing Tech University, Nanjing 210009, People's Republic of China
[4]International Joint Laboratory for Light Alloys (Ministry of Education), Department of Materials Science and Engineering, Chongqing University, Chongqing, People's Republic of China
[5]Department of Civil and Mechanical Engineering, Technical University of Denmark, DK-2800, Kgs. Lyngby, Denmark

E-mails: doje@dtu.dk; fengxiang.lin@kau.se; xclei@njtech.edu.cn

Abstract. In previous work [1], we analyzed the recrystallization kinetics in cold-rolled AA5657 and AA5182 based on measurements of the volume fraction of recrystallized material, the free (unimpinged) surface area density and the average size of the recrystallizing grains, plus the contiguity between recrystallizing grains, for a series of partially recrystallized samples. In that work, the calculated growth and nucleation rates were used as basis from the analysis. These two properties are derived from the measured parameters, which, as shown in the present work, adds to their uncertainty. In the present paper, we revisit the kinetics using improved stereological data and perform a full microstructural path analysis. The new analysis results are compared to the previous ones and the implications of using the different types of data sets and analytical approach for the interpretation of the kinetics are discussed.

1. Introduction

Primary recrystallization of metals and alloys is the thermally-activated microstructural evolution process whereby a population of new grains, which are generally assumed to be almost defect and strain free, emerges from within the volume of a deformed metal or alloy (nucleation) and with time these new grains increase in size by means of an interface migration process (growth) thereby consuming the deformed volumes [2-4]. The dynamics of recrystallization can generally be interpreted in terms of nucleation and growth kinetics using the science of stereology. The experimental data consist of measuring the time dependence of global microstructural properties, principally, the volume fraction recrystallized grains V_V, the grain boundary area density separating recrystallized grains from the deformed grains S_V, and the grain boundary area density of the recrystallized grains that have impinged upon one another during the course of annealing S_{Vrex} [5]. The dependence of such data on the annealing time allows microstructural path models (MPM) to be devised for the recrystallization process being studied [6,7].

In previous work [1], the recrystallization kinetics of two cold-rolled industrially processed alloys, namely AA5182 and AA5657 aluminum alloy sheets, have been investigated and compared. The work focused on effects of the different deformation microstructures of the two alloys on the recrystallization kinetics. In the initial 78% cold rolled, and 235-240°C batch annealed condition both alloys showed highly elongated grains. Large (1-10 µm) constituent phases e.g., $Al_x(Fe, Mn)_ySi$, $Al_6(Fe, Mn)$ were present in AA5182 (approx. 0.5 vol%), while AA5657 contained only a few Fe-containing particles (< 0.1%). Also, Mg_2Si particles and fine Mn-rich dispersoids of about 0.1~0.4 µm were present in AA5182. The low Mn content in AA5657 results in a negligible volume fraction of dispersoids in this alloy. In the initial condition the AA5182 specimen had an extremely high dislocation density with no evidence of recovery. By contrast, the AA5657 exhibited well-developed subgrains, 1-2 µm in size, typical of recovery in relatively high-purity aluminum alloys. For the experiment, specimens were cut from both alloy sheets and annealed isothermally at 245°C (AA5182) and 300°C (AA5657) for various durations, ranging from 5 minutes to 1440 minutes. Thereby a series of partially recrystallized specimens were obtained and characterized by electron backscattered diffraction to determine the key stereological parameters. For further details see [1].

The aim of the present paper is to revisit the data from the previous work and use an improved stereological analysis combined with advanced MPM to quantify the recrystallization kinetics. The present work will highlight issues related to the analysis in [1] of recrystallization kinetics, discuss effects of the grains not growing isotopically, present an advanced MPM methodology applicable to such case, and finally compare the recrystallization kinetics of the two industrial aluminum alloys as well as compare to the results obtained in [1].

2. Stereological analysis

Stereology involves the 2-dimensional interpretation of global microstructural properties from a 2-dimensional cross-sectional examination on surfaces cut through a material specimen. It is a method that utilizes random, systematic sampling to provide unbiased quantitative data [8]. The sectioning plane(s) should thus be at a randomly selected angle(s) relative to any major sample axis, and if lines are used for determining the microstructural parameters, they should also be in random directions.

Heavily cold rolling produces a strong directionality in the deformed microstructure, which gives rise to recrystallizing grains which are most often elongated along the rolling direction. This is also the case for the present aluminum alloys, see figure 1. The aspect ratio, i.e., the size of the recrystallizing grains measured along the rolling direction (RD) divided by that along the normal direction (ND), is around 1.8 for AA5657 while it is around 1.4 for AA5182 [1]. Furthermore the rolling will also affect the spatial distribution of the second phase particles and dispersoids, further adding to the directionality of the recrystallizing microstructures.

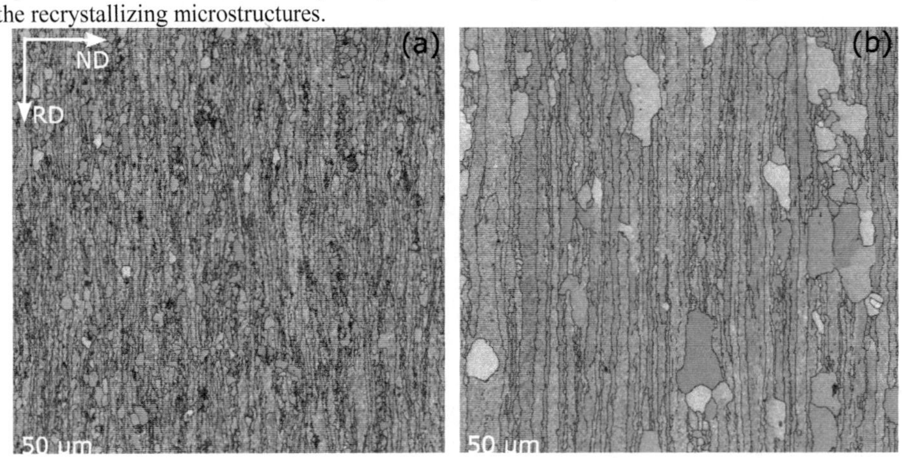

Figure 1. Boundary maps overlaid with band contrast maps of the partially recrystallized specimens. (a) AA5182 (annealed at 245 °C for 200 min) and (b) AA5657 (annealed at 300 °C for 60 min). The cyan and black lines represent low angle (2°-15°) and high angle (>15°) boundaries, respectively.

For studies of recrystallization kinetics in rolled samples, a specific plane, like the rolling plane (defined by RD and the transverse direction TD) or the longitudinal plane (defined by RD and ND,) is often used for the analysis, e.g. [1, 9-11]. This already violates the random principles of stereology. Furthermore, because of the marked directional character of the microstructure (see figure 1), we in [1], did choose to analyze the kinetics separately along the two dominating directions, namely RD and ND for both alloys. As expected, this gave rise to quite different S_V versus V_V curves for the two directions, while the V_V versus time data were unaffected by the line direction within the experimental uncertainty, see figure 2. The data thus gave a clear picture of the non-isotropic recrystallization, but the measured stereological properties were biased in RD or ND [1]. Averaging the microstructural properties measured along RD and ND may remove much of the bias and improve the ability to create viable microstructural models, as was done in e.g. [9]. This averaging approach is thus used here. Figure 2 shows the result. The unbiased V_V curves follow that found in the previous analysis while the S_V curves fall between the two curves measured along RD and ND for both alloys. These novel averaged data are in the following used as input for a MPM analysis of the recrystallization kinetics.

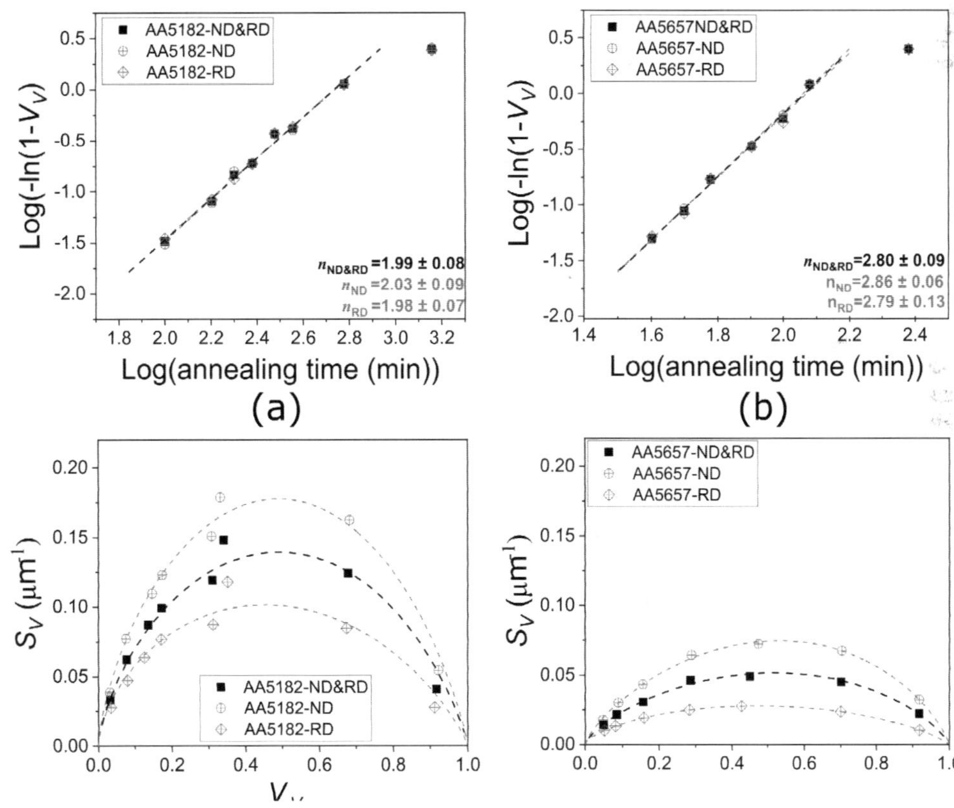

Figure 2. Microstructural properties describing recrystallization. (a) and (c) are for AA5182, (b) and (d) are for AA5657. In black is shown the data averaged over RD and ND, while grey lines and points are used for the biased data measured along RD or ND.

3. Microstructural Path Modelling

It was clear already from the early analysis [1], that the early KJMA models [12-14] were inappropriate for the present aluminum alloys. This can also be seen from figure 2 (a) and (b) showing that the data cannot be fitted well by a single straight line. We will thus here analyze the data based on the microstructural path approach, e.g. [15]. For a viable MPM to be devised, the kinetic character of the combined nucleation and growth behavior needs to be delineated and formalized.

Following a large body of prior studies, summarized in [15], a time-dependent nucleation rate, $\dot{N}$, is assumed to be of the form

$$\dot{N} = N_1 t^{\delta - 1} \quad (1)$$

where t is the annealing time and N_1 and δ are model constants. A growth rate, G, (grain boundary migration rate) is assumed to be given by

$$G = P t^{r-1} \quad (2)$$

where P and r are model constants.

An additional important assumption must be adopted regarding about how to account for the impingements between the recrystallizing grains. In the present work, it is assumed that the impingements occur uniformly throughout the microstructure. This 80-year-old uniform impingement concept has proven quite useful in recrystallization modelling studies and is expressed mathematically as

$$\frac{dV_V}{dV_{Vex}} = 1 - V_V \quad (3)$$

where V_{Vex} is the extended volume fraction, i.e. the imaginary volume fraction that would be estimated when the recrystallizing grains are conceptualized to grow unimpeded through one another and where already recrystallized grains would continue to nucleate new grains. That the uniform impingement criterion is appropriate for our purposes here, is demonstrated by the good agreement of the experimental data with characteristic time-dependent power-law functions (the simplest integrated form that equation 3 reduces to) for annealing times up to about 70% recrystallization. For example, the experimental data for $0 < V_V < 0.70$, for both alloys, are well described by the Avrami equation (see figure 2 a,b)

$$V_V = 1 - \exp(-Bt^k) \quad (4)$$

where B and k are constants. In this modelling work $V_{Vex} = Bt^k$. An important companion equation that is necessary to consider is the DeHoff equation for the interfacial density [5], namely

$$S_V = (1 - V_V) \cdot S_{Vex} \quad (5)$$

where the extended interfacial area, $S_{Vex} = Kt^m$, K and m are constants. Additionally, a microstructural path function may be described, relating S_V and V_V [15]. It is given by

$$S_V = (1 - V_V) \cdot C \cdot [-\ln(1 - V_V)]^q \quad (6)$$

where C and q are constants. Of the three equations 4-6, only two are independent.

For the present data, models such as the "relaxed" KJMA models [6] are not appropriate as they idealize the recrystallized grains as spheres, which they are clearly not [1], see also figure 1. Based on the directionally-measured average grain aspect ratios of 1.8 and 1.4 [1], it is proposed that the grains could be better modelled as shape-preserved spheroids of either oblate or prolate shapes. This requires an additional microstructural parameter to enter in the analysis, namely the grain eccentricity, ε, given by

$$\varepsilon^2 = 1 - \left(\frac{b}{a}\right)^2 \quad (7)$$

where a and b are the major and minor semi-axis radii, respectively. It is proposed that b/a equals the inverse of the measured average aspect ratios.

A generalized approach for modelling recrystallization kinetics in cases where the new grains nucleate and grow as spheroids has been detailed in [16]. The method utilizes the mathematics of LaPlace Transforms to analyze the time-dependence of the experimentally observed global microstructural properties. The recrystallized grains are assumed to be either oblate (o) or prolate (p) spheroids requiring shape factors K_V^i and K_S^i , where i refers to either (o) or (p), to describe their geometric volumes and interfaces, respectively. For the oblate grains

$$K_V^o = \frac{4\pi}{3}(1-\varepsilon^2)^{1/2} \tag{8a}$$

$$K_S^o = 2\pi\left\{1+\frac{1}{2}\left(\frac{1-\varepsilon^2}{\varepsilon}\right)\ln\left[\frac{1+\varepsilon}{1-\varepsilon}\right]\right\} \tag{8b}$$

and the prolate ones

$$K_V^p = \frac{4\pi}{3}(1-\varepsilon^2) \tag{8c}$$

$$K_S^p = 2\pi\left\{1-\varepsilon^2+(1-\varepsilon^2)^{1/2}\cdot\frac{\sin^{-1}\varepsilon}{\varepsilon}\right\} \tag{8d}$$

Following [16], the equations, equivalent to equations 4 and 5, are taken as

$$V_{Vex} = K_V^i P^3 N_1 \Gamma(B) t^{3r+\delta} = Bt^k \tag{9}$$

$$S_{Vex} = K_S^i P^2 N_1 \Gamma(K) t^{2r+\delta} = Kt^m \tag{10}$$

where the gamma functions are $\Gamma(B) = \Gamma(k+1-\delta)\Gamma(\delta)/\Gamma(k+1)$ and $\Gamma(K) = \Gamma(m+1-\delta)\Gamma(\delta)/\Gamma(m+1)$, which can be evaluated from the exponents given inside the parentheses.

Equations 9 and 10 are considered equations of the proposed model and the various quantities in them can be evaluated in the manner put forth in [15]. Thus, from the exponents, we calculate that $r = k-m$, the growth exponent term (see equation 2). Once r is known, δ can be estimated from the exponent k or m. By fitting the pre-exponential terms B and K, the model quantities N_1 and P can be calculated:

$$P = \frac{B}{K}\cdot\frac{\Gamma(K)}{\Gamma(B)}\cdot\frac{K_S^i}{K_V^i} \tag{11}$$

$$N_1 = \frac{B}{\Gamma(B)\cdot P^3\cdot K_V^i} \tag{12}$$

The calculated quantities, P and N_1, when inserted as the pre-exponential parameters with r and δ in equations 1 and 2, the result essentially describes the average growth path representing a spectrum of recrystallizing grains, i.e., a model of the recrystallization process in each of the two alloys.

4. Results on recrystallization kinetics of AA5657 and AA 5182 – revisited

The original analysis presented in [1] was based on calculations of the impingement-corrected Cahn Hagel growth rate [17] and the nucleation rate. These two parameters are derived mathematically from the measured microstructural parameters. For example, the Cahn-Hagel growth rate is determined from the time derivative of V_V

$$\langle v\rangle_{ch} = \left(\frac{1}{S_V}\right)\cdot\left(\frac{dV_V}{dt}\right) \tag{13}$$

A function for $V_V(t)$ is needed to calculate the derivative, and this function must be very accurately known, or else different results are obtained for the growth rates. This is illustrated in figure 3. A solid result requires many precise data points in particular for short and long annealing times. With this uncertainty in mind, the original data – now averaging RD and ND results - are re-analyzed by the above-described advanced MPM approach. Table 1 encapsulates the nucleation and growth characteristics obtained for the two alloys.

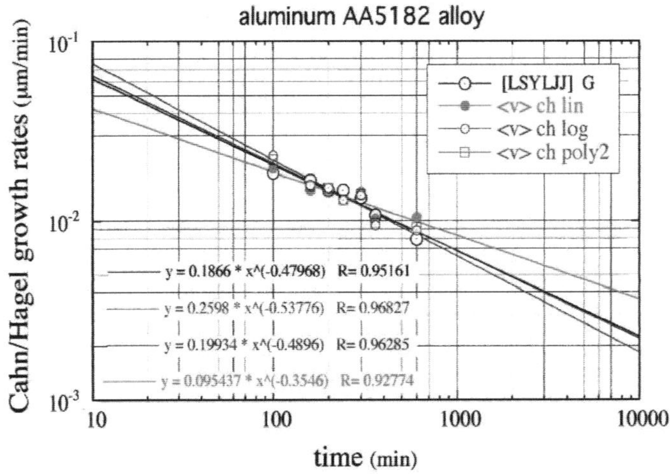

Figure 3. Figure illustrating issues related to determining precise growth rate results using equation 13. Four different functions are used for fitting the V_V (t) using the measured V_V data for AA5182. All functions gave an excellent fit to the experimental V_V data. (The one called LSYLJJ refers to that used in [1]).

Table 1. Measured and simulated recrystallization parameters, the aspect ratio and grain intercept length D are measured, while Nv, the growth rate and grain intercept length (λ) are simulated.

Alloy	Annealing time (min)	Annealing Temperature °C	Aspect ratio	D at V_V = 1 (μm)	N_V at V_V = 0.7	Growth rate at V_V = 0.7 μm/min	λ (μm) Underwood at $V_V = 0.7$
AA5182	600	245	1.4	8	3.7×10^{-4}	0.0173	9.2
AA5657	120	300	1.8	24	2.0×10^{-5}	0.292	24.4

Together the nucleation and growth behavior predicted by the models yield, upon application of Underwoods equation $\langle\lambda\rangle = \frac{3}{4}\left(\frac{V_V}{N_V}\right)^{\frac{1}{3}}$ [8], almost fully recrystallized grain sizes of 9.2 and 24.4 μm respectively for AA5182 and AA5657. These values are close to the experimentally observed values of 8 and 24 μm reported in [1], supporting the validity of the MPM model assumptions.

It is difficult to compare the recrystallization characteristics of the two alloys to one another directly because they were annealed at different temperatures and had different histories after cold deformation even though the deformation was approximately the same. The model results (see table 1), however clearly reveal that the AA5182 alloy was about 20 times more prolific in nucleating recrystallizing grains (see N_V in Table 1). The modelling data also show that nucleation in AA5182 is approximately "site-saturated", i.e., all nucleation happened at times short compared to the first annealing time, whereas the AA5657 demonstrated a somewhat reluctance to nucleation, requiring a detectable time-dependent nucleation component. The much more potent nucleation in AA5182 than in AA5657, is very likely to be a results of the many large second phase particles in this alloy leading to a substantial contribution from particle stimulated nucleation, e.g. [18,19]. The AA5657 alloy demonstrated a much faster growth path migration rate than did the AA5182. This may in part be due to the higher annealing temperature used for that alloy, but boundary pinning by the dispersoids in AA5182 is considered to be an important factor, as well. The observed the low Avrami exponent of 2 (see table 2) for this alloy relates to the significant decrease in growth rate with increasing annealing time (see figure 3). The pinning by dispersoids is likely to be an important factor leading to the decreasing growth rate, but other factors, such as orientation pinning, may also contribute.

In the previous analysis directional data and derived properties, like the growth rate, were used as the basis [1], whereas the present analysis is based on full MPM. In spite of this difference, reasonable agreement between the two sets of characteristic recrystallization exponents are obtained (see table 2). The previous analysis suggested that i) the nucleation density and rate are significantly higher in AA5182 than in AA5657, ii) the nucleation sites are not randomly distributed but clustered, and iii) the growth rate is faster and decreases less with time in AA5657 than in AA5182. The present much more advanced analysis confirms that many more nuclei develop, but the growth is slower, in AA5182 compared to the purer AA5657. For AA5657 the results related to the nucleation rates obtained by the two methods agree reasonably well, whereas a significant difference is found for AA5182 (see δ in table 2). In fact the present analysis suggests that site saturation is an adequate assumption for AA5182. The new analysis furthermore reveals that the importance of clustering, suggested in [1], may be weak. So far MPM approach can deal with effects of clustering in the case of site saturation, for the future, it would be of interest to extend the model to also be applicable for constant nucleation situations.

5. Conclusions

Stereology relies of random sampling of relevant microstructural parameters from 2-dimensional cross-sectional examinations. In most recrystallization cases, some directionality, i.e. variations from one principal sample axis to another, are typical. To understand how the underlying variations in deformation microstructures affect the recrystallization, it would be advantageous to have sampling methods which pick-up such directionalities and yet provide unbiased stereological data. If random sampling cannot be done, it is suggested to measure the microstructural parameters along all principal sample axes and then properly average the data for a stereological analysis.

From the measured relevant microstructural parameters, which for recrystallization includes the volume fraction recrystallized, and the grain boundary area densities separating recrystallized grains from the deformed matrix and from other recrystallizing grains, information about the nucleation and growth may be derived directly from the measured data; for example, the impingement corrected Cahn Hagel growth rate (see equation 13). The present work illustrates, that when doing so, very high quality data are needed, in particular, for short and long annealing times.

Microstructural path modelling is a very powerful tool to derive models quantifying the fundamentals of nucleation and growth processes. When used in its full, even directional results are achieved from proper stereological data. The present MPM suggests that the recrystallizing grains in AA5657 and AA5182 grow as shape preserved spheroids of either prolate or oblate shapes. Even though it is difficult to compare the recrystallization characteristics of the two alloys AA5182 and AA5657 to one another directly, it is clear that the many large particles and dispersoids in AA5182 leads to more and faster nucleation but much slower growth than that observed in the purer AA5657.

Table 2. Comparison the Avrami exponent (called k in this paper and n in [1]), and values for δ and r found in the present work and in [1]

Alloy		$n=k$	r	δ	$\delta+3r$	$\delta+3r-n$
AA5182	Present work	2.0	0.65	0.04	1.99	0
	[1]	2.0	0.57	0.41	2.12	0.12
AA5657	Present work	2.8	0.76	0.52	2.80	0
	[1]	2.8	0.82	0.49	2.95	0.15

Acknowledgement
DJJ and FL acknowledge the support from the European Research Council (ERC) under the European Union's Horizon 2020 research and innovation programme (Grant agreement No.788567).

References
[1] Lei X, Sanders R E, Yang X, Lin F, Juul Jensen D 2021 *Metallurgical and Materials Transactions A* **219** 012031
[2] Humphreys F J, Rohrer G S and Rollett A 2017. *Recrystallization and related annealing phenomena*. 3rd Ed., (Amsterdam, Elsevier)
[3] Doherty R D, Hughes D A, Humphreys F J, Jonas J J, Juul Jensen D, Kassner M E, King W E, McNelley T R, McQueen H J, Rollett A D 1998 *Materials Today* **1** 14-15
[4] Juul Jensen D, Zhang Y B 2020 *Current Opinion in Solid State and Materials Science* **24** 100821
[5] DeHoff R T 1972 *Treatise on Materials Science* **1** 247-92
[6] Vandermeer R A, Masumura R A, Rath B B. 1991 *Acta Metallurgica et Materialia* **39** 383-9
[7] Vandermeer R A 2005 *Acta Materialia* **53** 1449-57
[8] DeHoff R T and Rhines F N 1968 *Quantitative Microscopy* (New York, McGraw-Hill)
[9] Vandermeer R A and Juul Jensen D 2001 *Acta Materialia* **49** 2083-94
[10] Vandermeer R A and Juul Jensen D 1994 *Scripta Metallurgica et Materialia* **30** 1575-80
[11] Lu Y, Molodov D A and Gottstein G 2011 *Acta Materialia* **59** 3229-43
[12] Avrami M. 1939 *The Journal of Chemical Physics* **7** 1103-12
[13] Kolmogorov A N 1937 *Izv. Akad. Nauk SSSR, Ser. Mat.* **1** 355-9
[14] Johnson W A and Mehl R F 1939 *Transactions of AIME* **135** 396-415
[15] Vandermeer R A 1995 *Proc. 16th Risoe Int. Symp. on Materals Science* (September 1995, Risø, Denmark)193-213
[16] Vandermeer R A and Rath B B, *Proc. Recrystallization 90'* (January 1990, Wollongong,Australia) 48-58
[17] Cahn J W and Hagel WG 1963 *Acta Metallurgica* **11** 561-74
[18] Humphreys F J 1977 *Acta Metallurgica* **25** 1323-44
[19] Lei X, Zhang Y. Sun J, Bachmann F, Yang X, Sanders R E 2020 *Materials Research Letters* **9** 65-70

IOP Conf. Series: Materials Science and Engineering 1249 (2022) 012061 doi:10.1088/1757-899X/1249/1/012061

The Effect of Compositional Heterogeneity on the Martensite Start Temperature of a High Strength Steel During Rapid Austenitisation and Cooling

M Taylor[1], J W Fellowes[1], P O Hill[2], M J Rawson[2], T L Burnett[1] and E J Pickering[1,3]

[1]Department of Materials, University of Manchester, Manchester M13 9PL2
[2]Rolls-Royce plc., PO Box 31, Derby DE24 8BJ, UK 3UK
[3]Henry Royce Institute for Advanced Materials, The University of Manchester, Manchester M13 9PL, UK

E-mail: mark.taylor-5@manchester.ac.uk

Abstract. Many low alloy steels are compositionally heterogeneous due to the preferential segregation of alloying elements to the dendritic or interdendritic regions when the steel is first solidified (microsegregation). This segregation is often ignored when using CCT diagrams to predict the phase transformation behaviour of steels, but may be of importance in some industrially-relevant cases. This work focuses on the martensite start temperature (M_s) of Super CMV, a high strength aerospace steel, after rapid austenitisation treatment to various peak temperatures from 900 °C to 1300 °C. It was found that the average M_s temperature increased with increasing peak temperature (and prior austenite grain size) at peak temperatures of 1100°C and above, which is to be expected conventionally. However, at peak temperatures of below 1100°C, the M_s temperature increased with decreasing prior-austenite grain size. It is proposed that this was due to the presence of non-dissolved carbides in these conditions, particularly in enriched bands of material, which deplete the matrix and hence raise M_s.

1. Introduction

To predict the microstructures developed during cooling for a ferritic steel which has undergone a heat treatment above its austenitisation (Ac_3) temperature, we typically apply models of austenite grain growth alongside continuous cooling transformation (CCT) diagrams. This approach assumes the steel to be compositionally homogeneous. However, in most industrial situations the steel is chemically inhomogeneous owing to the preferential segregation of alloying elements to the dendritic or inter dendritic regions when the steel was first solidified. These regions of segregation often persist as bands, typically a few hundred microns in width, due to the subsequent thermomechanical processing of the steel, and may affect the phase transformation behaviour of the steel locally, both on heating and cooling [1, 2].

The martensite transformation start temperature (M_s) is a parameter commonly measured to construct CCT diagrams and to study/predict the transformation in many steels. It is known to be affected by prior austenite grain size (PAGS), alloy composition and mechanical deformation. When measured by quenching dilatometry, M_s varies with composition and increases with PAGS according to a logarithmic relationship [3, 4]. This relationship has been defined empirically for low-alloy steels by Lee and Park [5] as:

$$M_s = 475.9 - 335.1x_C - 34.5x_{Mn} - 1.3x_{Si} - 15.5x_{Ni} - 13.1x_{Mo} - 9.6x_{Cu} - 11.67ln\left(d_y\right) \quad (1)$$

Where M_s is the martensite start temperature in °C, x_m is the composition of respective element in wt% and d_y is the prior austenite grain (PAG) equivalent circle diameter in μm. The equation illustrates that, in some cases, chemical composition may have a greater effect on M_s than PAGS. It follows that local compositional changes could have a significant effect on phase transformation behaviour and various authors have shown that this is the case, although they often focus on situations where different microstructures/phases form in different bands within the same block of material [1, 6, 7, 8, 9, 10]. In cases where compositionally banded steel is rapidly quenched to a martensitic or bainitic microstructure in both enriched and depleted bands, evidence of the banding can usually still be found after etching [1, 11-13]. In most cases, the literature suggests that effect of banding appears to be reduced where the PAGS is greater than the width of the bands [1, 10]. It was reported by Basso et al. [12] that banding can create a bimodal PAGS distribution in a high carbon and silicon cast steel austenitising at 920 °C. This effect was attributed to the presence of undissolved carbides in one band.

The work presented here investigates the effect that extreme heat treatments, which involve rapid heating and cooling to/from the Ac_3 temperature, have on the martensitic transformation. Such conditions may be experienced in scenarios where welding or extreme frictional heating are involved. In these situations, the interplay between the chemical micro-segregation and the brevity of the event complicate the microstructures that are formed. Here, the martensitic transformation behaviour of a compositionally-banded specimen of Super CMV (SCMV), a high strength low alloy aerospace steel, is studied using quenching dilatometry after rapid austenitisation treatment to various peak temperatures from 900 °C to 1300 °C. The resulting microstructures are characterised using optical microscopy and high resolution scanning electron microscopy (SEM). Large area electron backscatter diffraction (EBSD) montage mapping with prior austenite grain (PAG) reconstructions is combined with electron probe micro analysis (EPMA), and micro-hardness to demonstrate the effect of micro-segregation on the resulting microstructures.

2. Experimental Procedure

SCMV of the composition 0.39C-3.25Cr-0.95Mo-0.55Mn-0.20V-0.23Si wt% was supplied as a block, but with the banding present elongated in a single direction (presumed to be the forging direction), such that it had a distinct top down and side on orientation (see figure S1 (a-b) in the Supplementary Information for examples). Dilatometry samples were electrical discharge machined from the block as 10 x 4 mm hollow cylinders with a 2 mm diameter hole through the centre. The longitudinal direction of the cylinders ran parallel to the bands in the block. A TA Instruments DIL805 A/D/T dilatometer was used in quenching mode to run at least 4 samples through each of the temperature profiles shown in figure 1 (a). To simulate conditions in a weld/frictional heating event heat affected zone (HAZ), a 150 °C s^{-1} heating rate was used and the samples were quenched immediately. The temperature profiles were kept identical save for the peak temperatures used, which were 900 °C, 1000 °C, 1100 °C, 1200 °C and 1300 °C. The 10 °C s^{-1} cooling rate was chosen to simplify the experiment and ensure that the samples transformed only to martensite upon cooling (not a martensite-bainite mix). The M_s and Ac_3 values were calculated using the offset method as described by Yang and Bhadeshia and demonstrated in figure 1 (b) [14]. Further details of this can be found in the Supplementary Information.

In preparation for microscopy and micro-hardness, at least one dilatometry sample from each heat treatment condition was cut in half and mounted in conductive Bakelite such that a cross section from the centre of the sample in either orientation could be imaged, although only the top down orientation underwent optical microscopy and single tile EBSD. The samples were ground, then polished with diamond paste to a quarter micron finish. For EPMA and EBSD the samples were polished to a 0.04

μm colloidal silica finish, and for optical microscopy and SEM and EDS, the samples were then etched in 5 % Nital. For the 1300 °C peak temperature sample, an entire cross section of the sample was EBSD mapped so that several bands were covered and enough grains were measured for statistical analysis after PAG reconstruction. For the finer grained 1000-1200 °C samples half a cross section was mapped, and for the even finer grained 900 °C sample a quarter cross section. Only a section of the 1000 °C sample underwent EPMA, and this area was correlated to a section of the EBSD montage for the same sample. More comprehensive experimental details can be found in the Supplementary Information and in Table S1.

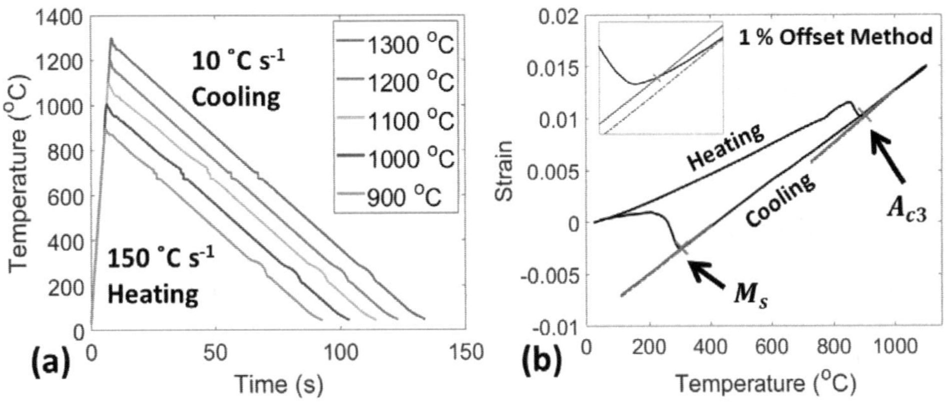

Figure 1. (a) The temperature profiles used for the dilatometry tests. (b) One of the 1100 °C peak temperature dilatometry curves with the M_s and Ac_3 temperatures calculated using the offset method.

3. Results

The microstructures in all dilatometry samples were identified as containing lath martensite and no bainite (figure S1 (c) in the Supplementary Information demonstrates this). The dilatometry curves confirmed this, since they only display one sharp transformation upon cooling and the measured transformation temperatures are close to the M_s temperatures predicted for the alloy using equation for the 1000-1300 °C tests. The average Ac_3 temperature based on seven 1100 °C tests was 898 ± 5 °C (one standard deviation). Since it is measured using the 1 % offset method we can say that the steel was 99 % austenitised by 898 °C.

An example PAG reconstruction from a single EBSD tile is shown in figure 2 (a-b). PAG size increased with peak temperature as expected. The average PAG equivalent circle diameters for each peak temperature can be found in the Supplementary Information (Table S2) alongside further information on the errors, reconstruction method and the EBSD maps from the samples not displayed in the main text (figures S2 and S3).

M_s temperatures measured using dilatometry are plotted against PAGS (from the EBSD) in figure 2 (c). The red, black and blue solid lines on figure 2 (c) are the M_s - PAGS curves predicted according to equation 1 calculated using the average EPMA compositions of the enriched band, the bulk alloy composition and the depleted band, respectively. They demonstrate the relationship that we would expect to see for this alloy. For 1300 °C and 1200 °C peak temperatures (larger PAGS) the measured M_s values roughly fit the trend predicted by the empirical equation: data points clustering around the solid line calculated from the average composition. However, from the 1100 °C peak temperature and below (≤ 8.90 μm average PAG diameter) the measured M_s values become increasingly higher than predicted by equation 1. This is especially true for the 900 °C tests where the average M_s value (312 °C) is 31 °C higher than the value predicted for the bulk composition, more than could be accounted for by either the compositional variation caused by banding or PAGS effects.

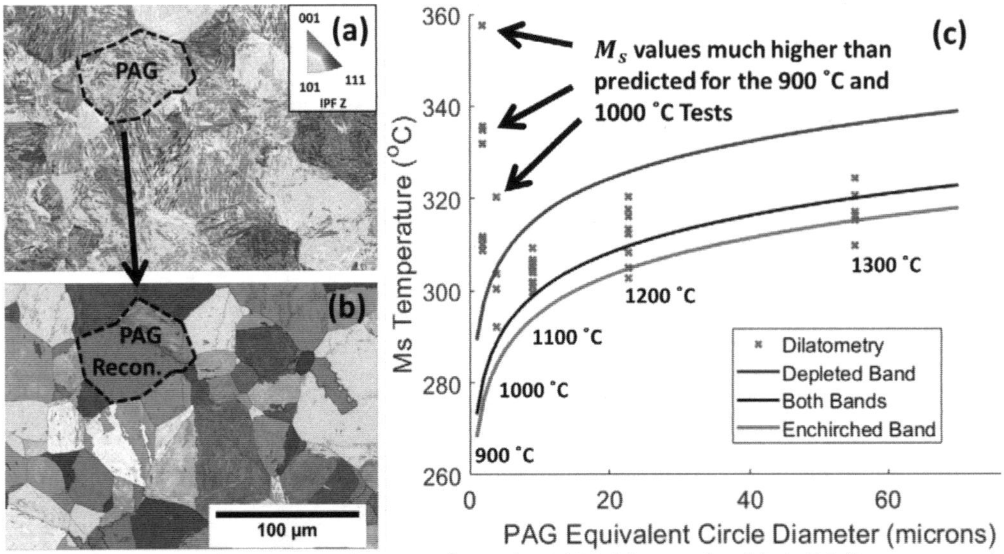

Figure 2. (a) An example IPF Z EBSD map from the 1200 °C sample. (b) A PAG reconstruction of (a) using AZtec Crystal 2.2. (c) The relationship between M_s and PAG size. The solid lines represent the M_s - PAGS predicted by equation 1. Errors in the M_s - PAGS measurement can be found in Table S2 in the Supplementary Information.

The EPMA results, an example of which is shown in figure 3 (a), confirms that the lighter etching bands contained higher concentrations of Cr, Mo, V and Mn. The areas belonging to either band were separated by thresholding using the Mo concentration. Regions with >0.95 wt% were considered to be enriched bands (<0.95 wt% considered to be depleted) and the average compositions of the enriched and depleted bands were calculated using the data from these areas - these values are displayed in Table S3 in the Supplementary Information. Mo was used for the thresholding because it has a low partition coefficient (k value) [13, 15-18] in steel when compared with the other alloying elements, and therefore was the most clearly segregated into distinct bands.

Figure 3 also displays the direct comparison between the distribution in PAGS measured by EBSD in (b) and the locations of compositional bands, both measured by EPMA within the 1000 °C sample (a) and highlighted in optical microscopy by Nital etching (c). It is clear that in this sample smaller PAGs are found in the regions with higher alloying content (light etching enriched bands).

Etching helped highlight spherodised particles which were seen with SE imaging in the enriched (light etching) bands of the 900-1100 °C samples. An example SE image is shown in figure 3 (d) alongside an EDS map of the same area (e-g). The EDS revealed that these particles were rich in Cr, Mo, V and C when compared with the matrix, although the carbon measurement from this EDS are only qualitative. Interestingly, these particles were rarely found in the depleted bands and almost never in samples that were heated to 1200 °C or 1300 °C. In the enriched bands of the 900 °C and 1000 °C samples, spherodised particles often appear on the more heavily etched PAG boundaries as well as in the inside or between the martensite laths (highlighted in figure 3). In the 1100 °C sample spherodised carbides on the PAG boundaries appear less common. More example micrographs can be found in figure S4 in the Supplementary Information.

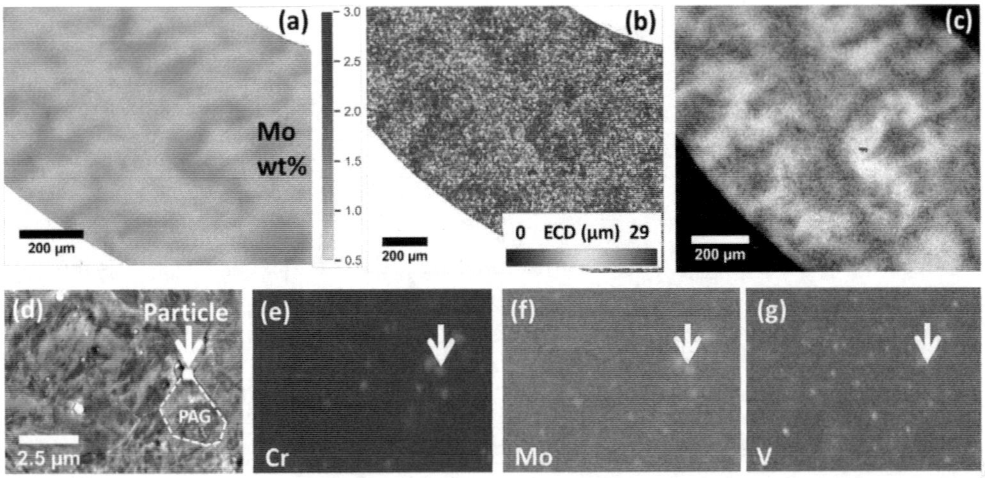

Figure 3. (a) EPMA map from the 1000 °C sample showing Mo distribution. (b) A PAG reconstruction from a montage EBSD of the same area, in which the grains are coloured according to equivalent circle diameter (ECD). (c) An optical image of the same area again after the sample was etched. (d) An SE image from one of the enriched bands in the 1000 °C sample after the etching. A PAG with what is presumed to be a large spherodised carbide on its boundary is highlighted. (e-f) An EDS map of (d), in which the same carbide is highlighted. The C, Mn and Si can be found in the figure S5 in the Supplementary Information. Intensity is relative to the number of counts and gamma brightness and contrast has been altered, the map is intended to be qualitative only.

In figure 4 (a) a single tile reconstructed EBSD map also from the 1000°C sample further illustrates the disparity in PAGS between compositional bands. The in-lens BSE images from either band, figure 4 (b) and (c), reveal spherodised particles that only appear in the enriched (smaller PAGS) bands. Similar particles could be seen in the enriched regions of the 900 °C and 1100 °C samples. As the 1200 °C sample reconstruction in figure 2 (b) demonstrated earlier, the distribution in PAGS appears to be even across both bands.

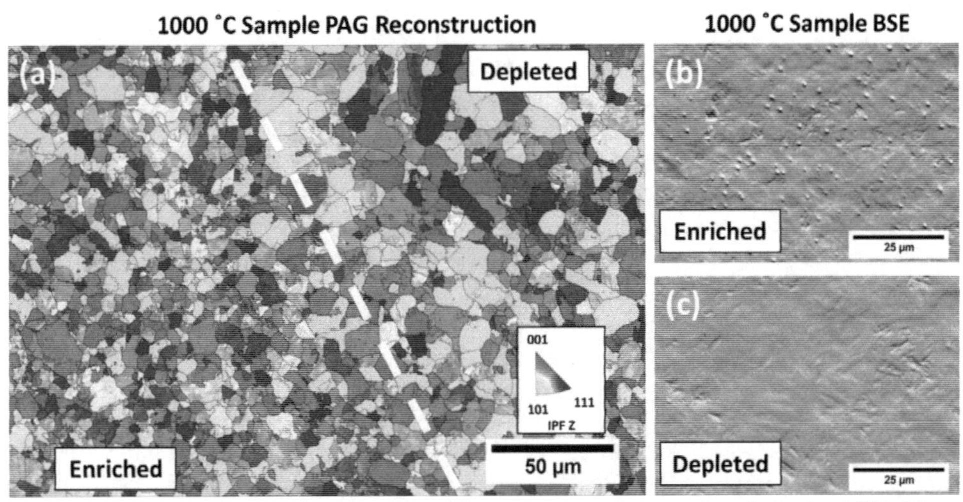

Figure 4. (a) An EBSD PAG reconstruction of an area covering two bands in the 1000 °C sample. The grains are smaller in the enriched band and when compared with the 1200 °C sample in figure 2 (b) are more irregular in shape than in both bands. (b-c) Higher magnification BSE images from either band in (a). Spherodised particles can be found in the enriched band but not the depleted band.

4. Discussion

Usually, it would be expected that M_s decreases for smaller PAG sizes, as shown by the solid lines in figure 2. However, the 900 °C sample shows a much higher M_s, and the 1000 °C and 1100 °C samples both have higher average M_s temperatures than would be predicted by equation 1 for the average composition of SCMV. As the dilatometry curve in figure 2 (c) demonstrates, the 1000 °C was fully austenitised (Ac_3 = 898 °C) so it is unlikely that this is an effect relating to untransformed ferrite. Also, the in M_s values for the 900 °C samples (and one 1000 °C sample) were much higher than was predicted for the depleted bands by the using equation 1, so the effect cannot be fully explained by the dilatometer measuring the depleted bands transforming before the enriched. Since the 900-1100 °C samples all contained spherodised particles in the enriched bands, it is reasonable to assume that they are related to the increase in M_s.

To investigate these particles, the average compositions of the enriched and depleted bands (as calculated by thresholding the EPMA for the 1000 °C sample) were used in ThermoCalc with the TCFE10 Steels/Fe-Alloys v10.1 database. ThermoCalc predicted that under equilibrium conditions SCMV should be fully austenitic in both bands above 811 °C and that it should contain various Cr, Mo and V carbides until 963 °C in the enriched band, and until 946 °C in the depleted. The volume fractions of the various carbides predicted by ThermoCalc to form in either band between 0-1000 °C are compared in figure 4. Since SCMV is a very clean steel and the particles in figure 3 (d-g) were shown to contain varying amounts of the aforementioned alloying elements, it is assumed that they are carbides.

The 10 °C s^{-1} cooling rate would have provided very little time for auto-tempering, yet some of the carbides in the 900-1100 °C samples are of large size and have spherodised as one might expect to see in a steel which has undergone prolonged tempering. This suggests that the carbides must have existed in the material prior to the temperature excursion, but that there was insufficient time to fully dissolve them all in the enriched bands, and that the surviving carbides were instead spheroidising during the heat treatment. Since the spherodised carbides are not present in the 1200 and 1300 °C samples, we can assume that they have dissolved in these cases. Thermocalc predicted that the MC, M_7C_6 and M_2C carbides would be thermodynamically stable at temperatures close to 900 °C in the enriched bands. It is therefore probable that the spherodised particles were some combination of the aforementioned alongside the $M_{23}C_6$ carbides, which are thought to form from M_7C_3 carbides upon cooling [19-22].

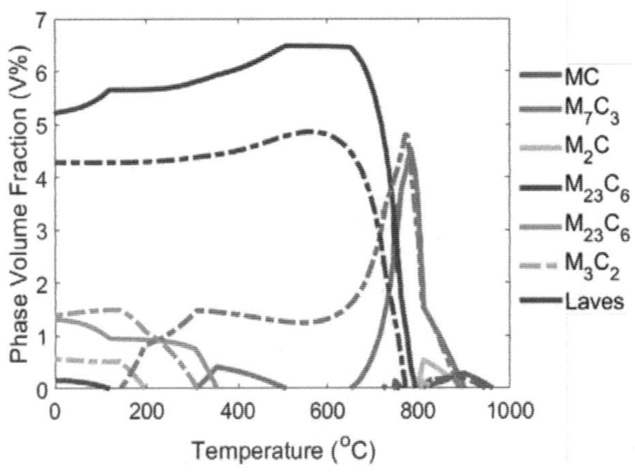

Figure 5. The ThermoCalc predictions for the carbides formed under equilibrium for the average composition of the enriched (solid lines) and depleted (dotted lines) bands, as measured by the EPMA. The carbides were composed of varying amounts of Cr, Mo, V, Mn and C. The compositions used for these calculations are in Table S3 in the Supplementary Information.

The correlated EPMA and EBSD PAG reconstruction in figure 4 demonstrated that the PAGs are smaller in the enriched regions of the 1000 °C samples and the SEM and EDS images displayed carbides on PAG boundaries. This is consistent with the notion that in the lower peak temperature samples the pre-existing alloy carbides from the parent material have a Zener pinning effect on the austenite boundaries, stifling grain growth at elevated temperatures. The effect is most prevalent in the enriched bands which retain the most pre-austenitision carbides and could be similar to the bimodal

PAGS affect found by Basso et al [12], albeit in a very different cast steel after an hour's austenitisation. Although carbides did appear in the 1100 °C samples, the grain boundaries appear to have enough time and/or enough mobility to move past the pinning carbides and form equiaxed grains in the duration of the experiment.

The most plausible explanation for the increase in M_s at lower peak temperatures is therefore that the alloy carbides deplete the surrounding matrix of alloying elements as they coarsen, most notably carbon, which has the most significant effect on M_s as demonstrated in equation 1. If they are not dissolved during the heat treatment the matrix remains depleted, significantly altering the transformation behaviour and raising the M_s temperature. This can be illustrated by inserting the composition of the matrix in the enriched bands, as predicted by ThermoCalc, at 850 °C into equation 1. The M_s value is found to be 316 °C for the 900 °C PAGS, over 40 °C higher than if carbides were not accounted for. Of course, this value is assuming equilibrium conditions, but it provides a useful illustration. This result is significant because the depletion of the matrix by un-dissolved carbides could change the entire CCT curve behaviour of the material depending on the peak temperature/time locally, and have a significant impact on the prediction of phase transformations. At the same time, the differential pinning of PAG boundaries between different bands could add another level of complexity to grain growth predictions in the HAZ. This is especially a problem for anyone trying to use the PAG size to help forensically determine the peak temperature experienced by a piece of material exposed to rapid heating/quenching. On the other hand, since the carbides only survive to certain temperatures, their presence in a HAZ microstructure could be used instead of PAG size to infer the maximum temperature that a section of material may have reached.

5. Conclusions and Summary

SCMV was rapidly austenitised to 900, 1000, 1100, 1200 and 1300 °C before quenching at 10 °C s^{-1} with no hold time. M_s temperatures were measured using a dilatometer and the microstructure was characterised using optical microscopy, SEM, EDS, EPMA and EBSD with PAG reconstructions. From the results the following conclusions have been drawn:

- When rapidly austenitising and quenching the steel, it was found that carbides did not fully dissolve at 1100 °C and below. In particular, carbides were seen to survive austenitisation most prevalently in enriched bands, which led to smaller PAGS in these bands.
- It was found that the measured M_s temperature increased significantly for the lowest PAGS, which is in contrast to the normally expected behaviour. This can be attributed to the presence of the undissolved carbides at low austenitisation temperatures, which 'lock up' carbon and mean that the remaining matrix is less hardenable.
- Normally, enriched bands would be expected to have a lower M_S. However, the preferential formation of carbides in enriched bands (and their increased stability at high temperatures in these bands) could have inverted this trend.

6. Supplementary Information

For the proceedings of the 42nd Risø International Symposium on Materials Science Supplementary Information and supporting data files can be found at https://doi.org/10.5281/zenodo.6665419. The files for the montaged EBSD datasets are not included but can be made available upon request by contacting the corresponding author.

7. Acknowledgments

The authors gratefully acknowledge financial support from Rolls-Royce plc, the EPSRC and SFI via the Centre for Doctoral Training (CDT) in Advanced Metallic Systems grant EP/S022635/1, and the EPSRC funded JEOL JXA-8530F FEG-EPMA, award EP/M028097/1. E. J. Pickering acknowledges support from the Henry Royce Institute through EPSRC grants EP/R00661X/1, EP/S019367/1, EP/P025021/1 and EP/P025498/1. The corresponding author would like to thank Kirstie Bruce for the contribution of their code used for the EPMA data analysis. Requests for access to the underlying research data should be directed to the corresponding author, and will be considered against commercial interests and data protection.

References

[1] Verhoeven J D 2000 Journal of Materials Engineering and Performance **9** 286–96
[2] Krauss G 2003 Metallurgical and Materials Transactions B **34** 781–92
[3] Yang H S and Bhadeshia H 2009 Scripta Materialia **60** 493–5
[4] Celada-Casero C, Sietsma J and Santofimia M J 2019 Materials & Design **167** 107625
[5] Lee S J and Park K S 2013 Metallurgical and Materials Transactions A **44** 3423–7
[6] Grange R A 1971 Metallurgical Transactions **2** 417–26
[7] Majka T F, Matlock D K and Krauss G 2002 Metallurgical and Materials Transactions A **33** 1627–37
[8] Krebs B, Germain L, Hazotte A and Goun´e M 2011 Journal of Materials Science **46** 7026 38
[9] Offerman S E, van Dijk N H, Rekveldt M T, Sietsma J and van der Zwaag S 2002 Materials Science and Technology **18** 297–303
[10] Thompson S W and Howell P R 1992 Materials Science and Technology **8** 777–84
[11] Tenaglia N E, Massone J M, Boeri R E and Speer J G 2020 Materials Science and Technology **36** 690–8
[12] Basso A D, Toda-Caraballo I, Eres-Castellanos A, San-Martin D, Jimenez J A and Caballero F G 2020 Metals **10** 574
[13] Jabar S, Siefert J A, Strangwood M and West G D 2021 Metallurgical and Materials Transactions A **52** 426–37
[14] Yang H S and Bhadeshia H K D H 2007 Materials Science and Technology **23** 556–60
[15] Kobayashi Y, Dobara K, Todoroki H, Nam C, Morishita K and Yasuda H 2020 ISIJ International **60** 276–85
[16] Pickering E J, Chesman C, Al-Bermani S, Holland M, Davies P and Talamantes-Silva J 2015 Metallurgical and Materials Transactions B **46** 1860–74
[17] Howe A A 1991 Modeling of Casting, Welding and Advanced Solidification Processes V **5** 461–6
[18] Fujii T, Poirier D R and Flemings M C 1979 Metallurgical Transactions B **10** 331–9
[19] Bhadeshia H and Honeycombe R 2017 Tempering of Martensite Steels: Microstructure and Properties (Elsevier) pp 237–70 4th ed
[20] Kroupa A, Výrostková A, Svoboda M and Janovec J 1998 Acta Materialia **46** 39–49
[21] Tunney R J, Lorimer G W and Ridley N 1978 Metal Science **12** 271–6
[22] Saroja S, Parameswaran P, Vijayalakshmi M and Raghunathan V 1995 Acta Metallurgica et Materialia **43** 2985–3000
[23] Huang C Y, Ni H C and Yen H W 2020 Materialia **9** 100554

Supplementary Information

When using the offset method to calculate the M_s and Ac_3 temperatures, the fitted line was taken at least 50 °C away from the transformation start and the offset strain was 1.099 51×10^{-4}, which is the strain predicted for the formation of 1 % martensite from ferrite for SCMV using Yang and Bhadeshia's equations [14]. This process was then repeated with the offset taken from at least 75 °C and then 100 °C, the final M_s was an average of the three values. Since the scales of the microstructures in each sample were different, different step sizes, exposure times and reconstruction parameters were used for the individual tiles of the EBSD montage maps are listed in Table S1. The montages were too big to be stitched in the AZtec 6 software in one without binning the information, so they were instead montaged in sections of up to 64 Mpx. The sections' PAGs were reconstructed using the Oxford Instruments Aztec Crystal 2.2 software based on the reconstruction method by Huang, Ni and Yen [23]. Prior to the reconstruction, the data was cleaned first by using the 'Remove Wild Spikes' tool which removes clusters of up to 4 pixels which have a completely different orientation to those surrounding them. This served the purpose of removing the stray clusters of indexed pixels that were found recorded in the Bakelite regions off the edge of the sample, preventing them from being included in the reconstruction. Clusters of non-indexed pixels left in the map can lead to false austenite islands appearing in the centre of reconstructed PAGs, since the algorithm detects them as regions with high misorientation relative to the surrounding pixels. For this reason the Remove Zero Solution tool was used iteratively to fill in non-indexed pixels until there was 100 % indexing in the parts of the map covered by the sample. The 'level 6' setting was used so that only non-indexed pixels with 3 or more neighbours of similar orientation were filled in an iteration. This prevented small clusters of indexed pixels found in the Bakelite regions from being grown into larger clusters that could affect the PAG size measurement. For the reconstruction the Nishiyama-Wassermann orientation relationship was used as a first approximation and no training regions were used. This was because the author found that they can sometimes make the reconstruction worse if selected poorly, and because for SCMV the reconstructions work well enough without them to measure PAGS. The following parameters were used: 5 ° Child Threshold, 20 pix Window Size, 25 % Step Width, 7 ° Voting threshold and 10 ° Parent Threshold. When extracting PAG statistics the data was thresholded by manually selecting the smallest grain that was considered to be real and removing any grains containing fewer pixels. The threshold for defining grain boundaries was a 10 ° misorientation angle.

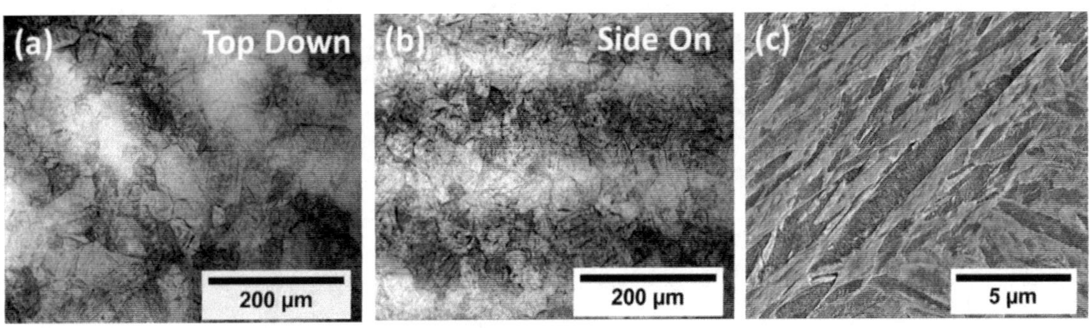

Figure S1. (a-b) Optical micrographs show banding in SCMV from the top down and side on orientations respectively. The lighter etching regions are those with higher alloy composition. (c) An SE SEM image of martensitic laths in the 1200 °C peak temperature samples.

Table S1. Details for the experimental techniques. All microscopy samples were mounted in conductive Bakelite, ground and polished to a quarter micron diamond finish prior to the final preparation methods listed here.

Technique	Sample Finish	Machine Model	Parameters
Dilatometry	S-Type thermocouples were spot welded to the samples.	DIL805 A/D/T in quenching mode	Silica or alumina pushrods were used for the tests that reached peak temperatures of 1100 °C or below, and the higher temperature capability alumina pushrods only were used for all tests that went higher.
Optical Microscopy	Etched in 5 % Nital until both bands start to etch (≃10 s)	BH2 series Olympus	Gamma, brightness and contrast were digitally altered after acquisition.
SE Imaging and EDS	Etched in 5 % Nital until both bands start to etch (≃10 s)	Tescan Mira 3 with an Oxford Instruments x-act	20 kV, 3 nA (measured using a Faraday cup), 15 mm working distance and 10 ms dwell time. Gamma, brightness and contrast were digitally altered after acquisition.
BSE Imaging and EBSD	0.04 μm neat colloidal silica for 3 minutes	ThermoScientific Apreo with an Oxford Instruments Symmetry 2 detector	20 kV, 51 nA , 21 mm working distance, 70 ° tilt. A 0.1 μm step size was used for the example single tile maps and the 900 °C montage map, 0.3 μm was used for the 1300°C montage map and 0.2 μm for everything else. Gamma, brightness and contrast were digitally altered for the BSE images after acquisition.
EPMA	0.04 μm neat colloidal silica for 3 minutes	JEOL JXA-8530F	15 kV, 100 nA, 2 μm step size and 100 ms dwell time.

Table S2. Equivalent circle diameter measurements calculated from PAG reconstructions of montage EBSD maps.

Peak Temperature (°C)	Average Equivalent Circle Diameter (μm)	Standard Deviation (μm)	Number of PAGs Counted	Standard Error (μm)
1300	55.24	36.77	2468	0.7401
1200	22.71	13.54	7648	0.1548
1100	8.90	6.28	52052	0.0275
1000	3.85	3.16	253241	0.0063
900	1.90	1.81	468631	0.0027

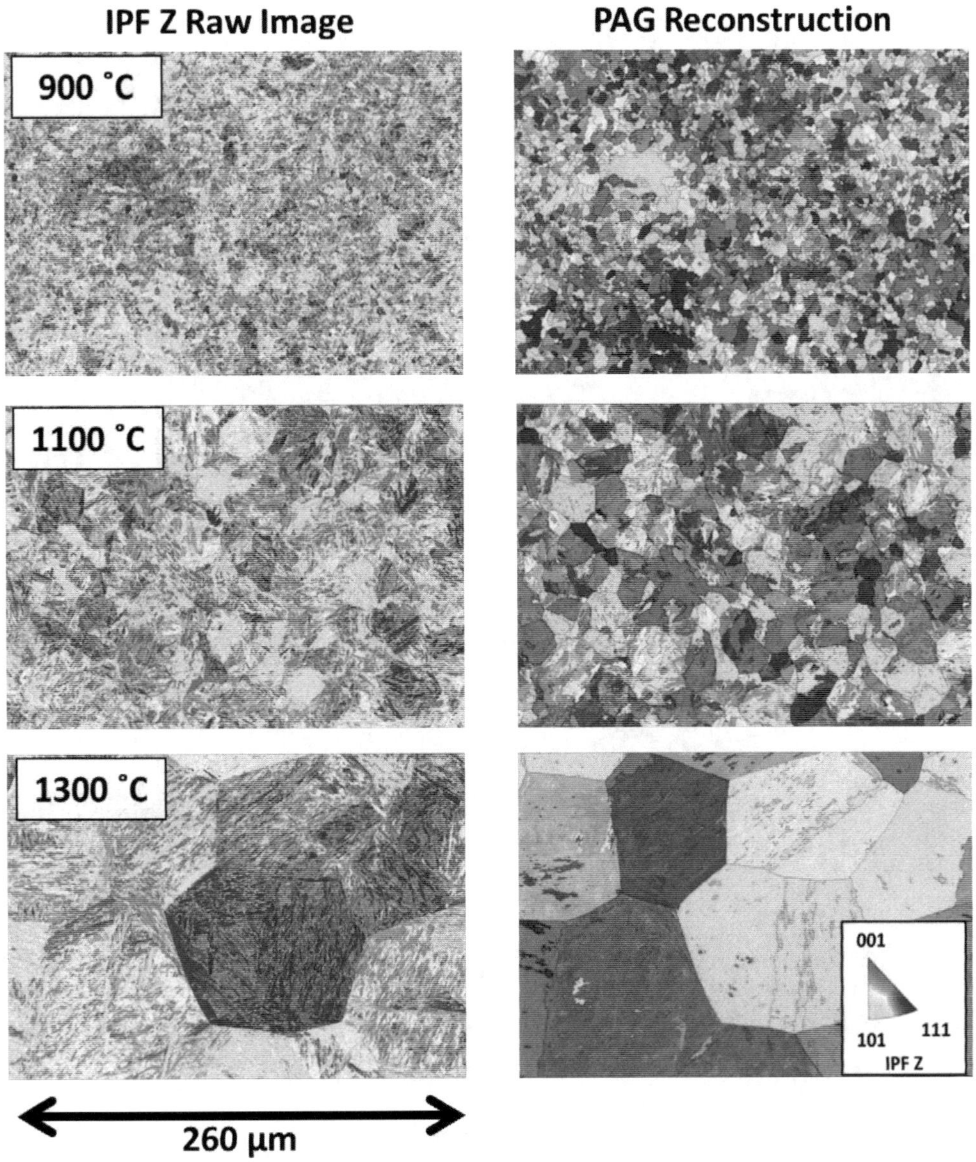

Figure S2. Single tile EBSD maps (left) with their corresponding PAG reconstructions (right) for the peak temperatures not shown in the main text. PAGS increases with peak temperature and the grains become more equiaxed after 1100 °C.

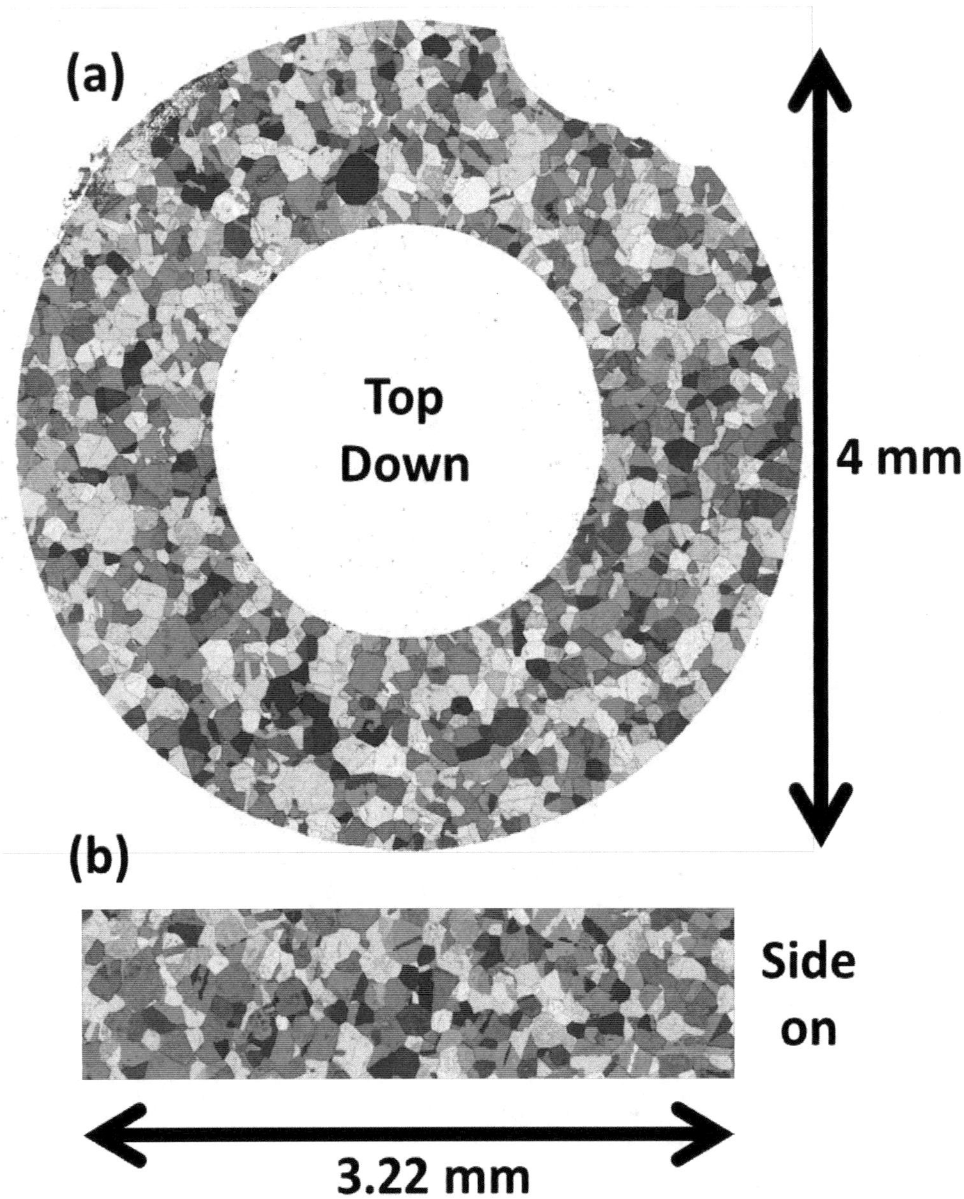

Figure S3. (a) A PAG reconstruction from a montage EBSD map of the entire 1300 °C sample. The reconstructions were performed in sections since the montage maps were so large. (b) A PAG reconstruction of the sample as in (a) from the side on orientation. Grain morphology was the same regardless of the sample orientation.

IOP Conf. Series: Materials Science and Engineering 1249 (2022) 012061 doi:10.1088/1757-899X/1249/1/012061

Table S3. The composition of SCMV (wt%) as measured in the 1000 °C by EPMA. The compositions of the enriched and depleted bands were calculated by thresholding the Mo greater or less than 0.95 wt% Mo respectively. Standard errors were less than ± 0.0011 for all measurements. The EPMA carbon measurement is not included since the experiment was not set up for accurate light element detection.

	Fe	C	Cr	Mo	Mn	V	Si	Ni
EPMA Overall	Bal	N/A	3.41	1.01	0.55	0.24	0.21	0.01
EPMA Enriched	Bal	N/A	3.50	1.12	0.56	0.26	0.22	0.01
EPMA Depleted	Bal	N/A	3.29	0.85	0.53	0.22	0.21	0.01

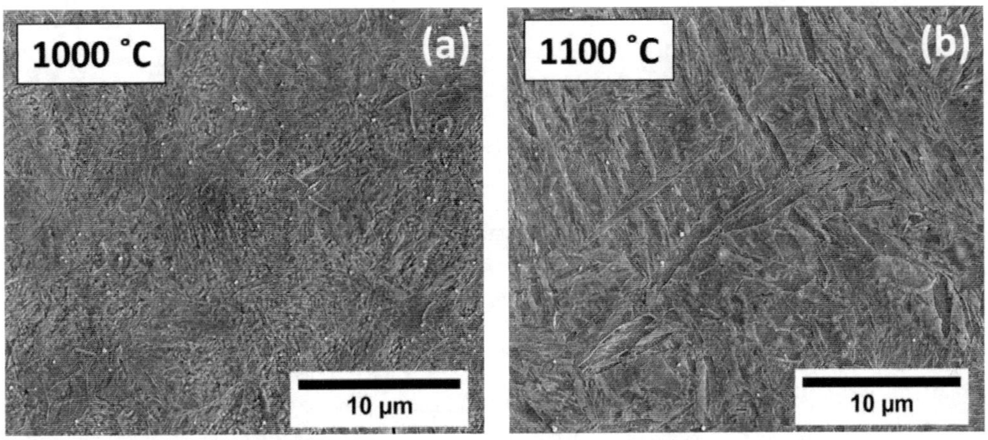

Figure S4. SE images of enriched bands in the (a) 1000 °C and (b) 1100 °C samples after etching in Nital. More carbides can be seen in the 1000 °C sample.

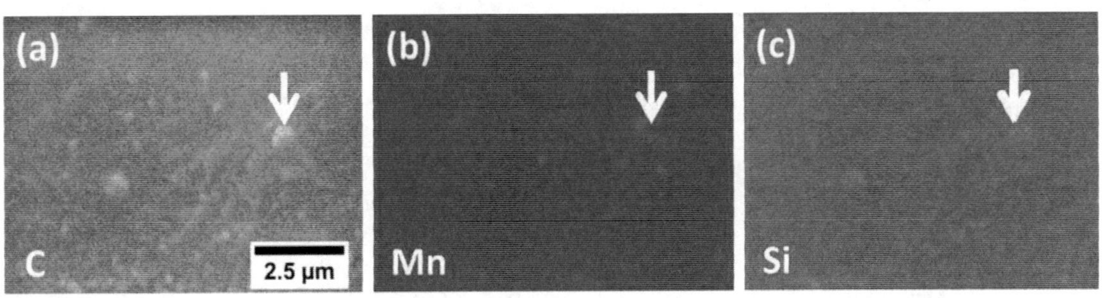

Figure S5. The EDS maps for the other elements omitted from figure 3. Intensity is left relative to the number of counts not the wt%, since wt% is not reliable for the EDS of C. The higher carbon counts around the particles in (a) support the theory that they are carbides. (b) There may be a small amount of Mn in some carbides. (c) The Si counts appear fairly diffuse.

IOP Conf. Series: Materials Science and Engineering 1249 (2022) 012062 doi:10.1088/1757-899X/1249/1/012062

Dislocation sub-structures, recovery and recrystallization in pure tantalum

C Moussa[1], J Baton[1] and W Geslin[2]

[1] MINES ParisTech, PSL Research University, CEMEF - Centre de mise en forme des matériaux, CNRS UMR 7635, CS 10207 rue Claude Daunesse, 06904 Sophia Antipolis Cedex, France
[2] CEA DAM Valduc, F-21120 Is-sur-Tille, France

E-mail: charbel.moussa@mines-paristech.fr

Abstract. Plastic deformation, static recovery and recrystallization in pure tantalum are analyzed in the present paper. Quantitative analysis of dislocation sub-structures were done. Sub-structures development is completely dependent on crystallographic orientation. No clear effect of recovery was observed when direct analysis was made. Hence, recovery was studied through its effect on recrystallization. Recovery heat treatments followed by a recrystallization heat treatment were applied to compare the recrystallized state of samples which were subject to prior recovery with those which were not. Results show that recrystallization is accelerated in the early stages of recovery (low temperature or low annealing time or low plastic strain). On the opposite, in the advanced stages of recovery, it slows down recrystallization. Results show again an influence of crystallographic orientation. Static recrystallization was also investigated and the influence of crystallographic orientation, most probably inherited from the deformed state through dislocation sub-structures, was observed. Based on experimental results, a discussion on the mechanisms of nucleation in the different orientation grains is presented. Finally, a discussion is made on the quantification of stored energy. The effect of the observation scale and the choices of considering dislocation density or dislocation sub-boundaries energies from EBSD data is discussed.

1. Introduction

Tantalum is a Body Centered Cubic (bcc) metal with three slip systems: {110}<111>, {112}<111> and {123}<111>. Even though tantalum slips mainly on {110} planes during cold deformation, it may also slip on planes with maximal resolved shear stress [1]. In bcc materials, cross slip can easily take place since no stable stacking faults were observed [2]. Many experimental evidence on dislocation sub-structures in bcc materials can be found [3,4]. Influence of crystallographic orientation on dislocation densities and sub-structures in bcc materials was observed [5]. More sub-structures and higher stored energy were observed in grains with <111> direction parallel to normal direction (ND) than in grains with <100> direction parallel to ND on cold rolled tantalum [6].

The influence of crystallographic orientation on the stored energy directly influences subsequent recrystallization. It was reported that nucleation is favored in grains with dislocation sub-structures in tantalum [7]. For estimating stored energy from Electron Back Scatter Diffraction (EBSD), several methods can be used [8]. Generally, the stored energy is calculated either as the sum of the energies of all individual dislocations or as the sum of the energies of the sub-boundaries.

IOP Conf. Series: Materials Science and Engineering 1249 (2022) 012062 doi:10.1088/1757-899X/1249/1/012062

Because of the ease of cross slip in bcc materials, recovery can take place. It is generally accepted that recovery, which is activated at lower temperatures, tends to slow down recrystallization kinetics. However, some studies showed that recovery may enhance recrystallization such in aluminum alloys [9]. Less studies were made on recovery in bcc materials.

The current paper is a synthesis of new findings on microstructure evolutions in pure tantalum during plastic deformation [10], recrystallization [11] and recovery [12]. Special focus is made on the effect of crystallographic orientation on sub-structures development and their effect on recrystallization and recovery through EBSD analysis at the polycrystalline and at the sub-structures' scales.

2. Material presentation and experimental conditions

High purity (>99.99% wt.) tantalum sheet was used in the present study. EBSD maps used for the analysis at the polycrystalline scale and at the substructures scale were made with a step size of 1.2µm and 90 nm respectively. Quantitative EBSD analysis was performed using homemade routines within the MTEX toolbox [13]. The microstructure of the as received material in the center of the sheet and in the quarter layer region are presented in figure 1(a). The fully recrystallized microstructure is very different in the two regions. In the center of the sheet, upper part of figure 1(a), a dominant γ-fiber (<111> direction parallel to ND) of around 70-80% of volume fraction can be observed. In the quarter layer region, lower part of figure 1(a), γ-fiber represents about 10% and θ-fiber (<100> direction parallel to ND) about 30%. Those textures are inherited from the last rolling steps with a strong shear component near the surface because of friction and mainly compressive components in the center of the sheet. A more detailed texture analysis with X-ray diffraction can be found in [10]. In this study, all grains (grain boundaries defined with a disorientation higher than 10°) were classified into three groups: γ-fiber, θ-fiber and "others" for the grains that do not belong to either of the first two classes. The material corresponding to the center, upper part of figure 1, and the one corresponding to the quarter layer, lower part of figure 1, are called AR1 and AR2 (for as-received).

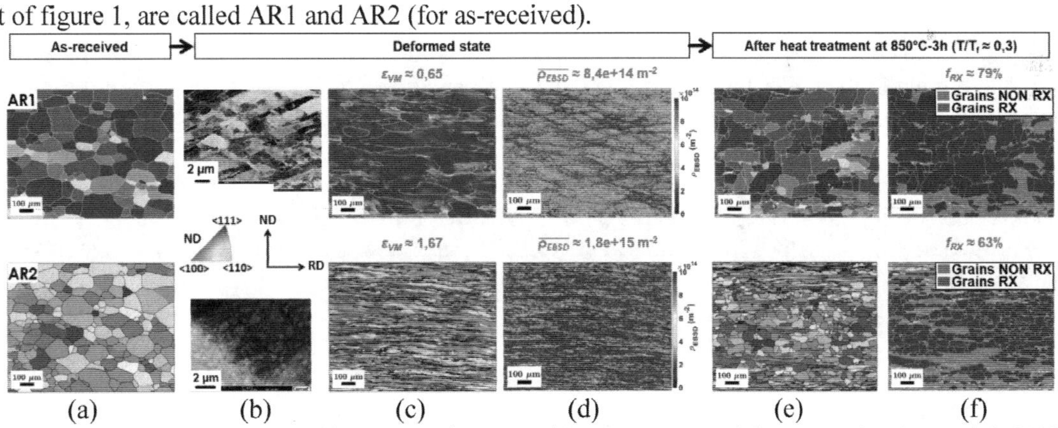

Figure 1. Upper part for AR1 and lower part for AR2. (a) EBSD maps of the as-received materials [10], (b) Electron Channeling Contrast Imaging (ECCI) micrographs of the deformed state [10], (c) EBSD maps after plastic deformation [10], (d) dislocation density ρ_{EBSD} maps [12], (e) EBSD maps after heat treatment [11] and (f) Grain Orientation Spread maps differentiating recrystallized and non-recrystallized grains (RD stands for Rolling direction).

3. Dislocation sub-structures in cold deformed tantalum

3.1. Polycrystalline scale analysis

To study sub-structures development during plastic deformation, AR1 was deformed by compression (parallel to ND) and AR2 by cold rolling [10]. For more details on deformation conditions and texture evolutions please report to [10]. Dislocation density ρ_{EBSD} was estimated from EBSD data using the Nye's tensor with the method proposed by Pantleon [14]. The deformed microstructures along with ρ_{EBSD} maps for $\varepsilon_{VM} = 0.65$ for the AR1 and $\varepsilon_{VM} = 1.67$ for the AR2 are presented in figures 1(c) and

1(d). As illustrated in figure 2(a), the evolution of the average dislocation density ($\overline{\rho_{EBSD}}$) in function of ε_{VM} is different for each class of orientations. $\overline{\rho_{EBSD}}$ is higher in the γ-fiber grains than in the θ-fiber grains of around 60% with a faster increase. $\overline{\rho_{EBSD}}$ in "other" grains is either between those of γ-fiber and θ-fiber grains or close to those of γ-fiber for others. This observation is consistent with the average Taylor factor of each class, 3.5, 2.4 and 3.0 for γ-fiber, θ-fiber and "other" grains respectively. However, no correlation was found between $\overline{\rho_{EBSD}}$ and the Taylor factor grain per grain. $\overline{\rho_{EBSD}}$ is higher in the γ-fiber grains of the AR1 than in those of the AR2 as illustrated in figure 2(b). This is probably due to interactions with neighboring grains as it will be discussed in the next section.

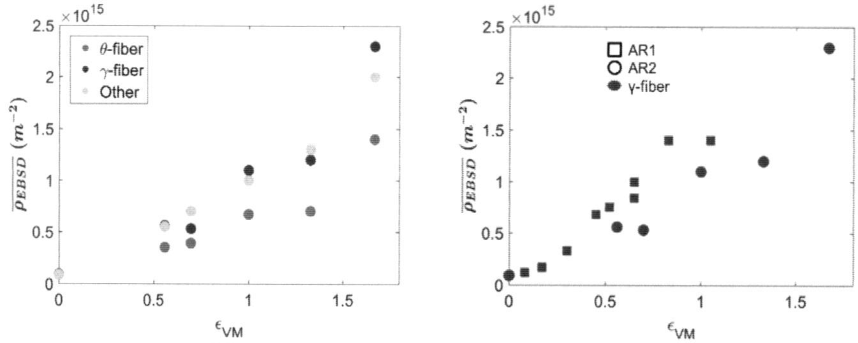

(a) (b)

Figure 2. Evolution of $\overline{\rho_{EBSD}}$ with ε_{VM} (a) for each class of orientations and (b) for the γ-fiber grains in AR1 and AR2 [10].

3.2. Dislocation substructures' scale analysis

To analyze dislocation substructures evolution in the grains of each class, Electron Channeling Contrast Imaging (ECCI) micrographs were made in the γ-fiber and θ-fiber grains as illustrated in figure 1(b). The micrograph in the upper part of figure 1(b) was obtained in a γ-fiber grain in AR1 and the one in the lower part of figure 1(b) was obtained in a θ-fiber grain in AR2. Figure 1(b) shows that sub-structures are formed in the γ-fiber grains even for small ε_{VM} (even at $\varepsilon_{VM} = 0.3$ [10]). As for θ-fiber grains, very few sub-structures were observed even for $\varepsilon_{VM} = 1.67$. This observation agrees with the observations made at the polycrystalline scale where $\overline{\rho_{EBSD}}$ is higher in the γ-fiber grains (figure 2(a)).

For the analysis of dislocation sub-structures EBSD maps were acquired with a step size of 90 nm. All segments with a disorientation between 1.5° and 10° were considered as sub-boundaries. They are quantified through their average disorientation $\overline{\theta_{sub}}$ and their density $\overline{\rho_{sub}}$ (total length of sub-boundaries over total area ratio). For statistical details please refer to [10].

The evolutions of $\overline{\theta_{sub}}$ and $\overline{\rho_{sub}}$ in function of ε_{VM} for the three orientation classes are presented in figures 3(a) and 3(b) respectively. As expected, $\overline{\theta_{sub}}$ and $\overline{\rho_{sub}}$ are higher in the γ-fiber grains than in the "other" oriented grains which in their turn are higher than in the θ-fiber grains. To describe the evolution of $\overline{\theta_{sub}}$ and $\overline{\rho_{sub}}$ in function of ε_{VM}, power laws were used for every case as illustrated in figure 3. The obtained laws are not in agreement with the ones proposed in [15] on aluminum for Geometrically Necessary Boundaries (GNB) and Incidental Dislocation Boundaries (IDB). One possible explanation of such differences is that the proposed laws on face centered cubic aluminum may not be suitable to describe sub-structures evolution in bcc tantalum.

From figure 3(b), it can be observed that for AR1, with a dominant γ-fiber ≈70-80%, the development of sub-boundaries occurs for small ε_{VM} and an early saturation appears (additional experiments for higher ε_{VM} would confirm this saturation at $\overline{\rho_{sub}} \approx 2\ \mu m^{-1}$). For AR2, with less γ-fiber ≈10%, figure 3(b) shows that $\overline{\rho_{sub}}$ keeps increasing even with high ε_{VM} with no saturation indicating a continuous sub-structures development with ε_{VM}. In addition, sub-structures development seems more homogenous (smaller error bars) in the case of AR1 than in the case of AR2. Those two observations illustrate the influence of the neighboring grains (majority of γ-fiber for AR1) on substructures development in γ-

IOP Conf. Series: Materials Science and Engineering 1249 (2022) 012062 doi:10.1088/1757-899X/1249/1/012062

fiber grains. Hence, even though the grain orientation is of prime importance, the neighboring grain orientations also affect the plastic deformation of grains.

No correlation between orientation long range gradients and crystallographic orientation was found [10].

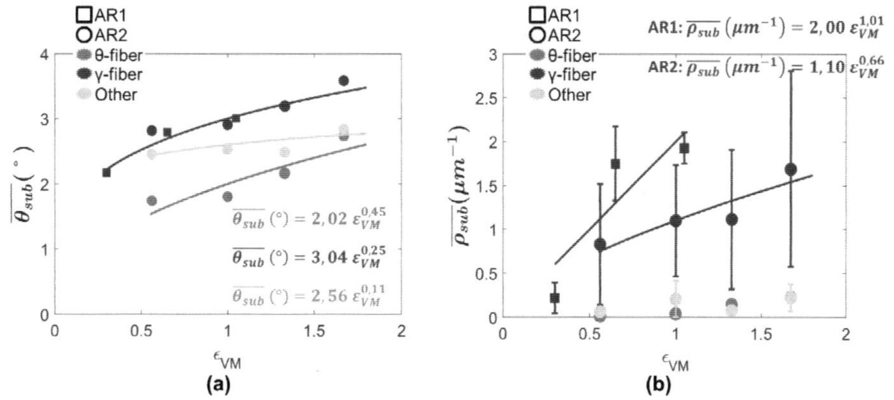

(a) (b)

Figure 3. Evolution of (a) $\overline{\theta_{sub}}$ and (b) $\overline{\rho_{sub}}$ in function of ε_{VM} for the three orientation classes. Power laws were used to fit experimental data [10].

4. Static recrystallization in pure tantalum

As illustrated in figures 1(e) and 1(f) recrystallization is much faster in AR1 than in AR2 even with smaller ε_{VM}. This faster recrystallization is due to the high amount of stored energy despite smaller ε_{VM}. As illustrated in figure 2(a) dislocation density is higher in the γ-fiber grains which explains the high $\overline{\rho_{EBSD}}$ in AR1 than in AR2 for a given ε_{VM}. The evolution of the recrystallized fraction for the three classes of orientation is presented in figure 4(a) and 4(b) for AR1 and AR2 respectively. Recrystallized grains mainly belong to the γ-fiber class. This crystallographic orientation of recrystallized grains is inherited from the deformed state. Because of well-formed sub-structures, figures 1(b) and 3, nucleation is enhanced in γ-fiber grains and diminished in θ-fiber grains. Hence, deformed θ-fiber grains are mainly consumed by recrystallized grains issued from neighboring γ-fiber grains than by those who nucleate in the deformed θ-fiber grains. This hypothesis is supported by the faster evolution of γ-fiber recrystallized grains volume fraction observed in figures 4(a) and 4(b). It can be also concluded that grains which recrystallize in the deformed γ-fiber grains are probably γ-fiber oriented.

The grain sizes values, presented in figures 4(a) and 4(b), show higher grain sizes for γ-fiber recrystallized grains for AR1 and AR2. A faster nucleation (shorter incubation time) as a result of the sub-structures in the deformed state explains this observation.

The comparison between figures 4(a) and (b) shows that recrystallized grain sizes in AR1 are bigger than in AR2. Two possible reasons may explain this difference. The first one is the faster nucleation in the γ-fiber deformed grains (dominant in AR1) leading to additional growth time and size advantage from the early stages of the annealing. The second one is the deformed grains morphology (figure 1(c)) were the deformed AR1 has almost three times less grain boundaries than the deformed AR2 (0.06μm/μm² and 0.175μm/μm² respectively). Considering that grains boundaries are potential nucleation sites, more nuclei may appear in the deformed AR2 leading to a smaller recrystallized grain size.

It is important to mention, that when $\overline{\rho_{EBSD}}$ estimated at the polycrystalline scale (step size of 1,2μm) was used to quantify stored energy, contradictory results were obtained. For example, the values of $\overline{\rho_{EBSD}}$ given in figure 1(d) are contradictory with those of recrystallized fraction in figure 1(f). However, when EBSD data at the sub-structures scale (step size of 90 nm) were used, $\overline{\rho_{EBSD}}$ gave consistent results. This observation is probably because Geometrically Necessary Dislocation (GND) density is correctly estimated with EBSD data acquired at the sub-structures scale [16]. Independently of the scale analysis, results showed that the quantification of the stored energy using the dislocation sub-structures [8] instead of $\overline{\rho_{EBSD}}$ gave more consistent results with recrystallized fractions [11].

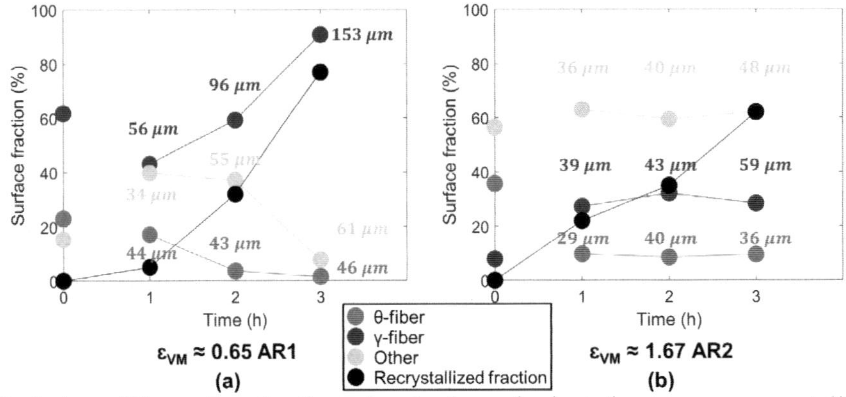

Figure 4. Evolution of the surface fraction of each class of orientations among recrystallized fraction (black curve) with annealing time for (a) AR1 and (b) AR2. The given values correspond the average recrystallized grain size [11].

5. Recovery in pure tantalum

Because of the ease of cross slip in tantalum, recovery is expected to take place and to influence recrystallization. Therefore, AR2 samples deformed up to $\varepsilon_{VM} = 0.56, 1.00\ and\ 1.67$ were submitted to a recovery heat treatment at 750°C for 4 hours. Then samples were analyzed through hardness measurements, $\overline{\rho_{EBSD}}$, $\overline{\theta_{sub}}$ and $\overline{\rho_{sub}}$ [12]. No clear evolution induced by recovery was observed.

Therefore, it was decided to analyze recovery indirectly through its effect on recrystallization. A one stage recrystallization heat treatment (950°C for 30 min.) and five two stages heat treatment of pre-recovery (700°C for 60 min., 750°C for 15 min., 750°C for 60 min., 750°C for 240 min. and 800°C for 60 min.) followed by a recrystallization heat treatment (950°C for 30 min.) were done without any cooling between the two treatments. The recrystallized fractions are reported in figures 5(a) and 5(b) in function of the temperature and the time of the pre-recovery treatment respectively. For all the cases, pre-recovery affects recrystallized fraction indicating that recovery is indeed activated.

Recovery is known to decrease the stored energy because of annihilation of redundant dislocations and therefore to slow-down recrystallization. In the present case, this was observed for the major cases with a smaller recrystallized fraction for samples subject to pre-recovery when compared to the samples which were not subject to pre-recovery, figure 5. However, for the three cases circled in figures 5(a) and 5(b), recovery enhanced recrystallization with a higher recrystallized fraction.

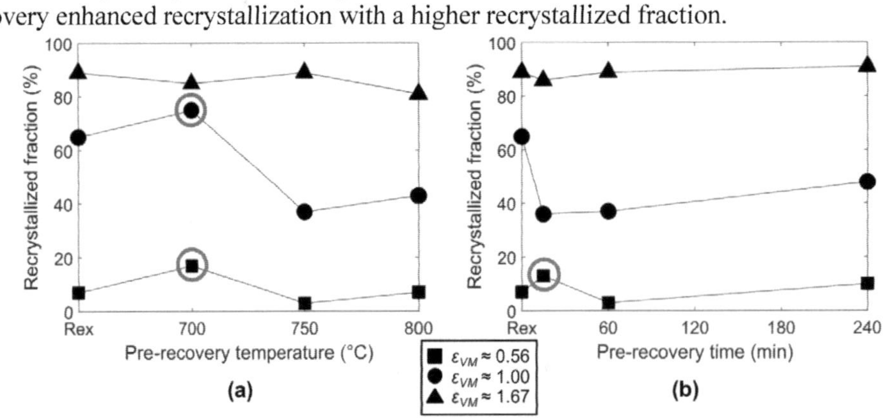

Figure 5. Recrystallized fraction in function of pre-recovery (a) temperature (for 60 min.) and (b) time (at 750°C) treatment followed by a recrystallization treatment at 950°C for 30 min. estimated from EBSD maps of 1.2 mm × 0.9 mm [12].

IOP Conf. Series: Materials Science and Engineering 1249 (2022) 012062 doi:10.1088/1757-899X/1249/1/012062

These observations illustrate two competitive mechanisms occurring during recovery. The first one, associated with annihilation of dislocations and a decrease of stored energy; slows down recrystallization. The second mechanism consists in the increase of the ability of sub-structures to migrate by decreasing dislocations tangles around them. This mechanism would make nucleation faster by decreasing incubation time. Therefore, the effect of recovery depends on the balance between those two mechanisms. Results show that at the early stages of recovery (low temperature or short time or low plastic strain) the second mechanism is favored and recrystallization is accelerated. On the opposite, when recovery advances the first mechanism is dominant, and recrystallization is slowed down.

Additional analysis of the effect of recovery on the recrystallized grains size and their crystallographic orientation can be found in [12].

6. Conclusions

Plastic deformation, static recovery and recrystallization in pure tantalum were analyzed in the present paper. Several main conclusions can be reported.

- Dislocations sub-structures develop in γ-fiber grains even at low plastic strains unlike in θ-fiber grains where very few sub-structures were observed even at high plastic strain.
- When γ-fiber is dominant, dislocation sub-boundaries develop homogenously from the early stages of deformation with a quick saturation and their disorientation continue to increase. When γ-fiber is not dominant, the appearance of new sub-boundaries with strain continues.
- The orientation dependance of plastic deformation influences recrystallization. Nucleation is enhanced in γ-fiber deformed grains giving place to mostly γ-fiber recrystallized grains. Hence, microstructure composed mainly of γ-fiber grains recrystallizes much faster.
- Estimation of stored energy using dislocation density estimated from EBSD at the polycrystalline scale gives contradictory results. For a more precise estimation, it is necessary to estimate stored energy from dislocation sub-boundaries from data acquired at sub-structures scale.
- Recovery gives place to two competitive mechanisms, one accelerating and one slowing down recrystallization. The first one is dominant at the early stages of recovery, the second one is dominant when recovery advances.

References

[1] Weinberger C R, Boyce B L and Battaile C C 2013 *Int. Mater.Rev.* **58** 296–314
[2] Hull D and Bacon D 2001 *Introduction to dislocations* (Oxford: Butterworth-Heinemann)
[3] Moussa C, Bernacki M, Besnard R and Bozzolo N 2017 *Ultramicroscopy* **179** 63–72
[4] Moussa C, Bernacki M, Besnard R and Bozzolo N 2015 *IOP Conf. Ser. Mater. Sci. Eng.* **89** 012038
[5] Wauthier-Monnin A, Chauveau T, Castelnau O, Réglé H and Bacroix B 2015 *Mater. Charact.* **104** 31–41
[6] Liu Y, Liu S, Deng C, Fan H, Yuan X and Liu Q 2018 *J. Mater. Sci. Technol.* **34** 2178–82
[7] Sandim H R Z, Padilha A F, Randle V and Blum W 1999 *Int. J. Refract. Hard Met.* **17** 431–5
[8] Godfrey A, Mishin O V and Yu T 2015 *IOP Conf. Ser. Mater. Sci. Eng.* **89** 012003
[9] Ren X, Huang Y, Zhou W, Zhao L, Wang Q and Min X 2019 *Mater. Sci. Eng.* A **764** 138159
[10] Baton J, Geslin W and Moussa C 2021 *Mater. Charact.* **171** 110789
[11] Baton J, Geslin W and Moussa C 2022 *Int. J. Refract.Hard Met.* **104** 105786
[12] Baton J, Geslin W and Moussa C 2021 *J. Mater. Sci.* **56** 15354–78
[13] Bachmann F, Hielscher R and Schaeben H 2010 *Solid State Phenom.* **160** 63–8
[14] Pantleon W 2008 *Scr. Mater.* **58** 994–7
[15] Pantleon W 2001 *Mater. Sci. Eng..* A **319–321** 211–5
[16] Ruggles T J, Rampton T M, Khosravani A and Fullwood D T 2016 *Ultramicroscopy* **164** 1–10

Stored energy calculation in AA 1050 based on deformation microstructure from electron backscatter diffraction data

Yingrui Yan[a], Hongjiang Pan[a,b,*], Andy Godfrey[c], Dorte Juul Jensen[b] and Xiaodan Zhang[b,*]

[a] Faculty of Materials Science and Engineering, Kunming University of Science and Technology, Kunming, Yunnan 650093, China
[b] Department of Civil and Mechanical Engineering, Technical University of Denmark, 2800 Kgs. Lyngby, Denmark
[c] Laboratory of Advanced Materials (MOE), School of Materials Science and Engineering, Tsinghua University, Beijing 100084, China

*E-mail: hj_pan@kust.edu.cn, xzha@mek.dtu.dk

Abstract. An AA1050 aluminum plate was cold rolled to a true strain of 1.9, and the stored energy was calculated based on microstructural parameters, namely the dislocation boundary spacing and the misorientation distribution, determined from large area electron back-scatter diffraction maps with an indexed fraction higher than 90% before any data cleaning operations. Three methods were used to determine the dislocation boundary spacing and the misorientation distribution. The results show significant differences in the dislocation boundary spacings and the misorientation distributions obtained by the different methods, especially for boundaries with misorientation angles below 2°. These deviations lead to significant differences in the calculated stored energy. A discussion of the three approaches to determine stored energies is presented.

1. Introduction

During the deformation of metals and alloys most of the energy associated with the process is converted into heat, while only a small amount of energy is retained in the form of crystal lattice defects, which is known as the stored energy of deformation. This stored energy is critical for the thermal response of metals and alloys, such as the microstructural evolution during recovery and recrystallization processes [1, 2].

Determination of stored energy is often based on X-ray diffraction measurements [3-8], or differential scanning calorimetry [9-12]. These methods are utilized to calculate the total stored energy of a whole specimen, are thereby often referred to as macroscopic testing methods. Calculation of stored energy can, however, also be done based on microstructure-based measurement, e.g. [13, 14]. Such measurements are normally done by spatially-resolved orientation measurements, recording and indexing diffraction patterns either in a scanning electron microscope (SEM) or a transmission electron microscope (TEM). Compared to TEM measurements, the electron backscatter diffraction (EBSD) characterization in a SEM is much easier and faster, and offers a higher statistical integrity based on the ability to readily inspect much larger areas [15].

Standard commercial EBSD post-processing software can automatically calculate misorientation distributions. However, the angular resolution of conventional EBSD measurements is around 1°. This may introduce errors during application of fully automated methods in the analysis of the misorientation distribution associated with dislocation boundaries even for moderately deformed metals and alloys. This aspect has, however, seldom been investigated.

The present study is designed to investigate different methods to determine the critical parameters, needed to calculate stored energies based on characterization of deformation microstructures. For this purpose the aluminum alloy AA1050 cold rolled to a medium strain has been chosen. Thereby, a microstructure with a substantial fraction of low misorientation angle dislocation boundaries is selected as the basis for the analysis. The stored energy values are compared and the reasons for the observed differences are discussed.

2. Experimental

The chemical composition of the aluminum AA1050 alloy used in this study is Al-0.056Si-0.17Fe-0.002Mn- 0.006Ni-0.006Zn-0.005Ti-0.003Zr-0.007V (wt. %). The material was received as a hot extruded cylindrical bar with an initial average grain size of 47 μm and a dislocation density estimated to be 1×10^{13} m^{-2}. Further detailed information regarding the initial microstructure is reported in [16, 17]. A plate with a cross section of 23 × 10 mm was cut out by wire-electrode cutting from the hot extruded bar. This plate was cold rolled to a thickness reduction of 85%, corresponding to a medium strain (ε) of 1.9 with the rolling direction parallel to the extrusion direction of the initial bar. The microstructure of the specimen after rolling was characterized by EBSD after careful electro-polishing to give an indexed fraction in the EBSD data of more than 90%. The step size of EBSD scanning is chosen to be 150 nm according to the fineness of deformation structure.

The stored energy was in this study calculated according to the following equation [13]:

$$E_s = S_v \cdot \gamma + E(\rho_0) \tag{1}$$

where S_v is boundary area per unit volume, γ is the energy per unit area of a dislocation boundary which is a function of the angular misorientation across the boundary (see below for further details), and $E(\rho_0)$ is a contribution from individual dislocations in the volume between dislocation boundaries. For high stacking fault energy metals such as Al, $E(\rho_0)$ is small and is neglected in the present work.

More precisely, the stored energy is calculated by a discrete summation of boundary energies for different angular misorientations based on the relationship:

$$E_s = S_v \int_{\theta_{start}}^{\theta_{end}} f(\theta/\theta_{av})\gamma(\theta)d\theta$$

$$= S_v f(\theta_0\ to\ \theta_1)\gamma(\theta_0\ to\ \theta_1) + \cdots + S_v f(\theta_{end-1}\ to\ \theta_{end})\gamma(\theta_{end-1}\ to\ \theta_{end}) \tag{2}$$

where $f(\theta/\theta_{av})$ is taken as the Rayleigh function, and $f(\theta_x\ to\ \theta_y)$ is the fraction of misorientation angles in the range from θ_x to θ_y°; $\gamma(\theta_x\ to\ \theta_y)$ is the value of γ in the range from θ_x to θ_y°, taken here as the value at the bin mid-point (this will be close to the "true" average value in each bin, as the function describing boundary energy is reasonably linear with misorientation angle when considering only a small angular range).

The energy per unit area of a dislocation boundary is related to the angular misorientation across the boundary using the Read-Shockley equation:

$$\gamma(\theta) = \gamma_m(\theta/\theta_m)[1 - ln(\theta/\theta_m)] \qquad \theta \le \theta_m \tag{3a}$$

$$\gamma(\theta) = \gamma_m \qquad \theta > \theta_m \tag{3b}$$

where γ_m is the energy per unit area of a high angle boundary, θ is the boundary misorientation angle, and θ_m is the misorientation angle above which the energy per unit area is independent of misorientation angle. For the present calculations we use pure Al values: γ_m=0.324Jm^{-2} [18] and θ_m=15°.

Generally, the boundary map (BM) and the inverse pole-figure (IPF) coloring superposed with a boundary map (IPF/BM) are the most usual ways to reveal the microstructure of data obtained by EBSD. However, in these cases, the values of critical misorientation angle dominate the imaging and data analysis of these plotting. For example, when the critical misorientation angle is smaller than 2°, the plotting of deformed pure Al may contain a large amount of incorrect orientation identifications or noise, and the boundaries are hard to be identified. When the critical misorientation angle is larger than 2°, the boundaries with smaller misorientation are missing. This leads to a significant error during data analysis. In contrast, pattern quality, is a measure of how sharp or fuzzy an EBSD diffraction pattern is [19]. Pattern quality parameters include band slope (BS), band contrast (BC) and pattern quality (PQ). Each of these measures can be plotted as a grey scale map. Pattern quality (PQ) maps are based on the sharpness of the Kikuchi band edges or the height of peaks in Hough space [20]. In commercial software, BS is the maximum gradient of intensity at the edge of the bands [21] and BC is determined by the software based on the contrast or sharpness difference between the Kikuchi bands and the background. For the present work, the BS values are plotted for each pixel to allow visual estimation of features that cause poor band contrast, including slip bands, grain boundaries, low angle dislocation boundaries (LABs), phase interfaces, surface imperfections such as pores, scratches, and small inclusions. From BS maps, it is thus easy to get an impression of the microstructure, which is generally clearer than backscatter electron (BSE) images, but also a lot more time-consuming to analyse properly. It should be noted also that modern EBSD systems perform diffraction pattern "quality" (e.g. BS) calculations, for every pattern analysed, regardless of the EBSD "indexability" or the state of the material beneath. BS maps generated are like microstructurally sensitive SEM images, collected at EBSD speeds and with a high grid-based spatial resolution [22].

In this work, the dislocation boundary spacings and misorientation distributions, required for a microstructure-based stored energy calculation, were obtained using three different approaches based on values generated automatically using built-in functions in the EBSD post-processing software or from values derived from manual inspection of the EBSD data. For comparison, the dislocation boundary spacing and misorientation distribution of three methods use automated, partially automated and manual work. Specifically, the following methods were used: i) dislocation boundary spacing values were calculated automatically as equal circle diameters following application of a grain reconstruction (GR) algorithm ("Detect Grains" in the Channel 5 software), while the misorientation distribution was obtained from an automatically generated list of adjacent pixel misorientations (designated in the following as the "GR" method), ii) since the IPF/BM is the most usual way to reveal microstructure, dislocation boundary spacings were measured manually using a line intersect (LI) method on an EBSD constructed using IPF/BM (plotting adjacent pixel misorientations) while the misorientation distribution was again obtained using from an automatically generated list of adjacent pixel misorientations (designated in the following as the "LI/BM" method), and iii) dislocation boundary spacing values were measured manually using a line-intersect method on an EBSD band slope (BS) map, while the misorientation distribution was calculated from measuring the cumulative misorientation across features identified as boundaries in the BS map (designated in the following as the "LI/BS" method).

For the first two methods, the critical misorientation angle definitions of 1° and 2° were used in a consistent manner (the same value for both boundary spacing and misorientation calculations) using the entire map area. For the third method, the misorientations were calculated using just 1/3 of the mapped area covering the whole distance of normal direction. For all line intersection measurements five equally spaced vertical and horizontal test lines were used. A summary of the methods used and their designation in this study are given in table 1.

Table 1. Summary of the three different methods used for boundary spacings and misorientations.

	Designation	Dislocation boundary spacing	Misorientation distribution
(i)	GR-1°	Grain reconstruction - automated	Adjacent pixel misorientations – automated
	GR-2°	Grain reconstruction - automated	Adjacent pixel misorientations – automated
(ii)	LI/BM -1°	Line intersection - manual	Adjacent pixel misorientations – automated
	LI/BM-2°	Line intersection - manual	Adjacent pixel misorientations – automated
(iii)	LI/BS	Line intersection - manual	Misorientations across BS map features - manual

3. Results and discussion

The microstructure and the lines used for the line-intersection measurement method to determine the average dislocation boundary spacing, D, and the boundary area per unit volume, S_v, are shown in figure 1. Five straight lines along the horizontal and vertical directions are used in these cases. The number of intersections, n, with boundaries along each measuring line is counted. S_v and D are then calculated as follows:

$$S_v = 2n/L \tag{4}$$

$$D = L\,/\,(n - 1) \tag{5}$$

where L is the total test line length.

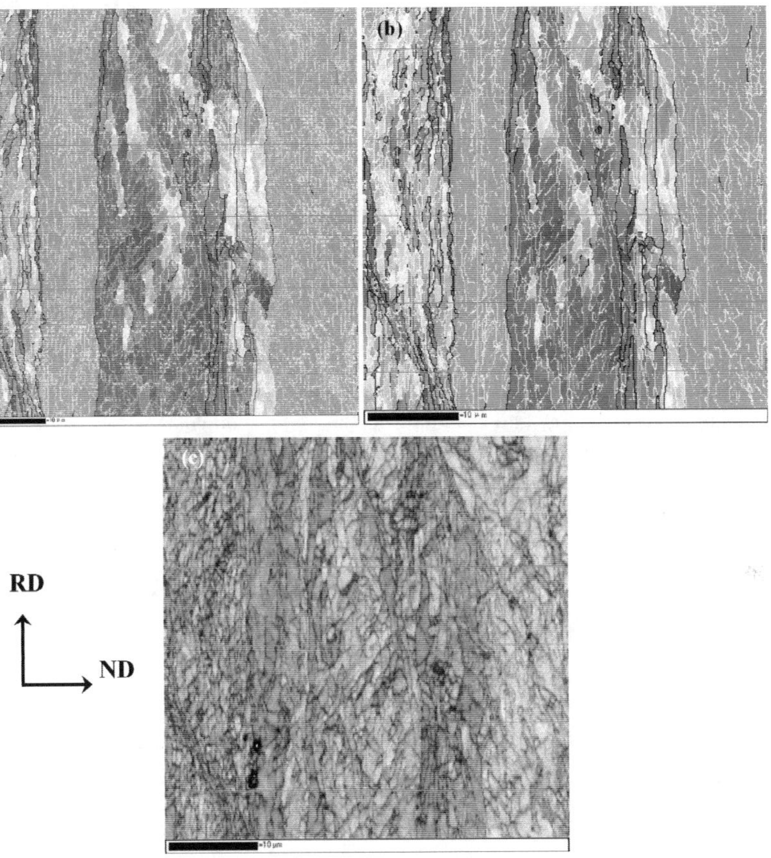

Figure 1. The microstructure of the cold-rolled AA1050 (ε = 1.9): (a) IPF/BM (1°), (b) IPF/BM (2°), (c) BS maps. Black lines show high angle boundaries (>15°) in both (a) and (b), while white lines correspond to misorientations of 1-15° and 2-15° in (a) and (b), respectively.

The average values of S_v and D values determined by the three methods are given in table 2. The average values of D using a critical misorientation angle of 1° obtained by the GR method and the LI/BM method are significantly smaller than the value obtained by the LI/BS method. However, when a critical misorientation angle of 2° is used, similar D values are obtained by all three methods. This may be directly related to the same line intersection methods applied in the present study, as these values are larger than the value (determined by TEM) of cold-rolled commercial purity aluminium at a similar strain [23]. The detailed comparison with the values determined by the common random line intersection method and the calculated ones from the measured geometrical necessary boundary (GNB) and incidental dislocation boundary (IDB) spacings will be carried out in near future.

As shown in figure 1(a), the IPF/BM using a critical angle of 1° shows a large number of incorrect orientation identifications, or noise, compared with that seen in figures 1(b) and 1(c). This phenomenon is also found and consistent with the comparison and analysis of deformed pure Al's data obtained by both EBSD and transmission electron microscope (TEM) methods in our previous work [24]. Thus, it is certainly the reason for the large deviations in estimation of the average boundary spacing when including 1° boundaries in determination.

Additionally, when the critical misorientation angle is 2°, the boundaries with misorientation smaller than 2° are missing, as shown in figure 1(b). This is considered to show certain deviations of *D* between GR-2°, LI/BM-2° and LI/BS, though the differences are small than our anticipation. The reasons are as following: i) when measuring along the normal direction, most boundaries are high angle boundaries and exhibit small deviations of *D* among GR-2°, LI/BM-2° and LI/BS methods, which leads to small total differences of average values, ii) the incorrect orientation identifications or noise of boundaries smaller than 2° are notably high, which remarkably magnify the deviations of *D*.

Table 2. Boundary spacings and boundary areas per unit volume as determined by the three methods.

Method	*n*	S_V (10^6 m^{-1})	*D* (μm)
GR-1°	-	5.56	0.362
GR-2°	-	2.80	0.729
LI/BM-1°	90 ± 13	4.00 ± 0.58	0.506 ± 0.27
LI/BM-2°	59 ± 7	2.62 ± 0.31	0.776 ± 0.09
LI/BS	62 ± 5	2.76 ± 0.22	0.738 ± 0.06

Figure 2 shows the misorientation distributions obtained by the two different approaches used, namely data obtained in an automated manner and based on manual measurements across features identified in the BS map. Significant differences in the two distributions can be observed: i) the misorientation distribution below the lower critical threshold (here 1°) is, by definition, not accurately identified from automated EBSD measurements; ii) the proportion of misorientations in the range 1 - 2° in (a) is much higher than that in (b). Thus, there is a significant error in the range of 0 - 2° for the data obtained in an automated manner from the post-processing software. This is consistent with the statistical results of the dislocation boundary spacing, as shown in table 2. It is thus clear that the automated methods cannot properly recognize dislocation boundaries with very low misorientations, e.g. 0 - 1°, as such misorientation angle values overlap with the typical EBSD orientation noise of around 1°. However, the BS map (see figure 1c) may be used to identify dislocation boundaries below 1°, and can as such provide a more accurate data set. This agrees with a previous study [25], which concluded that more accurate microstructural data can be obtained through manual measurement on BS maps, and can be used to supplement EBSD maps from indexing-derived data. For example, boundaries between two adjacent areas with a certain misorientation can be shown by markings superposed on an IPF map, but they do not in general reveal the actual microstructure, especially when the underlying misorientations are small.

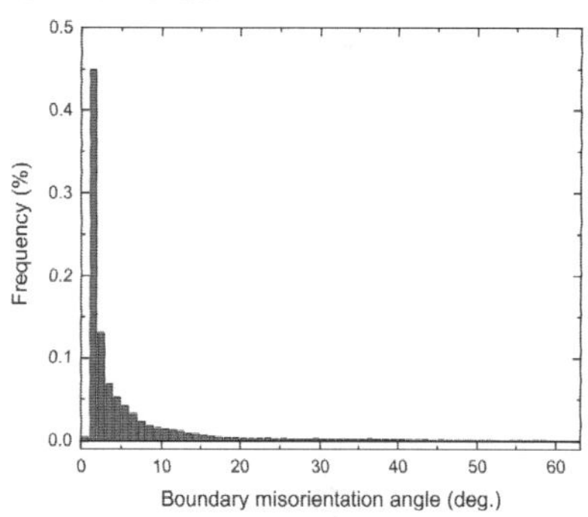

(a)

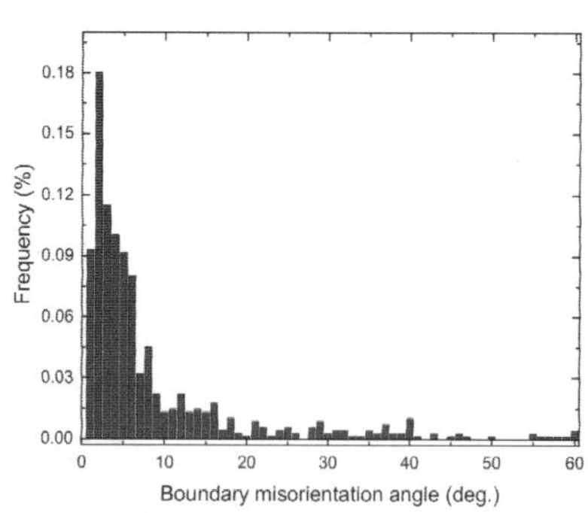

(b)

Figure 2. Misorientation distributions obtained by different methods: (a) data obtained in an automated manner from the Channel 5 software; (b) data based on manual measurement of misorientation across boundaries identified in the BS map.

The stored energy of the present sample has been calculated based on the microstructural data according to equation (2) and the obtained values are shown in table 3. It can be seen that the GR-1° and the LI/BM-1° methods, as used by many researchers, result in significantly higher estimates of stored energy compared to the LI/BS method, where boundaries are identified precisely in a manual manner, corresponding to relative deviations of 82% and 31%, respectively to the LI/BS method values. Incorrect orientation determinations, or orientation noise, are obviously very significant in these cases. In contrast, the deviations of stored energies calculated by the GR-2° and the LI/BM-2° methods are just 9% and 16% relative to the LI/BS value. As these two methods yield similar values of average dislocation boundary spacing to the LI/BS method, these deviations are dominated by the differences in the way the misorientation distribution is calculated.

Table 3. Calculated stored energies and deviations value determined by the LI/BS method.

Method	Stored energy (10^5 J m^{-3})	Deviation to LI/BS method value
GR-1°	10.82	82%
GR-2°	5.45	9%
LI/BM-1°	7.79 ± 0.11	31%
LI/BM-2°	5.11 ± 0.06	16%
LI/BS	5.95 ± 0.05	-

4. Conclusion

The effects on stored energy calculations of using data for boundary spacing and misorientation obtained by standard EBSD methods or through a manual (LI/BS) method have been investigated. The results reveal very significant differences between the data obtained by the standard methods and those obtained using the LI/BS method when a critical misorientation angle of 1° is used. The main reason for the difference is that the large number of dislocation boundaries with misorientations of 0 - 2°, which are formed in pure Al during cold-rolling, frequently are incorrectly identified by the standard EBSD software because of the overlap with the noise being in the same angular range. When a cut-off of 2° are used instead, much more consistent results are obtained. It is suggested that misorientation measurements based on BS maps, or other such non-indexing based patterns, result in good precision when determining stored energies, and a combination of the two types of maps and methods may be advantageous to use.

Acknowledgements

D. Juul Jensen and X. Zhang acknowledge the support from the European Research Council under the European Union Horizon 2020 research and innovation program (grant agreement No 788567-M4D). H. Pan acknowledges the support from the National Natural Science Foundation of China (NSFC) (No. 51901091), the China Scholarship Council (No. 201908530046), the Basic Research Project of Yunnan Science and Technology Program (No. 202001AU070081).

References

[1] Humphreys F J 2017 *Recrystallization and Related Annealing Phenomena* (Amsterdam, Netherlands: Elsevier).
[2] Doherty R D, Hughes D A, Humphreys F J, Jonas J J, Jensen D J, Kassner M E, King W E, McNelley T R, McQueen H J and Rollett A D 1997 *Mat. Sci. Eng. A* **238** 219-74.
[3] Essmann U, Rapp M, and Wilkens M 1968 *Acta Mater.* **16** 1275-87.
[4] Wilkens.M 1969 *Acta Mater.* **17** 1155-9.
[5] Kocks U F and Scattergood R O 1969 *Acta Mater.* **17** 1161-8.
[6] Mohamed G and Bacroix B 2000 *Acta Mater.* **48** 3295-302.
[7] Castelnau O, Ungar T, Miroux A, Chauveau T and Bacroix B 2000 *Mater. Sci. Forum* **321** 720-5.
[8] Gerber P, Tarasiuk J, Chauveau T and Bacroix B 2003 *Acta Mater.* **51** 6359-71.
[9] Clarebrough L M, Hargreaves M E and West G W 1955 *Proc. R. Soc. London. Ser. A* **7** 99-100.

[10] Bever M B, Holt D L and Titchener A L 1973 *The stored energy of cold work* (Oxford, New York: Pergamon Press).
[11] Woldt E and Jensen D J 1995 *Metall. Mater. Trans. A* **26** 1717-24.
[12] Knudsen T, Cao W Q, Godfrey A, Liu Q and Hansen N 2008 *Metall. Mater. Trans. A* **39** 430-40.
[13] Godfrey A, Cao W Q, Hansen N and Liu Q 2005 *Metall. Mater. Trans. A* **36** 2371-8.
[14] Godfrey A, Hansen N and Jensen D J 2007 *Metall. Mater. Trans. A* **38** 2329-39.
[15] Pan H, He Y and Zhang X 2021 *Materials (Basel)* **14** 1-47.
[16] Zhang X, Hansen N and Nielsen C V 2017 *IOP Conf. Ser. Mat. Sci.* **219** 1-7.
[17] Zhang X D, Nielsen C V, Hansen N, Silva C M A and Martins P A F 2019 *Int. J. Plasticity* **115** 93-110.
[18] Murr L E 1975 *Interfacial phenomena in metals and alloys* (Reading, Mass: Addison-Wesley Pub. Co. Advanced Book Program).
[19] Sitzman S D, Nolze G and Nowell M M 2010 *Microsc. Microanal.* **16** 698-9.
[20] Zhang F, Ruimi A and Field D P 2016 *Microsc. Microanal.* **22** 99-107.
[21] Afshar M 2017 *Mater. Charact.* **132** 405-12.
[22] Wright S I and Nowell M M 2006 *Microsc. Microanal.* **12** 72-84.
[23] Liu Q, Huang X, Lloyd D J, and Hansen N 2002 *Acta Mater.* **50** 3789-802.
[24] Cao W Q, Godfrey A and Liu Q 2003 *J. Microsc-Oxford* **211** 219-29.
[25] Zhang X D, Hansen N, Gao Y K and Huang X X 2012 *Acta Mater.* **60** 5933-43.

Cooperative enhancement of strength and ductility in metallic materials: The role of local stress/strain evolution on the plastic flow behavior

K S Miao[1], Y P Xia[2, 3], D Y Li[4], H Wu[1] and G H Fan[1*]

[1]Key Laboratory for Light-weight Materials, Nanjing Tech University, Nanjing, 211816, China
[2]School of Materials Science and Engineering, Harbin Institute of Technology, Harbin, 150001, China
[3]Department of Civil and Mechanical Engineering, Technical University of Denmark, DK-2800 Kgs. Lyngby, Denmark
[4]Laboratory for Space Environment and Physical Sciences, Harbin Institute of Technology, Harbin, 150001, China

*E-mail: ghfan@njtech.edu.cn

Abstract. Simultaneous enhancement of strength and ductility is one of the eternal topics for the community of metallic materials. The heterostructure architecture, e. g. layered structure, is one of the potential methods to evade the strength-ductility trade-off dilemma in recent years. However, there is still a lack of a deeper understanding to excellent strength-ductility combinations. In the present study, we focused on the local stress, which drives local plastic deformation. The local represents the affected scale of the stress. The evolution processes of local stress/strain upon loading of typical model materials of Ti/Ti, Ti/Al, and Al/Al layered metals (LMs) were quantitatively characterized by *in situ* electron microscope, synchrotron X-ray diffraction, and neutron diffraction. We found that the deformation characteristics were closely associated with the local stress/strain evolution process, and highlighted the key role of local stress/strain in revealing the deformation mechanisms. This work therefore may open a new route to design new metallic materials through the tailoring of local stress/strain evolution.

1. Introduction

Cooperative enhancement of strength and ductility is crucial for metallic materials. Over the past century, employing solid solution strengthening, work hardening, grain refinement strengthening, and second phase strengthening, the mechanical properties of metallic materials have been well modified. However, there is still a persistent trade-off between the strength and ductility of metallic materials, that is, the increase of strength is general at the expense of ductility.

Recently, a considerable attraction has grown around the theme of heterostructure architecture, which refers to intentionally arranging domains in metallic materials into a specific structure, such as layered structure, gradient nanostructure, harmonic structure, bimodal structure, etc [1]. As one of the heterostructures, layered structure has shown a promising perspective for enhancing work hardening and evading the strength-ductility trade-off dilemma [2]. By means of *in situ* high-resolution strain mapping, the existence of interface affected zone has been quantitatively evaluated in Cu/bronze layered metals (LMs), which settled the basement for mechanical property optimization of LMs [3].

IOP Conf. Series: Materials Science and Engineering 1249 (2022) 012065 doi:10.1088/1757-899X/1249/1/012065

The macroscopic mechanical performance is integrated from local stress/strain response of each domain within LMs, which emphasizes the importance of local stress/strain evolution in LMs. It has been reported that local stress/strain regulated the deformation mechanisms of metallic materials. For example, the local stress concentrations played critical roles in both twin nucleation and the growth process of twins [4]. It was also suggested that variant selection for twinning is jointly determined by local stress and local strain [5]. Although the deformation mechanism-based constitutive model has been established [6] to predict the tensile response of LMs considering local deformation heterogeneities, much uncertainty of the relation between the local stress/strain and deformation mechanisms still exists due to the deficient experimental validation.

The developments of synchrotron radiation X-ray diffraction and neutron diffraction shed light on local stress/strain characterization with high spatial resolution. In the present study, we quantitatively characterized the evolution processes of local stress/strain upon loading of typical model materials of Ti/Ti, Ti/Al, and Al/Al LMs by *in situ* electron microscopy, synchrotron radiation X-ray diffraction, and neutron diffraction. The differences between local stress/strain in domains with macroscopic applied stress/strain were addressed, while its effects on deformation mechanism regulation were evaluated. These findings provided insights into the design of high-performance materials via tailoring local stress/strain.

2. Microstructures and mechanical properties of the LMs

The LMs in the present study were fabricated via alternatively stacking of raw material foils (AA1060 commercially pure Al or TA1 commercially pure Ti), followed by thermal processing comprising hot-pressing, hot-rolling, annealing, etc. Figure 1 includes the microstructures and mechanical properties of the LMs, where the coordinate is denoted as normal direction (ND), transverse direction (TD), and loading direction (LD) hereafter. One prominent feature of the LMs is the drastic differences between constituent layers. For the Ti/Ti and Al/Al LMs, with the assistance of pre-annealing on raw material foils, the grain size of coarse-grained (CG) layer is general 5 times higher than that of fine-grained (FG) layer, that is, 50.2 μm versus 9.8 μm in Figure 1(a) and 21.6 μm versus 4.6 μm in Figure 1(c), respectively. As shown in Figure 1(b), the Ti/Al LM consists of heterophases that are at the variance of grain size, yield strength, elastic modulus, crystal structure, deformation mechanism, etc. In comparison with their homogeneous counterparts, attractive mechanical properties are achieved in the LMs. Figure 1(d) illustrates that the yield strength of Ti/Ti LM is beyond the predictions of both the Hall-Petch relationship (HP) and the rule of mixture (ROM). The average grain size for HP calculation is measured by the line-intercept method, which is grain boundary area weighted. In contrast, the strengthening contribution in the ROM is grain volume weighted. As suggested in Figure 1(e), the fracture strain of Ti/Al LM is superior to that of both constituent layers. The Al/Al LM also possesses an enhanced combination of yield strength and uniform elongation compared with homogeneous Al (Figure 1(f)).

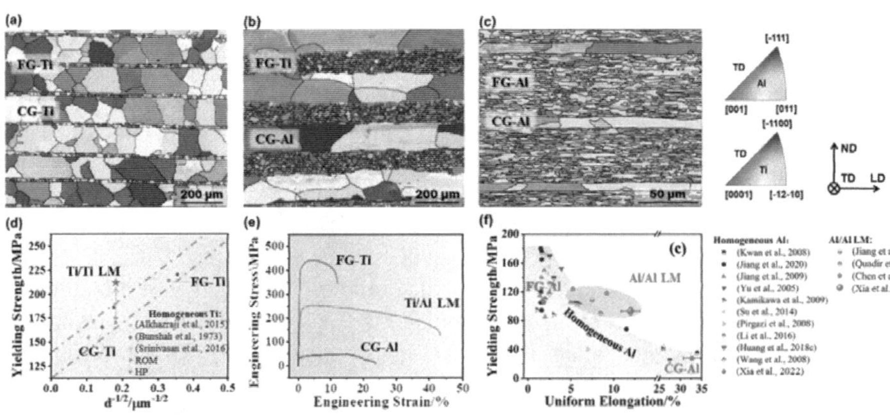

Figure 1. The microstructures of the (a) Ti/Ti LM, (b) Ti/Al LM, and (c) Al/Al LM. The mechanical property comparisons of (d) Ti/Ti LM, (e) Ti/Al LM, and (f) Al/Al LM with their homogeneous counterparts, respectively. Image reproduced with permission from Ref. [7-9].

3. Effects of local stress on deformation in the LMs

The high spatial resolution of synchrotron radiation Laue X-ray diffraction makes it available for detailed analysis at micron-scale in the LM. Laue diffraction adopts the incident beam with a wide wavelength range (polychromatic), enabling the acquisition of diffraction peaks from multiple crystalline planes in materials via single exposure (Laue pattern). The deviatoric lattice stress/strain could be calculated by comparing relative positions of diffraction peaks in experimental and ideal Laue patterns. Here, an *in situ* deviatoric lattice strain analysis of Ti/Ti LM was conducted on the Beamline 12.3.2 at the Advanced Light Source (ALS) in Lawrence Berkeley National Laboratory (LBNL) with a beam size of ~1 μm × 1 μm [7]. As shown in Figure 2(a), grain A is embedded in the CG-Ti layer of Ti/Ti LM, whose crystalline orientation is schematically illustrated in Figure 2(b). Figure 2(c-h) demonstrate the evolution of deviatoric lattice strain in grain A. It is worth noting that ε_{xx}, ε_{yy}, and ε_{zz} here do not include the hydrostatic components since the present experiments are not sensitive to lattice dilation. Before the yielding of Ti/Ti LM (212 MPa), there is an early local yielding through dislocation activities near the layer interface, indicated by the changes of shear strain components ε_{xy} and ε_{yz} at an applied stress of 160 MPa. Considering Laue diffraction is sensitive to defects, local plastic deformation leads to a drop in the indexing ratio for Laue patterns, reducing the visible area of grain A when applied stress exceeds 160 MPa. Meanwhile, the diffraction peak shapes in Laue patterns streak along various directions during the "interface-interior-interface" line scanning as shown in Figure 2(i), indicative of the distinguish deformation mechanisms in local areas. It is suggested that the layered structure induces local stress concentration at the layered interfaces, promoting the additional deformation mechanisms in Ti.

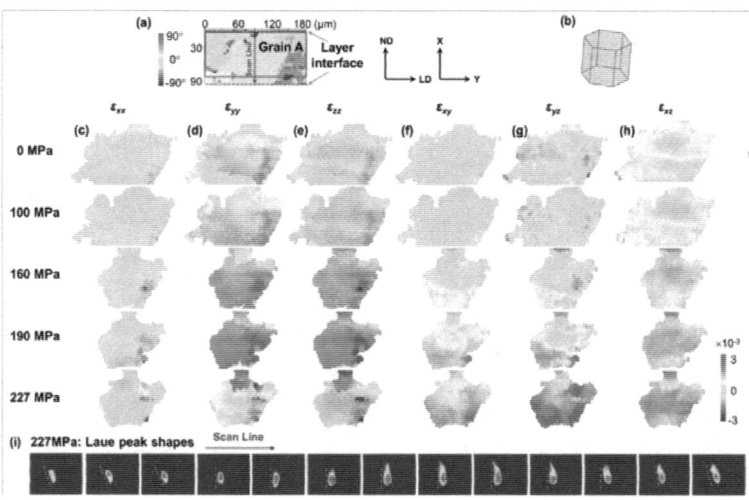

Figure 2. The analysis of Ti/Ti LM via synchrotron radiation Laue X-ray diffraction. (a) Euler angle (ϕ_1) map of Ti/Ti LM, where grain A is a grain in the CG-Ti layer. (b) The orientation of the grain A. The evolution of the deviatoric strain tensor in the grain A as the applied stress increased from 0 MPa to 227MPa: (c) ε_{xx} , (d) ε_{yy}, (e) ε_{zz} , (f) ε_{xy}, (g) ε_{yz} and (h) ε_{xz}. (i) The Laue peak shapes along scan line in (a) at an applied stress of 227 MPa. Image reproduced with permission from Ref. [7].

In addition to the local stress induced by interfaces, there is also an intrinsic local stress partition during deformation due to the incompatibility of constituent layers. The neutron diffraction, which permits high penetration and statistically reliable characterization, provides an important opportunity to advance the understanding of the local stress partition between constituent layers. For example, neutron diffraction experiment was employed to Ti/Al LM on the materials science diffractometer STRESS-SPEC at the neutron source FRM II [8]. A monochromatic neutron beam with a wavelength of ~1.68 Å and a gauge volume of 5×5×5 mm^3 were utilized. The diffraction vectors of selected {hkl} planes were parallel to the tensile direction. By analyzing the shifting of diffraction peak, the lattice strain evolutions of {111} for Al layer and {101} for Ti layer upon increasing applied stress are presented in Figure 3(a). It is found that deformation of Ti/Al LM before yielding (applied stress reaches 175 MPa) could be classified into two stages. During stage I, both Ti and Al deform elastically, following the condition of iso-strain. However, the different elastic modulus leads to the unequal local stress partition between Ti and Al, considering Hooke's law. Then the Al yields prior to Ti due to the lower yield strength, the further increase of applied stress is dominantly carried by Ti rather than Al. Moreover, the mismatched Poisson's ratio of plastic deformation for Al and elastic deformation for Ti generates local stress perpendicular to the LD [10]. In Figure 3(b), red and black solid lines or dotted lines indicate the measured or theoretical elastic modulus of Ti {101} and Al {111} planes, respectively. The deviation between measured and theoretical values also confirms the local stress partition in Ti/Al LM during deformation.

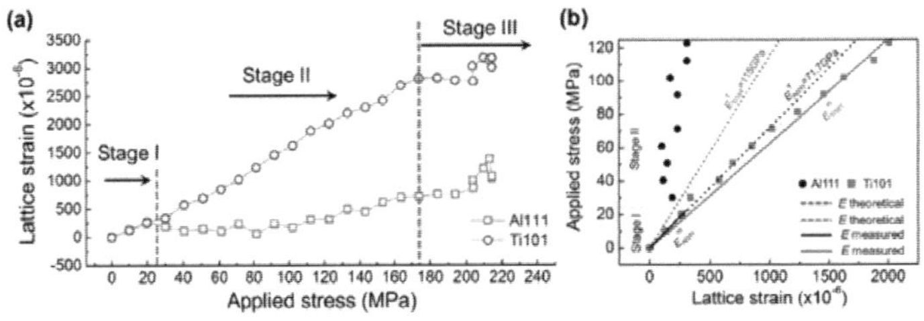

Figure 3. The evolution of lattice strain of Ti/Al LM measured by neutron diffraction. (a) Three deformation stages of Ti/Al LM with increasing applied stress, which were classified by the evolution of lattice strain of {111} for Al layer and {101} for Ti layer. (b) The measured elastic modulus (E_m) versus the theoretical elastic modulus (E_T). Image reproduced with permission from Ref. [8].

It is worth mentioning that the "local" in the above discussion, i. e. local stress induced by the interfaces and local stress partition during deformation, correspond to different spatial scales. That is, the stress state for a given area is the local stress superimposition of multiple effects (e. g. interface effect, elastic modulus, Poisson's effect, etc.). The existence of a multiaxial local stress state is expected in LMs even though only uniaxial stress is applied externally, which could drive the activation of deformation mechanisms that deviated from the prediction considering applied stress merely [11].

4. Effects of local strain on deformation in the LMs

The activations of different deformation mechanisms are known to contribute to strain tensor to various extent. The local strain serves as crucial role in regulating deformation mechanisms. For instance, a detailed investigation is conducted on 16 grains in the CG-Al layers of Al/Al LM by combining the *in situ* SEM and EBSD [9]. As is shown in Figure 4, the potential slip traces are deduced from the grain orientation and plot on SEM images. Meanwhile, the primary slip traces observed by SEM images are highlighted by red circles. These insert values alongside slip traces are Schmid factors (SFs) calculated from the applied stress. A useful misorientation mapping method, scaled Rodrigues vector mapping, was performed to highlight the formation of dislocation boundaries and deformation microbands [12]. Many deformation microbands formed with sharp boundaries aligned close to the slip planes with the first or second highest SFs, which also correspond to the obvious slip traces observed by the SEM images (donated as Schmid microbands, marked by red arrows). However, it can also be found that the microbands correspond to lower SFs (donated as non-Schmid microbands, marked by yellow arrows). The formation of microbands implies the predominant activation of corresponding slip systems. Accordingly, it was suggested that different groups of slip systems with lower SFs are activated dominantly in the CG-Al layers.

For rationalizing the deformation mechanisms regulated by the layered structure, we combined the local strain result with the Taylor ambiguity method [13,14], which can tell the plastic strains from different slip systems. The digital image correlation (DIC) was utilized coupling with SEM and *in situ* straining for measuring local strain, as shown in Figure 5(a-c). There is an obvious ε_{yy} strain partition in different constituent layers of Al/Al LM in spite of an iso-strain condition in the aspect of ε_{xx}. More specific, the CG-Al layers bear much more compressive strain than FG-Al layers, which could result from interface effects [3]. We quantitatively extracted the local strain tensor in CG-Al with strain incompressible condition and ignoring the shear components. The results are input into the Taylor ambiguity method [9]. The theoretical uniaxial strain tensor with a Poisson's ratio of 0.5, is also imported for comparison. As exemplified by two grains in Figure 5(d), the dominant slip systems predicted by Taylor ambiguity method can explain the microbands formation well. The results from theoretical uniaxial strain found that dominant slip systems have the first or second highest SFs, which agree well with the slip traces as well as the Schmid microbands. However, the formation of non-Schmid microbands can only be explained with considering the local strain tensor extracted from DIC. For instance, the [$\bar{1}$10](111) in G3 predicted by local strain case can correspond to the non-Schmid microbands near (111) plane. The slip system [110](1$\bar{1}$1) in G12 with much lower SF, is consistent with dark-colored non-Schmid microbands along (1$\bar{1}$1). In this work, the relation between local strain and activation of unexpected slip systems is revealed.

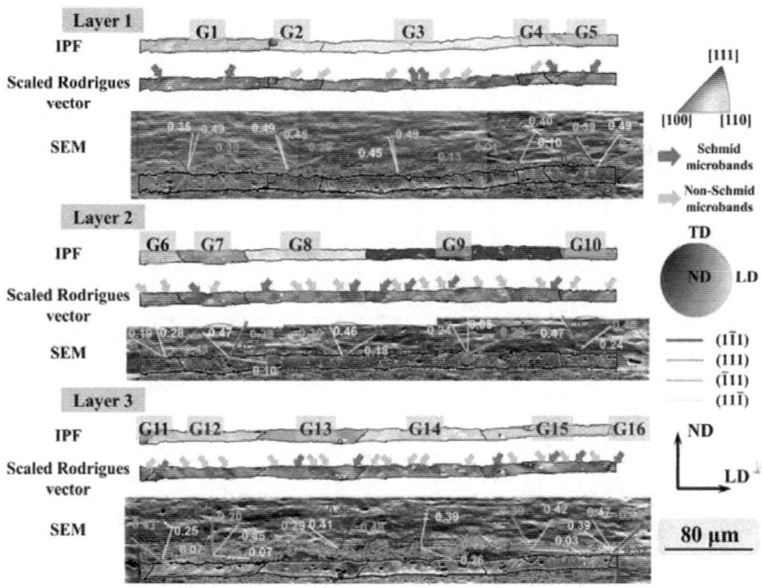

Figure 4. The IPF maps, scaled pixel-to-grain-average Rodrigues vector maps, and SEM images of CG-Al layers in Al/Al LM at a plastic strain of 10%. Image reproduced with permission from Ref. [9].

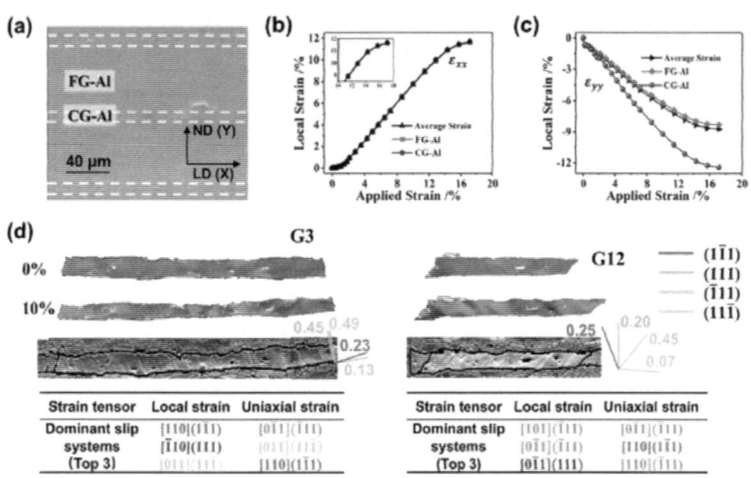

Figure 5. Analysis of the Al/Al LM based on local strain. (a) Schematic diagram for the microstructure and coordinate. The average strain of (b) ε_{xx} and (c) ε_{yy} in the whole region, the FG-Al layers, and the CG-Al layers, reflecting the strain partition along ND. (d) Pixel-to-grain-average Rodrigues vector maps coupled with the SEM micrograph showing slip traces a plastic strain of ~10%

for grains G3 and G12. The activation propensities of slip systems are predicted considering local strain and uniaxial strain and list in the tables. Image reproduced with permission from Ref. [9].

5. Conclusions

In summary, we characterized the local stress/strain evolution processes of typical model materials of Ti/Ti, Ti/Al, and Al/Al LMs upon deformation. It was found that the deformation characteristics were closely associated with the local stress/strain evolution process and highlighted the crucial role of local stress/strain in revealing deformation mechanisms. This work, therefore, opens a new window to design new metallic materials through the tailoring of local stress/strain evolution.

Acknowledgments

This work was financially supported by the National Key Research & Development Plan (Grant No. 2020YFA0405900), the National Natural Science Foundation of China (Grant No. 51927801), and Natural Science Foundation of Jiangsu Province (Grant No. BK20202010).

References

[1] Wu X L and Zhu Y T 2021 Gradient and lamellar heterostructures for superior mechanical properties *MRS Bull.* **46** 244-9
[2] Misra A, Göken M, Mara N A and Beyerlein I J 2021 Hierarchical and heterogeneous multiphase metallic nanomaterials and laminates *MRS Bull.* **46** 236-43
[3] Huang C X, Wang Y F, Ma X L, Yin S, Höppel H W, Göken M, Wu X L, Gao H J and Zhu Y T 2018 Interface affected zone for optimal strength and ductility in heterogeneous laminate *Mater. Today* **21** 713-9
[4] McCabe R J, Kumar M A, Liu W, Tomé C N and Capolungo L 2021 Revealing the effect of local stresses on twin growth mechanisms in titanium using synchrotron X-ray diffraction *Acta Mater.* **221** 117359
[5] Xu S, Schuman C and Lecomte J S 2016 Accommodative 10-12 twins at high angle grain boundaries in rolled pure titanium *Scripta Mater.* **116** 152-6
[6] Zhao J, Zaiser M, Lu X, Zhang B, Huang C, Kang G and Zhang X 2021 Size-dependent plasticity of hetero-structured laminates: A constitutive model considering deformation heterogeneities *Int. J. Plast.* **145** 103063
[7] Li D Y, Fan G H, Huang X X, Jensen D J, Miao K S, Xu C, Geng L, Zhang Y B and Yu T B 2021 Enhanced strength in pure Ti via design of alternating coarse- and fine-grain layers *Acta Mater.* **206** 116627
[8] Huang M, Xu C, Fan G H, Maawad E, Gan W M, Geng L, Lin F X, Tang G Z, Wu H, Du Y, Li D Y, Miao K S, Zhang T T, Yang X S, Xia Y P, Cao G J, Kang H J, Wang T M, Xiao T Q and Xie H L 2018 Role of layered structure in ductility improvement of layered Ti-Al metal composite *Acta Mater.* **153** 235-49
[9] Xia Y P, Miao K S, Wu H, Geng L, Xu C, Ku C S and Fan G H 2022 Superior strength-ductility synergy of layered aluminum under uniaxial tensile loading: The roles of local stress state and local strain state *Int. J. Plast.* **152** 103240
[10] Jiang S, Peng R L, Hegedűs Z, Gnäupel-Herold T, Moverare J J, Lienert U, Fang F, Zhao X, Zuo L and Jia N 2021 Micromechanical behavior of multilayered Ti/Nb composites processed by accumulative roll bonding: An in-situ synchrotron X-ray diffraction investigation *Acta Mater.* **205** 116546
[11] Miao K S, Huang M, Xia Y P, Wu H, Liu Q and Fan G H 2022 Unexpected de-twinning of strongly-textured Ti mediated by local stress *J. Mater. Sci. Technol.* **125** 231-7
[12] Yu T B, Du Y, Fan G H, Xu R Q, Barabash R, Hansen N, Huang X X and Zhang Y B 2021 In-situ synchrotron X-ray micro-diffraction investigation of ultra-low-strain deformation microstructure in laminated Ti-Al composites *Acta Mater.* **202** 149-58
[13] Mánik T and Holmedal B 2014 Review of the Taylor ambiguity and the relationship between rate-independent and rate-dependent full-constraints Taylor models *Int. J. Plast.* **55** 152-81
[14] Bunge H J 1970 Some applications of the Taylor theory of polycrystal plasticity *Kristall und Technik* **5** 145-75

IOP Conf. Series: Materials Science and Engineering 1249 (2022) 012066 doi:10.1088/1757-899X/1249/1/012066

Importance of Microstructure on Precipitation in Tempering of Martensitic Steels

Z Hou[1,2,3], **R P Babu**[3] **and L Zhang**[1,2,3]

[1] International Joint Laboratory for Light Alloys (MOE), College of Materials Science and Engineering, Chongqing University, Chongqing, 400044, China

[2] Shenyang National Laboratory for Materials Science, Chongqing University, Chongqing 400044, P.R. China

[3] Department of Materials Science and Engineering, KTH Royal Institute of Technology, SE-100 44 Stockholm, Sweden

[4] Electron Microscopy Center of Chongqing University, Chongqing University, Chongqing 400044, P.R. China

E-mail: houzy@cqu.edu.cn

Abstract. Precipitation hardening is one of most effective strengthening mechanisms in steels, and much research has been performed in the past. To evaluate the contribution of precipitates, the quantitative features of precipitates including mean size and particle size distribution etc., are vital and needed. However, the predictive modeling of precipitation is still a challenge so far, especially on a quantitative level. Thus, in the present work, precipitation of carbides after tempering of martensitic Fe-Cr-C alloys, consisting of hierarchically arranged substructures within the prior-austenite grains, namely packets and blocks of individual laths, up to 1000h has been investigated. Experimental measurements using electron microscopy and modeling using a Langer-Schwartz theory with the Kampmann-Wagner -Numerical (KWN) method have been conducted. The importance of a proper definition of the initial microstructure for predictive modeling is discussed, in terms of the comparison of calculated and experimental results.

IOP Conf. Series: Materials Science and Engineering 1249 (2022) 012066 doi:10.1088/1757-899X/1249/1/012066

1. Introduction

Since precipitation hardening is one of the most strengthening mechanisms in metallic alloys, further understanding and modelling of precipitation behavior has attractive numerous attentions in the past decades [1-4]. Martensitic steels have been widely used in many industrial fields such as tool steel, after subjected to heating treatment e.g. tempering to improve its toughness or/and strength [1]. During tempering, release of internal stresses generated from the martensitic transformation, diffusion of saturated solid solution atoms and annihilation of dislocations together with movement of various boundaries cause softening of steels [5]. While, dispersion precipitates form in the matrix as carbide-formed alloying elements added and strengthen the steels after tempering. To evaluate the contribution of precipitates on the strength, parameters like size, volume fraction, size distribution etc of the precipitates are needed [6-8]. By measuring and controlling these precipitation parameters, it is possible to tune and optimize properties of the steels, not only the mechanical properties such as strength and ductility but also creep resistance, wear resistance and degradation properties [8]. However, due to their very fine size and variation with the tempering conditions, quantitative parameters only from the experimental investigations using transmission electron microscopy (TEM), atom probe tomography (APT) and small-angle scattering (SAS) are always time-consuming and costly [9-11]. With recent progresses in both theoretical and experimental analysis of steels, inspired by the trends in materials design towards integrated computational materials engineering (ICME) [7], modelling to predict e.g. the mechanical property evolution from the microstructural evolution during tempering is becoming more tractable. This can provide an efficient way for a new alloy design and process optimization. Further improvements of the predictive capabilities of models correlating microstructure with nucleation, growth and coarsening mechanism of precipitates depend on an improved understanding of the microstructure and chemical composition evolution, as well as implies the control mechanism of precipitation at the interface of matrix/precipitates [12,13].

The purpose of the present work is to investigate the microstructure evolution and precipitation behavior in a Fe–0.15C–1Cr model alloy tempered at 700 °C. This alloy was chosen since it provides an excellent model for systems where it is necessary to handle diffusion of both interstitial elements and substitutional elements simultaneously. Meanwhile, the Langer–Schwartz–Kampmann–Wagner (LSKW) modeling [14-15] scheme as implemented in TC-PRISMA for predictive modeling of precipitation is assessed by comparing with experimental results from TEM analysis on carbon extraction replica samples and thin-foil samples.

2. Experimental procedures

The chemical composition of the investigated alloy is given in table 1. An isopleth phase diagram for Fe–0.14 wt% C–xCr and for the investigated steel, calculated using Thermo-Calc [16] and the TCFE12 database [17] are shown in figure 1. It can be seen in figure 1 that for the alloy the stable phases at 700 °C are ferrite, cementite(M_3C) and M_7C_3, and that below 524°C ferrite and M_3C_2 are the stable phases. It should be noted that the chemical composition of the present alloy is located nearby the border of ferrite and M_3C two-phase region, and contains a very small fraction of M_7C_3 (0.0017 V.%) at equilibrium accordingly.

Small specimens (10×10×1 mm3) were austenitized at 1100 °C for10 min and quenched in brine. For the as-quenched specimen, figure 2 shows a fully lath martensite as observed in low carbon low alloy steels. The as-quenched samples were thereafter tempered at 700 °C for 5 s, 5 min, 30 min, 5 h, 100 h and 1000 h in the Sn-Bi metal bath (5 s and 5 min), a tube furnace with Ar gas (30 min) and in evacuated quartz tubes (5, 100 and 1000 h). To study the as-quenched microstructure and its evolution during tempering, an optical microscope(OM), a JEOL JEM-2100F microscope operating at 200 kV and a JEOL JSM-7800F FEG scanning electron microscope (SEM) equipped with an electron backscatter diffraction (EBSD) detector and electron channeling contrast imaging (ECCI) detector operated 15 kV were applied. TEM thin-foil and carbon extraction replicas were prepared for identification and quantification of precipitates. Quantification of the precipitates was performed in a scanning transmission electron microscope (STEM). For each tempered sample, approximately 1000 precipitates were imaged from several random regions and their mean size was calculated.

Table 1. Nominal chemical composition of the investigated alloy (in wt. %).

Alloy	C	Cr	Si	Mn	S	Al	Cu	Ni
0.14C-0.98Cr	0.14	0.983	0.021	0.07	0.06	0.014	0.011	0.017

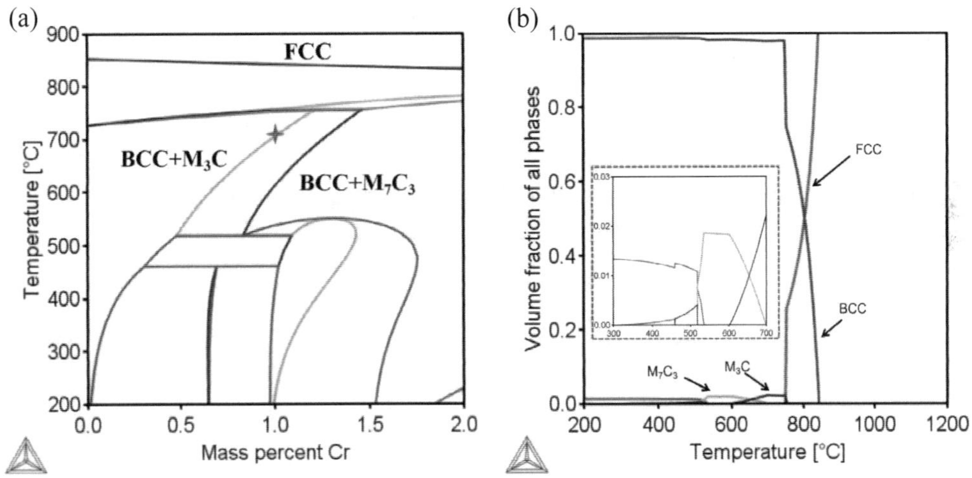

Figure 1. isopleth phase diagram for Fe–0.14 wt% C–*x*Cr and for 0.14C–98Cr steel

Figure 2. LOM microstructure of the as-quenched specimen

3. Modeling

In the present study the precipitation module (TC-PRISMA) as implemented in the Thermo-Calc software was used for modeling the precipitation during tempering. The details of the models can be found in [3] but for the convenience of the reader a short overview will now be given. All models assume that the particles are spherical with no diffusions inside them. Different types of growth models are implemented, for example, simplified, advanced, Para equilibrium and Neglect partitioning local equilibrium (NPLE). The difference between each model exists in the tie-line taken and the interface condition. In the simplified treatment a constant tie-line taken across the bulk composition is assumed, whereas for the advanced treatment a new tie-line is determined at each time step [16]. In the present work the Para and simplified treatment are used for quantitative predictions and comparison with experiments, to generate a better understanding of the operating growth mechanism. The thermodynamic and kinetic data for the simulations were taken from the TCFE12 and MOBFE7 databases. As mentioned earlier, M_3C and M_7C_3 are stable carbides at 700°C, thus both precipitates are included in the modeling. The interfacial energy between the precipitates and the matrix is critical for quantitative predictions and here the interfacial energy (σ_c) calculated based on a Becker's model [18] in combination with the thermodynamic data from TCFE12 was adopted directly, that are 0.22019J/m^2 for M_3C and 0.29691J/m^2 for M_7C_3, respectively. Furthermore, the number of potential nucleation sites as another critical parameters is selected as the dislocation density 2.13×10^{15} m^{-2} in the martensitic structure since most nucleation events occur at dislocations [10].

4. Results and discussion

The stability diagram of the investigated steel is presented in figure 3. In addition to the austenite and ferrite phases, the equilibrium carbides, M_3C, M_7C_3 and M_3C_2, are represented in the temperature range of 500 ~ 900°C. By which, the relative stability of all carbides is possible to calculate when these stable carbides are suspended. The dashed line in figure 3 show the metastable diagram after the stable carbides are suspended. Furthermore, it is shown that M3C carbides are not stable at temperatures higher than 782 °C and that M_7C_3 carbides are not expected to be observed, as they are only stable at temperatures below 702 °C.

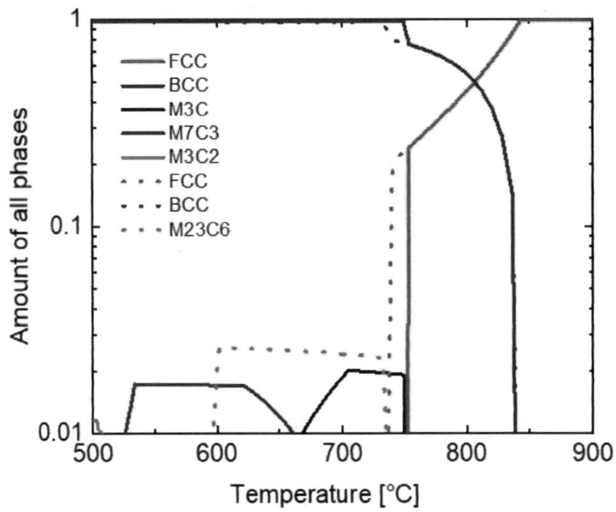

Figure 3. The stability diagram of 0.14C-0.98Cr steel

The tempered microstructure of specimens at 700°C for 5s up to 1000 h are presented in figure 4. After tempering for 5s, lath martensite is similar to that in the as-quenched one, with features of long and thin lath with straight boundaries. Noted that, plenty of carbides, showing as black or white points due to channeling contrast, are clear seen at intra- and inter-lath. This is consistent with the TEM observation (figure 5(a)). With the prolonged tempering, the width of lath as well as the size of precipitates are increased gradually, see figure4 (b~f) and figure 5(b). After tempered for 1000 h, large carbides are most located at various grain boundaries. Meanwhile, the specimen still keeps its hierarchical structure, i.e. lath, sub/-block packet, even after 1000 h, see figure 4(h). Interestingly, fine elongated ferrite with equivalent circle diameter (ECD) of 1.9±0.5 μm was reserved after such long time tempering, see figure 4 (f, g, h), which provide one simple way to obtain fine grain size microstructure only by quenching and tempering, as reported earlier [10]. During tempering, migration of high angle grain boundaries is pinned by the formed carbides, result in a slow movement of grain boundaries and coarsening rate of substructure. This type information of microstructure evolution together with its pinning effect are important to acquire as the mechanical property modelling aims to consider the grain boundary strengthening as an important factor.

Figure 5 shows the typical TEM microstructure for thin foils and carbon replicas. After identified by selected area diffraction patterns (SADPs) by TEM, not shown here, all the precipitates in the present steels are M_3C, no other carbides are found, although M_7C_3 carbides are expected to form. The absence of M_7C_3 carbides in the present work maybe mainly due to: 1) their very small volume fraction or 2) uncertainty of the database, the chemical composition is located nearby the region with only M_3C carbides as the stable carbide at 700°C.

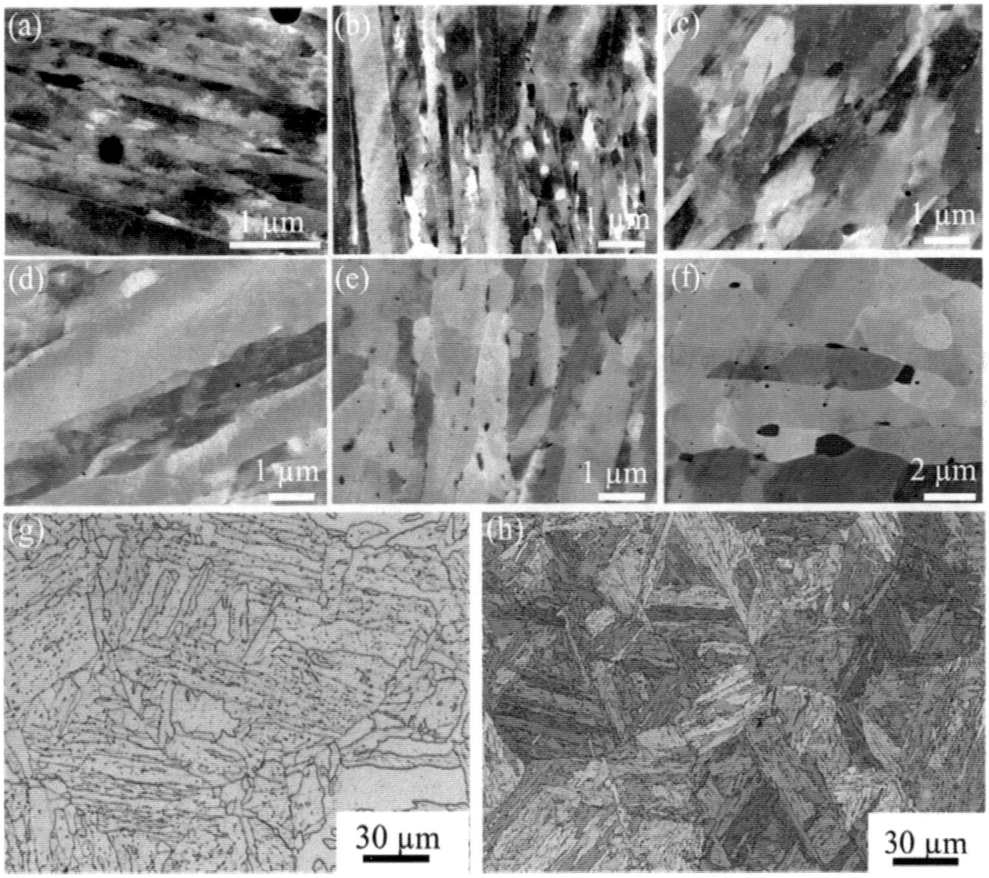

Figure 4. ECC images of as-quenched specimens after tempering at 700°C for: (a) 5s; (b) 5 min; (c) 30 min; (d) 5h; (e) 100 h; (f) 1000 h; OM (g) and IPF (h) maps for 1000 h

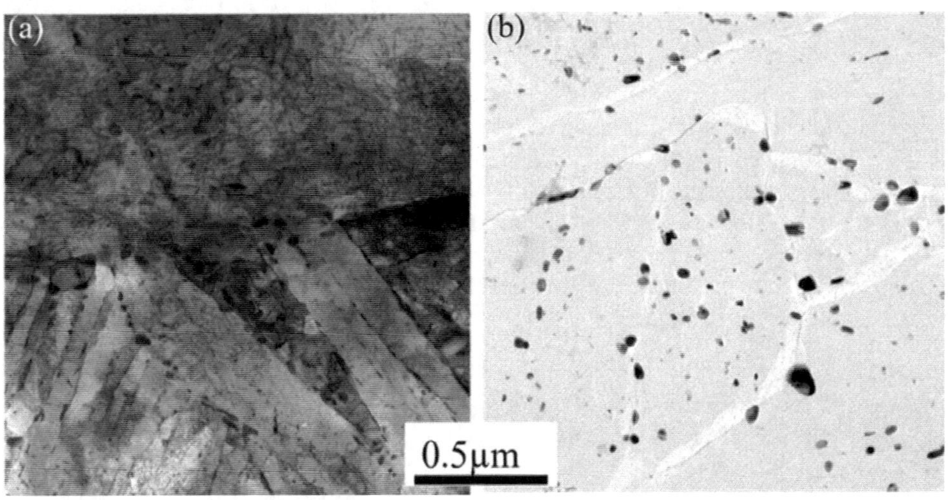

Figure 5. Typical TEM for thin foil and carbon replicas samples: (a) 5s; (b)30 min

The mean radius and volume fraction of M_3C from TC-Prisma simulations and experiments are presented in figure 6. For the simulations, both simplified and PARA growth models are plotted to give insight of growth mechanism. The mean radius of M_3C increases from 25±5 nm at 5 s to 682 ±45nm at 1000 h, while the volume fraction keeps constant at the range of 0.015±0.04 v.% indicating a coarsening stage even after 5 s at 700 °C. It is clearly that the volume fraction of carbides can be predicted very well using both growth models with default input data from misstructure and chemical composition. However, the PARA growth model significantly overestimates but the Simplified growth model underestimate the mean radius of M_3C carbides, especially at the very earlier stage, see figure 6. In general, there is no Cr diffusion occurs at the nucleation of M_3C carbides, which is more like the PARA condition [10,11]. As the tempering continuous Cr starts to partitioning to M_3C carbides from the matrix, that is why the agreement between the calculated values by Simplified growth model and experiments is good at the late stage. Other parameters as input data should be considered for a better agreement, such as grain boundaries as active nucleation sites. As seen from figure 4 and figure 5, in addition to at dislocations the precipitates are also nucleated at various boundaries. Furthermore, carbon trapped by dislocation or others defects are need to escape first prior to form M_3C, this could delay the start point of PARA growth model [19]. At the other hand, a higher number of nucleation sites refine the mean size of precipitates at a given volume fraction. Thereafter, to improve the predictive modelling of precipitation quantitative data of such as various boundaries and carbon suggestion etc., are needed.

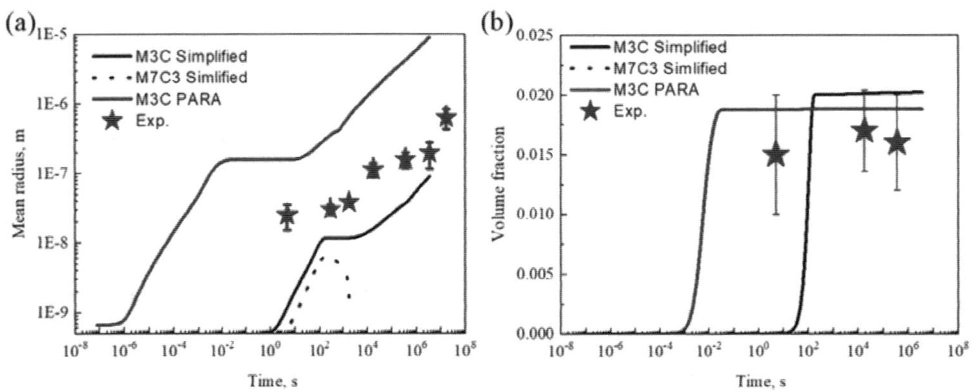

Figure 6. Comparison of calculation and experiments for (a) Mean radius and (b) Volume fraction of carbide evolution during tempering

5. Summary

The precipitation behavior in a 0.14C-0.98Cr steel during tempering at 700°C for 5 s up to 1000 h has been investigated together with microstructure evolution. Only M_3C carbides is observed through the tempering, and forms at both intra- and inter- lath. The mean radius of M_3C increases from 25±5 nm at 5 s to 682 ±45 nm at 1000 h, while the volume fraction keeps constant at the range of 0.015±0.04 v.%. By comparison, the PARA growth model significantly overestimates but the simplified growth model underestimates the mean radius of M_3C carbides, especially at the very earlier stage.

Acknowledgements

This work was supported by the National State Key Research and Development Program of China (2021YFB3702101), the "111" Project (B16007) by the Ministry of Education and the State Administration of Foreign Experts Affairs of China and Research Foundation for Advanced Talents.

References

[1] Miyamoto G, Oh J, Hono K, Furuhara T and Maki T 2007 Acta Mater. 55 5027–38
[2] Hou Z, Babu R, Hedström P and Odqvist J 2019 J. Mater. Sci. 54 9222–34.
[3] Hou Z, Hedström P, Chen Q, Xu Y, Wu D and Odqvist J 2016 Calphad 53 39–48
[4] Hou Z, Babu P, Hedström P and Odqvist J 2020 Mater. Sci Techn. 36 887–93
[5] Krauss G 2017 Steel Res Int.
[6] Olson G B 2000 Science 288 993–8
[7] Allison J 2011 JOM. 63 15–8
[8] Xiong W and Olson G 2016 NPJ Comput. Mater. 2 15009
[9] Zhou T, Babu P, Hou Z and Hedström P 2021 Critical Rev Solid State Mater. Sci.
[10] Hou Z, Hedström P, Xu Y, Di W and Odqvist J 2014 ISIJ Inter. 54 2649–56
[11] Hou Z, Babu P, Hedström P and Odqvist J 2018 J. Mater. Sci. 53 6939–50
[12] Jiang S, Wang H, Wu Y, Liu X, Chen H, Yao M, Gault B, Ponge D, Raabe D and Hirata A, 2017 Nature 544 460–4
[13] Wagner R, Kampmann R and Voorhees P, 2001Phase Transformations in Materials; Wiley-VCH: Weinheim
[14] Langer J and Schwartz K 1980 Phys. Rev. A 21 948–58
[15] Kampmann R and Wagner R 1984 Kinetics of Precipitation in Metastable Binary Alloys: Theory and Application to Copper–1. 9 at.% Titanium and Nickel–14 at.%Aluminium. In: Decomposition of Alloys: The Early Stages, edited by Haasen P, Gerold V and Wagner R
[16] Thermo-Calc Software TCFE12, Steels/Fe-based Alloys Database, version 12, Thermo-Calc Software AB, Stockholm, (2022).
[17] Thermo-Calc Software MOBFE7, Steels/Fe-based Alloys Database, version 7, Thermo-Calc Software AB, Stockholm, (2022)
[18] Becker R 1938 Ann. Phys. 32 128–32
[19] Maalekian M and Kozeschnik E 2008 Calphad 32 650–4

AUTHOR INDEX